T0092155

Mathematical Physics with Differential Equations

Mathematical Physics with Differential Equations

Yisong Yang
Professor, Courant Institute of Mathematical Sciences,
New York University

OXFORD
UNIVERSITY PRESS

Great Clarendon Street, Oxford, OX2 6DP,
United Kingdom

Oxford University Press is a department of the University of Oxford.
It furthers the University's objective of excellence in research, scholarship,
and education by publishing worldwide. Oxford is a registered trade mark of
Oxford University Press in the UK and in certain other countries

Published in the United States of America by Oxford University Press
198 Madison Avenue, New York, NY 10016, United States of America

British Library Cataloguing in Publication Data
Data available

Library of Congress Control Number: 2022943690

ISBN 978–0–19–287261–6
ISBN 978–0–19–287262–3 (pbk.)

DOI: 10.1093/oso/9780192872616.001.0001

Printed and bound by
CPI Group (UK) Ltd, Croydon, CR0 4YY

For Sheng,
Peter, Anna, and Julia

Contents

Preface

This book aims to present a broad range of fundamental topics in theoretical and mathematical physics in a thorough and transparent manner based on the viewpoint of differential equations. The subject areas covered include classical and quantum many-body problems, thermodynamics, electromagnetism, magnetic monopoles, special relativity, gauge field theories, general relativity, superconductivity, vortices and other topological solitons, and canonical quantization of fields, for which, differential equations are essential for comprehension and have played, and will continue to play important roles. Over the past decade, the author has used most of these topics at several universities, domestically and internationally, as courses and seminars mainly for mathematical graduate students and researchers trained and interested in differential equations. These activities and experiences convinced the author that many of the concepts, construction, structures, ideas, and insights of fundamental physics can be taught and learned effectively and productively, with emphasis on what are offered by or demanded from differential equations.

With this in mind, the book has several goals to accomplish. Firstly, the style of the presentation hopefully provides a handy and direct access to approach the subjects discussed. Secondly, it serves to render a fairly wide selection of themes that may further be tailored for a graduate-level mathematical physics curriculum out of the individual preference of the instructor or reader. Thirdly, it supplies a balanced pool of topics for upper-level or honors undergraduate seminars. Fourthly, it offers guidance and stimulation to the related contemporary research frontiers and literature.

Except for knowledge on differential equations, the prerequisite for the reader of the book is kept minimal, although certain levels of acquaintance with undergraduate general physics is helpful for the reader to proceed smoothly. Thus, the book begins with classical mechanics in canonical formalism and moves on to various advanced subjects. However, unless needed, the book excludes specialized topics of traditional classical mechanics such as fluids and elasticity theories, since they are treated extensively elsewhere in the literature. The book may be used for self-study, as a textbook, or as a supplemental source book for a course in mathematical physics with concentration and interests in quantum mechanics, field theory, and general relativity, emphasizing insights from differential equations.

While the book holds fifteen chapters, each chapter may be studied or presented separately in a more or less self-contained manner, depending on interests and readiness of the reader or audience.

In Chapter 1, we start with a presentation of the canonical formalism of classical mechanics. We then consider the classical many-body problems in three-, two-, and one-dimensional settings, subsequently. Specifically, in three dimensions, we discuss the many-body problem governed by Newton's gravity, consolidated by a thorough study of Kepler's laws of planetary motion and a derivation of Newton's law of gravitation, as a by-product; in two dimensions, we introduce the Helmholtz–Kirchhoff point-vortex model; and, in one dimension, we present a dynamical system problem in biophysics known as the DNA denaturation. For this third subject we also explain how to implement ideas of thermodynamics to study a temperature-dependent mechanical system. The goal of this chapter is to lay a Lagrangian field-theoretical foundation for field theory and enlighten the study with some exemplary applications.

In Chapter 2, we consider quantum many-body problems. We explain how the Schrödinger equation is conceptualized and the statistical interpretation of the wave function. Then, we formulate the quantum many-body problem describing an atomic system and discuss the hydrogen model as an illustration. Next, we show how the Hartree–Fock method, Thomas–Fermi approach, and density functional theory may be utilized in various situations as computational tools to find the ground state solution of a quantum many-body problem. An initial goal of this chapter is to illustrate a monumental transition from classical to quantum mechanics based on the Schrödinger equation realization of the photoelectric effect. A second goal of this chapter is to introduce some mathematical challenges presented by quantum many-body problems. Here the study of the hydrogen model serves as a motivating starting point of the quantum many-body problem, which naturally leads to the development of subsequent analytic methods of computational significance when the dimension of the problem goes up. In particular, we show that the quantum-mechanical description of a many-body problem, whose classical-mechanical behavior is governed by nonlinear ordinary differential equations, is now given by a linear partial differential equation, and that appropriate approximations of such a linear equation necessitate the formulation of various nonlinear equation problems in respectively specialized situations.

Chapter 3 is a study of the Maxwell equations and some distinguished consequences. First, we present the equations and discuss the associated electromagnetic duality phenomenon. We next formulate the Dirac monopole and Dirac strings and show how to use a gauge field to resolve the Dirac string puzzle and obtain Dirac's charge quantization formula. We demonstrate how this idea inspired Schwinger to derive a generalized quantization formula for a particle carrying both electric and magnetic charges, known as dyon. We then present the Aharonov–Bohm effect for wave interference, which demonstrates the significant roles played by the gauge field and topology of a system at quantum level. The goal of this chapter is to appreciate how some of the fundamental and rich contents of electromagnetic interaction may be investigated productively through exploring the structures of the differential equations governing the interaction.

Specifically, by complexifying the wave equation, we obtain a novel derivation of the Maxwell equations, which also embodies a clear and natural revelation of the electromagnetic duality, and, by considering the topological properties of the solutions to the Maxwell equations, we arrive at the findings of Dirac, Schwinger, and Aharonov–Bohm.

Chapter 4 is a succinct introduction to special relativity. Since most of the subjects covered in this text are concerned with relativistic field equations, some solid knowledge on special relativity is necessary. Thus we carry out a study of special relativity in this chapter. We first discuss spacetime, inertial frames, and the Lorentz transformations. We present topics that include spacetime line element, proper time, and a series of notions, including length contraction, time dilation, and simultaneity of events. Then, we study relativistic mechanics. Although this chapter is short, its goal is to serve as the foundation for many following chapters, including those on the Dirac equations, gauge field theory, general relativity, and topological solitons. In particular, the Lagrangian action for the motion of a relativistic particle will be the starting point for the formulation of the Nambu–Goto string action and the Born–Infeld theory.

In Chapter 5, we present the Abelian gauge field theory. We start with an introduction to the notions of covariance, contravariance, and invariance for quantities defined over a spacetime. We formulate the Klein–Gordon equation, which is a relativistic extension of the Schrödinger equation. We then show that a gauge field is brought up again naturally in order to promote the internal symmetry of the system from global to local such that the Maxwell equations are deduced as a consequence. Furthermore, we discuss various concepts of symmetry breaking and illustrate the ideas of the Higgs mechanism as another important application of gauge field equations. A goal achieved in this chapter is that a vista of important physical consequences may be obtained from examining some basic structural aspects of the equations of motion without sufficient knowledge about their solutions.

Chapter 6 centers around the Dirac equation. We first show how to obtain the classical Dirac equation and what immediate consequences the equation offers in contrast to the Klein–Gordon equation. We next consider the Dirac equation coupled with a gauge field and present its Schrödinger equation approximations in electrostatic and magnetostatic limits, respectively. In particular, we derive the Stern–Gerlach term, whose presence is essential for the explanation of the Zeeman effect. We then review some nonlinear Dirac equations. This study shows that sometimes profound physics may be unveiled unexpectedly through an exploration of some deeply hidden internal structures of the governing equations.

Chapter 7 covers the Ginzburg–Landau theory for superconductivity. We begin with a discussion of perfect conductors, superconductors, the Meissner effect, the London equations, and the Pippard equation. We next present the Ginzburg–Landau equations for superconductivity and show how to come up with the London equations in the uniform order-parameter limit and demonstrate the Meissner effect. We then study the classification of superconductivity in view of surface energy and discuss the appearance of mixed states in type II superconductors. We end the chapter with a review of some generalized

Ginzburg–Landau equations. This study retraces the historical path regarding how differential equations of varied subtleties have been exploited in line with real-world observations to advance the understanding of superconductivity. In particular, it also describes an unsolved two-point boundary value problem, arising in the Ginzburg–Landau theory, for the classification of superconductivity.

Chapter 8 grows out of the subjects covered in Chapter 5 and Chapter 7. Specifically, in this chapter, we focus on the static Abelian Higgs theory or the Ginzburg–Landau theory in two dimensions, which possesses a distinctive class of mixed-state solutions of a topological characteristic known as vortices. We describe such solutions in detail in view of several important facets including energy concentration, vortex-line distribution, quantization of magnetic flux or charge, and exponential decay properties. We also discuss the use of such vortex-line solutions in a linear confinement mechanism for magnetic monopoles, a topic actively pursued in quark confinement research in recent years. This study shows again the applications of solutions of gauge field equations, of topological characteristics, to fundamental physics, of both quantitative and conceptual values.

In Chapter 9, we move onto the subject of non-Abelian gauge field theory. We first present the theory on a general level, and then specialize on the Yang–Mills–Higgs theory. We discuss a series of concrete formalisms including the Georgi–Glashow model and the Weinberg–Salam electroweak theory. We also illustrate some important families of solutions such as the 't Hooft–Polyakov monopole, Julia–Zee dyon, and Bogomol'nyi–Prasad–Sommerfield explicit solution. The main goal of this chapter is to present a broad family of nonlinear partial differential equations of importance in elementary particle physics.

In Chapter 10, we study the Einstein equations of general relativity and related subjects. We begin with an introduction to the basics of Riemannian geometry and then present the Einstein tensor and the Einstein equations for gravitation. Subsequently, we unfold our discussion mainly around special solutions of the Einstein equations, categorized into time-dependent space-uniform solutions and time-independent space-symmetric solutions. In the former category, we elaborate on the cosmological consequences and implications richly contained in various solutions of the Friedmann type equations under the Robertson–Walker metric, which include the Big Bang cosmological scenario, patterns of expansion of the universe, and an estimate of the age of the universe. In the latter category, we begin with a presentation of the Schwarzschild solution and a discussion of several notions unveiled, such as the event horizon and black hole. We then present a derivation of the Reissner–Nordström solution for a black hole carrying both electric and magnetic charges and discuss its consequences. We will also discuss the Kerr solution describing a rotating black hole. Afterwards, we consider the gravitational mass problem and the Penrose bounds as additional themes. We next present a discussion of gravitational waves in the weak-field limit. We conclude the chapter with a study of the cosmological expansion of an isotropic and homogeneous universe propelled by a scalar-wave matter known as quintessence. The main goal of this chapter is to use the Einstein equations as a key to access a broad range of gravity-related research directions of contemporary interests.

Chapter 11 is about charged vortices and the Chern–Simons equations. For conciseness, we focus on the simplest Abelian situations. We first present the Julia–Zee theorem and its proof, which states that finite-energy electrically charged vortices, which are static solutions in two dimensions, do not exist in the usual Yang–Mills–Higgs theory. Thus, some modification of the theory is to be made in order to accommodate charged vortices, and the addition of a Chern–Simons topological term to the Lagrangian action density will serve the purpose. In this chapter, our goal is to present a brief introduction to the Chern–Simons vortex equations. Besides the motivation for allowing electrically charged vortices, other applications of the Chern–Simons theory include anyon physics of condensed matters, gravity theory, and high-temperature superconductivity, where non-Abelian structures are also abundantly utilized. It is hoped that this introduction will serve to spark interest and inspiration in the study of an enormous family of partial differential equation problems of challenges, under the shared title of the Chern–Simons vortex equations.

In Chapter 12, we consider the Skyrme model and some related topics. We begin with an exploration of the well-known dimensionality constraints brought forth by the Derrick theorem and the Pohozaev identity. We then introduce the Skyrme model to maneuver around the dimensionality constraints. As a related topic, we will also discuss the Faddeev model, which may be viewed as a descent of the Skyrme model and which brings about knot-like solutions characterized by fractionally-powered growth laws relating energy to topology. In addition, we present a discussion of Coleman's Q-ball model, which also has no dimensionality restriction. Due to the difficulties associated with the structures of the nonlinearities and topological characteristics of these field-theoretical models, it has been a daunting task to consider the equations of motion directly, except in numerical studies, and one needs to focus on their variational solutions. In either situation, hopefully this chapter serves as an invitation to many related research topics.

Chapter 13 is a short discussion on strings and branes. We first revisit the relativistic motion of a free particle and subsequently formulate the Nambu–Goto string equations. We then extend the study to consider branes and their governing equations. We next present the Polyakov strings and branes and their equations of motion. Thus, our goal of this chapter is to emphasize the challenges and difficulties encountered in these highly geometric and nonlinear partial differential equations as classical field-theory equations. Except in some extremely simplified or reduced limits, these equations are not yet well understood, regarding their solutions.

In Chapter 14, we present the Born–Infeld theory of electromagnetism and some of the associated mathematical problems. To start, we recall the energy divergence problem of the point-charge model of the electron and the idea of Born and Infeld in tackling the problem based on a revision of the action density motivated by special relativity, sometimes referred to as the first formulation of

Born and Infeld. Within this formalism, we consider some interesting illustrative calculations around the electric and dyonic point charge problems. We next present the second formulation of Born–Infeld based on invariance consideration and show how to resolve the energy divergence problem associated with a dyonic point charge encountered in the first formulation of Born and Infeld. We then relate the Born–Infeld equations to the minimal surface equations and propose some generalized Bernstein problems. Subsequently, we conduct a discussion of an integer-squared law of a universal nature regarding the global vortex solutions of the Born–Infeld equations in two dimensions. Furthermore, we also present a series of electrically and dyonically charged black hole solutions of the Einstein equations coupled with the Born–Infeld equations. Thereafter, we consider the generalized Born–Infeld theories and present some interesting applications, including a nonlinear mechanism for an exclusion of monopoles as finite-energy magnetically charged point particles, relegation and removal of curvature singularities of charged black holes of the Reissner–Nordström type, and theoretical realizations of cosmological expansion and equations of state of cosmological fluids through appropriate Born–Infeld scalar-wave matters in the form of k-essence. In some sense, this chapter may be regarded as a gauge-field or scalar-field extension of the subjects discussed in Chapter 13. Therefore, the difficulties we encounter here are similar to those there. On the other hand, within the limitation of the Born–Infeld theory, here we are able to see how real progress is made for many important issues of concern, such as the resolution of an electric point charge of divergent energy, electromagnetic asymmetry, singularity relegation for charged black holes, and k-essence scalar field cosmology, all based on pursuing special solutions of the governing equations in various situations.

Chapter 15 is the final chapter and provides some taste of field quantization and a further view expansion. For clarity and conciseness, our discussion will be clustered around harmonic oscillators. We start with a study of the quantum mechanics of harmonic oscillators based on canonical quantization. We next consider the Hamiltonian formalism of general field equations in terms of functional derivative and commutators. We then show how to quantize the Klein–Gordon equation and the Schrödinger equation. In doing so, we encounter the well-known infinity problem arising from a divergent zero-point energy, which gives us an opportunity to explain the concept of renormalization. We then move on to quantize the Maxwell equations that govern electromagnetic fields propagating in free space. We focus our attention on the quantization of energy, momentum, and spin angular momentum directly, rather than the electromagnetic fields themselves, and derive the Planck–Einstein and Compton–Debye formulas for the photoelectric effect and photon spin in the context of quantum field theory. We conclude the chapter with a discussion of the thermodynamics of a harmonic oscillator, both classically and quantum mechanically, such that we are able to come up with a picture about the relation, ranges of applicability and limitation, and transition with regard to temperature, of classical and quantum-mechanical descriptions of a physical system, in general. Thus, part of the goal of this chapter is to show in view of quantum field theory what may be expected beyond classical field equations both in sense of differential equations and meaning of quantum physics.

Exercises appear at the end of each chapter. These mostly straightforward problems serve either to supplement the details or expand the scope of the materials of the text. Working out some of the problems may be useful for checking the understanding of the subjects covered but omitting this process should not compromise the quality of learning too much since throughout the text the materials are presented in sufficient details and elaboration.

An ideal reader of this book is a person well versed in college-level differential equations who is motivated by physical applications and is interested in gaining insights into field-theoretical physics through differential equations. In order to keep the volume of the book to a reasonable size, we leave out introductory materials about basic physical concepts commonly covered in an undergraduate course in general physics. For example, when we discuss the Ginzburg–Landau theory of superconductivity, we assume the reader knows what a superconductor is and how it behaves. Thus, if this book is used as a textbook for a short or extensive course, it will serve the purpose better if it is supplemented with some extra conceptual nontechnical reading materials, which should be easily available.

In addition to serving for self-study, the materials covered in the book are planned in such a way that each of the chapters may be used for a short concentrated topic course ranging from two to seven weeks or longer, with about two to three hours of lectures per week. Specifically, Chapters 1, 3–9, 11, and 13 may be candidates for a two-week course, Chapters 2, 12, and 15 for a three- to four-week course, and Chapters 10 and 14 for a six- to seven-week course. For a one-semester course, the author suggests picking a collection of about six to seven chapters depending on the interests of the instructor and students. At an elementary level, a choice may be Chapters 1–4, Chapter 7, Chapter 8, and Chapter 11, supplemented with Section 5.1 if necessary. At a more advanced level, a choice may be Chapters 5–10 and Chapter 15. The materials of the full book are more than enough for a year-long course. Moreover, except for Chapters 4 and 15, all other chapters may be studied for research topics and projects of differential equations and nonlinear analysis in theoretical and mathematical physics.

We supplement the book with an Appendices chapter of six sections, which cover some concepts and subjects encountered and used elsewhere in the main text. In the first section, we give a full introduction to the notions of indices of vector fields and topological degrees of maps, in the context of the Euclidean spaces. We begin our discussion from the argument principle in complex analysis and then extend the construction to real situations, highlighted with some applications as examples, including a proof of the fundamental theorem of algebra and a study of the issue of existence and non-existence of periodic orbits of some dynamical systems. Subsequently, we develop the concepts in higher dimensions and conclude the discussion with a proof of the Brouwer fixed-point theorem. In the second section, we consider the concepts of linking number and the Hopf invariant based on our knowledge on the topological degree of a map. We then consider these constructions in view of the concepts of the helicity of a vector field, the Chern–Simons invariant, and the classical integral representation of the Hopf invariant by Whitehead. In the third section, we present a comprehensive discussion of the Noether theorem, which associates continuous symmetries of a Lagrangian mechanical or field-theoretical system with its conserved quantities schematically. As illustrations, we first consider the motion of a point mass and derive its energy, linear momenta, and angular momenta, as consequences of time- and space-translation invariance and rotation-invariance. We then develop the formalism in the setting of a general Lagrangian field-theoretical framework and show how to construct the associated energy-momentum tensor and various Noether charges and currents. In the fourth section, we describe the possible eigenvalues of the angular momentum operators of a particle in non-relativistic quantum-mechanical motion based on the associated commutation relations of these operators. As a by-product, we explain how to deduce Dirac's charge quantization formula using Saha's method without resorting to a treatment of the Dirac strings. In the fifth section, we show how the concept of the intrinsic spin of a particle in quantum-mechanical motion arises as a result of "correcting" a "deficiency" in the spectra of orbital quantum momentum operators. As a consequence, we are naturally led to the introduction of spin matrices and spinors. In particular, we show how the Pauli spin matrices are called upon, and then explain how the particle spins are related to particle statistics and classification by virtue of the spin-statistics theorem. In the sixth section, we present a comprehensive discussion on the problem of gravitational deflection of light near a massive celestial body. We begin by considering the light deflection problem in the context of Newtonian gravity and derive the associated bending angle. We then study the geodesic equations for the motion of a photon subject to the Schwarzschild black hole metric and deduce Einstein's deflection angle, that exactly doubles that of Newton and was famously confirmed by Dyson, Eddington, and Davidson in 1919.

Thus, these sections may be clustered into four subgroups. The first subgroup consists of the first two sections and concerns with some topological concepts and constructions; the second subgroup is made of the subsequent section that focuses on conservation laws in relation to continuous symmetries in a system; the third subgroup is comprised of the next two sections and addresses issues around the eigenvalues of angular momentum operators and spins of particles; the fourth subgroup, which is the last section of this chapter and supplements Chapter 1 loosely and Chapter 10 tightly, is a study of the gravitational light deflection phenomenon. Each of these four subgroups of subjects may be of independent interest to some readers. As in the rest of the book, exercises appear at the end of the chapter.

The author thanks Sven Bjarke Gudnason, Luciano Medina, Wenxuan Tao, and Tigran Tchrakian for some constructive comments and suggestions, and Dan Taber of Oxford University Press for valuable editorial suggestions and advice, which helped improve the presentation and organization of the contents of the book.

Author
West Windsor, New Jersey

Notation and Convention

We use $\mathbb{N}$ to denote the set of all natural numbers,

$$\mathbb{N} = \{0, 1, 2, \dots\},$$

and $\mathbb{Z}$ the set of all integers,

$$\mathbb{Z} = \{\dots, -2, -1, 0, 1, 2, \dots\}.$$

We use $\mathbb{R}$ and $\mathbb{C}$ to denote the sets of real and complex numbers, respectively.

We use the roman type letter i to denote the imaginary unit $\sqrt{-1}$. For a complex number $c = a + ib$ where a and b are real numbers we use

$$\overline{c} = a - ib$$

to denote the complex conjugate of c. We use $\mathrm{Re}\{c\}$ and $\mathrm{Im}\{c\}$ to denote the real and imaginary parts of the complex number $c = a + ib$. That is,

$$\mathrm{Re}\{c\} = a, \quad \mathrm{Im}\{c\} = b.$$

The signature of an $(n+1)$-dimensional Minkowski spacetime is always $(+-\cdots-)$. The $(n+1)$-dimensional flat Minkowski spacetime is denoted by $\mathbb{R}^{n,1}$ and is equipped with the scalar product

$$xy = x^0y^0 - x^1y^1 - \cdots - x^ny^n,$$

where $x = (x^0, x^1, \dots, x^n)$ and $y = (y^0, y^1, \dots, y^n) \in \mathbb{R}^{n,1}$ are spacetime vectors.

Unless otherwise stated, we always use the Greek letters α, β, μ, ν to denote the spacetime indices,

$$\alpha, \beta, \mu, \nu = 0, 1, 2, \dots, n,$$

and the Latin letters i, j, k, l to denote the space indices,

$$i, j, k, l = 1, 2, \dots, n.$$

We use t to denote the variable in a polynomial or a function or the transpose operation on a vector or a matrix.

When an expression, say X or Y, is given, we use $X \equiv Y$ to denote that Y, or X, is defined to be X, or Y, respectively.

Occasionally, we use the symbol $\forall$ to express "for all", and $\exists$, to express "there exists".

We use $[,]$ to denote the commutators operated on suitable "quantities" so that

$$[A, B] = AB - BA, \quad A_{[a}B_{b]} = A_aB_b - A_bB_a,$$

and so on.

We observe the summation convention over repeated indices unless otherwise stated. For example,

$$A_iB_i = \sum_{i=1}^{n} A_iB_i, \quad A_iB^i = \sum_{i=1}^{n} A_iB^i,$$

$$E_{ij}F^{ij} = \sum_{i,j=1}^{n} E_{ij}F^{ij}, \quad E_{ij}^2 = \sum_{i,j=1}^{n} E_{ij}^2.$$

The roman type letter e is reserved to denote the Euler number or the base of the natural logarithmic system and the italic type letter e to denote an irrelevant physical coupling constant such as the charge of the positron ($-e$ will then be the charge of the electron). The roman type letter d denotes the differential and the italic type letter d denotes a "quantity", in mathematical display mode.

The references in bibliography and their citations in text follow alphabetic orders by the last names of the authors.

Although the Greek letters μ, ν, etc., denote the indices of the spacetime coordinates, occasionally, they are also used to denote the Radon measures or some parameters in other contexts, when there is no risk of confusion but there is a need to be consistent with literature. Furthermore, sometimes ν is used to denote the outnormal to the boundary of a bounded domain.

The unit sphere centered at the origin, in $\mathbb{R}^n$ ($n = 2, 3, \dots$), is denoted by S^{n-1}.

The area element of a surface such as the boundary of a spatial domain is often denoted by $\mathrm{d}\sigma$. However, the Lebesgue measure of a domain for integration is sometimes omitted to save space when there is no risk of confusion.

Let S be a set of finitely many points. We use $|S|$ or $\#S$ to denote the number of points in S.

Let S be a subset of the set T in a certain space. We use $T \setminus S$ to denote the complement of S in T, or simply S^c when T is the full space.

We use the roman type abbreviation supp to denote the support of a function.

The letter C will be used to denote a positive constant which may assume different values at different places.

For a complex matrix A, we use $A^\dagger$ to denote its Hermitian conjugate, which consists of a matrix transposition and a complex conjugation.

The symbol $W^{k,p}$ denotes the Sobolev space of functions whose distributional derivatives up to the kth order are all in the space L^p.

By convention, various matrix Lie algebras are denoted by lowercase letters. For example, the Lie algebras of the Lie groups $SO(N)$ and $SU(N)$ are denoted by $so(N)$ and $su(N)$, respectively.

The notation for various derivatives is as follows,

$$\partial_\mu = \frac{\partial}{\partial x^\mu}, \quad \partial_\pm = \partial_1 \pm \mathrm{i}\partial_2, \quad \partial = \frac{1}{2}(\partial_1 - \mathrm{i}\partial_2), \quad \overline{\partial} = \frac{1}{2}(\partial_1 + \mathrm{i}\partial_2).$$

Besides, with the complex variable $z = x^1 + \mathrm{i}x^2$, we always understand that $\partial_z = \frac{\partial}{\partial z} = \partial, \partial_{\overline{z}} = \frac{\partial}{\partial \overline{z}} = \overline{\partial}$. Thus, for any function f that only has partial derivatives with respect to x^1 and x^2, the quantities $\partial_z f = \frac{\partial f}{\partial z}$ and $\partial_{\overline{z}} = \frac{\partial f}{\partial \overline{z}}$ are well defined.

Vectors and tensors are often simply denoted by their general components, respectively, following physics literature. For example, it is understood that

$$A_\mu \equiv (A_\mu) = (A_0, A_1, A_2, A_3), \quad g_{\mu\nu} \equiv (g_{\mu\nu}).$$

In a volume of this scope, it is inevitable to have a letter to carry different but standard meanings in different contexts, although such a multiple usage of letters has been kept to a minimum. Here are some examples. The Greek letter ν usually denotes a spacetime coordinate index but also stands for a unit normal vector to a surface; δ may stand for a small positive number, variation of a functional, or the Dirac distribution, and δ_{ij} is the Kronecker symbol; g may stand for a coupling constant such as a magnetic charge, a metric tensor or its determinant, or a function; x may denote the coordinate of a point in the real axis or a point in space or spacetime; P may denote a magnetic charge or a momentum vector; G usually stands for Newton's universal gravitational constant but may also denotes a Lie group or a function; ρ usually stands for a charge, mass, probability, or energy density, but may also denotes a radial variable or radial coordinate under consideration; the lower case letter c usually denotes the speed of light in vacuum but also occasionally a constant that should be made clear in the context.

For convenience, we sometimes use $\mathbf{x}$ to denote a point in $\mathbb{R}^3$ or $\mathbb{R}^n$ in general. We use $\nabla \times \mathbf{F}$ and $\mathrm{curl}\,\mathbf{F}$, and, $\nabla \cdot \mathbf{F}$ and $\mathrm{div}\,\mathbf{F}$, interchangeably for the curl and divergence operations, respectively, on a vector field $\mathbf{F}$ over $\mathbb{R}^3$. For an $\mathbb{R}^n$-valued vector, say $\mathbf{A}$, we use $\|\mathbf{A}\|$ and $|\mathbf{A}|$ alternatively to denote the length or norm of $\mathbf{A}$ with respect to the Euclidean scalar product $\cdot$ of $\mathbb{R}^n$ such that $\mathbf{A} \cdot \mathbf{A} = |\mathbf{A}|^2$ is also rewritten as $\mathbf{A}^2$ concisely.

For a positive quantity or variable, say, r, we use $r \ll 1$ or $r \gg 1$ to denote the assumption that r is sufficiently small or large, respectively.

We use the overdot $\dot{}$ to denote differentiation with respect to a "time variable", t, and the prime $'$ differentiation with respect to some other variable, which should be self-evident from the context. Alternatively, we also use $'$ to denote a quantity that is a variation of an original one following a specific rule or understanding.

All displayed mathematical expressions are numbered regardless whether they are referred to in the text for the sake of convenience of the reader in case any need of their reference is called on while using the book.

In some chapters, the first sections also serve to briefly survey the subjects to be covered in the subsequent sections.

1

Hamiltonian systems and applications

This chapter first introduces the Hamiltonian or Lagrangian formalism of classical mechanics, which is the conceptual foundation of all later developments. As illustrations of applications, it then presents a few important many-body problems. Among these, it first discusses the classical N-body problem in $\mathbb{R}^3$ and next considers Kepler's laws of planetary motion as an important application of the formalism and derive Newton's law of gravitation as a by-product. It then presents the Helmholtz–Kirchhoff point vortex problem, which may be regarded as an N-body problem in $\mathbb{R}^2$. The chapter ends with a study of an N-body problem in $\mathbb{R}$ modeling an over-simplified DNA system. In order to understand a thermodynamical phenomenon of the system known as DNA denaturation, it takes this opportunity to make a short introduction to some basic concepts of statistical mechanics.

1.1 Motion of massive particle

The Hamiltonian or Lagrangian formalism of classical mechanics lays the foundation of classical and quantum field theories and grows out of Newtonian mechanics describing the interaction of point masses. In this section, we formulate the Hamilton–Lagrange mechanics from Newton's law of motion.

Equations of motion of Newton

Consider the motion of a point particle of mass m and coordinates $(q^i) = q$ in a potential field $V(q,t)$ described by Newtonian mechanics in the n-dimensional

Mathematical Physics with Differential Equations. Yisong Yang, Oxford University Press.
 DOI: 10.1093/oso/9780192872616.003.0001

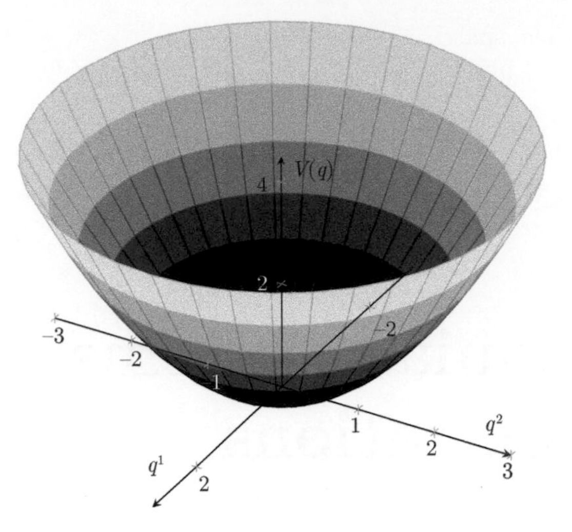

Figure 1.1 An illustrative example of a potential well of that of a two-dimensional harmonic oscillator defined by the quadratic potential function $V(q) = \frac{1}{2}([q^1]^2 + [q^2]^2)$ for which the equations of motion are seen to "drive" the particle to the equilibrium state given by $q^1 = q^2 = 0$, which minimizes the potential energy.

Euclidean space $\mathbb{R}^n$. The equations of motion are

$$m\ddot{q}^i = -\frac{\partial V}{\partial q^i}, \quad i = 1, 2, \dots, n, \tag{1.1.1}$$

where the (double) overdot $(\ddot{\ })\ \dot{}$ denotes (second) first-order time derivative. Since

$$-\nabla V = -\nabla_q V = -\left(\frac{\partial V}{\partial q^i}\right) \tag{1.1.2}$$

defines the direction along which the potential energy V decreases most rapidly, the equation (1.1.1) says that the particle is accelerated along the direction of the flow of the *steepest descent* of V. Figure 1.1 shows a typical profile of the potential energy function V in the form of a "potential well."

Lagrangian formalism

With the *Lagrangian function*

$$L(q, \dot{q}, t) = \frac{1}{2} m \sum_{i=1}^{n} (\dot{q}^i)^2 - V(q, t), \tag{1.1.3}$$

which is simply the difference of the kinetic and potential energies of the moving particle, the equations in (1.1.1) are the *Euler–Lagrange equations* of the action

$$\int_{t_1}^{t_2} L(q(t), \dot{q}(t), t)\, \mathrm{d}t, \tag{1.1.4}$$

over the admissible space of trajectories $\{q(t)\,|\,t_1 < t < t_2\}$ starting and terminating at fixed points at $t = t_1$ and $t = t_2$, respectively.

Hamiltonian formalism

The *Hamiltonian function* or *energy* at any time t is the sum of kinetic and potential energies given by

$$H = \frac{1}{2}m\sum_{i=1}^{n}(\dot{q}^i)^2 + V(q,t) = m\sum_{i=1}^{n}(\dot{q}^i)^2 - L. \tag{1.1.5}$$

Introduce the *momentum vector* $p = (p_i)$,

$$p_i = m\dot{q}^i = \frac{\partial L}{\partial \dot{q}^i}, \quad i = 1, 2, \ldots, n. \tag{1.1.6}$$

Then, in view of (1.1.5), H is defined from L, through a *Legendre transformation*, by

$$H(q,p,t) = \sum_{i=1}^{n} p_i\dot{q}^i - L(q,\dot{q},t), \tag{1.1.7}$$

with $p = (p_i)$, and the equations of motion, (1.1.1), are a *Hamiltonian system*,

$$\dot{q}^i = \frac{\partial H}{\partial p_i}, \quad \dot{p}_i = -\frac{\partial H}{\partial q^i}, \quad i = 1, 2, \ldots, n. \tag{1.1.8}$$

In a compressed fashion, the system (1.1.8) reads

$$\frac{\mathrm{d}}{\mathrm{d}t}\begin{pmatrix} q \\ p \end{pmatrix} = \begin{pmatrix} 0 & I_n \\ -I_n & 0 \end{pmatrix}\begin{pmatrix} \nabla_q H \\ \nabla_p H \end{pmatrix} = J\nabla_{q,p}H, \tag{1.1.9}$$

where I_n denotes the $n \times n$ identity matrix and J is a *symplectic matrix* satisfying $J^2 = -I_{2n}$.

General formalism

For general applications, it is important to consider when the Lagrangian function L is an arbitrary function of $q, \dot{q}$, and t. The equations of motion are the Euler–Lagrange equations of (1.1.4),

$$\frac{\mathrm{d}}{\mathrm{d}t}\left(\frac{\partial L}{\partial \dot{q}^i}\right) = \frac{\partial L}{\partial q^i}, \quad i = 1, 2, \ldots, n. \tag{1.1.10}$$

To make a similar Hamiltonian formulation, we are motivated from (1.1.6) to introduce the *generalized momentum vector* $p = (p_i)$ by setting

$$p_i = \frac{\partial L}{\partial \dot{q}^i}, \quad i = 1, 2, \ldots, n. \tag{1.1.11}$$

We still use the Legendre transformation (1.1.7) to define the corresponding Hamiltonian function H. A direct calculation shows that the system (1.1.10) is now equivalent to the Hamiltonian system (1.1.8) in the present general framework.

We note that an important property of a Hamiltonian function is that it is independent of the variables $\dot{q}^i$ $(i = 1, 2, \ldots, n)$. In fact, from the definition of the generalized momentum vector given by (1.1.11), we have

$$\frac{\partial H}{\partial \dot{q}^i} = p_i - \frac{\partial L}{\partial \dot{q}^i} = 0, \quad i = 1, 2, \ldots, n. \tag{1.1.12}$$

This fact justifies our notation of $H(q, p, t)$ in (1.1.7) instead of $H(q, p, \dot{q}, t)$.

Evolution equation

Let F be a dynamical quantity that is an arbitrary function depending on q^i, p_i $(i = 1, 2, \ldots, n)$ and time t. We see that F varies its value along a trajectory of the equations of motion, (1.1.8), according to

$$\begin{aligned} \frac{\mathrm{d}F}{\mathrm{d}t} &= \frac{\partial F}{\partial t} + \frac{\partial F}{\partial q^i}\dot{q}^i + \frac{\partial F}{\partial p_i}\dot{p}_i \\ &= \frac{\partial F}{\partial t} + \frac{\partial F}{\partial q^i}\frac{\partial H}{\partial p_i} - \frac{\partial F}{\partial p_i}\frac{\partial H}{\partial q^i}, \end{aligned} \tag{1.1.13}$$

where, and in the sequel, we observe the *summation convention* over repeated indices, although occasionally we also spell out the summation explicitly. Thus, we are motivated to use the *Poisson bracket* $\{\cdot, \cdot\}$ defined by

$$\begin{aligned} \{f, g\} &= \frac{\partial f}{\partial q^i}\frac{\partial g}{\partial p_i} - \frac{\partial f}{\partial p_i}\frac{\partial g}{\partial q^i} \\ &= (\nabla_q f, \nabla_p f)\begin{pmatrix} 0 & I_n \\ -I_n & 0 \end{pmatrix}\begin{pmatrix} \nabla_q g \\ \nabla_p g \end{pmatrix}, \end{aligned} \tag{1.1.14}$$

to rewrite the rate of change of F with respect to time t as

$$\frac{\mathrm{d}F}{\mathrm{d}t} = \frac{\partial F}{\partial t} + \{F, H\}. \tag{1.1.15}$$

In particular, when the Hamiltonian H does not depend on time t explicitly, $H = H(q, p)$, then (1.1.15) implies that

$$\frac{\mathrm{d}H}{\mathrm{d}t} = 0, \tag{1.1.16}$$

which gives the fact that energy is conserved and the mechanical system is thus called *conservative*.

Use of complexified coordinates

It will be useful to "*complexify*" our formulation of classical mechanics. For this purpose we introduce the complex variables

$$u_i = \frac{1}{\sqrt{2}}(q^i + \mathrm{i}p_i), \quad \overline{u}_i = \frac{1}{\sqrt{2}}(q^i - \mathrm{i}p_i), \quad i = 1, 2, \ldots, n, \quad \mathrm{i} = \sqrt{-1}. \tag{1.1.17}$$

Here the normalization factor $\frac{1}{\sqrt{2}}$ is introduced in order to make the transformation $(q^i, p_i) \to (u_i, \overline{u}_i)$ *isometric* or *unitary*, $|u_i|^2 + |\overline{u}_i|^2 = |q^i|^2 + |p_i|^2$, in the domain of complex quantities. Then the Hamiltonian function H depends only on $u = (u_i)$ and $\overline{u} = (\overline{u}_i)$,

$$H = H(u, \overline{u}, t). \tag{1.1.18}$$

Hence, in terms of differential operators, there hold

$$\frac{\partial}{\partial u_i} = \frac{1}{\sqrt{2}}\left(\frac{\partial}{\partial q^i} - \mathrm{i}\frac{\partial}{\partial p_i}\right), \quad \frac{\partial}{\partial \overline{u}_i} = \frac{1}{\sqrt{2}}\left(\frac{\partial}{\partial q^i} + \mathrm{i}\frac{\partial}{\partial p_i}\right), \tag{1.1.19}$$

and the Hamiltonian system (1.1.8) takes the concise form

$$\mathrm{i}\dot{u}_i = \frac{\partial H}{\partial \overline{u}_i}, \quad i = 1, 2, \ldots, n. \tag{1.1.20}$$

Again, let F be a function depending on $u, \overline{u}$, and t. Then (1.1.20) gives us

$$\begin{aligned} \frac{\mathrm{d}F}{\mathrm{d}t} &= \frac{\partial F}{\partial t} + \frac{\partial F}{\partial u_i}\dot{u}_i + \frac{\partial F}{\partial \overline{u}_i}\dot{\overline{u}}_i \\ &= \frac{\partial F}{\partial t} - \mathrm{i}\frac{\partial F}{\partial u_i}\frac{\partial H}{\partial \overline{u}_i} + \mathrm{i}\frac{\partial F}{\partial \overline{u}_i}\frac{\partial H}{\partial u_i}. \end{aligned} \tag{1.1.21}$$

Thus, with the notation

$$\{f, g\} = \frac{\partial f}{\partial u_i}\frac{\partial g}{\partial \overline{u}_i} - \frac{\partial f}{\partial \overline{u}_i}\frac{\partial g}{\partial u_i} \tag{1.1.22}$$

for the Poisson bracket, we have

$$\frac{\mathrm{d}F}{\mathrm{d}t} = \frac{\partial F}{\partial t} + \frac{1}{\mathrm{i}}\{F, H\}. \tag{1.1.23}$$

In particular, the complexified Hamiltonian system (1.1.20) becomes

$$\dot{u}_i = \frac{1}{\mathrm{i}}\{u_i, H\}, \quad i = 1, 2, \ldots, n, \tag{1.1.24}$$

which closely resembles the *Heisenberg equation*, in the *Heisenberg representation* of quantum mechanics, which we discuss later.

1.2 Many-body problem

The *many-body*, or more precisely, *N-body problem* stems from *celestial mechanics*, which treats celestial bodies as point particles interacting through Newton's law of gravitation.

We start from considering the gravitational force between a point mass M fixed at the origin of $\mathbb{R}^3$ and another point mass m at $\mathbf{x} \in \mathbb{R}^3 \setminus \{\mathbf{0}\}$. The potential field that induces the gravitational force is

$$U(\mathbf{x}) = -G\frac{Mm}{|\mathbf{x}|}, \quad \mathbf{x} \in \mathbb{R}^3 \setminus \{\mathbf{0}\}, \tag{1.2.1}$$

where $G > 0$ is Newton's universal gravitational constant, so that it exerts the force

$$-\nabla_{\mathbf{x}} U = -G\frac{Mm}{|\mathbf{x}|^3}\mathbf{x} \tag{1.2.2}$$

to the point mass m at $\mathbf{x}$, leading to the equation of motion,

$$m\ddot{\mathbf{x}} = -G\frac{Mm}{|\mathbf{x}|^3}\mathbf{x}. \tag{1.2.3}$$

Equations of motion of N-body problem

Now consider N point particles, each of mass m_i, located at $\mathbf{x}_i \in \mathbb{R}^3$, $i = 1, \dots, N$. Then the equations governing the motion of these masses are

$$m_i\ddot{\mathbf{x}}_i = -G\sum_{j\neq i}^{N} \frac{m_i m_j(\mathbf{x}_i - \mathbf{x}_j)}{|\mathbf{x}_i - \mathbf{x}_j|^3} = -\nabla_{\mathbf{x}_i} U, \quad i = 1, \dots, N, \tag{1.2.4}$$

where

$$U(\mathbf{x}_1, \dots, \mathbf{x}_n) = -G \sum_{1\leq i<j\leq N}^{N} \frac{m_i m_j}{|\mathbf{x}_i - \mathbf{x}_j|}, \tag{1.2.5}$$

is the total gravitational potential of the N-particle system. In particular, the motion follows the principle that the particles are accelerated along the directions of the fastest descendants that would lower the gravitational potential.

Hamiltonian system

In order to recast the system into a Hamiltonian system, we relabel the coordinate variables and masses according to

$$\begin{aligned} (\mathbf{x}_1, \dots, \mathbf{x}_N) &\mapsto (q^1, q^2, \dots, q^{3N}), \\ (m_1, m_1, m_1, \dots, m_N, m_N, m_N) &\mapsto (m_1, m_2, \dots, m_{3N}), \end{aligned} \tag{1.2.6}$$

which allows us to introduce the momentum variables

$$p_i = m_i \dot{q}^i, \quad i = 1, 2, \dots, 3N. \tag{1.2.7}$$

It can be seen that the equations of motion now take the form of a Hamiltonian system

$$\dot{q}^i = \frac{\partial H}{\partial p_i}, \quad \dot{p}_i = -\frac{\partial H}{\partial q^i}, \quad i = 1, 2, \ldots, 3N, \tag{1.2.8}$$

where the Hamiltonian function H is defined by

$$H = \sum_{i=1}^{3N} \frac{p_i^2}{2m_i} + U = \sum_{i=1}^{3N} \frac{1}{2}\dot{q}^i p_i + U. \tag{1.2.9}$$

Conserved quantities

As a general discussion, we begin with considering the first-order differential equations

$$\dot{x}^i = f^i(x), \quad x = (x^i) \in \mathbb{R}^n, \quad i = 1, \ldots, n, \tag{1.2.10}$$

where the functions $f^1(x), \ldots, f^n(x)$ do not depend on time t explicitly such that the equations are referred to as autonomous. A *first integration* of (1.2.10) is a function depending on $x^1, \ldots, x^n$, say $F(x)$, which is constant along any solution of (1.2.10). That is,

$$\frac{\mathrm{d}F(x(t))}{\mathrm{d}t} = \frac{\partial F}{\partial x^i}\dot{x}^i = f^i \partial_i F = 0, \tag{1.2.11}$$

where $x = x(t) = (x^i(t))$ is a solution to (1.2.10). In other words, a first integral is a conserved quantity of time t with respect to the equations of motion. On the other hand, since any solution to (1.2.10) is considered as a curve in $\mathbb{R}^n$ which may in turn be interpreted as the intersection of $n-1$ hypersurfaces, thus the general solution of (1.2.10) may assume the form

$$F_1(x) = C_1, \quad \ldots, \quad F_{n-1}(x) = C_{n-1}, \tag{1.2.12}$$

where $F_1, \ldots, F_n$ are *functionally independent* first integrals of (1.2.10), satisfying the condition that the Jacobian matrix

$$J(F, x) = (\partial_i F_j) \tag{1.2.13}$$

is of full rank. That is, $J(F, x)$ is of rank $n-1$. In this situation, we say that the autonomous system (1.2.10) is *integrable* or *completely integrable* or has a *complete integration.* Therefore, a system is integrable if and only if it possesses the maximum possible number of independent conserved quantities. In particular, an autonomous Hamiltonian system is integrable if and only if it possesses the maximum number of functionally independent quantities, each of which is commutative with respect to the underlying Hamiltonian function of the system and the induced Poisson bracket.

Since the system (1.2.8) consists of $6N$ first-order autonomous equations, its complete integration (solution) requires obtaining $6N-1$ independent integrals. By exploring the mechanical properties of the system, we have the following immediate integrals (or conserved quantities), namely, the center of masses $\mathbf{x}_0$ determined by

$$\mathbf{x}_0 \sum_{i=1}^{N} m_i = \sum_{i=1}^{N} m_i \mathbf{x}_i, \tag{1.2.14}$$

which moves at a constant velocity; the total (linear) momentum $\mathbf{L}_0$ given by

$$\mathbf{L}_0 = \sum_{i=1}^{N} m_i \dot{\mathbf{x}}_i = \sum_{i=1}^{N} \mathbf{p}_i; \tag{1.2.15}$$

the total angular momentum $\mathbf{a}_0$ expressed as

$$\mathbf{a}_0 = \sum_{i=1}^{N} m_i \mathbf{x}_i \times \dot{\mathbf{x}}_i = \sum_{i=1}^{N} \mathbf{x}_i \times \mathbf{p}_i; \tag{1.2.16}$$

and the conserved total energy H stated in (1.2.9). Thus, we have a total of ten obvious first integrals. This number count indicates that the N-body problem quickly becomes highly nontrivial when N increases. Indeed, the $N = 3$ situation is already notoriously hard, since its integration requires a total of $6 \cdot 3 - 1 = 17$ independent integrals, and is known as the three-body problem. In general, it is believed that the N-body problem is not integrable. The best understood situation is the two-body problem [394, 453], also known as *Kepler's problem* [20]. Here we only mention that the two-body problem is completely integrable for the following reasons:

(i) Using the center of mass coordinate frame it can be shown that the two-body problem is actually planar.

(ii) Notice that, as a planar problem, an N-body Hamiltonian system consists of $4N$ first-order equations.

(iii) The complete integration of an autonomous Hamiltonian system of $4N$ equations requires $4N - 1$ independent integrals.

(iv) The same collection of mechanical quantities give us $2 + 2 + 2 + 1 = 7$ independent integrals.

(v) When $N = 2$ these 7 integrals render the required number of independent integrals for a complete solution of the problem.

When the masses are replaced by charges so that Newton's gravitation is placed by Coulomb's law of electrostatics, we can study the N-body problem of charged particles. The quantum mechanical version of this is called the quantum N-body problem [303], which has important applications in the theory of atoms and molecules and is of contemporary research interest [357]. We consider this problem in Chapter 2.

1.3 Kepler's laws of planetary motion

As an important example of applications of Hamiltonian systems, this section presents a thorough study of Kepler's laws, which describe the motion of a planet around the sun.

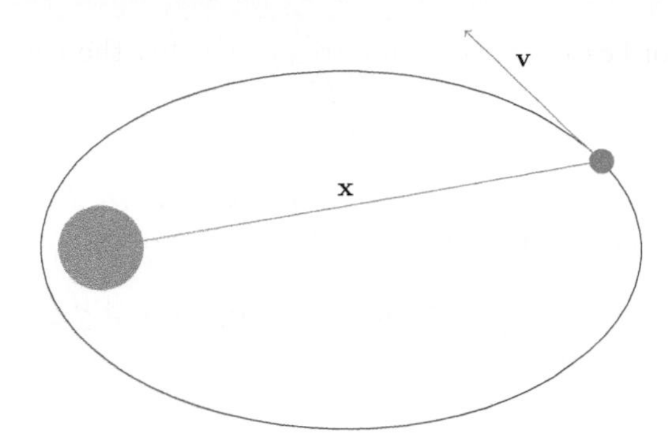

Figure 1.2 The planar motion of a planet around the sun. The orbit is an ellipse with the sun sitting at one of the foci of the ellipse.

Based on his study of then-available observed astronomical data for the motion of planets around the sun, Kepler published between 1609 and 1619 three fundamental laws, known as *Kepler's laws of planetary motion*, which may be stated as follows.

(i) The motion of a planet about the sun is planar and the orbit is an ellipse with the sun sitting at one of the foci of the ellipse. This is *Kepler's first law.*

(ii) The line segment connecting the sun and planet sweeps out equal areas during equal time lapses. This is *Kepler's second law.*

(iii) The square of the time period of the orbit of ellipse is proportional to the cube of the length of the semi-major axis of the ellipse and the proportionality constant is independent of the mass of the planet. This is *Kepler's third law.*

Figure 1.2 illustrates such planetary motion with the sun resting at a focus of the elliptical trajectory orbited by a planet with the position vector $\mathbf{x}$, measured from the focus, and the velocity vector $\mathbf{v} = \dot{\mathbf{x}}$.

Note that, strictly speaking, this is not a two-body problem since the sun is *fixed.*

Polar-variable representation of ellipse

In Cartesian coordinates x, y, the ellipse of semi-major axis $a > 0$ and semi-minor axis $b > 0$ with $a \geq b$ and foci at $(0, 0)$ and $(2c, 0)$ where

$$c = \sqrt{a^2 - b^2}, \tag{1.3.1}$$

centered at $(c, 0)$, is given by the equation

$$\frac{(x-c)^2}{a^2} + \frac{y^2}{b^2} = 1, \tag{1.3.2}$$

such that the quantity

$$e = \frac{c}{a} = \sqrt{1 - \frac{b^2}{a^2}}, \tag{1.3.3}$$

is referred to as the eccentricity of the ellipse. Of course, $0 \leq e < 1$. Thus, in terms of the polar coordinates r, θ, or

$$x = r\cos\theta, \quad y = r\sin\theta, \quad r > 0, \quad 0 \leq \theta \leq 2\pi, \tag{1.3.4}$$

we can recast (1.3.2) into

$$r^2 = \left([1 - e^2]a + er\cos\theta\right)^2. \tag{1.3.5}$$

Therefore we may resolve (1.3.5) to arrive at the polar-variable form of the ellipse (1.3.2):

$$r = r(\theta) = \frac{a(1 - e^2)}{1 - e\cos\theta}. \tag{1.3.6}$$

More generally, if the major axis of the ellipse is tilted up from the x-axis at an angle θ_0 about the origin, then (1.3.6) assumes the modified form

$$r = r(\theta) = \frac{a(1 - e^2)}{1 - e\cos(\theta - \theta_0)}. \tag{1.3.7}$$

Planar motion

Let $\mathbf{x}$ denote the position vector of a planet of mass m under consideration which is attracted toward the sun of mass M resting at the origin. Thus the second law of Newton leads to the equation of motion

$$m\mathbf{a} = -F(r)\mathbf{u}, \quad \mathbf{a} = \dot{\mathbf{v}}, \quad \mathbf{v} = \dot{\mathbf{x}}, \quad \mathbf{u} = \frac{\mathbf{x}}{r}, \quad r = |\mathbf{x}|, \quad \mathbf{x} \neq \mathbf{0}. \tag{1.3.8}$$

An immediate consequence of (1.3.8) is that the vector

$$\mathbf{w} = \mathbf{x} \times \mathbf{v} \tag{1.3.9}$$

is a constant vector. If $\mathbf{w} = \mathbf{0}$, then the planet moves along the radial direction, away or toward the sun. Eventually, a collision will occur, and the motion is not sustainable. Such a situation should be excluded. Thus we may now assume $\mathbf{w} \neq \mathbf{0}$. In this case, we conclude that the motion is confined to a plane that is perpendicular to $\mathbf{w}$.

Proof of Kepler's first law

To determine the orbit of the motion of the planet, we need to integrate the second-order differential equation (1.3.8). We have seen that (1.3.9) is an integral.

So we are to obtain another one. For this purpose, we form the vector $\mathbf{v} \times \mathbf{w}$ and investigate how this vector evolves with time. Hence, with $\mathbf{v} = \dot{r}\mathbf{u} + r\dot{\mathbf{u}}$, we have

$$\begin{aligned}\frac{\mathrm{d}}{\mathrm{d}t}(\mathbf{v} \times \mathbf{w}) &= \mathbf{a} \times \mathbf{w} \\ &= -\frac{F(r)}{m}\mathbf{u} \times (r\mathbf{u} \times [\dot{r}\mathbf{u} + r\dot{\mathbf{u}}]) \\ &= -\frac{F(r)r^2}{m}\mathbf{u} \times (\mathbf{u} \times \dot{\mathbf{u}}).\end{aligned} \tag{1.3.10}$$

Now recall the vector cross-product identity

$$\mathbf{A} \times (\mathbf{B} \times \mathbf{C}) = (\mathbf{A} \cdot \mathbf{C})\mathbf{B} - (\mathbf{A} \cdot \mathbf{B})\mathbf{C}. \tag{1.3.11}$$

Using (1.3.11) in (1.3.10) and applying the properties $\mathbf{u} \cdot \mathbf{u} = 1$ and $\mathbf{u} \cdot \dot{\mathbf{u}} = 0$, we obtain

$$\frac{\mathrm{d}}{\mathrm{d}t}(\mathbf{v} \times \mathbf{w}) = \frac{F(r)r^2}{m}\dot{\mathbf{u}}. \tag{1.3.12}$$

The left-hand side of (1.3.12) is a total derivative of t. Therefore, in order to render a total derivative for the right-hand side of (1.3.12) to maintain consistency with its left-hand side, it suffices to take

$$F(r)r^2 = K = \text{constant}. \tag{1.3.13}$$

In other words, we can draw the conclusion that the function $F(r)$ in (1.3.8) may be taken to follow an inverse-square law,

$$F(r) = \frac{K}{r^2}. \tag{1.3.14}$$

Moreover, consistency in (1.3.12) indicates that K should contain m as a factor such that the right-hand side of (1.3.12) is independent of m as its left-hand side. Besides, by reciprocal symmetry, we then conclude that K should contain M as a factor as well. Consequently we have

$$K = GmM, \tag{1.3.15}$$

where $G > 0$ is a proportionality constant independent of m and M which is in fact Newton's universal gravitational constant.

In view of (1.3.8), (1.3.14), and (1.3.15), we have somehow derived Newton's law for gravitation. However, this derivation is heuristic or plausible and its justification is yet to be made through a full examination of Kepler's laws, discussed next.

We now proceed to prove Kepler's first law.

Inserting (1.3.14) and (1.3.15) into (1.3.12), we have

$$\frac{\mathrm{d}}{\mathrm{d}t}(\mathbf{v} \times \mathbf{w} - GM\mathbf{u}) = \mathbf{0}. \tag{1.3.16}$$

Hence

$$\mathbf{v} \times \mathbf{w} - GM\mathbf{u} = \text{constant} = -\mathbf{A} \text{ (say)}, \tag{1.3.17}$$

which is a second integral of the equation (1.3.8) as desired, in addition to the first integral (1.3.9). Thus the equation (1.3.8) is integrated and it remains to see what its integration or solution looks like.

In fact, from (1.3.9) and (1.3.17), we get

$$\begin{aligned} w^2 \equiv |\mathbf{w}|^2 &= (\mathbf{x} \times \mathbf{v}) \cdot \mathbf{w} = \mathbf{x} \cdot (\mathbf{v} \times \mathbf{w}) \\ &= r\mathbf{u} \cdot (GM\mathbf{u} - \mathbf{A}) \\ &= GMr - r\mathbf{A} \cdot \mathbf{u}. \end{aligned} \tag{1.3.18}$$

Note that, since $\mathbf{v} \times \mathbf{w}$ lies in the plane of the motion of the planet, so does the constant vector $\mathbf{A}$ in view of (1.3.17). Besides, in view of the relation (1.3.18) and the condition $\mathbf{w} \neq \mathbf{0}$, we have

$$\mathbf{A} \cdot \mathbf{u} < GM. \tag{1.3.19}$$

Thus, if the orbit of the planet is a closed curve so that the unit vector $\mathbf{u}$ may assume all possible directions in the plane of the motion, then (1.3.19) is equivalent to

$$|\mathbf{A}| < GM. \tag{1.3.20}$$

If (1.3.20) is violated, the orbit of the planet will not be a closed curve. This situation is not our interest here.

Assume the condition (1.3.20). There are two cases to consider.

(i) $\mathbf{A} = \mathbf{0}$. Then (1.3.18) gives us

$$r = \frac{w^2}{GM}. \tag{1.3.21}$$

Thus the orbit is a circle. This is clearly an exceptional case.

(ii) $\mathbf{A} \neq \mathbf{0}$. Let θ be the angle between $\mathbf{A}$ and $\mathbf{u}$. Then (1.3.18) leads to

$$r = \frac{w^2}{GM - |\mathbf{A}| \cos\theta} \equiv r(\theta). \tag{1.3.22}$$

Comparing (1.3.22) with (1.3.6), we see that the orbit of the planet is an ellipse with eccentricity and semi-major axis given to be

$$e = \frac{|\mathbf{A}|}{GM}, \quad a = \frac{w^2}{GM\left(1 - \left[\frac{|\mathbf{A}|}{GM}\right]^2\right)}, \tag{1.3.23}$$

respectively, and a focus at $r = 0$. This is clearly a generic case.

Thus we have established Kepler's first law.

From (1.3.22), we see that the distance r is maximized at $\theta = 0$, indicating that the planet is at *aphelion*, and minimized at $\theta = \pi$, at *perihelion*, with the values

$$r_a = r(0) = \frac{w^2}{GM - |\mathbf{A}|}, \quad r_p = r(\pi) = \frac{w^2}{GM + |\mathbf{A}|}, \tag{1.3.24}$$

respectively. As a consequence of (1.3.23) and (1.3.24), we get

$$e = \frac{r_a - r_p}{r_a + r_p}. \tag{1.3.25}$$

For the earth, this quantity is about 0.016710218.

Proof of Kepler's second law

Use (r, θ) to denote the polar variables given in (1.3.22) which describe the elliptical motion of a planet about the sun. The area swept out by the line segment connecting the sun and the planet over the span of the angle θ between $\theta = 0$ (say) and $\theta > 0$ is given by the integral

$$\mathcal{A}(\theta) = \int_0^\theta \frac{1}{2} r^2(\varphi)\,\mathrm{d}\varphi. \tag{1.3.26}$$

Hence, we have

$$\frac{\mathrm{d}\mathcal{A}(\theta)}{\mathrm{d}t} = \frac{1}{2} r^2(\theta)\dot{\theta}. \tag{1.3.27}$$

On the other hand, let $\mathbf{e}_1$ and $\mathbf{e}_2$ be two fixed orthonormal vectors in the plane of the orbit of the planet and write the radial unit vector $\mathbf{u}$ as $\mathbf{u} = \cos\theta\,\mathbf{e}_1 + \sin\theta\,\mathbf{e}_2$. Then we have

$$\begin{aligned} \mathbf{w} = \mathbf{x} \times \mathbf{v} = r\mathbf{u} \times \dot{\mathbf{x}} &= r\mathbf{u} \times (\dot{r}\mathbf{u} + r\dot{\mathbf{u}}) = r^2\mathbf{u} \times \dot{\mathbf{u}} \\ &= (r^2\dot{\theta})\,(\cos\theta\,\mathbf{e}_1 + \sin\theta\,\mathbf{e}_2) \times (-\sin\theta\,\mathbf{e}_1 + \cos\theta\,\mathbf{e}_2) \\ &= (r^2\dot{\theta})(\mathbf{e}_1 \times \mathbf{e}_2), \end{aligned} \tag{1.3.28}$$

resulting in $w^2 = r^4(\theta)\,\dot{\theta}^2$. We may assume that θ increases with respect to time t. Thus we have

$$\dot{\theta} = \frac{w}{r^2(\theta)}. \tag{1.3.29}$$

Inserting (1.3.29) into (1.3.27), we arrive at

$$\frac{\mathrm{d}\mathcal{A}(\theta)}{\mathrm{d}t} = \frac{w}{2} = \text{constant}, \tag{1.3.30}$$

which establishes Kepler's second law.

A by-product of (1.3.29) is that it gives a description how the angular velocity of the planet depends on its distance from or location with respect to the sun. Substituting (1.3.22) into (1.3.29), we have

$$\dot{\theta} = \frac{(GM - |\mathbf{A}|\cos\theta)^2}{w^3}. \tag{1.3.31}$$

Furthermore, using $\mathbf{u}\cdot\dot{\mathbf{u}}=0$ in $\mathbf{v}=\dot{r}\mathbf{u}+r\dot{\mathbf{u}}$, we obtain the linear speed of the motion of the planet to be

$$\begin{aligned} v=|\mathbf{v}| &= \sqrt{\dot{r}^2+r^2\dot{\theta}^2}\\ &= \frac{1}{w}\sqrt{(|\mathbf{A}|-GM)^2+2GM|\mathbf{A}|(1-\cos\theta)}, \end{aligned} \tag{1.3.32}$$

by virtue of (1.3.22) and (1.3.29). In particular, we see that the linear speed v is maximized or minimized wherever the angular velocity $\dot{\theta}$ of the motion is.

Proof of Kepler's third law

Use $T>0$ to denote the time period for the orbiting planet. When it completes one round of its trip along its elliptical orbit of semi-major axis a and semi-minor axis $b=a\sqrt{1-e^2}$ determined by (1.3.23) such that the area swept out by the line segment connecting the planet to the sun equals πab, we have by applying (1.3.30) the relation

$$\pi ab=\int_0^T \frac{\mathrm{d}A(\theta)}{\mathrm{d}t}\,\mathrm{d}t=\frac{wT}{2}. \tag{1.3.33}$$

Using (1.3.23) in (1.3.33), we may eliminate the eccentricity to obtain the following neat expression,

$$\begin{aligned} T^2 &= \frac{4\pi^2a^2b^2}{w^2}\\ &= \left(\frac{4\pi^2}{GM}\right)a^3, \end{aligned} \tag{1.3.34}$$

which establishes Kepler's third law.

The successful establishment of Kepler's three laws justifies the earlier plausible argument leading to the expression (1.3.15) and, as a consequence, also renders Newton's law for gravitation.

Note that, when Kepler's problem is treated as a two-body system such that the planet is considered to orbit around the center of mass, instead, of a joint or effective total mass, $M+m$, the details of the results obtained are to be correspondingly modified. For example, in the two-body context, the formula (1.3.34) is updated by the analogous expression

$$\frac{a^3}{T^2}=\frac{G(M+m)}{4\pi^2}. \tag{1.3.35}$$

1.4 Helmholtz–Kirchhoff vortex model

In this section, we consider the motion of the Helmholtz–Kirchhoff point vortices in a planar fluid, which may naturally be modeled by an N-body problem in the plane for which a quantity called the *vortex strength* or *vortex charge* serves the role of mass as in the classical N-body problem in the Euclidean space discussed in Section 1.2. At first sight, the study of fluid motion is concerned with the dynamical properties of continuous media and does not seem to be a subject that would lend itself to a Hamiltonian description of a discrete N-body system. After all, it is a very different phenomenon from the study of orbits of planets moving in empty space subject to gravity as seen in Section 1.3. However, in this section, we will show that, in the context of the formalism by Helmholtz [286] and Kirchhoff [332, 349], the motion of point vortices in a planar fluid can indeed be described by a very simple Hamiltonian N-body system in the plane.

Vorticity field and strength of vorticity

It will be instructive to start from a general discussion. Let $\mathbf{v}$ be the velocity field of a fluid. Then

$$\mathbf{w} = \nabla \times \mathbf{v} \tag{1.4.1}$$

describes the tendency that the fluid swirls itself, which is commonly called the *vorticity field.* Imagine that we form a vortex tube by vortex lines, similar to streamlines induced from the velocity field. Then cut off two cross-sections, say S_1 and S_2, to form a cylindrically shaped finite vortex tube, say T. Then the divergence theorem says that

$$\int_T \nabla \cdot \mathbf{w}\, \mathrm{d}x = \int_T \nabla \cdot (\nabla \times \mathbf{v})\, \mathrm{d}x = 0, \tag{1.4.2}$$

which then implies

$$\int_{S_1} \mathbf{w} \cdot \mathrm{d}\mathbf{S} = \int_{S_2} \mathbf{w} \cdot \mathrm{d}\mathbf{S}, \tag{1.4.3}$$

where the orientations on S_1 and S_2 are chosen in an obviously compatible way. In other words, the flux of vortex lines across the vortex tube is constant along the tube. This common flux is called the *strength* or *tension of the vortex tube.* On the other hand, the circulation of a vector field $\mathbf{v}$ along a closed curve C is defined to be

$$\oint_C \mathbf{v} \cdot \mathrm{d}\mathbf{s}. \tag{1.4.4}$$

Thus, if C is the boundary curve of a cross-section of a vortex tube of the fluid with velocity field $\mathbf{v}$, the above discussion indicates that the strength of the vortex tube may be expressed as the circulation of the fluid around the vortex tube.

Planar situation

Now consider the motion confined in a horizontal plane so that $\mathbf{v} = (v_1, v_2, 0)$. Then the vorticity field $\mathbf{w}$ is always along the vertical direction so that we may express it as a scalar field given by

$$w = \partial_1 v_2 - \partial_2 v_1. \tag{1.4.5}$$

Of course, vortex lines are all vertical to the plane.

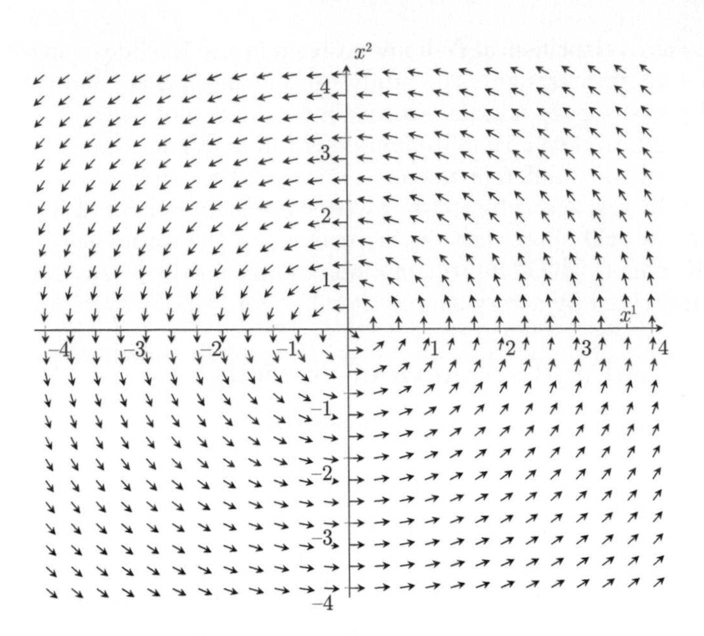

Figure 1.3 A vector field that generates concentric flow-lines plotted in the (x^1, x^2)-coordinate plane.

Helmholtz–Kirchhoff point vortices

A *Helmholtz–Kirchhoff point vortex* centered at the origin of $\mathbb{R}^2$ is an idealized situation where the velocity field is centrally generated from a specified scalar potential function according to the relations

$$v_j = \epsilon_{jk}\partial_k U, \quad j,k=1,2, \quad U(\mathbf{x}) = -\frac{\gamma}{2\pi}\ln|\mathbf{x}|, \quad \mathbf{x}=(x^1,x^2)\in\mathbb{R}^2, \quad (1.4.6)$$

where ϵ_{jk} is the standard skew-symmetric Kronecker symbol with $\epsilon_{12}=1$ and $\gamma>0$ is a parameter. It is clear that the flow-lines are concentric circles around the origin (Figure 1.3).

Let C_r be any one of such circles of radius $r>0$. Then, we have

$$\oint_{C_r} \mathbf{v}\cdot \mathrm{d}\mathbf{s} = \gamma, \quad (1.4.7)$$

which says the circulation along any flow line or the strength of any vortex tube containing the center of the vortex takes the constant value γ. Here we verify these facts.

Indeed, along C_r, the line element in terms of the polar coordinates r,θ with

$$x^1 = r\cos\theta, \quad x^2 = r\sin\theta, \quad r>0, \quad 0\le\theta\le 2\pi, \quad (1.4.8)$$

is

$$\mathrm{d}\mathbf{s} = \mathrm{d}(r\cos\theta, r\sin\theta) = (-r\sin\theta, r\cos\theta)\,\mathrm{d}\theta, \quad (1.4.9)$$

and the velocity vector field $\mathbf{v}$ assumes the form

$$\mathbf{v} = (\partial_2 U, -\partial_1 U) = \left(U_r \frac{x^2}{r}, -U_r \frac{x^1}{r}\right) = U_r(\sin\theta, -\cos\theta), \tag{1.4.10}$$

which is seen to generate concentric flowlines. Therefore, we have

$$\oint_{C_r} \mathbf{v}\cdot \mathrm{d}\mathbf{s} = -\int_0^{2\pi} U_r\, r\, \mathrm{d}\theta = -2\pi r U_r. \tag{1.4.11}$$

Hence, the constant circulation assumption leads to the differential equation

$$U_r = -\frac{\gamma}{2\pi}\frac{1}{r}, \tag{1.4.12}$$

which may be integrated to yield the result

$$U(r) = -\frac{\gamma}{2\pi}\ln r, \tag{1.4.13}$$

as anticipated.

The quantity γ gives the r-independent circulation or strength of the vortex centered at the origin. Furthermore, we can also compute the vorticity field directly,

$$w = -\Delta U = -(\partial_1^2 + \partial_2^2)U = \gamma\delta(\mathbf{x}), \tag{1.4.14}$$

in the sense of *distributions* or *weak derivatives*, which clearly reveals a *point vortex* at the origin given by the Dirac function and justifies again the quantity γ as the *strength of the point vortex.*

It may be instructive to make a note on how (1.4.13) is constructed. First recall that it is clear that a *radial vector field*, $\mathbf{u}$, can be realized as the gradient of a radially symmetric scalar field, say $U(r)$. That is,

$$\mathbf{u} = \nabla_{\mathbf{x}} U = \frac{U'(r)}{r}\mathbf{x}, \tag{1.4.15}$$

as shown in Figure 1.4. Next we can obtain a vector field that generates concentric flow-lines by a 90° counterclockwise rotation of the vector field $\mathbf{u}$ given in (1.4.15). Hence, we arrive at (1.4.10) and all the rest follows naturally.

Planar N-body problem of point vortices

Following the vortex model of Helmholtz [286] and the Hamiltonian formalism of Kirchhoff [332, 349], the dynamical interaction of N point vortices located at $\mathbf{x}_i = \mathbf{x}_i(t) \in \mathbb{R}^2$ of respective strengths γ_i's ($i = 1, \ldots, N$) at time t is governed by the interaction potential

$$U(\mathbf{x}_1, \ldots, \mathbf{x}_N) = -\frac{1}{2\pi}\sum_{1\le i<j\le N} \gamma_i\gamma_j \ln|\mathbf{x}_i - \mathbf{x}_j|, \tag{1.4.16}$$

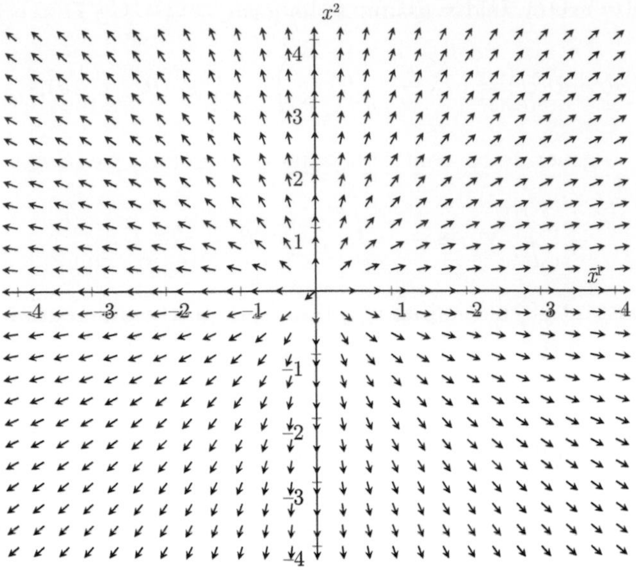

Figure 1.4 A radial vector field that generates co-centered ray-like flow-lines plotted in the (x^1, x^2)-coordinate plane.

and the equations of motion

$$\gamma_i \dot{\mathbf{x}}_i = J \nabla_{\mathbf{x}_i} U, \quad i = 1, \dots, N, \quad J = \begin{pmatrix} 0 & 1 \\ -1 & 0 \end{pmatrix}. \tag{1.4.17}$$

Rewriting $\mathbf{x}$ in the coordinate form with $\mathbf{x} = (x^1, x^2) \in \mathbb{R}^2$ and setting

$$x_i^1 = q^i, \quad \gamma_i x_i^2 = p_i, \quad i = 1, \dots, N, \tag{1.4.18}$$

we arrive at

$$\dot{q}^i = \frac{\partial U}{\partial p_i}, \quad \dot{p}_i = -\frac{\partial U}{\partial q^i}, \quad i = 1, \dots, N, \tag{1.4.19}$$

which is a Hamiltonian system. Note that the "momenta" p_i's appear in the Hamiltonian function U in a "non-standard" way. The reason for this odd appearance is that the p_i's actually do not have a mechanical meaning as momenta and are artificially identified as the momentum variables. However, it is interesting to note how the circulations γ_i's are being absorbed into these momenta so consistently.

See [64, 65, 360, 361, 415, 564] and references therein for further developments of the subject of the Helmholtz–Kirchhoff point vortices.

1.5 Partition function and thermodynamics

This section considers some *thermodynamical properties* of a Hamiltonian system in the formalism of *statistical mechanics*. The key to our study is the notion of *partition function*.

Partition function

For simplicity, consider a closed system that can occupy a countable set of states indexed by $s \in \mathbb{N}$ (the set of non-negative integers) and of distinct energies E_s ($s \in \mathbb{N}$). Then the partition function of the system is defined by

$$Z = \sum_{s=0}^{\infty} e^{-\beta E_s}, \tag{1.5.1}$$

where

$$\beta = \frac{1}{k_B T} \tag{1.5.2}$$

is the *inverse temperature* for which $k_B > 0$ is the *Boltzmann constant* and T the *absolute temperature*. Thus, in order that (1.5.1) makes sense, the sequence $\{E_s\}$ cannot have a limiting point and has to diverge sufficiently rapidly as $s \to \infty$.

Boltzmann factor

Assuming all conditions are valid so that $Z < \infty$ in (1.5.1), we see that

$$P_s = \frac{1}{Z} e^{-\beta E_s}, \quad s \in \mathbb{N}, \tag{1.5.3}$$

may naturally be interpreted as the probability that the system occupies the state s so that its energy is $E = E_s$ ($s \in \mathbb{N}$). The quantity $e^{-\beta E_s}$ is also called the *Boltzmann factor*. With such an understanding, the partition function Z may be regarded as the normalization factor of the sequence of the Boltzmann factors which give rise to the probability distribution of the random energy, E, of the system.

Thermodynamic quantities

We now illustrate how to use Z to obtain statistical information of the system.

First, the expected value of the energy (the *thermodynamic value* of the energy) is

$$\begin{aligned}\langle E \rangle &= \sum_{s=0}^{\infty} E_s P_s = \frac{1}{Z} \sum_{s=0}^{\infty} E_s e^{-\beta E_s} \\ &= -\frac{\partial \ln Z}{\partial \beta} = k_B T^2 \frac{\partial \ln Z}{\partial T},\end{aligned} \tag{1.5.4}$$

which is also commonly denoted as U. Next, in view of (1.5.4), the variance is

$$\begin{aligned}\sigma_E^2 &= \langle (E-\langle E\rangle)^2\rangle = \langle E^2\rangle - \langle E\rangle^2 \\ &= \frac{1}{Z}\sum_{s=0}^{\infty} E_s^2 \mathrm{e}^{-\beta E_s} - \left(\frac{\partial \ln Z}{\partial \beta}\right)^2 \\ &= \frac{1}{Z}\frac{\partial^2 Z}{\partial \beta^2} - \frac{1}{Z^2}\left(\frac{\partial Z}{\partial \beta}\right)^2 = \frac{\partial^2 \ln Z}{\partial \beta^2},\end{aligned} \tag{1.5.5}$$

which, in view of (1.5.4) and (1.5.5), gives rise to the *heat capacity*

$$C_v = \frac{\partial U}{\partial T} = \frac{\partial \langle E\rangle}{\partial T} = \frac{1}{k_{\mathrm{B}}T^2}\frac{\partial^2 \ln Z}{\partial \beta^2} = \frac{1}{k_{\mathrm{B}}T^2}\sigma_E^2. \tag{1.5.6}$$

Besides, the *entropy* or the *Gibbs entropy*, also often called the *Shannon entropy*, of the system, S, which measures the *disorder* or *uncertainty* of the system, is given in view of (1.5.4) by

$$\begin{aligned}S &= -k_{\mathrm{B}}\sum_{s=0}^{\infty} P_s \ln P_s = -k_{\mathrm{B}}\sum_{s=0}^{\infty} P_s \ln \frac{\mathrm{e}^{-\beta E_s}}{Z} \\ &= k_{\mathrm{B}}\sum_{s=0}^{\infty}(P_s \ln Z + \beta E_s P_s) \\ &= k_{\mathrm{B}}(\ln Z + \beta\langle E\rangle) = k_{\mathrm{B}}\ln Z + \frac{1}{T}\langle E\rangle \\ &= \frac{\partial}{\partial T}(k_{\mathrm{B}}T\ln Z) \equiv -\frac{\partial A}{\partial T},\end{aligned} \tag{1.5.7}$$

where

$$\begin{aligned}A &= -k_{\mathrm{B}}T\ln Z \\ &= U - TS \quad (U = \langle E\rangle),\end{aligned} \tag{1.5.8}$$

in view of the second line in (1.5.7), is the *Helmholtz free energy*. These examples show the usefulness of the partition function.

Derivation of partition function

After seeing the importance of the partition function (1.5.1), here we show how it arises naturally from the laws of thermodynamics.

First, when a system is at its thermodynamic equilibrium, the *second law of thermodynamics* asserts a *maximized entropy*. Thus, mathematically, the system follows the probability distribution

$$P_s = P(\{E = E_s\}), \tag{1.5.9}$$

that maximizes the entropy

$$S = -k_{\mathrm{B}} \sum_{s=0}^{\infty} P_s \ln P_s, \tag{1.5.10}$$

subject to the constraints

$$\sum_{s=0}^{\infty} P_s = 1, \quad \sum_{s=0}^{\infty} E_s P_s = U, \tag{1.5.11}$$

imposed to the total probability and average energy. Therefore, we are to extremize the Lagrange function

$$L = -k_{\mathrm{B}} \sum_{s=0}^{\infty} P_s \ln P_s + \lambda_1 \left(\sum_{s=0}^{\infty} P_s - 1 \right) + \lambda_2 \left(\sum_{s=0}^{\infty} E_s P_s - U \right), \tag{1.5.12}$$

where λ_1, λ_2 are the Lagrange multipliers. So we are led to setting up the equations

$$\frac{\partial L}{\partial P_s} = -k_{\mathrm{B}} (\ln P_s + 1) + \lambda_1 + \lambda_2 E_s = 0, \quad \forall s \in \mathbb{N}, \tag{1.5.13}$$

which render the solution

$$P_s = a \mathrm{e}^{\frac{\lambda_2 E_s}{k_{\mathrm{B}}}}, \quad a \equiv \mathrm{e}^{-1+\frac{\lambda_2}{k_{\mathrm{B}}}}. \tag{1.5.14}$$

Next, multiplying (1.5.13) by P_s, summing up, and using (1.5.11), we have

$$S - k_{\mathrm{B}} + \lambda_1 + \lambda_2 U = 0, \tag{1.5.15}$$

which yields the differential relation

$$\mathrm{d}S + \lambda_2 \mathrm{d}U = 0. \tag{1.5.16}$$

On the other hand, recall the *first law of thermodynamics*, which says the increment of heat to the system, $\mathrm{d}Q = T\mathrm{d}S$, is the result of increment of average energy, $\mathrm{d}U$, and the extra mechanical work done to the system $\mathrm{d}W$, namely, $T\mathrm{d}S = \mathrm{d}U + \mathrm{d}W$. However, in the present equilibrium situation, $\mathrm{d}W = 0$. Hence, $T\mathrm{d}S = \mathrm{d}U$. Inserting this into (1.5.16), we find

$$\lambda_2 = -\frac{1}{T}. \tag{1.5.17}$$

Then, substituting (1.5.17) into (1.5.14), we obtain

$$P_s = a \mathrm{e}^{-\beta E_s}, \quad \beta = \frac{1}{k_{\mathrm{B}} T}. \tag{1.5.18}$$

Finally, summing up P_s in (1.5.18) and using (1.5.11), we have

$$\frac{1}{a} = \sum_{s=0}^{\infty} \mathrm{e}^{-\beta E_s}, \tag{1.5.19}$$

which gives us the partition function defined in (1.5.1).

Hamiltonian system

For a classical Hamiltonian system with generalized coordinates $q = (q_1, \ldots, q_n)$ and momenta $p = (p_1, \cdots, p_n)$, governed by the Hamiltonian function $H(q, p)$, the partition function is expressed by

$$Z = \int \mathrm{e}^{-\beta H(q,p)}\, \mathrm{d}q\mathrm{d}p, \tag{1.5.20}$$

where q, p take over the role of the state index s and the integral replaces the summation in our earlier discussion. Therefore, a similar collection of knowledge can be gathered as before. For example, if $F(q,p)$ is a mechanical quantity of interest, then its expected or thermodynamic value is given by

$$\langle F \rangle = \frac{1}{Z} \int F(q,p)\mathrm{e}^{-\beta H(q,p)}\, \mathrm{d}q\mathrm{d}p. \tag{1.5.21}$$

A fairly thorough treatment of statistical mechanics may be found in [252, 297, 350, 391]. The next section applies these ideas to study a thermodynamic property of DNA.

1.6 Dynamic modeling of DNA denaturation

DNA, the short name for *deoxyribonucleic acid*, is a nucleic acid that contains the genetic instructions used in the development and functioning of all known living organisms. Chemically, a DNA consists of two long polymers of simple units called nucleotides, with backbones made of sugars and phosphate groups. These two strands run in parallel and form a double helix. Attached to each sugar is one of four types of nucleotide molecules, also called *bases*, named by letters A (adenine), C (cytosine), G (guanine), and, T (thymine), so that only A and T, C and G, from opposite strands may bind to form pairs. During the last four decades, biologists and physicists have carried out extensive research on the dynamics of DNA, using mathematical modeling and computer simulation, and obtained profound knowledge about DNA and its function.

Mathematical modeling of dynamics of DNA was initiated in 1980 by Englander and colleagues [192], who presented a discrete *sine-Gordon soliton interpretation of the DNA* of n pairs of bases and used the solitary wave in the continuous limit as an approximation in the limit $n \to \infty$ to obtain some qualitative behavior of DNA. In 1989, Peyrard and Bishop [446] published their pioneering work on DNA dynamical modeling in which the base pairing due to hydrogen bonding is recognized, the discreteness of the model is maintained, and a statistical mechanics study is fully carried out which describes the inter-strand separation in the double helix as a function of temperature, leading to a mathematical formulation of *DNA denaturation*, that is, the phenomenon that the two DNA strands in the double helix become separated when heated.

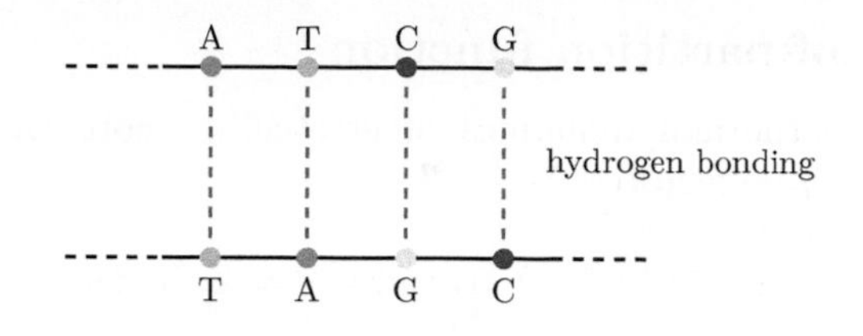

Figure 1.5 An oversimplified DNA chain model.

Chain model

Following [446], here we initially allow two degrees of freedom for each pair of bases and use u_i and v_i to denote the transverse displacements of the bases from their equilibrium positions along the direction of the hydrogen bonds that connect the two bases in a pair. The governing Hamiltonian for the *double helix model* containing a longitudinal *harmonic coupling* between neighboring bases due to *stacking* and n pairs of bases is given as

$$H = \sum_{i=1}^{n} \left(\frac{1}{2} m(\dot{u}_i^2 + \dot{v}_i^2) + \frac{1}{2}\kappa([u_i - u_{i-1}]^2 + [v_i - v_{i-1}]^2) + V(u_i - v_i) \right), \tag{1.6.1}$$

where a common mass m is taken for all bases, a uniform "elastic" (*stacking force*) constant κ is assumed for simplicity, and the potential energy V is defined by

$$V(u) = D(\mathrm{e}^{-au} - 1)^2, \tag{1.6.2}$$

which accounts for the hydrogen bonding, with a, D some positive constants, and is of the *Morse type* [401]. Figure 1.5 illustrates just such an oversimplified model. Here we note that the definitions of u_0 and v_0 depend on the specific boundary condition to be considered.

In terms of the new variables x_i, y_i and the associated momenta p_i, P_i, defined by

$$x_i = \frac{(u_i + v_i)}{\sqrt{2}}, \quad y_i = \frac{(u_i - v_i)}{\sqrt{2}}, \quad p_i = m\dot{x}_i, \quad P_i = m\dot{y}_i, \tag{1.6.3}$$

the Hamiltonian (1.6.1) is normalized into the form

$$\begin{aligned} H &= \sum_{i=1}^{n} \left(\frac{p_i^2}{2m} + \frac{1}{2}\kappa(x_i - x_{i-1})^2 \right) \\ &\quad + \sum_{i=1}^{n} \left(\frac{P_i^2}{2m} + \frac{1}{2}\kappa(y_i - y_{i-1})^2 + D(\mathrm{e}^{-a\sqrt{2}y_i} - 1)^2 \right). \end{aligned} \tag{1.6.4}$$

It is important to realize that the variable y_i measures the stretching distance between the bases in a pair of bases.

Factorization of partition function

To understand the thermal dynamics of stretching, note that the partition function Z is seen to be factored as

$$Z = \int \mathrm{e}^{-\beta H(p,x,P,y)}\, \mathrm{d}x\mathrm{d}y\mathrm{d}p\mathrm{d}P = Z_p Z_x Z_P Z_y, \tag{1.6.5}$$

where $\beta = (k_{\mathrm{B}}T)^{-1}$, with T the absolute temperature and k_{B} the Boltzmann constant, and, to save space, we use $\mathrm{d}x$ (say) to denote $\mathrm{d}x^1 \cdots \mathrm{d}x^n$ and use x to denote the vector coordinates (x^i) or a single variable interchangeably if there is no risk of confusion in the context. From this, Peyrard and Bishop [446] recognized that the mean stretching $\langle y_\ell \rangle$ of the bases at the position $\ell = 1, \dots, n$, due to the hydrogen bonding, is given by

$$\langle y_\ell \rangle = \frac{1}{Z} \int y_\ell \mathrm{e}^{-\beta H}\, \mathrm{d}x\mathrm{d}y\mathrm{d}p\mathrm{d}P = \frac{1}{Z_y} \int y_\ell \mathrm{e}^{-\beta \sum_{i=1}^n f(y_i, y_{i-1})}\, \mathrm{d}y, \tag{1.6.6}$$

where the factors involving x, p, P are dropped as a consequence of the decomposed Hamiltonian (1.6.4) and $f(y, y')$ is the reduced potential given by the y-dependent terms in (1.6.4) as

$$f(y, y') = \frac{1}{2}\kappa(y - y')^2 + D\left(\mathrm{e}^{-a\sqrt{2}y} - 1\right)^2. \tag{1.6.7}$$

It is still rather difficult to analyze the quantity (1.6.6) as a function of the temperature T without further simplification. In [446], Peyrard and Bishop take $n \to \infty$ in (1.6.6) to arrive at the position-independent mean stretching

$$\langle y \rangle = \langle \varphi_0 | y | \varphi_0 \rangle = \int \varphi_0^2(y) y\, \mathrm{d}y, \tag{1.6.8}$$

where

$$\varphi_0(y) = \frac{(\sqrt{2}a)^{\frac{1}{2}}(2d)^{d-\frac{1}{2}}}{\Gamma(2d-1)^{\frac{1}{2}}} \exp(-d\mathrm{e}^{-\sqrt{2}ay})\mathrm{e}^{-(d-\frac{1}{2})\sqrt{2}ay}, \tag{1.6.9}$$

$$d = \frac{1}{a}\beta(\kappa D)^{\frac{1}{2}} > \frac{1}{2}. \tag{1.6.10}$$

Based on this formalism, Peyrard and Bishop [446] succeeded in finding a thermodynamical description of the DNA denaturation phenomenon. Using (1.6.9) with (1.6.10) and numerical evaluation, they showed that the base mean stretching $\langle y_\ell \rangle$ increases significantly as the temperature climbs to a particular level which is an unambiguous indication of DNA denaturation. Another interesting by-product of such a calculation is that, since the dependence of the ground state on the absolute temperature $T = (k_{\mathrm{B}}\beta)^{-1}$ is through the parameter d given earlier, a greater value of the elastic constant κ leads to a higher DNA denaturation temperature, which is what was observed [214, 446] in the laboratory.

Hamiltonian of out-of-phase motion

In particular, we see that the dynamics of the DNA molecule is effectively described by the reduced Hamiltonian that contains the "out-of-phase" motion of the bases only given in terms of the y-variables as

$$H = \sum_{i=1}^{n} \left(\frac{1}{2} m \dot{y}_i^2 + \frac{1}{2} \kappa (y_i - y_{i-1})^2 + D \left(\mathrm{e}^{-a\sqrt{2} y_i} - 1 \right)^2 \right). \tag{1.6.11}$$

See [445] for a review of related topics and directions. This example shows how a simple system of ordinary differential equations may be used to investigate a fundamental problem in biophysics.

Exercises

1. Let $q = (q^1, \ldots, q^n) \in \mathbb{R}^n$ be the position coordinate vector of a particle that passes the points $q = q_1$ and $q = q_2$ at the times $t = t_1$ and $t = t_2$, respectively. Assume that the motion of such a particle is governed by the action
$$\mathcal{A}(q) = \int_{t_1}^{t_2} L(q(t), \dot{q}(t), t) \, \mathrm{d}t \tag{1.E.1}$$
so that the trajectory of the motion $q = q(t)$ is a critical point of the action. Show that $q(t)$ solves the equation of motion
$$\frac{\mathrm{d}}{\mathrm{d}t} \left(\frac{\partial L}{\partial \dot{q}^i} \right) = \frac{\partial L}{\partial q^i}, \quad i = 1, 2, \ldots, n. \tag{1.E.2}$$
2. Let (x, y) be a point on the ellipse defined by the equation (1.3.2). Show that the sum of the distances from (x, y) to the two foci, $(0, 0)$ and $(2c, 0)$, is $2a$.
3. Consider the equation of an ellipse given by (1.3.6). It is clear that the maximum $r_{\max}$ and minimum $r_{\min}$ of the radial distance $r = r(\theta)$ from a point on the ellipse to its focus at the origin are attained at the points where $\theta = 0$ and $\theta = \pi$, called the aphelion and perihelion, respectively. Show that the eccentricity e of the ellipse can be represented in terms of $r_{\max}$ and $r_{\min}$ by
$$e = \frac{r_{\max} - r_{\min}}{r_{\max} + r_{\min}}. \tag{1.E.3}$$
4. Formulate the equation of motion (1.3.8) in the form of a Hamiltonian system.
5. Consider the gravitational interaction of two point masses μ and m, fixed at the origin and moving around the origin, respectively, in spherical coordinates
$$x = r \sin\theta \cos\phi, \quad y = r \sin\theta \sin\phi, \quad z = r \cos\theta. \tag{1.E.4}$$

(a) Show that the kinetic energy of the moving point mass m is given by

$$\mathcal{K} = \frac{1}{2} m \left(\dot{r}^2 + r^2 \dot{\theta}^2 + r^2 \sin^2 \theta \, \dot{\phi}^2 \right), \tag{1.E.5}$$

and obtain the associated Lagrangian function.

(b) Show that the momenta of the moving mass associated with the cooordinates r, θ, ϕ are

$$p_r = m\dot{r}, \quad p_\theta = mr^2 \dot{\theta}, \quad p_\phi = mr^2 \sin^2 \theta \, \dot{\phi}. \tag{1.E.6}$$

(c) Use (b) to show that the Hamiltonian of the problem is

$$H(r, \theta, \phi, p_r, p_\theta, p_\phi) = \frac{1}{2m} \left(p_r^2 + \frac{p_\theta^2}{r^2} + \frac{p_\phi^2}{r^2 \sin^2 \theta} \right) - \frac{\mu m}{r}, \tag{1.E.7}$$

(d) Derive the Hamiltonian equation from (c) governing the mechanical variables $r, \theta, \phi, p_r, p_\theta, p_\phi$.

6. Consider the classical central-force motion of a particle of mass m governed by the equation

$$m\ddot{\mathbf{x}} = f(x, y, z)\mathbf{x}, \quad \mathbf{x} = (x, y, z) \in \mathbb{R}^3 \setminus \{\mathbf{0}\}, \tag{1.E.8}$$

where f is a real-valued continuous function.

(a) Establish the law of conservation for the angular momentum

$$\frac{\mathrm{d}}{\mathrm{d}t}(m\mathbf{x} \times \dot{\mathbf{x}}) = \mathbf{0}. \tag{1.E.9}$$

(b) Use (a) to show that the motion of the particle is planar. That is, its orbit is confined to a plane, say P.

(c) Prove that the areas in P swept out by the position vector $\mathbf{x}$ during equal time intervals are equal. In other words, Kepler's second law is valid for central-force motion, in general, governed by the equation (1.E.8).

7. Derive (1.3.32).

8. Consider the time-dependent Hamiltonian

$$H(q, p, t) = \frac{p^2}{2m} + \frac{kq^2}{2} - \mu \, q \cos(\omega t), \tag{1.E.10}$$

where $m, k, \mu, \omega > 0$ are constants.

(a) Write the Hamiltonian equations.

(b) Find the general solution of the Hamiltonian equations.

(c) Obtain the Lagrangian function and Lagrange equation.

9. Consider the nonlinear equation

$$\ddot{x} = -x + \frac{\lambda}{(1-x)^2}, \tag{1.E.11}$$

where $\lambda > 0$ is a constant, governing the dynamics of the moving plate of an electrostatic actuator in a microelectromechanical system (MEMS).

   (a) Find the Lagrangian and Hamiltonian functions of the equation.

   (b) Show that when $\lambda > 0$ is small the solution with the initial condition

$$x(0) = 0, \quad \dot{x}(0) = 0 \tag{1.E.12}$$

   is periodic. See [609].

10. Consider the motion of a point particle of mass m and electric charge Q in an electromagnetic field of electric potential V and magnetic potential $\mathbf{A}$. If the spatial position vector $\mathbf{x}$ of the particle at time t is $\mathbf{x}(t)$, then $\mathbf{x}(t)$ is governed by the Lagrangian function

$$L(\mathbf{x}, \dot{\mathbf{x}}, t) = \frac{1}{2} m\dot{\mathbf{x}}^2 - QV(\mathbf{x}, t) + Q\dot{\mathbf{x}} \cdot \mathbf{A}(\mathbf{x}, t). \tag{1.E.13}$$

   (a) Obtain the equation of motion or the Lagrange equation of the particle.

   (b) Use (a) to show that the equation of motion of the charged particle is of the form of the Newton law $m\ddot{\mathbf{x}} = \mathbf{F}$ where $\mathbf{F}$ is the *Lorentz force* given by

$$\mathbf{F} = Q(\mathbf{E} + \dot{\mathbf{x}} \times \mathbf{B}), \tag{1.E.14}$$

   with

$$\mathbf{E} = -\nabla V - \frac{\partial \mathbf{A}}{\partial t}, \quad \mathbf{B} = \nabla \times \mathbf{A}, \tag{1.E.15}$$

   being the electric and magnetic fields induced from the potential fields V and $\mathbf{A}$.

   (c) Find the Hamiltonian function of the system.

   (d) Recover the result in (a) by obtaining the Hamiltonian equations for the motion of the particle.

11. As in the discussion of the Kirchhoff vortex model in $\mathbb{R}^2$, construct a centralized (radial) vector field $\mathbf{F}$ in $\mathbb{R}^3$ such that the flux

$$\Phi = \int_{\partial\Omega} \mathbf{F} \cdot \mathrm{d}\mathbf{S} \tag{1.E.16}$$

is independent of the choice of the bounded domain Ω around the origin of $\mathbb{R}^3$. (You may start your study with the situation when Ω is an arbitrary ball.)

12. Let the energies of a system be $E_s = \varepsilon s$, $s = 0, 1, 2, \ldots$, where $\varepsilon > 0$ is fixed.

(a) Find the partition function Z.

(b) Compute the Helmholtz free energy.

(c) Compute the entropy of the system.

13. Let the energy spectrum of a thermodynamic system be a continuum given by $\{E_s = \varepsilon s \,|\, s > 0\}$ where $\varepsilon > 0$ is fixed.

(a) Find the partition function Z.

(b) Find the distribution function of the energy.

(c) Find the thermodynamic energy U.

14. Let the partition function of a system be $Z = 1 + \mathrm{e}^{-\beta\varepsilon}$ where $\varepsilon > 0$.

(a) Compute the thermodynamic energy U and find how it depends on the absolute temperature T. In particular, find

$$U(0) = \lim_{T\to 0} U, \quad U(\infty) = \lim_{T\to\infty} U. \tag{1.E.17}$$

(b) Use (a) to find the heat capacity of the system.

15. Derive the Hamiltonian equations of the model (1.6.11) assuming $y_0 = 0$ and $y_0 = y_n$, respectively.

16. Consider an over-simplified model of one-base DNA dynamics governed by the equation

$$\ddot{x} + kx + D\mathrm{e}^{x}(\mathrm{e}^{x} - 1) = f(t), \tag{1.E.18}$$

where $k, D > 0$ are constants and $f(t)$ is a forcing term.

(a) Find the Lagrangian and Hamiltonian functions of the equation.

(b) Let $f(t)$ be of period $T > 0$. Investigate whether the equation has a solution of period T under some appropriate conditions.

2

Schrödinger equation and quantum mechanics

Quantum mechanics, developed at the beginning of the twentiety century, attempted to explain a broad range of physical phenomena in microscopic scales based upon a series of celebrated experiments that could not be explained within the conceptual framework of classical physics. This chapter focuses on the Schrödinger equation, which is the foundation of quantum mechanics, and aims to understand some fundamental features of it. It then presents a few methodological approaches to analyzing the quantum many-body problem, which is of contemporary research interest. In doing so, it shows that the study of the quantum aspects of an N-body system, which is governed by a set of nonlinear ordinary differential equations, renders the problem linear through the Schrödinger equation, which is a partial differential equation, governing the state of the system. Furthermore, the search for approximate solutions of the Schrödinger equation leads us to considering nonlinear equations of combined differential-integral type and problems of calculus of variations.

2.1 Path to quantum mechanics

The core or essence of quantum mechanics is the Schrödinger equation. Thus, the path to quantum mechanics lies in understanding how the Schrödinger equation came about. The goal of the first two sections of this chapter is to understand how the Schrödinger equation is introduced. To this goal, in this section, we first recall some milestone early-day discoveries which revealed the wave-particle duality nature of matter interaction in microscopic world and how these discoveries were given precise mathematical perceptions or formulations successfully by the pioneers of quantum mechanics. Then, the next section shows

Mathematical Physics with Differential Equations. Yisong Yang, Oxford University Press.
 DOI: 10.1093/oso/9780192872616.003.0002

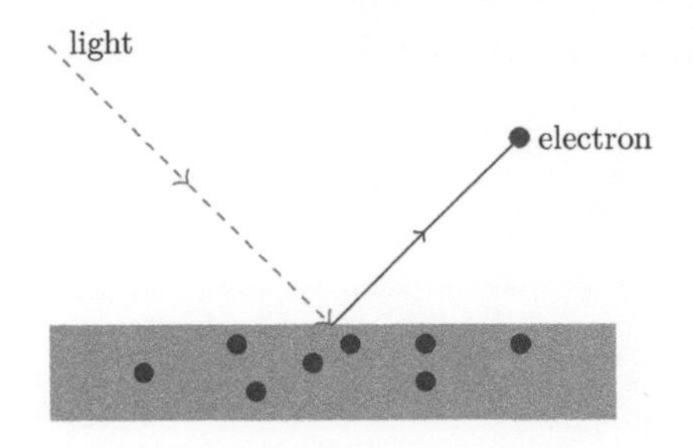

Figure 2.1 An illustration of the photoelectric effect that electrons in a piece of metal may be energized by a light beam to escape from the metal.

how the Schrödinger equation may be "derived," such that its distinguished wave-equation characteristics serve to unify these formulations in a single framework.

Photoelectric effect

Place a piece of metal in a vacuum tube and shoot a beam of light onto it. The electrons in the metal may become sufficiently energized to be emitted from the metal (Figure 2.1). This is known as the *photoelectric effect*, which has found a wide range of applications in today's electronics. Now measure the energy carried by an emitted electron and denote it by E_e. It is known that E_e may be written as the difference of two quantities, one is proportional to the frequency, ν, of the light beam so that the proportionality constant, h, is universal and independent of the metal, the other, ϕ, depends on the metal but is independent of the light frequency. Therefore, we have

$$E_e = h\nu - \phi. \tag{2.1.1}$$

Einstein's postulate

In 1905, Einstein postulated that light, a special form of electromagnetic waves, is composed of particles called photons. Each photon carries an amount of energy equal to $h\nu$. That is,

$$E = h\nu. \tag{2.1.2}$$

When the photon hits an electron in a metal, the electron receives this amount of energy, consumes the amount of the binding energy of the metal to the electron to escape from the metal, and becomes an emitted electron of the energy given by (2.1.1). The equation (2.1.2) is also known as the *Planck–Einstein relation.* Section 15.6 presents a derivation of the formula (2.1.2) as a consequence of quantization of electromagnetic fields.

In terms of angular frequency

In physics, frequency ν is measured in hertz with unit second^{-1} (times per second), and angular frequency ω is related to ν by $\omega = 2\pi\nu$ (radians per second). Hence, in terms of ω, the Einstein formula becomes

$$E = \hbar\omega, \quad \hbar = \frac{h}{2\pi}. \tag{2.1.3}$$

Recall that energy is measured in unit of Joules and one Joule is equal to one Newton×meter. The constant $\hbar$ in (2.1.3), called the *Planck constant*, is a tiny number of the unit of Joules × second and accepted to be

$$\hbar = 1.05457 \times 10^{-34}. \tag{2.1.4}$$

Historically, h is called the Planck constant, and $\hbar$ the *Dirac constant* or the *extended Planck constant.*

Compton effect

After Einstein's 1905 postulate that light is composed of photons, physicists began to wonder whether a photon might exhibit its kinetic momentum in interaction, that is, when colliding with another particle. In 1922, Compton and Debye came with a very simple mathematical description of this behavior, which was then experimentally observed by Compton himself in 1923 and further confirmed by Woo, then Compton's graduate student. In simple terms, when a photon hits an electron, it behaves like a particle when it collides with another particle so that one observes energy as well as momentum conservation relations, which is evidenced by a wavelength shift after the collision.

Intuitively and mathematically, we may write the energy of a photon by the *Einstein formula*, $E = mc^2$, where c is the speed of light in vacuum and m is the "virtual rest mass" of photon (note that a photon in fact has no rest mass so that the connection made this way is only intuitive, and a completely rigorous treatment along the same line may be formulated with the full relativistic energy-mass-momentum formula (4.3.22) to be derived in Chapter 4). Thus, in view of (2.1.3), we have

$$E = mc^2 = \hbar\omega. \tag{2.1.5}$$

On the other hand, recall that the wavenumber (also called the *angular wavenumber*) k, wavelength λ, frequency ν, angular frequency ω, and speed c of a photon are related by

$$k = \frac{2\pi}{\lambda}, \quad c = \lambda\nu = \lambda\frac{\omega}{2\pi} = \frac{\omega}{k}. \tag{2.1.6}$$

Consequently, the momentum of the photon is given by

$$\begin{aligned} p = mc &= \frac{E}{c} = \hbar\,\frac{\omega}{c} \\ &= \hbar\,k. \end{aligned} \tag{2.1.7}$$

De Broglie's wave-particle duality hypothesis

In 1924, de Broglie formulated his celebrated *wave-particle duality* hypothesis in his PhD thesis, which equalizes waves and particles, takes the Einstein formula (2.1.3) and the *Compton–Debye formula* (2.1.7) as two axioms, and reiterates the wave and particle characteristics of all interactions in nature:

$$E = \hbar\,\omega, \tag{2.1.8}$$

$$p = \hbar\,k. \tag{2.1.9}$$

In other words, a particle of energy E and momentum p behaves like a wave of wavenumber k and a wave of wavenumber k behaves like a particle of energy E and momentum p such that E, p, and k are related through (2.1.8) and (2.1.9). In Section 15.6, we show how the Compton–Debye formula (2.1.7) may be derived when we quantize electromagnetic fields.

2.2 Schrödinger equation

Based on de Broglie's wave-particle duality, we now derive the *Schrödinger equation*, first published by Schrödinger in 1926. The wave-equation features of this equation and the statistical interpretation of its solution enable us to perceive and understand some of the most profound physical properties of nature only available or observable at microscopic scales. Interestingly, we will see how classical and quantum mechanics are statistically linked, also through the Schrödinger equation.

Wave motion in terms of angular wavenumber and frequency

Consider a stationary wave distributed over the x-axis of wavenumber k (the wave has k repeated cycles over the standard angular (cell) interval $[0, 2\pi]$) whose simplest form is given by

$$C\mathrm{e}^{\mathrm{i}kx}. \tag{2.2.1}$$

Switch on time-dependence so that the wave moves to right (say) at velocity $v > 0$. We see from (2.2.1) that the wave is represented by

$$\phi(x,t) = C\mathrm{e}^{\mathrm{i}k(x-vt)}. \tag{2.2.2}$$

Notice that we can extend (2.1.6) as

$$k = \frac{2\pi}{\lambda}, \quad v = \lambda\nu = \lambda\,\frac{\omega}{2\pi}. \tag{2.2.3}$$

Combining (2.2.2) and (2.2.3), we have

$$\phi(x,t) = C\mathrm{e}^{\mathrm{i}(kx-\omega t)}. \tag{2.2.4}$$

Momentum and energy as eigenvalues

Formally, in view of (2.2.4), the de Broglie momentum (2.1.9) can be read off as an eigenvalue of the operator $-\mathrm{i}\hbar\frac{\partial}{\partial x}$. That is,

$$\left(-\mathrm{i}\hbar\frac{\partial}{\partial x}\right)\phi = (\hbar\, k)\phi = p\phi, \tag{2.2.5}$$

so that

$$\left(-\mathrm{i}\hbar\frac{\partial}{\partial x}\right)^2\phi = (\hbar\, k)^2\phi = p^2\phi. \tag{2.2.6}$$

Similarly, the de Broglie energy (2.1.8) can be read off as an eigenvalue of the operator $\mathrm{i}\hbar\frac{\partial}{\partial t}$. That is,

$$\left(\mathrm{i}\hbar\frac{\partial}{\partial t}\right)\phi = (\hbar\,\omega)\phi = E\phi. \tag{2.2.7}$$

Schrödinger equation

For a free particle of mass $m > 0$, we know that there holds the classical relation

$$E = \frac{p^2}{2m}. \tag{2.2.8}$$

In view of (2.2.6)–(2.2.8), we arrive at the *free Schrödinger equation*

$$\mathrm{i}\hbar\frac{\partial\phi}{\partial t} = -\frac{\hbar^2}{2m}\frac{\partial^2\phi}{\partial x^2}. \tag{2.2.9}$$

For a particle moving in a potential field $V = V(x,t)$, the energy-momentum relation (2.2.8) becomes

$$E = \frac{p^2}{2m} + V. \tag{2.2.10}$$

Therefore the Schrödinger equation (2.2.9) for a free particle is modified into form

$$\mathrm{i}\hbar\frac{\partial\phi}{\partial t} = -\frac{\hbar^2}{2m}\frac{\partial^2\phi}{\partial x^2} + V\phi. \tag{2.2.11}$$

This is called the *Schrödinger wave equation* whose solution, ϕ, is called a *wave function.*

Born's statistical interpretation of wave function

Consider the Schrödinger equation (2.2.11) describing a particle of mass m and assume that ϕ is a "normalized" solution of (2.2.11) which satisfies the normalization condition

$$\int |\phi(x,t)|^2\,\mathrm{d}x = 1, \tag{2.2.12}$$

and characterizes the "state" of the particle. According to Born, the mathematical meaning of such a wave function is that $\rho(x,t) = |\phi(x,t)|^2$ gives the *probability density* for the location of the particle at time t. In other words, the probability of finding the particle in an interval (a,b) at time t is

$$P(\{a < x(t) < b\}) = \int_a^b |\phi(x,t)|^2 \,\mathrm{d}x. \tag{2.2.13}$$

With this interpretation, we see that the expected location of the particle at time t is

$$\langle x\rangle(t) = \int x|\phi(x,t)|^2 \,\mathrm{d}x = \int \overline{\phi}(x,t)\, x\, \phi(x,t)\,\mathrm{d}x. \tag{2.2.14}$$

Operator representations of physical quantities

Naturally, the expected value of the momentum of the particle should be equal to the product of the particle mass and the expected value of the particle velocity. Therefore, in view of (2.2.11), we have

$$\begin{aligned}
\langle \hat{p}\rangle(t) &= m\frac{\mathrm{d}\langle x\rangle(t)}{\mathrm{d}t} \\
&= m\int (\overline{\phi}_t(x,t)\, x\, \phi(x,t) + \overline{\phi}(x,t)\, x\, \phi_t(x,t))\,\mathrm{d}x \\
&= \mathrm{i}\frac{\hbar}{2}\int (x\overline{\phi}\phi_{xx} - x\overline{\phi}_{xx}\phi)\,\mathrm{d}x \\
&= \int \overline{\phi}(x,t)\Big(-\mathrm{i}\hbar\frac{\partial}{\partial x}\Big)\phi(x,t)\,\mathrm{d}x.
\end{aligned} \tag{2.2.15}$$

Hence, formally, the expected value of the momentum is the "expected value" of the operator

$$\hat{p} = -\mathrm{i}\hbar\frac{\partial}{\partial x}. \tag{2.2.16}$$

In other words, within the framework of Born's statistical interpretation of the wave function, momentum has its elegant operator representation (2.2.16).

In this manner, we have the trivial operator representations

$$\hat{x} = x, \quad \hat{f}(x) = f(x), \tag{2.2.17}$$

for the particle coordinate x and its functions. Besides, (2.2.10) gives us the energy representation

$$\hat{E} = \frac{1}{2m}\hat{p}^2 + V, \tag{2.2.18}$$

which is the quantum-mechanical Hamiltonian. Thus,

$$\langle \hat{E}\rangle = \int \overline{\phi}(x,t)\Big(-\frac{\hbar^2}{2m}\frac{\partial^2}{\partial x^2} + V\Big)\phi(x,t)\,\mathrm{d}x. \tag{2.2.19}$$

Using (2.2.11) in (2.2.19), we have

$$\langle \hat{E} \rangle = \int \overline{\phi}(x,t) \left(\mathrm{i}\hbar \frac{\partial}{\partial t} \right) \phi(x,t) \, \mathrm{d}x. \tag{2.2.20}$$

In other words, energy should be represented by the operator

$$\hat{E} = \mathrm{i}\hbar \frac{\partial}{\partial t}. \tag{2.2.21}$$

These fundamental *operator* representations of various physical quantities composed from their *classical counterparts*, collectively carried out in a procedure known as the *first quantization*, form the foundation of quantum mechanics.

Conservation law and probability current

It is easily checked that the normalization condition is well posed because

$$\frac{\mathrm{d}}{\mathrm{d}t} \int |\phi(x,t)|^2 \, \mathrm{d}x = 0, \tag{2.2.22}$$

by virtue of the equation (2.2.11) so that it suffices to require the condition

$$\int |\phi(x,0)|^2 \, \mathrm{d}x = 1, \tag{2.2.23}$$

initially. Here, we look for some additional consequences from the *global conservation law* (2.2.22). For this purpose, we differentiate the probability density ρ to get

$$\begin{aligned} \rho_t &= \overline{\phi}_t \phi + \overline{\phi} \phi_t \\ &= -\mathrm{i} \frac{\hbar}{2m} (\phi \overline{\phi}_x - \overline{\phi} \phi_x)_x, \end{aligned} \tag{2.2.24}$$

where we have used (2.2.11) again. It is interesting to view ρ as a "*charge*" density and rewrite (2.2.24) in the form of a *conservation law*,

$$\frac{\partial}{\partial t} \rho + \frac{\partial}{\partial x} j = 0, \tag{2.2.25}$$

where j may be viewed as a "*current*" density, which is identified to be

$$j = \mathrm{i} \frac{\hbar}{2m} (\phi \overline{\phi}_x - \overline{\phi} \phi_x). \tag{2.2.26}$$

We note that it is crucial that ϕ is complex-valued: if it is real-valued, the current density will be zero identically and ρ will be time-independent.

Furthermore, differentiating (2.2.13) and using (2.2.26), we have

$$\begin{aligned} \frac{\mathrm{d}}{\mathrm{d}t} P(\{a < x(t) < b\}) &= \frac{\mathrm{d}}{\mathrm{d}t} Q(a,b)(t) \\ &= \frac{\mathrm{d}}{\mathrm{d}t} \int_a^b \rho(x,t) \, \mathrm{d}x \\ &= j(a,t) - j(b,t), \end{aligned} \tag{2.2.27}$$

where $Q(a,b)$ may be interpreted as the charge contained in the interval (a,b) at time t so that its rate of change is equal to the net current following into such an interval. Or more correctly, we may call Q the "*probability charge*" and j the "*probability current.*"

Ehrenfest theorem

Differentiating (2.2.15) and using (2.2.11), we have

$$\begin{aligned}\frac{\mathrm{d}\langle\hat{p}\rangle}{\mathrm{d}t} &= \int \left(\overline{\phi}_t\left[-\mathrm{i}\hbar\frac{\partial}{\partial x}\right]\phi + \overline{\phi}\left[-\mathrm{i}\hbar\frac{\partial}{\partial x}\right]\phi_t\right)\mathrm{d}x \\ &= -\mathrm{i}\hbar\int\left(\left[-\mathrm{i}\frac{\hbar}{2m}\overline{\phi}_{xx}+\mathrm{i}\frac{V}{\hbar}\overline{\phi}\right]\phi_x + \overline{\phi}\frac{\partial}{\partial x}\left[\mathrm{i}\frac{\hbar}{2m}\phi_{xx}-\mathrm{i}\frac{V}{\hbar}\phi\right]\right)\mathrm{d}x \\ &= -\frac{\hbar^2}{2m}\int([\overline{\phi}_x\phi_x]_x - [\overline{\phi}\phi_{xx}]_x)\,\mathrm{d}x - \int V_x|\phi|^2\,\mathrm{d}x \\ &= -\langle V_x\rangle,\end{aligned} \tag{2.2.28}$$

which may be compared with the equation of motion in the classical Newtonian mechanics,

$$m\frac{\mathrm{d}^2x}{\mathrm{d}t^2} = \frac{\mathrm{d}p}{\mathrm{d}t} = -V_x. \tag{2.2.29}$$

In other words, in quantum mechanics, in sense of expected value, quantum operators obey the equation of motion of Newtonian mechanics. This statement is known as the *Ehrenfest theorem.*

Complex potential and unstable particles

The profound meaning of the conservation law (2.2.12) is that a particle can never disappear once it is present. Here we show that a small modification may be made so that we are able to describe unstable particles which may disappear after some time lapse. We will not justify whether such a modification is physically correct but will only be content to know that there is room in the Schrödinger equation to accommodate theoretical explorations. To this end, we assume that the potential energy V in (2.2.11) is perturbed by an imaginary quantity,

$$V = V_1 + \mathrm{i}V_2, \quad V_1 \text{ and } V_2 \text{ are both real-valued,} \tag{2.2.30}$$

which is allowed in the mathematical setting of the equation. Hence, (2.2.11) becomes

$$\mathrm{i}\hbar\frac{\partial\phi}{\partial t} = -\frac{\hbar^2}{2m}\frac{\partial^2\phi}{\partial x^2} + (V_1+\mathrm{i}V_2)\phi. \tag{2.2.31}$$

In view of (2.2.31), we see that the probability that there is a particle present at time t, that is,

$$P(t) = \int |\phi(x,t)|^2\,\mathrm{d}x, \tag{2.2.32}$$

satisfies the equation

$$P'(t) = \frac{2}{\hbar} \int V_2(x,t)|\phi(x,t)|^2 \, \mathrm{d}x. \tag{2.2.33}$$

This clearly indicates that it is the presence of V_2 that breaks down the probability conservation law (2.2.22).

For simplicity, we further assume that there is a constant $K > 0$ such that

$$V_2(x,t) \leq -K, \quad \forall x, t. \tag{2.2.34}$$

Then (2.2.33) and (2.2.34) lead us to

$$P'(t) \leq -\frac{2K}{\hbar} P(t). \tag{2.2.35}$$

Thus, if a particle is present initially, then $P(0) = 1$. Consequently, we can integrate (2.2.35) to infer that

$$P(t) \leq \mathrm{e}^{-\frac{2K}{\hbar} t}, \quad t > 0. \tag{2.2.36}$$

In other words, in a bulk situation, we will observe loss of particles as time elapses, suggesting that we encounter *unstable particles.*

Equation in higher dimensions

Our discussion about the one-dimensional Schrödinger equations can be extended to arbitrarily high dimensions. For this purpose, we consider the spacetime of dimension $(n+1)$ with coordinates $t = x^0, \mathbf{x} = (x^1, \dots, x^n)$, for time and space, respectively. We use the Greek letters μ, ν, etc., to denote the spacetime indices, $\mu, \nu = 0, 1, \dots, n$, the Latin letters i, j, k, etc., the space indices, $i, j, k = 1, \dots, n$, and ∇ the gradient operator on functions depending on $x^1, \dots, x^n$.

The Schrödinger equation that quantum-mechanically governs a particle of mass m in $\mathbb{R}^n$ is given by

$$\mathrm{i}\hbar \frac{\partial \phi}{\partial t} = -\frac{\hbar^2}{2m} \Delta \phi + V\phi, \tag{2.2.37}$$

since the energy and momentum operators are, respectively, given by

$$\hat{E} = \mathrm{i}\hbar \frac{\partial}{\partial t}, \quad \hat{\mathbf{p}} = -\mathrm{i}\hbar \nabla, \tag{2.2.38}$$

and the total energy operator, or the Hamiltonian, is

$$\hat{H} = \frac{1}{2m} \hat{\mathbf{p}}^2 + V. \tag{2.2.39}$$

Subsequently, the associated probability current $j = (j^\mu) = (j^0, \mathbf{j}) = (\rho, j^i)$ is defined by

$$\rho = |\phi|^2, \quad j^i = \mathrm{i}\frac{\hbar}{2m}(\phi \partial_i \overline{\phi} - \overline{\phi} \partial_i \phi), \quad i = 1, \dots, n, \tag{2.2.40}$$

such that the conservation law relating probability density ρ or "charge" and probability current $\mathbf{j}$ or "current" reads

$$\partial_\mu j^\mu = 0, \quad \text{or} \quad \frac{\partial}{\partial t}\rho + \nabla \cdot \mathbf{j} = 0. \tag{2.2.41}$$

Moreover, the Ehrenfest theorem says

$$\frac{\mathrm{d}\langle \hat{\mathbf{p}} \rangle}{\mathrm{d}t} = -\langle \nabla V \rangle. \tag{2.2.42}$$

Note also that, in applications, the potential function V may also be self-induced by the wave function ϕ. For example, if $V = |\phi|^2$ in (2.2.37), it is the classical *cubic Schrödinger equation*.

Steady state and energy spectrum

Here we revisit the one-dimensional situation and look for a solution of (2.2.11), where $V = V(x)$, in the separable form

$$\phi(x,t) = T(t)u(x). \tag{2.2.43}$$

Following a standard procedure, we have

$$\mathrm{i}\hbar \frac{T'(t)}{T(t)} = -\frac{\hbar^2}{2m}\frac{u''(x)}{u(x)} + V(x) = E, \tag{2.2.44}$$

where E is a constant. In terms of E, we have

$$T(t) = \mathrm{e}^{-\mathrm{i}\frac{E}{\hbar}t} \tag{2.2.45}$$

and u and E satisfy the relation

$$\left(-\frac{\hbar^2}{2m}\frac{\mathrm{d}^2}{\mathrm{d}x^2} + V \right) u = \hat{H}u = Eu. \tag{2.2.46}$$

Suppose that a quantum state of the particle is described by the separable solution (2.2.43). Then the normalization condition (2.2.12) becomes

$$\int |u|^2 \, \mathrm{d}x = 1. \tag{2.2.47}$$

Using (2.2.46), (2.2.47), and the fact that E must be real otherwise ϕ is not normalizable, we get

$$\langle \hat{H} \rangle = \int \overline{\phi} \hat{H} \phi \, \mathrm{d}x = E \int |u|^2 \, \mathrm{d}x = E, \tag{2.2.48}$$

which should not come as a surprise. Furthermore, since the second moment of $\hat{H}$ is simply

$$\langle \hat{H}^2 \rangle = \int \overline{\phi} \hat{H}^2 \phi \, \mathrm{d}x = E^2 \int |u|^2 \, \mathrm{d}x = E^2, \tag{2.2.49}$$

we see that the variance $\sigma^2_{\hat{H}}$ of the measurements of $\hat{H}$ is zero because

$$\sigma^2_{\hat{H}} = \langle \hat{H}^2 \rangle - \langle \hat{H} \rangle^2 = E^2 - E^2 = 0. \tag{2.2.50}$$

Such a result implies that, when a particle occupies a separable state, its energy is a definite value.

Now consider the general situation. For simplicity, we assume that $\hat{H}$ has $\{E_n\}$ as its complete energy spectrum so that the corresponding energy eigenstates $\{u_n\}$ form an orthonormal basis with respect to the natural L^2-inner product. Therefore, an arbitrary solution $\phi(x,t)$ of (2.2.11) can be represented as

$$\phi(x,t) = \sum_{n=1}^{\infty} c_n \mathrm{e}^{-\mathrm{i}\frac{E_n}{\hbar}t} u_n(x). \tag{2.2.51}$$

Therefore, in view of the normalization condition (2.2.12), we arrive at

$$\sum_{n=1}^{\infty} |c_n|^2 = 1. \tag{2.2.52}$$

This result is amazing since it reminds us of the total probability of a discrete random variable whose probability mass density function is given by the sequence $\{|c_n|^2\}$. In quantum mechanics, indeed, such a random variable is *postulated* as the measured value of energy for the particle that occupies the state given by (2.2.51). In other words, if we use $\mathcal{E}$ to denote the random reading of the energy of the particle occupying the state (2.2.51), then $\mathcal{E}$ may only take $E_1, E_2, \dots, E_n, \dots,$ as possible values. Furthermore, if these values are distinct, then

$$P(\{\mathcal{E} = E_n\}) = |c_n|^2, \quad n = 1, 2, \dots. \tag{2.2.53}$$

In quantum mechanics, this statement appears as a major postulate, which is also referred to as the "*generalized statistical interpretation*" of eigenstate representation. In particular, when $\phi(x,t)$ itself is separable as given in (2.2.43), since E is an eigenvalue itself, we see that $\mathcal{E}$ takes the single value E with probability one. Thus, we recover the earlier observation made on a separable state.

Finally, using (2.2.51), we can compute the expected value of the energy operator $\hat{H}$ immediately:

$$\begin{aligned} \langle \hat{H} \rangle &= \int \overline{\phi} \hat{H} \phi \, \mathrm{d}x \\ &= \sum_{n=1}^{\infty} E_n |c_n|^2. \end{aligned} \tag{2.2.54}$$

It is interesting to note that (2.2.54) is consistent with the postulate (2.2.53). In fact, we can see that the interpretation (2.2.53) is motivated or supported by (2.2.54).

Note also that the right-hand side of (2.2.54) is independent of time t. Thus, (2.2.54) may be viewed as a quantum-mechanical version of the energy conservation law.

Heisenberg equation

Let $\phi(t) = \phi(\mathbf{x}, t)$ be a time-dependent state evolving from its initial state $\phi_0 = \phi_0(\mathbf{x})$ and governed by the Hamiltonian $\hat{H}$. Then it satisfies

$$\mathrm{i}\hbar\frac{\partial\phi}{\partial t} = \hat{H}\phi, \quad \phi(0) = \phi_0, \tag{2.2.55}$$

over $\mathbb{R}^n$. Formally the solution $\phi(t)$ may be rewritten as

$$\phi(t) = \mathrm{e}^{-\mathrm{i}\frac{t}{\hbar}\hat{H}}\phi_0. \tag{2.2.56}$$

Thus, if $\hat{A}$ is an observable, then its expected value in the state ϕ at time t is given by

$$\begin{aligned}\langle\hat{A}\rangle(t) &= \int \overline{\phi(t)}\hat{A}\phi(t)\,\mathrm{d}\mathbf{x} \\ &= \int \overline{\phi_0}\left(\mathrm{e}^{\mathrm{i}\frac{t}{\hbar}\hat{H}}\hat{A}\mathrm{e}^{-\mathrm{i}\frac{t}{\hbar}\hat{H}}\right)\phi_0\,\mathrm{d}\mathbf{x}.\end{aligned} \tag{2.2.57}$$

On the other hand, we may require that the observable evolve according to a certain law so that at time t it becomes $\hat{A}(t)$, but keep the state unchanging so that its expected value in the state ϕ_0 is

$$\langle\hat{A}(t)\rangle = \int \overline{\phi_0}\hat{A}(t)\phi_0\,\mathrm{d}\mathbf{x}. \tag{2.2.58}$$

Of course, such a new description should give us the same result as that given by the evolution of the state governed by the Schrödinger equation. So we have

$$\langle\hat{A}\rangle(t) = \langle\hat{A}(t)\rangle. \tag{2.2.59}$$

Combining (2.2.57)–(2.2.59) and noting that ϕ_0 is arbitrary, we arrive at

$$\hat{A}(t) = \mathrm{e}^{\mathrm{i}\frac{t}{\hbar}\hat{H}}\hat{A}\mathrm{e}^{-\mathrm{i}\frac{t}{\hbar}\hat{H}}, \tag{2.2.60}$$

which is the solution of the operator-valued evolution equation

$$\frac{\mathrm{d}}{\mathrm{d}t}\hat{A}(t) = \dot{\hat{A}}(t) = \frac{1}{\mathrm{i}\hbar}[\hat{A}(t), \hat{H}], \quad \hat{A}(0) = \hat{A}, \tag{2.2.61}$$

known as the *Heisenberg equation*, which gives an equivalent description of quantum mechanics, in terms of evolving observables instead of states, often referred to as the *Heisenberg picture*.

We can also compare the Heisenberg equation (2.2.61) with the Hamiltonian equation (1.1.23).

2.3 Quantum many-body problem

Consider a quantum-mechanical description of N particles of respective masses m_i and electric charges Q_i, $i = 1, \ldots, N$, interacting solely through the Coulomb force. The locations of these N particles are at $\mathbf{x}_i \in \mathbb{R}^3$, $i = 1, \ldots, N$. Thus we can express their respective momenta as

$$\hat{\mathbf{p}}_i = -\mathrm{i}\hbar\nabla_{\mathbf{x}_i}, \quad i = 1, \ldots, N. \tag{2.3.1}$$

The potential function is given by

$$V(\mathbf{x}_1, \ldots, \mathbf{x}_N) = \sum_{1 \leq i < j \leq N}^{N} \frac{Q_i Q_j}{|\mathbf{x}_i - \mathbf{x}_{i'}|}. \tag{2.3.2}$$

So the N-particle system Hamiltonian reads

$$\hat{H} = \sum_{i=1}^{N} \frac{1}{2m_i}\hat{\mathbf{p}}_i^2 + V(\mathbf{x}_1, \ldots, \mathbf{x}_N), \tag{2.3.3}$$

which forms the foundation of a *quantum N-body problem.*

Atom model

An important special situation of the quantum N-body problem is the classical atom model in which Z electrons, each of electric charge $-e$ and mass m, orbit around a heavy nucleus of electric charge Ze resting at the origin. In this case, the Hamiltonian becomes

$$\hat{H} = -\sum_{i=1}^{Z} \frac{\hbar^2}{2m}\nabla_{\mathbf{x}_i}^2 - \sum_{i=1}^{Z} \frac{Ze^2}{|\mathbf{x}_i|} + \sum_{1 \leq i < j \leq Z}^{Z} \frac{e^2}{|\mathbf{x}_i - \mathbf{x}_j|}, \tag{2.3.4}$$

where the second term describes the Coulomb interaction of the electrons with the nucleus and the third term that between the electrons.

Hydrogen atom

In the case of a hydrogen atom, $Z = 1$, and the third term disappears. Thus, we arrive at the simplest possible Hamiltonian

$$\hat{H} = -\frac{\hbar^2}{2m}\nabla_{\mathbf{x}}^2 - \frac{e^2}{|\mathbf{x}|}. \tag{2.3.5}$$

The spectrum of (2.3.5), say $\{E\}$, consists of two different portions: $E > 0$, which happens to be continuous, and $E < 0$, which happens to be discrete. Since the Coulomb potential vanishes at infinity, the state with $E < 0$ indicates that the electron is in a state that lies inside the "potential well" of the nucleus, the proton, which is called a *bound state*, and describes a situation when the electron and proton "bind" to form a composite particle, the hydrogen. Likewise, the state with $E > 0$ indicates that the electron is in a state that lies outside the potential well of the proton, which is called a *scattering state*, and describes a situation when the electron and proton interact as two charged "free" particles, which do not appear to have the characteristics of a composite particle, namely, a hydrogen atom. Hence we are interested in the bound state situation only.

Restricting to spherically symmetric configurations, the texts [255, 464] show that the bound-state energy spectrum of (2.3.5) is given by

$$E_n = -\frac{me^4}{2\hbar^2 n^2} = \frac{E_1}{n^2}, \quad n = 1, 2, \dots, \tag{2.3.6}$$

known as the *Bohr formula*. The ground-state energy,

$$E_1 = -\frac{me^4}{2\hbar^2}, \tag{2.3.7}$$

is about -13.6 eV, which is what is needed to ionize a hydrogen atom.

Suppose that the hydrogen atom absorbs or emits an amount of energy, E_δ, so that the initial and final energies are $E_{\rm i}$ and $E_{\rm f}$, corresponding to $n = n_{\rm i}$ and $n = n_{\rm f}$ in (2.3.6), respectively. Then

$$\begin{aligned} E_\delta &= E_{\rm i} - E_{\rm f} \\ &= E_1 \left(\frac{1}{n_{\rm i}^2} - \frac{1}{n_{\rm f}^2} \right). \end{aligned} \tag{2.3.8}$$

It will be instructive to examine in some detail that the hydrogen atom is made to emit energy through the form of light. The Einstein formula (2.1.2) states that the frequency ν of the light obeys

$$E_\delta = h\nu = h\frac{c}{\lambda}, \tag{2.3.9}$$

where c is the speed and λ is the wavelength of light. Substituting (2.3.9) into (2.3.8), we arrive at the celebrated *Rydberg formula*

$$\frac{1}{\lambda} = \frac{me^4}{4\pi c\hbar^3} \left(\frac{1}{n_{\rm f}^2} - \frac{1}{n_{\rm i}^2} \right), \quad n_{\rm i} > n_{\rm f}. \tag{2.3.10}$$

Specifically, transitions to the ground state $n_{\rm f} = 1$ give rise to ultraviolet (higher-frequency) lights with

$$\frac{1}{\lambda} = \frac{me^4}{4\pi c\hbar^3} \left(1 - \frac{1}{n^2} \right), \quad n = 2, 3, \dots, \tag{2.3.11}$$

called the Lyman series; transitions to the first excited state $n_{\rm f} = 2$ lead to visible (medium-frequency) lights with $n = 3, 4, \dots,$ called the Balmer series; transitions to the second excited state $n_{\rm f} = 3$ correspond to infrared (lower-frequency) lights with $n = 4, 5, \dots,$ called the Paschen series. The series with $n_{\rm f} = 4, 5, 6$ are named under Brackett, Pfund, and Humphreys, respectively. Rydberg presented his formula in 1888, many years before the formulation of the Schrödinger equation and quantum mechanics.

Helium model and beyond

The model for helium, with $Z = 2$, immediately becomes more difficult because the Hamiltonian takes the form

$$\hat{H} = -\left(\frac{\hbar^2}{2m}\nabla^2_{\mathbf{x}_1} + \frac{2e^2}{|\mathbf{x}_1|}\right) - \left(\frac{\hbar^2}{2m}\nabla^2_{\mathbf{x}_2} + \frac{2e^2}{|\mathbf{x}_2|}\right) + \frac{e^2}{|\mathbf{x}_1 - \mathbf{x}_2|}, \tag{2.3.12}$$

in which the last term renders the problem non-separable. The model for lithium, with $Z = 3$, shares the same difficulty.

Thus, we see that the quantum N-body problem is important for particle physics and quantum chemistry but difficult to deal with when $N \geq 2$. A way out of this is to develop approximation methods. Along this direction, three well-known approaches are the Hartree–Fock method, the Thomas–Fermi model, and density functional theory, all based on variational techniques. The subsequent sections briefly discuss these approaches.

We note that, while the classical N-body problem is nonlinear, its quantum-mechanical version, which asks about the spectrum of the N-body Hamiltonian, becomes linear.

For comprehensive textbook treatments of quantum mechanics, see [154, 169, 251, 255, 393, 464]. For some additional reading on the mathematical foundation and history of quantum mechanics, see [378, 412].

2.4 Hartree–Fock method

To motivate our discussion, we use $\{E_n\}$ to denote the complete sequence of eigenvalues of the Hamiltonian $\hat{H}$ so that

$$E_1 \leq E_2 \leq \cdots \leq E_n \leq \cdots, \tag{2.4.1}$$

and $\{u_n\}$ the corresponding eigenstates which form an orthonormal basis. Let ϕ be any normalized function, satisfying

$$\langle\phi|\phi\rangle = \int \overline{\phi}\phi \, \mathrm{d}x = \int |\phi|^2 \, \mathrm{d}x = 1, \tag{2.4.2}$$

where x denotes a generic spatial point of the problem. Thus, the expansion

$$\phi = \sum_{n=1}^{\infty} c_n u_n, \tag{2.4.3}$$

gives us

$$\sum_{n=1}^{\infty} |c_n|^2 = 1. \tag{2.4.4}$$

Consequently, we have

$$\langle\phi|\hat{H}|\phi\rangle = \int \overline{\phi}\hat{H}\phi \, \mathrm{d}x = \sum_{n=1}^{\infty} E_n|c_n|^2 \geq E_1 \sum_{n=1}^{\infty} |c_n|^2 = E_1. \tag{2.4.5}$$

Minimization problem

In other words, the lowest eigenpair (E_1, u_1) may be obtained from solving the minimization problem

$$\min\left\{\langle\phi|\hat{H}|\phi\rangle \mid \langle\phi|\phi\rangle = 1\right\}. \tag{2.4.6}$$

In practice, it is often hard to approach (2.4.6) directly due to lack of compactness. Instead, one may come up with a reasonably good wave-function configuration, a trial approximation, depending on finitely many parameters, say $\alpha_1, \dots, \alpha_m$, of the form

$$\phi(x) = \phi(\alpha_1, \dots, \alpha_m)(x). \tag{2.4.7}$$

Then one solves the minimization problem

$$\min\left\{\langle\phi(\alpha_1, \dots, \alpha_m|\hat{H}|\psi(\alpha_1, \dots, \alpha_m)\rangle \mid \langle\phi(\alpha_1, \dots, \alpha_m)|\phi(\alpha_1, \dots, \alpha_m)\rangle = 1\right\}, \tag{2.4.8}$$

involving multivariable functions of $\alpha_1, \dots, \alpha_m$ only.

Many-body problem

We now consider the Z electron system with $x = (\mathbf{x}_1, \dots, \mathbf{x}_Z)$ and rewrite the Hamiltonian (2.3.4) as

$$\hat{H} = \sum_{i=1}^{Z} \hat{H}_i + \frac{1}{2}\sum_{i \neq j} V_{ij}, \tag{2.4.9}$$

where

$$\hat{H}_i = -\frac{\hbar^2}{2m}\nabla^2_{\mathbf{x}_i} - \frac{Ze^2}{|\mathbf{x}_i|}, \quad V_{ij} = \frac{e^2}{|\mathbf{x}_i - \mathbf{x}_j|}, \quad i \neq j, \quad i, j = 1, \dots, Z, \tag{2.4.10}$$

are the ith electron Hamiltonian, without inter-electron interaction, and the inter-electron Coulomb potential between the ith and jth electrons, respectively.

Reduced energy functional

Since the non-interacting Hamiltonian $\sum_{i=1}^{Z} \hat{H}_i$ allows separation of variables, we are prompted to use the trial configuration

$$\phi(\mathbf{x}_1, \dots, \mathbf{x}_Z) = \phi_1(\mathbf{x}_1) \cdots \phi_Z(\mathbf{x}_Z), \quad \mathbf{x}_1, \dots, \mathbf{x}_Z \in \mathbb{R}^3, \tag{2.4.11}$$

known as the *Hartree product*, where $\phi_1, \dots, \phi_Z$ are unknowns. In order to implement the normalization condition $\langle\phi|\phi\rangle = 1$, we impose

$$\langle\phi_i|\phi_i\rangle = \int |\phi_i|^2(\mathbf{x}_i)\,\mathrm{d}\mathbf{x}_i = \int |\phi_i|^2(\mathbf{x})\,\mathrm{d}\mathbf{x} = 1, \quad i = 1, \dots, Z. \tag{2.4.12}$$

Inserting (2.4.11) and using (2.4.12), we arrive at

$$
\begin{aligned}
I(\phi_1,\dots,\phi_Z) &= \int \overline{\phi}\hat{H}\phi \,\mathrm{d}\mathbf{x}_1\cdots \mathrm{d}\mathbf{x}_Z \\
&= \sum_{i=1}^{Z}\int \overline{\phi}_i \hat{H}_i \phi_i \,\mathrm{d}\mathbf{x}_i + \frac{1}{2}\sum_{i\neq j}\int \overline{\phi}_i\overline{\phi}_j V_{ij}\phi_i\phi_j \,\mathrm{d}\mathbf{x}_i\,\mathrm{d}\mathbf{x}_j \\
&= \sum_{i=1}^{Z}\int \left(\frac{\hbar^2}{2m}|\nabla\phi_i|^2 - \frac{Ze^2}{|\mathbf{x}|}|\phi_i|^2\right)\mathrm{d}\mathbf{x} + \frac{e^2}{2}\sum_{i\neq j}\int \frac{|\phi_i(\mathbf{x})|^2|\phi_j(\mathbf{y})|^2}{|\mathbf{x}-\mathbf{y}|}\,\mathrm{d}\mathbf{x}\mathrm{d}\mathbf{y},
\end{aligned}
\tag{2.4.13}
$$

where we have renamed the dummy variables with $\mathbf{x},\mathbf{y}\in\mathbb{R}^3$.

Integro-differential equations

Consequently, we are led to considering the reduced constrained minimization problem

$$
\min\{I(\phi_1,\dots,\phi_Z)\,|\,\langle\phi_i|\phi_i\rangle = 1,\ i=1,\dots,Z\}, \tag{2.4.14}
$$

whose solutions may be obtained by solving the following system of nonlinear integro-differential equations

$$
\frac{\hbar^2}{2m}\Delta\phi_i + \frac{Ze^2}{|\mathbf{x}|}\phi_i + \lambda_i\phi_i = e^2\left(\sum_{j\neq i}^{Z}\int\frac{|\phi_j(\mathbf{y})|^2}{|\mathbf{x}-\mathbf{y}|}\,\mathrm{d}\mathbf{y}\right)\phi_i, \quad i=1,\dots,Z, \tag{2.4.15}
$$

with the Lagrange multipliers $\lambda_1,\dots,\lambda_Z$ appearing as eigenvalues. Thus, in particular, we see that, in order to solve a linear problem with interacting potential, we are offered a highly nontrivial nonlinear problem to tackle instead.

Slater determinant

Note that the discussion of the *Hartree–Fock method* is over-simplified. Since electrons are *fermions* that obey the *Pauli exclusion principle*, the wave function should be considered as skew-symmetric, with respect to permutations of $\mathbf{x}_1,\dots,\mathbf{x}_Z$, which give rise to their joint wave function $\phi(\mathbf{x}_1,\dots,\mathbf{x}_Z)$. Thus, practically, we need to consider the problem with the redesigned skew-symmetric wave function

$$
\phi(\mathbf{x}_1,\dots,\mathbf{x}_Z) = \frac{1}{\sqrt{Z!}}\begin{vmatrix} \phi_1(\mathbf{x}_1) & \phi_1(\mathbf{x}_2) & \dots & \phi_1(\mathbf{x}_Z) \\ \phi_2(\mathbf{x}_1) & \phi_2(\mathbf{x}_2) & \dots & \phi_2(\mathbf{x}_Z) \\ \dots & \dots & \dots & \dots \\ \phi_Z(\mathbf{x}_1) & \phi_Z(\mathbf{x}_2) & \dots & \phi_Z(\mathbf{x}_Z) \end{vmatrix}, \tag{2.4.16}
$$

subject to the orthonormal condition

$$
\langle\phi_i|\phi_j\rangle = \int\overline{\phi}_i(\mathbf{x})\phi_j(\mathbf{x})\,\mathrm{d}\mathbf{x} = \delta_{ij}, \quad i,j=1,\dots,Z. \tag{2.4.17}
$$

This formalism is known as the *Slater determinant* [154, 464, 517], which makes the problem more complicated.

2.5 Thomas–Fermi approach

The Hartree–Fock method is effective when the atom number Z is small. When Z is large, the problem quickly becomes difficult and alternative methods are needed. The Thomas [545] and Fermi [207] approach treats electrons as a *static electron gas cloud* surrounding a nucleus and subject to a continuously distributed electrostatic potential.

Self-consistency equation

The electron at $\mathbf{x}$ assumes the maximum energy, say $-eA$, where A is a constant, otherwise the electrons will not remain in the static state. Let the electrostatic potential be $\phi(\mathbf{x})$. Then $-e\phi(\mathbf{x})$ will be the potential energy carried by the electron. Thus, if we use $p(\mathbf{x})$ to denote the maximum momentum of the electron, we have the relation

$$-eA = \frac{p^2(\mathbf{x})}{2m} - e\phi(\mathbf{x}). \tag{2.5.1}$$

On the other hand, let $n(\mathbf{x})$ be the number of electrons over a tiny domain in space, say $\delta\Omega$, centered around $\mathbf{x}$ and of volume $\mathrm{d}\mathbf{x}$. That is, $n(\mathbf{x})$ is the electron number density. Then $p(\mathbf{x})$ is approximately a constant over $\delta\Omega$. We assume that all states in the momentum space are occupied by the electrons which take up a volume

$$\frac{4\pi}{3}p^3(\mathbf{x}) \tag{2.5.2}$$

in the momentum space. Since each state can be occupied by exactly one electron, due to Pauli's exclusion principle, we arrive at the electron number count (in $\delta\Omega$)

$$n(\mathbf{x}) = 2\frac{\frac{4\pi}{3}p^3(\mathbf{x})}{h^3}, \tag{2.5.3}$$

where h is the unbarred Planck constant so that the angular wave number in the de Broglie wave–particle duality hypothesis is replaced with the usual wave number and the factor 2 takes account of the two possible spins of the electrons. Inserting (2.5.1) into (2.5.3), we have

$$n(\mathbf{x}) = \frac{8\pi}{3h^3}\left(2me[\phi(\mathbf{x}) - A]\right)^{\frac{3}{2}}. \tag{2.5.4}$$

On the other hand, we know that the electrostatic potential function ϕ and the electron number density n are related through the Poisson equation

$$\Delta\phi = 4\pi e n, \tag{2.5.5}$$

where $-ne = \rho$ is the charge density (it may be helpful to compare (2.5.5) with some of its concrete realizations in Section 2.3). In view of (2.5.4) and (2.5.5), we obtain the *self-consistency equation*

$$\Delta\phi = \alpha(\phi - A)^{\frac{3}{2}}, \quad \alpha = \frac{32\pi^2 e}{3h^3}(2me)^{\frac{3}{2}}, \tag{2.5.6}$$

which serves as the governing equation of the Thomas–Fermi method, also called the *Thomas–Fermi equation*. Of course, a meaningful solution must satisfy $\phi \geq A$.

Since the electron cloud surrounds a nucleus of charge Ze, we see that ϕ should behave like a central Coulomb potential, $Ze/|\mathbf{x}|$, near the origin. Hence, we have the singular boundary condition

$$\lim_{|\mathbf{x}|\to 0} |\mathbf{x}|\phi(\mathbf{x}) = Ze. \tag{2.5.7}$$

Besides, if we assume the electron cloud is concentrated in a bounded domain, say Ω, then $n = 0$ on $\partial\Omega$. Thus, (2.5.4) leads to the boundary condition

$$\phi(\mathbf{x}) = A, \quad \mathbf{x} \in \partial\Omega. \tag{2.5.8}$$

Variational method

We next pursue a pure variational formalism.

Use $n(\mathbf{x})$ to denote the electron density at $\mathbf{x}$. Then the total potential energy of the electron cloud is clearly

$$\mathcal{V} = -Ze^2 \int \frac{n(\mathbf{x})}{|\mathbf{x}|}\,\mathrm{d}\mathbf{x} + \frac{1}{2}e^2 \int \frac{n(\mathbf{x})n(\mathbf{y})}{|\mathbf{x}-\mathbf{y}|}\,\mathrm{d}\mathbf{x}\mathrm{d}\mathbf{y}. \tag{2.5.9}$$

On the other hand, if we use $\tau(\mathbf{x})$ to denote the kinetic energy density, then the total kinetic energy of the electron cloud is

$$\mathcal{T} = \int \tau(\mathbf{x})\,\mathrm{d}\mathbf{x}. \tag{2.5.10}$$

We now aim to approximate $\mathcal{T}$. To this end, we are to compute the kinetic energy carried by the electrons in $\delta\Omega$. First, note that the number of electrons filling the momentum shell between the concentric spheres of radius p and $p+\mathrm{d}p$ in the momentum space may be computed by

$$\frac{2}{h^3}(4\pi p^2\,\mathrm{d}p). \tag{2.5.11}$$

This formula follows from the assumption that a shell of radius p, of thickness $\mathrm{d}p$, in the momentum space is filled with electrons in the unit h in each of the three momentum directions, with two extra degrees of freedom, that is spin up and down. Hence, using (2.5.11) and applying (2.5.3), we obtain

$$\begin{aligned} \tau(\mathbf{x}) &= \int_0^{p(\mathbf{x})} \frac{p^2}{2m}\frac{2}{h^3}4\pi p^2\,\mathrm{d}p \\ &= \frac{4\pi}{5mh^3}p^5(\mathbf{x}) \\ &= \frac{3h^2}{10m}\left(\frac{3}{8\pi}\right)^{\frac{2}{3}} n^{\frac{5}{3}}(\mathbf{x}) \\ &\equiv \gamma n^{\frac{5}{3}}(\mathbf{x}). \end{aligned} \tag{2.5.12}$$

Consequently, we arrive at the total energy of the electron cloud:

$$
\begin{aligned}
E &= \mathcal{T} + \mathcal{V} \\
&= \gamma \int n^{\frac{5}{3}}(\mathbf{x})\,\mathrm{d}\mathbf{x} - Ze^2 \int \frac{n(\mathbf{x})}{|\mathbf{x}|}\,\mathrm{d}\mathbf{x} + \frac{e^2}{2}\int \frac{n(\mathbf{x})n(\mathbf{y})}{|\mathbf{x}-\mathbf{y}|}\,\mathrm{d}\mathbf{x}\mathrm{d}\mathbf{y},
\end{aligned} \tag{2.5.13}
$$

known as the *Thomas–Fermi energy*, which is to be minimized subject to the constraint of the total number of electrons:

$$
\int n(\mathbf{x})\,\mathrm{d}\mathbf{x} = Z, \tag{2.5.14}
$$

resulting in the Euler–Lagrange equation

$$
\frac{5}{3}\gamma n^{\frac{2}{3}}(\mathbf{x}) = \frac{Ze}{|\mathbf{x}|} - e^2 \int \frac{n(\mathbf{y})}{|\mathbf{x}-\mathbf{y}|}\,\mathrm{d}\mathbf{y} + \mu, \tag{2.5.15}
$$

where $\mu \in \mathbb{R}$ is a Lagrangian multiplier. This equation is also known as the Thomas–Fermi equation.

Modified energy functional

For greater applicability, people often modify the Thomas–Fermi energy by adding the *Weizsäcker correction term* [575]:

$$
\mathcal{T}_W = \frac{\hbar^2}{8m}\int \frac{1}{n(\mathbf{x})}|\nabla n(\mathbf{x})|^2\,\mathrm{d}\mathbf{x}. \tag{2.5.16}
$$

Hence, with the substitution

$$
w = n^{\frac{1}{2}}, \tag{2.5.17}
$$

we may consider the following *Thomas–Fermi–Weizsäcker energy*

$$
\begin{aligned}
E(w) = {} & \frac{\hbar^2}{2m}\int |\nabla w(\mathbf{x})|^2\,\mathrm{d}\mathbf{x} + \gamma \int w^{\frac{10}{3}}(\mathbf{x})\,\mathrm{d}\mathbf{x} \\
& - Ze^2 \int \frac{w^2(\mathbf{x})}{|\mathbf{x}|}\,\mathrm{d}\mathbf{x} + \frac{1}{2}e^2 \int \frac{w^2(\mathbf{x})w^2(\mathbf{y})}{|\mathbf{x}-\mathbf{y}|}\,\mathrm{d}\mathbf{x}\mathrm{d}\mathbf{y},
\end{aligned} \tag{2.5.18}
$$

subject to the quadratic constraint

$$
\int w^2(\mathbf{x})\,\mathrm{d}\mathbf{x} = Z. \tag{2.5.19}
$$

See [2, 193] for more extended models with the Weizsäcker corrections.

It is interesting to note that the Thomas–Fermi semi-classical treatment, and its extensions, of the quantum many-body that is linear turns the problem back into a nonlinear problem. See [355, 356, 357, 358] for the mathematical work on the Thomas–Fermi model and its many extensions and variations.

2.6 Density functional theory

Another important variational method, known as the *density functional theory*, nicknamed DFT, was developed by Hohenberg and Kohn [294], which significantly reduces the number of degrees of freedom of the quantum many-body problem so that computation becomes tangible. See [46, 173, 313] for surveys and extensions. Here we present a concise description of this formalism.

Reduction with density functions

Consider the Z-electron energy

$$E = \langle \phi | \hat{H} | \phi \rangle, \tag{2.6.1}$$

where $\hat{H}$ is as defined in (2.4.9) and (2.4.10) and $\phi(x) = \phi(\mathbf{x}_1, \dots, \mathbf{x}_Z)$ is a general normalized wave function satisfying

$$\int |\phi|^2(x)\,\mathrm{d}x = \int |\phi(\mathbf{x}_1, \dots, \mathbf{x}_Z)|^2\,\mathrm{d}\mathbf{x}_1 \cdots \mathrm{d}\mathbf{x}_Z = 1. \tag{2.6.2}$$

In view of this, we may naturally write the "marginal" density function for the ith electron as

$$n_i(\mathbf{x}) = \int |\phi(\mathbf{x}_1, \dots, \underset{i}{\mathbf{x}}, \dots, \mathbf{x}_Z)|^2\,\mathrm{d}\mathbf{x}_1 \cdots \widehat{\mathrm{d}\mathbf{x}_i} \cdots \mathrm{d}\mathbf{x}_Z, \tag{2.6.3}$$

where $\widehat{\mathrm{d}\mathbf{x}_i}$ indicates the absence of the factor $\mathrm{d}\mathbf{x}_i$ from the measure, and define the Z-electron density function $n(\mathbf{x})$ is by

$$n(\mathbf{x}) = \sum_{i=1}^{Z} n_i(\mathbf{x}). \tag{2.6.4}$$

As a consequence, the normalization condition (2.6.2) leads to

$$\int n(\mathbf{x})\,\mathrm{d}\mathbf{x} = Z, \tag{2.6.5}$$

which is identical to (2.5.14) as imposed in the Thomas–Fermi formalism.

To proceed further, assume that ϕ is either symmetric or anti-symmetric with respect to the permutation of its arguments:

$$\phi(\dots, \mathbf{x}_i, \dots, \mathbf{x}_j, \dots) = \pm\phi(\dots, \mathbf{x}_j, \dots, \mathbf{x}_i, \dots), \quad i \neq j. \tag{2.6.6}$$

Then there holds the compressed formula

$$n(\mathbf{x}) = Z \int |\phi(\mathbf{x}, \mathbf{x}_2, \dots, \mathbf{x}_Z)|^2\,\mathrm{d}\mathbf{x}_2 \cdots \mathrm{d}\mathbf{x}_Z. \tag{2.6.7}$$

Similarly, we may use (2.6.7) to calculate the nucleus-electron energy to be

$$\begin{aligned}\langle\phi|\hat{V}|\phi\rangle &= \int \overline{\phi}(x)\left(-\sum_{i=1}^{Z}\frac{Ze^2}{|\mathbf{x}_i|}\right)\phi(x)\,\mathrm{d}x \\ &= -Ze^2\int\frac{n(\mathbf{x})}{|\mathbf{x}|}\,\mathrm{d}\mathbf{x},\end{aligned} \tag{2.6.8}$$

given as the first part of (2.5.9).

We next calculate the electron-electron energy. Using (2.6.6) again, we have

$$\begin{aligned}\langle\phi|\hat{U}|\phi\rangle &= \int \overline{\phi}(x)\left(\frac{1}{2}\sum_{i,j=1,i\neq j}^{Z}\frac{e^2}{|\mathbf{x}_i-\mathbf{x}_j|}\right)\phi(x)\,\mathrm{d}x \\ &= \frac{e^2}{2}\int\frac{\rho(\mathbf{x},\mathbf{y})}{|\mathbf{x}-\mathbf{y}|}\,\mathrm{d}\mathbf{x}\,\mathrm{d}\mathbf{y},\end{aligned} \tag{2.6.9}$$

where

$$\rho(\mathbf{x},\mathbf{y}) = Z(Z-1)\int|\phi(\mathbf{x},\mathbf{y},\mathbf{x}_3,\ldots,\mathbf{x}_Z)|^2\,\mathrm{d}\mathbf{x}_3\cdots\mathrm{d}\mathbf{x}_Z, \tag{2.6.10}$$

is an order-2 density function called the *pair density*. It is clear that the density function $n(\mathbf{x})$ and pair density $\rho(\mathbf{x},\mathbf{y})$ are related through the equation

$$n(\mathbf{x}) = \frac{1}{Z-1}\int\rho(\mathbf{x},\mathbf{y})\,\mathrm{d}\mathbf{y}. \tag{2.6.11}$$

We see that (2.6.9) resembles but is not quite as given as the second part of (2.5.9).

We then consider the kinetic energy

$$\begin{aligned}\langle\phi|\hat{T}|\phi\rangle &= \int \overline{\phi}(x)\left(-\frac{\hbar^2}{2m}\sum_{i=1}^{Z}\nabla^2_{\mathbf{x}_i}\right)\phi(x)\,\mathrm{d}x \\ &= -\frac{Z\hbar^2}{2m}\int\overline{\phi}(\mathbf{x},\mathbf{x}_2,\ldots,\mathbf{x}_Z)\nabla^2_{\mathbf{x}}\phi(\mathbf{x},\mathbf{x}_2,\ldots,\mathbf{x}_Z)\,\mathrm{d}\mathbf{x}\,\mathrm{d}\mathbf{x}_2\cdots\mathrm{d}\mathbf{x}_Z \\ &= -\frac{\hbar^2}{2m}\int\left(\nabla^2_{\mathbf{y}}P_1(\mathbf{x},\mathbf{y})\right)_{\mathbf{y}=\mathbf{x}}\,\mathrm{d}\mathbf{x},\end{aligned} \tag{2.6.12}$$

where

$$P_1(\mathbf{x},\mathbf{y}) = Z\int\overline{\phi}(\mathbf{x},\mathbf{x}_2,\ldots,\mathbf{x}_Z)\phi(\mathbf{y},\mathbf{x}_2,\ldots,\mathbf{x}_Z)\,\mathrm{d}\mathbf{x}_2\cdots\mathrm{d}\mathbf{x}_Z, \tag{2.6.13}$$

whose "diagonal" part, $P_1(\mathbf{x},\mathbf{x})$, coincides with the density function $n(\mathbf{x})$.

The construction of $P_1(\mathbf{x},\mathbf{y})$ motivates a more general order-2 *density matrix* given as

$$\begin{aligned}&P_2(\mathbf{x},\mathbf{y};\mathbf{u},\mathbf{v}) \\ &= Z(Z-1)\int\overline{\phi}(\mathbf{x},\mathbf{y},\mathbf{x}_3,\ldots,\mathbf{x}_Z)\phi(\mathbf{u},\mathbf{v},\mathbf{x}_3,\ldots,\mathbf{x}_Z)\,\mathrm{d}\mathbf{x}_3\cdots\mathrm{d}\mathbf{x}_Z,\end{aligned} \tag{2.6.14}$$

resulting in the simple relation

$$\rho(\mathbf{x},\mathbf{y}) = P_2(\mathbf{x},\mathbf{y};\mathbf{x},\mathbf{y}), \tag{2.6.15}$$

for the pair density.

In summary, we see that the energy of the state ϕ is given by

$$\begin{aligned} E &= T + V + U = \langle\phi|\hat{T}|\phi\rangle + \langle\phi|\hat{V}|\phi\rangle + \langle\phi|\hat{U}|\phi\rangle \\ &= -\frac{\hbar^2}{2m}\int \left(\nabla_{\mathbf{y}}^2 P_1(\mathbf{x},\mathbf{y})\right)_{\mathbf{y}=\mathbf{x}} \,\mathrm{d}\mathbf{x} - Ze^2\int \frac{n(\mathbf{x})}{|\mathbf{x}|}\,\mathrm{d}\mathbf{x} + \frac{e^2}{2}\int \frac{\rho(\mathbf{x},\mathbf{y})}{|\mathbf{x}-\mathbf{y}|}\,\mathrm{d}\mathbf{x}\,\mathrm{d}\mathbf{y}. \end{aligned} \tag{2.6.16}$$

This energy involves interrelated densities $n(\mathbf{x}), \rho(\mathbf{x},\mathbf{y}), P_1(\mathbf{x},\mathbf{y})$ such that the dimension of the quantum Z-electron problem is reduced from $3Z$ to 6, which may be regarded as a theoretical progress.

Density functional theory

The idea of the density functional theory is to seek for a further reduction of the degrees of freedom of the problem. To this end, we rewrite the quantum-mechanical Hamiltonian in a more general manner as

$$\hat{H} = \hat{F} + \hat{V}, \tag{2.6.17}$$

where

$$\hat{F} = -\frac{\hbar^2}{2m}\sum_{i=1}^{Z}\nabla_i^2 + \frac{e^2}{2}\sum_{i\neq j}\frac{1}{|\mathbf{x}_i - \mathbf{x}_i|}, \tag{2.6.18}$$

denotes the "internal" or "universal" part of the system, which is independent of the model, and

$$\hat{V} = \sum_{i=1}^{Z} V(\mathbf{x}_i), \tag{2.6.19}$$

represents the "external" part of the system, which is dependent on the model.

As a preliminary fact, we first observe that, up to an additive constant, the potential function V is uniquely determined by an eigenstate of the Hamiltonian $\hat{H}$ it generates. Indeed, if $\hat{H}$ and $\hat{H}' = \hat{F} + \hat{V}'$ share the same eigenstate, $|\phi\rangle$, then we have

$$\hat{H}|\phi\rangle = E|\phi\rangle, \tag{2.6.20}$$

$$\hat{H}'|\phi\rangle = E'|\phi\rangle, \tag{2.6.21}$$

for some $E, E' \in \mathbb{R}$. From $\hat{H}' = \hat{F} + \hat{V} + (\hat{V}' - \hat{V}) = \hat{H} + (\hat{V}' - \hat{V})$ and using (2.6.20)–(2.6.21), we get $(E' - E)|\phi\rangle = (\hat{V}' - \hat{V})|\phi\rangle$, which leads to

$$\hat{V}' - \hat{V} = \sum_{i}^{Z} V'(\mathbf{x}_i) - \sum_{i=1}^{Z} V(\mathbf{x}_i) = E' - E. \tag{2.6.22}$$

Setting $\mathbf{x}_1 = \cdots = \mathbf{x}_Z = \mathbf{x}$ in (2.6.22), we see that $V'(\mathbf{x}) = V(\mathbf{x})$ modulo a constant.

Earlier, (2.6.7) indicated that the density function $n(\mathbf{x})$ is uniquely determined by the wave function ϕ. We next observe that the converse is also true under the assumption that the groundstate energy of the problem is non-degenerate. That is, the eigenspace associated with the minimum energy of the corresponding Hamiltonian is one-dimensional. In fact, for this a stronger statement has been established, known as the first *Hohenberg–Kohn theorem*: The density function uniquely determines the external potential, up to an additive constant, and the ground state, of the Hamiltonian.

To prove this theorem, use $|\phi\rangle$ and $|\phi'\rangle$ to denote two normalized ground states of $\hat{H}$ and $\hat{H}'$ generated from the potentials $\hat{V}$ and $\hat{V}'$, associated with the minimum eigenvalues E_1 and E_1' of $\hat{H}$ and $\hat{H}'$, respectively, such that the difference of $\hat{V}$ and $\hat{V}'$ is not a constant. Hence, $|\phi\rangle$ and $|\phi'\rangle$ are linearly independent. Assume otherwise that they give rise to the same density function $n(\mathbf{x})$. Then we have

$$\begin{aligned} E_1 &= \langle\phi|\hat{H}|\phi\rangle < \langle\phi'|\hat{H}|\phi'\rangle \\ &= \langle\phi'|\hat{H}'|\phi'\rangle + \langle\phi'|\hat{H} - \hat{H}'|\phi'\rangle \\ &= E_1' + \int n(\mathbf{x})\,(V(\mathbf{x}) - V'(\mathbf{x}))\,\mathrm{d}\mathbf{x}. \end{aligned} \tag{2.6.23}$$

Likewise, we have

$$E_1' < E_1 + \int n(\mathbf{x})\,(V'(\mathbf{x}) - V(\mathbf{x}))\,\mathrm{d}\mathbf{x}. \tag{2.6.24}$$

Adding (2.6.23) and (2.6.24), we arrive at the false statement $E_1 + E_1' < E_1' + E_1$. Thus the proof follows.

Consequently, the groundstate wave function may be regarded as a density function dependent quantity, denoted as $\phi = \phi(n)$, leading to the relation

$$\begin{aligned} E_1 = \langle\phi(n)|\hat{H}|\phi(n)\rangle &= \langle\phi(n)|\hat{F}|\phi(n)\rangle + \int n(\mathbf{x})V(\mathbf{x})\,\mathrm{d}\mathbf{x} \\ &\equiv F(n) + \int n(\mathbf{x})V(\mathbf{x})\,\mathrm{d}\mathbf{x}. \end{aligned} \tag{2.6.25}$$

Thus, if $n'(\mathbf{x})$ is a trial density function associated with its correspondingly determined potential function V', resulting in a normalized groundstate wave function $\phi' = \phi'(n')$, then there holds $\langle\phi'|\hat{H}|\phi'\rangle \geq \langle\phi|\hat{H}|\phi\rangle = E_1$, or equivalently,

$$\begin{aligned} E_1 &= F(n) + \int n(\mathbf{x})V(\mathbf{x})\,\mathrm{d}\mathbf{x} \\ &\leq \langle\phi'|\hat{F}|\phi'\rangle + \int n'(\mathbf{x})V(\mathbf{x})\,\mathrm{d}\mathbf{x} \\ &= F(n') + \int n'(\mathbf{x})V(\mathbf{x})\,\mathrm{d}\mathbf{x}, \end{aligned} \tag{2.6.26}$$

which spells out a variational principle known as the *second Hohenberg–Kohn theorem*. Thus we see, at least theoretically, that the determination of the groundstate of the system amounts to solving the constrained minimization problem

$$\min\left\{E(n) = F(n) + \int n(\mathbf{x})V(\mathbf{x})\,\mathrm{d}\mathbf{x} \,\middle|\, \int n(\mathbf{x})\,\mathrm{d}\mathbf{x} = Z\right\}, \tag{2.6.27}$$

over $\mathbb{R}^3$. This formalism is the essence of the density functional theory.

To proceed further, we now rewrite the density functional in view of (2.5.13) as

$$E(n) = T(n) + E_{\mathrm{H}}(n) + E_{\mathrm{xc}}(n) + \int n(\mathbf{x})V(\mathbf{x})\,\mathrm{d}\mathbf{x}, \tag{2.6.28}$$

where

$$T(n) = \langle \phi(n)|\hat{T}|\phi(n)\rangle \tag{2.6.29}$$

is the free kinetic energy,

$$E_{\mathrm{H}}(n) = \frac{e^2}{2}\int \frac{n(\mathbf{x})n(\mathbf{y})}{|\mathbf{x}-\mathbf{y}|}\,\mathrm{d}\mathbf{x}\mathrm{d}\mathbf{y}, \tag{2.6.30}$$

the *Hartree energy* or *Coulomb energy*, and $E_{\mathrm{xc}}(n)$ the *exchange-correlation energy* whose exact form expressed in terms of the density function n is usually unknown.

Thus, with the Slater determinant (2.4.16) and the normalization condition (2.4.17), we have the Z-electron density function given by

$$n(\mathbf{x}) = \sum_{i=1}^{Z} |\phi_i(\mathbf{x})|^2. \tag{2.6.31}$$

Besides, we also have

$$T(n) = -\frac{\hbar^2}{2m}\sum_{i=1}^{Z}\int \overline{\phi}_i(\mathbf{x})\nabla^2\phi_i(\mathbf{x})\,\mathrm{d}\mathbf{x}. \tag{2.6.32}$$

Governing equations

Using (2.6.31) and (2.6.32) in (2.6.28), and applying the constraints (2.4.17), we see that a ground state must satisfy the variational equations

$$\left(-\frac{\hbar^2}{2m}\nabla^2 + V_{\mathrm{eff}}(\mathbf{x})\right)\phi_i(\mathbf{x}) = \sum_{j=1}^{Z} c_{ij}\phi_j(\mathbf{x}), \quad i = 1,\ldots,Z, \tag{2.6.33}$$

for some constant, Z by Z, matrix $C = (c_{ij})$, consisting of the Lagrange multipliers, where the effective potential function V_{eff} is given as

$$V_{\mathrm{eff}}(\mathbf{x}) = V(\mathbf{x}) + e^2\int \frac{n(\mathbf{y})}{|\mathbf{x}-\mathbf{y}|}\,\mathrm{d}\mathbf{y} + \frac{\delta E_{\mathrm{xc}}(n)}{\delta n}(\mathbf{x}). \tag{2.6.34}$$

Since the left-hand operator of (2.6.33) is Hermitian, we see that the matrix C is also Hermitian.

Now consider the column vector $\Phi = (\phi_i)$ and recast (2.6.33) into the matrix form

$$\left(-\frac{\hbar^2}{2m}\nabla^2 + V_{\text{eff}}(\mathbf{x})\right)\Phi = C\Phi. \tag{2.6.35}$$

Let Q be a unitary matrix such that

$$Q^\dagger C Q = \text{diag}\{\varepsilon_1, \dots, \varepsilon_Z\} \equiv D, \tag{2.6.36}$$

where $\varepsilon_1, \dots, \varepsilon_Z \in \mathbb{R}$. Set $\Psi = Q^\dagger\Phi$. Thus, in terms of Ψ, the equation (2.6.35) takes a "diagonal" form:

$$\left(-\frac{\hbar^2}{2m}\nabla^2 + V_{\text{eff}}(\mathbf{x})\right)\Psi = D\Psi. \tag{2.6.37}$$

Consequently, with $\Psi = (\psi_i)$, we may rewrite (2.6.37) componentwise as

$$\left(-\frac{\hbar^2}{2m}\nabla^2 + V_{\text{eff}}(\mathbf{x})\right)\psi_i(\mathbf{x}) = \varepsilon_i\psi_i(\mathbf{x}), \quad i = 1, \dots, Z. \tag{2.6.38}$$

These equations are often referred to as the *Kohn–Sham equations*, and ψ_i is called the ith orbital and the Lagrange multiplier ε_i the ith orbital energy, $i = 1, \dots, Z$. The effective potential V_{eff} is also called the *Kohn–Sham potential.*

It is worth noting that the density function $n(\mathbf{x})$ is invariant with respect to the unitary change of variables into ψ_i ($i = 1, \dots, Z$). That is,

$$n(\mathbf{x}) = \sum_{i=1}^{Z} |\phi_i(\mathbf{x})|^2 = \sum_{i=1}^{Z} |\psi_i(\mathbf{x})|^2. \tag{2.6.39}$$

Moreover, in view of (2.6.36), the orthonormality condition (2.4.17) is preserved:

$$\langle\psi_i|\psi_j\rangle = \delta_{ij}, \quad i, j = 1, \dots, Z. \tag{2.6.40}$$

Furthermore, since there holds the matrix equation

$$(\Phi(\mathbf{x}_1), \dots, \Phi(\mathbf{x}_Z)) = Q(\Psi(\mathbf{x}_1), \dots, \Psi(\mathbf{x}_Z)), \tag{2.6.41}$$

we see that the Slater determinant

$$\psi(\mathbf{x}_1, \dots, \mathbf{x}_Z) = \frac{1}{\sqrt{Z!}}\begin{vmatrix} \psi_1(\mathbf{x}_1) & \psi_1(\mathbf{x}_2) & \dots & \psi_1(\mathbf{x}_Z) \\ \psi_2(\mathbf{x}_1) & \psi_2(\mathbf{x}_2) & \dots & \psi_2(\mathbf{x}_Z) \\ \dots & \dots & \dots & \dots \\ \psi_Z(\mathbf{x}_1) & \psi_Z(\mathbf{x}_2) & \dots & \psi_Z(\mathbf{x}_Z) \end{vmatrix}, \tag{2.6.42}$$

generated from $\psi_1, \dots, \psi_Z$, differs from the Slater determinant (2.4.16), generated from $\phi_1, \dots, \phi_Z$, only by a global phase factor, $\mathrm{e}^{\mathrm{i}\theta}$ ($\theta \in \mathbb{R}$), by the unitarity of the matrix Q, which establishes the invariance of the density function again.

Thus, we are led to a well-formulated nonlinear spectrum problem given in terms of the Kohn–Sham equations involving Z orbitals and the density function composed from these orbitals.

Exchange-correlation energy

Various approximate forms of the exchange-correlation energy E_{xc} are of practical interests, among which the *local density approximation* model [336] may be defined by

$$E_{\mathrm{xc}}(n) = \int n(\mathbf{x})(f_{\mathrm{x}}(n(\mathbf{x})) + f_{\mathrm{c}}(n(\mathbf{x})))\,\mathrm{d}\mathbf{x}, \tag{2.6.43}$$

where f_{x} and f_{c} are specifically defined functions taking accounts of the exchange and correlation energies, respectively and separately [379], for example,

$$f_{\mathrm{x}}(n) = \frac{a}{r_s}, \quad f_{\mathrm{c}}(n) = \frac{b}{r_s + c}, \tag{2.6.44}$$

where r_s is the dimensionless *Wigner–Seitz radius* given by the relation

$$\frac{1}{n} = \frac{4\pi r_s^3}{3}, \tag{2.6.45}$$

and the *generalized gradient approximation* model assumes the form

$$E_{\mathrm{xc}}(n) = \int n(\mathbf{x}) f(n(\mathbf{x}), \nabla n(\mathbf{x}))\,\mathrm{d}\mathbf{x}, \tag{2.6.46}$$

with, for example,

$$f(n, \nabla n) = n^{\frac{1}{3}} \left(1 + a\sigma^2 + b\sigma^4 + c\sigma^6\right)^m, \quad \sigma = \frac{|\nabla n|}{2p_{\mathrm{F}}(n) n}, \tag{2.6.47}$$

where $p_{\mathrm{F}}(n)$ is the local Fermi momentum determined through n by (2.5.3) which gives us

$$p_{\mathrm{F}}^3(n) = 3\pi^2 \hbar^3 n. \tag{2.6.48}$$

See [46] and references therein for more discussion on the generalized gradient approximation model.

Kinetic energy

Moreover, in implementing the computation in density functional theory, it is also practical to approximate the kinetic energy $T(n)$ in $E(n)$ by putting together the Thomas–Fermi kinetic energy consisting of (2.5.10) and (2.5.12) and the Weizsäcker correction term (2.5.16) to come up with the functional [108]

$$T_{\mathrm{TFW}}(n) = \frac{3\hbar^2}{10m} \left(3\pi^2\right)^{\frac{2}{3}} \int n^{\frac{5}{3}}(\mathbf{x})\,\mathrm{d}\mathbf{x} + \frac{\hbar^2}{8m} \int \frac{|\nabla n(\mathbf{x})|^2}{n(\mathbf{x})}\,\mathrm{d}\mathbf{x}. \tag{2.6.49}$$

See [39, 46, 108, 312, 432] and references therein for some comprehensive presentations of the density functional theory and its approximations in broad ranges of applications and modeling situations.

Exercises

1. Consider the Hamiltonian

$$\hat{H} = -\frac{\hbar^2}{2m}\nabla^2 + V(\mathbf{x}), \quad \mathbf{x} \in \mathbb{R}^n, \tag{2.E.1}$$

where V is a real-valued continuous function, and the eigenstate ϕ satisfies $\hat{H}\phi = E\phi$ for some $E \in \mathbb{R}$. Show that under the assumption $\phi(\mathbf{x}) = 0$ as $|\mathbf{x}| \to \infty$ (the probability of finding the particle near infinity is zero) there must hold

$$E > V(\mathbf{x}) \quad \text{for some } \mathbf{x} \in \mathbb{R}^n. \tag{2.E.2}$$

2. Consider the model of a one-dimensional infinite-square well given by the potential

$$V(x) = \begin{cases} 0, & -1 < x < 1, \\ \infty, & x < -1 \text{ or } x > 1, \end{cases} \tag{2.E.3}$$

so that the probability of finding the particle away from the well $-1 < x < 1$ is zero.

(a) Find all eigenstates and eigenvalues of the one-particle Hamiltonian

$$\hat{H} = -\frac{\hbar^2}{2m}\frac{\mathrm{d}^2}{\mathrm{d}x^2} + V(x). \tag{2.E.4}$$

In particular, show that the lowest possible energy level of the particle, called the *zero-point energy*, is actually nonzero, and becomes significant when the mass of the particle is small.

(b) Use (a) to determine the wave function $\phi(x,t)$ governing the particle at "rest" initially at the center of the well:

$$\phi(x,0) = \phi_0(x) = \begin{cases} \frac{1}{\sqrt{2\varepsilon}}, & -\varepsilon < x < \varepsilon, \\ 0, & \text{elsewhere}, \end{cases} \tag{2.E.5}$$

where $0 < \varepsilon < 1$ is a small constant. Plot a sample graph of the wave function to see that the particle cannot stay resting but is actually being "boiled" to rattle around (for simplicity, you may choose a concrete value for ε, say $\varepsilon = 1/32$, and make a finite-term truncation for the wave function, say keeping 3–4 modes).

(c) What happens when $\varepsilon \to 0$ in (b)?

3. Derive (2.2.33).
4. Establish (2.2.42).
5. Explain why E in (2.2.46) must be real.

6. Let the Hermitian operator $\hat{O}$ be an observable in the Schrödinger picture and $\hat{H}$ the Hamiltonian operator. Assume that the system lies in the state ϕ, which satisfies the Schrödinger equation

$$\mathrm{i}\hbar\frac{\partial\phi}{\partial t} = \hat{H}\phi. \tag{2.E.6}$$

   (a) Establish the formula

$$\frac{\mathrm{d}}{\mathrm{d}t}\langle\hat{O}\rangle = \frac{1}{\mathrm{i}\hbar}\langle[\hat{O},\hat{H}]\rangle + \left\langle\frac{\partial\hat{O}}{\partial t}\right\rangle. \tag{2.E.7}$$

   This formula is a more general realization of the Ehrenfest theorem.

   (b) Use (a) to derive (2.2.42).

7. Consider the Thomas–Fermi problem

$$\Delta\phi = \alpha(\phi - A)^{\frac{3}{2}},\ x \in \Omega;\ \phi(\mathbf{x}) = A,\ \mathbf{x} \in \partial\Omega;\ \lim_{|\mathbf{x}|\to 0}|\mathbf{x}|\phi(\mathbf{x}) = Ze, \tag{2.E.8}$$

   where α, A, Z, e are positive constant and Ω is a ball in $\mathbb{R}^3$ centered at the origin. Show that the problem has a radially symmetric solution. Is such a solution unique?

8. Show that the two Thomas–Fermi equations, (2.5.6) subject to (2.5.7) and (2.5.15), are in fact equivalent over $\mathbb{R}^3$ by noting that the *Newton potential*

$$u(\mathbf{x}) = \int \frac{f(\mathbf{y})}{|\mathbf{x}-\mathbf{y}|}\,\mathrm{d}\mathbf{y} \tag{2.E.9}$$

   leads to the relation [236]:

$$\Delta u = -4\pi f. \tag{2.E.10}$$

9. Consider the Hartree product given in (2.4.11) subject to the normalization condition (2.4.12). Show that the ith particle density defined by (2.6.3) is given as

$$n_i(\mathbf{x}) = |\phi_i(\mathbf{x})|^2, \quad i = 1, \ldots, Z. \tag{2.E.11}$$

10. For $Z = 2$, show that the functions

$$\phi_\pm(\mathbf{x}_1, \mathbf{x}_2) = \frac{1}{\sqrt{2}}(\phi_1(\mathbf{x}_1)\phi_2(\mathbf{x}_2) \pm \phi_1(\mathbf{x}_2)\phi_2(\mathbf{x}_1)), \tag{2.E.12}$$

    subject to the condition (2.4.17), are normalized wave functions, which are symmetric and skew-symmetric, respectively.

11. With the symmetric and skew-symmetric wave function defined in (2.E.12), obtain the corresponding Hartree–Fock energy and governing equations, as given in (2.4.13) and (2.4.15), respectively.

12. Verify (2.6.31).
13. Verify (2.6.32).
14. Let P be the set of all permutations of the set $\{1, 2, \ldots, Z\}$ so that each $p \in P$ is of the form $p = \{p_1, p_2, \ldots, p_Z\}$. Let $\phi_1, \phi_2, \ldots, \phi_Z$ be Z normalized wave functions satisfying (2.4.17) and formulate the symmetrized Z-particle wave function by

$$\phi(\mathbf{x}_1, \mathbf{x}_2, \ldots, \mathbf{x}_Z) = \frac{1}{\sqrt{Z!}} \sum_{p \in P} \phi_1(\mathbf{x}_{p_1})\phi_2(\mathbf{x}_{p_2}) \cdots \phi_Z(\mathbf{x}_{p_Z}). \tag{2.E.13}$$

With (2.E.13), work out the corresponding Z-particle density function $n(\mathbf{x})$ and the kinetic energy $\langle \phi | \hat{T} | \phi \rangle$, in terms of $\phi_1, \phi_2, \ldots, \phi_Z$.

3

Maxwell equations, Dirac monopole, and gauge fields

This chapter embarks on an exploration of a series of profound features offered by the Maxwell equations for electromagnetism. It begins with a discussion on electromagnetic duality and a mathematical derivation of the Maxwell equations. Next, it presents the Dirac magnetic monopole and the associated Dirac strings. It then shows how to resolve the puzzle of the Dirac strings through considering the motion of a charged massive particle in an electromagnetic field and by introducing gauge fields in a natural way. As a result, it obtains Dirac's charge quantization formula. It also demonstrates how to derive Schwinger's charge quantization formula as an extension of the discussion. Finally it presents the Aharonov–Bohm effect, which tells that some refined quantum-mechanical properties of a system may be contained in gauge fields themselves, rather than the force fields induced by the gauge fields.

3.1 Maxwell equations and electromagnetic duality

The Maxwell equations are the foundation of electromagnetism theory, and also form the foundation of gauge field theory describing fundamental interactions in nature. In this section, we present a mathematical derivation of the Maxwell equations based on exploring the wave nature of the propagation of electromagnetic fields and electromagnetic duality.

Maxwell equations

Let the vector fields $\mathbf{E}$ and $\mathbf{B}$ denote the electric and magnetic fields, respectively, which are induced from the presence of an electric charge density distribution, ρ,

Mathematical Physics with Differential Equations. Yisong Yang, Oxford University Press.
 DOI: 10.1093/oso/9780192872616.003.0003

and a current density, $\mathbf{j}$, in a vacuum background space, also called a *free space*. Then these fields are governed by the Maxwell equations

$$\nabla \cdot \mathbf{E} = \frac{\rho}{\varepsilon_0}, \tag{3.1.1}$$

$$\nabla \times \mathbf{B} = \mu_0 \left(\mathbf{j} + \varepsilon_0 \frac{\partial \mathbf{E}}{\partial t} \right), \tag{3.1.2}$$

$$\nabla \cdot \mathbf{B} = 0, \tag{3.1.3}$$

$$\nabla \times \mathbf{E} = -\frac{\partial \mathbf{B}}{\partial t}, \tag{3.1.4}$$

where ε_0 and μ_0 are the electric permittivity and magnetic permeability of the free space, respectively, such that

$$c = \frac{1}{\sqrt{\varepsilon_0 \mu_0}} \tag{3.1.5}$$

renders the speed of light in free space. Among these equations, (3.1.1) is the Coulomb law, which directly relates the electric field to an electric charge distribution, (3.1.2) the Ampére law, which indicates that magnetism may be switched on by the onset of an electric current or a changing electric field, (3.1.3) the Gauss law, which states that magnetic field is solenoidal or source-free, and (3.1.4) the Faraday law, which implies that electricity is generated by a changing magnetic field.

General situation

Alternatively, in terms of the electric displacement field $\mathbf{D}$ and auxiliary magnetic field, also called magnetizing field, $\mathbf{H}$, associated with $\mathbf{E}$ and $\mathbf{B}$ by $\mathbf{D} = \varepsilon_0 \mathbf{E}$ and $\mu_0 \mathbf{H} = \mathbf{B}$, respectively, the equations (3.1.1)–(3.1.4) are

$$\nabla \cdot \mathbf{D} = \rho, \tag{3.1.6}$$

$$\nabla \times \mathbf{H} = \mathbf{j} + \frac{\partial \mathbf{D}}{\partial t}, \tag{3.1.7}$$

$$\nabla \cdot \mathbf{B} = 0, \tag{3.1.8}$$

$$\nabla \times \mathbf{E} = -\frac{\partial \mathbf{B}}{\partial t}, \tag{3.1.9}$$

so that the last two equations remain intact. In more general situations, the pairs $\mathbf{E}$, $\mathbf{B}$ and $\mathbf{D}$, $\mathbf{H}$ may be related through the *constitutive equations*

$$\mathbf{D} = \varepsilon \mathbf{E}, \quad \mathbf{B} = \mu \mathbf{H}, \tag{3.1.10}$$

in which the electric permittivity ε and magnetic permeability μ are constant for homogeneous medium, coordinate-dependent for non-homogeneous medium, and field-dependent for nonlinear medium. In other words, in practical applications, the Maxwell equations (3.1.6)–(3.1.9) are the fundamental governing equations for electromagnetism described in terms of $\mathbf{D}, \mathbf{E}, \mathbf{H}, \mathbf{B}$.

For our theoretical study, we will now focus on (3.1.1)–(3.1.4).

For convenience, we assume the Heaviside–Lorentz rationalized units in which $\varepsilon_0 = 1$ and $\mu_0 = 1$. Hence (3.1.1)–(3.1.4) become

$$\nabla \cdot \mathbf{E} = \rho, \tag{3.1.11}$$

$$\nabla \times \mathbf{B} - \frac{\partial \mathbf{E}}{\partial t} = \mathbf{j}, \tag{3.1.12}$$

$$\nabla \cdot \mathbf{B} = 0, \tag{3.1.13}$$

$$\nabla \times \mathbf{E} + \frac{\partial \mathbf{B}}{\partial t} = \mathbf{0}, \tag{3.1.14}$$

Electromagnetic duality

In absence of charge and current when $\rho = 0$, $\mathbf{j} = \mathbf{0}$, the equations (3.1.11)–(3.1.14) are invariant under the *dual correspondence*

$$\mathbf{E} \mapsto \mathbf{B}, \quad \mathbf{B} \mapsto -\mathbf{E}. \tag{3.1.15}$$

That is, in another world "dual" to the original, electricity and magnetism are seen as magnetism and electricity, and vice versa. This property is called *electromagnetic duality* or *dual symmetry of electricity and magnetism.* However, such a symmetry is *broken* in the presence of electric charge and current, ρ and $\mathbf{j}$. Dirac proposed a procedure that may be used to symmetrize (3.1.11)–(3.1.14), which starts from the extended equations

$$\nabla \cdot \mathbf{E} = \rho_e, \tag{3.1.16}$$

$$\nabla \times \mathbf{B} - \frac{\partial \mathbf{E}}{\partial t} = \mathbf{j}_e, \tag{3.1.17}$$

$$\nabla \cdot \mathbf{B} = \rho_m, \tag{3.1.18}$$

$$\nabla \times \mathbf{E} + \frac{\partial \mathbf{B}}{\partial t} = -\mathbf{j}_m, \tag{3.1.19}$$

where ρ_e, $\mathbf{j}_e$ and ρ_m, $\mathbf{j}_m$ denote the electric and magnetic source terms, respectively. These equations are invariant under the transformation (3.1.15) if the source terms are transformed accordingly,

$$\rho_e \mapsto \rho_m, \quad \mathbf{j}_e \mapsto \mathbf{j}_m, \quad \rho_m \mapsto -\rho_e, \quad \mathbf{j}_m \mapsto -\mathbf{j}_e. \tag{3.1.20}$$

Hence, with the new magnetic source terms in (3.1.18) and (3.1.19), duality between electricity and magnetism is again achieved, as in the source-free case.

Note that, although electromagnetic duality prompted Dirac to propose the concept of monopoles, or magnetic point charges (discussed in the next three sections), an isolated monopole has not been observed so far, excepting some of its simulated forms in laboratory settings [235, 465, 594]. Thus, with this regard, electromagnetic duality remains a pure theoretical curiosity of mathematical elegance.

In (3.1.16)–(3.1.19), ρ_e and $\mathbf{j}_e$ are the usual electric charge and current density functions, but ρ_m and $\mathbf{j}_m$, which are the magnetic charge and current density functions, introduce completely new ideas, both technically and physically.

Derivation of Maxwell equations

Before we move on we present a mathematical derivation of the Maxwell equations (3.1.1)–(3.1.4). For this purpose, we use $\mathbf{F}$ to denote a vector field that generically describes an electromagnetic field propagating in space. To start, we assume a sourceless situation,

$$\nabla \cdot \mathbf{F} = 0, \tag{3.1.21}$$

and understand that $\mathbf{F}$ travels like light of speed c in free space such that it satisfies the wave equation

$$\left(\frac{1}{c^2}\frac{\partial^2}{\partial t^2} - \nabla^2\right)\mathbf{F} = \mathbf{0}. \tag{3.1.22}$$

On the other hand, using the operator identity

$$\nabla \times \nabla \times \mathbf{F} = \mathrm{curl}^2\mathbf{F} = \nabla(\nabla \cdot \mathbf{F}) - \nabla^2\mathbf{F} \tag{3.1.23}$$

and (3.1.21), we may rewrite (3.1.22) as

$$\left(\frac{1}{c^2}\frac{\partial^2}{\partial t^2} + \mathrm{curl}^2\right)\mathbf{F} = \mathbf{0}. \tag{3.1.24}$$

We now proceed to further reduce this equation. In order to do so, we factor the differential operator in (3.1.24) as follows,

$$\left(\frac{1}{c^2}\frac{\partial^2}{\partial t^2} + \mathrm{curl}^2\right) = \left(\frac{1}{c}\frac{\partial}{\partial t} - \mathrm{i}\,\mathrm{curl}\right)\left(\frac{1}{c}\frac{\partial}{\partial t} + \mathrm{i}\,\mathrm{curl}\right) \equiv D^\dagger D, \tag{3.1.25}$$

such that (3.1.24) becomes $D^\dagger D\mathbf{F} = \mathbf{0}$, which may be fulfilled by solving either

$$D\mathbf{F} = \left(\frac{1}{c}\frac{\partial}{\partial t} + \mathrm{i}\,\mathrm{curl}\right)\mathbf{F} = \mathbf{0}, \tag{3.1.26}$$

or

$$D^\dagger\mathbf{F} = \left(\frac{1}{c}\frac{\partial}{\partial t} - \mathrm{i}\,\mathrm{curl}\right)\mathbf{F} = \mathbf{0}, \tag{3.1.27}$$

since $D^\dagger$ and D are commutative. For example, we may consider (3.1.26).

The complex form of (3.1.26) suggests that we should assume $\mathbf{F}$ to be a complex-valued vector field,

$$\mathbf{F} = \mathbf{P} + \mathrm{i}\,\mathbf{Q}, \tag{3.1.28}$$

where $\mathbf{P}$ and $\mathbf{Q}$ are real-valued vector fields. Inserting (3.1.28) into (3.1.26), we get

$$\frac{1}{c}\frac{\partial \mathbf{P}}{\partial t} = \operatorname{curl}\mathbf{Q}, \quad \frac{1}{c}\frac{\partial \mathbf{Q}}{\partial t} = -\operatorname{curl}\mathbf{P}. \tag{3.1.29}$$

On the other hand, in the sourceless situation, the equations (3.1.2) and (3.1.4) read

$$\operatorname{curl}\mathbf{B} = \frac{1}{c^2}\frac{\partial \mathbf{E}}{\partial t}, \quad \operatorname{curl}\mathbf{E} = -\frac{\partial \mathbf{B}}{\partial t}. \tag{3.1.30}$$

With this hint, we may compare (3.1.29) and (3.1.30) and subsequently take $\mathbf{P}, \mathbf{Q}$ to be proportional to $\mathbf{E}, \mathbf{B}$, respectively,

$$\mathbf{P} = a\mathbf{E}, \quad \mathbf{Q} = b\mathbf{B}, \quad a, b \in \mathbb{R}, \tag{3.1.31}$$

resulting in the equation

$$b = ac, \tag{3.1.32}$$

which indicates that there is one-degree of freedom in choosing a, b. For our purpose, we may set

$$a = \sqrt{\varepsilon_0}, \quad b = \frac{1}{\sqrt{\mu_0}}, \tag{3.1.33}$$

in view of (3.1.5). Thus, combining (3.1.21), (3.1.29), (3.1.31), and (3.1.33), we arrive at the source-free version of the equations (3.1.1)–(3.1.4).

Next, we consider the presence of source terms.

Let ρ be an electric charge density distributed over the space $\mathbb{R}^3$. Then the associated electric displacement field $\mathbf{D} = \varepsilon_0\mathbf{E}$ is given by $\nabla\cdot\mathbf{D} = \rho$, which leads to (3.1.1). Consequently, the electric charge q over a bounded domain Ω in $\mathbb{R}^3$ may be calculated as

$$q = \int_\Omega \rho \,\mathrm{d}x = \int_{\partial\Omega} \mathbf{D}\cdot \mathrm{d}\mathbf{S}. \tag{3.1.34}$$

On the other hand, since the rate of increase of electric charge in Ω is determined by the electric current, measured by a vector density $\mathbf{j}$, flowing through $\partial\Omega$ into Ω, we have

$$\frac{\mathrm{d}q}{\mathrm{d}t} = \int_\Omega \frac{\partial\rho}{\partial t}\mathrm{d}x = -\int_{\partial\Omega} \mathbf{j}\cdot \mathrm{d}\mathbf{S}. \tag{3.1.35}$$

Combining (3.1.34) and (3.1.35), we arrive at

$$\int_{\partial\Omega} \left(\frac{\partial \mathbf{D}}{\partial t} + \mathbf{j}\right)\cdot \mathrm{d}\mathbf{S} = 0. \tag{3.1.36}$$

The arbitrariness of Ω then leads to the conclusion that there is a vector field $\mathbf{M}$ such that

$$\frac{\partial \mathbf{D}}{\partial t} + \mathbf{j} = \operatorname{curl}\mathbf{M}. \tag{3.1.37}$$

However, in the sourceless limit, (3.1.2) reads

$$\operatorname{curl}\mathbf{H} = \frac{\partial \mathbf{D}}{\partial t}, \tag{3.1.38}$$

so that $\mathbf{M}$ is naturally identified with $\mathbf{H}$, the magnetizing field. Thus, the equation (3.1.37) should simply be

$$\frac{\partial \mathbf{D}}{\partial t} + \mathbf{j} = \operatorname{curl} \mathbf{H}, \tag{3.1.39}$$

which is (3.1.2). Of course, the equations (3.1.3) and (3.1.4) remain valid. So the full Maxwell equations (3.1.1)–(3.1.4) now follow.

3.2 Dirac monopole and strings

To motivate our study, consider the classical situation that electromagnetism is generated from an ideal *electric point charge* q lying at the origin,

$$\rho_e = 4\pi q\delta(\mathbf{x}), \quad \mathbf{j}_e = \mathbf{0}, \quad \rho_m = 0, \quad \mathbf{j}_m = \mathbf{0}. \tag{3.2.1}$$

It is clear that, inserting (3.2.1), the system (3.1.16)–(3.1.19) can be solved to yield $\mathbf{B} = \mathbf{0}$ and

$$\mathbf{E} = \frac{q}{|\mathbf{x}|^3}\mathbf{x}, \tag{3.2.2}$$

which is the well-known Coulomb law in static electricity.

Magnetic monopole

We now consider the case of a *magnetic point charge* g, or a monopole, often referred to as the *Dirac monopole* [167, 240, 386, 452, 460, 465, 466], resting at the origin,

$$\rho_e = 0, \quad \mathbf{j}_e = \mathbf{0}, \quad \rho_m = 4\pi g\delta(\mathbf{x}), \quad \mathbf{j}_m = \mathbf{0}. \tag{3.2.3}$$

Hence $\mathbf{E} = 0$ and

$$\mathbf{B} = \frac{g}{|\mathbf{x}|^3}\mathbf{x} = -g\nabla\left(\frac{1}{|\mathbf{x}|}\right). \tag{3.2.4}$$

Consequently, the magnetic flux through a sphere centered at the origin and of radius $r > 0$ is

$$\Phi = \int_{|\mathbf{x}|=r} \mathbf{B} \cdot \mathrm{d}\mathbf{S} = 4\pi g, \tag{3.2.5}$$

which is independent of r and is identical to the Gauss law for static electricity. Nevertheless, we show below through quantum mechanics that the introduction of a magnetic charge yields drastically new physics because electric and magnetic fields are induced differently from their respective potential fields.

We next evaluate the energy of a monopole. Recall that the total energy of an electromagnetic field with electric component $\mathbf{E}$ and magnetic component $\mathbf{B}$ is given by

$$E = \frac{1}{2}\int_{\mathbb{R}^3} (\mathbf{E}^2 + \mathbf{B}^2)\,\mathrm{d}x. \tag{3.2.6}$$

Inserting (3.2.4) into (3.2.6) and using $r = |\mathbf{x}|$, we have

$$E = 2\pi g^2 \int_0^\infty \frac{1}{r^2}\,\mathrm{d}r = \infty. \tag{3.2.7}$$

This energy blow-up seems to suggest that the idea of a magnetic monopole encounters an unacceptable obstacle. However, since the Coulomb law expressed in (3.2.2) for an electric point charge also leads to a divergent energy of the same form, (3.2.7), the infinite energy problem for a monopole is not a more serious one than that for an electric point charge which has been used effectively as good approximation for various particle models.

Dirac strings

We then study the magnetic field generated from a monopole more closely.

Recall that for the electric field generated from a point electric charge q, the Coulomb law (3.2.2) gives us a scalar potential function $\phi = -q/|\mathbf{x}|$ such that $\mathbf{E} = \nabla\phi$ holds everywhere away from the electric point charge.

Similarly, we consider the magnetic field $\mathbf{B}$ generated from a monopole of charge g, given in (3.2.4). Based on the classical knowledge on magnetic field, we know that $\mathbf{B}$ should be solenoidal. That is, there should exist a vector field $\mathbf{A}$, also referred to as a gauge potential of $\mathbf{B}$, such that

$$\mathbf{B} = \nabla \times \mathbf{A}, \tag{3.2.8}$$

except at the origin where the point monopole is placed. Unfortunately, using (3.2.5) and the Stokes theorem, it is easy to see that (3.2.8) cannot hold everywhere on any closed sphere centered at the origin. In other words, any such sphere would contain a singular point at which (3.2.8) fails. Shrinking a sphere to the origin would give us a continuous locus of singular points, which is a string that links the origin to infinity. Such a string is called a *Dirac string* (Figure 3.1).

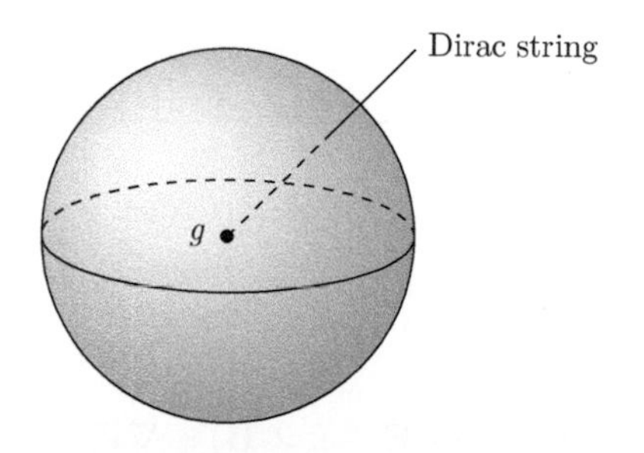

Figure 3.1 The generic presence of the Dirac strings as a consequence of the solenoidal nature of the magnetic field generated from a monopole.

Concretely, it may be checked directly that

$$\mathbf{A}^+ = (A_1^+, A_2^+, A_3^+),$$
$$A_1^+ = \frac{-x^2}{|\mathbf{x}|(|\mathbf{x}|+x^3)}g, \quad A_2^+ = \frac{x^1}{|\mathbf{x}|(|\mathbf{x}|+x^3)}g, \quad A_3^+ = 0, \tag{3.2.9}$$

satisfies (3.2.8) everywhere except on the negative x^3-axis, given by $x^1 = 0, x^2 = 0, x^3 \leq 0$. That is, with $\mathbf{A}^+$, the Dirac string, say S^-, is the negative x^3-axis. Similarly,

$$\mathbf{A}^- = (A_1^-, A_2^-, A_3^-),$$
$$A_1^- = \frac{x^2}{|\mathbf{x}|(|\mathbf{x}|-x^3)}g, \quad A_2^- = \frac{-x^1}{|\mathbf{x}|(|\mathbf{x}|-x^3)}g, \quad A_3^- = 0, \tag{3.2.10}$$

satisfies (3.2.8) everywhere except on the positive x^3-axis, given by $x^1 = 0$, $x^2 = 0, x^3 \geq 0$. That is, with $\mathbf{A}^-$, the Dirac string, say S^+, is the positive x^3-axis.

Gauge potential ambiguity

Note that $\mathbf{A}^+$ and $\mathbf{A}^-$ do not agree on the common part of their domains of definition because

$$\mathbf{A}^+ - \mathbf{A}^- = \mathbf{a} = (a_1, a_2, a_3), \quad \mathbf{x} \in \mathbb{R}^3 \setminus (S^+ \cup S^-), \tag{3.2.11}$$

where

$$a_1 = \frac{-2gx^2}{(x^1)^2 + (x^2)^2}, \quad a_2 = \frac{2gx^1}{(x^1)^2 + (x^2)^2}, \quad a_3 = 0. \tag{3.2.12}$$

Thus, the appearance of the Dirac strings gives rise to the gauge potential ambiguity (3.2.11).

In order to resolve this ambiguity, we need to consider the quantum-mechanical motion of an electric charge in an electromagnetic field. Such a consideration then introduces the gauged Schrödinger equation and the concept of gauge symmetry discussed in the next section.

3.3 Charged particle in electromagnetic field

Consider a point particle of mass m and electric charge Q moving in an electric field $\mathbf{E}$ and a magnetic field $\mathbf{B}$, in addition to a potential field V, in the Euclidean space $\mathbb{R}^3$ so that $\mathbf{x} = (x^1, x^2, x^3)$ gives the position coordinates of the particle. The equation of motion is

$$m\ddot{\mathbf{x}} = Q(\mathbf{E} + \dot{\mathbf{x}} \times \mathbf{B}) - \nabla V, \tag{3.3.1}$$

where $Q\mathbf{E}$ is the electric force and $Q\dot{\mathbf{x}} \times \mathbf{B}$ is the *Lorentz force* of the magnetic field $\mathbf{B}$ exerted on the particle of velocity $\dot{\mathbf{x}}$.

Gauge potential fields

Let the magnetic field $\mathbf{B}$ and electric field $\mathbf{E}$ be represented by a vector potential $\mathbf{A}$ and a scalar potential Ψ as follows:

$$\mathbf{B} = \nabla \times \mathbf{A}, \tag{3.3.2}$$

$$\mathbf{E} = \nabla\Psi - \frac{\partial \mathbf{A}}{\partial t}. \tag{3.3.3}$$

At the classical level, Ψ and $\mathbf{A}$ do not contribute directly to the underlying physics because they do not make their appearance, except their derivatives, in the governing equations (3.1.1)–(3.1.4) and (3.3.1). However, at the quantum-mechanical level, they do make observable contributions. This phenomenon, known as the Aharonov–Bohm effect, was predicted by Aharonov and Bohm [6, 7] and is discussed later in this chapter. Thus, in order to explore the meaning of the "ambiguity" of the vector potential associated with (3.2.11), we need to consider the quantum-mechanical description of the motion of a charged particle.

Lagrangian formalism

Using $\mathbf{y} = (y_i) = m\dot{\mathbf{x}} = m(\dot{x}^i)$ to denote the mechanical momentum vector, then we see that the equation (3.3.1) becomes

$$\begin{aligned} \dot{y}_i &= Q\left(\frac{\partial \Psi}{\partial x^i} - \frac{\partial A_i}{\partial t}\right) + Q\dot{x}^j\left(\frac{\partial A_j}{\partial x^i} - \frac{\partial A_i}{\partial x^j}\right) - \frac{\partial V}{\partial x^i} \\ &= -Q\frac{\mathrm{d}A_i}{\mathrm{d}t} + Q\frac{\partial \Psi}{\partial x^i} + Q\dot{x}^j\frac{\partial A_j}{\partial x^i} - \frac{\partial V}{\partial x^i}, \end{aligned} \tag{3.3.4}$$

which may be recast into the form

$$\frac{\mathrm{d}}{\mathrm{d}t}(y_i + QA_i) = \frac{\partial}{\partial x^i}(Q\Psi + Q\dot{x}^j A_j - V), \tag{3.3.5}$$

or

$$\frac{\mathrm{d}}{\mathrm{d}t}\left(\frac{\partial L}{\partial \dot{x}^i}\right) = \frac{\partial L}{\partial x^i}, \quad i = 1, 2, 3, \tag{3.3.6}$$

if we define the function L to be

$$\begin{aligned} L(\mathbf{x}, \dot{\mathbf{x}}, t) &= \frac{1}{2}m(\dot{x}^i)^2 + Q\Psi + Q\dot{x}^i A_i - V \\ &= \frac{1}{2}m\dot{\mathbf{x}}^2 + Q\Psi(\mathbf{x}, t) + Q\dot{\mathbf{x}} \cdot \mathbf{A}(\mathbf{x}, t) - V(\mathbf{x}, t). \end{aligned} \tag{3.3.7}$$

In other words, the formula (3.3.7) gives us the Lagrangian function of the problem. Note that the momentum vector has a correction due to the presence of the electromagnetic field through the *vector potential* $\mathbf{A}$,

$$p_i = \frac{\partial L}{\partial \dot{x}^i} = y_i + QA_i, \quad i = 1, 2, 3. \tag{3.3.8}$$

Hamiltonian

With (3.3.7), the Hamiltonian function becomes

$$\begin{aligned} H &= p_i\dot{x}^i - L = \frac{1}{2m}{y_i}^2 - Q\Psi + V \\ &= \frac{1}{2m}(p_i - QA_i)^2 - Q\Psi + V. \end{aligned} \tag{3.3.9}$$

Finally, if we use $A = (A_\mu)$ ($\mu = 0, 1, 2, 3$) to denote a vector field with four components, $A = (\Psi, \mathbf{A})$, the Hamiltonian function (3.3.9) takes the form

$$H = \frac{1}{2m}(p_i - QA_i)^2 - QA_0 + V. \tag{3.3.10}$$

Gauged Schrödinger equation

From the Hamiltonian (3.3.10) and the correspondence (2.2.38), we have

$$\mathrm{i}\hbar\frac{\partial\psi}{\partial t} = -\frac{1}{2m}(\hbar\partial_i - \mathrm{i}QA_i)^2\psi - QA_0\psi + V\psi. \tag{3.3.11}$$

Thus, if we introduce the *gauge-covariant derivatives*

$$D_\mu\psi = \partial_\mu\psi - \mathrm{i}\frac{Q}{\hbar}A_\mu\psi, \quad \mu = 0, 1, 2, 3, \tag{3.3.12}$$

then the *gauged Schrödinger equation* (3.3.11) assumes an elegant form,

$$\mathrm{i}\hbar D_0\psi = -\frac{\hbar^2}{2m}D_i^2\psi + V\psi. \tag{3.3.13}$$

Note that (3.3.11) or (3.3.13) is semi-quantum mechanical in the sense that the point particle of mass m is treated quantum mechanically by the Schrödinger equation but that the electromagnetic field is a classical field, through the coupling of the vector potential A_μ, also called the *gauge field*, which will be made more specific later.

Gauge symmetry

We can see that (3.3.13) is invariant under the transformation

$$\psi \mapsto \psi' = \mathrm{e}^{\mathrm{i}\omega}\psi, \quad A_\mu \mapsto A'_\mu = A_\mu + \frac{\hbar}{Q}\partial_\mu\omega, \tag{3.3.14}$$

which is also called the *gauge transformation*, *gauge equivalence*, or *gauge symmetry*, of the system. Such an invariance follows from the *gauge-covariance* property

$$D'_\mu\psi' = \mathrm{e}^{\mathrm{i}\omega}D_\mu\psi, \quad D'_\mu = \partial_\mu - \mathrm{i}\frac{Q}{\hbar}A'_\mu, \tag{3.3.15}$$

in view of (3.3.12) and (3.3.14).

Two gauge equivalent field configurations, (ψ, A_μ) and (ψ', A'_μ), describe identical physics, which may be understood within the formalism of a *vector bundle*, which we now elaborate briefly.

Bundle formalism

In the context of differential geometry, the gauge-field formalism here defines a *complex line bundle* ξ over the Minkowski spacetime $\mathbb{R}^{3,1}$ where the symmetry group is $U(1) = \{e^{i\omega} \mid \omega \in \mathbb{R}\}$ so that ψ is a *cross-section* and A_μ a *connection* which jointly obey the transformation property

$$\psi' = \Omega\psi, \quad A'_\mu = A_\mu - i\frac{\hbar}{Q}\Omega^{-1}\partial_\mu\Omega, \quad \Omega \in C^2(\mathbb{R}^{3,1}, U(1)). \tag{3.3.16}$$

It will be convenient to consider the problem in the framework of such a global transformation property.

3.4 Removal of Dirac strings and charge quantization

We now look at the relation between the vector potentials $\mathbf{A}^+$ and $\mathbf{A}^-$, given in (3.2.9) and (3.2.10), induced from a magnetic point charge g placed at the origin.

Consider (3.3.1) when the electric and magnetic fields are given by (3.2.2) and (3.2.4), respectively, and the potential field V is absent,

$$m\ddot{\mathbf{x}} = Q(\mathbf{E} + \dot{\mathbf{x}} \times \mathbf{B}), \quad \mathbf{E} = \frac{q}{|\mathbf{x}|^3}\mathbf{x}, \quad \mathbf{B} = \frac{g}{|\mathbf{x}|^3}\mathbf{x}. \tag{3.4.1}$$

Gauge fields

It is clear that $\mathbf{E}$ and $\mathbf{B}$ in (3.4.1) are produced by the gauge fields

$$A^\pm = (A^\pm_\mu) = (A^\pm_0, \mathbf{A}^\pm), \quad A^\pm_0 = -\frac{q}{|\mathbf{x}|}, \tag{3.4.2}$$

in their respective domains of definition. To avoid ambiguity in the overlapped part of these domains, we need to look for the possibility that makes A^-_μ and A^+_μ gauge equivalent there.

Gauge transformation

First, from (3.2.12), we see that if we use (r, θ, φ) to denote the spherical coordinates where θ is the azimuth angle and φ the inclination angle, then

$$a_1 = 2g\partial_1\theta, \quad a_2 = 2g\partial_2\theta, \quad \theta = \tan^{-1}\left(\frac{x^2}{x^1}\right). \tag{3.4.3}$$

Next, inserting $\Omega = e^{i\omega}$ into

$$A^+_\mu - A^-_\mu = -i\frac{\hbar}{Q}\Omega^{-1}\partial_\mu\Omega, \tag{3.4.4}$$

we have

$$A_\mu^+ - A_\mu^- = \frac{\hbar}{Q}\partial_\mu \omega. \tag{3.4.5}$$

Then, substituting $A^+ - A^- = (a_\mu) = (0, a_1, a_2, 0)$ into (3.4.5) and assuming that ω depends on x^1, x^2 only, we arrive at

$$a_1 = \frac{\hbar}{Q}\partial_1 \omega, \quad a_2 = \frac{\hbar}{Q}\partial_2 \omega, \tag{3.4.6}$$

which are consistent with (3.4.3) away from the strings.

In view of (3.4.3) and (3.4.6) we see that on $\mathbb{R}^3 \setminus (S^+ \cup S^-)$ the relation

$$\frac{\hbar}{Q}\omega - 2g\theta = c \tag{3.4.7}$$

holds, where $c \in \mathbb{R}$ is a constant.

Dirac quantization formula

Hence, ω depends on θ only and the single-valuedness of the quantity $\Omega = \mathrm{e}^{\mathrm{i}\omega}$ requires

$$\frac{\hbar}{Q}(2\pi n) = 2g(2\pi), \quad n \in \mathbb{N}, \tag{3.4.8}$$

which leads to the *Dirac charge quantization formula*

$$gQ = \frac{\hbar}{2}n, \quad n \in \mathbb{N}. \tag{3.4.9}$$

Alternatively, we may also rewrite (3.2.12) as

$$a_1 = -2g\frac{\sin\theta}{r}, \quad a_2 = 2g\frac{\cos\theta}{r}, \tag{3.4.10}$$

and solve the equation

$$\frac{\hbar}{Q}\partial_j \omega = a_j, \quad j = 1, 2, \tag{3.4.11}$$

directly for a solution depending on θ only, so that we arrive at the equation

$$\frac{\mathrm{d}\omega}{\mathrm{d}\theta} = \frac{2gQ}{\hbar}. \tag{3.4.12}$$

Hence, (3.4.7) follows again.

Consequently, when the condition (3.4.9) holds, the magnetic field away from a magnetic point charge g is well defined everywhere and is generated *piecewise* from suitable gauge potentials defined on their corresponding domains. In particular, the Dirac strings are seen to be artifacts and are removed. Consequently, like electric point charges, magnetic monopoles are also truly magnetic point charges, which henceforth we call monopoles.

The derivation here along the formalism of a complex line bundle (or a $U(1)$-principal bundle) was worked out by T. T. Wu and C. N. Yang [589, 590] in the mid 1970s and lays a firm mathematical foundation for monopoles in particular and topological solitons [386, 606] in general.

An immediate popular-science implication of the formula (3.4.9) is that the existence of a single monopole in the universe would predict that all electric charges are integer multiples of a basic unit charge. Indeed, this is what is observed in nature, since all electric charges are measured to be the multiples of the charge of the electron.

As mentioned earlier, although a monopole has never been found in nature, some recent studies by experimental physicists confirmed the occurrence of some monopole-like structures in condensed-matter systems [79, 114, 234, 400, 465, 594].

3.5 Schwinger dyons and extended charge quantization formula

The charge quantization formula (3.4.9) was originally derived by Dirac [167] through a computation of the quantum-mechanical eigenspectrum of the momentum operator associated with the motion (3.3.1), which was then further explored by Schwinger [501] to obtain an extended charged quantization formula for electrically *and* magnetically charged *hypothetical* particles, called *dyons*, or the *Schwinger dyons.*

Motion of point charge

Consider again the motion of a point particle of mass m and electric charge Q in the electric and magnetic fields $\mathbf{E}$ and $\mathbf{B}$ of a point charge and a monopole given by (3.2.2) and (3.2.4), respectively, governed by (3.4.1), which is rewritten

$$m\ddot{\mathbf{x}} = \frac{qQ}{|\mathbf{x}|^3}\mathbf{x} + \frac{gQ}{|\mathbf{x}|^3}(\dot{\mathbf{x}} \times \mathbf{x}). \tag{3.5.1}$$

First, we note that the classical orbital momentum $\mathbf{x} \times \mathbf{p} = \mathbf{x} \times m\dot{\mathbf{x}}$ is no longer a conserved quantity due to (3.5.1). In fact, we have, instead,

$$\begin{aligned} \frac{\mathrm{d}}{\mathrm{d}t}(\mathbf{x} \times m\dot{\mathbf{x}}) &= \mathbf{x} \times m\ddot{\mathbf{x}} \\ &= \frac{gQ}{|\mathbf{x}|^3}\mathbf{x} \times (\dot{\mathbf{x}} \times \mathbf{x}) = \frac{\mathrm{d}}{\mathrm{d}t}\left(gQ\frac{\mathbf{x}}{|\mathbf{x}|}\right), \end{aligned} \tag{3.5.2}$$

in view of the vector identity $\mathbf{a} \times (\mathbf{b} \times \mathbf{c}) = (\mathbf{a} \cdot \mathbf{c})\mathbf{b} - (\mathbf{a} \cdot \mathbf{b})\mathbf{c}$, which suggests that the modified quantity

$$\mathbf{J} = \mathbf{x} \times m\dot{\mathbf{x}} - gQ\frac{\mathbf{x}}{|\mathbf{x}|}, \tag{3.5.3}$$

not containing the electric charge q but the magnetic charge g, in the background source fields is defined as the angular momentum of the particle, whose quantum mechanical representation possesses eigenvalues in the multiples of $\frac{\hbar}{2}$, which led Dirac to conclude with the quantization formula (3.4.9).

Motion of point dyon

Next, we consider the motion of a dyon of mass m and electric charge Q and magnetic charge G in an electric $\mathbf{E}$ and a magnetic field $\mathbf{B}$. In this situation, the equation (3.3.1) becomes

$$m\ddot{\mathbf{x}} = Q(\mathbf{E} + \dot{\mathbf{x}} \times \mathbf{B}) + G(\mathbf{B} - \dot{\mathbf{x}} \times \mathbf{E}), \tag{3.5.4}$$

as a consequence of the electromagnetic duality (3.1.15). Thus, if $\mathbf{E}$ and $\mathbf{B}$ are generated from a *dyonically charged point particle* located at the origin with electric charge q and magnetic charge g, then we have

$$\mathbf{E} = q\frac{\mathbf{x}}{|\mathbf{x}|^3}, \quad \mathbf{B} = g\frac{\mathbf{x}}{|\mathbf{x}|^3}, \tag{3.5.5}$$

so that (3.5.4) assumes the form

$$m\ddot{\mathbf{x}} = (qQ + gG)\frac{\mathbf{x}}{|\mathbf{x}|^3} + (gQ - qG)\dot{\mathbf{x}} \times \frac{\mathbf{x}}{|\mathbf{x}|^3}. \tag{3.5.6}$$

Comparing (3.5.6) with (3.5.1), we obtain the conserved angular momentum

$$\mathbf{J} = \mathbf{x} \times m\dot{\mathbf{x}} - (gQ - qG)\frac{\mathbf{x}}{|\mathbf{x}|}, \tag{3.5.7}$$

which is of the same form as (3.5.3).

Schwinger quantization formula

Therefore, by analogue, we arrive at the *Schwinger charge quantization formula for dyons,*

$$gQ - qG = \frac{\hbar}{2}n, \quad n \in \mathbb{N}. \tag{3.5.8}$$

In particular, when $G = 0$ or $q = 0$, we recover the Dirac formula (3.4.9).

Of course, the Schwinger formula (3.5.8) may also be established directly by using a gauge transformation between the gauge potentials

$$A^+ = (A_0^+, A_1^+, A_2^+, A_3^+),$$

$$A_0^+ = -\frac{q'}{|\mathbf{x}|}, \; A_1^+ = \frac{-x^2}{|\mathbf{x}|(|\mathbf{x}| + x^3)}g', \; A_2^+ = \frac{x^1}{|\mathbf{x}|(|\mathbf{x}| + x^3)}g', \; A_3^+ = 0, \tag{3.5.9}$$

$$A^- = (A_0^-, A_1^-, A_2^-, A_3^-),$$

$$A_0^- = -\frac{q'}{|\mathbf{x}|}, \; A_1^- = \frac{x^2}{|\mathbf{x}|(|\mathbf{x}| - x^3)}g', \; A_2^- = \frac{-x^1}{|\mathbf{x}|(|\mathbf{x}| - x^3)}g', \; A_3^- = 0, \tag{3.5.10}$$

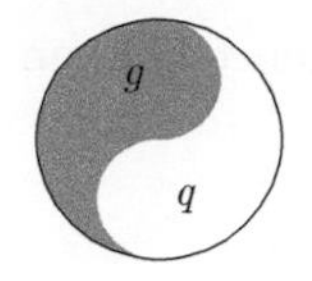

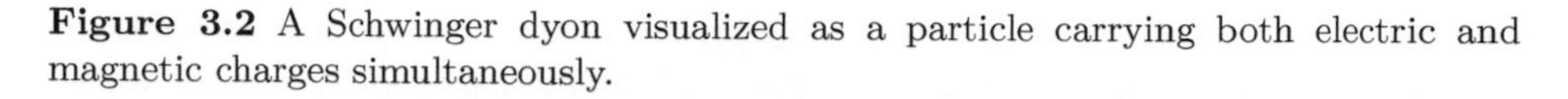

Figure 3.2 A Schwinger dyon visualized as a particle carrying both electric and magnetic charges simultaneously.

in $\mathbb{R}^3 \setminus S_-$ and $\mathbb{R}^3 \setminus S_+$, respectively, where

$$q' = qQ + gG, \quad g' = gQ - qG, \tag{3.5.11}$$

because (3.5.6) describes the motion of a unit-charge particle in a dyon field with effective electric charge q' and magnetic charge g', in view of (3.5.1).

Like that of a point Coulomb charge or Dirac monopole, a Schwinger dyon also carries infinite energy.

Figure 3.2 depicts a Schwinger dyon in the form of a hypothetical "yin-yang" particle.

3.6 Aharonov–Bohm effect

Following the study of the quantum-mechanical motion of a charged particle in an electromagnetic field embodied by a gauge field, here we elaborate on the Aharonov–Bohm effect, which states that a gauge field may measurably demonstrate its presence quantum mechanically.

Formalism and governing equations

Consider the motion of a free particle of mass m and electric charge Q in absence of an electromagnetic field, in such a spatial region that is the complement of a cylindrically-shaped infinitely long tubular subregion, T, governed by the wave function $\phi(\mathbf{x}, t)$ satisfying the Schrödinger equation

$$\mathrm{i}\hbar \frac{\partial \phi}{\partial t} = -\frac{\hbar^2}{2m}\nabla^2 \phi. \tag{3.6.1}$$

Now switch on a vector potential $\mathbf{A}$ in $\mathbb{R}^3$ such that

$$\mathbf{B} \equiv \nabla \times \mathbf{A} = \mathbf{0}, \tag{3.6.2}$$

outside T. Thus, classically, the charged massive particle is subject to the same physical environment.

Quantum mechanically, however, the presence of the vector potential $\mathbf{A}$ will be "felt" by the wave function, as a result, because in view of (3.3.11), the wave function satisfies the gauged Schrödinger equation

$$\mathrm{i}\hbar \frac{\partial \psi}{\partial t} = -\frac{\hbar^2}{2m}\left(\nabla - \mathrm{i}\frac{Q}{\hbar}\mathbf{A}\right)^2 \psi, \quad \mathbf{x} \in \mathbb{R}^3 \setminus T, \tag{3.6.3}$$

which clearly indicates the dependence of ψ on $\mathbf{A}$ itself, rather than $\mathbf{B}$, and will be addressed in detail as follows.

Interference of waves

Specifically, this section shows that the presence of a gauge field $\mathbf{A}$, which does not generate a magnetic field outside T, is demonstrated by way of wave interference registered by the *updated* wave function ψ governed by (3.6.3).

To see how, we consider a cross section of the spatial region and imagine how the matter waves of the particle initially placed at the position marked "emitter" are received at the position marked "receiver", in Figure 3.3.

To solve for the wave function, we note that what are received as matter waves at the receiver may be regarded as a result of superposition of waves propagating along the "upper" and "lower" paths, respectively (Figure 3.3), denoted by C_1 and C_2.

Use ϕ_1, ϕ_2 to denote the solutions of (3.6.1) representing two matter waves along upper and lower paths, respectively, in absence of a gauge field. Then $I_1 = |\phi_1|^2, I_2 = |\phi_2|^2$ measure the corresponding wave intensities. By the principle of superposition, the wave function ϕ observed at the receiver is the sum of ϕ_1 and ϕ_2: $\phi = \phi_1 + \phi_2$, such that the wave intensity $I = |\phi|^2$ enjoys the simple expression

$$I = |\phi_1 + \phi_2|^2 = |\phi_1|^2 + |\phi_2|^2 + 2\,\mathrm{Re}(\overline{\phi}_1\phi_2) = I_1 + I_2 + 2\,\mathrm{Re}(\overline{\phi}_1\phi_2). \quad (3.6.4)$$

In other words, the intensity I of the waves at the receiver is the sum of the intensities I_1 and I_2 of the waves coming along the upper and lower paths plus an *interference term.*

Now use ψ_1 to denote a solution of (3.6.3) representing the matter waves along an upper path. Note that, restricted to an upper region, which becomes simply connected, the path integral, say,

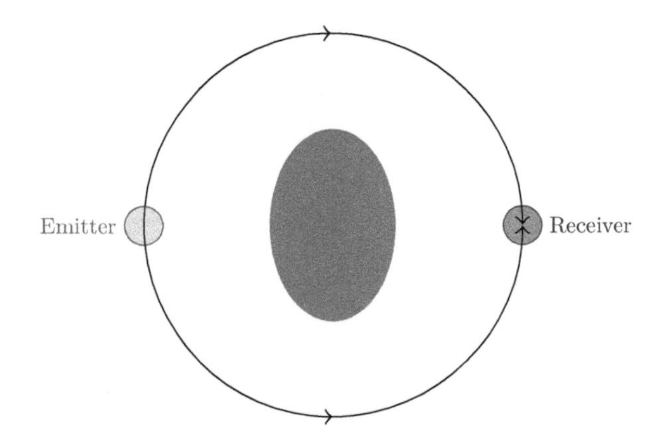

Figure 3.3 An illustration of the Aharonov–Bohm effect as a result of interference of matter waves.

$$\omega_1 \equiv \int_{C_1} \mathbf{A}(\mathbf{x}') \cdot \mathrm{d}\mathbf{x}', \tag{3.6.5}$$

is independent of the details of C_1 but only depends on the initial and terminal points of C_1, which are taken here to be $\mathbf{x}_0$, at the emitter, and $\mathbf{x}$, at the receiver. In particular, (3.6.5) defines ω_1 as a function of $\mathbf{x}$ such that

$$\nabla \omega_1(\mathbf{x}) = \mathbf{A}(\mathbf{x}). \tag{3.6.6}$$

Thus, from (3.6.6), we see that

$$\phi_1 = \mathrm{e}^{-\mathrm{i}\frac{Q}{\hbar}\omega_1} \psi_1 \tag{3.6.7}$$

satisfies (3.6.1) since ψ_1 solves (3.6.3).

Likewise, we can also define without ambiguity

$$\omega_2(\mathbf{x}) \equiv \int_{C_2} \mathbf{A}(\mathbf{x}') \cdot \mathrm{d}\mathbf{x}', \tag{3.6.8}$$

where C_2 initiates at $\mathbf{x}_0$ and terminates at $\mathbf{x}$ such that, with ψ_2, we see that

$$\phi_2 = \mathrm{e}^{-\mathrm{i}\frac{Q}{\hbar}\omega_2} \psi_2 \tag{3.6.9}$$

satisfies (3.6.1) as well.

In summary, we obtain

$$\psi_1(\mathbf{x}, t) = \mathrm{e}^{\mathrm{i}\frac{Q}{\hbar}\omega_1(\mathbf{x})} \phi_1(\mathbf{x}, t), \quad \psi_2(\mathbf{x}, t) = \mathrm{e}^{\mathrm{i}\frac{Q}{\hbar}\omega_2(\mathbf{x})} \phi_2(\mathbf{x}, t), \tag{3.6.10}$$

for the waves traveling along the upper and lower paths, respectively, where ϕ_1, ϕ_2 represent waves along the corresponding upper and lower paths and solve (3.6.1), for the situation of an absent gauge field.

Consequently, by the principle of superposition again, we may express the wave function observed at the receiver as

$$\psi(\mathbf{x}, t) = \psi_1(\mathbf{x}, t) + \psi_2(\mathbf{x}, t) = \mathrm{e}^{\mathrm{i}\frac{Q}{\hbar}\omega_1(\mathbf{x})} \phi_1(\mathbf{x}, t) + \mathrm{e}^{\mathrm{i}\frac{Q}{\hbar}\omega_2(\mathbf{x})} \phi_2(\mathbf{x}, t), \tag{3.6.11}$$

which indicates that the intensity of waves measured at the receiver follows the formula

$$\begin{aligned} I &= |\psi|^2 = |\psi_1 + \psi_2|^2 \\ &= |\phi_1|^2 + |\phi_2|^2 + 2\,\mathrm{Re}\left(\overline{\phi}_1 \phi_2 \mathrm{e}^{\mathrm{i}\Omega}\right), \end{aligned} \tag{3.6.12}$$

where

$$\Omega = \Omega(\mathbf{x}) = \frac{Q}{\hbar}(-\omega_1(\mathbf{x}) + \omega_2(\mathbf{x})) = \frac{Q}{\hbar} \int_C \mathbf{A}(\mathbf{x}') \cdot \mathrm{d}\mathbf{x}', \tag{3.6.13}$$

by (3.6.5) and (3.6.8), where C is the closed curve consisting of two pieces, C_1 and C_2, with appropriately adjusted orientations, and in view of the Stokes theorem, we may rewrite (3.6.13) as

$$\Omega = \Omega(\mathbf{x}) = \frac{Q}{\hbar}\Phi, \tag{3.6.14}$$

where Φ is the magnetic flux through a cross section of the tube T, which may be converted into a line integral along the boundary of the cross section by (3.6.13).

Aharonov–Bohm effect

Comparing (3.6.12) with (3.6.4), we see that the presence of a gauge field spells out a detectable variation in the interference term, measured by a phase angle shift, Ω. This phenomenon is the celebrated Aharonov–Bohm effect. The phase angle shift Ω given by (3.6.14) is known as the *Aharonov–Bohm phase shift* which is actually seen to be position independent in the present setting.

Apparently, when the magnetic flux contained in the tube satisfies the critical condition

$$\Phi = \frac{2\pi n\hbar}{Q}, \quad n \in \mathbb{N}, \tag{3.6.15}$$

the wave interference term in (3.6.12) returns to that in (3.6.4) such that the Aharonov–Bohm effect is switched off.

Magnetic field contained in tube

There are many different ways to construct a needed gauge field that produces a nontrivial magnetic field inside the tube T but nothing outside it.

For example, let $\mathbf{x} = (x^1, x^2, x^3)$ and assume that the tube T is a round cylinder of radius $R > 0$ and centered about the x^3-axis. With $r = \sqrt{(x^1)^2 + (x^2)^2} > 0$, set

$$U(r) = -\frac{\gamma}{2\pi} \ln r, \quad \mathbf{A} = (\partial_2 U, -\partial_1 U, 0), \tag{3.6.16}$$

in view of the *Kirchhoff vortex model*, where $\gamma > 0$ is a prescribed constant. Then

$$\mathbf{B} = \nabla \times \mathbf{A} = (0, 0, -(\partial_1^2 + \partial_2^2)U) = (0, 0, \gamma\delta(x)), \tag{3.6.17}$$

in which $\delta(x)$ is the planar Dirac distribution in $x = (x^1, x^2)$ concentrated at the origin of the (x^1, x^2)-plane. Therefore, a magnetic field contained in the tube T is obtained, whose flux through the cross section given by $r < R, x^3 = 0$ is $\Phi = \gamma$.

Mathematical features

Note that the occurrence of the Aharonov–Bohm effect is of a topological nature in that the containment of magnetic field in a cylindrical tube makes the space non-simply connected.

Furthermore, the refined role played by the gauge field, say $\mathbf{A}$, rather than the induced magnetic field, say $\mathbf{B}$, is an indication of a non-local dependence of the problem on the magnetic field. Specifically, from $\mathbf{B} = \nabla \times \mathbf{A}$, we have

$$\mathbf{A}(\mathbf{x}) = \frac{1}{4\pi} \int_{\mathbb{R}^3} \frac{\nabla_{\mathbf{y}} \times \mathbf{B}(\mathbf{y})}{|\mathbf{x} - \mathbf{y}|} \, \mathrm{d}\mathbf{y}, \tag{3.6.18}$$

in general, which indeed reveals how quantum physics depends on $\mathbf{B}$ *non-locally*, as a consequence.

The Aharonov–Bohm effect has long been confirmed experimentally [120, 304, 424, 500, 548]. See also [9, 427, 505, 596] for some development of the subject.

Exercises

1. Show that the charge density ρ and current density $\mathbf{j}$ in (3.1.1)–(3.1.4) satisfy the conservation law

$$\frac{\partial \rho}{\partial t} + \nabla \cdot \mathbf{j} = 0. \tag{3.E.1}$$

2. Show that in the source-free situation, $\rho = 0$, $\mathbf{j} = 0$, the electric and magnetic fields, $\mathbf{E}$ and $\mathbf{B}$, both propagate in the form of d'Alembertian waves of speed c. That is, they satisfy the equations

$$\frac{1}{c^2}\frac{\partial^2 \mathbf{E}}{\partial^2 t} - \nabla^2 \mathbf{E} = \mathbf{0}, \quad \frac{1}{c^2}\frac{\partial^2 \mathbf{B}}{\partial^2 t} - \nabla^2 \mathbf{B} = \mathbf{0}. \tag{3.E.2}$$

3. For the Maxwell equations (3.1.1)–(3.1.4), the energy density is given by

$$u = \frac{\varepsilon_0}{2}\mathbf{E}^2 + \frac{1}{2\mu_0}\mathbf{B}^2. \tag{3.E.3}$$

Establish the energy conservation law

$$\frac{\partial u}{\partial t} + \nabla \cdot \mathbf{S} + \mathbf{j} \cdot \mathbf{E} = 0, \tag{3.E.4}$$

where $\mathbf{S}$ is the *Poynting vector*:

$$\mathbf{S} = \frac{1}{\mu_0}\left(\mathbf{E} \times \mathbf{B}\right). \tag{3.E.5}$$

(Hint: Use the vector identity $\nabla \cdot (\mathbf{a} \times \mathbf{b}) = (\nabla \times \mathbf{a}) \cdot \mathbf{b} - \mathbf{a} \cdot (\nabla \times \mathbf{b})$.)

4. Establish the electromagnetic duality by showing that (3.1.16)–(3.1.19) are invariant under the transformation given jointly by (3.1.15) and (3.1.20).
5. Show that the equation (3.2.8) is valid in $\mathbb{R}^3 \setminus S^{\mp}$ for $\mathbf{A} = \mathbf{A}^{\pm}$ given in (3.2.9) and (3.2.10), respectively.
6. For a_1, a_2 defined in (3.2.12), verify that

$$\partial_1 a_2 - \partial_2 a_1 = 0, \quad \text{when } (x^1, x^2) \neq (0,0). \tag{3.E.6}$$

7. Verify (3.3.4).
8. Let ϕ, ψ be two complex-valued functions. For the gauge-covariant derivative (3.3.12) where A_μ is a real-valued gauge field, establish the identity

$$\partial_\mu(\phi\overline{\psi}) = (D_\mu\phi)\overline{\psi} + \phi(\overline{D_\mu\psi}). \tag{3.E.7}$$

9. Establish the gauge-covariance property (3.3.15).
10. Establish (3.4.3).

11. Consider the motion of a dyon of mass m and electric charge Q and magnetic charge G in an electric $\mathbf{E}$ and a magnetic field $\mathbf{B}$. In this situation, the equation (3.3.1) becomes

$$m\ddot{\mathbf{x}} = Q(\mathbf{E} + \dot{\mathbf{x}} \times \mathbf{B}) + G(\mathbf{B} - \dot{\mathbf{x}} \times \mathbf{E}) - \nabla V. \tag{3.E.8}$$

    (a) In view of (3.3.2) and (3.3.3), find the Lagrangian function of the equation (3.E.8).

    (b) Find the momenta of the motion of the dyon and use them to obtain the Hamiltonian function of the equation (3.E.8).

12. In view of the previous problem, find the gauged Schrödinger equation that describes the quantum mechanical motion of the dyon.

13. Identify the underlying gauge symmetry of the gauged Schrödinger equation obtained in Exercise 12 and use it to derive the Schwinger charge quantization formula (3.5.8) for dyons.

14. Consider the wave interference problem.

    (a) Show that the interference term in (3.6.4) may be rewritten as

$$2\sqrt{I_1 I_2}\cos\theta, \tag{3.E.9}$$

    and express the quantity θ in terms of the phase angles of the wave functions ϕ_1 and ϕ_2.

    (b) Show that the interference term in (3.6.12) has the form

$$2\sqrt{I_1 I_2}\cos(\theta + \Omega). \tag{3.E.10}$$

15. Use the Newton potential to explain the validity of (3.6.18).

4

Special relativity

This chapter presents a short introduction to the theory of special relativity. It begins with a discussion of spacetime, coordinates, events, and inertial frames, and then considers line element, length contraction, time dilation, and other distinct notions in special relativity. It next studies relativistic mechanics. As an excursion, it describes the relativistic Doppler effect and compares it with the classical one.

4.1 Inertial frames, Minkowski spacetime, and Lorentz boosts

Einstein proposed the special theory of relativity, or special relativity, in 1905 to reformulate and treat fundamental mechanical problems and concepts with the viewpoint that laws of physics and the speed of light are both independent of the choice of an inertial coordinate frame. This development completely reshaped physics. In fact, in subsequent chapters, our studies center around theories and constructions satisfying the required invariance or symmetry properties of special relativity. This section discusses special relativity and covers some basic key concepts, including spacetime, events, coordinates, coordinate transformations, and inertial frames. These concepts lay the foundation for special relativity.

Event, history, and inertial frames

We use t to denote time and (x, y, z) the Cartesian coordinates of a point in space. Then the 4-tuple (t, x, y, z) denote a spacetime point called an *event*, which may represent a particle located at (x, y, z) when time is t, with spacetime coordinates t, x, y, z or (t, x, y, z). A sequence of the events or coordinate vectors may represent the trajectory of a moving particle, called the *history* or *world line* of the particle. Use F, called a frame, to denote the coordinate system of the

Mathematical Physics with Differential Equations. Yisong Yang, Oxford University Press.
 DOI: 10.1093/oso/9780192872616.003.0004

spacetime with coordinates (t, x, y, z). If F' is another frame with coordinates (t', x', y', z') that moves with a constant velocity relative to F but without rotation, then F and F' are called *inertial frames.* Note that, within this formalism, the concept of an absolute time is abandoned, since in frame F' time is measured by t', which may (in fact, must if F' does move) be different from t in F. In such a terminology, an event is happening in a *Minkowski spacetime* coordinated with a given frame F at *coordinate time* t and *coordinate location* (x, y, z).

Fundamental postulates of special relativity

Based on real-world observations, Einstein formulated the following fundamental postulates of his special theory of relativity:

(i) Physical laws are the same in inertial frames.

(ii) Speed of light is the same in inertial frames.

Thus, if we use $\mathbf{r} = (x, y, z)$ to denote the location of a particle in free motion, that is, the particle is not subject to an applied force, then the particle remains in such a state without acceleration. Thus, in terms of a differential equation, its law of motion is

$$\frac{\mathrm{d}^2\mathbf{r}}{\mathrm{d}t^2} = \mathbf{0}. \tag{4.1.1}$$

Thus, for inertial frames F, F' where the coordinates of F' are t', x', y', z' with $\mathbf{r}' = (x', y', z')$, we obtain from the postulate (i) and the law (4.1.1) the law

$$\frac{\mathrm{d}^2\mathbf{r}'}{\mathrm{d}t'^2} = \mathbf{0}, \tag{4.1.2}$$

expressed in frame F'. Furthermore, if we use c to denote the speed of light in a vacuum (which is about 300,000 km/s) and $(t, \mathbf{r})$ and $(t', \mathbf{r}')$ the spacetime coordinates of a photon in frame F and F', respectively, then the postulate (ii) leads us to

$$c = \left|\frac{\mathrm{d}\mathbf{r}}{\mathrm{d}t}\right| = \frac{\mathrm{d}r}{\mathrm{d}t} = \left|\frac{\mathrm{d}\mathbf{r}'}{\mathrm{d}t'}\right| = \frac{\mathrm{d}r'}{\mathrm{d}t'}, \tag{4.1.3}$$

where $\mathrm{d}r = |\mathrm{d}\mathbf{r}|$ such that

$$\mathrm{d}r^2 = \mathrm{d}x^2 + \mathrm{d}y^2 + \mathrm{d}z^2, \tag{4.1.4}$$

and a similar expression for $\mathrm{d}r'$. In terms of (4.1.4), we may rewrite (4.1.3) as

$$c^2\mathrm{d}t^2 - \mathrm{d}x^2 - \mathrm{d}y^2 - \mathrm{d}z^2 = c^2\mathrm{d}t'^2 - \mathrm{d}x'^2 - \mathrm{d}y'^2 - \mathrm{d}z'^2 = 0, \tag{4.1.5}$$

for the history of a photon dictated by the postulate (ii).

Lorentz boosts

Let F and F' be inertial frames described earlier such that F' moves along the x-axis of F with constant speed, v, and that initially the origins of F and F' coincide. Hence, the coordinates (t,x,y,z) and (t',x',y',z') are related through the following general homogeneous linear transformation (illustrated in Figure 4.1)

$$t' = \alpha t + \beta x + \epsilon y + \delta z, \tag{4.1.6}$$
$$x' = \gamma(x - vt), \tag{4.1.7}$$
$$y' = y, \tag{4.1.8}$$
$$z' = z. \tag{4.1.9}$$

Inserting (4.1.6)–(4.1.9) into (4.1.5), we see that consistency gives us $\epsilon = \delta = 0$ and

$$c^2\alpha^2 - v^2\gamma^2 = c^2, \tag{4.1.10}$$
$$\gamma^2 - c^2\beta^2 = 1, \tag{4.1.11}$$
$$c^2\alpha\beta + v\gamma^2 = 0. \tag{4.1.12}$$

For convenience, we assume the transformation preserves the direction of time and space coordinate orientation. So we have $\alpha, \gamma > 0$. Substituting β given in (4.1.12) into (4.1.11), we get

$$\gamma^2 - \frac{v^2\gamma^4}{c^2\alpha^2} = 1. \tag{4.1.13}$$

Combining (4.1.10) and (4.1.13), we arrive at the relation

$$\alpha = \gamma. \tag{4.1.14}$$

Inserting this into (4.1.10), we find

$$(c^2 - v^2)\gamma^2 = c^2. \tag{4.1.15}$$

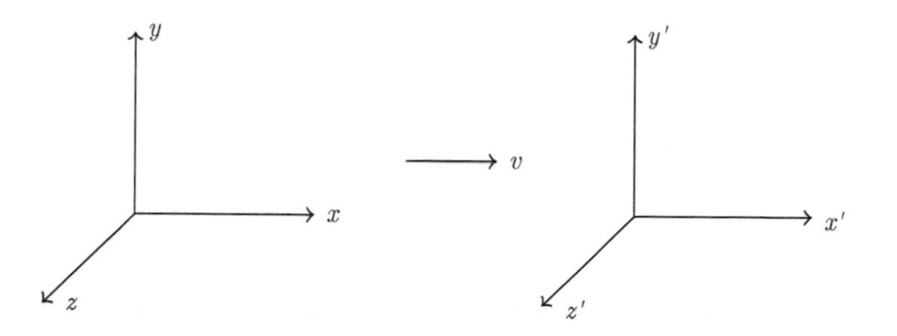

Figure 4.1 An illustration of the relative motion of two inertial frames.

In particular, from (4.1.15), we obtain the important restriction to the allowed value of velocity v:

$$|v| < c. \tag{4.1.16}$$

That is, the allowed velocity v between two inertial frames never exceeds the speed of light c. Consequently, from (4.1.10)–(4.1.16), we have

$$\alpha = \gamma = \left(1 - \frac{v^2}{c^2}\right)^{-\frac{1}{2}}, \quad \beta = -\frac{v}{c^2}\left(1 - \frac{v^2}{c^2}\right)^{-\frac{1}{2}}. \tag{4.1.17}$$

The quantity γ is called the *Lorentz factor.*

We can summarize our transformation, called the *Lorentz boost*, along the x-axis with velocity v, as follows:

$$t' = \gamma\left(t - \frac{v}{c^2}x\right), \tag{4.1.18}$$

$$x' = \gamma(x - vt), \quad y' = y, \quad z' = z, \tag{4.1.19}$$

$$\gamma = \left(1 - \frac{v^2}{c^2}\right)^{-\frac{1}{2}}. \tag{4.1.20}$$

Analogously, we may derive the Lorentz boosts along the y- and z-axis. These are collectively called the Lorentz boosts.

4.2 Line element, proper time, and consequences

Note that, in the derivation of the Lorentz boost (4.1.18)–(4.1.20), we actually did not use the fact that the right-hand side of (4.1.5) vanishes, but only the *invariance*

$$\mathrm{d}s^2 \equiv c^2\mathrm{d}t^2 - \mathrm{d}x^2 - \mathrm{d}y^2 - \mathrm{d}z^2 = c^2\mathrm{d}t'^2 - \mathrm{d}x'^2 - \mathrm{d}y'^2 - \mathrm{d}z'^2, \tag{4.2.1}$$

between two inertial frames. In other words, two inertial frames $F : (t, x, y, z)$ and $F' : (t', x', y', z')$ related by the Lorentz boosts follow the invariance law (4.2.1) such that $\mathrm{d}s^2 = 0$ for the history of a photon.

Line element

Imagine we observe a moving particle rather than a photon. Thus, (4.1.16) indicates that $\left(\frac{\mathrm{d}r}{\mathrm{d}t}\right)^2 < c^2$. So by (4.1.4) we have

$$\mathrm{d}s^2 = c^2\mathrm{d}t^2 - \mathrm{d}x^2 - \mathrm{d}y^2 - \mathrm{d}z^2 > 0. \tag{4.2.2}$$

When an event satisfies $\mathrm{d}s^2 > 0$, it is called *timelike*; $\mathrm{d}s^2 = 0$, *lightlike*; $\mathrm{d}s^2 < 0$, *spacelike*. From now on we will concentrate on timelike situations. Since $\mathrm{d}s^2$ has dimension of length squared, it is called the *line element* of the spacetime.

Proper time

Since $\mathrm{d}s^2$ is an invariance of dimension of length squared and c is an invariant speed indifferent to the choice of an inertial frame, we see that $\mathrm{d}\tau^2$ given by

$$\mathrm{d}s^2 = c^2\mathrm{d}\tau^2 \tag{4.2.3}$$

gives rise to an *invariant time*, τ, called *proper time*. In view of (4.2.1), we see that the lapse of proper time τ agrees with the lapse of coordinate time t in frame F for a particle at rest with $\mathrm{d}x = 0, \mathrm{d}y = 0, \mathrm{d}z = 0$. Besides, for a particle moving with velocity v along the x-axis (say), we may "follow" it with frame $F' : (t', x', y', z')$ as given in (4.1.18)–(4.1.20) such that the particle is at rest in frame F' with $\mathrm{d}x' = 0, \mathrm{d}y' = 0, \mathrm{d}z' = 0$. Thus, in view of (4.2.1) again, we see that the lapse of proper time τ now agrees with the lapse of coordinate time t' in frame F'. In other words, for a moving particle, proper time τ measures the time that "goes along" with the particle.

Length contraction

Consider a stick moving along the x-axis at velocity v, which is represented as a displacement vector, in space,

$$\mathbf{l} = (l_1, l_2, l_3) = (x_1, y_1, z_1) - (x_2, y_2, z_2), \tag{4.2.4}$$

in frame $F : (t, x, y, z)$, at a given time $t = t_0$. Thus, in frame $F' : (t', x', y', z')$, moving with the stick, given by (4.1.18)–(4.1.20), we see that the stick is represented by the displacement vector

$$\mathbf{l}' = (l_1', l_2', l_3') = (x_1', y_1', z_1') - (x_2', y_2', z_2'), \tag{4.2.5}$$

where

$$l_1' = \gamma l_1, \quad l_2' = l_2, \quad l_3' = l_3. \tag{4.2.6}$$

So there occurs a length variation in the direction of the motion. More precisely, since $L_0 = |\mathbf{l}'|$ is the length of the stick in the moving frame F' where the stick is at rest, it is called *rest length* or *proper length*. On the other hand, $L = |\mathbf{l}|$ is the length of the moving stick seen in frame F at time $t = t_0$. From (4.2.6), we observe that length variation occurs along the direction of motion but no length variation occurs in the directions that are transverse to the direction of motion. Thus, if we consider the motion of a stick placed along the x-axis, we obtain the following relation between the rest length L_0 and the length L observed in frame F of the moving stick:

$$L = \gamma^{-1} L_0 = \left(1 - \frac{v^2}{c^2}\right)^{\frac{1}{2}} L_0. \tag{4.2.7}$$

Thus, $L < L_0$ when $v \neq 0$. That is, we observe a *length contraction*, also known as the *Lorentz contraction*, for a moving stick.

Time dilation

Consider the motion of a particle at constant velocity v and use $F' : (t', x', y', z')$ to denote the inertial frame that moves with the particle. Then $\mathrm{d}x' = 0, \mathrm{d}y' = 0, \mathrm{d}z' = 0$ in (4.2.1). Thus, using T and T' to denote the time lapses recorded in frame F, where the particle is seen in motion and in frame F' where the particle is seen at rest, respectively, we are led by (4.2.1), or $(c^2 - v^2)\,\mathrm{d}t^2 = c^2 \mathrm{d}t'^2$, to the conclusion

$$T' = \left(1 - \frac{v^2}{c^2}\right)^{\frac{1}{2}} T. \tag{4.2.8}$$

Thus, $T > T'$ when $v \neq 0$. That is, we observe longer time lapse in frame F than that in frame F', which moves with the particle. This phenomenon is known as *time dilation.*

Simultaneity

Imagine we are to observe a stick set to glow light. Let $F : (t, x, y, z)$ and $F' : (t', x', y', z')$ be two inertial frames related by (4.1.18)–(4.1.20). Viewed in frame F, the stick is parametrized in space by

$$a \leq x \leq b, \quad y = 0, \quad z = 0, \tag{4.2.9}$$

and set to glow light at time $t = t_0$ (say). This is a sequence of events of glowing light at various spots of the stick that occur simultaneously. From (4.1.18)–(4.1.20), we see that, in frame F', the stick is parametrized in space by

$$\gamma(a - vt_0) \leq x' \leq \gamma(b - vt_0), \quad y' = 0, \quad z' = 0, \tag{4.2.10}$$

which is seen to glow light when time t' is in the interval with the corresponding end-points depending on a, b:

$$t'_a = \gamma\left(t_0 - \frac{v}{c^2}a\right), \quad t'_b = \gamma\left(t_0 - \frac{v}{c^2}b\right). \tag{4.2.11}$$

In particular, in F', the events of glowing light at various spots of the stick will not occur simultaneously. In other words, *simultaneity* fails. For example, if F' moves along the positive x-axis such that $v > 0$, then $t'_a > t'_b$ for $a < b$ and the interval for the occurrence of light glowing is given by

$$\gamma\left(t_0 - \frac{v}{c^2}b\right) \leq t' \leq \gamma\left(t_0 - \frac{v}{c^2}a\right). \tag{4.2.12}$$

Hence, the right-most end of the stick glows first and the left-most end of the stick glows last.

Superposition of velocities

Let F, F' and F', F'' be two pairs of inertial frames such that F' moves along the x-axis of F with velocity v and F'' moves along the x'-axis of F' with velocity w. Thus, we may relate F, F' and F', F'' by (4.1.18)–(4.1.20) so that the corresponding Lorentz factors are denoted specifically by γ_v and γ_w, respectively. Thus, with this notation, $F : (t, x, y, z), F' : (t', x', y', z')$ and $F', F'' : (t'', x'', y'', z'')$ are related through the equations

$$t' = \gamma_v \left(t - \frac{v}{c^2} x \right), \tag{4.2.13}$$

$$x' = \gamma_v (x - vt), \quad y' = y, \quad z' = z, \tag{4.2.14}$$

$$t'' = \gamma_w \left(t' - \frac{w}{c^2} x' \right), \tag{4.2.15}$$

$$x'' = \gamma_w (x' - wt'), \quad y'' = y', \quad z'' = z'. \tag{4.2.16}$$

Iterating (4.2.13)–(4.2.16), we obtain

$$t'' = \gamma_v \gamma_w \left(1 + \frac{vw}{c^2} \right) t - \gamma_v \gamma_w \frac{(v + w)}{c^2} x, \tag{4.2.17}$$

$$x'' = \gamma_v \gamma_w \left(1 + \frac{vw}{c^2} \right) x - \gamma_v \gamma_w (v + w) t, \quad y'' = y, \quad z'' = z. \tag{4.2.18}$$

To proceed further, we set

$$a = \gamma_v \gamma_w \left(1 + \frac{vw}{c^2} \right), \quad b = \gamma_v \gamma_w (v + w), \tag{4.2.19}$$

which allows us to suppress (4.2.17)–(4.2.18) into

$$t'' = at - \frac{b}{c^2} x, \tag{4.2.20}$$

$$x'' = ax - bt, \quad y'' = y, \quad z'' = z. \tag{4.2.21}$$

Moreover, from (4.2.19), we see that the quantity $u \equiv \frac{b}{a}$ is given by the formula

$$u = \frac{v + w}{1 + \frac{vw}{c^2}}. \tag{4.2.22}$$

With (4.2.22), we have

$$a = \gamma_v \gamma_w \left(1 + \frac{vw}{c^2} \right) = \gamma_u, \tag{4.2.23}$$

which may be combined with (4.2.20)–(4.2.22) to give us the joint Lorentz boost associated with the velocity u:

$$t'' = \gamma_u \left(t - \frac{u}{c^2} x \right), \tag{4.2.24}$$

$$x'' = \gamma_u (x - ut), \quad y'' = y, \quad z'' = z. \tag{4.2.25}$$

In other words, we have seen that the frames F and F'' are related through a Lorentz boost along the x-axis associated with the velocity u. This velocity is "superimposed" from the velocities v and w following the formula (4.2.22). This formula spells out the *law of superposition of velocities* in the present context.

Let F, F' be two inertial frames related through (4.1.18)–(4.1.20) with the associated relative velocity v. In special relativity, the ratio $\frac{v}{c}$ may uniquely be expressed in terms of a *hyperbolic angle* ϕ by the formula

$$\frac{v}{c} = \tanh\phi. \tag{4.2.26}$$

The hyperbolic angle ϕ is also called *rapidity* associated with the inertial frames F, F' with relative velocity v. With such a terminology, if we use ψ and θ to denote the rapidities associated with the inertial frames F', F'' and F, F'' with relative velocities w and u respectively, then (4.2.22) leads to

$$\tanh\theta = \frac{\tanh\phi + \tanh\psi}{1 + \tanh\phi\tanh\psi} = \tanh(\phi + \psi), \tag{4.2.27}$$

or equivalently,

$$\theta = \phi + \psi. \tag{4.2.28}$$

Thus, rapidities are additive in special relativity, which replaces additivity of velocities in Newtonian mechanics.

Note that, in view of (4.2.26), we see that the Lorentz factor associated with velocity v is

$$\gamma = \left(1 - \tanh^2\phi\right)^{-\frac{1}{2}} = \cosh\phi. \tag{4.2.29}$$

Thus, inserting (4.2.29) into (4.1.18)–(4.1.20), we get the following elegant matrix form of the Lorentz boost:

$$\begin{pmatrix} ct' \\ x' \\ y' \\ z' \end{pmatrix} = \begin{pmatrix} \cosh\phi & -\sinh\phi & 0 & 0 \\ -\sinh\phi & \cosh\phi & 0 & 0 \\ 0 & 0 & 1 & 0 \\ 0 & 0 & 0 & 1 \end{pmatrix} \begin{pmatrix} ct \\ x \\ y \\ z \end{pmatrix}, \tag{4.2.30}$$

which may be regarded as a "hyperbolic rotation." Note also that such a formalism puts the coordinate variables naturally in equal footing since ct, ct' are in dimension of length as well. By virtue of (4.2.30), the law of additivity of rapidities expressed in (4.2.27) or (4.2.28) becomes obvious.

4.3 Relativistic mechanics

It is convenient to recast the temporal and spatial coordinates, t and x, y, x, into an "equal footing" by setting

$$x^0 = ct, \quad x^1 = x, \quad x^2 = y, \quad x^3 = z, \tag{4.3.1}$$

and rewrite a spacetime point, or event, (t, x, y, z), collectively as $x \equiv (x^\mu) = (x^0, x^i)$ $(\mu = 0, 1, 2, 3, i = 1, 2, 3)$. We also use the notation $x = (x^0, \mathbf{x})$. Observe that x^μs are all of dimension of length.

Minkowski metric

With the 4×4 matrix

$$\eta = (\eta_{\mu\nu}) = \begin{pmatrix} 1 & 0 & 0 & 0 \\ 0 & -1 & 0 & 0 \\ 0 & 0 & -1 & 0 \\ 0 & 0 & 0 & -1 \end{pmatrix}, \tag{4.3.2}$$

known as the standard *Minkowski metric* of the Minkowski spacetime $\mathbb{R}^{3,1}$, and following the convention that summation is conducted over repeated upper and lower indices, we can express the line element (4.2.2) as

$$\mathrm{d}s^2 = \eta_{\mu\nu}\mathrm{d}x^\mu \mathrm{d}x^\nu. \tag{4.3.3}$$

Here and in the sequel, we observe the rule that the Greek letters $\mu, \nu = 0, 1, 2, 3$ are spacetime indices and Latin letters $i, j = 1, 2, 3$ space indices, unless otherwise stated.

Lorentz transformations

We can rewrite the matrix form of the spacetime coordinate transformation (4.2.30) as

$$x'^\mu = L^\mu_\nu x^\nu, \tag{4.3.4}$$

where $L = (L^\mu_\nu)$ is a 4×4 real matrix. The invariance stated in (4.2.1), or

$$\eta_{\mu\nu}\mathrm{d}x^\mu \mathrm{d}x^\nu = \eta_{\mu\nu}\mathrm{d}x'^\mu \mathrm{d}x'^\nu, \tag{4.3.5}$$

indicates in view of (4.3.4) that $L = (L^\mu_\nu)$ satisfies the equation

$$\eta_{\mu\nu} = \eta_{\alpha\beta} L^\alpha_\mu L^\beta_\nu. \tag{4.3.6}$$

Of course, the matrices given by the Lorentz boosts along the x^1, x^2, x^3 axes all satisfy (4.3.6). Besides, space rotations given by

$$x'^0 = x^0, \quad x^i = R^i_j x^j, \tag{4.3.7}$$

where $R = (R^i_j)$ is a 3×3 orthogonal matrix, also satisfy (4.3.6). The spacetime transformations satisfying (4.3.6) are called the *Lorentz transformations*. Thus, the Lorentz boosts, space rotations given in (4.3.7), and their compositions are Lorentz transformations. The set of all Lorentz transformations equipped with composition operation is a Lie group called the *Lorentz group*, which is generated from the Lorentz boosts and space rotations.

Velocity vectors

We have seen that coordinate time is not an invariant, but depends on the inertial frame chosen. We have also seen that proper time is invariant, which is frame-independent. Thus, in order to express frame-independent physical laws, we should use proper time instead of coordinate time.

As before, use τ to denote proper time as given in (4.2.3). We define the *velocity 4-vector*, or *world velocity*, to be given by the rates of change of spacetime coordinates with respect to proper time:

$$u = (u^\mu), \quad u^\mu = \frac{\mathrm{d}x^\mu}{\mathrm{d}\tau}. \tag{4.3.8}$$

Inserting (4.3.8) into (4.2.3) or (4.3.3), we arrive at the identity

$$\eta_{\mu\nu} u^\mu u^\nu = c^2, \tag{4.3.9}$$

which is a conservation law. As a comparison, the *coordinate velocity*, in terms of coordinate time t, is given by

$$(v^\mu), \quad v^\mu = \frac{\mathrm{d}x^\mu}{\mathrm{d}t}, \tag{4.3.10}$$

such that $v^0 = c$, resulting in $(v^\mu) = (c, \mathbf{v})$, where

$$\mathbf{v} = (v^i), \quad v^i = \frac{\mathrm{d}x^i}{\mathrm{d}t}, \tag{4.3.11}$$

is the Newtonian velocity vector. Thus, with the notation $v = |\mathbf{v}|$, we may relate (u^μ) and (v^μ) as follows:

$$(u^\mu) = (v^\mu)\frac{\mathrm{d}t}{\mathrm{d}\tau} = \gamma(c, \mathbf{v}), \quad \frac{\mathrm{d}t}{\mathrm{d}\tau} = \gamma = \left(1 - \frac{v^2}{c^2}\right)^{-\frac{1}{2}}. \tag{4.3.12}$$

Momentum vectors

Let $m > 0$ denote the mass of a moving particle. The Newtonian momentum vector is $m\mathbf{v}$ which may be viewed as the "spatial part" of the coordinate momentum

$$mv^\mu = (mc, m\mathbf{v}), \tag{4.3.13}$$

which is given in terms of coordinate time, rather than proper time, causing complication with regard to frame choice. Nevertheless, from (4.3.13), we are led to introducing the relativistic *momentum vector* defined by

$$p = (p^\mu) = (p^0, \mathbf{p}), \quad p^\mu = mu^\mu = m\frac{\mathrm{d}x^\mu}{\mathrm{d}\tau}. \tag{4.3.14}$$

We elaborate on some details of (4.3.14) as follows.

The temporal component of p is

$$p^0 = mc\frac{\mathrm{d}t}{\mathrm{d}\tau} = \gamma mc. \tag{4.3.15}$$

The spatial component of p is

$$\mathbf{p} = m\frac{\mathrm{d}\mathbf{x}}{\mathrm{d}\tau} = m\frac{\mathrm{d}\mathbf{x}}{\mathrm{d}t}\frac{\mathrm{d}t}{\mathrm{d}\tau} = \gamma m\mathbf{v}. \tag{4.3.16}$$

Both p^0 and $\mathbf{p}$ are in the form of classical momentum quantities, with the former associated with the speed of light c and the latter the coordinate velocity $\mathbf{v}$, and an *effective mass*

$$\gamma m = \left(1 - \frac{v^2}{c^2}\right)^{-\frac{1}{2}} m. \tag{4.3.17}$$

Consequently, in the coordinate frame, the moving particle appears more massive.

Furthermore, from (4.3.9), we arrive at the *law of energy-momentum conservation* in special relativity:

$$\eta_{\mu\nu}p^\mu p^\nu = m^2c^2, \tag{4.3.18}$$

also called *four-momentum relation* or *mass shell condition.*

Energy formula

We may rewrite (4.3.18) as

$$(p^0)^2 - |\mathbf{p}|^2 = m^2c^2. \tag{4.3.19}$$

To explore the mechanical meaning of this identity, we rewrite it as

$$(p^0)^2 = m^2c^2 + |\mathbf{p}|^2, \tag{4.3.20}$$

and note that the first term on the right-hand side is invariant, or motion independent, and the second term is motion dependent, or kinetic. Thus, the first term leads to a notion of *rest mass energy*, and the second, kinetic energy, of the moving particle of mass m. In other words, the left-hand side of (4.3.20) pertains to the total energy of the particle. Recall that energy is of dimension of mass times velocity squared, and momentum of mass times velocity. So we may write

$$p^0 = \frac{E}{c}, \tag{4.3.21}$$

in which the speed of light c is used simply to achieve correct dimensionality such that it is consistent that the quantity E has the dimension of energy.

Inserting (4.3.21) into (4.3.20), we arrive at the celebrated Einstein energy formula:

$$E^2 = m^2c^4 + |\mathbf{p}|^2c^2, \tag{4.3.22}$$

so that the rest mass energy reads

$$E = mc^2, \tag{4.3.23}$$

for a resting particle with $\mathbf{p} = \mathbf{0}$. Inserting (4.3.16) into (4.3.22) and choosing the positive radical root, we have

$$\begin{aligned} E &= mc^2\sqrt{1+\frac{\gamma^2 v^2}{c^2}} = \gamma mc^2 \\ &= mc^2 + \frac{1}{2}mv^2 + \frac{3}{8}mv^2\left(\frac{v}{c}\right)^2 + \frac{5}{16}mv^2\left(\frac{v}{c}\right)^4 + \cdots . \end{aligned} \tag{4.3.24}$$

Hence, the rest mass energy is the leading term and the usual Newtonian kinetic energy comes next.

Mass of photon

Quantum mechanically, when a particle of mass m is considered to behave like a wave of wave number k, wavelength λ, and angular frequency ω, then its energy E and momentum $p \equiv |\mathbf{p}|$ are given by the Planck–Einstein and Compton–Debye formulas, (2.1.8) and (2.1.9), respectively. On the other hand, if the particle is a photon that travels at the speed of light, then k, λ, and ω observe (2.1.6). Thus, inserting (2.1.6), (2.1.8), and (2.1.9) into (4.3.22), we arrive at $m = 0$ for the photon.

Equation of motion

In Newtonian mechanics, when a particle of mass $m > 0$ is subject to an external force $\mathbf{F}$, the rate of change of its momentum vector obeys the equation

$$\frac{\mathrm{d}}{\mathrm{d}t}(m\mathbf{v}) = m\frac{\mathrm{d}^2\mathbf{x}}{\mathrm{d}t^2} = \mathbf{F}, \tag{4.3.25}$$

with respect to coordinate time t, which is known as Newton's second law. In special relativity, this equation is replaced by the frame-free equation in terms of proper time τ:

$$\frac{\mathrm{d}p^\mu}{\mathrm{d}\tau} = m\frac{\mathrm{d}^2x^\mu}{\mathrm{d}\tau^2} = f^\mu, \tag{4.3.26}$$

where $f = (f^\mu) = (f^0, \mathbf{f})$ is an applied 4-vector force.

If the observer travels with the particle, then proper time coincides with coordinate time such that spatial part of (4.3.26) is the same as the Newton equation (4.3.25). Thus $\mathbf{f} = \mathbf{F}$. On the other hand, the temporal component of (4.3.26), in view of (4.3.21), becomes

$$\frac{\mathrm{d}p^0}{\mathrm{d}t} = \frac{1}{c}\frac{\mathrm{d}E}{\mathrm{d}t}. \tag{4.3.27}$$

Moreover, note that the rate of change of energy E equals to the rate of change of the work done to the particle by the applied force:

$$\frac{\mathrm{d}E}{\mathrm{d}t} = \frac{\mathrm{d}}{\mathrm{d}t}\int \mathbf{F}\cdot\mathrm{d}\mathbf{x} = \frac{\mathrm{d}}{\mathrm{d}t}\int \mathbf{F}\cdot\mathbf{v}\,\mathrm{d}t = \mathbf{F}\cdot\mathbf{v}, \tag{4.3.28}$$

resulting in $f^0 = \frac{1}{c}\mathbf{F}\cdot\mathbf{v}$ in the co-moving frame (so that proper time coincides with coordinate time). Consequently, since in a general coordinate frame the momentum vector $(p^0, \mathbf{p})$ is related to the rest-frame momentum vector $(mc, m\mathbf{v})$ through the relations (4.3.15)–(4.3.16) by a Lorentz factor γ, thus, by consistency, the force 4-vector (f^μ) should follow the same rule. Hence, we arrive at

$$(f^\mu) = (f^0, \mathbf{f}) = \gamma\left(\frac{\mathbf{F}\cdot\mathbf{v}}{c}, \mathbf{F}\right). \tag{4.3.29}$$

Inserting (4.3.29) into (4.3.26), using (4.3.16), and noting $\frac{\mathrm{d}t}{\mathrm{d}\tau} = \gamma$ again, we see that the temporal component of (4.3.26) is simply (4.3.28) in view of (4.3.27), of course, and the spatial part of (4.3.26) becomes

$$m\frac{\mathrm{d}}{\mathrm{d}t}\left(\frac{\mathbf{v}}{\sqrt{1-\frac{v^2}{c^2}}}\right) = \mathbf{F}, \tag{4.3.30}$$

which governs the relativistic motion of the particle. In particular, if the force $\mathbf{F}$ is conservative produced by a potential $V(\mathbf{x})$, that is,

$$\mathbf{F} = -\nabla_{\mathbf{x}} V, \tag{4.3.31}$$

then the equation (4.3.30) is the Euler–Lagrange equation of the Lagrange function

$$L(\mathbf{x}, \dot{\mathbf{x}}) = -mc^2\sqrt{1-\frac{\dot{\mathbf{x}}^2}{c^2}} - V(\mathbf{x}). \tag{4.3.32}$$

In the limit $c \to \infty$, (4.3.32) becomes the familiar Newtonian Lagrange function

$$L_{\text{Newton}}(\mathbf{x}, \dot{\mathbf{x}}) = \frac{1}{2}m\dot{\mathbf{x}}^2 - V(\mathbf{x}), \tag{4.3.33}$$

up to a constant translation, $-mc^2$. Thus, the rest mass energy, mc^2, may well be interpreted as a background energy that contributes to the potential energy. The associated momentum of the Lagrange function (4.3.32) is

$$\mathbf{p} = \nabla_{\dot{\mathbf{x}}} L = \frac{m\dot{\mathbf{x}}}{\sqrt{1-\frac{\dot{\mathbf{x}}^2}{c^2}}} = \gamma m\mathbf{v}, \tag{4.3.34}$$

which returns to the expression (4.3.16). Furthermore, the Hamiltonian in view of (4.3.32) and (4.3.34) is

$$\begin{aligned} H &= \mathbf{p}\cdot\dot{\mathbf{x}} - L(\mathbf{x}, \dot{\mathbf{x}}) \\ &= \gamma m v^2 + mc^2\sqrt{1-\frac{\dot{\mathbf{x}}^2}{c^2}} + V(\mathbf{x}) \\ &= \gamma mc^2 + V(\mathbf{x}), \end{aligned} \tag{4.3.35}$$

whose first term on the right-hand side coincides with the particle energy E stated in (4.3.24), and the second term being the potential energy stored in the potential field. Thus, H is the total energy. Hence, from (4.3.28), we have

$$\frac{\mathrm{d}H}{\mathrm{d}t} = \frac{\mathrm{d}E}{\mathrm{d}t} + \dot{\mathbf{x}} \cdot \nabla_{\mathbf{x}} V(\mathbf{x}) = 0. \tag{4.3.36}$$

In other words, the temporal component of (4.3.26) is simply that the total energy is conserved when applied force is conservative.

4.4 Doppler effects

For simplicity, we consider the situation of a stationary observer and a moving wave source. The source waves travel with speed w, which is independent of the speed v_s of the source. In general, we use $T > 0$ to denote the time period of the waves measuring time lapse between two consecutive crests of the waves. So, comparing T with unit time, we have the wave frequency

$$f = \frac{1}{T}, \tag{4.4.1}$$

measuring the number of wave crests in unit time. Thus, use λ, the wavelength, to denote the distance between two consecutive wave crests. Then $f\lambda$ gives the wave speed. That is,

$$f = \frac{w}{\lambda}. \tag{4.4.2}$$

We are now prepared to discuss the Doppler effects.

Classical Doppler effect

Consider two consecutive wave crests emitted from a wave source approaching the stationary observer at speed v_s, $v_s < w$. Over the time period T of the two emitted wave crests, the distance between the source and the observer, hence the distance between the two wave crests, is shortened by $v_s T$, so the distance traveled by the wave between the two wave crests becomes

$$\lambda' = \lambda - v_s T, \tag{4.4.3}$$

which is the wavelength of the waves observed by the stationary observer. On the other hand, combining (4.4.1) and (4.4.2), we find

$$T = \frac{\lambda}{w}. \tag{4.4.4}$$

Inserting (4.4.4) into (4.4.3), we arrive at

$$\lambda' = \lambda\left(1 - \frac{v_s}{w}\right). \tag{4.4.5}$$

Since the speed of waves is independent of moving source, using (4.4.5) and (4.4.2), we get the updated frequency of the waves observed by the observer:

$$f' = \frac{f}{1 - \frac{v_s}{w}}. \tag{4.4.6}$$

The situation of receding wave source is left as an exercise.

Relativistic Doppler effect

We continue to assume the wave velocity is constant in inertial frames. Let us consider waves emitted from the spatial origin of the frame $F : (t, x, y, z)$. Let $F' : (t', x', y', z')$ be another frame moving along the x-axis with velocity $v > 0$, which is related to frame F by the Lorentz boost (4.1.18)–(4.1.20) so that the two frames coincide at the origin and the observer resides at $x' = 0$, $y' = 0, z' = 0$. So, equivalently, the situation may also be regarded that the observer is stationary but the source moves away. Let $(0, 0, 0, 0)$ and $(T, 0, 0, 0)$ denote the events of emitting two consecutive wave crests over the wave period $T > 0$ in frame F. From (4.1.18)–(4.1.20), we see that these two events are recorded in F' as $(0, 0, 0, 0)$ and

$$t' = \gamma T, \quad x' = -\gamma v T, \quad y' = 0, \quad z' = 0. \tag{4.4.7}$$

In other words, in frame F', the second event happens at time $t' = \gamma T$ and at the negative x'-axis of distance $\gamma v T$ away. Since the wave velocity is w, it takes an additional time, say $\Delta t'$, for the second wave crest to arrive at the observer at the spatial origin of F'. Hence, we have

$$w \Delta t' = \gamma v T. \tag{4.4.8}$$

Therefore, from (4.4.7) and (4.4.8), we obtain the total "wait" time by the observer to observe the arrival of the second wave crest

$$\begin{aligned} T' &= t' + \Delta t' \\ &= \gamma T \left(1 + \frac{v}{w}\right), \end{aligned} \tag{4.4.9}$$

which gives the period formula of the relativistic Doppler effect for a receding source. For an approaching source, there occurs a sign change in front of the velocity factor v in the formula. Thus, as in the discussion of the classical Doppler effect, if we use v_s to denote the source velocity and update the Lorentz factor γ with $v = v_s$, then we may use (4.4.2) to obtain the frequency formula for the Doppler effect:

$$f' = \frac{f \gamma^{-1}}{1 \pm \frac{v_s}{w}} = \frac{f}{\left(1 \pm \frac{v_s}{w}\right)} \sqrt{1 - \frac{v_s^2}{c^2}}, \tag{4.4.10}$$

which is seen to differ from the formula in the classical Doppler effect, as stated in (4.4.6) for the approaching source situation, simply by a Lorentz factor term,

γ^{-1}. Here, the sign convention in (4.4.10) is taken such that the plus and minus signs account for receding and approaching wave sources, respectively.

In particular, if the wave source is that of light or electromagnetic waves, then $w = c$ and (4.4.10) becomes

$$f' = f\frac{\sqrt{c \mp v_s}}{\sqrt{c \pm v_s}}, \quad v_s > 0, \tag{4.4.11}$$

for receding and approaching sources, giving rise to *redshift* and *blueshift* of the signal waves, respectively.

Exercises

1. Find the inverse transformation of (4.1.18)–(4.1.20) and explain your results.
2. Consider a car moving along the x-axis at constant velocity u in a fixed frame $F : (t, x, y, z)$. Let $F' : (t', x', y', z')$ be another frame that moves along the x-axis of frame F at constant velocity v. Determine the velocity of the car in frame F'.
3. Let F, F' be inertial frames related by (4.1.18)–(4.1.20) and θ the angle between a rod and the x-axis in frame F. Find the angle θ', in terms of θ, between the rod and x'-axis in frame F'.
4. Using (4.3.16), derive the relativistic velocity-momentum formula

$$\mathbf{v} = \frac{c\mathbf{p}}{\sqrt{m^2c^2 + |\mathbf{p}|^2}}, \tag{4.E.1}$$

   and find the limits of the right-hand side of (4.E.1) as $c \to \infty$ and $m \to 0$, respectively.
5. Use the full Einstein energy formula (4.3.22) to derive the Compton–Debye formula (2.1.7).
6. With the Hamiltonian function (4.3.35), show that its coordinate-momentum form is

$$H(\mathbf{x}, \mathbf{p}) = c\sqrt{m^2c^2 + |\mathbf{p}|^2} + V(\mathbf{x}), \tag{4.E.2}$$

   and obtain the associated relativistic Hamiltonian equations in the form

$$\frac{\mathrm{d}\mathbf{x}}{\mathrm{d}t} = \nabla_{\mathbf{p}} H, \quad \frac{\mathrm{d}\mathbf{p}}{\mathrm{d}t} = -\nabla_{\mathbf{x}} H. \tag{4.E.3}$$

7. Using (4.3.30), show that the relativistic version of the second law of Newton, $ma = F$, is

$$ma\left(1 - \frac{v^2}{c^2}\right)^{-\frac{3}{2}} = F, \tag{4.E.4}$$

when the applied force is in the same direction of the velocity such that $\mathbf{F} = F\mathbf{u}$ and $\mathbf{v} = v\mathbf{u}$ where $\mathbf{u}$ is the unit vector in the direction of $\mathbf{v}$, and $a = \frac{dv}{dt}$ is the usual acceleration.

8. Apply Exercise 7 to the situation of the motion of an electrically charged particle of electric charge $q > 0$ and mass $m > 0$ in a constant electric field of the form $\mathbf{E} = E\mathbf{u}$ where $\mathbf{u}$ is a given unit vector. Then (4.E.4) becomes

$$ma\left(1 - \frac{v^2}{c^2}\right)^{-\frac{3}{2}} = qE. \tag{4.E.5}$$

   (a) Show that $a = a(t)$ decreases.
   (b) Find the limit of $v = v(t)$ as $t \to \infty$ with any initial velocity $v(0)$, without integrating (4.E.5).
   (c) Find $v = v(t)$ with $v(0) = 0$ explicitly by integrating (4.E.5).
   (d) Find the trajectory of the particle initially at the spatial origin by using part (c).

9. Show that, when the wave source recedes from the stationary observer at speed v_s, then the wavelength and wave frequency of the emitted waves observed by the observer are given by

$$\lambda' = \lambda\left(1 + \frac{v_s}{w}\right), \quad f' = \frac{f}{1 + \frac{v_s}{w}}, \tag{4.E.6}$$

respectively.

10. Use the *Galilean transformation*

$$t' = t, \quad x' = x - vt, \quad y' = y, \quad z' = z, \tag{4.E.7}$$

as in the study of the relativistic Doppler effect, to derive the classical Doppler effect.

11. Let v_s denote the velocity of a moving light source such that $v_s > 0$ or $v_s < 0$ indicates the source is approaching or receding, respectively. Use (4.4.11) to derive the general formula

$$v_s = c\left(\frac{\rho^2 - 1}{\rho^2 + 1}\right), \quad \rho = \frac{f'}{f}, \tag{4.E.8}$$

which determines v_s in terms of the blue- or redshift of the light received by an observer at rest.

12. Extend Exercise 11 to the situation of a general moving wave source not necessarily that of light. That is, use (4.4.10) to obtain a formula for v_s similar to (4.E.8), in terms of w in (4.4.10) and ρ in (4.E.8).

5

Abelian gauge field equations

This chapter studies the Abelian gauge field theory. It starts with a discussion of the notions of covariance, contravariance, and invariance, which provides the necessary foundations to develop relativistic field theory. Next, it introduces the Klein–Gordon equation and explores what the symmetry of the underlying theory offers. In particular, it is seen that gauge field and electromagnetism arise as a natural consequence from the need to preserve local internal symmetry. It then considers various notions of symmetry breaking, especially the notion of spontaneous symmetry breaking. It shows in such a context that gauge field again plays an important role through the celebrated Higgs mechanism.

5.1 Spacetime, covariance, and invariance

The Abelian gauge field theory is the simplest relativistic Lagrangian field theory containing the Maxwell theory of electromagnetism. Yet, this theory possesses all the essential ingredients of a general gauge field theory and lays the foundation for all subsequent extensions embracing more sophisticated and useful symmetry properties. Mathematically, gauge fields are described in terms of vector fields enjoying covariant and contravariant properties subject to coordinate transformations. This section looks at such properties.

Vector space and its dual space

Let $\mathcal{V}$ be an n-dimensional vector space over $\mathbb{R}$, with its dual space, that is, the vector space of all linear functionals over $\mathcal{V}$, denoted by $\mathcal{V}'$. For a given basis, $\mathcal{B} = \{u_1, \dots, u_n\}$, over $\mathcal{V}$, use $\mathcal{B}' = \{u_1', \dots, u_n'\} \subset \mathcal{V}'$ to denote the associated dual basis [267, 607] over $\mathcal{V}'$ satisfying

$$\langle u_i, u_j' \rangle \equiv u_j'(u_i) = \delta_{ij}, \quad i, j = 1, \dots, n. \tag{5.1.1}$$

Mathematical Physics with Differential Equations. Yisong Yang, Oxford University Press.
© Yisong Yang (2023). DOI: 10.1093/oso/9780192872616.003.0005

Covariance and contravariance

For an invertible linear transformation $T : \mathcal{V} \to \mathcal{V}$,

$$\mathcal{C} = \{v_1, \ldots, v_n\} = \{Tu_1, \ldots, Tu_n\}, \tag{5.1.2}$$

is another basis for $\mathcal{V}$. Suppose the dual basis corresponding to $\mathcal{C}$ is $\mathcal{C}' = \{v_1', \ldots, v_n'\}$ and the adjoint of T is $T' : \mathcal{V}' \to \mathcal{V}'$. Then we have

$$\delta_{ij} = \langle v_i, v_j' \rangle = \langle Tu_i, v_j' \rangle = \langle u_i, T'v_j' \rangle, \quad i, j = 1, \ldots, n, \tag{5.1.3}$$

which leads us to the conclusion that

$$T'v_i' = u_i' \quad \text{or} \quad (T')^{-1}u_i' = v_i', \quad i = 1, \ldots, n. \tag{5.1.4}$$

Thus, the dual basis transformation follows a reversed direction relative to the direction of the original transformation. These rules are illustrated by the commutative diagram

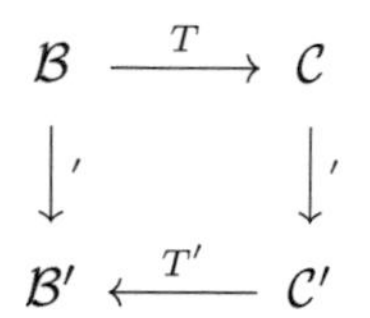

Thus, we may say that the bases in $\mathcal{V}$ are transformed in a *covariant* way, while the corresponding bases in $\mathcal{V}'$ are transformed in a *contravariant* way.

Matrix representation

If we express the transformation $T : \mathcal{V} \to \mathcal{V}$ explicitly in terms of a matrix $A = (A_{ij})$, called the basis transition matrix, by

$$v_j = Tu_j = \sum_{i=1}^{n} A_{ij} u_i, \quad j = 1, \ldots, n, \tag{5.1.5}$$

then we have

$$(v_1, \ldots, v_n) = (u_1, \ldots, u_n)A. \tag{5.1.6}$$

Thus, we see that the bases are directly related by the basis transition matrix. On the other hand, we can express a vector $u \in \mathcal{V}$ in terms of the bases $\mathcal{B}$ and $\mathcal{C}$ as

$$u = \sum_{i=1}^{n} x_i u_i = \sum_{i=1}^{n} y_i v_i, \tag{5.1.7}$$

with the corresponding coordinate vectors $x = (x_i), y = (y_i) \in \mathbb{R}^n$ taken to be column vectors. Inserting (5.1.5) into (5.1.7), we find

$$x = Ay \quad \text{or} \quad y = A^{-1}x, \tag{5.1.8}$$

indicating that the coordinate vectors are related inversely by the basis transition matrix. In other words, coordinate vectors are related in the opposite way to that of basis vectors, in $\mathcal{V}$, in terms of basis transition matrices. For this reason coordinate vectors for vectors in $\mathcal{V}$ are called contravariant.

Similarly, we rewrite (5.1.4) as

$$u'_j = T'v_j = \sum_{i=1}^n A'_{ij}v'_i, \tag{5.1.9}$$

where $A' = (A'_{ij})$ is the corresponding basis transition matrix from the basis $\mathcal{C}'$ into the basis $\mathcal{B}'$. Thus, in place of (5.1.6), we have

$$(u'_1, \ldots, u'_n) = (v'_1, \ldots, v'_n)A'. \tag{5.1.10}$$

So for any vector $u' \in \mathcal{V}'$ expressed with respect to bases $\mathcal{B}', \mathcal{C}'$ as

$$u' = \sum_{i=1}^n x'_i u'_i = \sum_{i=1}^n y'_i v'_i, \quad x' = (x'_i),\, y' = (y'_i) \in \mathbb{R}^n, \tag{5.1.11}$$

we have

$$y' = A'x', \tag{5.1.12}$$

for the coordinate vectors, which again follows the opposite direction to that for the bases, with regard to the basis transition matrix. Therefore, coordinate vectors for vectors in $\mathcal{V}'$ are called covariant.

Matrices of maps

Let $T : \mathcal{V} \to \mathcal{V}$ be a linear map and $T' : \mathcal{V}' \to \mathcal{V}'$ its adjoint. For a fixed basis $\mathcal{B} = \{u_1, \ldots, u_n\}$ over $\mathcal{V}$ and the corresponding dual one $\mathcal{B}' = \{u'_1, \ldots, u'_n\}$ over $\mathcal{V}'$, assume that $A = (A_{ij})$ and $A' = (A'_{ij})$ are their respective matrix representatives such that

$$Tu_j = \sum_{i=1}^n A_{ij}u_i, \quad T'u'_j = \sum_{i=1}^n A'_{ij}u'_i, \quad j = 1, \ldots, n. \tag{5.1.13}$$

Then, using (5.1.13), we have

$$A_{ij} = \langle Tu_j, u'_i\rangle = \langle u_j, T'u'_i\rangle = A'_{ji}. \tag{5.1.14}$$

That is, we arrive at the well-known relation, $A' = A^t$.

Einstein notation

With respect to the covariant basis $\{u_1, \ldots, u_n\}$ of $\mathcal{V}$, we may raise the indices of the coordinate vector of any vector u in $\mathcal{V}$ to emphasize that such a coordinate vector is contravariant:

$$u = \sum_{i=1}^n x^i u_i \equiv x^i u_i, \tag{5.1.15}$$

with the summation convention over repeated indices observed as well. Following this practice, the dual basis $\{u'_1, \dots, u'_n\}$ may be denoted contravariantly as $\{u^1, \dots, u^n\}$. Correspondingly, since the coordinate vectors of vectors in $\mathcal{V}'$ are covariant, we may consistently use lower-indexed vectors in $\mathbb{R}^n$ to denote covariant vectors such that any $u' \in \mathcal{V}'$ can be expressed as

$$u' = \sum_{i=1}^{n} x_i u^i \equiv x_i u^i. \tag{5.1.16}$$

This expression is in elegant company with the expression (5.1.15).

Moreover, in a compatible way, we may rewrite the matrix $A = (A_{ij})$ as $A = (A^i_j)$ such that the row and column indices of the entries correspond to upper and lower or contravariant and covariant indices, respectively. Hence, in view of (5.1.13) and (5.1.14), we get

$$Tu_j = A^i_j u_i, \quad T'u^j = A^j_i u^i. \tag{5.1.17}$$

Consequently, the coordinate vectors $(x^i), (y^i)$ and $(x_i), (y_i)$ of $u, v \in \mathcal{V}$ and $u', v' \in \mathcal{V}'$, linked by $v = Tu$ and $v' = T'u'$, are related by the matrix (A^i_j) following

$$y^i = A^i_j x^j, \quad y_i = A^j_i x_j, \tag{5.1.18}$$

which says in particular that the matrix (A^i_j) transforms contravariant and covariant vectors into contravariant and covariant vectors, respectively.

Invariance

We have seen that coordinate vectors and matrix representatives of maps are basis dependent and follow specific covariant or contravariant rules when there is a change of basis.

There are some important quantities that are independent of choice of basis. For example, we may consider the pairing

$$f(u', u) \equiv u'(u) = \langle u, u' \rangle, \quad u' \in \mathcal{V}', \quad u \in \mathcal{V}, \tag{5.1.19}$$

which is a bilinear map from $\mathcal{V}' \times \mathcal{V}$ into $\mathbb{R}$. More generally, for a linear map $T : \mathcal{V} \to \mathcal{V}$, we may construct the bilinear form $f_T : \mathcal{V}' \times \mathcal{V} \to \mathbb{R}$ by

$$f_T(u', u) = u'(Tu) = \langle Tu, u' \rangle, \quad u' \in \mathcal{V}', \quad u \in \mathcal{V}. \tag{5.1.20}$$

It is clear that both (5.1.19) and (5.1.20) are basis-independent. This basis-independent property is referred to as being *invariant*.

The discussion here motivates the following concept.

Tensor

A multilinear map of the form

$$\mathcal{T} : \underbrace{\mathcal{V}' \times \cdots \times \mathcal{V}'}_{p} \times \underbrace{\mathcal{V} \times \cdots \times \mathcal{V}}_{q} \to \mathbb{R} \tag{5.1.21}$$

is called a tensor of type (p, q). Thus, when the tensor $\mathcal{T}$ over a fixed basis $\{u_1, \dots, u_n\}$ for $\mathcal{V}$ and the associated dual basis $\{u^1, \dots, u^n\}$ for $\mathcal{V}'$, we have the quantities

$$\mathcal{T}(u^{i_1}, \dots, u^{i_p}, u_{j_1}, \dots, u_{j_q}) \equiv T^{i_1 \cdots i_p}_{j_1 \cdots j_q}, \tag{5.1.22}$$

which depend on p contravariant and q covariant indices, are called the $\binom{i_1 \cdots i_p}{j_1 \cdots j_q}$-th components of $\mathcal{T}$ with respect to the given bases, and completely determine the tensor. Commonly one refers to $(T^{i_1 \cdots i_p}_{j_1 \cdots j_q})$ or simply the generic component $T^{i_1 \cdots i_p}_{j_1 \cdots j_q}$ as a type (p, q) tensor without mentioning the specific basis upon which it depends.

As an example, (5.1.20) is a type $(1, 1)$ tensor.

If $\mathcal{T}$ is a tensor of either $(1, 1)$, $(0, 2)$, or $(2, 0)$ type, then, with the vector representation (5.1.15)–(5.1.16), we have

$$\mathcal{T}(u', u) = T^i_j x_i x^j, \quad \mathcal{T}(u, v) = T_{ij} x^i y^j, \quad \mathcal{T}(u', v') = T^{ij} x_i y_j, \tag{5.1.23}$$

where $v = y_i u^i \in \mathcal{V}, v' = y^i u_i \in \mathcal{V}'$. Consequently, invariants arise from summing over or *contracting* covariant and contravariant indices.

We see also that the concept of tensor extends those of bilinear forms and matrices.

We can see how the components of a tensor depend on choice of bases. For example, assume the bases $\{u_1, \dots, u_n\}$ and $\{\tilde{u}_1, \dots, \tilde{u}_n\}$ of $\mathcal{V}$ are related by

$$\tilde{u}_j = A^i_j u_i. \tag{5.1.24}$$

Then, for a type $(0, 2)$ tensor $\mathcal{T}$, we have with the notation (5.1.22),

$$\tilde{T}_{ij} = \mathcal{T}(\tilde{u}_i, \tilde{u}_j) = \mathcal{T}(A^k_i u_k, A^l_j u_l) = A^k_i A^l_j T_{kl}, \tag{5.1.25}$$

which may be useful to compare how the coordinate vectors for a vector $u \in \mathcal{V}$ are related in the present notation. In fact, with $u = x^i u_i = \tilde{x}^i \tilde{u}_i$, we have

$$x^i = A^i_j \tilde{x}^j. \tag{5.1.26}$$

Spacetime situation

Now we return to the spacetime setting for field theory.

Based on Section 4.3 and using the terminology here, the standard Minkowski spacetime $\mathbb{R}^{3,1}$ is chosen to consist of contravariant 4 vectors of the form $x = (x^\mu)$ $(\mu = 0, 1, 2, 3)$, which are the coordinate vectors of vectors of $\mathbb{R}^4$ under standard basis. These will no longer be explicitly mentioned so that basis transformations are simply invertible matrices, equipped with the inner product

$$(x, y) = x^t \eta y, \quad x = (x^\mu), y = (y^\mu), \quad \eta = (\eta_{\mu\nu}). \tag{5.1.27}$$

With this, a Lorentz transformation is a real 4×4 invertible matrix, of the form $L = (L^\mu_\nu)$, with μ, ν being again the row and column indices, respectively, which preserves the inner product (5.1.27), that is,

$$(x', y') = (x, y), \quad x' = Lx, \quad y' = Ly, \quad x, y \in \mathbb{R}^{3,1}, \tag{5.1.28}$$

and will be our *restricted family* of transformations, instead of general invertible transformations. Therefore, we arrive at

$$x^t \eta y = (x')^t \eta y' = x^t L^t \eta L y, \quad \forall x, y \in \mathbb{R}^{3,1}. \tag{5.1.29}$$

This invariance property has a few important consequences.

(i) Rewriting the left-hand side of (5.1.29) as

$$x^\mu \eta_{\mu\nu} y^\nu, \tag{5.1.30}$$

we see that $\eta_{\mu\nu}$ is a type $(0,2)$ tensor, or a covariant 2-tensor, since it contracts two contravariant vectors x^μ and y^ν, and

$$\eta_{\mu\nu} y^\nu \tag{5.1.31}$$

is a covariant vector, which may be denoted by y_μ, since it contracts the contravariant vector x^μ.

(ii) The equation (5.1.29) also renders the invariance of η in the sense that

$$\eta = L^t \eta L, \quad \text{or} \quad \eta_{\mu\nu} = L^\alpha_\mu \eta_{\alpha\beta} L^\beta_\nu, \tag{5.1.32}$$

which in view of (5.1.25) indicates that $\eta_{\mu\nu}$ is a fixed point under the Lorentz transformations.

(iii) From (5.1.32), we have $\eta^{-1} = (L^{-1})\eta^{-1}(L^{-1})^t$ or

$$\eta^{-1} = L \eta^{-1} L^t, \tag{5.1.33}$$

for any Lorentz transformation L. Since matrix transposition switches the indices of columns and rows of a matrix, or the indices of covariance and contravariance, we see that η^{-1} is a contravariant 2-tensor, written as $\eta^{\mu\nu}$, which happens to be identical to η as a matrix,

$$\eta^{-1} = (\eta^{\mu\nu}) = (\eta_{\mu\nu}) = \eta. \tag{5.1.34}$$

(iv) With (5.1.34), we may alternately raise or lower the indices of vectors and tensors following

$$A_\mu = \eta_{\mu\nu} A^\nu, \quad A^\mu = \eta^{\mu\nu} A_\nu, \quad T_{\mu\nu} = \eta_{\mu\alpha}\eta_{\nu\beta} T^{\alpha\beta}, \quad T^{\mu\nu} = \eta^{\mu\alpha}\eta^{\nu\beta} T_{\alpha\beta}, \tag{5.1.35}$$

and so on, consistently, since $\eta^{\mu\alpha}\eta_{\nu\alpha} = \delta^\mu_\nu$.

Lorentz invariants

We can form various Lorentz invariants by contracting covariant and contravariant indices, for example:

$$A_\mu B^\mu, \quad S_{\mu\nu} T^{\mu\nu}, \quad S^\mu_\nu T^\nu_\mu, \quad T^\mu_\mu. \tag{5.1.36}$$

Here, we examine the invariance of $S_{\mu\nu}T^{\mu\nu}$.

For $(x')^\mu = L^\mu_\nu x^\nu$, we have

$$(T')^{\mu\nu} = L^\mu_\alpha L^\nu_\beta T^{\alpha\beta}, \tag{5.1.37}$$

and the same rule is valid for $S^{\mu\nu}$. Consequently, we obtain

$$\begin{aligned}(S')_{\mu\nu}(T')^{\mu\nu} &= (S')^{\alpha\beta}\,\eta_{\mu\alpha}\eta_{\nu\beta}(T')^{\mu\nu} \\ &= S^{\alpha'\beta'}L^\alpha_{\alpha'}L^\beta_{\beta'}\eta_{\mu\alpha}\eta_{\nu\beta}L^\mu_{\mu'}L^\nu_{\nu'}T^{\mu'\nu'} \\ &= S^{\alpha'\beta'}\eta_{\alpha'\mu'}\eta_{\beta'\nu'}T^{\mu'\nu'} \\ &= S_{\mu\nu}T^{\mu\nu},\end{aligned} \tag{5.1.38}$$

as anticipated, where we have applied the fixed-point property (5.1.32) for $\eta_{\mu\nu}$.

We can also check that the differential operator $\partial_\mu = \frac{\partial}{\partial x^\mu}$ behaves covariantly and the differential $\mathrm{d}x^\mu$ contravariantly. Thus, in particular, the differential

$$\mathrm{d}f = \frac{\partial f}{\partial x^\mu}\,\mathrm{d}x^\mu = \partial_\mu f \mathrm{d}x^\mu, \tag{5.1.39}$$

is an invariant. In other words, it is independent of choice of coordinates.

We can extend the formalism to include complex-valued vectors and tensors. We omit such an obvious extension, but will use it throughout.

5.2 Relativistic field equations

First, recall that the Schrödinger equation for the motion of a free particle of mass m is derived from the Newtonian energy-momentum relation

$$E = \frac{\mathbf{p}^2}{2m}, \tag{5.2.1}$$

which is non-relativistic. Thus, in order to extend the Schrödinger equation to the relativistic realm, the most direct approach is to replace the Newtonian relation (5.2.1) by its relativistic extension.

In fact, recall that the relativistic energy-momentum relation for the motion of a free particle of mass m and momentum $\mathbf{p}$ is given by the equation (4.3.22) or

$$E^2 = c^2\mathbf{p}^2 + m^2c^4, \tag{5.2.2}$$

where c is the speed of light. Taking the positive root of the equation (5.2.2), we get

$$E = mc^2\sqrt{\frac{1}{m^2c^2}\mathbf{p}^2 + 1}. \tag{5.2.3}$$

Fractionally powered equation

Thus, by (2.2.38), quantization of (5.2.2) or (5.2.3) gives us the following *fractionally powered Schrödinger equation*:

$$i\hbar\frac{\partial\psi}{\partial t} = mc^2\left(\sqrt{-\frac{\hbar^2}{m^2c^2}\nabla^2 + 1}\right)\psi, \tag{5.2.4}$$

for the motion of a free particle.

Schrödinger equation

If we expand (5.2.3) in the Taylor series, we have

$$E = mc^2 + \frac{1}{2m}\mathbf{p}^2 + \mathrm{O}(m^{-3}c^{-2}), \tag{5.2.5}$$

which recovers the classical relation (5.2.1) after truncating the tail term $\mathrm{O}(m^{-3}c^{-2})$, except that a relativistic rest-energy term, mc^2, appears as an additional background term. Thus, with such a truncation, we arrive at the modified Schrödinger equation

$$i\hbar\frac{\partial\psi}{\partial t} = -\frac{\hbar^2}{2m}\nabla^2\psi + mc^2\psi, \tag{5.2.6}$$

with a massive term. Now setting

$$\psi = \mathrm{e}^{-\mathrm{i}\frac{mc^2}{\hbar}t}\phi, \tag{5.2.7}$$

we eliminate the massive term to recast (5.2.6) into the usual free-particle Schrödinger equation

$$i\hbar\frac{\partial\phi}{\partial t} = -\frac{\hbar^2}{2m}\nabla^2\phi. \tag{5.2.8}$$

Klein–Gordon equation

We now maintain the full relation (5.2.2) for consideration. Thus, using the quantization procedure (2.2.38) again, we arrive at the equation

$$\frac{1}{c^2}\frac{\partial^2\psi}{\partial t^2} = \nabla^2\psi - \frac{m^2c^2}{\hbar^2}\psi, \tag{5.2.9}$$

governing a complex scalar field ψ, which is relativistic and commonly called the free-particle *Klein–Gordon equation*. Although the derivation of this equation is based on the first quantization formalism such that the equation is quantum-mechanical by nature, its classical field interpretation is also of important physical applications, as will be seen in subsequent development.

Using $x^0 = ct, (x^i) = \mathbf{x}$, or $(x^\mu) = (x^0, \mathbf{x})$, to denote the contravariant coordinates of the position vector in the Minkowski spacetime $\mathbb{R}^{3,1}$ of the free particle, we may rewrite (5.2.9) as

$$\partial_\mu \partial^\mu \psi + \frac{m^2 c^2}{\hbar^2} \psi = 0, \quad \text{or} \quad \Box \psi + \frac{m^2 c^2}{\hbar^2} \psi = 0, \tag{5.2.10}$$

which is clearly Lorentz invariant and is the Euler–Lagrange equation of the invariant action

$$L = \int \left(\frac{1}{2} \partial_\mu \psi \overline{\partial^\mu \psi} - \frac{1}{2} \frac{m^2 c^2}{\hbar^2} |\psi|^2 \right) \mathrm{d}x. \tag{5.2.11}$$

Conservation laws

From (5.2.10), we have the conservation law

$$\partial_\mu j^\mu = 0, \quad j^\mu = \frac{\mathrm{i}}{2}(\psi \partial^\mu \overline{\psi} - \overline{\psi} \partial^\mu \psi) = \mathrm{Im}(\overline{\psi} \partial^\mu \psi), \tag{5.2.12}$$

which is of a local nature and gives rise to the charge density

$$\rho = \frac{\mathrm{i}}{2}(\psi \partial^0 \overline{\psi} - \overline{\psi} \partial^0 \psi) = \frac{\mathrm{i}}{2c}(\psi \overline{\psi}_t - \overline{\psi} \psi_t), \tag{5.2.13}$$

and the current density

$$\mathbf{j} = (j^i) = \frac{\mathrm{i}}{2}(\psi \partial^i \overline{\psi} - \overline{\psi} \partial^i \psi) = -\frac{\mathrm{i}}{2}(\psi \nabla \overline{\psi} - \overline{\psi} \nabla \psi). \tag{5.2.14}$$

Thus, with (5.2.13) and (5.2.14), we may rewrite (5.2.12) in the familiar form

$$\frac{1}{c} \frac{\partial \rho}{\partial t} + \nabla \cdot \mathbf{j} = 0. \tag{5.2.15}$$

A consequence of (5.2.15) is that the total charge

$$Q = \int \rho \, \mathrm{d}\mathbf{x} = \int \frac{\mathrm{i}}{2c} \left(\psi \overline{\psi}_t - \overline{\psi} \psi_t \right) \mathrm{d}\mathbf{x}, \tag{5.2.16}$$

is conserved or time-independent, which is of a global feature.

The quantities (5.2.13) and (5.2.14) are often interpreted as electric charge and current densities, respectively. With such an interpretation, a real-valued wave function is *electrically neutral*. In particular, a charged wave function may be viewed as composed from two neutral wave functions. Besides, the complex conjugate $\overline{\psi}$ of a complex-valued wave function ψ carries the opposite electric charge of that of ψ, $-Q$.

On the other hand, since the charge density (5.2.13) does not remain non-negative in general, it cannot be used as a probability density function. In other words, the Klein–Gordon equation formalism here lacks a usual statistical interpretation, unlike what has been seen for the Schrödinger equation in Chapter 2. In this regard, the Klein–Gordon equation is not a successful quantum-mechanical realization of the relativistic energy-momentum relation (5.2.2). Chapter 6 presents Dirac's method to quantize (5.2.2) successfully.

Exact conversion of Klein–Gordon equation to Schrödinger equations

Note that the Klein–Gordon equation (5.2.9) and the Schrödinger equation (5.2.6) are linked [154].

Let ψ satisfy (5.2.9) and set

$$\phi = \frac{1}{2}\left(\psi + \frac{\mathrm{i}\hbar}{mc^2}\frac{\partial \psi}{\partial t}\right), \quad \chi = \frac{1}{2}\left(\psi - \frac{\mathrm{i}\hbar}{mc^2}\frac{\partial \psi}{\partial t}\right). \tag{5.2.17}$$

In view of (5.2.9), (5.2.17), and

$$\psi = \phi + \chi, \quad \mathrm{i}\hbar\frac{\partial \psi}{\partial t} = mc^2(\phi - \chi), \tag{5.2.18}$$

we see that ϕ and χ satisfy the coupled Schrödinger equations

$$\mathrm{i}\hbar\frac{\partial \phi}{\partial t} = -\frac{\hbar^2}{2m}\nabla^2(\phi + \chi) + mc^2\phi, \tag{5.2.19}$$

$$\mathrm{i}\hbar\frac{\partial \chi}{\partial t} = \frac{\hbar^2}{2m}\nabla^2(\phi + \chi) - mc^2\chi, \tag{5.2.20}$$

both of the type of (5.2.6). Thus, we see from (5.2.18)–(5.2.20) that the Klein–Gordon equation enjoys a greater degree of freedom than a single Schrödinger equation. It seems that a particle-physics realization or interpretation of these coupled equations has not yet been discussed in the literature.

General setting

Based on (5.2.11), we now consider the generalized Lagrangian action density

$$\mathcal{L} = \frac{1}{2}\partial_\mu\phi\overline{\partial^\mu\phi} - V(|\phi|^2), \tag{5.2.21}$$

where ϕ is a complex-valued scalar field as before and V a potential density function. The Euler–Lagrange equation of this action is

$$\partial_\mu\partial^\mu\phi = -2V'(|\phi|^2)\phi, \tag{5.2.22}$$

which is the Klein–Gordon equation in a general setting.

Global and local symmetries

Clearly, the Lagrangian shown is invariant under the *phase change* for the field ϕ,

$$\phi(x) \mapsto \mathrm{e}^{\mathrm{i}\omega}\phi(x), \tag{5.2.23}$$

where ω is a real constant. Such a symmetry is called a *global symmetry* because a global identical phase shift for the field ϕ does not change anything. However, when this global symmetry is *enlarged* to a *local* one for which ω becomes a function of the spacetime coordinates, $\omega = \omega(x)$, the invariance is no longer valid. A way out of this is to replace the ordinary derivatives by gauge-covariant derivatives and modify the Lagrangian (5.2.21) into

$$\mathcal{L} = \frac{1}{2} D_\mu \phi \overline{D^\mu \phi} - V(|\phi|^2), \tag{5.2.24}$$

$$D_\mu \phi = \partial_\mu \phi - \mathrm{i} e A_\mu \phi, \tag{5.2.25}$$

where $e > 0$ is a coupling constant resembling an electric charge as in the study of gauged Schrödinger equation, and require that the vector field A_μ obey the transformation rule

$$A_\mu \mapsto A'_\mu = A_\mu + \frac{1}{e}\partial_\mu \omega, \tag{5.2.26}$$

and behave like a covariant vector field under the Lorentz transformations. We can check that, under a local (x-dependent) phase shift

$$\phi(x) \mapsto \phi'(x) = \mathrm{e}^{\mathrm{i}\omega(x)}\phi(x), \tag{5.2.27}$$

the gauge-covariant derivative (5.2.25) changes itself covariantly according to

$$D_\mu \phi \mapsto D'_\mu \phi' \equiv \partial_\mu \phi' - \mathrm{i} e A'_\mu \phi' = \mathrm{e}^{\mathrm{i}\omega(x)} D_\mu \phi, \tag{5.2.28}$$

so that the modified Lagrangian indeed becomes invariant under the *gauge transformation* consisting of the rules (5.2.26) and (5.2.27).

Dynamics of gauge field

It is clear that the Lagrangian (5.2.24) is incomplete because it cannot give rise to an equation of motion for the newly introduced gauge field A_μ, which is an additional dynamical variable. In order to derive a suitable dynamic law for the motion of A_μ, we compare A_μ with ϕ and require that any candidate Lagrangian should contain quadratic terms involving the first-order derivatives of A_μ. Since such terms are to be invariant under (5.2.26), we see that a minimal way to do this is to introduce the covariant 2-tensor

$$F_{\mu\nu} = \partial_\mu A_\nu - \partial_\nu A_\mu, \tag{5.2.29}$$

which is then to be contracted with its contravariant partner $F^{\mu\nu}$ in order to observe the Lorentz invariance. Thus, we arrive at the minimally modified complete Lagrange action density

$$\mathcal{L} = -\frac{1}{4} F_{\mu\nu} F^{\mu\nu} + \frac{1}{2} D_\mu \phi \overline{D^\mu \phi} - V(|\phi|^2), \tag{5.2.30}$$

where factor $\frac{1}{4}$ is for convenience. We then see that (5.2.30) is invariant under gauge and Lorentz transformations.

The Euler–Lagrange equations of the updated Lagrangian action density (5.2.30) are

$$D_\mu D^\mu \phi = -2V'(|\phi|^2)\phi, \tag{5.2.31}$$

$$\partial_\nu F^{\mu\nu} = -J^\mu, \tag{5.2.32}$$

which govern the evolution of the fields ϕ and A_μ, where

$$J^\mu = \frac{\mathrm{i}}{2e}(\phi\overline{D^\mu\phi} - \overline{\phi}D^\mu\phi) = e\,\mathrm{Im}(\overline{\phi}D^\mu\phi), \tag{5.2.33}$$

is the charge-current density in the present context given in terms of gauge-covariant derivative.

Electromagnetism

The equation (5.2.31) is a gauged wave equation that extends (5.2.9) or (5.2.10).

We now explore what is contained in the equation (5.2.32). The rest of the chapter uses the Heaviside–Lorentz rationalized units for convenience such that both the electric permittivity and magnetic permeability are unity such that $c = 1$ and $x^0 = t$.

First we apply (5.2.31) to check that J^μ is a conserved quantity satisfying

$$\partial_\mu J^\mu = 0. \tag{5.2.34}$$

Thus, if we introduce the charge density ρ and current density $\mathbf{j}$ by setting

$$\rho = J^0 = \frac{\mathrm{i}}{2e}(\phi\overline{D^0\phi} - \overline{\phi}D^0\phi), \tag{5.2.35}$$

$$\mathbf{j} = (J^i), \quad J^i = \frac{\mathrm{i}}{2e}(\phi\overline{D^i\phi} - \overline{\phi}D^i\phi), \tag{5.2.36}$$

then (5.2.34) is a conservation law balancing charge and current densities:

$$\frac{\partial\rho}{\partial t} + \nabla\cdot\mathbf{j} = 0, \tag{5.2.37}$$

which reveals that (5.2.32) may be identified with the Maxwell equations with a suitable interpretation of various components of $F^{\mu\nu}$.

For such a purpose, we next introduce the electric and magnetic fields $\mathbf{E} = (E^i)$ and $\mathbf{B} = (B^i)$ by setting

$$(F^{\mu\nu}) = \begin{pmatrix} 0 & -E^1 & -E^2 & -E^3 \\ E^1 & 0 & -B^3 & B^2 \\ E^2 & B^3 & 0 & -B^1 \\ E^3 & -B^2 & B^1 & 0 \end{pmatrix}, \tag{5.2.38}$$

or in a compressed way,

$$E^i = -F^{0i}, \quad B^i = -\frac{1}{2}\epsilon^{ijk}F^{jk}, \quad i,j,k = 1,2,3. \tag{5.2.39}$$

Then we can check that the temporal component of (5.2.32), with $\mu = 0$, is simply

$$\nabla \cdot \mathbf{E} = \rho, \tag{5.2.40}$$

which is the Coulomb law (3.1.11). The spatial components of (5.2.32), with $\mu = i = 1, 2, 3$, render the equation

$$\frac{\partial \mathbf{E}}{\partial t} + \mathbf{j} = \nabla \times \mathbf{B}, \tag{5.2.41}$$

which is the Ampère law (3.1.12). Hence, we have partially recovered the Maxwell equations, namely, the part with charge and current sources.

In order to recover the source-free part, namely, the equations (3.1.13) and (3.1.14), we note that the definition (5.2.29) implies the *Bianchi identity*

$$\partial^\gamma F^{\mu\nu} + \partial^\mu F^{\nu\gamma} + \partial^\nu F^{\gamma\mu} = 0, \tag{5.2.42}$$

which leads to (3.1.13) and (3.1.14) in view of (5.2.38). In other words, (3.1.13) and (3.1.14) automatically hold as a consequence of the definition (5.2.29).

In summary, (5.2.32) is indeed the full set of the Maxwell equations.

Therefore, we see that the Maxwell equations can be derived as a consequence of *imposing gauge symmetry.*

External and internal symmetries

Let us explore what we have just learned.

In the field-theoretical formalism considered, the spacetime where the fields are defined over is called the *external space* of the theory. In our theory the scalar field may be viewed as a cross-section of a principal bundle or complex line over the spacetime. The bundle into which the scalar field takes its values is called the *internal space* of the theory. The external space has the Lorentz group as the symmetry group as a result of accommodating special relativity. The internal space has the $U(1)$ group as the symmetry group, or phase symmetry, whose local invariance or gauge invariance leads to the introduction of a gauge field that generates electromagnetism. The external space symmetry is called *external symmetry* and the internal space symmetry *internal symmetry.* Local $U(1)$ internal symmetry requires the presence of electromagnetism. The Lorentz symmetry discussed is a global external symmetry giving rise to special relativity. When such a global external symmetry is elevated to local, new fields arise. This situation leads to general relativity and presence of gravitation, as shown by Einstein (studied in Chapter 10). Furthermore, it is foreseeable that, when the internal symmetry is modified to be given by larger gauge groups such as $SU(N)$ ($N \geq 2$), other physical forces may be generated. In fact, this is the case and the forces generated can be weak and strong forces for nuclear interactions. Such a process of realization of various symmetries in fact enables a comprehensive understanding of all four known forces in nature: gravitational, electromagnetic, weak, and strong forces. In short, we conclude that external symmetry leads to the presence of gravity, and internal symmetry leads to the presence of electromagnetic, weak, and strong interactions.

Roles of terms in action density

Consider (5.2.30). First, from (5.2.38), we have

$$-\frac{1}{4}F_{\mu\nu}F^{\mu\nu} = -\frac{1}{4}F^{\alpha\beta}\eta_{\mu\alpha}\eta_{\nu\beta}F^{\mu\nu} = \frac{1}{2}(\mathbf{E}^2 - \mathbf{B}^2). \tag{5.2.43}$$

Hence, (5.2.30) becomes

$$\mathcal{L} = \mathcal{K} - \mathcal{V}, \tag{5.2.44}$$

where

$$\mathcal{K} = \frac{1}{2}\mathbf{E}^2 + \frac{1}{2}|D_0\phi|^2, \tag{5.2.45}$$

$$\mathcal{V} = \frac{1}{2}\mathbf{B}^2 + \frac{1}{2}|D_i\phi|^2 + V(|\phi|^2), \tag{5.2.46}$$

are recognized as the kinetic and potential energy densities, respectively. Thus, the total energy or Hamiltonian density reads

$$\mathcal{H} = \mathcal{K} + \mathcal{V} = \frac{1}{2}\left(\mathbf{E}^2 + \mathbf{B}^2\right) + \frac{1}{2}\left(|D_0\phi|^2 + |D_i\phi|^2\right) + V(|\phi|^2). \tag{5.2.47}$$

5.3 Coupled nonlinear hyperbolic and elliptic equations

We need to express the equations (5.2.31)–(5.2.32) more explicitly and concretely. We treat the time-dependent and time-independent cases separately.

Hyperbolic equations

First we explore the time-dependent case. To proceed, we use the notation

$$\Box_D = D_\mu D^\mu, \tag{5.3.1}$$

to denote the gauged d'Alembertian operator with respect to gauge-covariant derivatives. Hence, $\Box_\partial = \Box$ is the standard one and we see that (5.2.31)–(5.2.32) become

$$\Box_D\phi = -2V'(|\phi|^2)\phi, \tag{5.3.2}$$

$$\Box_\partial A^\mu = J^\mu + \partial^\mu(\partial_\nu A^\nu), \tag{5.3.3}$$

where the 4-current density J^μ is given as in (5.2.33). More explicitly, these equations are

$$\Box\phi = \mathrm{i}e([\partial_\mu A^\mu]\phi + 2A^\mu\partial_\mu\phi) + e^2(A_\mu A^\mu)\phi - 2V'(|\phi|^2)\phi, \tag{5.3.4}$$

$$\Box A^\mu = \partial^\mu(\partial_\nu A^\nu) + \frac{\mathrm{i}}{2}e(\phi\partial^\mu\overline{\phi} - \overline{\phi}\partial^\mu\phi) - e^2A^\mu|\phi|^2. \tag{5.3.5}$$

Clearly, (5.3.2)–(5.3.3) or (5.3.4)–(5.3.5) are considerably simplified under the *Lorenz gauge* condition

$$\partial_\mu A^\mu = 0, \tag{5.3.6}$$

and the wave equation nature of the system becomes more transparent.

The existence and uniqueness of a solution to the initial value problem of (5.3.2)–(5.3.3) or (5.3.4)–(5.3.5) over $\mathbb{R}^2$ are studied in [95] subject to the Lorenz gauge condition (5.3.6) when the potential density function V is a quartic function of ϕ. Specifically, here, (5.3.6) and the temporal component, or the equation with $\mu = 0$, of (5.3.5), which now assumes the form

$$\Box A^0 = \frac{\mathrm{i}}{2} e(\phi \partial^0 \overline{\phi} - \overline{\phi} \partial^0 \phi) - e^2 A^0 |\phi|^2, \tag{5.3.7}$$

are treated as constraints and imposed initially at $t = 0$. These constraints are preserved by the solution of the equations for $t > 0$, thereby establishing the local existence and uniqueness of the solution. Global existence then follows from a series of *a priori* estimates based on the conservation law

$$\frac{\mathrm{d}}{\mathrm{d}t} \int \mathcal{H} \, \mathrm{d}\mathbf{x} = 0, \tag{5.3.8}$$

where $\mathcal{H}$ is as given in (5.2.47) and integration is taken over the full space.

Elliptic equations

Let us also consider the static or time-independent situation.

From (5.3.5), we see that it is consistent to impose the *temporal gauge* condition

$$A_0 = 0, \tag{5.3.9}$$

in the subsequent discussion.

For convenience, we use vector notation $\mathbf{A} = (A_1, A_2, A_3)$ and set $D_{\mathbf{A}}\phi = (D_i \phi)$. Then there hold

$$D_{\mathbf{A}}\phi = \nabla\phi - \mathrm{i}e\mathbf{A}\phi, \tag{5.3.10}$$

$$D_i D^i \phi = -D_{\mathbf{A}}^2 \phi = -\Delta_{\mathbf{A}}\phi, \tag{5.3.11}$$

where

$$\Delta_{\mathbf{A}}\phi = \Delta\phi - 2\mathrm{i}e\mathbf{A}\cdot\nabla\phi - \mathrm{i}e(\nabla\cdot\mathbf{A})\phi - \mathrm{e}^2|\mathbf{A}|^2\phi, \tag{5.3.12}$$

is the gauge-covariant Laplace operator. Thus, (5.3.2) and (5.3.3) become a system of nonlinear equations,

$$\Delta\phi = 2\mathrm{i}e\mathbf{A}\cdot\nabla\phi + \mathrm{i}e(\nabla\cdot\mathbf{A})\phi + \mathrm{e}^2|\mathbf{A}|^2\phi + 2V'(|\phi|^2)\phi, \tag{5.3.13}$$

$$\Delta\mathbf{A} = \nabla(\nabla\cdot\mathbf{A}) + \frac{\mathrm{i}}{2}e(\overline{\phi}\nabla\phi - \phi\nabla\overline{\phi}) + e^2\mathbf{A}|\phi|^2, \tag{5.3.14}$$

which are strictly elliptic and further simplified when $\mathbf{A}$ satisfies the condition

$$\nabla\cdot\mathbf{A} = 0, \tag{5.3.15}$$

known as the *Coulomb gauge* condition.

Note that the equations (5.3.13)–(5.3.14) are the Euler–Lagrange equations of the energy functional

$$E(\phi, \mathbf{A}) = \int \left\{ \frac{1}{2}|\nabla \times \mathbf{A}|^2 + \frac{1}{2}|D_{\mathbf{A}}\phi|^2 + V(|\phi|^2) \right\} \mathrm{d}\mathbf{x}. \tag{5.3.16}$$

The equations (5.3.13)–(5.3.14) and the energy (5.3.16) are actually known as the *Ginzburg–Landau equations* and the *Ginzburg–Landau energy* arising in the *theory of superconductivity* [155, 237, 326, 481, 543, 544], when the potential density function assumes a specific form, which is discussed in more detail later.

5.4 Symmetry breaking

Symmetry breaking is an essential concept in theoretical physics and has found rich applications in a broad range of subject areas such as elementary particles and interaction, solid-state and condensed-matter systems, and superconductivity. The concept of symmetry breaking may well be explained in the context of the Abelian gauge field theory. This section explains some notions of symmetry breaking without a gauge field, and focuses on the notion of spontaneously broken symmetry.

Temperature-dependent potential density function

We start from the Klein–Gordon equation (5.2.9), with $c = 1, \hbar = 1$, governing a complex scalar field ϕ, of mass m, so that the associated Lagrangian action density function is

$$\mathcal{L} = \frac{1}{2}\partial_\mu\phi\partial^\mu\overline{\phi} - \frac{1}{2}m^2|\phi|^2. \tag{5.4.1}$$

The Hamiltonian energy density of (5.4.1) reads

$$\mathcal{H} = \frac{1}{2}|\partial_0\phi|^2 + \frac{1}{2}|\nabla\phi|^2 + \frac{1}{2}m^2|\phi|^2. \tag{5.4.2}$$

An important departure from the simple model (5.4.1) is the generalized Lagrangian action density (5.2.21) where the potential density function V is taken [326, 481] to assume the prototype form

$$V(|\phi|^2) = \frac{1}{2}m^2(T)|\phi|^2 + \frac{1}{8}\lambda|\phi|^4, \tag{5.4.3}$$

in which m^2 is a function of the absolute temperature T given by

$$m^2(T) = a\left(\left[\frac{T}{T_c}\right]^2 - 1\right), \tag{5.4.4}$$

the quantities $\lambda, a > 0$ are suitable parameters, and $T_c > 0$ is a critical temperature. Thus, the updated energy density reads

$$\mathcal{H} = \frac{1}{2}|\partial_0\phi|^2 + \frac{1}{2}|\nabla\phi|^2 + V(|\phi|^2). \tag{5.4.5}$$

Ground states

Vacuum solutions, or ground states, are the lowest energy solutions. In view of (5.4.5), these solutions may be sought among constant states which minimize (5.4.3).

High-temperature situation

In high temperature, $T > T_c$, we have $m^2(T) > 0$. The only minimizer of (5.4.3) is

$$\phi_{\mathrm{V}} = 0, \tag{5.4.6}$$

which is the unique vacuum state of the problem. Of course, both the Lagrangian action density (5.4.1) and the vacuum state (5.4.6) are invariant under the $U(1)$-symmetry group $\phi \mapsto \mathrm{e}^{\mathrm{i}\omega}\phi$.

Symmetry and symmetry-breaking

In general, given a symmetry group G, the Lagrangian density should be invariant under G as well if the vacuum state is already invariant under G, as asserted in the *Coleman theorem*, which roughly says that the invariance of the vacuum state implies the invariance of the universe. If both the vacuum state and the Lagrangian density are invariant, we say that there is an *exact symmetry*. If the vacuum state is noninvariant, the Lagrangian density may be noninvariant or invariant. In both cases, we say that the symmetry as a whole is *broken*. The former case is referred to as *explicit symmetry-breaking* and the latter case is referred to as *spontaneous symmetry-breaking*, which is one of the fundamental phenomena in low-temperature physics.

To explain these concepts [118], we now assume $T < T_c$. We have $m^2(T) < 0$ and we see that there is a phase transition. Although $\phi = 0$ is still a solution, it is no longer stable. In fact, the minimum of (5.4.3) is attained instead at any of the configurations

$$\phi_{\mathrm{V},\theta} = \phi_0 \mathrm{e}^{\mathrm{i}\theta}, \quad \phi_0 = \sqrt{\left(1 - \left[\frac{T}{T_c}\right]^2\right)\frac{2a}{\lambda}} > 0, \quad \theta \in \mathbb{R}, \tag{5.4.7}$$

which give us a *continuous family of distinct vacuum states*, a *vacuum manifold*, which happens to be a circle labeled by θ. Since for $\omega \neq 2k\pi$ the map $\phi \mapsto \mathrm{e}^{\mathrm{i}\omega}\phi$ transforms any given vacuum state, $\phi_{\mathrm{V},\theta}$, to a different one, $\phi_{\mathrm{V},\theta+\omega}$, we observe the non invariance of vacuum states, although the Lagrangian density is still invariant. Consequently, the symmetry is now spontaneously broken.

The quantity ϕ_0 given in (5.4.7) measures the *scale of the broken symmetry*, which increases as temperature decreases.

As illustrations, we consider two concrete cases of the potential function (5.4.3), given specifically by

$$V(\phi) = \frac{\phi^2}{2} + \frac{\phi^4}{8}, \quad T > T_c; \quad V(\phi) = -\frac{\phi^2}{2} + \frac{\phi^4}{8}, \quad T < T_c, \tag{5.4.8}$$

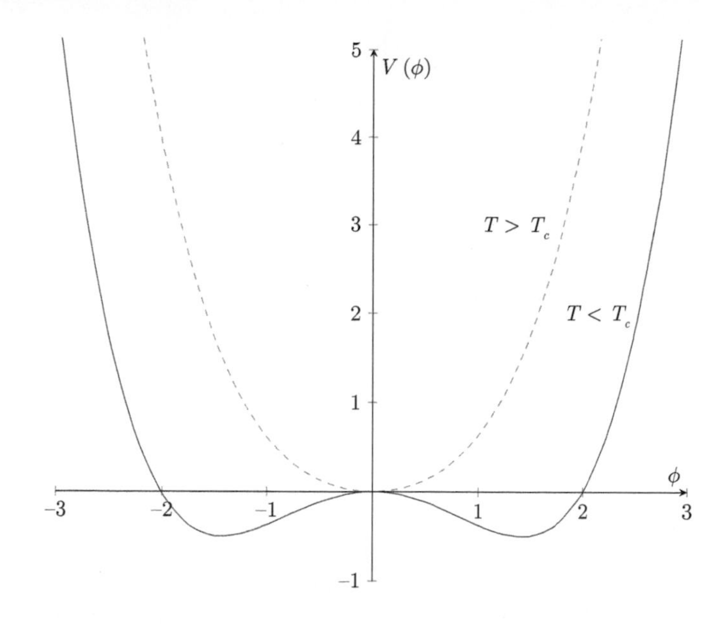

Figure 5.1 Plots of profiles of potential density functions versus temperature. At high temperature, the profile of the potential function is a dashed-line curve and the vacuum state presents itself as the unique minimum at the center. As a result, symmetry is unbroken and exact. At low temperature, the profile of the potential function is a solid-line curve and the vacuum states are minima deviating away from the center which fails to be a vacuum state. That is, the shifted minima become new vacuum states, and symmetry is spontaneously broken.

in terms of a real-valued scalar field ϕ for simplicity, corresponding to the situations $T > T_c$ and $T < T_c$, respectively. The theory enjoys a Z_2-*symmetry* characterized by the apparent invariant property under the transformation

$$\phi \leftrightarrow -\phi. \tag{5.4.9}$$

In the former situation, the potential function $V(\phi)$ is minimized at $\phi = \phi_V = 0$ so that there is an exact symmetry. In the latter situation, V is minimized at $\phi = \phi_V = \pm\sqrt{2}$ so that there is a spontaneously broken symmetry (Figure 5.1).

We can also illustrate the concept of spontaneous symmetry-breaking with an example from mechanics [477] as follows.

Place a cylindrically shaped thin stick up vertically on a flat surface. Without disturbance, the stick will stay in its present "straight-up" position. Initially, we may make a mark on a cross-section of the cylinder so that we can label its position in terms of a "twisting angle," θ, relative to a reference direction (Figure 5.2). Now apply a downward force, F, to the stick. If F is small, the stick will remain upright as if no force were applied and the position angle θ may assume any value. In other words, the system is rotationally symmetric and symmetry is exact.

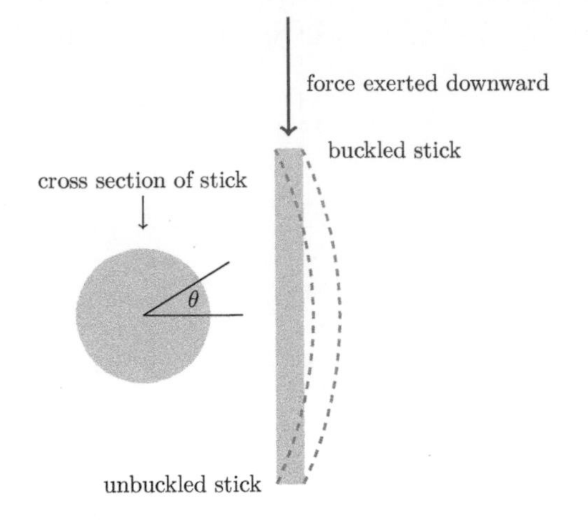

Figure 5.2 The buckling of a cylindrically shaped thin stick demonstrating spontaneous symmetry breaking. A force is applied downward to the top of the stick. When the force is small, there is rotational symmetry about the axis of the cylinder. When the applied force is beyond a critical value, the stick buckles to acquire a new equilibrium position and the rotational symmetry is spontaneously broken.

However, if we increase F beyond a critical value, F_c, the stick will buckle away from its original axis of symmetry and achieve a new equilibrium position with a definite but otherwise unpredictable position angle. Consequently, rotational symmetry is lost or *spontaneously broken.*

5.5 Higgs mechanism

This section explores some further implications of spontaneously broken symmetry and uses the gauge field in the resolution of a puzzle associated with the broken symmetry.

Here, we continue to consider the system governed by the Lagrangian action density (5.2.21) for which the potential density is given by (5.4.3)–(5.4.4). At low temperature, $T < T_c$, the symmetry of the system is spontaneously broken.

In this situation, from (5.4.4) and (5.4.7), we have

$$m^2(T) = -\frac{\lambda}{2}\phi_0^2 < 0. \tag{5.5.1}$$

On the other hand, comparing (5.2.21) under (5.4.3), that is,

$$\mathcal{L} = \frac{1}{2}\partial_\mu\phi\partial^\mu\overline{\phi} - \frac{1}{2}m^2(T)|\phi|^2 - \frac{\lambda}{8}|\phi|^4, \tag{5.5.2}$$

with (5.4.1), we see that (5.4.1) may be viewed as a leading-order approximation of (5.5.2) for ϕ fluctuating around the zero state, $\phi = 0$. Thus, if we maintain the massive-wave interpretation for the model (5.4.1) in the weak-wave limit around the zero state, it suggests that we would have particles of imaginary mass. However, this is not the case.

Goldstone particles

In fact, the Lagrangian action density (5.5.2) governs wave fluctuations around its vacuum state, which we now investigate.

For $T > T_c$, for which symmetry is unbroken, the vacuum state is the zero state as given in (5.4.6) and $m(T)$ clearly defines mass.

For $T < T_c$, for which symmetry is spontaneously broken, we need to consider fluctuations around a given *nonzero* vacuum state, say ϕ_0, as given in (5.4.7), represented by two real scalar fields ϕ_1 and ϕ_2:

$$\phi(x) = \phi_0 + \phi_1(x) + \mathrm{i}\phi_2(x). \tag{5.5.3}$$

In this case, the minimum of V is strictly negative,

$$V(\phi_0^2) = \frac{1}{2}m^2(T)\phi_0^2 + \frac{\lambda}{8}\phi_0^4 = -\frac{\lambda}{8}\phi_0^4. \tag{5.5.4}$$

Thus, in order to maintain finite energy in an unbounded space, we need to shift the potential energy density by the quantity given on the right-hand side of (5.5.4),

$$V \mapsto V(|\phi|^2) + \frac{\lambda}{8}\phi_0^4 = \frac{\lambda}{8}(|\phi|^2 - \phi_0^2)^2, \tag{5.5.5}$$

and the shifted minimum energy level is zero.

With the updated potential density function, we can rewrite the corresponding Lagrangian action density as

$$\begin{aligned}\mathcal{L} = &\frac{1}{2}\partial_\mu\phi_1\partial^\mu\phi_1 + \frac{1}{2}\partial_\mu\phi_2\partial^\mu\phi_2 - \frac{\lambda}{2}\phi_0^2\phi_1^2 \\ &- \frac{\lambda}{8}(\phi_1^4 + \phi_2^4 + 2\phi_1^2\phi_2^2 + 4\phi_0\phi_1^3 + 4\phi_0\phi_1\phi_2^2),\end{aligned} \tag{5.5.6}$$

which governs the scalar fields ϕ_1 and ϕ_2 fluctuating around the vacuum state, $\phi_1 = 0, \phi_2 = 0$. The coefficient of ϕ_1^2 defines the mass of the ϕ_1-waves or particles,

$$m_1 = \sqrt{\lambda}\phi_0 = \sqrt{2}|m(T)| > 0. \tag{5.5.7}$$

However, since the ϕ_2^2 term is absent and the higher-order terms describe cross-interactions between the ϕ_1- and ϕ_2-waves, the ϕ_2-waves or particles are seen to be massless and referred to as the *Goldstone particles.* Hence, we conclude that spontaneous symmetry-breaking leads to the presence of the Goldstone particles, namely, particles of zero mass instead of particles of imaginary mass. This statement is known as the *Goldstone theorem.*

Higgs mechanism

Being massless, the Goldstone particles appear curious. We will now see that these particles may be removed from the system when gauge fields are switched on.

For this purpose, we return to the locally invariant Lagrangian action density to get

$$\mathcal{L} = -\frac{1}{4}F_{\mu\nu}F^{\mu\nu} + \frac{1}{2}D_\mu\phi\overline{D^\mu\phi} - \frac{\lambda}{8}(|\phi|^2 - \phi_0^2)^2, \tag{5.5.8}$$

with the associated Hamiltonian density

$$\mathcal{H} = \frac{1}{2}F_{0i}^2 + \frac{1}{4}F_{ij}^2 + \frac{1}{2}|D_0\phi|^2 + \frac{1}{2}|D_i\phi|^2 + \frac{\lambda}{8}(|\phi|^2 - \phi_0^2)^2, \tag{5.5.9}$$

which leads us to consider wave fluctuations around the vacuum state

$$\phi_{\mathrm{V}} = \phi_0, \quad (A_\mu)_{\mathrm{V}} = 0, \tag{5.5.10}$$

where ϕ_0 is given in (5.4.7). Using (5.5.3), we obtain the Lagrangian action density governing the fields ϕ_1, ϕ_2, and A_μ as follows,

$$\begin{aligned}\mathcal{L} = &-\frac{1}{4}F_{\mu\nu}F^{\mu\nu} + \frac{1}{2}e^2\phi_0^2 A_\mu A^\mu \\ &+ \frac{1}{2}\partial_\mu\phi_1\partial^\mu\phi_1 + \frac{1}{2}\partial_\mu\phi_2\partial^\mu\phi_2 - \frac{\lambda}{2}\phi_0^2\phi_1^2 + \mathcal{L}_{\mathrm{inter}},\end{aligned} \tag{5.5.11}$$

where $\mathcal{L}_{\mathrm{inter}}$ is the interaction Lagrangian density that contains all off-diagonal terms involving the mixed products of the fields ϕ_1, ϕ_2, A_μ, and their derivatives:

$$\begin{aligned}\mathcal{L}_{\mathrm{inter}} = &\frac{1}{2}\,e^2 A_\mu A^\mu\left(\phi_1^2 + \phi_2^2 + 2\phi_0\phi_1\right) \\ &- e\,\mathrm{Im}\left(A_\mu\partial^\mu[\phi_1 + \mathrm{i}\phi_2][\phi_0 + \phi_1 - \mathrm{i}\phi_2]\right) \\ &- \frac{\lambda}{8}(\phi_1^4 + \phi_2^4 + 2\phi_1^2\phi_2^2 + 4\phi_0\phi_1^3 + 4\phi_0\phi_1\phi_2^2).\end{aligned} \tag{5.5.12}$$

From the Lagrangian action density (5.5.11), we see that ϕ_2 remains massless. Besides, if we ignore any interactions, the gauge field A_μ is now governed by the equation

$$\partial_\nu F^{\mu\nu} = e^2\phi_0^2 A^\mu, \quad \text{or} \quad \Box A^\mu + e^2\phi_0^2 A^\mu = \partial^\mu(\partial_\nu A^\nu), \tag{5.5.13}$$

known as the *Maxwell–Proca equation* [250, 477] describing massive electromagnetic or gauge fields of an acquired mass of the value $e\phi_0$.

Furthermore, from the gauge transformation (5.2.26)–(5.2.27), we have

$$\phi' = (\phi_0 + \phi_1' + \mathrm{i}\phi_2') = \mathrm{e}^{\mathrm{i}\omega}(\phi_0 + \phi_1 + \mathrm{i}\phi_2), \quad A_\mu' = A_\mu + \frac{1}{e}\partial_\mu\omega, \tag{5.5.14}$$

resulting in

$$\begin{aligned}\phi_1' &= \phi_1\cos\omega - \phi_2\sin\omega + \phi_0(\cos\omega - 1), \\ \phi_2' &= \phi_1\sin\omega + \phi_2\cos\omega + \phi_0\sin\omega.\end{aligned} \tag{5.5.15}$$

Now choose

$$\omega = -\arctan\frac{\phi_2}{\phi_0+\phi_1}. \tag{5.5.16}$$

Then, $\phi_2' = 0$. In other words, with this choice of gauge condition, the remaining field variables are ϕ_1', A_μ'. Thus, suppressing the prime sign $'$, we see that the governing Lagrangian action density is reduced into

$$\mathcal{L} = -\frac{1}{4}F_{\mu\nu}F^{\mu\nu} + \frac{1}{2}\partial_\mu\phi_1\partial^\mu\phi_1 + \frac{1}{2}e^2\phi_0^2A_\mu A^\mu - \frac{\lambda}{2}\phi_0^2\phi_1^2 + \mathcal{L}_{\text{inter}}, \tag{5.5.17}$$

where $\mathcal{L}_{\text{inter}}$ contains all off-diagonal interaction terms involving the mixed products of the fields ϕ_1, A_μ, and their derivatives. Therefore, we see that, in such a fixed gauge, what remain are a massive real scalar field and a massive gauge field, and the massless Goldstone particle is eliminated.

Consequently, spontaneous breaking of a continuous symmetry does not lead to the appearance of a massless Goldstone particle, but rather to the disappearance of a scalar field and the appearance of a massive gauge field. This statement is known as the *Higgs mechanism* and the remaining massive scalar particle is called the *Higgs particle.* The Lagrangian action density considered that contains a spontaneously broken symmetry is commonly referred to as the *Abelian Higgs model* and the complex scalar field ϕ is called the *Higgs field.* This fundamental mechanism is due to the independent work by three different groups of people: Englert and Brout [194], Higgs [290], and Guralnik, Hagen, and Kibble [262], published in the same journal at the same time.

Exercises

1. Establish (5.1.18).
2. With (5.1.17), show that the type $(1,1)$ tensor defined in (5.1.20) has the components
$$(f_T)^i_j = A^i_j. \tag{5.E.1}$$
3. Establish analogous relations to the formula (5.1.25) for tensors of types $(1,1)$ and $(2,0)$.
4. Establish the consistency in (5.1.35).
5. Using the same idea as in (5.1.38), obtain the invariance for the rest of the quantities listed in (5.1.36).
6. Show that one may use a refined truncation of the energy-momentum relation (5.2.3) to obtain the *4th-order free-particle Schrödinger equation*
$$\mathrm{i}\hbar\frac{\partial\phi}{\partial t} = -\frac{\hbar^2}{2m}\nabla^2\phi - \frac{\hbar^4}{8m^3c^2}\nabla^4\phi. \tag{5.E.2}$$
7. Show that ϕ and χ given in (5.2.17) both satisfy the Klein–Gordon equation (5.2.9).

8. For the coupled Schrödinger equations (5.2.19)–(5.2.20), derive the associated conservation laws involving charge and current densities generated from the wave functions ϕ and χ.
9. Show that the Euler–Lagrange equations of the action density (5.2.30) are (5.2.31) and (5.2.32).
10. Establish the conservation law (5.2.34).
11. Verify the Bianchi identity (5.2.42) and use (5.2.42) and (5.2.38) to check that (3.1.13) and (3.1.14) hold.
12. Use the covariant gauge field strength $F_{\mu\nu}$ to realize the electromagnetic field matrix as that given by (5.2.38).
13. Establish the conservation law (5.3.8).
14. Show that the equations (5.3.13)–(5.3.14) are the Euler–Lagrange equations of the energy functional (5.3.16).
15. Consider the Lagrangian action density

$$\mathcal{L} = \frac{1}{2}u_t^2 - \frac{1}{2}u_x^2 - \frac{\alpha}{2}u^2 - \frac{\beta}{4}u^4, \tag{5.E.3}$$

where α, β are constants, $\beta > 0$, and u is a real-valued scalar field.

(a) Show that the model (5.E.3) enjoys a Z_2-symmetry $u \mapsto \pm u$.

(b) Determine the values of the parameters α and β so that the model has either an exact symmetry or a spontaneously broken symmetry.

16. Consider the Lagrangian action density

$$\mathcal{L} = \frac{1}{2}u_t^2 - \frac{1}{2}u_x^2 - (1 - \cos u), \tag{5.E.4}$$

governing a real-valued scalar field u, known as the *sine-Gordon model.*

(a) Determine the full symmetry of this model.

(b) Is the symmetry spontaneously broken?

6

Dirac equations

This chapter presents a brief introduction to the Dirac equation, which lays the foundation for *quantum electrodynamics* and describes interaction of electrons and electromagnetic fields at quantum field theory level. It begins by deriving the Dirac equation and discussing some of its direct consequences. Next, it presents a few special solutions of the equation and considers the Dirac equation coupled with an electromagnetic gauge field. Finally, it reviews some nonlinear extensions.

6.1 Pauli matrices, spinor fields, and Dirac equation

Section 5.2 showed how the Klein–Gordon equation (5.2.9) was introduced to realize the relativistic energy-momentum relation (5.2.2). This equation suffers from a serious setback by a lack of statistical interpretation associated with the absence of a probability density function in such a formalism. This chapter presents Dirac's method [166, 169], which is used to arrive at a quantum-mechanical realization of (5.2.2) (the Dirac equation), successfully accommodating both special relativity and quantum mechanics and reestablishing a statistical interpretation. Originally, Dirac developed his theory as an attempt to resolve the discrepancy issue that "the observed number of stationary states for an electron in an atom being twice the number by theory" [166], a puzzle also (then) duly called the "duplexity" phenomenon. Dirac's idea was to work to reconcile special relativity and quantum mechanics. His success enabled him to interpret electron spin — quite a mystery at the time — and to achieve many other monumental accomplishments. Since then, the Dirac equation and its associated concepts have formed the very foundation of relativistic quantum mechanics in general, and quantum electrodynamics in particular.

Mathematical Physics with Differential Equations. Yisong Yang, Oxford University Press.
 DOI: 10.1093/oso/9780192872616.003.0006

Mathematically, Dirac's idea in finding a correct way to quantize (5.2.2) is to seek its appropriate factorization.

Factorization of relativistic energy-momentum relation

In view of (5.2.2) and the quantization procedure (2.2.38), or

$$\hat{E} = \mathrm{i}\hbar \frac{\partial}{\partial t}, \quad \hat{\mathbf{p}} = -\mathrm{i}\hbar\nabla \equiv (\hat{p}_i), \quad \hat{p}_i = -\mathrm{i}\hbar\,\partial_i, \quad i = 1,2,3, \tag{6.1.1}$$

we may rewrite the free-particle Klein–Gordon wave equation (5.2.9) as

$$\left(\hat{E}^2 - c^2\hat{\mathbf{p}}^2 - m^2c^4\right)\psi = D_- D_+ \psi = 0, \tag{6.1.2}$$

where

$$D_\pm = \hat{E} \pm \left(c\alpha^i \hat{p}_i + mc^2\beta\right), \tag{6.1.3}$$

with summation convention over repeated spatial index i and α^i ($i = 1,2,3$) and β being some coordinate-independent "quantities" to be determined. Expanding the product of D_- and D_+ and using the commutativity of $\hat{E}$ and $\hat{p}_i$, we have

$$D_- D_+ = \hat{E}^2 - c^2\alpha^i\alpha^j\hat{p}_i\hat{p}_j - mc^3(\alpha^i\beta + \beta\alpha^i)\hat{p}_i - m^2c^4\beta^2. \tag{6.1.4}$$

In particular, the operators D_- and D_+ are commutative,

$$D_- D_+ = D_+ D_-. \tag{6.1.5}$$

In order to recover the left-hand side of (6.1.2), we impose the following conditions among α^i ($i = 1,2,3$) and β:

$$(\alpha^i)^2 = 1, \quad i = 1,2,3, \quad \beta^2 = 1, \tag{6.1.6}$$

$$\alpha^i\alpha^j + \alpha^j\alpha^i = 0, \quad i \neq j, \quad i,j = 1,2,3, \tag{6.1.7}$$

$$\alpha^i\beta + \beta\alpha^i = 0, \quad i = 1,2,3. \tag{6.1.8}$$

In terms of the *anti-commutator* notation, these conditions may be summarized as

$$\{\alpha^i, \alpha^j\} \equiv \alpha^i\alpha^j + \alpha^j\alpha^i = 2\delta^{ij}, \quad \{\alpha^i, \beta\} = 0, \quad \beta^2 = 1. \tag{6.1.9}$$

Of course, the quantities α^i's and β cannot be scalars but matrices. In particular, the unity "1" in (6.1.6) is understood to be a unit matrix. Thus, the wave function ψ in (6.1.2) is a multi-component function.

Construction by Pauli matrices

In order to construct α^i's and β, consider the *Pauli spin matrices*:

$$\sigma^1 = \begin{pmatrix} 0 & 1 \\ 1 & 0 \end{pmatrix}, \quad \sigma^2 = \begin{pmatrix} 0 & -\mathrm{i} \\ \mathrm{i} & 0 \end{pmatrix}, \quad \sigma^3 = \begin{pmatrix} 1 & 0 \\ 0 & -1 \end{pmatrix}. \tag{6.1.10}$$

Then we see that the σ^i's satisfy the properties

$$(\sigma^i)^2 = 1, \quad \sigma^i = -\mathrm{i}\sigma^j\sigma^k, \quad [\sigma^i, \sigma^j] = 2\mathrm{i}\sigma^k, \tag{6.1.11}$$

as 2×2 matrices, where $\{i, j, k\}$ is taken to be all the cyclic permutations of $\{1, 2, 3\}$. Now construct the following 4×4 matrices from the Pauli matrices (6.1.10):

$$\alpha^i = \begin{pmatrix} 0 & \sigma^i \\ \sigma^i & 0 \end{pmatrix}, \quad i = 1, 2, 3, \quad \beta = \begin{pmatrix} 1 & 0 \\ 0 & -1 \end{pmatrix}. \tag{6.1.12}$$

Applying (6.1.11), we may show that the α^i's and β defined in (6.1.12) fulfill (6.1.6)–(6.1.8).

Consequently, with (6.1.12) and (6.1.5), we see that the free-particle relativistic wave equation (6.1.2) may be reduced into

$$\left(\hat{E} \pm \left(c\alpha^i\hat{p}_i + mc^2\beta\right)\right)\psi = 0, \tag{6.1.13}$$

which is of the first order, where ψ is a complex-valued four-component scalar function often referred to as the *Dirac spinor field*. Without loss of generality, we now focus on the minus-sign situation (say),

$$\left(\hat{E} - \left(c\alpha^i\hat{p}_i + mc^2\beta\right)\right)\psi = 0, \tag{6.1.14}$$

since the plus-sign situation may be recovered by flipping the sign of the matrices α^i's and β while maintaining (6.1.6)–(6.1.8).

Dirac equation

Multiplying (6.1.14) from the left by β, setting

$$\gamma^0 = \beta, \quad \gamma^i = \beta\alpha^i, \quad i = 1, 2, 3, \tag{6.1.15}$$

inserting (6.1.1), and using the updated temporal coordinate $x^0 = ct$ such that $\partial_t = c\partial_0$, we arrive at

$$\mathrm{i}\hbar\gamma^\mu\partial_\mu\psi - mc\psi = 0, \tag{6.1.16}$$

which is the celebrated *Dirac equation* for a free-moving particle of mass m. The matrices γ^μ's are called the *gamma matrices* or the *Dirac matrices*:

$$\gamma^0 = \begin{pmatrix} 1 & 0 \\ 0 & -1 \end{pmatrix}, \quad \gamma^i = \begin{pmatrix} 0 & \sigma^i \\ -\sigma^i & 0 \end{pmatrix}, \quad i = 1, 2, 3, \tag{6.1.17}$$

which satisfy the anti-commutator relation

$$\{\gamma^\mu, \gamma^\nu\} = 2\eta^{\mu\nu}I, \tag{6.1.18}$$

where $(\eta^{\mu\nu})$ is the Minkowski metric (4.3.2) and I denotes the 4×4 identity matrix, and

$$\gamma^{0\dagger} = \gamma^0, \quad \gamma^{i\dagger} = -\gamma^i, \quad i = 1, 2, 3, \tag{6.1.19}$$

which may be summarized as

$$\gamma^{\mu\dagger} = \gamma^0 \gamma^\mu \gamma^0. \tag{6.1.20}$$

These properties will be useful in calculation.

6.2 Action, probability, and current densities

For a Dirac spinor ψ, the quantity

$$\overline{\psi} = \psi^\dagger \gamma^0, \tag{6.2.1}$$

is referred to as the *Dirac adjoint spinor* or *Dirac conjugate spinor*. With this, we can check that (6.1.16) is the Euler–Lagrange equation of the Lagrangian action density

$$\mathcal{L} = \mathrm{i}\hbar c \overline{\psi} \gamma^\mu \partial_\mu \psi - mc^2 \overline{\psi} \psi. \tag{6.2.2}$$

Besides, taking the Hermitian conjugate of (6.1.16) and multiplying the result by γ^0 from the right, we see that $\overline{\psi}$ satisfies

$$\mathrm{i}\hbar \partial_\mu \overline{\psi} \gamma^\mu + mc \overline{\psi} = 0, \tag{6.2.3}$$

also known as the *adjoint Dirac equation*. In view of (6.1.16) and (6.2.3), we have

$$\partial_\mu \left(\overline{\psi} \gamma^\mu \psi \right) = 0, \tag{6.2.4}$$

which leads to the definition of the 4-current density

$$j^\mu = c \overline{\psi} \gamma^\mu \psi, \tag{6.2.5}$$

satisfying the conservation law

$$\partial_t \rho + \nabla \cdot \mathbf{j} = 0, \tag{6.2.6}$$

where ρ and $\mathbf{j}$ are given by

$$\rho = j^0 = c \overline{\psi} \gamma^0 \psi = c|\psi|^2, \quad \mathbf{j} = (j^i) = \left(c \overline{\psi} \gamma^i \psi \right), \tag{6.2.7}$$

respectively. Since ρ is positive definite, it may well be regarded as giving rise to the probability density of the Dirac spinor particle and $\mathbf{j}$ the associated probability current density.

6.3 Special solutions

It will be convenient to rewrite the 4-spinor field ψ as two "stacked" 2-spinor fields

$$\psi = \begin{pmatrix} \phi \\ \chi \end{pmatrix}. \tag{6.3.1}$$

In terms of this notation, the Dirac equation (6.1.16) becomes

$$\mathrm{i}\hbar\left(\partial_0\phi+\sigma^i\partial_i\chi\right)-mc\phi=0, \tag{6.3.2}$$

$$\mathrm{i}\hbar\left(\partial_0\chi+\sigma^i\partial_i\phi\right)+mc\chi=0, \tag{6.3.3}$$

which shows clearly how ϕ and χ are coupled with each other through the reduced equations.

We now seek plane-wave solutions of the form

$$\psi(x)=u(p)\mathrm{e}^{-\frac{\mathrm{i}}{\hbar}p_\mu x^\mu}, \tag{6.3.4}$$

where $p_\mu=\eta_{\mu\nu}p^\nu$, p^μ is the relativistic energy-momentum vector with

$$p^0=\frac{E}{c},\quad (p^i)=\mathbf{p}, \tag{6.3.5}$$

and $u(p)$ a 4-vector to be determined. Inserting (6.3.4) into (6.1.16), we arrive at the algebraic equation

$$(\gamma^\mu p_\mu-mc)\,u(p)=0. \tag{6.3.6}$$

Correspondingly, if we further use (6.3.1) to split the representation (6.3.4) with

$$\phi(x)=v(p)\mathrm{e}^{-\frac{\mathrm{i}}{\hbar}p_\mu x^\mu},\quad \chi(x)=w(p)\mathrm{e}^{-\frac{\mathrm{i}}{\hbar}p_\mu x^\mu}, \tag{6.3.7}$$

then we obtain from (6.3.6) the coupled system

$$\left(\frac{E}{c}-mc\right)v+\sigma^i p_i w=0, \tag{6.3.8}$$

$$\left(\frac{E}{c}+mc\right)w+\sigma^i p_i v=0, \tag{6.3.9}$$

whose solutions have some interesting properties.

Rest-particle solutions

For a particle at rest such that the momentum vector vanishes, $\mathbf{p}=\mathbf{0}$, the system (6.3.8)–(6.3.9) decouples to render the fundamental solutions

$$E=mc^2,\quad v=\begin{pmatrix}1\\0\end{pmatrix},\begin{pmatrix}0\\1\end{pmatrix},\quad w=0,\quad \phi=v\mathrm{e}^{-\frac{\mathrm{i}}{\hbar}Et},\quad \chi=0, \tag{6.3.10}$$

$$E=-mc^2,\quad v=0,\quad w=\begin{pmatrix}1\\0\end{pmatrix},\begin{pmatrix}0\\1\end{pmatrix},\quad \phi=0,\quad \chi=w\mathrm{e}^{-\frac{\mathrm{i}}{\hbar}Et}. \tag{6.3.11}$$

Thus, denoting the solutions given in (6.3.10) and (6.3.11) by ψ_+ and ψ_-, respectively, we may apply the energy operator $\hat{E}$ in (6.1.1) to get

$$\hat{E}\psi_\pm=\pm mc^2\psi_\pm. \tag{6.3.12}$$

Apparently, the solution ψ_+ represents the particle at rest with rest energy mc^2, as expected. On the other hand, the solution ψ_- describes a mysterious new particle at the rest with rest energy $-mc^2$, which prompted Dirac [166, 169] to formulate his vacuum theory, which states that the vacuum is occupied by all the negative-energy state particles, known as the *Dirac sea*, and a hole made out of it may be viewed as an *anti-particle* at rest. For example, in the situation of the electron, its antiparticle, produced as a hole in the Dirac sea called the *positron*, carries the same energy or mass but the opposite electric charge, momentum, and spin.

Moving-particle solutions

In this situation the momentum is not zero, $\mathbf{p} \neq \mathbf{0}$, which is arbitrary otherwise. Thus, to ensure a nontrivial solution in (6.3.8)–(6.3.9), we see that the 4×4 coefficient matrix of the system of equations, (6.3.8)–(6.3.9), must be singular, resulting in

$$\det \begin{pmatrix} \frac{E}{c} - mc & \sigma^i p_i \\ \sigma^i p_i & \frac{E}{c} + mc \end{pmatrix} = 0. \tag{6.3.13}$$

After some manipulation, we may reduce the 4×4 determinant equation (6.3.13) into

$$\left(\frac{E^2}{c^2} - m^2c^2\right)^2 - \left(\frac{E^2}{c^2} - m^2c^2\right) \mathrm{Tr}\left((\sigma^i p_i)^2\right) + \left(\det(\sigma^i p_i)\right)^2 = 0. \tag{6.3.14}$$

Inserting the Pauli matrices (6.1.10), we have

$$\mathrm{Tr}\left((\sigma^i p_i)^2\right) = 2|\mathbf{p}|^2, \quad \det(\sigma^i p_i) = -|\mathbf{p}|^2. \tag{6.3.15}$$

Therefore, we arrive at the equation

$$E^2 = m^2c^4 + |\mathbf{p}|^2c^2. \tag{6.3.16}$$

In particular, we see that the condition $\mathbf{p} \neq \mathbf{0}$ leads to

$$E^2 > m^2c^4. \tag{6.3.17}$$

In view of (6.3.16), we see that (6.3.8) and (6.3.9) are related: the validity of one equation implies that of the other. In fact, if (6.3.8) holds, then we can apply $\sigma^i p_i$ to it to get

$$\left(\frac{E}{c} - mc\right)(\sigma^i p_i)v + (\sigma^i p_i)(\sigma^j p_j)w = 0. \tag{6.3.18}$$

Besides, note that

$$\begin{aligned}(\sigma^i p_i)(\sigma^j p_j) &= \sigma^i \sigma^j p_i p_j \\ &= (\delta^{ij} + \mathrm{i}\epsilon^{ijk}\sigma^k)p_i p_j = |\mathbf{p}|^2.\end{aligned} \tag{6.3.19}$$

Inserting (6.3.19) into (6.3.18) and using (6.3.16), we obtain

$$\left(\frac{E}{c} - mc\right)(\sigma^i p_i)v + \left(\frac{E^2}{c^2} - m^2c^2\right)w = 0, \tag{6.3.20}$$

which establishes (6.3.9).

Furthermore, (6.3.16) has two roots:

$$E = E_\pm \equiv \pm\sqrt{m^2c^4 + |\mathbf{p}|^2c^2}. \tag{6.3.21}$$

Thus, for $E = E_+$, we may use (6.3.9) to obtain

$$w = -\left(\frac{E_+}{c} + mc\right)^{-1} \sigma^i p_i v, \quad v = \begin{pmatrix} 1 \\ 0 \end{pmatrix}, \begin{pmatrix} 0 \\ 1 \end{pmatrix}, \tag{6.3.22}$$

and for $E = E_-$, we may use (6.3.8) to get

$$v = -\left(\frac{E_-}{c} - mc\right)^{-1} \sigma^i p_i w, \quad w = \begin{pmatrix} 1 \\ 0 \end{pmatrix}, \begin{pmatrix} 0 \\ 1 \end{pmatrix}. \tag{6.3.23}$$

Therefore, if we use $\psi_\pm$ to denote the corresponding 4-spinor field given by (6.3.22) and (6.3.23), respectively, then the energy operator $\hat{E}$ in (6.1.1) and (6.3.21) lead to

$$\hat{E}\psi_\pm = E\psi_\pm = \pm\sqrt{m^2c^4 + |\mathbf{p}|^2c^2}\,\psi_\pm. \tag{6.3.24}$$

As before, the field ψ_+, associated with the positive energy

$$E_+ = \sqrt{m^2c^4 + |\mathbf{p}|^2c^2}, \tag{6.3.25}$$

represents a freely moving particle of mass m and momentum $\mathbf{p}$, and the field ψ_-, with the negative energy

$$E_- = -\sqrt{m^2c^4 + |\mathbf{p}|^2c^2}, \tag{6.3.26}$$

describes an anti-particle of the same mass and momentum, in company. Note that the rest-particle energy formula (6.3.12) may be recovered from (6.3.24) by setting $\mathbf{p} = \mathbf{0}$ there.

Anti-matter consists of anti-particles, a stunning revelation made by the Dirac equation.

6.4 Dirac equation coupled with gauge field

If we consider the motion of a charged particle of mass m and charge q in an Abelian gauge field A_μ, then we need to replace the usual derivatives by the gauge-covariant derivatives,

$$\partial_\mu \mapsto D_\mu = \partial_\mu + \frac{\mathrm{i}}{\hbar} q A_\mu. \tag{6.4.1}$$

In terms of (6.4.1), the Dirac equation (6.1.16) becomes

$$\mathrm{i}\hbar\gamma^\mu D_\mu \psi - mc\psi = 0. \tag{6.4.2}$$

Electrostatic equation

We first consider the electrostatic situation where $A_0 = \frac{V}{c}$, $V = V(\mathbf{x})$, and $A_i = 0$ $(i = 1,2,3)$. Inserting these into (6.4.2) and using the 2-spinor representation (6.3.1), we obtain the equations

$$\mathrm{i}\hbar\left(\partial_0\phi + \sigma^i\partial_i\chi\right) - mc\phi - q\frac{V}{c}\phi = 0, \tag{6.4.3}$$

$$\mathrm{i}\hbar\left(\partial_0\chi + \sigma^i\partial_i\phi\right) + mc\chi - q\frac{V}{c}\chi = 0, \tag{6.4.4}$$

which update (6.3.2)–(6.3.3). For a stationary-state solution of the form

$$\phi(x) = v(\mathbf{x})\mathrm{e}^{-\frac{\mathrm{i}}{\hbar}Et}, \quad \chi(x) = w(\mathbf{x})\mathrm{e}^{-\frac{\mathrm{i}}{\hbar}Et}, \tag{6.4.5}$$

we see that (6.4.3)–(6.4.4) become

$$\left(E - mc^2 - qV\right)v + \mathrm{i}\hbar c\sigma^i\partial_i w = 0, \tag{6.4.6}$$

$$\left(E + mc^2 - qV\right)w + \mathrm{i}\hbar c\sigma^i\partial_i v = 0. \tag{6.4.7}$$

Thus, we may formally substitute (6.4.7) into (6.4.6) to obtain the reduced equation

$$\left(E - mc^2\right)v = qVv - \hbar^2c^2\sigma^i\sigma^j\partial_i\left(\frac{\partial_j v}{E + mc^2 - qV}\right), \tag{6.4.8}$$

which may be more conveniently rewritten as

$$E'v = qVv - \hbar^2c^2\sigma^i\sigma^j\partial_i\left(\frac{\partial_j v}{E' + 2mc^2 - qV}\right), \quad E' = E - mc^2, \tag{6.4.9}$$

which is a *nonlinear eigenvalue problem* with E' appearing as an "eigenvalue."

We further note that, in the non-relativistic limit $c \gg 1$, we may take the leading-order truncation

$$\begin{aligned}\frac{c^2}{E' + 2mc^2 - qV} &= \frac{1}{2m}\frac{1}{\left(1 + \frac{E'-qV}{2mc^2}\right)}\\ &= \frac{1}{2m}\left(1 - \frac{E' - qV}{2mc^2}\right) + \mathrm{O}\left(\frac{1}{c^4}\right),\end{aligned} \tag{6.4.10}$$

in (6.4.9) to get the approximate equation

$$E'v = qVv - \frac{\hbar^2}{2m}\left(1 - \frac{E'}{2mc^2}\right)\sigma^i\sigma^j\partial_i\partial_j v - \frac{\hbar^2 q}{4m^2c^2}\sigma^i\sigma^j\partial_i(V\partial_j v), \tag{6.4.11}$$

which becomes linear in E'. Thus, interchanging the dummy indices i, j in (6.4.11), adding the resulting equation back to (6.4.11), and using the identities

$$\sigma^i\sigma^j = \delta^{ij} + \mathrm{i}\varepsilon^{ijk}\sigma^k, \quad \{\sigma^i, \sigma^j\} = 2\delta^{ij}, \quad [\sigma^i, \sigma^j] = 2\mathrm{i}\varepsilon^{ijk}\sigma^k, \tag{6.4.12}$$

with summation convention over repeated indices, we have

$$\begin{aligned}E'v =& qVv - \frac{\hbar^2}{2m}\Delta v \\ &+ \frac{\hbar^2}{4m^2c^2}\left([E' - qV]\Delta v - q\,\nabla_{\mathbf{x}}V\cdot\nabla_{\mathbf{x}}v - \mathrm{i}q\,\varepsilon^{ijk}\sigma^k\partial_i V\partial_j v\right).\end{aligned} \tag{6.4.13}$$

The first line of this equation is simply the non-relativistic Schrödinger equation governing the motion of a spin-free particle of mass m and charge q:

$$\mathrm{i}\hbar\frac{\partial\phi}{\partial t} = -\frac{\hbar^2}{2m}\nabla^2_{\mathbf{x}}\phi + qV(\mathbf{x})\phi, \tag{6.4.14}$$

in terms of the usual stationary-state ansatz

$$\phi(x) = v(\mathbf{x})\mathrm{e}^{-\frac{\mathrm{i}}{\hbar}E't}, \tag{6.4.15}$$

and the second line of the equation takes account of relativistic corrections and spin contribution for the motion of the particle.

Magnetostatic equation

We next consider the magnetostatic situation with $A_0 = 0$ and $A_i = A_i(\mathbf{x})$ $(i = 1,2,3)$ in the covariant derivatives in (6.4.1). Thus, (6.4.2) in terms of (6.3.1) becomes

$$\mathrm{i}\hbar\left(\partial_0\phi + \sigma^i\partial_i\chi\right) - mc\phi - q\,A_i\sigma^i\,\chi = 0, \tag{6.4.16}$$

$$\mathrm{i}\hbar\left(\partial_0\chi + \sigma^i\partial_i\phi\right) + mc\chi - q\,A_i\sigma^i\,\phi = 0. \tag{6.4.17}$$

Inserting the stationary-state ansatz (6.4.5) into (6.4.16)–(6.4.17), we have

$$(E - mc^2)v + \mathrm{i}\hbar c\sigma^i D_i w = 0, \tag{6.4.18}$$

$$(E + mc^2)w + \mathrm{i}\hbar c\sigma^i D_i v = 0. \tag{6.4.19}$$

Solving w from (6.4.19) and substituting it into (6.4.18), we obtain

$$E'v = -\frac{\hbar^2c^2}{E' + 2mc^2}\sigma^i\sigma^j D_i D_j v, \quad E' = E - mc^2, \tag{6.4.20}$$

which is again a nonlinear eigenvalue problem so that the eigenvalue E' appears nonlinearly. Taking the truncation (6.4.10) with $V = 0$ in (6.4.20), we arrive as before at the following approximate equation:

$$E'v = -\frac{\hbar^2}{2m}\left(1 - \frac{E'}{2mc^2}\right)\sigma^i\sigma^j D_i D_j v, \tag{6.4.21}$$

which is linear in E'. As in the electrostatic situation, we may interchange the dummy indices i, j in (6.4.21) and adding the resulting equation back to (6.4.21) to obtain

$$E'v = -\frac{\hbar^2}{2m}\left(1 - \frac{E'}{2mc^2}\right)D_i^2 v - \frac{\mathrm{i}q\hbar}{4m}\left(1 - \frac{E'}{2mc^2}\right)\sigma^i\sigma^j F_{ij}v, \tag{6.4.22}$$

where we have used the operator identity

$$(D_\mu D_\nu - D_\nu D_\mu) = [D_\mu, D_\nu] = \mathrm{i}\frac{q}{\hbar}F_{\mu\nu}, \quad F_{\mu\nu} = \partial_\mu A_\nu - \partial_\nu A_\mu, \tag{6.4.23}$$

and (6.4.12). With (6.4.12), we also have

$$\sigma^i \sigma^j F_{ij} = \sum_{i<j} [\sigma^i, \sigma^j] \, F_{ij}. \tag{6.4.24}$$

At this juncture, we recall that the magnetic field $\mathbf{B} = (B^i)$ induced from the gauge field A_μ is given by

$$\mathbf{B} = (B^i) = (-F_{23}, F_{13}, -F_{12}) = \nabla \times \mathbf{A}, \quad \mathbf{A} = (A^i) = -(A_i). \tag{6.4.25}$$

Inserting (6.4.24) and (6.4.25) into (6.4.22) and applying (6.4.12), we obtain the equation

$$E'v = -\frac{\hbar^2}{2m}\left(1 - \frac{E'}{2mc^2}\right) D_i^2 v - \frac{q\hbar}{2m}\left(1 - \frac{E'}{2mc^2}\right) \boldsymbol{\sigma} \cdot \mathbf{B} v, \quad \boldsymbol{\sigma} = (\sigma^i), \tag{6.4.26}$$

which may be rewritten as

$$\begin{aligned} E'v = &-\frac{\hbar^2}{2m}\left(\nabla_{\mathbf{x}} - \frac{\mathrm{i}}{\hbar} q\mathbf{A}\right)^2 v - \frac{q\hbar}{2m}\boldsymbol{\sigma} \cdot \mathbf{B} v \\ &+ \frac{E'}{2mc^2}\left(\frac{\hbar^2}{2m}\left[\nabla_{\mathbf{x}} - \frac{\mathrm{i}}{\hbar} q\mathbf{A}\right]^2 v + \frac{q\hbar}{2m}\boldsymbol{\sigma} \cdot \mathbf{B} v\right). \end{aligned} \tag{6.4.27}$$

The first line of this equation is the steady-state form of the gauged Schrödinger equation

$$\mathrm{i}\hbar\frac{\partial \phi}{\partial t} = -\frac{\hbar^2}{2m}\left(\nabla_{\mathbf{x}} - \frac{\mathrm{i}}{\hbar} q\,\mathbf{A}\right)^2 \phi - \frac{q\hbar}{2m}\boldsymbol{\sigma} \cdot \mathbf{B}\,\phi, \tag{6.4.28}$$

with an extra magnetic potential term

$$U_m = -\frac{q\hbar}{2m}\boldsymbol{\sigma} \cdot \mathbf{B}, \tag{6.4.29}$$

taking account of the *Zeeman effect* — a splitting of the energy spectrum of the gauged Hamiltonian due to the presence of an applied magnetic field, **B**, generated by the *intrinsic magnetic moment*

$$\boldsymbol{\mu} = \frac{q\hbar}{2m}\boldsymbol{\sigma}, \tag{6.4.30}$$

all correctly *predicted* by the Dirac equation. This equation governs the nonrelativistic motion of a particle of mass m and charge q in a magnetostatic field, with the *intrinsic spin* or *spin angular momentum*

$$\mathbf{S} = \frac{\hbar}{2}\boldsymbol{\sigma}, \tag{6.4.31}$$

as *required* by the classical non-relativistic theory in order to explain the Zeeman effect, within the stationary-state wave-function ansatz (6.4.15). The resulting term on the right-hand side of (6.4.28) is also known as the *Stern–Gerlach term.* The second line on the right-hand side of the equation (6.4.27) contains other more refined relativistic contributions.

6.5 Dirac equation in Weyl representation

It is clear that the 4×4 matrices

$$\alpha^i = \pm \begin{pmatrix} \sigma^i & 0 \\ 0 & -\sigma^i \end{pmatrix}, \quad \beta = \begin{pmatrix} 0 & 1 \\ 1 & 0 \end{pmatrix}, \tag{6.5.1}$$

also fulfill (6.1.6)–(6.1.8). With these, we obtain (6.1.13). Hence, setting (6.1.15) such that

$$\gamma^0 = \beta = \begin{pmatrix} 0 & 1 \\ 1 & 0 \end{pmatrix}, \quad \gamma^i = \beta\alpha^i = \pm \begin{pmatrix} 0 & -\sigma^i \\ \sigma^i & 0 \end{pmatrix}, \tag{6.5.2}$$

we arrive at (6.1.16) again, which is often referred to as the Dirac equation represented in the *Weyl* (or *chiral*) *basis*, given in terms of the *Weyl gamma matrices.* In fact, the minus-sign situation in (6.5.2) coincides with (6.1.17) for the gamma matrices γ^i $(i = 1, 2, 3)$. In the rest of this section, however, we stay with the plus-sign choice in (6.5.2).

With (6.5.2) and inserting (6.3.1) into (6.1.16), we have

$$\mathrm{i}\hbar \left(\partial_0\phi + \sigma^i\partial_i\phi\right) - mc\chi = 0, \tag{6.5.3}$$

$$\mathrm{i}\hbar \left(\partial_0\chi - \sigma^i\partial_i\chi\right) - mc\phi = 0, \tag{6.5.4}$$

whose obvious advantage over (6.3.2)–(6.3.3) is that the leading derivative parts of the equations decouple. In particular, for a massless particle, $m = 0$, and (6.5.3)–(6.5.4) become

$$\partial_0\phi + \sigma^i\partial_i\phi = \sigma^\mu\partial_\mu\phi = 0, \tag{6.5.5}$$

$$\partial_0\chi - \sigma^i\partial_i\chi = \overline{\sigma}^\mu\partial_\mu\chi = 0, \tag{6.5.6}$$

which are called the *Weyl equations* and ϕ, χ the *Wely* (or *chiral*) *spinors*, where

$$\sigma^0 = \begin{pmatrix} 1 & 0 \\ 0 & 1 \end{pmatrix}, \quad \overline{\sigma}^\mu = (\sigma^0, -\sigma^i). \tag{6.5.7}$$

This convention is sometimes convenient in facilitating computation. In terms of (6.5.7), the Lagrangian action density of the equations (6.5.5)–(6.5.6) is

$$\mathcal{L} = \mathrm{i}\hbar\phi^\dagger\sigma^\mu\partial_\mu\phi + \mathrm{i}\hbar\chi^\dagger\overline{\sigma}^\mu\partial_\mu\chi - mc(\phi\chi^\dagger + \phi^\dagger\chi). \tag{6.5.8}$$

6.6 Nonlinear Dirac equations

There are numerous nonlinear extensions of the Dirac equation with rich applications and mathematical features. Here we present a few of these. For simplicity in presentation, we work in suitable units such that $\hbar = 1$ and $c = 1$ in this section. Use (6.2.1) to denote the Dirac adjoint spinor with ψ a Dirac spinor field, $m > 0$ mass, and $g > 0$ a coupling constant.

Thirring model

The Thirring model [542] is defined in the $(1 + 1)$-dimensional Minkowski spacetime $\mathbb{R}^{1,1}$ with the Lagrangian action density

$$\mathcal{L} = \mathrm{i}\overline{\psi}\gamma^{\mu}\partial_{\mu}\psi - m\overline{\psi}\psi - \frac{g}{2}(\overline{\psi}\gamma^{\mu}\psi)(\overline{\psi}\gamma_{\mu}\psi), \tag{6.6.1}$$

where $\mu = 0, 1$ and ψ is a two-component Dirac spinor field, with $\overline{\psi} = \psi^{\dagger}\gamma^{0}$ the adjoint of ψ, and γ^{μ}s ($\mu = 0, 1$) are the *two-dimensional gamma matrices* given by

$$\gamma^{0} = \sigma^{1}, \quad \gamma^{1} = -\mathrm{i}\sigma^{2}, \tag{6.6.2}$$

following the formalism of Coleman [140] such that $\{\gamma^{\mu}, \gamma^{\nu}\} = 2\eta^{\mu\nu}$ $(\mu, \nu = 0, 1)$. The Euler–Lagrange equation of (6.6.1), or the *Thirring equation*, is

$$\mathrm{i}\gamma^{\mu}\partial_{\mu}\psi - m\psi - g\gamma^{\mu}\psi(\overline{\psi}\gamma_{\mu}\psi) = 0, \tag{6.6.3}$$

which is integrable [122].

Soler model

The Soler model [518] is formulated over the usual $(3+1)$-dimensional Minkowski spacetime $\mathbb{R}^{3,1}$ such that the Lagrangian action density in the Weyl basis reads

$$\mathcal{L} = \mathrm{i}\overline{\psi}\gamma^{\mu}\partial_{\mu}\psi - m\overline{\psi}\psi + \frac{g}{2}(\overline{\psi}\psi)^{2}, \tag{6.6.4}$$

where $\mu = 0, 1, 2, 3$. The Euler–Lagrange equation of (6.6.4), or the *Soler equation*, is

$$\mathrm{i}\gamma^{\mu}\partial_{\mu}\psi - m\psi + g(\overline{\psi}\psi)\psi = 0. \tag{6.6.5}$$

The work [518] presents some rest-particle solutions.

Gross–Neveu model

The Gross–Neveu model [259] is an N-degree of freedom or N-flavor extension of the Soler model, but formulated instead over the $(1+1)$-dimensional Minkowski spacetime $\mathbb{R}^{1,1}$ (as in the Thirring model) such that the Lagrangian action density assumes the form

$$\mathcal{L} = \mathrm{i}\overline{\psi}_{a}\gamma^{\mu}\partial_{\mu}\psi_{a} - m\overline{\psi}_{a}\psi_{a} + \frac{g}{2N}(\overline{\psi}_{a}\psi_{a})^{2}, \tag{6.6.6}$$

where $\mu = 0, 1$, and the summation convention is observed over the repeated indices $a = 1, \dots, N$. The Euler–Lagrange equations of (6.6.6), or the *Gross–Neveu equations*, are

$$\mathrm{i}\gamma^\mu \partial_\mu \psi_a - m\psi_a + \frac{g}{N}(\overline{\psi}_b \psi_b)\psi_a = 0, \quad a = 1, \dots, N. \tag{6.6.7}$$

Although the equations (6.6.3), (6.6.5), and (6.6.7) share the common feature of having cubic nonlinearities, other types of nonlinearities are also investigated. Furthermore, the massless limits with setting $m = 0$ and gauged extensions of the equations are of independent interest as well.

See [198] for a survey on the studies on stationary solutions to linear and nonlinear Dirac equation problems based on variational methods.

Exercises

1. Verify the properties stated in (6.1.11) for the Pauli matrices.
2. Show that the α^i's and β defined in (6.1.12) fulfill (6.1.6)–(6.1.8).
3. Use (6.1.18) to show that a solution of (6.1.16) satisfies the massive Klein–Gordon equation

$$\hbar^2 \partial_\mu \partial^\mu \psi + m^2 c^2 \psi = 0 \quad \text{or} \quad \Box\psi + \frac{m^2 c^2}{\hbar^2}\psi = 0. \tag{6.E.1}$$

   In this way the components of ψ decouple into freely propagating massive Klein–Gordon waves.

4. For $a, b \in \mathbb{C}$ and A any 2×2 Hermitian matrix, show that there holds the identity

$$\det \begin{pmatrix} aI_2 & A \\ A & bI_2 \end{pmatrix} = (ab)^2 - ab\,\mathrm{Tr}(A^2) + (\det(A))^2, \tag{6.E.2}$$

   where I_2 is the 2×2 identity matrix. Then use (6.E.2) to derive (6.3.14).

5. Verify (6.3.15).
6. Use (6.3.14) and (6.3.15) to derive (6.3.16).
7. Consider the static situation when the gauge field A_μ in (6.4.1) is time-independent and use (6.4.5) to derive a steady-state Dirac equation such that its leading-order part governs the steady-state solution of the gauged, non-relativistic, Schrödinger equation of the form

$$\mathrm{i}\hbar\frac{\partial \phi}{\partial t} = -\frac{\hbar^2}{2m}\left(\nabla_{\mathbf{x}} - \frac{\mathrm{i}}{\hbar}q\,\mathbf{A}\right)^2 \phi + qV(\mathbf{x})\phi - \frac{q\hbar}{2m}\boldsymbol{\sigma}\cdot\mathbf{B}\,\phi, \tag{6.E.3}$$

   containing both (6.4.14) and (6.4.28) as limiting cases and accommodating the Zeeman effect.

8. Show that the gamma matrices in the Weyl basis given in (6.5.2) also obey the anti-commutator relation (6.1.18).

9. Use the gauge-covariant derivative (6.4.1) to obtain the gauged Dirac equation in the Weyl representation of the form (6.5.3)–(6.5.4) in terms of 2-spinors ϕ, χ governing the motion of a particle with mass m and charge q. Obtain both electrostatic and magnetostatic situations of the equation as special cases with the steady-state ansatz (6.4.5).

10. By looking for a plane-wave solution to (6.5.5) of the form

$$\phi(x) = v(p)\mathrm{e}^{-\frac{\mathrm{i}}{\hbar}p_\mu x^\mu}, \quad p^0 = \frac{E}{c}, \quad (p^i) = \mathbf{p}, \tag{6.E.4}$$

show that there holds the energy-momentum formula $E^2 = |\mathbf{p}|^2 c^2$, indeed resulting in the motion of a particle of vanishing rest mass.

11. With the notation $\boldsymbol{\sigma} = (\sigma^i)$ and ϕ a 2-spinor field, the equation

$$\mathrm{i}\hbar\frac{\partial \phi}{\partial t} = -\frac{\hbar^2}{2m}\left(\boldsymbol{\sigma}\cdot\left[\nabla_{\mathbf{x}} - \frac{\mathrm{i}}{\hbar}q\,\mathbf{A}\right]\right)^2\phi + qV(\mathbf{x})\phi \tag{6.E.5}$$

is known as the *Pauli equation*. Show that the Pauli equation (6.E.5) is exactly the gauged Schrödinger equation (6.E.3) with the extra Stern–Gerlach term to take account of the Zeeman effect.

12. Obtain conserved charge and current densities of the Thirring, Soler, and Gross–Neveu equations.

7

Ginzburg–Landau equations for superconductivity

This chapter presents the Ginzburg–Landau equations, which are the core of Ginzburg and Landau's phenomenological theory of superconductivity. First it follows the historical path to describe the formalism of F. London and H. London and derive their equations. It also gives a brief introduction to the non-local generalization of one of the London equations by Pippard. Next, it presents the Ginzburg–Landau equations, and discusses the classification of superconductivity by surface energy by virtue of a two-point boundary value problem. It then studies mixed states. The chapter ends with a brief overview of some extended Ginzburg–Landau equations along several directions of research.

7.1 Perfect conductors, superconductors, and London equations

The phenomenon of superconductivity refers to the onset of a state of zero electric resistance in a certain material cooled below a critical temperature, and was first observed by Onnes in 1911 in solid mercury kept at a temperature below 4.2°K. In general, when superconductivity occurs, the material also exhibits the distinctive behavior that it rejects from it magnetic field, an associated phenomenon known as the Meissner effect, observed by Meissner and Ochsenfeld in 1933. In correspondence, when an external magnetic field is made strong such that it starts to penetrate the material, superconductivity will then be turned off and the material behaves like a normal conductor. In other words, an applied magnetic field may play the role of temperature as a switch to turn superconductivity on and off, at low temperature. Progress on theoretical understanding of superconductivity was pioneered by F. and H. London [376] who, in 1935, succeeded in using modified Maxwell equations, known as the London equations,

Mathematical Physics with Differential Equations. Yisong Yang, Oxford University Press.
© Yisong Yang (2023). DOI: 10.1093/oso/9780192872616.003.0007

to interpret the Meissner effect. In 1950, Ginzburg and Landau proposed their theory of superconductivity [237] based on a system of nonlinear partial differential equations, containing a steady form of the Schrödinger equation and the Maxwell equations, naturally coupled together to preserve the underlying gauge symmetry, subsequently known as the Ginzburg–Landau equations. In 1959, Gorkov [242] showed that the Ginzburg–Landau equations may be derived from the quantum theory of superconductivity of Bardeen, Cooper, and Schrieffer [41, 42], which was conceived and formulated two years earlier. Besides being a successful phenomenological theory of superconductivity, the Ginzburg–Landau equations form the basis for the Abelian gauge field theory with a spontaneous broken symmetry, known as the Abelian Higgs theory [118, 298], which was later developed to embrace non-Abelian gauge symmetries and enable the conceptualization of the non-Abelian Yang–Mills–Higgs theory, thereby laying the foundation for the theory of fundamental interactions of elementary particles [570, 571, 572, 614]. The research and development brought forth by the Ginzburg–Landau equations, both mathematical and physical, are rich and active up to the present time. In this chapter, we introduce the Ginzburg–Landau equations following the key "landmarks" described above.

Motion of electron

Consider the motion of an electron of mass $m > 0$ and electric charge $-e$ subject to an applied electric field $\mathbf{E}$ and an applied magnetic field $\mathbf{B}$. Newton's law then gives us the equation of motion

$$m\dot{\mathbf{v}} = -e(\mathbf{E} + \mathbf{v} \times \mathbf{B}) - \frac{m}{\tau}\mathbf{v}, \tag{7.1.1}$$

where the right-hand side of the equation represents the force acting on the electron of velocity $\mathbf{v}$ consisting of the electric force, the Lorentz force, and a dragging force resembling "viscosity" or "friction," giving rise to electric resistance, with $\tau > 0$ a *dragging constant.*

Some special cases of are of interest to us as illustrations.

First, assume the weak magnetic field limit, $\mathbf{B} = \mathbf{0}$. Then becomes

$$m\dot{\mathbf{v}} = -e\mathbf{E} - \frac{m}{\tau}\mathbf{v}. \tag{7.1.2}$$

Use n to denote the density of conductive electrons in the system. Thus,

$$\mathbf{j} = -en\mathbf{v}, \tag{7.1.3}$$

is the current density. Multiplying (7.1.2) by $-en$, we have

$$\frac{\partial \mathbf{j}}{\partial t} = \frac{e^2 n}{m}\mathbf{E} - \frac{1}{\tau}\mathbf{j}. \tag{7.1.4}$$

Consequently, in steady-state situation when the current flows at constant velocity, $\frac{\partial \mathbf{j}}{\partial t} = \mathbf{0}$, the equation (7.1.4) leads to the relation

$$\mathbf{j} = \frac{e^2 n\tau}{m}\mathbf{E}, \tag{7.1.5}$$

which simply says that the current density is proportional to the electric field, a statement known as *Ohm's law*, such that $\frac{1}{\tau}$ naturally gives rise to *electric resistance* and τ then is related to *electric conductance*.

Perfect conductor and first London equation

Next, assume vanishing electric resistance or infinite conductance, $\tau = \infty$. Then (7.1.4) takes the form

$$\frac{\partial \mathbf{j}}{\partial t} = \frac{e^2 n}{m} \mathbf{E}. \tag{7.1.6}$$

This equation is known as the *first London equation*, which implies that the motion of electrons in a perfect conductor is accelerated due to vanishing resistance.

Balance of motion of electrons and second London equation

Electromagnetism is a coupled phenomenon such that an electric current, an electric field, and a magnetic field affect each other in interaction, and are governed by the Maxwell equations, among which the one that couples with the current density is

$$\nabla \times \mathbf{B} - \frac{\partial \mathbf{E}}{\partial t} = \mathbf{j}, \tag{7.1.7}$$

where $\mathbf{B}$ is the magnetic field. In F. and H. London's original work [376], the *displacement current density* $\frac{\partial \mathbf{E}}{\partial t}$ in (7.1.7) is neglected. Here, however, we keep it for the moment for the sake of completeness. Differentiating (7.1.7) with respect to time and using (7.1.6), we have

$$\nabla \times \frac{\partial \mathbf{B}}{\partial t} - \frac{\partial^2 \mathbf{E}}{\partial t^2} = \frac{e^2 n}{m} \mathbf{E}. \tag{7.1.8}$$

On the other hand, recall another equation in the Maxwell equations that is similar to (7.1.7) and of the form

$$\nabla \times \mathbf{E} + \frac{\partial \mathbf{B}}{\partial t} = \mathbf{0}. \tag{7.1.9}$$

Thus, applying curl to (7.1.8) and using (7.1.9), we arrive at

$$\nabla \times \nabla \times \left(\frac{\partial \mathbf{B}}{\partial t} \right) + \frac{e^2 n}{m} \left(\frac{\partial \mathbf{B}}{\partial t} \right) + \frac{\partial^2}{\partial t^2} \left(\frac{\partial \mathbf{B}}{\partial t} \right) = \mathbf{0}, \tag{7.1.10}$$

governing the magnetic field in a perfect conductor, which may be integrated to yield

$$\nabla \times \nabla \times (\mathbf{B} - \mathbf{B}_0) + \frac{e^2 n}{m} (\mathbf{B} - \mathbf{B}_0) + \frac{\partial^2}{\partial t^2} (\mathbf{B} - \mathbf{B}_0) = \mathbf{0}, \tag{7.1.11}$$

where $\mathbf{B}_0$ is the magnetic field at initial time $t = 0$, called the *frozen-in magnetic field* by London and London [376], representing a "permanent memory" of the field when the superconductor was cooled below the transition temperature. Consistency with the *Meissner effect* that a superconductor screens the magnetic field when cooled below the transition led London and London [376] to choose $\mathbf{B}_0 = \mathbf{0}$. The major distinction between a perfect conductor and a superconductor is that the magnetic field in a perfect conductor may be zero or a nonzero constant, but in a superconductor it may only be zero. See Henyey [264]. Thus, the frozen-in magnetic field $\mathbf{B}_0$ may be a nonzero constant. Now, using the zero frozen-in field assumption in (7.1.11), we arrive at

$$\nabla \times \nabla \times \mathbf{B} + \frac{e^2 n}{m} \mathbf{B} + \frac{\partial^2 \mathbf{B}}{\partial t^2} = \mathbf{0}, \tag{7.1.12}$$

governing the magnetic field in a superconductor, which is the *second London equation*. In [376], under the zero displacement current density assumption, the time-derivative term in (7.1.12) is absent and the equation assumes a steady-state form:

$$\nabla \times \nabla \times \mathbf{B} + \frac{e^2 n}{m} \mathbf{B} = \mathbf{0}, \tag{7.1.13}$$

which is popularly referred to as the London equation. In the Coulomb gauge

$$\nabla \cdot \mathbf{B} = 0, \tag{7.1.14}$$

also called the *London gauge* (which is simply one of the Maxwell equations stating that magnetic field is solenoidal), the equation (7.1.13) assumes the transparent form

$$-\nabla^2 \mathbf{B} + \frac{e^2 n}{m} \mathbf{B} = 0. \tag{7.1.15}$$

Meissner effect and penetration depth

Use (x, y, z) to denote a space point and consider a superconducting slab placed parallel to the (x, y)-plane, with boundary at $z = 0$ and interior $z > 0$. Let

$$\mathbf{B}_{\text{ext}} = (0, 0, M) \tag{7.1.16}$$

be an externally applied magnetic field distributed in the direction of the z-axis. By symmetry, we see that $\mathbf{B} = (B^i)$ assumes the same form, $B^1 = B^2 = 0$ and B^3 depends on $z > 0$ only. Using these results in (7.1.14), we see that B^3 is a constant. Inserting these into (7.1.15), we find $B^3 = 0$ as well. Thus, we see that the magnetic field perpendicular to the surface of a superconductor is completely screened. In other words, a perpendicularly applied magnetic field cannot penetrate a superconductor.

We now consider an applied magnetic field that is parallel to the surface of the superconducting slab. After a rotation if necessary, we may assume that the field is along the direction of the x-axis such that

$$\mathbf{B}_{\text{ext}} = (M, 0, 0). \tag{7.1.17}$$

By symmetry again we may let the responding magnetic field $\mathbf{B} = (B^i)$ assume the same form such that $B^1 = B^1(z)$ and $B^2 = B^3 = 0$. Of course, (7.1.14) is automatically fulfilled. Then subjecting the equation (7.1.15) in $z > 0$ to (7.1.17) as a boundary condition at $z = 0$, we get the reduced boundary-value problem

$$-\frac{\mathrm{d}^2 B^1}{\mathrm{d}z^2} + \frac{e^2 n}{m} B^1 = 0, \quad z > 0; \quad B^1(0) = M, \tag{7.1.18}$$

whose bounded solution is

$$B^1(z) = M \mathrm{e}^{-\frac{z}{\lambda_{\mathrm{L}}}}, \quad z \geq 0, \tag{7.1.19}$$

where

$$\lambda_{\mathrm{L}} = \sqrt{\frac{m}{e^2 n}}, \tag{7.1.20}$$

is called the *London penetration depth*, which provides a measurement about how rapidly an applied magnetic field parallel to the surface of a superconductor that has "entered" the interior of the superconductor will vanish.

This discussion may be regarded as a mathematical confirmation of the Meissner effect, which generally concludes that magnetic field is exponentially screened in a superconductor.

Inserting the result (7.1.19) into (7.1.7), we obtain the superconductive current density

$$\mathbf{j} = \left(0, -\frac{M}{\lambda_{\mathrm{L}}} \mathrm{e}^{-\frac{z}{\lambda_{\mathrm{L}}}}, 0\right), \tag{7.1.21}$$

indicating that *supercurrent* flows along the surface of a superconductor in a direction perpendicular to the direction of the applied magnetic field and "congests" to the surface of the superconductor.

Pippard equation

The Pippard equation [447] is an interesting non-local generalization of the London equation (7.1.13) at gauge field level. We maintain the zero displacement current density assumption so that (7.1.7) reads

$$\nabla \times \mathbf{B} = \mathbf{j}. \tag{7.1.22}$$

Taking curl in (7.1.22) and using (7.1.13), we have

$$\nabla \times \mathbf{j} = -\frac{e^2 n}{m} \mathbf{B}. \tag{7.1.23}$$

On the other hand, we may assume that the magnetic field $\mathbf{B}$ is generated from a gauge field, $\mathbf{A}$, such that $\mathbf{B} = \nabla \times \mathbf{A}$. From this and (7.1.23) we may write down a simple relation between $\mathbf{j}$ and $\mathbf{A}$:

$$\mathbf{j} = -\frac{e^2 n}{m} \mathbf{A}, \tag{7.1.24}$$

which relates **j** to **A** locally or pointwise. This relation is very much in the same spirit of Ohm's law (7.1.5), or $\mathbf{j} = \sigma\mathbf{E}$, which has been generalized by Chambers [447] into the following non-local form:

$$\mathbf{j}(\mathbf{x}) = \frac{3\sigma}{4\pi\ell}\int \frac{\mathbf{r}(\mathbf{r}\cdot\mathbf{E}(\mathbf{x}'))}{r^4}\mathrm{e}^{-\frac{r}{\ell}}\,\mathrm{d}\mathbf{x}', \quad r = |\mathbf{r}|, \quad \mathbf{r} = \mathbf{x} - \mathbf{x}', \tag{7.1.25}$$

where $\ell > 0$ is a constant. Comparing (7.1.24) with (7.1.25), we obtain

$$\mathbf{j}(\mathbf{x}) = -\frac{3e^2 n}{4\pi m\xi}\int \frac{\mathbf{r}(\mathbf{r}\cdot\mathbf{A}(\mathbf{x}'))}{r^4}\mathrm{e}^{-\frac{r}{\xi}}\,\mathrm{d}\mathbf{x}', \quad r = |\mathbf{r}|, \quad \mathbf{r} = \mathbf{x} - \mathbf{x}', \tag{7.1.26}$$

where $\xi > 0$ is a coupling constant replacing ℓ in (7.1.25). Inserting this into (7.1.22), we arrive at

$$\nabla\times\nabla\times\mathbf{A} + \frac{3e^2 n}{4\pi m\xi}\int \frac{\mathbf{r}(\mathbf{r}\cdot\mathbf{A}(\mathbf{x}'))}{r^4}\mathrm{e}^{-\frac{r}{\xi}}\,\mathrm{d}\mathbf{x}' = \mathbf{0}, \quad r = |\mathbf{r}|, \quad \mathbf{r} = \mathbf{x} - \mathbf{x}', \tag{7.1.27}$$

which is the *Pippard equation*, a non-local generalization of the London equation (7.1.13).

7.2 Superconductors and Ginzburg–Landau equations

The onset of superconductivity is described by a quantity called the *order parameter* so that when it is zero the conductor behaves normally and when it is positive the conductor is in its superconducting phase. The microscopic theory of superconductivity is the celebrated *BCS theory* of Bardeen, Cooper, and Schrieffer [41, 42], in which superconducting current consists of electron pairs called the *Cooper pairs*. Thus, the density of the Cooper pairs may serve the purpose of an order parameter. Indeed in the earlier phenomenological study of Ginzburg and Landau [237], the order parameter is taken to be given by a complex-valued quantum-mechanical wave function ϕ such that $|\phi|^2$ gives rise to the density of the Cooper pairs. In other words, in terms of the wave function ϕ, we have the following characterization of the phases of a superconductor:

$$|\phi|^2 = 0, \quad \text{occurrence of normal phase,} \tag{7.2.1}$$

$$|\phi|^2 > 0, \quad \text{onset of superconducting phase.} \tag{7.2.2}$$

In the context of the Ginzburg–Landau theory, the complex-valued wave function ϕ is often directly referred to as the order parameter as well.

Variational principle of normal and superconducting phase transition

Superconductivity is a low-temperature phenomenon in that there is a critical temperature $T_c > 0$ such that when $T > T_c$ the normal phase prevails and when $T < T_c$ the superconducting phase occurs, where $T \geq 0$ denotes absolute temperature. In the Ginzburg–Landau theory, the characterization consisting of (7.2.1)–(7.2.2) is realized by a T-dependent "free-energy" density function, $F = F(T, |\phi|^2)$, through a variational principle so that when $T > T_c$, the function F is minimized by the normal phase (7.2.1), and when $T < T_c$, it is minimized by the superconducting phase (7.2.2). Mathematically, we require that for given T the global minimum point of F in terms of $\rho^2 = |\phi|^2$ is a function of T, say $\rho^2 = \rho^2(T)$, satisfying the properties

$$\rho^2(T) = 0, \quad T > T_c; \quad \rho^2(T) > 0, \quad T < T_c. \tag{7.2.3}$$

Such features of F are qualitatively plotted in Figure 7.1 and share similarities with Figure 5.1. So, in terms of the Taylor expansion around $\rho = 0$, we have

$$F = F|_{\rho=0} + \alpha(T)\rho^2 + \frac{1}{2}\beta(T)\rho^4 + \frac{1}{6}\gamma(T)\rho^6 + \cdots, \tag{7.2.4}$$

whose leading-order truncation neglecting terms beyond quartic ones in ρ reads

$$F = F|_{\rho=0} + \alpha(T)\rho^2 + \frac{1}{2}\beta(T)\rho^4. \tag{7.2.5}$$

Thus, solving the equation $\frac{\partial F}{\partial \rho} = 0$ in view of (7.2.5), we have

$$\rho^2 = -\frac{\alpha(T)}{\beta(T)} = \rho^2(T). \tag{7.2.6}$$

Guided by the required properties of F with respect to T as depicted in Figure 7.1 and (7.2.6), we impose the condition

$$\begin{array}{ll} \alpha(T) = 0, & T > T_c, \\ \alpha(T) < 0, & T < T_c; \end{array} \quad \beta(T) > 0, \quad T \geq 0. \tag{7.2.7}$$

In [237], it is assumed that in the vicinity of T_c, $T \approx T_c$, the quantities $F|_{\rho=0}$ and $\beta(T)$ are approximately constant and $\alpha(T)$ is linear in $T - T_c$, namely,

$$F|_{\rho=0} = F_0, \quad \alpha(T) = \begin{cases} -\alpha_0(T_c - T), & T < T_c, \\ 0, & T > T_c, \end{cases} \quad \beta(T) = \beta_0, \tag{7.2.8}$$

where $F_0, \alpha_0 > 0, \beta_0 > 0$ are constants. Consequently, inserting (7.2.8) into (7.2.6), we see that the anticipated picture of normal and superconducting phase transition is specifically described in terms of the order parameter $|\phi|^2$ as follows:

$$|\phi|^2 = 0, \quad T > T_c; \quad |\phi|^2 = \frac{\alpha_0}{\beta_0}(T - T_c) \equiv \phi_0^2, \quad T < T_c. \tag{7.2.9}$$

In the present formalism, however, we only assume the condition (7.2.7), which gives us the general expression (7.2.6) for $\phi_0^2 = \rho^2(T)$, or

$$\phi_0^2 = -\frac{\alpha(T)}{\beta(T)}. \tag{7.2.10}$$

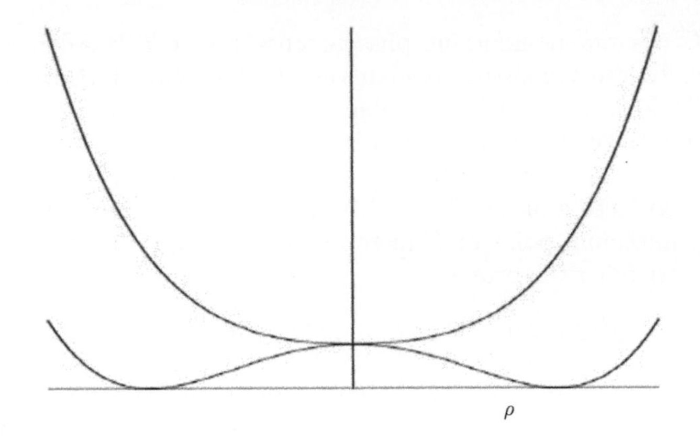

Figure 7.1 Profiles of the free-energy density function $F(T, \rho)$ with regard to the absolute temperature T such that when $T > T_c$ the function F attains its minimum at $\rho = 0$, realizing the normally conductive phase by the point $(\rho, F) = (0, F(T, 0))$ on the upper curve, and when $T > T_c$ the function F is minimized at $\rho = \pm\rho(T)$ for some $\rho(T) > 0$, realizing the superconductive phase by the point $(\rho, F) = (\pm\rho(T), F(T, \rho^2(T)))$ on the lower curve.

Inserting (7.2.10) into (7.2.5), we obtain (after a suitable T-dependent shift if necessary) the normalized form of the free-energy density function

$$V = \frac{\lambda}{8}(|\phi|^2 - \phi_0^2)^2, \quad \lambda = \lambda(T) > 0, \tag{7.2.11}$$

taken to be a potential density function as in the study of the Abelian Higgs model.

Ginzburg–Landau equations

To take account of non-uniformity of superconductivity, the order parameter ϕ is assumed to be a location-dependent function, $\phi = \phi(\mathbf{x})$. Thus, in addition to the potential energy part (7.2.11), we must consider the kinetic energy part as well where $m > 0$ is the mass of a Cooper pair, which is given in view of quantum mechanics by

$$K = \frac{\hbar^2}{2m}|\nabla_{\mathbf{x}}\phi|^2, \tag{7.2.12}$$

which may be combined with (7.2.11) to render us the expanded free-energy density

$$F = K + V = \frac{\hbar^2}{2m}|\nabla_{\mathbf{x}}\phi|^2 + \frac{\lambda}{8}(|\phi|^2 - \phi_0^2)^2, \tag{7.2.13}$$

which enjoys the global $U(1)$-symmetry as already shown. Thus, again, to enlarge the global $U(1)$-symmetry into a local one, we need to replace the conventional derivative by the gauge-covariant one,

$$D_{\mathbf{A}}\phi = \nabla_{\mathbf{x}}\phi - \mathrm{i}\frac{q}{\hbar}\,\mathbf{A}\phi, \tag{7.2.14}$$

where $q = 2e$ counts for the charge of a Cooper pair and $\mathbf{A}$ a gauge vector field, resulting in the minimally expanded locally invariant *Ginzburg–Landau free-energy density*:

$$F_{\text{GL}} = \frac{1}{2}|\nabla \times \mathbf{A}|^2 + \frac{\hbar^2}{2m}|D_{\mathbf{A}}\phi|^2 + \frac{\lambda}{8}(|\phi|^2 - \phi_0^2)^2. \tag{7.2.15}$$

The Euler–Lagrange equations of (7.2.15) are

$$\frac{\hbar^2}{2m} D_{\mathbf{A}}^2\phi = \frac{\lambda}{4}(|\phi|^2 - \phi_0^2)\phi, \tag{7.2.16}$$

$$\nabla \times \mathbf{B} = \mathbf{j} = \mathrm{i}\,\frac{\hbar q}{2m}\,(\phi\overline{D_{\mathbf{A}}\phi} - \overline{\phi}D_{\mathbf{A}}\phi), \tag{7.2.17}$$

where $\mathbf{B} = \nabla \times \mathbf{A}$ is the induced magnetic field in the superconductor, which are the *Ginzburg–Landau equations.* These equations are of a semi-quantum-mechanical character in that the order parameter ϕ is a quantum-mechanical wave function describing the density of the Cooper pairs microscopically, but the gauge field $\mathbf{A}$ or the magnetic field $\mathbf{B}$ follows the Maxwell equations, which determine the magnetic response of the superconductor macroscopically.

London equation limit

Assume low temperature such that $\phi_0 > 0$ in (7.2.10). At the uniform order parameter limit, $\phi \approx \phi_0$, the equation (7.2.17) gives us

$$\nabla \times \mathbf{B} = -\frac{q^2\phi_0^2}{m}\mathbf{A}. \tag{7.2.18}$$

Taking curl in (7.2.18), we get

$$\nabla \times \nabla \times \mathbf{B} = -\frac{q^2\phi_0^2}{m}\mathbf{B}, \tag{7.2.19}$$

which is identical to the London equation (7.1.13) except that the electron charge $-e$ and mass in (7.1.13) are now replaced by the Cooper pair charge $-q = -2e$, and mass and the electron density n by the Cooper pair density ϕ_0^2. Thus, in place of the London penetration depth, we have the following *Ginzburg–Landau penetration depth*:

$$\lambda_{\text{GL}} = \sqrt{\frac{m}{q^2\phi_0^2}} = \sqrt{\frac{m\beta(T)}{q^2|\alpha(T)|}} \equiv \lambda_{\text{GL}}(T), \tag{7.2.20}$$

where we have inserted (7.2.10), which allows us to see how the penetration depth depends on temperature. In particular, when we observe the approximation (7.2.8), we have

$$\lambda_{\text{GL}}(T) = \sqrt{\frac{m\beta_0}{q^2\alpha_0(T_c - T)}}, \quad T < T_c, \tag{7.2.21}$$

which blows up as $T \to T_c$, since magnetic field can penetrate a normal conductor.

Meissner effect

The free-energy density of the Ginzburg–Landau theory in the presence of an external magnetic field, $\mathbf{B}_{\text{ext}} \neq \mathbf{0}$, may now be written as

$$F(\phi, \mathbf{A}) = \frac{1}{2}\mathbf{B}^2 + \frac{\hbar^2}{2m}|D_{\mathbf{A}}\phi|^2 + \frac{\lambda}{8}(|\phi|^2 - \phi_0^2)^2 - \mathbf{B}\cdot\mathbf{B}_{\text{ext}}, \quad T < T_c. \quad (7.2.22)$$

If $\mathbf{B}_{\text{ext}}$ is constant, then the equations (7.2.16)–(7.2.17) are the same as the Euler–Lagrange equations of (7.2.22) and have two apparent solutions,

$$\phi^{\text{n}} = 0, \quad \nabla \times \mathbf{A}^{\text{n}} = \mathbf{B}^{\text{n}} = \mathbf{B}_{\text{ext}}; \quad \phi^{\text{s}} = \phi_0, \quad \mathbf{A}^{\text{s}} = \mathbf{0}, \quad (7.2.23)$$

representing the normal and completely superconducting phases, respectively. Inserting these into (7.2.22), we have

$$F(\phi^{\text{n}}, \mathbf{A}^{\text{n}}) = \frac{\lambda}{8}\phi_0^4 - \frac{1}{2}\mathbf{B}_{\text{ext}}^2, \quad F(\phi^{\text{s}}, \mathbf{A}^{\text{s}}) = 0. \quad (7.2.24)$$

Consequently, when $\mathbf{B}_{\text{ext}}$ satisfies

$$|\mathbf{B}_{\text{ext}}| < \frac{1}{2}\phi_0^2\sqrt{\lambda}, \quad (7.2.25)$$

we have $F(\phi^{\text{n}}, \mathbf{A}^{\text{n}}) > F(\phi^{\text{s}}, \mathbf{A}^{\text{s}})$ and $(\phi^{\text{s}}, \mathbf{A}^{\text{s}})$ is *energetically favored* over $(\phi^{\text{n}}, \mathbf{A}^{\text{n}})$. Thus, the superconductor is in the superconducting phase described by $(\phi^{\text{s}}, \mathbf{A}^{\text{s}})$ and the magnetic field is completely expelled from the superconductor, $\mathbf{B}^{\text{s}} = \mathbf{0}$. On the other hand, when

$$|\mathbf{B}_{\text{ext}}| > \frac{1}{2}\phi_0^2\sqrt{\lambda}, \quad (7.2.26)$$

we have $F(\phi^{\text{n}}, \mathbf{A}^{\text{n}}) < F(\phi^{\text{s}}, \mathbf{A}^{\text{s}})$ and $(\phi^{\text{n}}, \mathbf{A}^{\text{n}})$ is energetically favored over $(\phi^{\text{s}}, \mathbf{A}^{\text{s}})$. Thus, the superconductor is in the normal phase described by $(\phi^{\text{n}}, \mathbf{A}^{\text{n}})$ and the externally applied magnetic field penetrates the superconductor completely, $\mathbf{B}^{\text{n}} = \mathbf{B}_{\text{ext}}$. Such a picture depicts the Meissner effect described earlier. Furthermore, it spells out a critical magnetic field

$$B_c = \frac{1}{2}\phi_0^2\sqrt{\lambda}, \quad (7.2.27)$$

for the onset and destruction of superconductivity, untouched in the context of the London theory.

Mixed states

If $(\phi, \mathbf{A})$ is a solution of (7.2.16)–(7.2.17) such that $\phi(\mathbf{x}) = 0$ somewhere and $\phi(\mathbf{x}) \neq 0$ somewhere else, then, according to (7.2.1)–(7.2.2), the solution represents a *mixed state* embracing both normal and superconducting phases at corresponding locations.

7.3 Classification of superconductivity by surface energy

For greater generality, we consider the Ginzburg–Landau theory without assuming the normalized form (7.2.11) for the potential energy density. Thus, with (7.2.5), we rewrite the Ginzburg–Landau free-energy density as

$$F_{\mathrm{GL}} = \frac{1}{2}|\nabla \times \mathbf{A}|^2 + \frac{\hbar^2}{2m}|D_{\mathbf{A}}\phi|^2 + F_0 + \alpha(T)|\phi|^2 + \frac{1}{2}\beta(T)|\phi|^4, \qquad (7.3.1)$$

so that the associated Euler–Lagrange equations are

$$\frac{\hbar^2}{2m}\,D_{\mathbf{A}}^2\phi = \alpha\phi + \beta|\phi|^2\phi, \qquad (7.3.2)$$

$$\nabla \times \mathbf{B} = \mathbf{j} = \mathrm{i}\,\frac{\hbar q}{2m}\,(\phi\overline{D_{\mathbf{A}}\phi} - \overline{\phi}D_{\mathbf{A}}\phi), \qquad (7.3.3)$$

replacing (7.2.16)–(7.2.17). Since (7.3.3) is the same as (7.2.17), we are led to the same Ginzburg–Landau penetration depth (7.2.20). Besides, although the equation (7.3.2) contains numerous parameters, it may be recast into the following suppressed form

$$\frac{\hbar^2}{2m\beta}\,D_{\mathbf{A}}^2\phi = (|\phi|^2 - \phi_0^2)\phi, \qquad (7.3.4)$$

where ϕ_0^2 is given as in (7.2.10) for $T < T_c$.

Coherence length

With $\phi_0 > 0$ and $\psi = \frac{\phi}{\phi_0}$, the equation assumes the form

$$\xi^2 D_{\mathbf{A}}^2\psi = (|\psi|^2 - 1)\psi, \qquad (7.3.5)$$

where

$$\xi = \sqrt{\frac{\hbar^2}{2m\beta\phi_0^2}} = \sqrt{\frac{\hbar^2}{2m|\alpha(T)|}}. \qquad (7.3.6)$$

Since the factor ξ^2 in front of the left-hand side of (7.3.5) may be transformed away with a modified length scale

$$\mathbf{x} \mapsto \frac{1}{\xi}\,\mathbf{x}, \quad \mathbf{A} \mapsto \frac{1}{\xi}\,\mathbf{A}, \qquad (7.3.7)$$

the quantity ξ is called the *Ginzburg–Landau coherence length.*

Ginzburg–Landau parameter

We saw that both the Ginzburg–Landau penetration depth λ_{GL} and coherence length ξ are temperature-dependent quantities and blow up as $T \to T_c$. However, their ratio

$$\kappa = \frac{\lambda_{\text{GL}}}{\xi} = \frac{m}{q\hbar}\sqrt{2\beta(T)}, \tag{7.3.8}$$

as a consequence of (7.2.20) and (7.3.6), cancels out such a blow-up behavior. In particular, under the assumption (7.2.8) for T near T_c, the parameter κ is independent of T:

$$\kappa = \frac{m}{q\hbar}\sqrt{2\beta_0}. \tag{7.3.9}$$

The parameter κ plays a crucial role in the classification of superconductivity.

Critical magnetic field

In a superconductor, superconductivity may be maintained when an externally applied magnetic field is kept sufficiently weak. However, superconductivity will be quenched when the applied field becomes strong. Thus, there should be a critical value for the applied field at which a mixed state embracing both normal and superconducting phases occurs. In this situation, we may imagine an external magnetic field applied along the surface of a superconductor. At the critical value, the applied magnetic field destroys superconductivity near the surface of the superconductor but fails to switch off superconductivity in regions distant from the surface of the superconductor. Such a situation may be realized in the following simplified manner.

Let (x, y, z) denote a coordinate point of $\mathbb{R}^3$. Assume an applied critical magnetic field, $\mathbf{B}_{\text{ext}}$, is along the direction of the y-axis such that $\mathbf{B}_{\text{ext}} = (0, B_c, 0)$, where B_c is a certain constant. Let the surface of the superconductor in contact with exterior world be at $z = -\infty$ and the region distant from the surface of the superconductor at $z = \infty$. The excited magnetic field inside the superconductor will be given by

$$\mathbf{B} = (0, B(z), 0), \quad -\infty < z < \infty. \tag{7.3.10}$$

The form of the magnetic field $\mathbf{B}$ in (7.3.10) suggests that the order parameter ϕ and gauge field $\mathbf{A}$ may be assumed to be functions depending on z only such that

$$\mathbf{A} = (A(z), 0, 0), \quad -\infty < z < \infty, \tag{7.3.11}$$

where $A(z)$ is a real-valued function, to realize the relation $\mathbf{B} = \nabla \times \mathbf{A}$, resulting in

$$A'(z) = B(z), \quad -\infty < z < \infty. \tag{7.3.12}$$

Within this fixed gauge field configuration, it is consistent to assume that the order parameter ϕ is real-valued. Therefore, we have

$$|D_{\mathbf{A}}\phi|^2 = (\phi'(z))^2 + \frac{q^2}{\hbar^2}A^2(z)\phi^2(z). \tag{7.3.13}$$

Using (7.3.11) and (7.3.13), we see that the Ginzburg–Landau free-energy density in the presence of the critical magnetic field $\mathbf{B}_{\text{ext}} = (0, B_c, 0)$ becomes

$$\begin{aligned} F =& \frac{1}{2}|\nabla \times \mathbf{A}|^2 + \frac{\hbar^2}{2m}|D_{\mathbf{A}}\phi|^2 + F_0 + \alpha|\phi|^2 + \frac{1}{2}\beta|\phi|^4 - \mathbf{B}\cdot\mathbf{B}_{\text{ext}} \\ =& \frac{1}{2}(A')^2 + \frac{1}{2m}\left(\hbar^2(\phi')^2 + q^2A^2\phi^2\right) + F_0 + \alpha\phi^2 + \frac{1}{2}\beta\phi^4 - A'B_c. \end{aligned} \tag{7.3.14}$$

The associated Euler–Lagrange equations of (7.3.14) are

$$\frac{\hbar^2}{2m}\phi'' = \frac{q^2}{2m}A^2\phi + \alpha\phi + \beta\phi^3, \tag{7.3.15}$$

$$A'' = \frac{q^2}{m}\phi^2 A, \tag{7.3.16}$$

which are the one-dimensional, "thin-film" reduction of the full Ginzburg–Landau equations (7.3.2)–(7.3.3).

Recall that the normal phase is at $z = -\infty$ and the superconducting phase is at $z = \infty$. Thus, we are led to the boundary condition

$$\phi(-\infty) = 0, \quad A'(-\infty) = B_c, \tag{7.3.17}$$

$$\phi(\infty) = \phi_0, \quad A'(\infty) = 0, \tag{7.3.18}$$

as illustrated in Figure 7.2.

Inserting (7.3.17) into (7.3.14) and imposing $F = 0$ at $z = -\infty$ due to the finite-energy condition, we obtain

$$F_0 = \frac{1}{2}B_c^2. \tag{7.3.19}$$

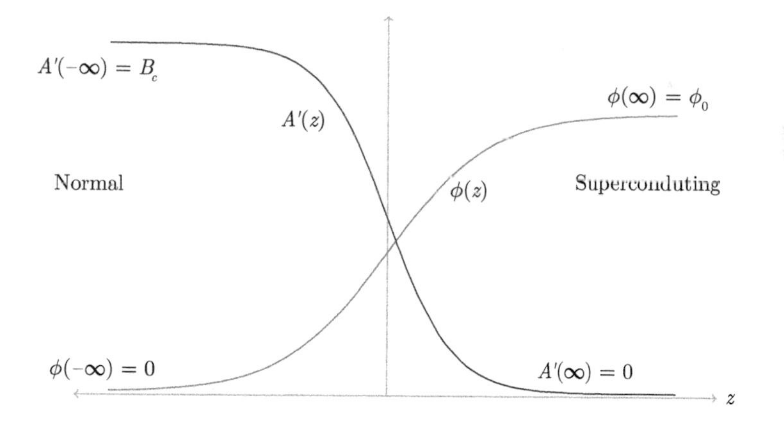

Figure 7.2 Behavior of the excited magnetic field $A'(z)$ and order parameter $\phi(z)$ in a superconductive film, with depth coordinate z, undergoing a phase transition subject to an applied magnetic field at a critical level, B_c, such that at $z = -\infty$, the film is in the normal phase, with $\phi(-\infty) = 0$ and $A'(-\infty) = B_c$, and at $z = \infty$, the film is in the full superconducting phase, with $\phi(\infty) = \phi_0$ and $A'(\infty) = 0$, as a result of the Meissner effect.

Besides, since $\phi_0 > 0$ in (7.3.18), we have $A(\infty) = 0$. In view of this fact, $\phi'(\infty) = 0$, and (7.3.18), we have by $F = 0$ at $z = \infty$, due to the finite-energy condition again, the relation

$$F_0 = -\alpha\phi_0^2 - \frac{1}{2}\beta\phi_0^4 = \frac{\alpha^2}{2\beta}, \tag{7.3.20}$$

using (7.2.10). Combining (7.3.19) and (7.3.20), we arrive at the formula

$$B_c = \frac{|\alpha(T)|}{\sqrt{\beta(T)}}, \quad T < T_c, \tag{7.3.21}$$

relating the critical magnetic field B_c to absolute temperature T. In particular, with the near T_c approximation (7.2.8), we obtain the classical expression

$$B_c = \frac{\alpha_0 T_c}{\sqrt{\beta_0}}\left(1 - \frac{T}{T_c}\right), \quad T < T_c, \tag{7.3.22}$$

in the context of the Ginzburg–Landau theory.

Surface energy and its normalization

Observing (7.3.15)–(7.3.21), we may evaluate the total energy

$$\begin{aligned}\sigma_s =& \int_{-\infty}^{\infty} F \,\mathrm{d}z \\ =& \frac{1}{2}\int_{-\infty}^{\infty}\left((A' - B_c)^2 + \frac{1}{m}\left(\hbar^2(\phi')^2 + q^2A^2\phi^2\right) + 2\alpha\phi^2 + \beta\phi^4\right)\mathrm{d}z, \end{aligned} \tag{7.3.23}$$

which may be regarded as the energy carried by the mixed state realizing a transition between the normal and superconducting phases across the superconducting film. Thus, σ_s is also called the *surface energy*.

Although (7.3.23) looks complicated in that it contains many parameters at various spots, it can be miraculously simplified in terms of the rescaled variables

$$z = \lambda_{\mathrm{GL}}\,\eta, \quad A = \lambda_{\mathrm{GL}} B_c\, a, \quad \phi = \phi_0\, f, \tag{7.3.24}$$

to yield

$$\sigma_s = \frac{\lambda_{\mathrm{GL}} B_c^2}{2}\int_{-\infty}^{\infty}\left((a' - 1)^2 + \frac{2}{\kappa^2}(f')^2 + a^2 f^2 - 2f^2 + f^4\right)\mathrm{d}\eta, \tag{7.3.25}$$

where we have applied (7.2.10), (7.2.20), (7.3.6), (7.3.8), and (7.3.21) in the computation to absorb the parameters and we continue to use the prime $'$ to denote differentiation with respect to the new independent variable η. The associated Euler–Lagrange equations of (7.3.25) are

$$\frac{1}{\kappa^2} f'' = \frac{1}{2}\, a^2 f + (f^2 - 1) f, \tag{7.3.26}$$

$$a'' = f^2 a, \tag{7.3.27}$$

which are the normalized equations of (7.3.15)–(7.3.16), and the boundary condition consisting of (7.3.17)–(7.3.18) becomes

$$f(-\infty) = 0, \quad a'(-\infty) = 1; \quad f(\infty) = 1, \quad a'(\infty) = 0. \tag{7.3.28}$$

This reduction recasts the problem into a much more transparent form since the energy density under integration in (7.3.25) contains only the Ginzburg–Landau parameter, κ.

Classification of superconductivity

A physically relevant solution to the one-dimensional Ginzburg–Landau equations (7.3.26)–(7.3.27) subject to the boundary condition (7.3.28) would be the one that minimizes the surface energy (7.3.25). Here, we assume such a solution exists for any $\kappa > 0$ and we study its significance.

First, we note that the integrand in (7.3.25) satisfies

$$\begin{aligned}(a'-1)^2 &+ \frac{2}{\kappa^2}(f')^2 + a^2 f^2 - 2f^2 + f^4 \\ &= (a' - 1 + f^2)^2 + (2f' + af)^2 + \left(\frac{2}{\kappa^2} - 4\right)(f')^2 - 2(af^2)'. \end{aligned} \tag{7.3.29}$$

Next, we see that the boundary asymptotics in (7.3.28) are achieved sufficiently rapidly such that the boundary terms arising from the integration of the last term on the right-hand side of (7.3.29) disappear, resulting in

$$\int_{-\infty}^{\infty} (af^2)' \,\mathrm{d}\eta = 0. \tag{7.3.30}$$

Consequently, at the critical situation

$$\kappa = \frac{1}{\sqrt{2}}, \tag{7.3.31}$$

we have by using (7.3.29) and (7.3.30) that there holds the lower bound

$$\sigma_s|_{\kappa=\frac{1}{\sqrt{2}}} = \frac{\lambda_{\mathrm{GL}} B_{\mathrm{c}}^2}{2} \int_{-\infty}^{\infty} \left((a' - 1 + f^2)^2 + (2f' + af)^2\right) \mathrm{d}\eta \geq 0, \tag{7.3.32}$$

and that the lower bound is saturated when (f, a) satisfies (7.3.28) and the equations

$$a' - 1 + f^2 = 0, \quad 2f' + af = 0, \quad -\infty < \eta < \infty. \tag{7.3.33}$$

Then we can check that, if (f, a) satisfies (7.3.33), it satisfies (7.3.26)–(7.3.27) under the condition (7.3.31).

If (f, a) is a solution of (7.3.33) subject to (7.3.28), then $f > 0$. Hence, the second equation in (7.3.33) may be integrated to yield $a = -2(\ln f)'$, which is then inserted into the first equation in (7.3.33) to render

$$u'' = \mathrm{e}^u - 1, \quad u = \ln f^2, \tag{7.3.34}$$

which is a one-dimensional Liouville type nonlinear equation. Multiplying the differential equation in (7.3.34) by u', integrating, and using the condition

$$u(\infty) = 0, \quad u'(\infty) = 0, \tag{7.3.35}$$

we arrive at

$$(u')^2 = 2(\mathrm{e}^u - u - 1), \quad -\infty < \eta < \infty. \tag{7.3.36}$$

Since the behavior of f implies $u' > 0$, we are led by (7.3.36) to the equation

$$u' = \sqrt{2}\,\sqrt{\mathrm{e}^u - u - 1}, \quad -\infty < \eta < \infty. \tag{7.3.37}$$

Although this equation denies further integration, we see that up to a translation it has a unique solution satisfying the desired boundary condition

$$u(-\infty) = -\infty, \quad u(\infty) = 0. \tag{7.3.38}$$

See [109] for detail. Returning to the original variables f and a, we see that (7.3.28) is fulfilled. In other words, the system (7.3.33) subject to (7.3.28) has a solution.

We are prepared to discuss the classification problem for superconductivity.

Recall that we have identified the surface energy in (7.3.25) as a function of the Ginzburg–Landau parameter κ given by

$$\sigma_s(\kappa) = \frac{\lambda_{\mathrm{GL}} B_c^2}{2} \min\left\{ \int_{-\infty}^{\infty} \left((a'-1)^2 + \frac{2}{\kappa^2}(f')^2 + a^2 f^2 - 2f^2 + f^4 \right) \mathrm{d}\eta \,\middle|\, (f,a) \text{ satisfies the boundary condition (7.3.28)} \right\}. \tag{7.3.39}$$

Then it is apparent from the structure of (7.3.39) that $\sigma_s(\kappa)$ is monotone decreasing in $\kappa > 0$. Therefore, since we have just shown that

$$\sigma_s\left(\frac{1}{\sqrt{2}}\right) = 0, \tag{7.3.40}$$

we see that there holds the conclusion

$$\sigma_s(\kappa) > 0, \quad \kappa < \frac{1}{\sqrt{2}}; \quad \sigma_s(\kappa) < 0, \quad \kappa > \frac{1}{\sqrt{2}}. \tag{7.3.41}$$

In other words, we have arrived at the following picture for the classification of superconductivity or superconductors:

(i) When $\kappa < \frac{1}{\sqrt{2}}$, the surface energy σ_s is positive and the *superconductivity or superconductor is of type I.*

(ii) When $\kappa > \frac{1}{\sqrt{2}}$, the surface energy σ_s is negative and the *superconductivity or superconductor is of type II.*

(iii) When $\kappa = \frac{1}{\sqrt{2}}$, the surface energy σ_s is zero and it is a nonphysical *borderline* situation regarding superconductivity.

Importance of existence of solution

The study in this section illustrates the importance of the knowledge on the existence of solutions to various nonlinear differential equation problems: The existence of a minimizer of the optimization problem (7.3.39) provides that the surface energy $\sigma_s = \sigma_s(\kappa)$ is well defined and monotonically depends on the parameter $\kappa > 0$. Besides, the existence of a solution to (7.3.33) establishes $\sigma_s = 0$ for the borderline value of κ spelled out in (7.3.31). As a consequence, the celebrated classification picture (i)–(iii) stated earlier for superconductivity stands out. Note also that, in order to achieve this understanding, the existence of a solution is crucial but the detailed properties and uniqueness of the solution are not.

7.4 Mixed state and its magnetic characterizations

Mixed states occur in type-II superconductors in which coexisting superconducting and normal regions are simultaneously accommodated. This section aims to present some characteristic descriptions of such states.

Quantization of magnetic flux

Consider a superconductor and use S to denote a piece of surface spanned over a bounded boundary simple closed curve Γ inside the superconductor. Assume that Γ is chosen so that superconductivity is maintained on it so that the order parameter ϕ, governed by the Ginzburg–Landau equations (7.3.2)–(7.3.3), is nonvanishing on Γ. Thus, along Γ, we can express ϕ in terms of its amplitude and phase angle as follows:

$$\phi(\mathbf{x}) = |\phi(\mathbf{x})|\mathrm{e}^{\mathrm{i}\theta(\mathbf{x})}. \tag{7.4.1}$$

Hence, on Γ, the current density $\mathbf{j}$ on the right-hand side of (7.3.3) becomes

$$\mathbf{j} = \frac{\hbar q}{m}|\phi(\mathbf{x})|^2\left(\nabla_{\mathbf{x}}\theta - \frac{q}{\hbar}\mathbf{A}\right), \tag{7.4.2}$$

in view of (7.2.14). On the other hand, the Meissner effect implies that the magnetic field $\mathbf{B}$ vanishes around Γ where superconductivity prevails, and we conclude from (7.3.3) that $\mathbf{j} = 0$ along Γ. Inserting this result into (7.4.2), we have

$$\nabla_{\mathbf{x}}\theta = \frac{q}{\hbar}\mathbf{A}, \quad \mathbf{x} \in \Gamma. \tag{7.4.3}$$

Consequently, by the Stokes theorem and (7.4.3), we see that the magnetic flux through S obeys

$$\Phi = \int_S \mathbf{B}\cdot \mathrm{d}\mathbf{S} = \oint_\Gamma \mathbf{A}\cdot \mathrm{d}\mathbf{s} = \frac{\hbar}{q}\Delta\theta|_\Gamma = 2\pi n\,\frac{\hbar}{q}, \quad n \in \mathbb{Z}, \tag{7.4.4}$$

where $\Delta\theta|_\Gamma = 2\pi n$ is the phase angle gain along Γ whose orientation is compatible with that of S. Hence, we see that Φ is *quantized* over S. For a mixed state, the magnetic field **B** may be nonvanishing where superconductivity is absent or weak. In other words, there, a magnetic field may penetrate the superconductor.

The property stated in (7.4.4), known as *quantization of magnetic flux*, is an outstanding feature of a type-II superconductor.

If we take

$$\Phi_0 = \frac{2\pi\hbar}{q} \tag{7.4.5}$$

as the unit flux, which is a universal quantity, then (7.4.4) assumes the more pronounced form

$$\Phi = n\Phi_0, \tag{7.4.6}$$

indicating that the magnetic flux through S is an integral multiple of a universal unit flux and is insensitive to the geometric conformation of S.

Secondary critical magnetic field

If we subject a type-II superconductor to an externally applied magnetic field $\mathbf{B}_{\text{ext}}$, the superconductor will be in a pure superconducting phase when the applied field is weak. Next, it will lose its pure superconducting phase when the applied field passes a critical level — the *first critical magnetic field* — at which a transition occurs between the pure superconducting phase and a mixed superconducting-normal phase. Then, it will stay in a mixed superconducting-normal phase until the applied field passes another critical level — the *second critical magnetic field* — beyond which the superconductor completely loses its superconductivity and is in its normal conductive phase. Here we discuss such a transition process.

For simplicity, we assume the applied magnetic field is in the vertical z-axis direction and constant, B. In view of the previous section, we see that the first critical magnetic field B_{c_1} is given by (7.3.21):

$$B_{c_1} = \frac{|\alpha(T)|}{\sqrt{\beta(T)}}, \quad T < T_c. \tag{7.4.7}$$

On the other hand, since the applied field is constant, the governing equations for the superconductor are still (7.3.2)–(7.3.3). We suppose B is slightly below the second critical magnetic field B_{c_2} such that the induced magnetic field is approximately that of the applied field,

$$\mathbf{B} = \nabla \times \mathbf{A} = (0, 0, B). \tag{7.4.8}$$

Hence, the gauge field **A** may be written as

$$\mathbf{A} = (0, Bx, 0). \tag{7.4.9}$$

With (7.4.9), it is consistent to take the ansatz $\phi = f(x) = \text{real}$ in (7.3.2), which enables us to arrive at the following variable-coefficient nonlinear equation

$$f'' - \frac{q^2B^2}{\hbar^2}x^2 f = \frac{2m}{\hbar^2}\left(\alpha f + \beta f^3\right), \quad -\infty < x < \infty. \tag{7.4.10}$$

Since we consider the situation that the applied field is slightly below the second critical field, we may imagine that superconductivity is residual and weak such that the order parameter f in (7.4.10) is nearly zero and the nonzero regions of f are local in x. So (7.3.3) is also fulfilled approximately within first-order terms in f. Besides, the structure of the left-hand side of (7.4.10) suggests that we consider the Hamiltonian

$$-\frac{1}{2}\frac{\mathrm{d}^2}{\mathrm{d}x^2}+\frac{x^2}{2}, \tag{7.4.11}$$

whose eigenvalues and eigenvectors are well known:

$$E_n = n+\frac{1}{2}, \quad u_n(x) = a_n H_n(x)\mathrm{e}^{-\frac{x^2}{2}}, \quad n = 0,1,2,\dots, \tag{7.4.12}$$

where $H_n(x)$'s are Hermite polynomials [149] and a_n's are normalization constants such that

$$\int_{-\infty}^{\infty} u_m(x)u_n(x)\,\mathrm{d}x = \delta_{mn}, \quad m,n = 0,1,2,\dots. \tag{7.4.13}$$

Thus, by rescaling

$$x \mapsto \sqrt{\frac{\hbar}{qB}}x, \tag{7.4.14}$$

we see that the Hamiltonian

$$\hat{H} = -\frac{1}{2}\frac{\mathrm{d}^2}{\mathrm{d}x^2}+\frac{q^2B^2x^2}{2\hbar^2}, \tag{7.4.15}$$

given by the left-hand side of (7.4.10) has the full set of eigenvalues

$$\lambda_n = \frac{qB}{\hbar}\left(n+\frac{1}{2}\right), \quad n = 0,1,2,\dots. \tag{7.4.16}$$

Use $\{v_n\}$ to denote the corresponding set of orthonormal set of eigenfunctions, generated from $\{u_n\}$. Represent the solution f of (7.4.10) by

$$f = \sum_{n=0}^{\infty} c_n v_n. \tag{7.4.17}$$

Then we have

$$\begin{aligned}\int_{-\infty}^{\infty}\left(\frac{1}{2}(f')^2+\frac{q^2B^2x^2}{2\hbar^2}f^2\right)\mathrm{d}x &= \int_{-\infty}^{\infty} f\hat{H}f\,\mathrm{d}x\\ &= \sum_{n=0}^{\infty}\lambda_n c_n^2 \geq \lambda_0\sum_{n=0}^{\infty}c_n^2 = \lambda_0\int_{-\infty}^{\infty}f^2\,\mathrm{d}x = \frac{qB}{2\hbar}\int_{-\infty}^{\infty}f^2\,\mathrm{d}x.\end{aligned} \tag{7.4.18}$$

Thus, multiplying (7.4.10) by f, integrating, and using (7.4.18), we obtain

$$\begin{aligned}\frac{m}{\hbar^2}\beta\int_{-\infty}^{\infty}f^4\,\mathrm{d}x &= \int_{-\infty}^{\infty}\left(-\frac{m}{\hbar^2}\alpha f^2 - f\hat{H}f\right)\mathrm{d}x\\ &\leq \int_{-\infty}^{\infty}\left(\frac{m}{\hbar^2}|\alpha| - \frac{qB}{2\hbar}\right)f^2\,\mathrm{d}x,\end{aligned} \tag{7.4.19}$$

leading to $f = 0$, that is, the prevalence of the normal phase everywhere in the superconductor, whenever B satisfies

$$B \geq \frac{2m}{q\hbar}|\alpha|. \tag{7.4.20}$$

We can use the threshold value given in (7.4.20) as an estimate for the secondary critical magnetic field. That is,

$$B_{c_2} = \frac{2m}{q\hbar}|\alpha|. \tag{7.4.21}$$

Consequently, in view of (7.3.8) and (7.4.7), we arrive from (7.4.21) at

$$B_{c_2} = \sqrt{2}\kappa B_{c_1}. \tag{7.4.22}$$

As described earlier, a type-II superconductor possesses a secondary critical magnetic field beyond the first critical magnetic, $B_{c_2} > B_{c_1}$. Inserting this into (7.4.22), we obtain $\kappa > \frac{1}{\sqrt{2}}$, as already encountered and spelled out through a study of surface energy.

It is between the first and secondary critical magnetic fields that a mixed state in a type-II superconductor takes place.

7.5 Some generalized Ginzburg–Landau equations

There is a wealth of generalizations of the Ginzburg–Landau equations introduced for various applications. This section covers a few of them to conclude our discussion of the subject.

Time-dependent Ginzburg–Landau equations

Gorkov and Eliashberg [241] introduced a time-dependent extension of the system of the Ginzburg–Landau equations accommodating both electric and magnetic fields and modeling superconductive alloys with paramagnetic impurities which, in normalized units, reads

$$\eta\left(\frac{\partial\phi}{\partial t} + \mathrm{i}V\phi\right) = D_{\mathbf{A}}^2\phi - \left(1 - \frac{T}{T_c}\right)(|\phi|^2 - 1)\phi + f, \tag{7.5.1}$$

$$\begin{aligned}\frac{\partial\mathbf{A}}{\partial t} + \nabla V + \kappa^2\nabla\times\nabla\times\mathbf{A} = &\left(1 - \frac{T}{T_c}\right)\frac{\mathrm{i}}{2}(\phi\overline{D_{\mathbf{A}}\phi} - \overline{\phi}D_{\mathbf{A}}\phi)\\ &+ \gamma\nabla\times\mathbf{B}_{\mathrm{ext}},\end{aligned} \tag{7.5.2}$$

where $\eta, \gamma > 0$ are coupling constants, κ is the Ginzburg–Landau parameter, V an electric potential, f and $\mathbf{B}_{\mathrm{ext}}$ are some externally applied electric and magnetic fields, respectively, and the temperature T is close to its critical level $T_c > 0$ and stays subcritical, $T < T_c$. These equations are also known as the *Gorkov–Eliashberg equations.*

Two-flavor Ginzburg–Landau equations

Suhl, Matthias, and Walker [531] pioneered the quantum theory of two-band superconductivity [354], which extends the Bardeen–Cooper–Schrieffer theory. In its Ginzburg–Landau equations limit, there naturally arises an extended system of differential equations governing two flavors of the order parameters ϕ_1 and ϕ_2 with inter-flavor interaction and interaction through an induced magnetic potential $\mathbf{A}$. Following Babaev [30, 31], these *two-flavor Ginzburg–Landau equations* are the Euler–Lagrange equations of the free energy density

$$F = \frac{1}{2}|\nabla \times \mathbf{A}|^2 + \frac{1}{2m_1}|D_{\mathbf{A}}\phi_1|^2 + \frac{1}{2m_2}|D_{\mathbf{A}}\phi_2|^2 + \eta\left(\overline{\phi}_1\phi_2 + \phi_1\overline{\phi}_2\right) + V_1(|\phi_1|^2) + V_2(|\phi_2|^2), \tag{7.5.3}$$

where m_1, m_2 are two effective masses of the superconductive particles of respective flavors, $D_{\mathbf{A}}\phi_{1,2} = \nabla\phi_{1,2} - iq\mathbf{A}\phi_{1,2}$, q an effective electric charge, η a positive constant, and

$$V_{1,2}(|\phi|^2) = \alpha_{1,2}|\phi|^2 + \frac{\beta_{1,2}}{2}|\phi|^4, \tag{7.5.4}$$

are the potential densities of the two order parameters with $\alpha_{1,2} < 0$ and $\beta_{1,2} > 0$ some temperature-dependent coupling constants as in the Ginzburg–Landau theory.

Ginzburg–Landau equations away from critical temperature

Gorkov's celebrated work [242] shows that the Ginzburg–Landau equations may be derived microscopically from the Bardeen–Cooper–Schrieffer quantum theory of superconductivity for an isotropic superconductor when the temperature T is slightly below the critical temperature T_c. Later this derivation is also carried out for anisotropic superconductors [243] in the same vicinity of T_c. This vicinity requirement restrains the range of applications of the Ginzburg–Landau equations. Thus, it will be important to be able to go far from T_c with some modified and justified forms of the Ginzburg–Landau equations. In [593], this task is taken by Xu, Shu, and Wang who present derivations of some extended Ginzburg–Landau equations for T in the range

$$\frac{\Delta(T)}{\pi k} < T < T_c, \tag{7.5.5}$$

where k is the Boltzmann constant and $\Delta(T)$ the *energy gap* function defined by the *Bardeen–Cooper–Schrieffer self-consistent equation* [41, 42], which, in the isotropic situation, may be seen to take the oversimplified form

$$1 = 2K\int_0^{\Lambda} \frac{\tanh\left(\frac{1}{2}\beta\sqrt{\eta^2 + \Delta^2(T)}\right)}{\sqrt{\eta^2 + \Delta^2(T)}}\,\mathrm{d}\eta, \tag{7.5.6}$$

where $K > 0$ is a T-independent constant responsible for the pairing of electron pairs, $\beta = \frac{1}{kT}$ the inverse temperature, and $\Lambda > 0$ an energy cut-off level. We see that there is a $T_c > 0$ such that (7.5.6) has a unique positive gap solution, $\Delta(T)$, for $0 \leq T < T_c$, which depends on T monotonically such that

$$\lim_{T \to T_c^-} \Delta(T) = 0. \tag{7.5.7}$$

See [604] and references therein for a general mathematical study. This description renders some idea about what condition (7.5.5) spells out, although the situation $\Delta(T) > \pi kT$ that supplements (7.5.5) is also studied in [593].

The two cases, isotropic and anisotropic, are established in [593] separately. In the isotropic case, the free energy density is

$$F = \frac{1}{2}|\nabla \times \mathbf{A}|^2 + \frac{\hbar^2}{2m}|D_{\mathbf{A}}\phi|^2 + \alpha|\phi|^2 + \sum_{n=2}^{\infty} \frac{\beta_n}{n}|\phi|^{2n}, \tag{7.5.8}$$

where $D_{\mathbf{A}}$ is defined by (7.2.14), and $\alpha < 0$, $\beta_n > 0$ $(n = 2, 3, \dots)$ are some T-dependent constants. Thus, the associated Euler–Lagrange equations are

$$\frac{\hbar^2}{2m} D_{\mathbf{A}}^2 \phi = \alpha\phi + \sum_{n=2}^{\infty} \beta_n |\phi|^{2n-2}\phi, \tag{7.5.9}$$

$$\nabla \times \nabla \times \mathbf{A} = \mathbf{j} = \mathrm{i}\frac{\hbar q}{2m}(\phi\overline{D_{\mathbf{A}}\phi} - \overline{\phi}D_{\mathbf{A}}\phi), \tag{7.5.10}$$

among which the magnetic field equation, (7.5.10), is unaltered from (7.3.3).

In the anisotropic case, the free energy density is modified in such a way that anisotropy is realized by prescribing effective mass components along the corresponding coordinate directions, $m \mapsto m_1, m_2, m_3$, giving rise to

$$\frac{\hbar^2}{2m}|D_{\mathbf{A}}\phi|^2 \mapsto \sum_{i=1}^{3} \frac{\hbar^2}{2m_i}|D_i\phi|^2, \quad D_i\phi = \partial_i\phi - \mathrm{i}\frac{q}{\hbar}A_i\phi, \quad i = 1, 2, 3, \tag{7.5.11}$$

where $\mathbf{A} = (A_i)$. Thus, the free energy density is now

$$F = \frac{1}{2}|\nabla \times \mathbf{A}|^2 + \sum_{i=1}^{3} \frac{\hbar^2}{2m_i}|D_i\phi|^2 + \alpha|\phi|^2 + \sum_{n=2}^{\infty} \frac{\beta_n}{n}|\phi|^{2n}, \tag{7.5.12}$$

and the associated Euler–Lagrange equations are

$$\sum_{i=1}^{3} \frac{\hbar^2}{2m_i} D_i^2 \phi = \alpha\phi + \sum_{n=2}^{\infty} \beta_n |\phi|^{2n-2}\phi, \tag{7.5.13}$$

$$(\nabla \times \nabla \times \mathbf{A})_i = j_i = \mathrm{i}\frac{\hbar q}{2m_i}(\phi\overline{D_i\phi} - \overline{\phi}D_i\phi), \quad i = 1, 2, 3. \tag{7.5.14}$$

Since the sets of equations (7.5.9)–(7.5.10) and (7.5.13)–(7.5.14) are derived without assuming T stays close to T_c, hence taking truncation at $n = 2$, they are referred to in [593] as the *complete Ginzburg–Landau equations*.

It will be interesting to investigate what modifications these extensions will introduce regarding various important issues considered earlier in the classical Ginzburg–Landau equations, including penetration depth, coherence length, critical magnetic fields, and classification of superconductivity.

For mathematical research on the Ginzburg–Landau equations, see [61, 62, 85, 86, 87, 174, 183, 320, 363, 364, 429, 483, 597] for progress and development regarding many of their analytic aspects related to and motivated from the physical contents of the equations.

Exercises

1. Show that the "energy"

$$E(\mathbf{B}) = \frac{1}{2}\int \left(|\partial_t \mathbf{B}|^2 + |\nabla \times \mathbf{B}|^2 + \frac{e^2 n}{m}|\mathbf{B}|^2\right) \mathrm{d}\mathbf{x}, \tag{7.E.1}$$

when neglecting boundary terms in integration, for a solution of the second London equation (7.1.12) is conserved.

2. Obtain an action functional whose Euler–Lagrange equation is the London equation (7.1.13).
3. Use the London equation to investigate the magnetic response in the interior of a superconducting slab occupying $\{(x, y, z) \mid z \geq 0\}$ such that the applied magnetic field parallel to the surface of the superconductor assumes the general form

$$\mathbf{B}_{\text{ext}} = (M_1, M_2, 0). \tag{7.E.2}$$

Use your result to compute the supercurrent and show that its direction is perpendicular to $\mathbf{B}_{\text{ext}}$.

4. Find the gauge transformation of the Ginzburg–Landau equations.
5. Show that (7.2.16)–(7.2.17) are the Euler–Lagrange equations of the free-energy density (7.2.15).
6. Derive the Euler–Lagrange equations of (7.2.22) and show that the equations have two apparent solutions given in (7.2.23) when $\mathbf{B}_{\text{ext}}$ is constant.
7. Consider (7.3.4) over the full space $\mathbb{R}^3$ with $\mathbf{A} = \mathbf{0}$. For a solution that is asymptotically superconducting such that $|\phi|^2 = \phi_0^2$ at infinity, show that $|\phi|^2 - \phi_0^2$ decays at infinity exponentially fast and determine the decay rate in terms of the Ginzburg–Landau coherence length.
8. Assume that the order parameter ϕ is real-valued and depends on z only and the gauge field $\mathbf{A}$ is of the form (7.3.11). Show that in this situation the equations (7.3.2)–(7.3.3) become (7.3.15)–(7.3.16).
9. Let (ϕ, A) be a solution of (7.3.15)-(7.3.16) satisfying (7.3.18). Show that $A(\infty) = 0$.

10. Let (f, a) be a solution to (7.3.26)–(7.3.28). Show that f, a enjoy the asymptotic behavior

$$f(x) = \mathrm{O}(\mathrm{e}^{-b\eta^2}), \quad a(x) = x + \mathrm{O}(\mathrm{e}^{-b\eta^2}), \quad x \to -\infty, \tag{7.E.3}$$

where $b > 0$ is a constant.

11. Show that if (f, a) satisfies (7.3.33), it satisfies (7.3.26)–(7.3.27) when $\kappa = \frac{1}{\sqrt{2}}$.

12. Show that there is a unique $u_0 \in (-\infty, 0)$ such that the local solution of the equation (7.3.37) satisfying $u(0) = u_0$ is globally defined over $(-\infty, \infty)$ and fulfills the boundary condition (7.3.38).

13. Derive the two-flavor Ginzburg–Landau equations as the Euler–Lagrange equations associated with the energy density (7.5.3).

14. Consider the Bardeen–Cooper–Schrieffer equation (7.5.6) and show that there is a critical temperature $T_c > 0$ such that the equation has a unique positive solution $\Delta(T)$ for $T \in [0, T_c)$. Show that $\Delta(T)$ decreases in $T \in [0, T_c)$ and satisfies the property (7.5.7). Use this result and take limit as $T \to T_c^-$ to show that T_c is determined implicitly by the equation

$$\frac{1}{2K} = \int_0^\Lambda \frac{1}{\eta} \tanh\left(\frac{\eta}{2kT_c}\right) \mathrm{d}\eta. \tag{7.E.4}$$

Finally, use (7.E.4) to obtain the critical temperature formula [41, 42]

$$T_c \approx 1.13 \frac{\Lambda}{k} \mathrm{e}^{-\frac{1}{2K}}. \tag{7.E.5}$$

(See [604] for a thorough study of this and other more general issues.)

8

Magnetic vortices in Abelian Higgs theory

We have seen that in the static limit and the temporal gauge the Abelian Higgs theory reduces itself into the Ginzburg–Landau theory for superconductivity. In this situation, the scalar Higgs field is an order parameter describing the onset of superconductivity. This chapter focuses on the two-dimensional situation and presents an interesting class of solutions to the Abelian Higgs equations — the Abrikosov vortices, or the Nielsen–Olesen dual strings. Such solutions are typical realizations of the mixed states in type-II superconductors and offer insight into a linear confinement mechanism between a magnetic monopole and anti-monopole pair, whose non-Abelian extensions in recent research endeavors help gain new levels of understanding of the puzzle of quark confinement.

8.1 Energy partition, flux quantization, and topological properties

In modern physics, the notion of vortices plays an important role in a vast range of applications and theoretical constructions where the field configurations confine themselves in a two-dimensional setting. Mathematically, a vortex may be generated from a vector field whose vorticity field has a localized lump, or from a wave function that vanishes somewhere, resulting in a phase singularity, or from both. In all situations, the associated energy density becomes concentrated at the center of the vortex, or vortex core, so that the vortex configuration is sometimes referred to as a soliton. Such a structure often owes its presence to the quantum nature of the problem and is dictated by the underlying topological characterization of its vacuum manifold with a broken symmetry. Thus, it is also called a *quantum vortex* or *topological defect*, first by Onsager in 1949 in his

Mathematical Physics with Differential Equations. Yisong Yang, Oxford University Press.
 DOI: 10.1093/oso/9780192872616.003.0008

study of superfluidity of helium, and then predicted by Abrikosov [1] in 1957 in his study of classification of superconductivity in terms of mixed states. It was later realized by Chiao, Garmire, and Townes [133] in 1964 in their study of laser optics in which a vortex core serves to guide the propagation of a highly concentrated light beam following a mechanism called self-trapping. For this latter subject, see also [162, 321, 608] and references therein.

This chapter presents Abrikosov's vortices in the two-dimensional static Abelian Higgs model because they are characterized by both the concentration of the induced vorticity field and vanishing of the wave function so that their topological nature and broken vacuum symmetry are well exhibited and fully appreciated. To this goal, it begins with a discussion of some basic features of the problem. In particular, in the first section, it derives the relation between the boundary condition for a finite-energy field configuration and its topological characterization. The next section discusses the importance of the presence of zeros of the wave function with respect to having a nontrivial vortex charge for a finite-energy field configuration. It also obtains some energy estimates that enable us to conclude that, in sense of energy-minimizing solutions, the presence of zeros in the wave function and the non-triviality of the vortex charge of the solution given in terms of the flux of the associated magnetic field, are equivalent. Subsequently, it presents a family of N-vortex solutions and explains how such vortex solutions may be used to provide conceptual ideas for an understanding of a linear confinement mechanism.

Static planar Abelian Higgs model

For convenience of discussion, in this chapter we normalize the parameters so that in the two-dimensional static temporal-gauge setting with the gauge field $A_\mu = (0, A_1, A_2, 0)$ and the Higgs field ϕ, the Hamiltonian density (5.5.9) assumes the form

$$\mathcal{H} = \frac{1}{2}F_{12}^2 + \frac{1}{2}|D_1\phi|^2 + \frac{1}{2}|D_2\phi|^2 + \frac{\lambda}{8}(|\phi|^2 - 1)^2, \tag{8.1.1}$$

over $\mathbb{R}^2$ with the coordinates x^1, x^2, where $D_i\phi = \partial_i\phi - \mathrm{i}A_i\phi$ and $x = (x^i)$, $i = 1, 2$. The Euler–Lagrange equations of (8.1.1), or the static Abelian Higgs equations over $\mathbb{R}^2$, are

$$D_iD_i\phi = \frac{\lambda}{2}(|\phi|^2 - 1)\phi, \tag{8.1.2}$$

$$\partial_j F_{ij} = \frac{\mathrm{i}}{2}(\phi\overline{D_i\phi} - \overline{\phi}D_i\phi), \tag{8.1.3}$$

where $i, j = 1, 2$. It is clear that these equations coincide with the Ginzburg–Landau equations over $\mathbb{R}^2$.

Ginzburg–Landau parameter

Comparing the equations (8.1.2)–(8.1.3) with the Ginzburg–Landau equations, we find that λ is related to the Ginzburg–Landau parameter κ by

$$\lambda = 2\kappa^2. \tag{8.1.4}$$

Hence, in view of the classification of superconductivity studied earlier, we see that $\lambda = 1$ borders the two types of superconductors such that

(i) When $\lambda < 1$, the superconductor is of type I.

(ii) When $\lambda > 1$, the superconductor is of type II.

The parameter λ is sometimes also called the Higgs self-coupling parameter or the Ginzburg–Landau parameter.

Boundary conditions and appearance of vortices

Since the total energy of the field-configuration pair, ϕ and $A = (A_i)$, is given by

$$\begin{aligned} E(\phi, A) &= \int_{\mathbb{R}^2} \mathcal{H}\, \mathrm{d}x \\ &= \int_{\mathbb{R}^2} \left(\frac{1}{2} F_{12}^2 + \frac{1}{2}|D_1\phi|^2 + \frac{1}{2}|D_2\phi|^2 + \frac{\lambda}{8}(|\phi|^2 - 1)^2 \right) \mathrm{d}x, \end{aligned} \tag{8.1.5}$$

the finite-energy condition implies that a solution (ϕ, A) must satisfy the boundary condition

$$F_{12} \to 0, \quad |D_i\phi| \to 0, \quad |\phi| \to 1 \quad \text{as } |x| \to \infty. \tag{8.1.6}$$

In fact, it can be shown that this decay may be achieved exponentially fast.

Note that the magnetic field F_{12} may be viewed as the vorticity field if the vector field $A = (A_1, A_2)$ is viewed as the velocity field in a two-dimensional fluid. Therefore, wherever $F_{12} \neq 0$, nontrivial vorticity is present. For this reason, nontrivial solutions of two-dimensional gauge field equations are also called "vortices" or "vortex-lines."

A general virial theorem

For mathematical interest, we consider a solution of (8.1.2) and (8.1.3) over $\mathbb{R}^n$, with $i, j = 1, \dots, n$, so that the energy

$$E(\phi, A) = \int_{\mathbb{R}^n} \left(\frac{1}{4} F_{ij}^2 + \frac{1}{2}|D_i\phi|^2 + \frac{\lambda}{8}(|\phi|^2 - 1)^2 \right) \mathrm{d}x \tag{8.1.7}$$

of the field-configuration pair, ϕ and $A = (A_i)$, stays finite. That is, we consider a finite-energy critical point (ϕ, A) of the energy functional (8.1.7).

For a real parameter σ, we introduce

$$\phi^\sigma(x) = \phi(\sigma x), \quad A_i^\sigma(x) = \sigma A_i(\sigma x), \quad i = 1, \ldots, n; \quad x_\sigma = \sigma x \in \mathbb{R}^n. \tag{8.1.8}$$

Therefore

$$(D_i\phi^\sigma)(x) = \sigma \left(\frac{\partial \phi}{\partial x_\sigma^i} - \mathrm{i}A_i\phi \right)(\sigma x), \tag{8.1.9}$$

$$(\partial_i A_j^\sigma - \partial_j A_i^\sigma)(x) = \sigma^2 \left(\frac{\partial A_j}{\partial x_\sigma^i} - \frac{\partial A_i}{\partial x_\sigma^j} \right)(\sigma x). \tag{8.1.10}$$

Thus, we have

$$\begin{aligned} E(\phi^\sigma, A^\sigma) &= \int_{\mathbb{R}^n} \left(\frac{\sigma^4}{4} \left(\frac{\partial A_j}{\partial x_\sigma^i} - \frac{\partial A_i}{\partial x_\sigma^j} \right)^2 \right. \\ &\quad \left. + \frac{\sigma^2}{2} \left| \frac{\partial \phi}{\partial x_\sigma^i} - \mathrm{i}A_i\phi \right|^2 + \frac{\lambda}{8}(|\phi|^2 - 1) \right)(\sigma x)\sigma^{-n}\,\mathrm{d}x^\sigma \\ &= \int_{\mathbb{R}^n} \left(\frac{\sigma^{4-n}}{4} F_{ij}^2 + \frac{\sigma^{2-n}}{2}|D_i\phi|^2 + \frac{\lambda\sigma^{-n}}{8}(|\phi|^2 - 1)^2 \right) \mathrm{d}x. \end{aligned} \tag{8.1.11}$$

Since (ϕ, A) is a critical point of (8.1.7), we have

$$\frac{\mathrm{d}}{\mathrm{d}\sigma} E(\phi^\sigma, A^\sigma) \bigg|_{\sigma=1} = 0. \tag{8.1.12}$$

Inserting (8.1.11) into (8.1.12), we arrive at

$$\frac{(4-n)}{4} \int_{\mathbb{R}^n} F_{ij}^2\,\mathrm{d}x + \frac{(2-n)}{2} \int_{\mathbb{R}^n} |D_i\phi|^2\,\mathrm{d}x = \frac{n\lambda}{8} \int_{\mathbb{R}^n} (|\phi|^2 - 1)^2\,\mathrm{d}x. \tag{8.1.13}$$

From (8.1.13), it is clear that, when $n \geq 5$, all solutions are gauge-equivalent to the trivial one, $\phi = 1, A = 0$; when $n = 4$, a nontrivial solution may only exist under the condition $\lambda = 0$ so that $\phi \equiv 0$; when $n = 2, 3$, solutions with $\lambda > 0$ may be permitted. Thus, we may conclude that, in view of static solutions of finite energies, physics prefers low spatial dimensions, $n = 2, 3, 4$.

Partition identity and consequence

Consider the case when $\lambda > 0$. The work [206] argues that, when $n = 3$, all finite-energy critical points of (8.1.7) are gauge-equivalent to the trivial one $\phi \equiv 1$, $A_j \equiv 0$ $(j = 1, 2, 3)$. Thus, here we concentrate on the case when $n = 2$ so that (8.1.13) assumes the special form

$$4 \int_{\mathbb{R}^2} F_{12}^2\,\mathrm{d}x = \lambda \int_{\mathbb{R}^2} (|\phi|^2 - 1)^2\,\mathrm{d}x. \tag{8.1.14}$$

This identity says that there holds an exact partition between the magnetic energy and the potential energy of the Higgs particle. In particular, the absence of the magnetic field, $F_{12} \equiv 0$ implies the triviality of the Higgs scalar field, $|\phi| \equiv 1$, and vice versa. In particular, the equations do not allow a reduction into

$$A_1 = A_2 \equiv 0, \quad \Delta\phi = \frac{\lambda}{2}(|\phi|^2 - 1)\phi, \quad x \in \mathbb{R}^2, \tag{8.1.15}$$

among finite-energy solutions. Such a property may be regarded as a demonstration of the Meissner effect from a different angle: A finite-energy solution has a nontrivial scalar field sector if and only if the solution has a nontrivial gauge field sector.

Flux quantization and topological invariants

Since $|\phi(x)| \to 1$ as $|x| \to \infty$, we see that

$$\Gamma = \frac{\phi}{|\phi|} : \quad S^1_R \to S^1 \tag{8.1.16}$$

is well defined when $R > 0$ is large enough, where S^1_R denotes the circle in $\mathbb{R}^2$ centered at the origin and of radius R. Therefore, Γ may be viewed as an element in the fundamental group

$$\pi_1(S^1) = \mathbb{Z} \tag{8.1.17}$$

and represented by an integer N. In fact, this integer N is the winding number of ϕ around S^1_R and may be expressed by the integral

$$N = \frac{1}{2\pi\mathrm{i}} \int_{S^1_R} \mathrm{d}\ln\phi. \tag{8.1.18}$$

Note that the continuous dependence of the right-hand side of (8.1.18) with respect to R implies that it is actually independent of R. An important consequence of such an observation is the *flux quantization condition*

$$\Phi = \int_{\mathbb{R}^2} F_{12}\,\mathrm{d}x = 2\pi N \tag{8.1.19}$$

which follows from

$$\begin{aligned}
&\left| \int_{|x|\leq R} F_{12}\,\mathrm{d}x + \mathrm{i}\int_{|x|=R} \mathrm{d}\ln\phi \right| \\
&= \left| \int_{|x|=R} A_i\,\mathrm{d}x_i + \mathrm{i}\int_{|x|=R} \phi^{-1}\partial_i\phi\,\mathrm{d}x_i \right| \\
&\leq \int_{|x|=R} |\phi^{-1}||D_A\phi|\,\mathrm{d}s \\
&\leq C\mathrm{e}^{-\delta R}\int_{|x|=R} \mathrm{d}s = 2\pi R C\mathrm{e}^{-\delta R} \to 0 \quad \text{as } R \to \infty.
\end{aligned} \tag{8.1.20}$$

Note that, when the theory is formulated in the language of a complex line bundle, say ξ, so that ϕ is a cross section, A is a connection 1-form, $F = \mathrm{d}A$ is the curvature, and D_A is the bundle connection, then the integer N is nothing but the *first Chern class* $c_1(\xi)$, which completely classifies the line bundle up to an isomorphism. That is,

$$\frac{\Phi}{2\pi} = N = c_1(\xi). \tag{8.1.21}$$

An important open question is whether for any given $N \in \mathbb{Z}$ there is a solution to the constrained minimization problem

$$E_N \equiv \inf\left\{E(\phi, A) \,\middle|\, \int_{\mathbb{R}^2} F_{12}\,\mathrm{d}x = 2\pi N\right\}. \tag{8.1.22}$$

The problem is solvable only when $\lambda = 1$ (see Taubes [310, 540, 541]).

Topology, vacuum manifold, and spontaneous broken symmetry

The boundary condition (8.1.6) simply indicates that a finite-energy field configuration stays in the vacuum manifold, $\mathcal{V}$, at infinity, which is

$$\mathcal{V} = \{(\phi, A) \,|\, \phi \in S^1, A = 0\}, \tag{8.1.23}$$

and may well be identified with $U(1) = S^1$ so that $U(1)$ acts on $\mathcal{V}$ transitively, resulting in a full spontaneously broken $U(1)$-symmetry whose topological richness provided by (8.1.17) or $\pi_1(\mathcal{V}) = \mathbb{Z}$ forebodes the existence of vortices with respective integer representations or topological charges.

8.2 Vortex-lines, solitons, and particles

Recall that in the Ginzburg–Landau theory of superconductivity, the complex scalar field ϕ is an order parameter that characterizes the two phases (superconducting and normal states) of a solid. Mathematically, $|\phi|^2$ gives rise to the density of superconducting electron pairs (the Cooper pairs) so that $\phi \neq 0$ indicates the presence of electron pairs and onset of superconductivity and $\phi = 0$ indicates the absence of electron pairs and the dominance of normal state. As a consequence, when the order parameter ϕ is such that it is nonvanishing somewhere but vanishing elsewhere, we are then having a mixed state. Recall also that, according to the Meissner effect, a superconductor screens the magnetic field. So the presence of superconductivity prevents the penetration of a magnetic field. Therefore, in a mixed state, the magnetic field, F_{12}, always has its maximum penetration at the spots where $\phi = 0$. Or equivalently, $|F_{12}|$ assumes its local maximum values at the zeros of ϕ. Since F_{12} may be interpreted as a vorticity field, the zeros of ϕ give rise to centers of vortices or locations of vortex-lines distributed over $\mathbb{R}^2$. Hence, we may expect that the energy density (8.1.1) as a function of $x \in \mathbb{R}^2$ attains its local maxima at the zeros of ϕ since the potential energy density and the magnetic energy density both peak there. However, since energy and mass are equivalent, we observe *mass concentration* centered at the zeros of ϕ. In other words, we have produced a distribution of *solitons* realized as mass lumps that may also be identified with *"particles"* in quantum field theory [206, 306, 386, 466, 477, 614].

Role of zeros of wave function

We now show that the presence of zeros of ϕ is essential for a solution to be nontrivial: *If (ϕ, A) is a finite-energy solution of the Abelian Higgs equations so that ϕ never vanishes, then (ϕ, A) is gauge-equivalent to the trivial solution* $A \equiv 0, \phi \equiv 1$.

Here is a quick proof.

Since ϕ never vanishes, we may rewrite ϕ as

$$\phi = \varphi e^{i\omega} \tag{8.2.1}$$

for globally defined real-valued smooth functions φ and ω over $\mathbb{R}^2$. In fact, we may assume $\varphi > 0$. Using the gauge transformation

$$\phi \mapsto \phi e^{-i\omega}, \quad A_i \mapsto A_i - \partial_i \omega, \tag{8.2.2}$$

we see that (ϕ, A) becomes "unitary" in the sense that ϕ is real valued,

$$\phi = \varphi, \tag{8.2.3}$$

and we say that we have chosen a "unitary gauge." In unitary gauge, the equations of motion decompose significantly to take the form

$$\Delta\varphi = |A|^2\varphi + \frac{\lambda}{2}(\varphi^2 - 1)\varphi, \tag{8.2.4}$$

$$2A_i\partial_i\varphi + (\partial_i A_i)\varphi = 0, \tag{8.2.5}$$

$$\partial_1 F_{12} = \varphi^2 A_2, \tag{8.2.6}$$

$$\partial_2 F_{12} = -\varphi^2 A_1. \tag{8.2.7}$$

From (8.2.6) and (8.2.7), we have

$$\int_{\mathbb{R}^2} (A_2\partial_1 F_{12} - A_1\partial_2 F_{12})\,dx = \int_{\mathbb{R}^2} |A|^2\varphi^2\,dx. \tag{8.2.8}$$

Integrating further by parts and dropping the boundary terms, we obtain

$$\int_{\mathbb{R}^2} (F_{12}^2 + |A|^2\varphi^2)\,dx = 0. \tag{8.2.9}$$

Since φ never vanishes, we have $A \equiv 0$. Returning to (8.2.4), we have

$$\Delta\varphi = \frac{\lambda}{2}\varphi(\varphi^2 - 1), \tag{8.2.10}$$

which may be rewritten as

$$\Delta(\varphi - 1) = \frac{\lambda}{2}\varphi(\varphi + 1)(\varphi - 1). \tag{8.2.11}$$

Using the boundary condition $\varphi - 1 \to 0$ as $|x| \to \infty$ and the maximum principle [199, 165], we deduce $\varphi \equiv 1$ as claimed.

Estimate of energy from below — topological lower bound

A useful identity involving gauge-covariant derivatives is

$$|D_i\phi|^2 = |D_1\phi \pm \mathrm{i}D_2\phi|^2 \pm F_{12}|\phi|^2 \pm \mathrm{i}(\partial_1[\phi\overline{D_2\phi}] - \partial_2[\phi\overline{D_1\phi}]). \tag{8.2.12}$$

Therefore, the total energy satisfies

$$\begin{aligned} E(\phi, A) &\geq \frac{1}{2}\min\{\lambda, 1\}\int_{\mathbb{R}^2}\left(F_{12}^2 + |D_i\phi|^2 + \frac{1}{4}(|\phi|^2 - 1)^2\right)\mathrm{d}x \\ &= \frac{1}{2}\min\{\lambda, 1\}\int_{\mathbb{R}^2}\left(\left[F_{12} \pm \frac{1}{2}(|\phi|^2 - 1)\right]^2 + |D_1\phi \pm \mathrm{i}D_2\phi|^2 \pm F_{12}\right)\mathrm{d}x \\ &\geq \min\{\lambda, 1\}\pi|N|, \end{aligned} \tag{8.2.13}$$

where $\Phi = 2\pi N$ and the signs $\pm$ follow $N = \pm|N|$. In the sequel, we focus on $N \geq 0$ for convenience. When $\lambda = 1$, we have

$$E(\phi, A) \geq \pi N, \tag{8.2.14}$$

and such an energy lower bound is saturated if and only if (ϕ, A) satisfies the *self-dual* system of equations

$$D_1\phi + \mathrm{i}D_2\phi = 0, \tag{8.2.15}$$

$$F_{12} = \frac{1}{2}(1 - |\phi|^2), \tag{8.2.16}$$

which is a reduction of the original equations of motion and is also often called a *BPS system* after Bogomol'nyi [67] and Prasad and Sommerfield [459], who first derived these equations.

Structure of BPS system

It is convenient to complexify our variables and use

$$A = A_1 + \mathrm{i}A_2, \quad A_1 = \frac{1}{2}(A + \overline{A}), \quad A_2 = \frac{1}{2\mathrm{i}}(A - \overline{A}), \quad z = x^1 + \mathrm{i}x^2, \tag{8.2.17}$$

$$\partial = \frac{1}{2}(\partial_1 - \mathrm{i}\partial_2),\ \overline{\partial} = \frac{1}{2}(\partial_1 + \mathrm{i}\partial_2),\ \partial_1 = \partial + \overline{\partial},\ \partial_2 = \mathrm{i}(\partial - \overline{\partial}), \tag{8.2.18}$$

$$\Delta = \partial_1^2 + \partial_2^2 = 4\partial\overline{\partial} = 4\overline{\partial}\partial. \tag{8.2.19}$$

Hence, (8.2.15) assumes the form

$$\overline{\partial}\phi = \frac{\mathrm{i}}{2}A\phi. \tag{8.2.20}$$

To understand this relation, recall the $\overline{\partial}$-*Poincaré lemma*, which states that the equation

$$\overline{\partial}\omega(z) = \mathrm{i}\alpha(z) \tag{8.2.21}$$

over a disk $B \subset \mathbb{C}$ always has a solution [310]. In fact, alternatively, we may understand this as follows: With the notation

$$\omega = u + \mathrm{i}v, \quad \mathrm{i}\alpha = f + \mathrm{i}g, \tag{8.2.22}$$

for some real-valued functions u, v, f, g, we may rewrite (8.2.21) as

$$\partial_1 u - \partial_2 v = f, \quad \partial_2 u + \partial_1 v = g, \tag{8.2.23}$$

resulting in the integrability condition

$$0 = \partial_1\partial_2 u - \partial_2\partial_1 u = -\Delta v - (\partial_2 f - \partial_1 g), \tag{8.2.24}$$

in which the right-hand-side equation, that is, $\Delta v = \partial_1 g - \partial_2 f$, can always be solved for any f, g. Hence, the lemma follows.

Now, let ψ solve

$$\overline{\partial}\psi = \frac{\mathrm{i}}{2}A \tag{8.2.25}$$

locally. Then we see that the complex-valued function $f = \phi\mathrm{e}^{-\psi}$ satisfies the Cauchy–Riemann equation

$$\overline{\partial} f = \overline{\partial}(\phi\mathrm{e}^{-\psi}) = \mathrm{e}^{-\psi}(\overline{\partial}\phi - \phi\overline{\partial}\psi) = 0. \tag{8.2.26}$$

Therefore, $f(z)$ is analytic. In particular, f (and hence ϕ) may only have isolated zeros with integer multiplicities. In other words, if z_0 is a zero of ϕ, then

$$\phi(z) = (z - z_0)^n h(z) \tag{8.2.27}$$

for z near z_0 and the function $h(z)$ never vanishes, where n is a positive integer that is also the local winding number of ϕ around z_0. From (8.2.16), we see clearly that the vorticity field F_{12} achieves it maximum value at z_0 as well,

$$\max\{F_{12}\} = F_{12}(z_0) = \frac{1}{2}, \tag{8.2.28}$$

so that the point z_0 defines the center of a (magnetic) vortex. The integer n is also called the local vortex charge. Besides, since $|\phi| \to 1$ as $|x| \to \infty$, we see that ϕ can only have a finite number of zeros over $\mathbb{C}$. Assume that the zeros of ϕ and their respective multiplicities are

$$z_1, n_1, \quad z_2, n_2, \quad \ldots\ldots, \quad z_k, n_k. \tag{8.2.29}$$

Counting multiplicities of these zeros (i.e. a zero of multiplicity m is counted as m zeros), the total vortex charge is the total number of zeros of ϕ, say $N(\phi)$,

$$N(\phi) = \sum_{s=1}^{k} n_s. \tag{8.2.30}$$

On the other hand, away from the zeros of ϕ, (8.2.15) may be rewritten

$$A = -\mathrm{i}2\overline{\partial}\ln\phi. \tag{8.2.31}$$

Therefore, there, we can represent F_{12} as

$$F_{12} = \partial_1 A_2 - \partial_2 A_1 = -\mathrm{i}(\partial A - \overline{\partial A}) = -2\partial\overline{\partial} \ln |\phi|^2 = -\frac{1}{2}\Delta \ln |\phi|^2. \quad (8.2.32)$$

Inserting (8.2.32) into (8.2.16), we have

$$\Delta \ln |\phi|^2 + 1 - |\phi|^2 = 0 \quad \text{(away from the zeros of } \phi). \quad (8.2.33)$$

Now define

$$u = \ln |\phi|^2. \quad (8.2.34)$$

Then, near z_s ($s = 1, 2, \ldots, k$), we have

$$u(z) = 2n_s \ln |z - z_s| + \text{ a regular term.} \quad (8.2.35)$$

Consequently, we arrive at the Liouville-type equation [375]

$$\Delta u = \mathrm{e}^u - 1 + 4\pi \sum_{s=1}^{k} n_s \delta(z - z_s) \quad \text{in } \mathbb{C} = \mathbb{R}^2, \quad (8.2.36)$$

subject to the boundary condition

$$u \to 0 \quad \text{as } |x| = |z| \to \infty \quad (8.2.37)$$

(since $|\phi| \to 1$ as $|x| \to \infty$), which may be solved by various techniques, such as the calculus of variations [310] and monotone iterations [565, 606].

Conversely, for any given data $\{(z_s, n_s)\}$, the solution of the elliptic equation (8.2.36) gives rise to a solution pair (ϕ, A), which represents multiply distributed vortices at $\{z_s\}$ with the corresponding local vortex charges $\{n_s\}$.

Formally, we see that (see Exercise 6)

$$\int_{\mathbb{R}^2} \Delta u \, \mathrm{d}x = 0. \quad (8.2.38)$$

Hence, from (8.2.36), we obtain

$$\begin{aligned} \int_{\mathbb{R}^2} (1 - |\phi|^2) \, \mathrm{d}x &= \int_{\mathbb{R}^2} (1 - \mathrm{e}^u) \, \mathrm{d}x \\ &= 4\pi \sum_{s=1}^{k} n_s = 4\pi N(\phi). \end{aligned} \quad (8.2.39)$$

Integrating (8.2.16) and inserting (8.2.39), we get the beautiful result

$$\begin{aligned} N = c_1(\xi) &= \frac{1}{2\pi} \int_{\mathbb{R}^2} F_{12} \, \mathrm{d}x \\ &= \frac{1}{4\pi} \int_{\mathbb{R}^2} (1 - |\phi|^2) \, \mathrm{d}x = N(\phi). \end{aligned} \quad (8.2.40)$$

In other words, the total vortex number is nothing but the first Chern class of the solution we have seen before, which also determines the total magnetic flux, $\Phi = 2\pi N$.

Let us document the important conclusion that the solution constructed from solving (8.2.36) carries the minimum energy,

$$E_N = \pi N. \tag{8.2.41}$$

We emphasize that such an exact result of the energy value is only known for $\lambda = 1$ [310, 540, 541], but unknown for $\lambda \neq 1$. The parameter λ here classifies superconductivity so that $\lambda < 1$ corresponds to type-I and $\lambda > 1$ corresponds to type-II superconductivity, respectively.

Estimate of energy from above — topological upper bound

Using (ϕ, A) to denote a solution of (8.2.15)–(8.2.16) with $N = \pm|N|$ as a trial field configuration pair, we have

$$\begin{aligned} E(\phi, A) &\leq \max\{\lambda, 1\} \int_{\mathbb{R}^2} \left(\frac{1}{2} F_{12}^2 + \frac{1}{2} |D_i \phi|^2 + \frac{1}{8} (|\phi|^2 - 1)^2 \right) \mathrm{d}x \\ &= \max\{\lambda, 1\} \pi |N|. \end{aligned} \tag{8.2.42}$$

Therefore, we may combine (8.2.13) and (8.2.42) to obtain the following lower and upper bounds

$$\min\{\lambda, 1\} \pi |N| \leq E_N \leq \max\{\lambda, 1\} \pi |N|, \tag{8.2.43}$$

which implies that the energy E grows in proportion to the total vortex number N and suggests that these vortices may be viewed as particles. We also see that a nonvanishing vortex number N is essential for the existence of a nontrivial solution.

The work [94] demonstrates that E_N is asymptotically like $\frac{\pi}{2} N^2 \ln \lambda$ for λ large.

Energy gap

An interesting fact contained in (8.2.43) is that the Ginzburg–Landau equations have no nontrivial energy-minimizing solution with an energy in the open interval

$$I = (0, \min\{\lambda, 1\} \pi). \tag{8.2.44}$$

Such a result may be viewed as an *energy* or *mass gap theorem* at classical level.

As a conclusion, we see that, in sense of an energy-minimizing solution, (ϕ, A), the presence of zeros of the wave function ϕ and the appearance of a nontrivial vortex charge N giving rise to the total magnetic flux induced from A, are in fact, equivalent, such that a vortex solution of the problem may either refer to ϕ possessing zeros or A inducing a nontrivial magnetic field.

8.3 Radially symmetric solutions

We saw that a finite-energy solution to (8.1.2)–(8.1.3) carries a quantized vortex charge determined by an integer N as stated in (8.1.19). Thus, we ask whether each such integer may be realized by a finite-energy solution. This section shows that radially symmetric solutions to (8.1.2)–(8.1.3) may be used to realize (8.1.19) for any given integer N.

Radial ansatz

Using polar coordinates (r, θ) for $\mathbb{R}^2$, a radially symmetric field configuration pair (ϕ, A) representing N vortices nested at the origin may be defined by the "ansatz"

$$\phi(x) = u(r)\mathrm{e}^{\mathrm{i}N\theta}, \tag{8.3.1}$$

$$A_i(x) = Nv(r)\varepsilon_{ij}\frac{x^j}{r^2}, \quad i, j = 1, 2. \tag{8.3.2}$$

It is clear that regularity at the origin imposes the condition

$$\lim_{r\to 0} u(r) = 0, \quad \lim_{r\to 0} v(r) = 0. \tag{8.3.3}$$

Governing equations

From the radial ansatz (8.3.1)–(8.3.2), we can write down the expressions

$$F_{12} = N\frac{v'(r)}{r}, \tag{8.3.4}$$

$$D_1\phi = u'(r)\frac{x^1}{r}\mathrm{e}^{\mathrm{i}N\theta} + \mathrm{i}N(v(r) - 1)u(r)\frac{x^2}{r^2}\mathrm{e}^{\mathrm{i}N\theta}, \tag{8.3.5}$$

$$D_2\phi = u'(r)\frac{x^2}{r}\mathrm{e}^{\mathrm{i}N\theta} - \mathrm{i}N(v(r) - 1)u(r)\frac{x^1}{r^2}\mathrm{e}^{\mathrm{i}N\theta}, \tag{8.3.6}$$

which allow us to reduce the original equations of motion into a system of ordinary differential equations

$$u'' + \frac{1}{r}u' = \frac{N^2}{r^2}(v - 1)^2 u + \frac{\lambda}{2}u(u^2 - 1), \tag{8.3.7}$$

$$v'' - \frac{1}{r}v' = (v - 1)u^2. \tag{8.3.8}$$

Note that the energy functional (8.1.5) reduces itself into the following one-dimensional energy functional

$$\begin{aligned} E(\phi, A) &= I(u, v) \\ &= \pi\int_0^\infty \left(N^2\frac{(v')^2}{r} + r(u')^2 + \frac{N^2}{r}u^2(v - 1)^2 + \frac{\lambda}{4}r(u^2 - 1)^2\right)\mathrm{d}r, \end{aligned} \tag{8.3.9}$$

which leads us to impose the boundary condition

$$\lim_{r\to\infty} u(r) = 1, \quad \lim_{r\to\infty} v(r) = 1. \tag{8.3.10}$$

The existence of a solution gives rise to an N-vortex solution, as can be readily verified by computing the flux

$$\Phi = \int_{\mathbb{R}^2} F_{12}\,\mathrm{d}x = 2\pi N \int_0^\infty v'(r)\,\mathrm{d}r = 2\pi N, \tag{8.3.11}$$

using (8.3.4) with (8.3.3) and (8.3.10). In fact, such a solution may be found by solving the minimization problem

$$\inf\{I(u,v) \,|\, I(u,v) < \infty, u(0) = v(0) = 0, u(\infty) = v(\infty) = 1\}, \tag{8.3.12}$$

as done in [61, 448], but it is not clear whether the solution gives rise to a global energy minimizer saturating E_N. On the other hand, Jaffe and Taubes [310] conjecture that, for $\lambda \neq 1$, radially symmetric N-vortex solutions are the only possible solutions of finite energies and, in particular, there are no multiply distributed zeros of ϕ for a finite-energy solution pair (ϕ, A) unless $\lambda = 1$.

8.4 From monopole confinement to quark confinement

This section comments briefly on how magnetic vortices may help us to conceptually understand a particle confinement mechanism. To illustrate the ideas, it first recalls a non-confinement situation in the context of Newtonian gravity. It next discusses what could happen in a type-II superconductor when a magnetic monopole and an anti-monopole are put in interaction subject to the Meissner effect. It therefore arrives at a linear confinement mechanism following Mandelstam [381, 382], Nambu [406], and 't Hooft [534, 536]. It then relates this discussion to the literature of monopole-quark confinement research [25, 338, 339, 506, 507].

Non-confinement subject to Newtonian gravity

Consider two masses, m_1 and m_2, initially placed r distance away. The gravitational attractive force between the masses is then

$$F(r) = G\frac{m_1 m_2}{r^2}, \tag{8.4.1}$$

following Newton's law. Thus, the work or energy needed to completely separate them, when they are initially placed at a distance $R > 0$, so that they eventually stay away as non-interacting "free masses" with an infinite distance between them is

$$W = \int_R^\infty F(r)\,\mathrm{d}r = G\frac{m_1 m_2}{R}, \tag{8.4.2}$$

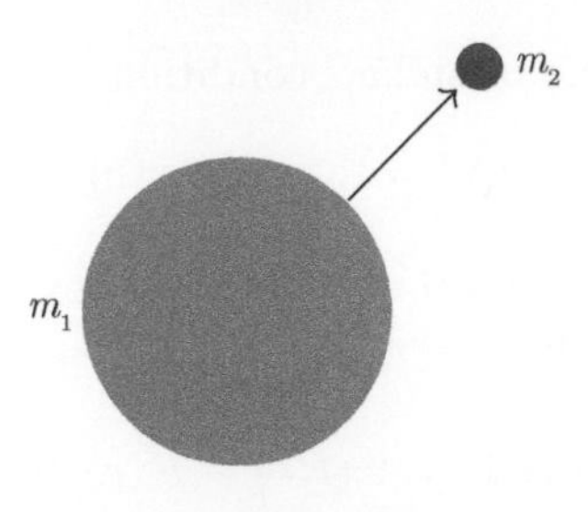

Figure 8.1 Attraction of a pair of masses, m_1 and m_2, through Newton's gravity. Due to the inverse-square law decay of gravitational force, the work needed to separate the two masses is calculated to be finite, rendering a non-confinement phenomenon.

which is finite. In other words, it consumes a *finite* amount of energy to separate the two masses to replace them at an infinite distance apart so that, gravitationally, they no longer "feel" the presence of each other. Consequently we observe a *non-confinement* phenomenon (Figure 8.1).

Of course, we can do the same thing for a pair of magnetic monopole and anti-monopole because the magnetic force between them obeys the same inverse-square law.

Monopole confinement in type-II superconductor

However, when the monopole and anti-monopole are placed in a type-II superconductor, we will encounter an entirely different situation. Due to the presence of superconducting Cooper pairs formed at low temperature as a result of electron condensation, the magnetic field is repelled from the bulk region of the superconductor due to the Meissner effect and squeezed into thin tubes of force lines, in the form of vortex-lines, which penetrate through the spots where superconductivity is destroyed, manifested by the vanishing of the order parameter. Since the strengths of these vortex-lines are constant with respect to the separation distance of the monopole and anti-monopole, the binding force between them remains constant. Consequently, the work needed to place the monopole and monopole, initially separated at a distance r_0 away, at a further distance r away is

$$W(r) = K(r - r_0), \quad r > r_0, \tag{8.4.3}$$

which is a linear function of r. In particular, the work needed to completely split the monopole pair so that the monopole and anti-monopole are seen as isolated free entities will be infinite. Thus, it will be practically impossible to separate the monopole and anti-monopole immersed in a type-II superconductor. In other words, the monopole and anti-monopole are confined. More precisely, we see that, due to the Meissner effect, the monopole and anti-monopole placed in a type-II superconductor interact with each other through narrowly formed vortex-lines that give rise to a constant inter-monopole binding force and the linear law of potential expressed in (8.4.3). As a result, it is impossible to separate

Figure 8.2 A pair of monopole and anti-monopole of charges g and $-g$, respectively, immersed in a type-II superconductor. Due to the Meissner effect, the pair attract each other through thin vortex lines whose strengths stay constant with respect to the separation distance, thus demonstrating a linear confinement phenomenon.

such a pair of monopole and anti-monopole, and the monopole confinement phenomenon takes place (Figure 8.2). Such a confinement picture is also called *linear confinement.*

From monopole confinement to quark confinement

A fundamental puzzle in physics, known as the *quark confinement* [249], is that quarks, which make up elementary particles like mesons and baryons, cannot be observed in isolation. A well-accepted confinement mechanism, exactly known as the linear confinement model, interestingly states that, when attempting to separate a pair of quarks, such as a quark and an anti-quark constituting a meson, the energy consumed would grow linearly with respect to the the separation distance between the quarks and it would require an infinite amount of energy in order to split the pair. The quark and anti-quark may be regarded as a pair of source and sink of color-charged force fields. The source and sink interact through color-charged fluxes that are screened in the bulk of space but form thin tubes in the form of color-charged vortex-lines so that the strength of the force remains constant over arbitrary distance. This results in a linear dependence relation for the potential energy with regard to the separation distance, which is similar to the described magnetic monopole and anti-monopole pair immersed in a type-II superconductor. In this situation, the magnetic fluxes mediating the interacting monopoles are not governed by the Maxwell equations, which would otherwise give rise to an inverse-square-power law type of decay of the forces and lead to non-confinement, but rather by the Ginzburg–Landau equations or the static Abelian Higgs model in the temporal gauge, which produce narrowly distributed vortex-lines, known as the Abrikosov vortices [1] or the Nielsen–Olesen strings [416], leading to a linear confinement result, as given in (8.4.3).

Inspired by the described monopole confinement in a type-II superconductor, in the 1970s, Mandelstam [381, 382], Nambu [406], and 't Hooft [534, 536] proposed that the ground state of quantum-chromodynamics (QCD) is a condensate of chromomagnetic (color-charged) monopoles, causing the chromoelectric fluxes between quarks to be squeezed into narrowly formed tubes or vortex-lines, similar to the electron condensation in the bulk of a superconductor, in the form of the Cooper pairs, resulting in the formation of color-charged flux-tubes or vortex-lines. These mediate the interaction between quarks, following a non-Abelian version of the Meissner effect (*dual Meissner effect*), which is responsible for the screening of chromoelectric fluxes [25, 338, 339, 506, 507]. For some mathematical constructions of such color-charged multiple vortices using techniques of nonlinear partial differential equations and other related studies, see [127, 359, 362].

Thus, we have seen that the notion of magnetic monopoles is not only a theoretical construction as a result of the electromagnetic duality but also supplies as a useful "*theoretical phenomenon*" that provides a crucial hint to a hopeful solution of one of the greatest puzzles of modern physics: quark confinement.

Exercises

1. Establish the relation (8.1.4).
2. Consider the domain wall energy functional

$$E(u) = \int_{-\infty}^{\infty} \left(\frac{1}{2}(u')^2 + \frac{\lambda}{8}(u^2 - 1)^2 \right) \mathrm{d}x, \quad \lambda > 0, \tag{8.E.1}$$

where u is a real-valued function satisfying the finite-energy boundary condition

$$u(\pm\infty) = \pm 1. \tag{8.E.2}$$

(a) Derive the Euler–Lagrange equation of (8.E.1).

(b) Derive the BPS equation of the model.

(c) Prove that the Euler–Lagrange equation and the BPS equation are equivalent for finite-energy solutions.

(d) Show that the energy of a solution of the Euler–Lagrange equation of the model is given by

$$E = \frac{2}{3}\sqrt{\lambda}. \tag{8.E.3}$$

(e) Find all finite-energy solutions of the Euler–Lagrange equation of the model.

(f) For a solution u obtained in (e), plot the energy density

$$\mathcal{E}(x) = \left(\frac{1}{2}(u')^2 + \frac{\lambda}{8}(u^2 - 1)^2 \right)(x), \quad -\infty < x < \infty \tag{8.E.4}$$

and explain your finding.

(g) If a finite-energy solution obtained in (e) is denoted by u_λ, find the limits

$$\lim_{\lambda\to\infty} u_\lambda, \quad \lim_{\lambda\to\infty} u'_\lambda. \tag{8.E.5}$$

(h) Can you solve (g) without using the explicit solution formula obtained in (e)?

3. Consider the details of the proof of the $\overline{\partial}$-Poincaré lemma by furnishing the following steps in solving (8.2.24).

(a) Show that the equation

$$\Delta v = \partial_1 g - \partial_2 f \tag{8.E.6}$$

has a local solution for any smooth functions f, g.

(b) Show that, with v, f, g given in (a), the system (8.2.23), namely,

$$\partial_1 u = f + \partial_2 v, \quad \partial_2 u = g - \partial_1 v, \tag{8.E.7}$$

has a solution over any small disk.

4. Consider a one-dimensional reduction for the Hamiltonian density (8.1.1) with

$$\phi = f = \text{real}, \quad A_1 = 0, \quad A_2 = A = \text{real}, \tag{8.E.8}$$

such that f and A are functions of $x^1 = x$ only.

(a) Show that (8.1.1) becomes

$$\mathcal{H} = \frac{1}{2}(A')^2 + \frac{1}{2}(f')^2 + \frac{1}{2}A^2 f^2 + \frac{\lambda}{8}(f^2 - 1)^2. \tag{8.E.9}$$

(b) Derive the Euler–Lagrange equations of (8.E.9) and obtain the corresponding boundary conditions for a finite-energy solution.

(c) Show that the equations obtained in (b) are the equations (8.1.2)–(8.1.3) when the Higgs and gauge fields are given by the one-dimensional ansatz (8.E.8).

(d) Show that finite energy of a solution, (f, A), of the equations implies $A \equiv 0$. Thus, all finite-energy solutions can be obtained explicitly.

5. Consider (8.2.36) over a boundary domain in $\mathbb{R}^2$ containing all vortex points and show that the equation has a unique solution under the homogeneous boundary condition.

6. For (8.2.36), consider the background function

$$u_0 = -\sum_{s=1}^{k} n_s \ln(1 + |z - z_s|^{-2}). \tag{8.E.10}$$

(a) Establish that

$$\Delta u_0 = 4\pi \sum_{s=1}^{k} n_s \delta(z - z_s) - g_0(z), \quad g_0(z) = 4\sum_{s=1}^{k} \frac{n_s}{(1 + |z - z_s|^2)^2}. \tag{8.E.11}$$

(b) Calculate to obtain the result

$$\int_{\mathbb{R}^2} \Delta u_0 \, \mathrm{d}x = 0. \tag{8.E.12}$$

(c) Rewrite the solution of (8.2.36) subject to (8.2.37) as $u = u_0 + v$, where v is a smooth function, to obtain (8.2.38).

7. (The Tong–Wong model for magnetic vortices with impurities [547]). Consider the energy density

$$\begin{aligned}\mathcal{E}(q, p, \hat{A}, \tilde{A}) =& \frac{1}{2}\hat{F}_{12}^2 + \frac{1}{2}\tilde{F}_{12}^2 + |D_i q|^2 + |D_i p|^2 \\ &+ \frac{1}{2}(|q|^2 - \zeta)^2 + \frac{1}{2}(-|q|^2 + |p|^2 - \tilde{\zeta})^2,\end{aligned} \tag{8.E.13}$$

over $\mathbb{R}^2$, where q, p are complex-valued scalar fields, $\hat{A} = (\hat{A}_i), \tilde{A} = (\tilde{A}_i)$ $(i = 1, 2)$ are real-valued gauge fields, $\zeta, \tilde{\zeta} > 0$ are constants, and

$$D_i q = \partial_i q - \mathrm{i}(\hat{A}_i - \tilde{A}_i)q, \quad D_i p = \partial_i p - \mathrm{i}\tilde{A}_i p, \quad i = 1, 2, \tag{8.E.14}$$

are gauge-covariant derivatives.

(a) Derive the Euler–Lagrange equations of (8.E.13).

(b) Derive a virial theorem.

(c) Determine the gauge transformations of the model.

(d) Let $(q, p, \hat{A}, \tilde{A})$ be a finite-energy solution of the Euler–Lagrange equations of (8.E.13) over $\mathbb{R}^2$. Show that if q, p are nowhere vanishing then the solution is gauge-equivalent to the trivial one,

$$|q|^2 = \zeta, \quad |p|^2 = \zeta + \tilde{\zeta}, \quad \hat{A} = 0, \quad \tilde{A} = 0. \tag{8.E.15}$$

(e) For a finite-energy solution of the Euler–Lagrange equations of (8.E.13), show that there are integers $\hat{N}, \tilde{N}$ such that

$$\frac{1}{2\pi}\int_{2} \hat{F}_{12} \, \mathrm{d}x = \hat{N}, \quad \frac{1}{2\pi}\int_{\mathbb{R}^2} \tilde{F}_{12} \, \mathrm{d}x = \tilde{N}. \tag{8.E.16}$$

Determine how $\hat{N}, \tilde{N}$ are related to the winding numbers of q, p near infinity of $\mathbb{R}^2$.

(f) Show that the model possesses the following BPS reduction

$$D_1 q \pm \mathrm{i} D_2 q = 0, \tag{8.E.17}$$

$$D_1 p \pm \mathrm{i} D_2 p = 0, \tag{8.E.18}$$

$$\hat{F}_{12} \pm (|q|^2 - \zeta) = 0, \tag{8.E.19}$$

$$\tilde{F}_{12} \pm (-|q|^2 + |p|^2 - \tilde{\zeta}) = 0. \tag{8.E.20}$$

(g) Show that the zero sets of q, p are finite and each zero has an integer multiplicity.

(h) Let the zero sets of q, p be denoted by $\mathscr{Z}(q, p)$ in which repeated appearances of points account for multiplicities. Show that the subsitutions $u = \ln|q|^2, v = \ln|p|^2$ recast the BPS equations in (f) into the following governing nonlinear elliptic equations

$$\Delta u = 4\mathrm{e}^u - 2\mathrm{e}^v - 2(\zeta - \tilde{\zeta}) + 4\pi \sum_{z \in \mathscr{Z}(q)} \delta_z(x), \tag{8.E.21}$$

$$\Delta v = -2\mathrm{e}^u + 2\mathrm{e}^v - 2\tilde{\zeta} + 4\pi \sum_{z \in \mathscr{Z}(p)} \delta_z(x). \tag{8.E.22}$$

(i) Show that u, v approach their asymptotic values at infinity exponentially fast.

(See [274] for a subsequent mathematical development of the problem.)

8. Introduce a radial ansatz for the Tong–Wong model that gives rise to the quantized fluxes (8.E.16) and find the associated governing equations along with the induced boundary conditions within this ansatz.
9. Continuing from Exercise 8, formulate a variational energy functional for the radial governing equations derived.
10. Obtain a one-dimensional reduction for the Tong–Wong model in the same spirits as those for the classical Abelian Higgs model discussed earlier.
11. Assume that Newton's law (8.4.1) is replaced by the modified law

$$F(r) = G\frac{m_1 m_2}{r^\alpha}, \quad r > 0, \tag{8.E.23}$$

for some constant $\alpha > 0$. Estimate the range of α such that such a modified law would lead to confinement.

9

Non-Abelian gauge field equations

Chapter 5 discussed a gauge field theory with the Abelian gauge group $U(1)$. This chapter presents a study of the gauge field theory with a general non-Abelian gauge group also known as the Yang–Mills theory. It begins with a general formulation, and then discusses the Georgi–Glashow model. It next presents the 't Hooft–Polyakov monopole and the Julia–Zee dyon arising in the Georgi–Glashow model. It complements this study with an introduction to the classical Bogomol'nyi–Prasad–Sommerfield solutions. Finally, it presents the Weinberg–Salam equations that describe the unified field theory of electromagnetic and weak interaction.

9.1 Yang–Mills theory

Chapter 5 introduced the Abelian gauge field theory based on the minimal Abelian gauge group $U(1)$. Although such an Abelian theory is limited mathematically and physically as a result of its simplicity, it contains almost all the essential ingredients of a general gauge field theory. These ingredients include the notions of global and local symmetries, symmetry breaking and symmetry restoration, gauge transformation, and the need of switching on a gauge field in order to preserve local symmetry and to eliminate a massless scalar particle, called the Goldstone particle, arising in the bare theory with a broken symmetry. The non-Abelian gauge field theory, or the Yang–Mills theory, is a natural extension of the Abelian gauge field theory in which the classical $U(1)$-gauge group is replaced by a compact non-Abelian Lie group, G, usually taken to be a matrix group. In fact, in 1954, Yang and Mills [595] initially considered $G = SU(2)$. Physically, the gauge field generated from the single generator of $U(1)$ gives rise to photons or bosons mediating electromagnetic interactions and the gauge fields generated from the three generators of $SU(2)$ give rise to three

Mathematical Physics with Differential Equations. Yisong Yang, Oxford University Press.
 DOI: 10.1093/oso/9780192872616.003.0009

species of weak gauge bosons known as the charged W- and neutral Z-particles mediating weak interactions [239, 482, 568]. Analogously, the eight generators of $SU(3)$ give rise to eight species of strong gauge bosons known as gluons mediating strong interactions between elementary particles such as quarks. Thus, in view of using gauge field theory to describe fundamental interactions, nature has been rather "sparing" in choosing its mathematical sophistication. In this spirit, the gauge field theory that unifies electromagnetic and weak interactions, called the *electroweak theory*, is based on $G = SU(2) \times U(1)$, and the theory that unifies electromagnetic, weak, and strong interactions, called the *standard model*, are based on $G = SU(3) \times SU(2) \times U(1)$, giving rise to a count of a total of twelve gauge bosons for the three fundamental interactions, of course. The *grand unified theory*, known also as GUT, is a gauge field theory based on $G = SU(5)$ [229] since $SU(5)$ is the smallest simple Lie group containing the standard model Lie group $SU(3) \times SU(2) \times U(1)$. (A simple Lie group is a connected non-Abelian Lie group without nontrivial connected normal subgroups. Since gauge fields take values in the Lie algebra of the Lie group, what is actually used in non-Abelian gauge field theory with assuming the gauge group being simple is that its Lie algebra is simple. See [302, 309] for details of mathematics.) In such a formalism, the gauge bosons representing the three fundamental forces (electromagnetic, weak, and strong) are realized as appropriate components of a single $SU(5)$-Lie algebra valued gauge field, which unifies these forces in one setting. However, since $SU(5)$ has twenty-four generators, GUT now contains twelve additional gauge bosons, contributing to and participating in this "grand unified" picture. See [118, 298, 316, 526, 570, 571, 614] for development and treatment of GUT in the broad context of quantum field theory.

This section presents a short introduction to the Yang–Mills theory. For notational convenience, we shall concentrate on the specific case where the gauge group G is either the orthogonal matrix group, $O(n)$, or unitary matrix group, $U(n)$, which is sufficient for all physical applications.

Rise of non-Abelian gauge fields

Let ϕ be a scalar field over the Minkowski spacetime and take values in either $\mathbb{R}^n$ or $\mathbb{C}^n$, which is the representation space of G (with $G = O(n)$ or $U(n)$). The field ϕ is also said to be in the *fundamental representation* of G. We use $\dagger$ to denote the Hermitian transpose or Hermitian conjugate in $\mathbb{R}^n$ or $\mathbb{C}^n$. Then $|\phi|^2 = \phi^\dagger \phi$. We may start from the Lagrangian density

$$\mathcal{L} = \frac{1}{2}(\partial_\mu \phi)^\dagger (\partial^\mu \phi) - V(|\phi|^2). \tag{9.1.1}$$

It is clear that $\mathcal{L}$ is invariant under the global symmetry group G,

$$\phi \mapsto \Omega \phi, \quad \Omega \in G. \tag{9.1.2}$$

However, as in the Abelian case, if the group element Ω is replaced by a local one depending on spacetime points,

$$\Omega = \Omega(x), \tag{9.1.3}$$

the invariance of $\mathcal{L}$ is no longer valid and a modification is needed. Thus, we are again motivated to consider the derivative

$$D_\mu \phi = \partial_\mu \phi + A_\mu \phi, \tag{9.1.4}$$

where we naturally choose A_μ to be an element in the Lie algebra $\mathcal{G}$ of G which has an obvious representation over the space of ϕ. The dynamical term in $\mathcal{L}$ becomes

$$\frac{1}{2}(D_\mu \phi)^\dagger (D^\mu \phi). \tag{9.1.5}$$

Of course, the invariance of this under the local transformation

$$\phi(x) \mapsto \phi'(x) = \Omega(x)\phi(x) \tag{9.1.6}$$

is ensured if $D_\mu \phi$ transforms itself covariantly according to

$$D_\mu \phi \mapsto D'_\mu \phi' = \partial_\mu \phi' + A'_\mu \phi' = \partial_\mu(\Omega\phi) + A'_\mu \Omega \phi = \Omega(D_\mu \phi). \tag{9.1.7}$$

Comparing the results, we conclude that A_μ should obey the following rule of transformation,

$$A_\mu \mapsto A'_\mu = \Omega A_\mu \Omega^{-1} - (\partial_\mu \Omega)\Omega^{-1}. \tag{9.1.8}$$

The $U(1)$ gauge field theory presented in Chapter 5 is contained here as a special case (the Lie algebra is the imaginary axis $\mathrm{i}\mathbb{R}$). Thus, in general, the Lie algebra valued field A_μ is also a gauge field.

Field strength tensor or curvature

In order to include dynamics for the gauge field A_μ, we need to introduce invariant quadratic terms involving derivatives of A_μ. For this purpose, recall that there is a standard inner product, the *Hilbert–Schmidt inner product* [607], over the space of $n \times n$ matrices, given by

$$(A, B) = \mathrm{Tr}(A^\dagger B), \ \ A, B \in \mathcal{G}. \tag{9.1.9}$$

For any $A \in \mathcal{G}$, since A is anti-Hermitian or $A^\dagger = -A$, we see that

$$|A|^2 = (A, A) = -\mathrm{Tr}\ (A^2). \tag{9.1.10}$$

In complete analogy with the electromagnetic field in the Abelian case, we can examine the non-commutativity of the gauge-covariant derivatives to get

$$D_\mu D_\nu \phi - D_\nu D_\mu \phi = (\partial_\mu A_\nu - \partial_\nu A_\mu + [A_\mu, A_\nu])\phi, \tag{9.1.11}$$

where $[\cdot,\cdot]$ is the Lie bracket (or commutator) of $\mathcal{G}$. Hence, we are motivated to define the skew-symmetric Yang–Mills field (curvature) 2-tensor $F_{\mu\nu}$ as

$$F_{\mu\nu} = \partial_\mu A_\nu - \partial_\nu A_\mu + [A_\mu, A_\nu]. \tag{9.1.12}$$

We can see that $F_{\mu\nu}$ transforms itself according to

$$F_{\mu\nu} \mapsto F'_{\mu\nu} = \partial_\mu A'_\nu - \partial_\nu A'_\mu + [A'_\mu, A'_\nu] = \Omega F_{\mu\nu} \Omega^{-1}. \tag{9.1.13}$$

Hence, we obtain the analogous invariant term,

$$\frac{1}{4}\mathrm{Tr}\ (F_{\mu\nu} F^{\mu\nu}). \tag{9.1.14}$$

Yang–Mills theory

Hence, we arrive at the final form of our locally gauge-invariant Lagrangian action density

$$\mathcal{L} = \frac{1}{4}\text{Tr}\,(F_{\mu\nu}F^{\mu\nu}) + \frac{1}{2}(D_\mu\phi)^\dagger(D^\mu\phi) - V(|\phi|^2), \tag{9.1.15}$$

which defines a non-Abelian gauge field theory, known as the *Yang–Mills theory*.

To proceed further, we use $\{t^a\}$ ($a = 1, \ldots, \dim(G)$) to denote a basis of $\mathcal{G}$ such that

$$[t^a, t^b] = f^{abc}t^c, \quad a, b, c = 1, \ldots, \dim(G), \tag{9.1.16}$$

where f^{abc}'s are *structural constants* of G and summation convention is observed for the repeated group indices a, b, c. With this notation, the $\mathcal{G}$-valued gauge field A_μ may be rewritten as

$$A_\mu = A_\mu^a t^a, \tag{9.1.17}$$

where A_μ^a are real-valued vector fields, $a = 1, \ldots, \dim(G)$.

With the shown preparation, we may vary ϕ and A_μ^a in (9.1.15) to obtain the associated Euler–Lagrange equations:

$$D_\mu D^\mu\phi = -2V'(|\phi|^2)\phi, \tag{9.1.18}$$

$$D_\mu F^{\mu\nu} = j^\nu, \tag{9.1.19}$$

which are generically called the *Yang–Mills equations*, where

$$D_\mu A = \partial_\mu A + [A_\mu, A], \tag{9.1.20}$$

is the gauge-covariant derivative for a Lie-algebra valued quantity A and, by using the fact that t^as are anti-Hermitian, the Lie-algebra valued current density j^μ reads

$$(j^\mu)^a = -\frac{1}{2}\left(\phi^\dagger t^a[D^\mu\phi] - [D^\mu\phi]^\dagger t^a\phi\right), \quad a = 1, \ldots, \dim(G). \tag{9.1.21}$$

In the situation where the potential density V introduces a spontaneously broken symmetry, of the prototype form

$$V(|\phi|^2) = \frac{\lambda}{4}(|\phi|^2 - v^2)^2, \quad \lambda, v > 0, \tag{9.1.22}$$

the theory is called the *Yang–Mills–Higgs theory* and the associated equations of motion are the *Yang–Mills–Higgs equations.*

When the matter component (containing ϕ) is neglected, the action density becomes

$$\mathcal{L} = \frac{1}{4}\text{Tr}\,(F_{\mu\nu}F^{\mu\nu}), \tag{9.1.23}$$

which is simply called the (pure) Yang–Mills theory. The Euler–Lagrange equations of such a Lagrangian, namely,

$$D_\mu F^{\mu\nu} = 0, \tag{9.1.24}$$

are called the (pure) Yang–Mills equations, which are non-Abelian extension of the Maxwell equations in vacuum. Note that (9.1.24) may be rewritten in the following Maxwell equation form:

$$\partial_\mu F^{\mu\nu} = j^\nu, \tag{9.1.25}$$

where

$$j^\nu = -[A_\mu, F^{\mu\nu}], \tag{9.1.26}$$

is formerly the associated current density, which is self-induced from the gauge field due to the nonlinearity caused internally by the non-Abelian nature of the gauge symmetry, rather than the usual source term in the classical Maxwell electromagnetism prescribed externally. In other words, "electromagnetism" may be self-sustained in a non-Abelian theory.

Self-dual Yang–Mills equations

To elaborate a little more on the Yang–Mills equations (9.1.24), we note that the covariant derivative (9.1.20) follows the *Jacobi identity*

$$[D_\alpha, [D_\beta, D_\gamma]] + [D_\beta, [D_\gamma, D_\alpha]] + [D_\gamma, [D_\alpha, D_\beta]] = 0. \tag{9.1.27}$$

Combining this with the Jacobi identity valid for the Lie-algebra valued quantities

$$[A, [B, C]] + [B, [C, A]] + [C, [A, B]] = 0, \quad A, B, C \in \mathcal{G}, \tag{9.1.28}$$

we can verify the *Bianchi identity*

$$D_\alpha F_{\beta\gamma} + D_\beta F_{\gamma\alpha} + D_\gamma F_{\alpha\beta} = 0, \tag{9.1.29}$$

for the Yang–Mills field $F_{\mu\nu}$ defined in (9.1.12). Now, define the *Hodge dual* tensor

$$(*F)^{\mu\nu} = \frac{1}{2}\epsilon^{\mu\nu\alpha\beta} F_{\alpha\beta}. \tag{9.1.30}$$

Then, the Bianchi identity (9.1.29) assumes the suppressed form

$$D_\mu (*F)^{\mu\nu} = 0. \tag{9.1.31}$$

In view of (9.1.24) and (9.1.31), we obtain a first integral of (9.1.24):

$$(*F)^{\mu\nu} = c\, F^{\mu\nu}, \tag{9.1.32}$$

where $c \neq 0$ is a constant, so that any solution of (9.1.32) fulfills (9.1.24) as well. In order to determine the constant c, note that the signature of the Minkowski metric leads to the anti-reflectivity relation

$$(*(*F))^{\mu\nu} = -F^{\mu\nu}. \tag{9.1.33}$$

Applying the *Hodge star map* $*$ to (9.1.32) and using (9.1.33), we get $c^2 = -1$. Hence, (9.1.32) becomes

$$(*F)^{\mu\nu} = \pm \mathrm{i}\, F^{\mu\nu}, \tag{9.1.34}$$

known as the *self-dual* and *anti-self-dual* Yang–Mills equations in the Minskowski spacetime, which makes sense only in suitable complex non-compact situations, for example, when $G = SL(n, \mathbb{C})$. See [407].

9.2 Georgi–Glashow model

After a brief introduction to the general non-Abelian gauge field theory and its governing equations, this section considers the theory with the simplest non-Abelian Lie group, $G = SO(3)$, known as the Georgi–Glashow model [228]. We concentrate on the partial differential equation aspect of this model and study the associated algebraic and topological characteristics as well as some direct physical properties of the model.

Lie algebra structure

Recall that $SO(3)$ has a set of generators $\{t^a\}$ $(a = 1, 2, 3)$ satisfying

$$[t^a, t^b] = \epsilon^{abc} t^c. \tag{9.2.1}$$

Consequently, for two $so(3)$-valued quantities $A = A^a t^a$ and $B = B^a t^a$, we have the commutator relation

$$[A, B] = \epsilon^{abc} A^a B^b t^c. \tag{9.2.2}$$

For convenience, we may view A and B as two 3-vectors, $\mathbf{A} = (A^a)$ and $\mathbf{B} = (B^a)$. Then, by (9.2.2), $[A, B]$ corresponds to the vector cross-product, $\mathbf{A} \times \mathbf{B}$. With these in mind, we make the following introduction to the $SO(3)$, or $SU(2)$, since $SO(3)$ and $SU(2)$ have identical Lie algebras, Yang–Mills–Higgs theory, specifically known as the Georgi–Glashow model [240], which serves as a foundation for the theory of weak interaction [228].

Georgi–Glashow model

Following the described formalism, let $\mathbf{A}_\mu = (A^a_\mu)$ $(\mu = 0, 1, 2, 3)$ and $\phi = (\phi^a)$ $(a = 1, 2, 3)$ be a gauge and matter Higgs fields, respectively, interacting through the normalized action density [228, 240]

$$\mathcal{L} = -\frac{1}{4}\mathbf{F}^{\mu\nu} \cdot \mathbf{F}_{\mu\nu} + \frac{1}{2}D^\mu\phi \cdot D_\mu\phi - \frac{\lambda}{4}(|\phi|^2 - 1)^2, \tag{9.2.3}$$

where the field strength tensor $\mathbf{F}_{\mu\nu}$ is defined by

$$\mathbf{F}_{\mu\nu} = \partial_\mu \mathbf{A}_\nu - \partial_\nu \mathbf{A}_\mu - e\mathbf{A}_\mu \times \mathbf{A}_\nu, \tag{9.2.4}$$

and the gauge-covariant derivative D_μ is defined by

$$D_\mu\phi = \partial_\mu\phi - e\mathbf{A}_\mu \times \phi, \tag{9.2.5}$$

where $e > 0$ is a coupling constant. Thus, we have

$$D_\gamma \mathbf{F}_{\alpha\beta} = \partial_\gamma \mathbf{F}_{\alpha\beta} - e\mathbf{A}_\gamma \times \mathbf{F}_{\alpha\beta}, \tag{9.2.6}$$

which results in the Bianchi identity

$$D_\alpha \mathbf{F}_{\beta\gamma} + D_\beta \mathbf{F}_{\gamma\alpha} + D_\gamma \mathbf{F}_{\alpha\beta} = \mathbf{0}. \tag{9.2.7}$$

The Hamiltonian energy density of (9.2.3) assumes the form

$$\mathcal{H} = \frac{1}{2}|\mathbf{F}_{0i}|^2 + \frac{1}{4}|\mathbf{F}_{ij}|^2 + \frac{1}{2}|D_0\phi|^2 + \frac{1}{2}|D_i\phi|^2 + \frac{\lambda}{4}(|\phi|^2 - 1)^2. \tag{9.2.8}$$

Furthermore, the equations of motion of (9.2.3), or the Georgi–Glashow equations, can be derived as

$$D_\mu \mathbf{F}^{\mu\nu} = e\phi \times D^\nu \phi, \tag{9.2.9}$$

$$D_\mu D^\mu \phi = -\lambda\phi(|\phi|^2 - 1). \tag{9.2.10}$$

Induced electromagnetism

In analogy with the theory of electromagnetism, we may define *vector-valued* electric and magnetic fields by

$$\mathbf{E}_i = \mathbf{F}_{0i}, \quad \mathbf{B}_i = \frac{1}{2}\,\epsilon_{ijk}\mathbf{F}_{jk}, \quad i, j, k = 1, 2, 3, \tag{9.2.11}$$

which mediate weak interaction. On the other hand, based on consideration on interactions, there are several proposals for the definition of conventional real-valued electric and magnetic fields, among which the simplest one may be that of Bogomol'nyi [67] and Faddeev [202] given by

$$f_{\mu\nu} = \phi \cdot \mathbf{F}_{\mu\nu}, \tag{9.2.12}$$

and the more popular one by 't Hooft [532, 535] given by

$$f_{\mu\nu} = \frac{1}{|\phi|}\phi \cdot \mathbf{F}_{\mu\nu} + \frac{1}{e|\phi|^3}\phi \cdot (D_\mu\phi \times D_\nu\phi), \tag{9.2.13}$$

also called the *'t Hooft electromagnetic tensor*. See [19, 240, 477] for detailed discussion.

Electric and magnetic charges

Following (9.2.12) or (9.2.13), we consider the real-valued electric and magnetic fields $\mathbf{e} = (e_i)$ and $\mathbf{b} = (b_i)$ defined by

$$e_i = f_{0i}, \quad b_i = \frac{1}{2}\,\epsilon_{ijk}f_{jk}. \tag{9.2.14}$$

Then the electric and magnetic charges, q, g, associated with (9.2.14) are

$$q = \lim_{R\to\infty}\int_{|\mathbf{x}|=R} \mathbf{e}\cdot \mathrm{d}\mathbf{S}, \tag{9.2.15}$$

$$g = \lim_{R\to\infty}\int_{|\mathbf{x}|=R} \mathbf{b}\cdot \mathrm{d}\mathbf{S}. \tag{9.2.16}$$

In view of the finite-energy condition

$$E = \int_{\mathbb{R}^3} \mathcal{H}\,\mathrm{d}\mathbf{x} < \infty, \tag{9.2.17}$$

where $\mathcal{H}$ is given by (9.2.8), we see that $|D_\mu\phi|^2$ vanishes at infinity sufficiently rapidly, which implies that (9.2.12) and (9.2.13) give rise to the same electric and magnetic charges defined in (9.2.15) and (9.2.16), respectively. Consequently, we may use (9.2.12) for simplicity, without loss of generality.

To compute the electric charge q, we first note that there holds

$$\begin{aligned} \phi\cdot D_i\mathbf{E}_i &= \phi\cdot D_i\mathbf{F}_{0i} \\ &= \phi\cdot D_i\mathbf{F}^{i0} \\ &= \phi\cdot(\phi\times D^0\phi) = 0, \end{aligned} \tag{9.2.18}$$

in view of (9.2.9). Thus, using the divergence theorem in (9.2.15) and (9.2.18), we obtain

$$\begin{aligned} q &= \int_{\mathbb{R}^3} \partial_i e_i\,\mathrm{d}x \\ &= \int_{\mathbb{R}^3} \partial_i(\phi\cdot\mathbf{F}_{0i})\,\mathrm{d}x \\ &= \int_{\mathbb{R}^3} \{(\partial_i\phi)\cdot\mathbf{F}_{0i} + \phi\cdot(e\mathbf{A}_i\times\mathbf{F}_{0i}) + \phi\cdot(\partial_i\mathbf{F}_{0i} - e\mathbf{A}_i\times\mathbf{F}_{0i})\}\,\mathrm{d}x \\ &= \int_{\mathbb{R}^3} (\partial_i\phi - e\mathbf{A}_i\times\phi)\cdot\mathbf{F}_{0i}\,\mathrm{d}x \\ &= \int_{\mathbb{R}^3} D_i\phi\cdot\mathbf{E}_i\,\mathrm{d}x. \end{aligned} \tag{9.2.19}$$

On the other hand, in view of the Bianchi identity (9.2.7), we have

$$D_i\mathbf{B}_i = \frac{1}{2}\epsilon_{ijk}D_i\mathbf{F}_{jk} = \mathbf{0}. \tag{9.2.20}$$

Hence, using the divergence theorem in (9.2.16) and (9.2.20), we find

$$\begin{aligned} g &= \int_{\mathbb{R}^3} \partial_i b_i\,\mathrm{d}x \\ &= \int_{\mathbb{R}^3} \partial_i(\phi\cdot\mathbf{B}_i)\,\mathrm{d}x \\ &= \int_{\mathbb{R}^3} \{(\partial_i\phi)\cdot\mathbf{B}_i + \phi\cdot(e\mathbf{A}_i\times\mathbf{B}_i) + \phi\cdot(\partial_i\mathbf{B}_i - e\mathbf{A}_i\times\mathbf{B}_i)\}\,\mathrm{d}x \\ &= \int_{\mathbb{R}^3} (\partial_i\phi - e\mathbf{A}_i\times\phi)\cdot\mathbf{B}_i\,\mathrm{d}x \\ &= \int_{\mathbb{R}^3} D_i\phi\cdot\mathbf{B}_i\,\mathrm{d}x. \end{aligned} \tag{9.2.21}$$

These elegant charge formulas may be used to obtain an energy lower bound estimate for a finite-energy solution of (9.2.9)–(9.2.10) as done in [67, 240].

An energy lower bound

From (9.2.8), we have

$$\begin{aligned} E &= \frac{1}{2}\int_{\mathbb{R}^3}\left(|\mathbf{E}_i|^2 + |\mathbf{B}_i|^2 + |D_0\phi|^2 + |D_i\phi|^2 + \frac{\lambda}{2}(|\phi|^2-1)^2\right)\,\mathrm{d}\mathbf{x} \\ &\geq \frac{1}{2}\int_{\mathbb{R}^3}\left(|\mathbf{E}_i|^2 + |\mathbf{B}_i|^2 + |D_i\phi|^2\right)\,\mathrm{d}\mathbf{x} \\ &= \frac{1}{2}\int_{\mathbb{R}^3}\left(|\mathbf{E}_i - \sin\theta D_i\phi|^2 + |\mathbf{B}_i - \cos\theta D_i\phi|^2\right)\,\mathrm{d}\mathbf{x} + q\sin\theta + g\cos\theta \\ &\geq q\sin\theta + g\cos\theta, \end{aligned} \tag{9.2.22}$$

where θ is an undetermined parameter. Maximizing the right-hand side of (9.2.22), we arrive at the estimate

$$E \geq \sqrt{q^2+g^2}. \tag{9.2.23}$$

Monopole

In particular, for static solutions in the temporal gauge, $\mathbf{A}_0 = \mathbf{0}$, the electric charge (9.2.19) vanishes, $q = 0$, and the energy lower bound (9.2.23) becomes

$$E \geq |g|, \tag{9.2.24}$$

which is the classical result for magnetic monopoles in the Georgi–Glashow model. In particular, if $\lambda = 0$, the lower bound in (9.2.24) is attained when

$$\mathbf{B}_i = \pm D_i\phi, \quad g = \pm|g|. \tag{9.2.25}$$

This is the well-known Bogomol'nyi monopole equation [67, 240, 310, 459] whose solutions are presented later. Here we examine that the equation (9.2.25) implies (9.2.9)–(9.2.10), which are

$$D_i\mathbf{F}_{ij} = -e\phi\times D_j\phi, \tag{9.2.26}$$

$$D_iD_i\phi = \mathbf{0}. \tag{9.2.27}$$

In fact, applying D_i to (9.2.25), we see that (9.2.27) follows from the Bianchi identity (9.2.20). Furthermore, using the Jacobi identity for vector cross product,

$$\mathbf{A}\times(\mathbf{B}\times\mathbf{C}) + \mathbf{B}\times(\mathbf{C}\times\mathbf{A}) + \mathbf{C}\times(\mathbf{A}\times\mathbf{B}) = \mathbf{0}, \tag{9.2.28}$$

we can get the commutator relation

$$D_\mu D_\nu\phi - D_\nu D_\mu\phi = -e\mathbf{F}_{\mu\nu}\times\phi. \tag{9.2.29}$$

On the other hand, by (9.2.11), we have

$$D_j \mathbf{F}_{ji} = -\epsilon_{ijk} D_j \mathbf{B}_k. \tag{9.2.30}$$

Thus, inserting (9.2.25) into the right-hand side of (9.2.30), we get

$$\begin{aligned} D_j \mathbf{F}_{ji} &= \mp \epsilon_{ijk} D_j D_k \phi \\ &= \mp \frac{1}{2} \epsilon_{ijk} (D_j D_k \phi - D_k D_j \phi) \\ &= \pm \frac{e}{2} \epsilon_{ijk} \mathbf{F}_{jk} \times \phi \quad \text{(using (9.2.29))} \\ &= \pm e \mathbf{B}_i \times \phi \quad \text{(using (9.2.11))} \\ &= \pm e(\pm D_i \phi) \times \phi \quad \text{(using (9.2.25) again)} \\ &= -e\phi \times D_i \phi, \end{aligned} \tag{9.2.31}$$

which establishes (9.2.26) as well.

Topology

Following the treatment in [147, 240] we explore the topological characterization of a finite-energy solution $(\phi, \mathbf{A}_\mu)$ to the Georgi–Glashow equations (9.2.9)–(9.2.10). For this purpose, we first note that finite-energy solution implies $D_\mu \phi$ and $|\phi|^2 - 1$ vanish at infinity exponentially fast [240] as a consequence of the spontaneously broken symmetry so that

$$D_\mu \phi = \partial_\mu \phi - e\mathbf{A}_\mu \times \phi = 0, \tag{9.2.32}$$

$$\phi \cdot \phi = 1, \tag{9.2.33}$$

near infinity of $\mathbb{R}^3$, modulo exponentially small errors terms that are neglected without harming the subsequent computation, near infinity. Taking the cross-product of (9.2.32) with ϕ, we get

$$\phi \times \partial_\mu \phi - e\phi \times (\mathbf{A}_\mu \times \phi) = 0. \tag{9.2.34}$$

Thus, using the vector identity

$$\mathbf{A} \times (\mathbf{B} \times \mathbf{C}) = (\mathbf{A} \cdot \mathbf{C})\mathbf{B} - (\mathbf{A} \cdot \mathbf{B})\mathbf{C}, \tag{9.2.35}$$

and (9.2.33), we obtain from (9.2.34) the resolution of $\mathbf{A}_\mu$ as follows:

$$\mathbf{A}_\mu = \frac{1}{e} (\phi \times \partial_\mu \phi) + (\phi \cdot \mathbf{A}_\mu)\phi. \tag{9.2.36}$$

At the first sight, the second term on the right-hand side of (9.2.36) depends on $\mathbf{A}_\mu$. However, since this term is in the direction of ϕ, which is actually allowed to be arbitrary otherwise due to the form of the equation (9.2.34) or (9.2.32), we see that (9.2.36) may be generalized into the form

$$\mathbf{A}_\mu = \frac{1}{e} (\phi \times \partial_\mu \phi) + A_\mu \phi, \tag{9.2.37}$$

where A_μ is an arbitrary real-valued vector field that may well be assumed to be defined globally for convenience. Now inserting (9.2.37) into (9.2.4) and using (9.2.33) such that $\phi \cdot \partial_\mu \phi = 0$ and (9.2.35) to simplify the expression, we have

$$\begin{aligned}\mathbf{F}_{\mu\nu} &= \frac{2}{e}(\partial_\mu \phi \times \partial_\nu \phi) - \frac{1}{e}(\phi \times \partial_\mu \phi) \times (\phi \times \partial_\nu \phi) + F_{\mu\nu}\phi \\ &= \frac{2}{e}(\partial_\mu \phi \times \partial_\nu \phi) - \frac{1}{e}(\phi \cdot [\partial_\mu \phi \times \partial_\nu \phi])\phi + F_{\mu\nu}\phi, \end{aligned} \tag{9.2.38}$$

where $F_{\mu\nu} = \partial_\mu A_\nu - \partial_\nu A_\mu$ as usual. Moreover, since $\partial_\mu \phi \times \partial_\nu \phi$ is parallel to ϕ, we have

$$\partial_\mu \phi \times \partial_\nu \phi = (\phi \cdot [\partial_\mu \phi \times \partial_\nu \phi])\phi. \tag{9.2.39}$$

In view of (9.2.38) and (9.2.39), we finally get the following much simplified result:

$$\mathbf{F}_{\mu\nu} = \frac{1}{e}(\phi \cdot [\partial_\mu \phi \times \partial_\nu \phi])\phi + F_{\mu\nu}\phi, \tag{9.2.40}$$

valid near infinity. Therefore, by virtue of (9.2.12), (9.2.14), and (9.2.16), we have

$$eg = a_1 + a_2, \tag{9.2.41}$$

where

$$a_1 = \lim_{R\to\infty} \int_{|\mathbf{x}|=R} \frac{1}{2}\, \epsilon_{ijk}\, \phi \cdot (\partial_j \phi \times \partial_k \phi)\, \mathrm{d}S^i = 4\pi N, \tag{9.2.42}$$

with N being the *Brouwer degree* of the map induced from ϕ, which maps the sphere S^2_∞ surrounding the infinity of $\mathbb{R}^3$ onto the unit sphere S^2 [19], and

$$\begin{aligned} a_2 &= \lim_{R\to\infty} \int_{|\mathbf{x}|=R} \frac{1}{2}\, \epsilon_{ijk}\, F_{jk}\, \mathrm{d}S^i \\ &= \int_{\mathbb{R}^3} \partial_i \left(\frac{1}{2}\, \epsilon_{ijk} F_{jk}\right) \mathrm{d}\mathbf{x} = 0, \end{aligned} \tag{9.2.43}$$

by the divergence theorem and the Bianchi identity. Inserting (9.2.42) and (9.2.43) into (9.2.41), we arrive at

$$g = \left(\frac{4\pi}{e}\right) N, \tag{9.2.44}$$

which says that the total magnetic charge g is quantized as an integer multiple of the unit quantity $\frac{4\pi}{e}$.

9.3 't Hooft–Polyakov monopole and Julia–Zee dyon

Chapter 3 discussed monopoles and dyons in terms of the Maxwell equations for electromagnetism, which is a theory of Abelian gauge fields. In fact, it is more natural for monopoles and dyons to exist in non-Abelian gauge field-theoretical models because nonvanishing commutators themselves are now present as electric and magnetic source terms. In other words, non-Abelian monopoles and dyons are self-induced and inevitable. This section presents the simplest non-Abelian dyons, known as the Julia–Zee dyons [315], which also contain as special solutions the 't Hooft–Polyakov monopoles [454, 532]. Specifically, we are interested in static solutions of (9.2.9)–(9.2.10). This is a difficult partial differential equation problem in general. Here we can only consider radially symmetric solutions so that the equations are reduced into ordinary differential equations and the problem becomes tractable.

Radial equations

Set $r = |\mathbf{x}|$. Following Julia and Zee [315], consider radially symmetric solutions of (9.2.9)–(9.2.10) given by the expressions

$$A_0^a = -\frac{x^a}{er^2} J(r), \tag{9.3.1}$$

$$A_i^a = -\epsilon^{abi} \frac{x^b}{er^2}(K(r) - 1), \tag{9.3.2}$$

$$\phi^a = -\frac{x^a}{er^2} H(r), \tag{9.3.3}$$

where $a, b, c = 1, 2, 3$. Inserting (9.3.1)–(9.3.3) into (9.2.9)–(9.2.10) and using prime to denote differentiation with respect to the radial variable r, we have

$$r^2 J'' = 2JK^2, \tag{9.3.4}$$

$$r^2 H'' = 2HK^2 - \lambda r^2 H\left(1 - \frac{1}{e^2 r^2} H^2\right), \tag{9.3.5}$$

$$r^2 K'' = K(K^2 - J^2 + H^2 - 1). \tag{9.3.6}$$

We need to specify boundary conditions for these equations. Setting

$$u(r) = \frac{H(r)}{er}, \quad v(r) = \frac{J(r)}{er}, \quad K(r) = K(r), \tag{9.3.7}$$

we see from (9.3.1)–(9.3.3) and regularity requirement that u, v, K must satisfy

$$\lim_{r \to 0} (u(r), v(r), K(r)) = (0, 0, 1). \tag{9.3.8}$$

Correspondingly, the equations (9.3.4)–(9.3.6) become

$$v'' + \frac{2}{r} v' = \frac{2}{r^2} K^2 v, \tag{9.3.9}$$

$$u'' + \frac{2}{r} u' = \frac{2}{r^2} K^2 u + \lambda(u^2 - 1)u, \tag{9.3.10}$$

$$K'' = \frac{1}{r^2}(K^2 - 1)K + e^2(u^2 - v^2)K. \tag{9.3.11}$$

Radial action and energy

In terms of (9.3.1)–(9.3.3), with (9.3.7), the action and energy densities of the Georgi–Glashow model are

$$\mathcal{L}(u,v,K) = -\frac{1}{e^2r^2}(K')^2 - \frac{1}{2}(u')^2 + \frac{1}{2}(v')^2 - \frac{1}{2e^2r^4}(K^2-1)^2 \\ - \frac{1}{r^2}K^2(u^2-v^2) - \frac{\lambda}{4}(u^2-1)^2, \tag{9.3.12}$$

$$\mathcal{H}(u,v,K) = \frac{1}{e^2r^2}(K')^2 + \frac{1}{2}(u')^2 + \frac{1}{2}(v')^2 + \frac{1}{2e^2r^4}(K^2-1)^2 \\ + \frac{1}{r^2}K^2(u^2+v^2) + \frac{\lambda}{4}(u^2-1)^2, \tag{9.3.13}$$

respectively. The equations (9.3.9)–(9.3.11) are the Euler–Lagrange equations of the action

$$L(u,v,K) = 4\pi \int_0^\infty \mathcal{L}(u,v,K)\, r^2 \mathrm{d}r. \tag{9.3.14}$$

Furthermore, the total energy is

$$E(u,v,K) = 4\pi \int_0^\infty \mathcal{H}(u,v,K)\, r^2 \mathrm{d}r. \tag{9.3.15}$$

Hence, finite-energy condition implies that $u(r) \to 1$ and $K(r) \to 0$ as $r \to \infty$. Besides, the structure of (9.3.13) indicates that $v(r) \to$ some constant C_0 as $r \to \infty$. However, C_0 cannot be determined completely. We record these results as follows,

$$\lim_{r\to\infty} (u(r), v(r), K(r)) = (1, C_0, 0). \tag{9.3.16}$$

The challenge here is that, in order to obtain a regular finite-energy solution to the radial version of the Georgi–Glashow equations, (9.3.9)–(9.3.11), we need to find a critical point of the *indefinite action functional* (9.3.14), instead of the positive-definite energy functional (9.3.15). Such a task is carried out in [485] by Schechter and Weder adopting a constrained optimization method developed in [57]. For a simplified approach to this problem, see also [606]. The existence theorem obtained in [485, 606] states that, for any $0 < C_0 < 1$, the equations (9.3.9)–(9.3.11) have a finite-energy solution (u, v, K) that satisfies the boundary conditions (9.3.8) and (9.3.16), and $u, v, K > 0$. The subsequent discussion describes such a solution.

Electromagnetism and occurrence of dyon

We now follow [315] to compute the electric and magnetic charges of a radially symmetric solution. First, insert (9.3.1)–(9.3.3), with (9.3.7), into (9.2.13) to get the electric field $\mathbf{e} = (e_i)$ to be

$$e_i = f_{0i} = \frac{x^i}{r} v'. \tag{9.3.17}$$

Thus, by the divergence theorem we have

$$\begin{aligned} q &= \lim_{R\to\infty} \int_{|\mathbf{x}|=R} \mathbf{e}\cdot \mathrm{d}\mathbf{S} \\ &= \int_{\mathbb{R}^3} \partial_i e_i \,\mathrm{d}\mathbf{x} \\ &= 4\pi \int_0^\infty \left(v'' + \frac{2}{r} v'\right) r^2 \,\mathrm{d}r \\ &= 8\pi \int_0^\infty K^2(r) v(r)\,\mathrm{d}r, \end{aligned} \tag{9.3.18}$$

in view of (9.3.9). On the other hand, in order to assess the magnetic charge, we consider the representation (9.2.12), since (9.2.12) and (9.2.13) only differ in exponentially vanishing error terms that do not affect our flux computation near infinity. With such a choice, we obtain, similar to (9.3.17), the expression

$$b_i = \frac{x^i}{er^3} u(r)(1 - K^2(r)). \tag{9.3.19}$$

Consequently, we can compute the associated magnetic charge g as follows:

$$\begin{aligned} g &= \lim_{R\to\infty} \int_{|\mathbf{x}|=R} \mathbf{b}\cdot \mathrm{d}\mathbf{S} \\ &= \frac{4\pi}{e}, \end{aligned} \tag{9.3.20}$$

using the boundary condition (9.3.16). Comparing with (9.2.44), we see that the solution obtained is of unit topological charge, $N = 1$.

Besides, the asymptotic properties of u, v, K near infinity and (9.3.17) and (9.3.19) indicate that the electric field $\mathbf{e}$ and magnetic field $\mathbf{b}$ both obey the inverse-square law, as in the Maxwell electromagnetism situation.

The static radially symmetric particle-like solution to the Georgi–Glashow equations described here which is both electrically and magnetically charged, $q > 0$ and $g > 0$, is called a *dyon*, or a *Julia–Zee dyon*, first studied in [315], which extends the 't Hooft–Polyakov monopole construction work [454, 532], which we now describe.

Monopole

From (9.3.17) and (9.3.7), we see that, if $J = 0$ or $v = 0$, then $\mathbf{e} = \mathbf{0}$ and $q = 0$ so that there is no electricity. In other words, the solution is *electrically neutral.* Now, the equations of motion (9.3.4)–(9.3.6) are reduced into the form

$$r^2 H'' = 2HK^2 - \lambda r^2 H\left(1 - \frac{1}{e^2 r^2} H^2\right), \tag{9.3.21}$$

$$r^2 K'' = K(K^2 + H^2 - 1), \tag{9.3.22}$$

subject to the boundary conditions (9.3.8) and (9.3.16) with $C_0 = 0$. The solutions of this problem cannot be obtained explicitly for general $\lambda > 0$, but an existence theorem has been established by using functional analysis [554]. See also [93, 615]. Such a particle-like solution, carrying a nontrivial magnetic charge given by (9.3.20) and a unit topological charge, with a locally concentrated energy profile, presents a magnetic monopole, known as the 't Hooft–Polyakov monopole [240, 477], which is clearly a limiting case of the Julia–Zee dyon.

9.4 Monopoles and dyons in BPS limit

We now present a family of explicit dyon and monopole solutions of the Georgi–Glashow equations at the BPS limit $\lambda = 0$ due to Bogomol'nyi [67] and Prasad–Sommerfield [459]. In this case, the action density becomes

$$\mathcal{L} = -\frac{1}{4}\mathbf{F}^{\mu\nu} \cdot \mathbf{F}_{\mu\nu} + \frac{1}{2}D^\mu\phi \cdot D_\mu\phi, \tag{9.4.1}$$

such that the associated energy density assumes the form

$$\mathcal{H} = \frac{1}{2}|\mathbf{F}_{0i}|^2 + \frac{1}{4}|\mathbf{F}_{ij}|^2 + \frac{1}{2}|D_0\phi|^2 + \frac{1}{2}|D_i\phi|^2. \tag{9.4.2}$$

Since the Higgs potential density is absent, there is a freedom in choosing the asymptotic behavior of ϕ near infinity. In order to accommodate spontaneously broken symmetry as before, we *impose* the asymptotic limit

$$\lim_{|\mathbf{x}|\to\infty} |\phi(\mathbf{x})| = \phi_0 > 0. \tag{9.4.3}$$

Consequently, the electromagnetic field (9.2.12) needs to be replaced by the rescaled expression [67, 202, 240]

$$f_{\mu\nu} = \frac{\phi}{\phi_0} \cdot \mathbf{F}_{\mu\nu}, \tag{9.4.4}$$

which may also be present as leading-term contribution in the 't Hooft tensor (9.2.13). Hence, we should now work with (9.4.4) to compute electric and magnetic charges. Note that, in terms of these correspondingly updated electric and magnetic charges, q and g, both being rescaled by the factor $\frac{1}{\phi_0}$, the energy lower bound (9.2.23) now becomes

$$E \geq \phi_0\sqrt{q^2 + g^2}. \tag{9.4.5}$$

We first consider the monopole equations and then use their solution to solve the dyon equations.

Monopole equations in BPS limit

In the BPS limit $\lambda = 0$, the equations (9.3.21)–(9.3.22) become

$$r^2 H'' = 2HK^2, \tag{9.4.6}$$

$$r^2 K'' = K(K^2 + H^2 - 1), \tag{9.4.7}$$

with the associated energy

$$\begin{aligned} E &= \int_{\mathbb{R}^3} \mathcal{H}\, \mathrm{d}x \\ &= \frac{4\pi}{e^2} \int_0^\infty \left((K')^2 + \frac{1}{2r^2}(rH' - H)^2 + \frac{1}{2r^2}(K^2 - 1)^2 + \frac{1}{r^2} K^2 H^2 \right) \mathrm{d}r. \end{aligned} \tag{9.4.8}$$

Note that the equations (9.4.6)–(9.4.7) are the Euler–Lagrange equations of (9.4.8). Besides, using the boundary conditions (9.3.8) and (9.3.16), we have

$$\begin{aligned} E = {} & \frac{4\pi}{e^2} \int_0^\infty \left(\left[K' + \frac{1}{r} KH \right]^2 + \frac{1}{2r^2} \left[rH' - H + (K^2 - 1) \right]^2 \right) \mathrm{d}r \\ & + \frac{4\pi}{e} \int_0^\infty \left(\frac{H}{er} - \frac{K^2 H}{er} \right)' \mathrm{d}r \geq \frac{4\pi}{e}. \end{aligned} \tag{9.4.9}$$

Hence, we have the energy lower bound, $E \geq \frac{4\pi}{e}$, which is attained when (H, K) satisfies

$$rK' = -KH, \tag{9.4.10}$$

$$rH' = H - (K^2 - 1). \tag{9.4.11}$$

Any solution of (9.4.10)–(9.4.11) also satisfies (9.4.6)–(9.4.7). These equations are the radially symmetric version of the Bogomol'nyi equation (9.2.25). In [380], it is shown that (9.4.6)–(9.4.7) and (9.4.10)–(9.4.11) are equivalent under finite-energy condition.

The energy carried by the monopole solution to the BPS equations (9.4.10)–(9.4.11),

$$E = \frac{4\pi}{e}, \tag{9.4.12}$$

is exactly what expressed on the right-hand side of (9.4.5) when $q = 0$, since $\phi_0 = 1$ and $g = \frac{4\pi}{e}$. Based on educated guesswork, an explicit solution of (9.4.10)–(9.4.11) is obtained in [459]:

$$H(r) = er \coth(er) - 1, \tag{9.4.13}$$

$$K(r) = \frac{er}{\sinh(er)}. \tag{9.4.14}$$

Here, we show how to solve (9.4.10)–(9.4.11) systematically by relating the equations to a one-dimensional Liouville-type equation.

Link to Liouville equation

Returning to the variable $u = \frac{H}{er}$, (9.4.10)–(9.4.11) become

$$K' + euK = 0, \tag{9.4.15}$$

$$e r u' - \frac{(1-K^2)}{r} = 0, \tag{9.4.16}$$

subject to the corresponding boundary condition, which we rewrite for convenience:

$$u(0) = 0, \quad K(0) = 1, \quad u(\infty) = 1, \quad K(\infty) = 0. \tag{9.4.17}$$

From (9.4.15)–(9.4.16) and (9.4.17), we see from the uniqueness of the solution to the initial value problem of an ordinary differential equation that any solution of our interest satisfies $0 < K(r) < 1$ and $0 < u(r) < 1$ for $r > 0$. Thus, we may convert (9.4.15) into

$$(\ln K)' = -eu. \tag{9.4.18}$$

Inserting (9.4.18) into (9.4.16), we arrive at

$$r^2(\ln K)'' = K^2 - 1. \tag{9.4.19}$$

Furthermore, we take $K = rw$ to convert (9.4.19) (rather miraculously) into

$$(\ln w)'' = w^2, \tag{9.4.20}$$

which is a one-dimensional Liouville-type equation [109, 375, 606] and should be integrable.

Integration

To this end, set $\ln w = f$ in (9.4.20) and use the boundary condition of K to arrive at

$$f'' = \mathrm{e}^{2f}, \quad 0 < r < \infty; \quad f(0) = \infty, \quad f(\infty) = -\infty. \tag{9.4.21}$$

Multiplying both sides of the differential equation in (9.4.21) by f', integrating, and applying the fact that $f'(r)$ increases, we obtain

$$f'(r) = -\sqrt{\mathrm{e}^{2f(r)} + a^2}; \tag{9.4.22}$$

where $a = \lim_{r\to\infty} |f'(r)| \geq 0$. It may be seen that $a > 0$. In fact, if $a = 0$, then (9.4.22) gives us the solution $f = \ln \frac{1}{r}$ leading to $w = \frac{1}{r}$, which does not provide the desired solution for the function K described. Hence, we have $a > 0$ in (9.4.22), which leads us to the integral

$$r = -\int_0^r \frac{f'(\rho)}{\sqrt{\mathrm{e}^{2f(\rho)} + a^2}} \mathrm{d}\rho = \int_0^r \frac{\mathrm{d}\mathrm{e}^{-f(\rho)}}{\sqrt{1 + a^2 \mathrm{e}^{-2f(\rho)}}}. \tag{9.4.23}$$

For $f \in (-\infty, \infty)$, set $\mathrm{e}^{-f} = \frac{1}{a}\tan\theta$ with $0 < \theta < \frac{\pi}{2}$. So (9.4.23) gives the result

$$\mathrm{e}^{ar} = \frac{1+\sin\theta(r)}{\cos\theta(r)} = \frac{1-\cos(\theta(r)+\frac{\pi}{2})}{\sin(\theta(r)+\frac{\pi}{2})} = \tan\left(\frac{\theta(r)}{2}+\frac{\pi}{4}\right), \tag{9.4.24}$$

rendering

$$\begin{aligned} a\mathrm{e}^{-f(r)} &= \tan\theta(r) = -\cot(2\arctan\mathrm{e}^{ar}) \\ &= -\frac{(\cot\arctan\mathrm{e}^{ar})^2-1}{2\cot(\arctan\mathrm{e}^{ar})} = -\frac{\mathrm{e}^{-2ar}-1}{2\mathrm{e}^{-ar}}. \end{aligned} \tag{9.4.25}$$

Therefore, we obtain

$$f(r) = \ln\frac{a}{\sinh ar}, \quad a > 0, \quad r > 0. \tag{9.4.26}$$

Thus, we see that the only relevant solution to the Liouville-type equation (9.4.20) is

$$w(r) = \frac{a}{\sinh ar}, \quad a > 0. \tag{9.4.27}$$

Inserting (9.4.27) into (9.4.18) with $K = rw$, we obtain

$$eu(r) = a\coth(ar) - \frac{1}{r}, \quad r > 0. \tag{9.4.28}$$

Hence, using the boundary condition $u(\infty) = 1$, we get the matching condition $a = e$. In summary, we arrive at the unique solution to the equations (9.4.15)–(9.4.16) subject to the boundary condition (9.4.17) given explicitly by the formulas

$$K(r) = \frac{er}{\sinh er}, \quad u(r) = \coth er - \frac{1}{er}, \quad r > 0, \tag{9.4.29}$$

recovering (9.4.13)–(9.4.14). The main advantage of linking the problem to the Liouville type equation (9.4.20) instead of relying on guesswork is that a full integration of the equations becomes available.

Dyons

We next present a continuous family of explicit dyon solutions at the BPS limit, $\lambda = 0$, so that the equations (9.3.4)–(9.3.6) become

$$r^2 J'' = 2JK^2, \tag{9.4.30}$$

$$r^2 H'' = 2HK^2, \tag{9.4.31}$$

$$r^2 K'' = K(K^2 - J^2 + H^2 - 1). \tag{9.4.32}$$

Since (9.4.30) and (9.4.31) are of the form (9.4.6), we may use the substitution

$$H \mapsto (\cosh\alpha)H, \quad J \mapsto (\sinh\alpha)H, \tag{9.4.33}$$

where α is a constant, to reduce (9.4.30)–(9.4.32) into (9.4.6)–(9.4.7). Therefore, using (9.4.13)–(9.4.14), we have

$$H(r) = \cosh\alpha(er\coth(er) - 1), \tag{9.4.34}$$

$$J(r) = \sinh\alpha(er\coth(er) - 1), \tag{9.4.35}$$

$$K(r) = \frac{er}{\sinh(er)}. \tag{9.4.36}$$

Thus, with (9.3.7), we see that the boundary condition at $r = 0$, (9.3.8), is maintained but that at $r = \infty$, (9.3.16), is updated with

$$u(\infty) = \cosh\alpha, \quad v(\infty) = \sinh\alpha, \quad K(\infty) = 0, \tag{9.4.37}$$

which alters the normalized asymptotic condition for the Higgs field ϕ, assuming $\lim_{|\mathbf{x}|\to\infty} |\phi(\mathbf{x})| = 1$, used earlier.

Thus, in view of (9.3.3), resulting in (9.4.3) with $\phi_0 = \cosh\alpha$, we may apply (9.4.4) to modify (9.3.17) into

$$\mathbf{e} = (e_i), \quad e_i = \frac{x^i}{r\cosh\alpha}v'. \tag{9.4.38}$$

Consequently, from (9.4.38), we have

$$\begin{aligned} q &= \lim_{R\to\infty} \int_{|\mathbf{x}|=R} \mathbf{e}\cdot \mathrm{d}\mathbf{S} \\ &= \frac{4\pi}{e\cosh\alpha} \lim_{R\to\infty} (RJ'(R) - J(R)) = \frac{4\pi\tanh\alpha}{e}. \end{aligned} \tag{9.4.39}$$

Similarly, (9.4.4) leads to an updated expression for the induced magnetic field, replacing (9.3.19), as follows:

$$\mathbf{b} = (b_i), \quad b_i = \frac{x^i}{er^3\cosh\alpha}u(r)(1 - K^2(r)). \tag{9.4.40}$$

Using (9.4.40), we obtain the associated magnetic charge

$$\begin{aligned} g &= \lim_{R\to\infty} \int_{|\mathbf{x}|=R} \mathbf{b}\cdot \mathrm{d}\mathbf{S} \\ &= \frac{4\pi}{e\cosh\alpha} \lim_{R\to\infty} u(R)(1 - K^2(R)) \\ &= \frac{4\pi}{e}, \end{aligned} \tag{9.4.41}$$

which is identical to that given by (9.3.20) since the asymptotic value of $|\phi|$ or u cancels out eventually.

Of course, the solution becomes electrically neutral, $q = 0$, when $\alpha = 0$ and (9.4.34)–(9.4.36) reduce to the monopole solution (9.4.13)–(9.4.14).

Electric and magnetic charges

Since α in (9.4.39) is arbitrary, the electric charge q given in (9.4.39) is *not quantized* but assumes its value in a continuum. The main reason for the discrepancy with what is expressed by the Dirac charge quantization formula is that the electric charge q here is the total charge induced from a continuously distributed electric field but it is not a pure point charge presented as a coupling parameter in gauge-covariant derivatives. In particular, the Higgs vacuum level $\phi_0 = \cosh\alpha$ is not fixed but continuously parametrized by α instead.

In contrast, the magnetic charge g is fixed, on the other hand, by the topology of the dyon solution.

Higgs field may generate charges

Like mass, both electric and magnetic charges may be *generated* from the Higgs fields.

Dyon energy and its saturation

First, following [459], we use (9.4.30)–(9.4.31) to get

$$\frac{1}{2r^2}(rH' - H)^2 + \frac{H^2K^2}{r^2} = \frac{1}{2}\left(H\left[H' - \frac{H}{r}\right]\right)', \tag{9.4.42}$$

$$\frac{1}{2r^2}(rJ' - J)^2 + \frac{J^2K^2}{r^2} = \frac{1}{2}\left(J\left[J' - \frac{J}{r}\right]\right)'. \tag{9.4.43}$$

Thus, the energy (9.3.15) becomes

$$\begin{aligned} E &= \frac{4\pi}{e^2}\int_0^\infty \left((K')^2 + \frac{1}{2r^2}(rH' - H)^2 + \frac{1}{2r^2}(rJ' - J)^2 \right. \\ &\qquad \left. + \frac{K^2}{r^2}(H^2 + J^2) + \frac{1}{2r^2}(K^2 - 1)^2\right) \mathrm{d}r \\ &= \frac{2\pi}{e^2}\left(\left(H\left[H' - \frac{H}{r}\right]\right)\Big|_0^\infty + \left(J\left[J' - \frac{J}{r}\right]\right)\Big|_0^\infty\right) \\ &\quad + \frac{4\pi}{e^2}\int_0^\infty \left((K')^2 + \frac{1}{2r^2}(K^2 - 1)^2\right) \mathrm{d}r \\ &= \frac{4\pi}{e}\cosh^2\alpha, \end{aligned} \tag{9.4.44}$$

as obtained in [459].

On the other hand, it will be interesting to compare (9.4.44) with the energy lower bound (9.2.23), which in terms of (9.4.39) and (9.4.41), reads

$$E \geq \frac{4\pi}{e}\cosh\alpha\sqrt{\tanh^2\alpha + 1}. \tag{9.4.45}$$

The right-hand side of (9.4.45) is lower than that given in (9.4.44). Thus, we can investigate whether the lower bound in (9.4.45) may be saturated. To study this question, in view of (9.4.33), and setting

$$\sigma^2 = \cosh^2\alpha + \sinh^2\alpha, \tag{9.4.46}$$

for convenience, we see that the energy of a BPS dyon is

$$\begin{aligned} E &= \frac{4\pi}{e^2}\int_0^\infty \left((K')^2 + \frac{\sigma^2}{2r^2}(rH'-H)^2 + \frac{1}{2r^2}(K^2-1)^2 + \frac{\sigma^2}{r^2}K^2H^2 \right) \mathrm{d}r \\ &= \frac{4\pi}{e^2}\int_0^\infty \left(\left[K' + \frac{\sigma}{r}KH\right]^2 + \frac{1}{2r^2}(\sigma[rH'-H] + [K^2-1])^2 \right) \mathrm{d}r \\ &\quad + \frac{4\pi\sigma}{e}\int_0^\infty \left(\frac{H}{er} - \frac{K^2H}{er} \right)' \mathrm{d}r \\ &\geq \frac{4\pi\sigma}{e} \\ &= \frac{4\pi}{e}\cosh\alpha\sqrt{\tanh^2\alpha + 1}, \end{aligned} \tag{9.4.47}$$

which coincides with (9.4.45), where we have inserted the normalized boundary condition for H and K:

$$\lim_{r\to 0}\frac{H(r)}{er} = 0, \quad \lim_{r\to\infty}\frac{H(r)}{er} = 1, \quad K(0) = 1, \quad K(\infty) = 0. \tag{9.4.48}$$

In order to saturate the lower bound in (9.4.47), we impose the equations

$$K' = -\frac{\sigma}{r}KH, \tag{9.4.49}$$

$$\sigma(rH' - H) = -(K^2 - 1), \tag{9.4.50}$$

which are similar to (9.4.10)–(9.4.11) and may be recast into the same Liouville-type equation (9.4.20), resulting the unique solution

$$H(r) = er\coth(\sigma er) - \frac{1}{\sigma}, \tag{9.4.51}$$

$$K(r) = \frac{\sigma er}{\sinh(\sigma er)}, \tag{9.4.52}$$

observing the boundary condition (9.4.48).

Hence, in terms of (9.4.51)–(9.4.52), we obtain from the ansatz (9.4.33) the following field profile

$$H(r) = \cosh\alpha\left(er\coth(\sigma er) - \frac{1}{\sigma}\right), \tag{9.4.53}$$

$$J(r) = \sinh\alpha\left(er\coth(\sigma er) - \frac{1}{\sigma}\right), \tag{9.4.54}$$

$$K(r) = \frac{\sigma er}{\sinh(\sigma er)}, \tag{9.4.55}$$

which minimizes dyon energy given in its radial form by

$$E = \frac{4\pi}{e^2}\int_0^\infty \left((K')^2 + \frac{r^2}{2}\left(\left[\frac{H}{r}\right]'\right)^2 + \frac{r^2}{2}\left(\left[\frac{J}{r}\right]'\right)^2 \right.$$

$$\left. + \frac{1}{2r^2}(K^2-1)^2 + \frac{1}{r^2}K^2(H^2+J^2)\right) \mathrm{d}r, \tag{9.4.56}$$

whose Euler–Lagrange equations are

$$H'' = \frac{2}{r^2}K^2 H, \tag{9.4.57}$$

$$J'' = \frac{2}{r^2}K^2 J, \tag{9.4.58}$$

$$K'' = \frac{K}{r^2}(K^2-1) + \frac{1}{r^2}(H^2+J^2)K. \tag{9.4.59}$$

It is clear that the structure of these equations accommodates (9.4.33), leading to the reduced equations

$$H'' = \frac{2}{r^2}K^2 H, \tag{9.4.60}$$

$$K'' = \frac{K}{r^2}(K^2-1) + \frac{\sigma^2}{r^2}H^2 K, \tag{9.4.61}$$

which are the Euler–Lagrange equations of the energy functional given in (9.4.47). In other words, we saw that (9.4.53)–(9.4.55) give us an energy-minimizing solution of the Euler–Lagrange equations (9.4.57)–(9.4.59) of the radial version of the Georgi–Glashow energy functional (9.4.56). Of course, this solution does not solve the dyon equations (9.4.30)–(9.4.32), which are the Euler–Lagrange equations of the associated indefinite action functional instead.

In particular, since the lower bound (9.4.45) is attained by (9.4.53)–(9.4.55), we see that it *can never* be attained by a dyon solution owing to the difference between (9.4.57)–(9.4.59) and (9.4.30)–(9.4.32). This also explains why the Prasad–Sommerfield solution given by (9.4.34)–(9.4.36) fails to saturate the energy minimum given on the right-hand side of (9.4.45).

Monopole mass

In the electrically neutral situation, $J = 0$, we see that the monopole equations (9.4.6)–(9.4.7) or (9.4.10)–(9.4.11) have a unique solution (H, K) satisfying the boundary condition

$$\lim_{r\to 0}\frac{H(r)}{er} = 0, \quad \lim_{r\to\infty}\frac{H(r)}{er} = \phi_0, \quad K(0) = 1, \quad K(\infty) = 0, \tag{9.4.62}$$

where $\phi_0 > 0$ is arbitrarily prescribed, which may be obtained as a solution to the BPS equations (9.4.10)–(9.4.11), carrying the minimum energy

$$E = g\phi_0, \quad g = \frac{4\pi}{e}. \tag{9.4.63}$$

In quantum field theory, this energy is also identified as the mass [55, 240, 310] of the monopole. Since the magnetic charge g is fixed by the topology of the solution, contributing to the total mass in the form of a universal factor, the mass is proportional to the quantity ϕ_0, often simply referred to as the *monopole mass*.

Dyon mass

For the BPS dyon given by (9.4.34)–(9.4.36), we have

$$\phi_0 = \lim_{r\to\infty} \frac{H(r)}{er} = \cosh\alpha. \tag{9.4.64}$$

Hence, the result (9.4.44) leads to the associated dyon energy or mass as follows:

$$E = \left(\frac{4\pi}{e}\right)\phi_0^2, \tag{9.4.65}$$

which is in sharp contrast with the formula (9.4.63) in that the BPS monopole mass is proportional to ϕ_0, but the *dyon mass* to ϕ_0^2, instead.

Dyon action

The action of a BPS dyon assumes the form

$$\begin{aligned} L = \frac{4\pi}{e^2}\int_0^\infty \bigg(&-(K')^2 - \frac{1}{2r^2}(rH' - H)^2 + \frac{1}{2r^2}(rJ' - J)^2 \\ &- \frac{1}{2r^2}(K^2 - 1)^2 - \frac{1}{r^2}K^2(H^2 - J^2)\bigg)\,\mathrm{d}r, \end{aligned} \tag{9.4.66}$$

which, in view of the substitution (9.4.33) and (9.4.8), (9.4.47), (9.4.10)–(9.4.11), and (9.4.13)–(9.4.14), renders the result

$$\begin{aligned} L &= -\frac{4\pi}{e^2}\int_0^\infty \left((K')^2 + \frac{1}{2r^2}(rH' - H)^2 + \frac{1}{2r^2}(K^2 - 1)^2 + \frac{1}{r^2}K^2H^2\right)\mathrm{d}r \\ &= -\frac{4\pi}{e}, \end{aligned} \tag{9.4.67}$$

which, unlike the dyon energy (9.4.44), is independent of the parameter α in (9.4.33). More generally, if (H, K) is a solution to (9.4.10)–(9.4.11) satisfying the boundary condition (9.4.62), then the shown calculation demonstrates that the corresponding dyon action takes the value

$$L = -\left(\frac{4\pi}{e}\right)\phi_0, \tag{9.4.68}$$

which is linear in ϕ_0 and *coincides* with that of the BPS monopole, $L = -E$, where E is as given in (9.4.63). These results are surprising.

For a rather comprehensive survey on the solutions to the classical $SU(2)$ Yang–Mills equations, see [3].

9.5 Weinberg–Salam electroweak equations

The Weinberg–Salam electroweak theory, also known as the Glashow–Weinberg–Salam theory, is a Yang–Mills–Higgs theory describing in a unified framework the interaction of weak force, which is an $SU(2)$ gauge field theory, and electromagnetic force, which is a $U(1)$ gauge field theory, through a scalar Higgs field, a complex doublet in the fundamental representation of $SU(2)\times U(1)$. This section covers this topic in some detail, regarding especially its governing partial differential equations.

Algebraic structure

To proceed, let σ^a $(a = 1,2,3)$ denote the Pauli spin matrices defined in Chapter 6. Then

$$t^a = \frac{\sigma^a}{2}, \quad a = 1,2,3, \tag{9.5.1}$$

give us a set of generators of $SU(2)$ satisfying the commutation relation

$$[t^a, t^b] = \mathrm{i}\epsilon^{abc}t^c, \tag{9.5.2}$$

and the normalization condition

$$\mathrm{Tr}(t^a t^b) = \frac{\delta^{ab}}{2}. \tag{9.5.3}$$

A group element Ω of $SU(2)$ is then expressed as

$$\Omega = \exp(-\mathrm{i}\lambda^a t^a); \quad \lambda^a \in \mathbb{R}, \quad a = 1,2,3, \tag{9.5.4}$$

such that any associated Lie-algebra- or $su(2)$-valued gauge potential field A_μ may be represented in the matrix form

$$A_\mu = A_\mu^a t^a. \tag{9.5.5}$$

Thus, we may alternatively use the matrix-valued quantity A_μ or isovector-valued quantity $\mathbf{A}_\mu = (A_\mu^a)_{a=1,2,3}$ to express an $SU(2)$-gauge potential field. This gauge field generates weak force described by the field strength tensor

$$F_{\mu\nu} = \partial_\mu A_\nu - \partial_\nu A_\mu + \mathrm{i}g[A_\mu, A_\nu], \tag{9.5.6}$$

where $g > 0$ is a coupling parameter.

As before, we may use B_μ to denote a real-valued gauge potential field, the $U(1)$-gauge potential field, such that electromagnetic force is described by the field strength tensor

$$G_{\mu\nu} = \partial_\mu B_\nu - \partial_\nu B_\mu. \tag{9.5.7}$$

The Higgs field is a complex doublet taking values in $\mathbb{C}^2$ and of the form

$$\phi = \begin{pmatrix} \phi_1 \\ \phi_2 \end{pmatrix}, \tag{9.5.8}$$

which is transformed by the gauge group $SU(2) \times U(1)$ following the rule

$$\phi \mapsto \exp(-\mathrm{i}\lambda^a t^a)\phi, \quad \lambda^a \in \mathbb{R}, \quad a = 1,2,3; \tag{9.5.9}$$

$$\phi \mapsto \exp(-\mathrm{i}\lambda^0 t^0)\phi, \quad \lambda^0 \in \mathbb{R}, \quad t^0 = \frac{\sigma^0}{2} = \frac{1}{2}\begin{pmatrix} 1 & 0 \\ 0 & 1 \end{pmatrix}. \tag{9.5.10}$$

Consequently, we arrive at the gauge-covariant derivative

$$D_\mu \phi = \partial_\mu \phi + \mathrm{i}gA^a_\mu t^a \phi + \mathrm{i}g' B_\mu t^0 \phi, \tag{9.5.11}$$

where $g' > 0$ is another coupling parameter responsible for electromagnetism.

Weinberg–Salam model

With this preparation, the bosonic Lagrangian action density of the *Weinberg–Salam electroweak model* may be written as

$$\mathcal{L} = -\frac{1}{2}\mathrm{Tr}(F_{\mu\nu}F^{\mu\nu}) - \frac{1}{4}G_{\mu\nu}G^{\mu\nu} + \frac{1}{2}(D_\mu\phi)^\dagger(D^\mu\phi) - \frac{\lambda}{4}(|\phi|^2 - v^2)^2, \tag{9.5.12}$$

where $\lambda, v > 0$ are constants. The Euler–Lagrange equations of (9.5.12) are

$$D_\mu D^\mu \phi = \lambda(|\phi|^2 - v^2)\phi, \tag{9.5.13}$$

$$D_\mu F^{\mu\nu a} = \frac{\mathrm{i}}{2} g \left(\phi^\dagger t^a [D^\nu \phi] - [D^\nu \phi]^\dagger t^a \phi\right), \tag{9.5.14}$$

$$\partial_\mu G^{\mu\nu} = \frac{\mathrm{i}}{2} g' \left(\phi^\dagger [D^\nu \phi] - [D^\nu \phi]^\dagger \phi\right), \tag{9.5.15}$$

where

$$D_\gamma F^{\mu\nu} = \partial_\gamma F^{\mu\nu} + \mathrm{i}g[A_\gamma, F^{\mu\nu}]. \tag{9.5.16}$$

The equations (9.5.13)–(9.5.15) are the full (bosonic) *Weinberg–Salam electroweak equations.*

Broken symmetry and unitary gauge

In order to isolate physics, we consider a fixed vacuum state. In doing so, the symmetry is spontaneously broken. Without loss of generality, it is customary to choose the vacuum state

$$\phi_0 = \begin{pmatrix} 0 \\ v \end{pmatrix}, \tag{9.5.17}$$

which breaks the full $SU(2) \times U(1)$ symmetry but enjoys a residual $U(1)$-symmetry that serves to generate electromagnetism in the unified electroweak theory. To find this residual symmetry, assume it is generated by the Lie-algebra element

$$t = c^0 t^0 + c^a t^a, \quad c^0, c^1, c^2, c^3 \in \mathbb{R}, \tag{9.5.18}$$

such that $\exp(-\mathrm{i}\omega t)\phi_0 = \phi_0$ ($\omega \in \mathbb{R}$). Differentiating this equation with respect to ω, we have

$$t\phi_0 = \begin{pmatrix} c^0 + c^3 & c^1 - \mathrm{i}c^2 \\ c^1 + \mathrm{i}c^2 & c^0 - c^3 \end{pmatrix} \begin{pmatrix} 0 \\ v \end{pmatrix} = 0, \tag{9.5.19}$$

resulting in the solution

$$c^1 = 0, \quad c^2 = 0, \quad c^0 = c^3. \tag{9.5.20}$$

As a consequence, we see that the residual $U(1)$-symmetry is generated by the matrix

$$Q \equiv t^3 + t^0, \tag{9.5.21}$$

such that eQ is referred to as the *charge operator*. In other words, we expect to derive electromagnetism arising from a massless gauge field associated with the generator eQ. The structure of Q indicates that such an electromagnetic gauge field should be generated as a linear combination the t^0- and t^3-components of the full gauge field. Thus, we then rotate the vector fields A^3_μ and B_μ in (9.5.11) to reveal a new pair of vector fields P_μ and Z_μ as follows:

$$P_\mu = B_\mu \cos\theta + A^3_\mu \sin\theta, \tag{9.5.22}$$

$$Z_\mu = -B_\mu \sin\theta + A^3_\mu \cos\theta. \tag{9.5.23}$$

Inserting (9.5.22)–(9.5.23) into the gauge-covariant derivative given in (9.5.11), we obtain

$$\begin{aligned} D_\mu = \partial_\mu &+ \mathrm{i}g(A^1_\mu t^1 + A^2_\mu t^2) + \mathrm{i}P_\mu(g\sin\theta\, t^3 + g'\cos\theta\, t^0) \\ &+ \mathrm{i}Z_\mu(g\cos\theta\, t^3 - g'\sin\theta\, t^0). \end{aligned} \tag{9.5.24}$$

We now set

$$eQ = e(t^3 + t^0) = g\sin\theta\, t^3 + g'\cos\theta\, t^0. \tag{9.5.25}$$

Then, we see that the coupling constants e, g, g' and the angular parameter θ are confined by the relations

$$e = g\sin\theta = g'\cos\theta, \tag{9.5.26}$$

$$e = \frac{gg'}{(g^2 + g'^2)^{\frac{1}{2}}}, \tag{9.5.27}$$

$$\cos\theta = \frac{g}{(g^2 + g'^2)^{\frac{1}{2}}}. \tag{9.5.28}$$

The so-determined angle θ is called the *Weinberg mixing angle*, which we now assume. As a consequence, the gauge-covariant derivative (9.5.24) becomes

$$D_\mu = \partial_\mu + \mathrm{i}g(A^1_\mu t^1 + A^2_\mu t^2) + \mathrm{i}P_\mu eQ + \mathrm{i}Z_\mu eQ', \tag{9.5.29}$$

where

$$Q' = \cot\theta\, t^3 - \tan\theta\, t^0, \tag{9.5.30}$$

is called the *neutral charge operator.*

In the unitary gauge, the Higgs doublet ϕ is taken to fluctuate around the ground state (9.5.17) in the form

$$\phi = \begin{pmatrix} 0 \\ \varphi \end{pmatrix}, \quad \varphi \text{ is real-valued.} \tag{9.5.31}$$

Then we have

$$D_\mu \phi = \begin{pmatrix} \frac{\mathrm{i}g}{2}(A^1_\mu - \mathrm{i}A^2_\mu)\varphi \\ \partial_\mu \varphi - \frac{\mathrm{i}g}{2\cos\theta} Z_\mu \varphi \end{pmatrix}. \tag{9.5.32}$$

Introduce a complex-valued vector field

$$W_\mu = \frac{1}{\sqrt{2}}(A^1_\mu + \mathrm{i}A^2_\mu), \tag{9.5.33}$$

and set

$$\mathcal{D}_\mu = \partial_\mu - \mathrm{i}gA^3_\mu = \partial_\mu - \mathrm{i}g(\sin\theta\, P_\mu + \cos\theta\, Z_\mu). \tag{9.5.34}$$

Then the Lagrange action density (9.5.12) reads

$$\begin{aligned}\mathcal{L} = &-\frac{1}{2}(\mathcal{D}_\mu W_\nu - \mathcal{D}_\nu W_\mu)\overline{(\mathcal{D}^\mu W^\nu - \mathcal{D}^\nu W^\mu)} - \frac{1}{4}Z_{\mu\nu}Z^{\mu\nu} - \frac{1}{4}P_{\mu\nu}P^{\mu\nu} \\ &+ \frac{1}{2}g^2\left([W_\mu \overline{W}^\mu]^2 - [W_\mu W^\mu]\overline{[W_\nu W^\nu]}\right) \\ &- \mathrm{i}g\,(Z^{\mu\nu}\cos\theta + P^{\mu\nu}\sin\theta)\,\overline{W}_\mu W_\nu + \frac{1}{4}g^2\varphi^2 W^\mu \overline{W}_\mu + \frac{1}{2}\partial_\mu\varphi\partial^\mu\varphi \\ &+ \frac{g^2}{8\cos^2\theta}\varphi^2 Z_\mu Z^\mu - \frac{\lambda}{4}(\varphi^2 - v^2)^2,\end{aligned} \tag{9.5.35}$$

which appears rather complicated. Nevertheless, this reduction allows us to identify and compute various particle masses quickly.

Particle masses

First, with $\varphi(x) = v + u(x)$, we may expand the Higgs potential density as

$$\frac{\lambda}{4}(\varphi^2 - v^2)^2 = \lambda v^2 u^2 + \lambda\left(v + \frac{u}{4}\right)u^3. \tag{9.5.36}$$

In view of (9.5.35) and (9.5.36), we see from reading off the quadratic term that the mass of the Higgs scalar particle is

$$m_{\mathrm{H}} = \sqrt{2\lambda}v. \tag{9.5.37}$$

Similarly, from the expansion

$$\frac{g^2}{8\cos^2\theta}\varphi^2 Z_\mu Z^\mu = \frac{g^2 v^2}{8\cos^2\theta}Z_\mu Z^\mu + \frac{g^2}{8\cos^2\theta}(2v+u)uZ_\mu Z^\mu, \tag{9.5.38}$$

we see that the Z-field or Z-*particle* mass is

$$m_{\mathrm{Z}} = \frac{gv}{2\cos\theta} = \frac{v(g^2 + g'^2)^{\frac{1}{2}}}{2}, \tag{9.5.39}$$

by applying (9.5.28).

The expression (9.5.35) also clearly indicates that the P-field or P-*photon* mass is zero:

$$m_{\mathrm{P}} = 0. \tag{9.5.40}$$

In other words, the vector field P_μ describes electromagnetism, as anticipated.

Finally, the coefficient of the quadratic term of W_μ is

$$\frac{1}{4}g^2\varphi^2 = \frac{1}{4}g^2v^2 + \frac{g^2}{4}(2v + u)u, \tag{9.5.41}$$

which gives rise to the W-field or W-*particle* mass

$$m_{\mathrm{W}} = \frac{gv}{2}. \tag{9.5.42}$$

With (9.5.39) and (9.5.42), we relate the Weinberg mixing angle θ with the celebrated W- *and* Z-*particle mass ratio* as follows:

$$\frac{m_{\mathrm{W}}}{m_{\mathrm{Z}}} = \cos\theta. \tag{9.5.43}$$

Since by (9.5.33) the field W_μ is complex-valued given in terms of two real-valued gauge fields, A^1_μ and A^2_μ, it actually represents two particles—the W^+- and W^--particles—governed by the fields

$$W^+_\mu = \frac{1}{\sqrt{2}}(A^1_\mu + \mathrm{i}A^2_\mu), \quad W^-_\mu = \frac{1}{\sqrt{2}}(A^1_\mu - \mathrm{i}A^2_\mu), \tag{9.5.44}$$

where W^+_μ coincides with W_μ, of course. Besides, since

$$W^+_\mu \overline{W}^{+\mu} = W_\mu \overline{W}^\mu, \quad W^-_\mu \overline{W}^{-\mu} = W_\mu \overline{W}^\mu, \tag{9.5.45}$$

the W^+- and W^--particles are of the same mass given by (9.5.42).

In conclusion, we see that the Weinberg–Salam electroweak theory describes in a unified formalism the interaction of a massless, long-range, P-photon field and three massive, short-range, W^+-, W^--, and Z-particle fields.

Electroweak equations in unitary gauge

Varying φ, P_μ, Z_μ, and $\overline{W}_\mu$ in (9.5.35), we obtain the associated Euler–Lagrange equations as follows:

$$\partial_\mu \partial^\mu \varphi = \frac{g^2}{2}(W_\mu \overline{W}^\mu)\varphi + \frac{g^2}{4\cos^2\theta}(Z_\mu Z^\mu)\varphi - \lambda(\varphi^2 - v^2)\varphi, \tag{9.5.46}$$

$$\begin{aligned}\partial_\mu P^{\mu\nu} &= \mathrm{i}g \sin\theta\, \partial_\mu (W^\mu \overline{W}^\nu - \overline{W}^\mu W^\nu) \\ &\quad + \mathrm{i}g\sin\theta \left(W_\mu \overline{(\mathcal{D}^\mu W^\nu - \mathcal{D}^\nu W^\mu)} - \overline{W}_\mu (\mathcal{D}^\mu W^\nu - \mathcal{D}^\nu W^\mu) \right),\end{aligned} \tag{9.5.47}$$

$$\begin{aligned}\partial_\mu Z^{\mu\nu} &= -\frac{g^2}{4\cos^2\theta}\varphi^2 Z^\nu + \mathrm{i}g\cos\theta\, \partial_\mu (W^\mu \overline{W}^\nu - \overline{W}^\mu W^\nu) \\ &\quad + \mathrm{i}g\cos\theta \left(W_\mu \overline{(\mathcal{D}^\mu W^\nu - \mathcal{D}^\nu W^\mu)} - \overline{W}_\mu (\mathcal{D}^\mu W^\nu - \mathcal{D}^\nu W^\mu) \right),\end{aligned} \tag{9.5.48}$$

$$\begin{aligned}&\mathcal{D}_\mu (\mathcal{D}^\mu W^\nu - \mathcal{D}^\nu W^\mu) \\ &\quad = -\frac{1}{4} g^2 \varphi^2 W^\nu - g^2 \left([W_\mu \overline{W}^\mu] W^\nu - [W_\mu W^\mu] \overline{W}^\nu \right) \\ &\quad\quad - \mathrm{i}g(Z^{\mu\nu}\cos\theta + P^{\mu\nu}\sin\theta) W_\mu,\end{aligned} \tag{9.5.49}$$

which are the Weinberg–Salam electroweak equations in a fixed unitary gauge. It is clear that (9.5.46)–(9.5.48) are real wave equations, while (9.5.49) is complex. In view of (9.5.44), the equation (9.5.49) governs the dynamics of the W^+-field and its complex conjugate, the W^--field.

Solutions

For these electroweak equations, previous studies have focused on static solutions. In [11, 12, 13, 14], it is shown that when the coupling parameters fulfill a critical condition, W-condensed multivortex solutions of the BPS-type appear. Some existence theorems for such vortices are established in [521, 522]. See [47, 539] for some further development and progress of this subject. Furthermore, the studies in [137, 600] develop an existence theory for dyon solutions. The construction of such solutions relies on the methods of calculus of variations.

See [184, 185, 323, 341, 420, 421, 502] for some studies on the well-posedness of the Yang–Mills equations as nonlinear wave equations on the Minkowski spaces.

Exercises

1. Show that the special unitary group $SU(N)$ $(N \geq 2)$ satisfying $\Omega^\dagger \Omega = \Omega\Omega^\dagger = I$ and $\det(\Omega) = 1$ for any $\Omega \in SU(N)$ depends on $N^2 - 1$ real parameters.
2. Show that the Yang–Mills field defined in (9.1.12) follows the transformation rule (9.1.13).

3. Consider the Yang–Mills equations (9.1.25). Show that there holds the necessary condition
$$\partial_\nu j^\nu = 0, \tag{9.E.1}$$
that is, a charge-current conservation law, which directly follows from (9.1.25).

4. Consider the current density j^ν defined by (9.1.26), which is formally the non-homogeneous term in (9.1.25). Establish the identity
$$\partial_\nu j^\nu = \frac{1}{2}[F_{\mu\nu}, F^{\mu\nu}] = 0, \tag{9.E.2}$$
hence, the self-consistency of (9.1.25).

5. Show that the Yang–Mills field defined in (9.1.12) and the gauge-covariant derivative (9.1.20) obey the identity
$$D_\mu D_\nu F^{\mu\nu} = \frac{1}{2}[F_{\mu\nu}, F^{\mu\nu}] = 0, \tag{9.E.3}$$
which implies the non-Abelian conservation law
$$D_\nu j^\nu = 0, \tag{9.E.4}$$
for the current density given in (9.1.21) through (9.1.19). In particular, the usual conservation law, $\partial_\nu j^\nu = 0$, is no longer valid.

6. In the special case when $G = U(1)$, show how (9.1.18), (9.1.19), and (9.1.21) recover (5.2.31), (5.2.32), and (5.2.33) in the Abelian Higgs theory.

7. Verify the Jacobi identity (9.1.27) and the Bianchi identity (9.1.29).

8. Establish the identity (9.1.33).

9. Develop the Yang–Mills theory when the scalar field ϕ is in the *adjoint representation* of the symmetry group G (with $G = O(n)$ or $U(n)$) so that ϕ lies in the Lie algebra of G and the global symmetry (9.1.2) is replaced by
$$\phi \mapsto \Omega^{-1}\phi\Omega. \tag{9.E.5}$$

10. Verify the Bianchi identity (9.2.7).

11. Establish the identity (9.2.20).

12. Show that the angle θ that maximizes the right-hand side of (9.2.22) is given by that which makes the vector $(\sin\theta, \cos\theta)$ lie parallel to the vector (q, g), which establishes the lower bound estimate (9.2.23).

13. Prove the commutation relation (9.2.29).

14. Establish (9.2.30).

15. Assume that $(\phi, \mathbf{A}_\mu)$ is a static radially symmetric solution of the Georgi–Glashow equations (9.2.9)–(9.2.10) given by (9.3.1)–(9.3.3) and (9.3.7). Show that
$$|D_\mu\phi|^2 = (u')^2 + \frac{2}{r^2}K^2u^2. \tag{9.E.6}$$

Use the boundary condition (9.3.16) in (9.3.9)–(9.3.11) and assume $0 < C_0 < 1$ in (9.E.6) to show that $D_\mu\phi$ vanishes at infinity exponentially fast.

16. With the radial ansatz (9.3.1)–(9.3.3), establish the results

$$(D_i\phi)^a = \frac{\delta^{ai}}{er^2}HK + \frac{x^a x^i}{er^4}(rH' - H - HK), \tag{9.E.7}$$

$$\mathbf{B}_i^a = -\frac{\delta^{ai}}{er^2}(rK') + \frac{x^a x^i}{er^4}(rK' + 1 - K^2). \tag{9.E.8}$$

Then use these to derive from (9.2.25) the following radially symmetric Bogomol'nyi equations

$$rH' - H = \pm(1 - K^2), \tag{9.E.9}$$

$$rK' = \mp HK, \tag{9.E.10}$$

which contain (9.4.10)–(9.4.11).

17. Show that a solution to (9.4.49)–(9.4.50) also fulfills (9.4.60)–(9.4.61).
18. Find a monopole solution to (9.4.6)–(9.4.7) subject to the boundary condition (9.4.62), which carries the minimum energy/mass as given in (9.4.63).
19. Supply details to derive the formula (9.4.68) for the action of a BPS dyon.
20. Verify the formula (9.5.24) for the electroweak gauge-covariant derivative.
21. With the relations (9.5.22)–(9.5.23) and (9.5.33), establish the expression

$$\begin{aligned} F^3_{\mu\nu}F^{3\mu\nu} &= \sin^2\theta P_{\mu\nu}P^{\mu\nu} + \cos^2\theta Z_{\mu\nu}Z^{\mu\nu} + 2\cos\theta\sin\theta P_{\mu\nu}Z^{\mu\nu} \\ &\quad + 2g^2\left([\overline{W}_\mu\overline{W}^\mu][W_\nu W^\nu] - [W_\mu\overline{W}^\mu]^2\right) \\ &\quad + \mathrm{i}\,2g\,(\sin\theta P_{\mu\nu} + \cos\theta Z_{\mu\nu})(\overline{W}^\mu W^\nu - W^\mu\overline{W}^\nu). \end{aligned} \tag{9.E.11}$$

22. With the relations (9.5.22)–(9.5.23), (9.5.33), and (9.5.34), establish the expression

$$F^1_{\mu\nu}F^{1\mu\nu} + F^2_{\mu\nu}F^{2\mu\nu} = 2(\mathcal{D}_\mu W_\nu - \mathcal{D}_\nu W_\mu)\overline{(\mathcal{D}^\mu W^\nu - \mathcal{D}^\nu W^\mu)}. \tag{9.E.12}$$

23. Use (9.5.32) where φ is real to obtain the relation

$$(D_\mu\phi)^\dagger(D^\mu\phi) = \frac{g^2\varphi^2}{2}W_\mu\overline{W}^\mu + \partial_\mu\varphi\partial^\mu\varphi + \frac{g^2\varphi^2}{4\cos^2\theta}Z_\mu Z^\mu. \tag{9.E.13}$$

24. Use (9.E.11)–(9.E.13) to derive the electroweak action density (9.5.35) in the unitary gauge.
25. Derive the electroweak equations (9.5.46)–(9.5.49) as the Euler–Lagrange equations of the action density (9.5.35).

10

Einstein equations and related topics

This chapter introduces the Einstein equations of general relativity and some of their direct consequences. It begins with a brief discussion on Riemannian geometry, a calculation of the metric energy-momentum tensor, and a derivation of the Einstein equations for gravitation. It next elaborates on some cosmological consequences of the Einstein equations and presents the Schwarzschild black-hole solution and the Reissner–Nordström charged black-hole solution. It then covers gravitational mass and the Penrose bounds, and considers gravitational waves in weak gravity limit. Finally, it discusses the cosmological expansion of an isotropic and homogeneous universe propelled by a scalar-wave matter known as quintessence.

10.1 Einstein field equations

Based on Riemannian geometry, in 1915, Einstein formulated the theory of gravitation know as general relativity. In this theory, the fundamental equations governing gravity interacting with and determined by the underlying physical matters in spacetime, namely, the Einstein equations, are at the center of the theory and expressed in terms of coordinate-covariant quantities. These quantities are generically referred to as curvature tensors, describing spacetime geometry, so that gravitation demonstrates its presence through affecting the trajectory of a particle in "free motion." Therefore, we present Einstein's theory by starting with a discussion about the free motion of a particle in spacetime with a location-dependent, or local, line element. We show that minimization of the action functional formed from integrating the proper time results in the equations of motion of the particle and leads to the introduction of the Christoffel symbols. These allow the formulation of various types of Riemann and Ricci tensors and

Mathematical Physics with Differential Equations. Yisong Yang, Oxford University Press.
 DOI: 10.1093/oso/9780192872616.003.0010

the introduction of covariant derivatives. Finally, we formulate or derive the Einstein equations at the end of this section.

For convenience, throughout this chapter and unless otherwise stated, we set $c = 1$ for the speed of light.

Action principle and equations of motion

Let $(g_{\mu\nu})$ be the metric tensor of spacetime. The spacetime line element or the first fundamental form is defined by

$$\mathrm{d}s^2 = g_{\mu\nu}\mathrm{d}x^\mu \mathrm{d}x^\nu, \tag{10.1.1}$$

which is also a measurement of the *proper time* (see (4.2.2) or (4.3.3) for its flat-spacetime version). A freely moving particle in spacetime follows a *timelike* curve (with $\mathrm{d}s^2 > 0$) that stationarizes the action

$$I = \int \sqrt{g_{\mu\nu}(x^\gamma(s))\frac{\mathrm{d}x^\mu}{\mathrm{d}s}\frac{\mathrm{d}x^\nu}{\mathrm{d}s}}\,\mathrm{d}s. \tag{10.1.2}$$

We now derive the equations of motion from this action functional. For this purpose, we recall that the Euler–Lagrange equations of the action of the general form

$$I(x^\mu) = \int \varphi(x^\mu(s), \dot{x}^\mu(s))\,\mathrm{d}s, \quad \dot{x}^\mu = \frac{\mathrm{d}x^\mu}{\mathrm{d}s}, \tag{10.1.3}$$

in terms of an arbitrary variable parameter s, are

$$\frac{\mathrm{d}}{\mathrm{d}s}\left(\frac{\partial\varphi}{\partial\dot{x}^\mu}\right) - \frac{\partial\varphi}{\partial x^\mu} = 0, \quad \mu = 0, 1, 2, 3. \tag{10.1.4}$$

In the following we use the notation $x^\mu(s)$ to denote the stationarized (desired) curve, trajectory of the particle, and s its proper time parameter or the "arclength" of the trajectory. Then we have

$$\sqrt{g_{\mu\nu}(x^\gamma(s))\frac{\mathrm{d}x^\mu}{\mathrm{d}s}\frac{\mathrm{d}x^\nu}{\mathrm{d}s}} = 1. \tag{10.1.5}$$

On the other hand, if we use $x^\mu(s)$ to denote a general field configuration expressed in terms of the proper time s defined here and use the notation

$$\varphi(x^\mu(s), \dot{x}^\mu(s)) = \sqrt{g_{\alpha\beta}(x^\gamma(s))\dot{x}^\alpha(s)\dot{x}^\beta(s)}, \tag{10.1.6}$$

we obtain the results

$$\begin{aligned}\frac{\partial\varphi}{\partial\dot{x}^\mu} &= \frac{1}{2\sqrt{g_{\alpha\beta}\dot{x}^\alpha\dot{x}^\beta}}\left(g_{\mu\gamma}\dot{x}^\gamma + g_{\gamma\mu}\dot{x}^\gamma\right)\\ &= \frac{1}{\sqrt{g_{\alpha\beta}\dot{x}^\alpha\dot{x}^\beta}}\left(g_{\mu\gamma}\dot{x}^\gamma\right)\\ &= g_{\mu\alpha}\dot{x}^\alpha, \qquad (10.1.7)\\ \frac{\partial\varphi}{\partial x^\mu} &= \frac{1}{2\sqrt{g_{\alpha\beta}\dot{x}^\alpha\dot{x}^\beta}}\left(\frac{\partial g_{\alpha\beta}}{\partial x^\mu}\dot{x}^\alpha\dot{x}^\beta\right)\\ &= \frac{1}{2}g_{\alpha\beta,\mu}\dot{x}^\alpha\dot{x}^\beta, \qquad (10.1.8)\end{aligned}$$

at the stationarized configuration so that (10.1.5) holds, where and in the sequel we use the notation

$$f_{,\alpha}, \quad A_{\mu,\alpha}, \quad F_{\mu\nu,\alpha}, \quad T^{\mu\nu}{}_{,\alpha}, \qquad (10.1.9)$$

etc., to denote the conventional partial derivative with respect to the variable x^α of various quantities.

Christoffel symbols

Inserting (10.1.7) and (10.1.8) into (10.1.4), we arrive at

$$g_{\mu\alpha}\ddot{x}^\alpha + g_{\mu\alpha,\beta}\dot{x}^\alpha\dot{x}^\beta - \frac{1}{2}g_{\alpha\beta,\mu}\dot{x}^\alpha\dot{x}^\beta = 0. \qquad (10.1.10)$$

Or more collectively, we have

$$\begin{aligned}&g_{\mu\alpha}\ddot{x}^\alpha + \frac{1}{2}\left(g_{\mu\alpha,\beta} + g_{\mu\beta,\alpha} - g_{\alpha\beta,\mu}\right)\dot{x}^\alpha\dot{x}^\beta\\ &\equiv g_{\mu\alpha}\ddot{x}^\alpha + \Gamma_{\mu\alpha\beta}\dot{x}^\alpha\dot{x}^\beta = 0, \qquad (10.1.11)\end{aligned}$$

where

$$\Gamma_{\mu\alpha\beta} = \frac{1}{2}\left(g_{\mu\alpha,\beta} + g_{\mu\beta,\alpha} - g_{\alpha\beta,\mu}\right) \qquad (10.1.12)$$

are called the *Christoffel symbols of the first kind*, which are obviously symmetric with respect to the interchange of the last two indices:

$$\Gamma_{\mu\alpha\beta} = \Gamma_{\mu\beta\alpha}. \qquad (10.1.13)$$

Note that the definition of $\Gamma_{\mu\nu\alpha}$ gives us the identity

$$\Gamma_{\mu\nu\alpha} + \Gamma_{\nu\mu\alpha} = g_{\mu\nu,\alpha}. \qquad (10.1.14)$$

Furthermore, use $(g^{\mu\nu})$ to denote the inverse of $(g_{\mu\nu})$ such that

$$g_{\mu\alpha}g^{\alpha\nu} = \delta^\nu_\mu, \qquad (10.1.15)$$

etc. Thus, multiplying (10.1.11) by $g^{\mu\nu}$ and summing up, we obtain the following explicit form of the stationary equations

$$\ddot{x}^\mu + \Gamma^\mu_{\alpha\beta}\dot{x}^\alpha\dot{x}^\beta = 0, \tag{10.1.16}$$

where

$$\Gamma^\mu_{\alpha\beta} = g^{\mu\nu}\Gamma_{\nu\alpha\beta} \tag{10.1.17}$$

are called the *Christoffel symbols of the second kind.*

Geodesics

More conveniently, using u^μ to denote the components of the *4-velocity*, we have

$$u^\mu(s) = \frac{\mathrm{d}x^\mu(s)}{\mathrm{d}s}. \tag{10.1.18}$$

Consequently, the equations of motion (10.1.11) or (10.1.16) become

$$g_{\mu\alpha}\frac{\mathrm{d}u^\alpha}{\mathrm{d}s} + \Gamma_{\mu\alpha\beta}\,u^\alpha u^\beta = 0 \quad \text{or} \quad \frac{\mathrm{d}u^\mu}{\mathrm{d}s} + \Gamma^\mu_{\alpha\beta}u^\alpha u^\beta = 0. \tag{10.1.19}$$

The curves that are solutions of (10.1.19) are called *geodesics.*

Covariant derivative

One of the most important applications of the Christoffel symbols is their role in the definition of covariant derivatives for covariant and contravariant quantities,

$$A_{\mu;\alpha} = A_{\mu,\alpha} - \Gamma^\beta_{\mu\alpha}A_\beta, \tag{10.1.20}$$

$$T_{\mu\nu;\alpha} = T_{\mu\nu,\alpha} - \Gamma^\beta_{\mu\alpha}T_{\beta\nu} - \Gamma^\beta_{\nu\alpha}T_{\mu\beta}, \tag{10.1.21}$$

$$A^\mu{}_{;\alpha} = A^\mu{}_{,\alpha} + \Gamma^\mu_{\beta\alpha}A^\beta, \tag{10.1.22}$$

$$T^{\mu\nu}{}_{;\alpha} = T^{\mu\nu}{}_{,\alpha} + \Gamma^\mu_{\beta\alpha}T^{\beta\nu} + \Gamma^\nu_{\beta\alpha}T^{\mu\beta}. \tag{10.1.23}$$

We will sometimes use ∇_α to denote covariant derivative. Since

$$\Gamma^\alpha_{\mu\nu} = \Gamma^\alpha_{\nu\mu}, \tag{10.1.24}$$

we have $f_{,\mu;\nu} = f_{,\nu;\mu}$ or $\nabla_\mu\nabla_\nu f = \nabla_\nu\nabla_\mu f$, which says the covariant derivative is *torsion-free.* Another direct consequence of this definition and the identity (10.1.14) is that

$$\begin{aligned} g_{\mu\nu;\alpha} &= g_{\mu\nu,\alpha} - \Gamma^\beta_{\mu\alpha}g_{\beta\nu} - \Gamma^\beta_{\nu\alpha}g_{\mu\beta} \\ &= g_{\mu\nu,\alpha} - \Gamma_{\mu\alpha\nu} - \Gamma_{\nu\alpha\mu} = 0. \end{aligned} \tag{10.1.25}$$

Similarly, $g^{\mu\nu}{}_{;\alpha} = 0$. Therefore, we have seen that the covariant and contravariant metric tensors, $g_{\mu\nu}$ and $g^{\mu\nu}$, behave like constants under covariant differentiation. This property is very useful in computation when combined with the Leibniz rule operated over tensor products.

Riemann curvature tensor

Let A_μ be a test covariant vector. Following (10.1.20)–(10.1.23), we obtain through an easy calculation the commutator

$$A_{\mu;\alpha;\beta} - A_{\mu;\beta;\alpha} = [\nabla_\alpha, \nabla_\beta]A_\mu = R^\nu_{\mu\alpha\beta}A_\nu, \tag{10.1.26}$$

where

$$R^\nu_{\mu\alpha\beta} = \Gamma^\nu_{\mu\beta,\alpha} - \Gamma^\nu_{\mu\alpha,\beta} + \Gamma^\gamma_{\mu\beta}\Gamma^\nu_{\gamma\alpha} - \Gamma^\gamma_{\mu\alpha}\Gamma^\nu_{\gamma\beta} \tag{10.1.27}$$

is a mixed 4-tensor called the *Riemann curvature tensor*. There hold the simple properties

$$R^\nu_{\mu\alpha\beta} = -R^\nu_{\mu\beta\alpha}, \tag{10.1.28}$$

$$R^\nu_{\mu\alpha\beta} + R^\nu_{\alpha\beta\mu} + R^\nu_{\beta\mu\alpha} = 0. \tag{10.1.29}$$

Furthermore, similar to (10.1.26), for covariant 2-tensors, we have

$$T_{\mu\nu;\alpha;\beta} - T_{\mu\nu;\beta;\alpha} = R^\gamma_{\mu\alpha\beta}T_{\gamma\nu} + R^\gamma_{\nu\alpha\beta}T_{\mu\gamma}. \tag{10.1.30}$$

Therefore, in particular, for a covariant vector field A_μ, we have

$$A_{\mu;\nu;\alpha;\beta} - A_{\mu;\nu;\beta;\alpha} = R^\gamma_{\mu\alpha\beta}A_{\gamma;\nu} + R^\gamma_{\nu\alpha\beta}A_{\mu;\gamma}. \tag{10.1.31}$$

We now make cyclic permutations of the indices ν, α, β and add the three resulting equations. In view of (10.1.26), the left-hand side of (10.1.31) gives us

$$\begin{aligned}
&(A_{\mu;\alpha;\beta;\nu} - A_{\mu;\beta;\alpha;\nu}) + \text{cyclic permutations}\\
&= (R^\gamma_{\mu\alpha\beta}A_\gamma)_{;\nu} + \text{cyclic permutations}\\
&= (R^\gamma_{\mu\alpha\beta}A_{\gamma;\nu} + R^\gamma_{\mu\alpha\beta;\nu}A_\gamma) + \text{cyclic permutations}.
\end{aligned} \tag{10.1.32}$$

In view of (10.1.29), the right-hand side of (10.1.31) gives us

$$R^\gamma_{\mu\alpha\beta}A_{\gamma;\nu} + \text{cyclic permutations}. \tag{10.1.33}$$

Equating (10.1.32) and (10.1.33), we arrive at

$$R^\gamma_{\mu\alpha\beta;\nu}A_\gamma + \text{cyclic permutations} = 0. \tag{10.1.34}$$

Since A_μ is arbitrary, we find that

$$R^\gamma_{\mu\nu\alpha;\beta} + R^\gamma_{\mu\alpha\beta;\nu} + R^\gamma_{\mu\beta\nu;\alpha} = 0. \tag{10.1.35}$$

This result is also known as the *Bianchi identity*, which is sometimes compactly rewritten as [563]

$$R^\gamma_{\mu[\nu\alpha;\beta]} = 0. \tag{10.1.36}$$

Ricci tensor

The *Ricci tensor* $R_{\mu\nu}$ is defined from $R^{\nu}_{\mu\alpha\beta}$ through contraction [168],

$$R_{\mu\nu} = R^{\alpha}_{\mu\nu\alpha}. \tag{10.1.37}$$

It is clear that $R_{\mu\nu}$ is symmetric. The *scalar curvature* R is then defined by

$$R = g^{\mu\nu} R_{\mu\nu}. \tag{10.1.38}$$

In the Bianchi identity (10.1.35), set $\gamma = \nu$ and then contract by multiplying by $g^{\mu\beta}$. We obtain

$$(g^{\mu\beta} R^{\nu}_{\mu\nu\alpha})_{;\beta} + (g^{\mu\beta} R^{\nu}_{\mu\alpha\beta})_{;\nu} + (g^{\mu\beta} R^{\nu}_{\mu\beta\nu})_{;\alpha} = 0, \tag{10.1.39}$$

or

$$-(g^{\mu\beta} R^{\nu}_{\mu\alpha\nu})_{;\beta} + (g^{\mu\beta} R^{\nu}_{\mu\alpha\beta})_{;\nu} + (g^{\mu\beta} R^{\nu}_{\mu\beta\nu})_{;\alpha} = 0. \tag{10.1.40}$$

The first and third terms on the left-hand side of (10.1.40) are simply $R^{\beta}_{\alpha;\beta}$, where of course $R^{\beta}_{\alpha} = g^{\mu\beta} R_{\mu\alpha} = g^{\mu\beta} R^{\nu}_{\mu\alpha\nu}$, and $R_{;\alpha}$, respectively. The second term, on the other hand, is not as clear. To clarify, we may use the Riemann tensor

$$R_{\alpha\beta\mu\nu} = g_{\alpha\gamma} R^{\gamma}_{\beta\mu\nu} \tag{10.1.41}$$

and its properties

$$R_{\alpha\beta\mu\nu} = R_{\mu\nu\alpha\beta} = -R_{\beta\alpha\mu\nu} = -R_{\alpha\beta\nu\mu} \tag{10.1.42}$$

to get

$$\begin{aligned} g^{\mu\beta} R^{\nu}_{\mu\alpha\beta} &= g^{\mu\beta} g^{\nu\gamma} R_{\gamma\mu\alpha\beta} \\ &= -g^{\nu\gamma} g^{\mu\beta} R_{\beta\alpha\gamma\mu} \\ &= -g^{\nu\gamma} R^{\mu}_{\alpha\gamma\mu} \\ &= -g^{\nu\gamma} R_{\alpha\gamma} \\ &= -R^{\nu}_{\alpha}. \end{aligned} \tag{10.1.43}$$

Collecting these results, we arrive at

$$2R^{\beta}_{\alpha;\beta} - R_{;\alpha} = 0. \tag{10.1.44}$$

Einstein tensor

Multiplying (10.1.44) by $g^{\mu\alpha}$, we have the following very important result,

$$G^{\mu\nu}{}_{;\nu} = 0, \tag{10.1.45}$$

where

$$G^{\mu\nu} = R^{\mu\nu} - \frac{1}{2} g^{\mu\nu} R, \tag{10.1.46}$$

or its covariant partner, $G_{\mu\nu}$, is called the *Einstein tensor*.

Energy-momentum tensor

We next consider physics over the curved spacetime of metric $(g_{\mu\nu})$ governed by a matter field u, which is either a scalar field or a vector field and governed by the action

$$S = \int \mathcal{L}(x, Du, g)\sqrt{|g|}\,\mathrm{d}x, \tag{10.1.47}$$

where we have emphasized the influence of the metric tensor $g = (g_{\mu\nu})$ and used the *canonical volume element* $\sqrt{|g|}\,\mathrm{d}x$. Here, $|g|$ is the absolute value of the determinant of the metric g. Without loss of clarity, we may also use g to denote the determinant of $(g_{\mu\nu})$ when there is no risk of confusion. With this notation, we have $|g| = -g$ because the signature of the metric is taken to be $(+---)$ such that $g < 0$. Since physics is independent of coordinates, $\mathcal{L}$ must be a real scalar. For example, the Klein–Gordon action density governing a real scalar field u now reads

$$\mathcal{L}(x, u, Du, g) = \frac{1}{2}g^{\mu\nu}\partial_\mu u \partial_\nu u - V(u), \tag{10.1.48}$$

which is dependent on the metric $g = (g_{\mu\nu})$. In other words, physics can no longer be purely material.

We see that the Euler–Lagrange equations, or the equations of motion, of (10.1.47) are now

$$\frac{1}{\sqrt{|g|}}\partial_\mu\left(\sqrt{|g|}\frac{\partial\mathcal{L}}{\partial(\partial_\mu u)}\right) = \frac{\partial\mathcal{L}}{\partial u}. \tag{10.1.49}$$

Using the translation invariance of the action and (10.1.49), we can derive the *energy-momentum tensor*,[1] also called the *stress tensor*, $T^{\mu\nu}$, given as

$$T^{\mu\nu} = 2\frac{\partial\mathcal{L}}{\partial g_{\mu\nu}} - g^{\mu\nu}\mathcal{L} = 2g^{\alpha\mu}g^{\beta\nu}\frac{\partial\mathcal{L}}{\partial g^{\alpha\beta}} - g^{\mu\nu}\mathcal{L}, \tag{10.1.50}$$

which obeys the *pointwise conservation law*

$$T^{\mu\nu}{}_{;\nu} = 0. \tag{10.1.51}$$

Einstein equations

The basic principle that led Einstein to write down his fundamental equations for gravitation states that the geometry of a spacetime is determined by the matter it contains. Mathematically, Einstein's idea was to consider the equation

$$Q^{\mu\nu} = -\kappa T^{\mu\nu}, \tag{10.1.52}$$

where $Q^{\mu\nu}$ is a 2-tensor generated from the spacetime metric $(g_{\mu\nu})$, which is purely geometric, $T^{\mu\nu}$ is the energy-momentum tensor, which is purely material,

[1] The foundational framework that allows us to derive conserved quantities as a result of symmetry properties of the action is called the *Noether theorem*, which is discussed in Section A.3 in Appendices.

κ is a constant called the Einstein gravitational constant, and the negative sign in front of κ is inserted for convenience. This equation imposes severe restriction to the possible form of the 2-tensor $Q^{\mu\nu}$. For example, $Q^{\mu\nu}$ should also satisfy the same conservation law (or the divergence-free condition),

$$Q^{\mu\nu}{}_{;\nu} = 0, \tag{10.1.53}$$

as $T^{\mu\nu}$ (see (10.1.51)). The simplest candidate for $Q^{\mu\nu}$ could be $g^{\mu\nu}$. However, since $g^{\mu\nu}$ is non-degenerate, this choice is incorrect because it makes $T^{\mu\nu}$ non-degenerate, which is absurd in general. The next candidate could be the Ricci curvature $R^{\mu\nu}$. Since $R^{\mu\nu}$ does not satisfy the required identity (10.1.53), it is abandoned. Consequently, based on both the compatibility condition (10.1.53) and simplicity consideration, we are naturally led to the choice of the Einstein tensor, $G^{\mu\nu}$, defined in (10.1.46). Therefore, we obtain the *Einstein equations*,

$$G^{\mu\nu} = -\kappa T^{\mu\nu} \quad \text{or} \quad G_{\mu\nu} = -\kappa T_{\mu\nu}. \tag{10.1.54}$$

Interestingly, if we multiply (10.1.54) by $g_{\mu\nu}$ or $g^{\mu\nu}$ and sum up, we obtain

$$R = \kappa T, \quad T = g_{\mu\nu} T^{\mu\nu} = g^{\mu\nu} T_{\mu\nu}, \tag{10.1.55}$$

which relates the scalar curvature R and the *trace of energy-momentum tensor* $T = T^{\mu}_{\mu}$ in a simple way. In view of (10.1.55), the Einstein tensor (10.1.46) becomes "half geometric" and "half material" of the form

$$G^{\mu\nu} = R^{\mu\nu} - \frac{\kappa}{2} g^{\mu\nu} T, \tag{10.1.56}$$

such that (10.1.54) becomes a system of the *Ricci tensor equations*

$$R^{\mu\nu} = -\kappa \left(T^{\mu\nu} - \frac{1}{2} g^{\mu\nu} T \right) \quad \text{or} \quad R_{\mu\nu} = -\kappa \left(T_{\mu\nu} - \frac{1}{2} g_{\mu\nu} T \right). \tag{10.1.57}$$

In Section 10.2, we show that the equation (10.1.54) or (10.1.57) recovers Newton's law of gravitation,

$$F = -G \frac{m_1 m_2}{r^2}, \tag{10.1.58}$$

which gives the magnitude of an attractive force between two point particles of masses m_1 and m_2 with a distance r apart, in the static spacetime and slow motion limit, when

$$\kappa = 8\pi G. \tag{10.1.59}$$

Recall that the constant G is called the Newton universal gravitational constant, which is extremely small compared to other quantities.

In summary, we have just derived the Einstein gravitational field equations,

$$G_{\mu\nu} = -8\pi G \, T_{\mu\nu}. \tag{10.1.60}$$

The determination of the constant κ as given in (10.1.59) is presented in the next section.

In the rest of this chapter, the Einstein equations (10.1.60) are the foundational equations of all our subsequent studies of a broad range of interesting problems involving gravitation. Specifically, we begin our study on some of the cosmological consequences of (10.1.60) based on the hypothesis that the universe is modeled with a homogeneous and isotropic perfect fluid. We next consider various black hole solutions of the equations, including massive, charged, and rotating ones. We then discuss a few associated gravitational mass and energy problems. We also present a study of gravitational waves in view of wave equations in weak gravity limit. We end the chapter with a discussion about using scalar-wave matters as a driving force to realize accelerated cosmological expansions of the universe, under the title "*quintessence*" by cosmologists.

10.2 Cosmological consequences

In modern cosmology, the universe is thought to be *homogeneous* (the number of stars per unit volume is uniform throughout large regions of space) and *isotropic* (the number of stars per unit solid angle is the same in all directions). This basic property is known as the *Cosmological Principle* and is supported by astronomical observations. A direct implication of such a principle is that synchronized clocks may be placed throughout the universe to give a uniform measurement of time, often referred to as the *cosmic time*. Another is that the space curvature, say sectional curvature, K, is constant at any fixed cosmic time t. Hence, we have the following simple mathematical descriptions for the space.

Space scenarios

(i) If $K = K(t) > 0$, the space is closed and may be defined as a 3-sphere

$$x^2 + y^2 + z^2 + w^2 = a^2, \quad a = a(t) = \frac{1}{\sqrt{K(t)}}, \tag{10.2.1}$$

embedded in the flat Euclidean space with the line element

$$\mathrm{d}\ell^2 = \mathrm{d}x^2 + \mathrm{d}y^2 + \mathrm{d}z^2 + \mathrm{d}w^2. \tag{10.2.2}$$

(ii) If $K = K(t) < 0$, the space is open and may be defined similarly by the equation

$$x^2 + y^2 + z^2 - w^2 = -a^2, \quad a = a(t) = \frac{1}{\sqrt{|K(t)|}}, \tag{10.2.3}$$

which is embedded in the flat Minkowski space with the line element

$$\mathrm{d}\ell^2 = \mathrm{d}x^2 + \mathrm{d}y^2 + \mathrm{d}z^2 - \mathrm{d}w^2, \tag{10.2.4}$$

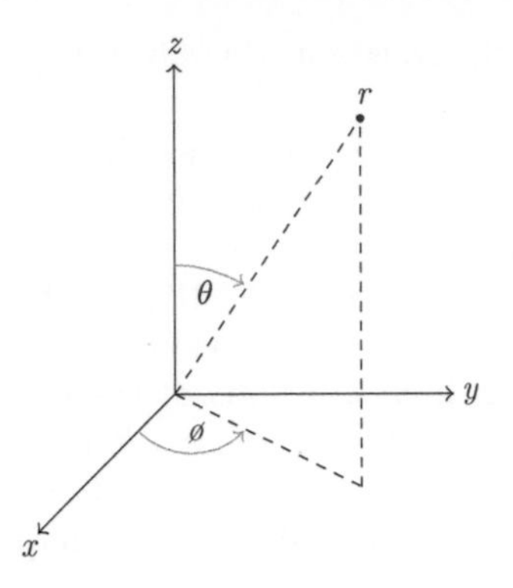

Figure 10.1 An illustration of the spherical coordinates r, θ, ϕ for a point in space, where r is the radial distance, θ the polar angle, and ϕ the azimuthal angle.

known as the *anti-de Sitter (adS) space*,[2] and may well be regarded as being obtained from the closed-space case after making the radius of the 3-sphere and the variable w imaginary:

$$a \mapsto \mathrm{i}a, \quad w \mapsto \mathrm{i}w. \tag{10.2.5}$$

(iii) If $K = K(t) = 0$, the space is the Euclidean space $\mathbb{R}^3$. In particular, the space is flat and its line element is given by

$$\mathrm{d}\ell^2 = \mathrm{d}x^2 + \mathrm{d}y^2 + \mathrm{d}z^2. \tag{10.2.6}$$

For simplicity, we work out the flat case first. Use the conventional spherical coordinates (r, θ, ϕ) to replace the Cartesian coordinates (x, y, z) (Figure 10.1). We have

$$x = r\cos\phi\sin\theta, \quad y = r\sin\phi\sin\theta, \quad z = r\cos\theta. \tag{10.2.7}$$

Thus, we have

$$\begin{aligned}\mathrm{d}\ell^2 &= \mathrm{d}r^2 + r^2\mathrm{d}\theta^2 + r^2\sin^2\theta\,\mathrm{d}\phi^2 \\ &\equiv \mathrm{d}r^2 + r^2\mathrm{d}\Omega_2^2,\end{aligned} \tag{10.2.8}$$

where $\mathrm{d}\Omega_2^2 = \mathrm{d}\theta^2 + \sin^2\theta\,\mathrm{d}\phi^2$ denotes the line element of the unit sphere S^2.

In the closed-space case (i) the line element is formally given by (10.2.2), subject to the constraint (10.2.1). That is, $r\mathrm{d}r + w\mathrm{d}w = 0$. Hence,

$$\mathrm{d}w^2 = \frac{r^2}{w^2}\mathrm{d}r^2 = \frac{r^2}{a^2 - r^2}\mathrm{d}r^2. \tag{10.2.9}$$

[2]In the Minkowski spacetime $\mathbb{R}^{n,1}$ with the metric $\mathrm{d}s^2 = -(\mathrm{d}x^0)^2 + \sum_{i=1}^n(\mathrm{d}x^i)^2$, the de Sitter and anti-de Sitter spaces are the hyperbolic submanifolds defined by $-(x^0)^2 + \sum_{i=1}^n(x^i)^2 = \pm a^2$ $(a > 0)$, denoted by dS_n and adS_n, whose sectional and scalar curvatures are $K = \pm\frac{1}{a^2}$ and $R = \pm\frac{n(n-1)}{a^2}$, respectively.

Consequently,

$$d\ell^2 = \frac{a^2}{a^2 - r^2} dr^2 + r^2 d\Omega_2^2. \tag{10.2.10}$$

Similarly, for the open-space case (ii), in view of the constraint (10.2.3), we have $r dr - w dw = 0$ so that

$$dw^2 = \frac{r^2}{w^2} dr^2 = \frac{r^2}{r^2 + a^2} dr^2. \tag{10.2.11}$$

Substituting (10.2.11) into (10.2.4), we find

$$d\ell^2 = \frac{a^2}{a^2 + r^2} dr^2 + r^2 d\Omega_2^2. \tag{10.2.12}$$

In summary, the line element of a closed or open space is given by

$$d\ell^2 = \frac{a^2}{(a^2 \mp r^2)} dr^2 + r^2 d\theta^2 + r^2 \sin^2\theta \, d\phi^2, \tag{10.2.13}$$

respectively.

Robertson–Walker metric

Finally, inserting (10.2.13) into the spacetime line element

$$ds^2 = dt^2 - d\ell^2, \tag{10.2.14}$$

and making the rescaling $r \mapsto ar$, we have

$$ds^2 = dt^2 - a^2(t) \left(\frac{1}{(1 - kr^2)} dr^2 + r^2 d\theta^2 + r^2 \sin^2\theta \, d\phi^2 \right), \tag{10.2.15}$$

where $k = \pm 1$ or $k = 0$ according to whether the space is assumed to be closed, open, or flat, respectively. This is the most general line element of a homogeneous and isotropic spacetime and is known as the *Robertson–Walker metric.*

Perfect fluid

In cosmology, the large-scale viewpoint allows us to treat stars or galaxies as particles of a perfect "fluid" or "gas" that fills the universe and is characterized by its mass-energy density ρ, counting both rest mass and kinetic energy per unit volume, and pressure p, which appear as two scalar fields, so that the associated energy-momentum tensor $T_{\mu\nu}$ is given by

$$T_{\mu\nu} = (\rho + p) u_\mu u_\nu - p g_{\mu\nu} \quad \text{or} \quad T^{\mu\nu} = (\rho + p) u^\mu u^\nu - p g^{\mu\nu}, \tag{10.2.16}$$

where u_μ is the 4-velocity of the gas particles and $(g_{\mu\nu})$ is the spacetime metric.

Energy-momentum tensor digression

At this moment, we pause to discuss some features of the energy-momentum tensor (10.2.16). For simplicity, we assume a flat spacetime with the global Minkowski metric

$$\mathrm{d}s^2 = c^2\mathrm{d}\tau^2 = c^2\mathrm{d}t^2 - \mathrm{d}(x^1)^2 - \mathrm{d}(x^2)^2 - \mathrm{d}(x^3)^2 = \eta_{\mu\nu}\mathrm{d}x^\mu\mathrm{d}x^\nu, \tag{10.2.17}$$

where we restore the speed of light c in order to see the role played by relativity more explicitly, τ denotes proper time, $x^0 = ct$, and $(\eta_{\mu\nu}) = \mathrm{diag}(1, -1, -1, -1)$, as in Chapter 4. Since in a co-moving frame with $u^0 = c$ and $u^i = 0$ the quantity T^{00} as rest energy density is $c^2\rho$ (other components of $T^{\mu\nu}$ are seen to enjoy the properties $T^{0i} = T^{i0} = 0$ and $T^{ij} = p\delta^{ij}$, indicating absence of heat conduction or momentum density and vanishing viscosity or shear stresses but an isotropic fluid pressure, respectively), we see that the energy-momentum tensor (10.2.16) should assume the updated form

$$T^{\mu\nu} = \left(\rho + \frac{p}{c^2}\right) u^\mu u^\nu - p\eta^{\mu\nu}, \tag{10.2.18}$$

in general. In terms of the mixed tensor $T^\mu_\nu = \eta_{\nu\alpha}T^{\mu\alpha}$, we have the associated Hamiltonian energy density $\mathcal{H}$ and trace T given by

$$\mathcal{H} = T^0_0 = c^2\rho, \tag{10.2.19}$$

$$T = T^\mu_\mu = c^2\rho - 3p. \tag{10.2.20}$$

For convenience, we rewrite (10.2.18) in the matrix form

$$(T^{\mu\nu}) = \begin{pmatrix} \left(\rho + \frac{p}{c^2}\right)(u^0)^2 - p & \left(\rho + \frac{p}{c^2}\right) u^0 u^j \\ \left(\rho + \frac{p}{c^2}\right) u^i u^0 & \left(\rho + \frac{p}{c^2}\right) u^i u^j + p\delta^{ij} \end{pmatrix}_{i,j=1,2,3}. \tag{10.2.21}$$

Thus, noting that the conservation law (10.1.51) is now replaced with the flat-space one:

$$\partial_\nu T^{\mu\nu} = T^{\mu\nu}{}_{,\nu} = 0, \tag{10.2.22}$$

inserting (10.2.21) into (10.2.22) with $\partial_t = c\partial_0$, and using the low-velocity approximation

$$(u^\mu) = (v^\mu)\frac{\mathrm{d}t}{\mathrm{d}\tau} \approx (c, \mathbf{v}), \quad \mathbf{v} = (v^i), \quad v^i = \frac{\mathrm{d}x^i}{\mathrm{d}t}, \tag{10.2.23}$$

we see that the component of (10.2.22) at $\mu = 0$ becomes

$$\partial_t\rho + \partial_i(\rho v^i) + \frac{1}{c^2}\partial_i(pv^i) = 0, \tag{10.2.24}$$

which appears in the form of a law of matter conservation:

$$\partial_t\rho + \nabla\cdot\left(\left[\rho + \frac{p}{c^2}\right]\mathbf{v}\right) = 0, \tag{10.2.25}$$

and the component of (10.2.22) at each $\mu = i = 1, 2, 3$ reads

$$\partial_t(\rho v^i) + \partial_j(\rho v^i)v^j + \frac{1}{c^2}\left(\partial_t(pv^i) + \partial_j(pv^i)v^j\right)$$
$$+ \left(\rho + \frac{p}{c^2}\right)(\partial_j v^j)v^i + \delta^{ij}\partial_j p = 0, \tag{10.2.26}$$

which with the *total* or *convective derivative*

$$\frac{D}{dt} = \partial_t + v^j\partial_j \equiv \partial_t + \mathbf{v}\cdot\nabla, \tag{10.2.27}$$

may be rewritten as

$$\frac{D(\rho\mathbf{v})}{dt} + \frac{1}{c^2}\frac{D(p\mathbf{v})}{dt} + \left(\rho + \frac{p}{c^2}\right)(\nabla\cdot\mathbf{v})\,\mathbf{v} = -\nabla p. \tag{10.2.28}$$

In particular, in the classical fluid limit, $c \to \infty$, (10.2.25) and (10.2.28) become

$$\partial_t\rho + \nabla\cdot(\rho\mathbf{v}) = 0, \tag{10.2.29}$$

$$\frac{D(\rho\mathbf{v})}{dt} + \rho\,(\nabla\cdot\mathbf{v})\,\mathbf{v} = -\nabla p. \tag{10.2.30}$$

Substituting (10.2.29) into (10.2.30), the equation (10.2.30) is simplified into

$$\rho\frac{D\mathbf{v}}{dt} + \rho(\nabla\cdot\mathbf{v})\mathbf{v} = -\nabla p, \tag{10.2.31}$$

known as the *Euler equation for inviscid flow*. When the fluid is *viscous*, the equation (10.2.31) is modified into

$$\rho\frac{D\mathbf{v}}{dt} + \rho(\nabla\cdot\mathbf{v})\mathbf{v} = \mu\nabla^2\mathbf{v} - \nabla p, \tag{10.2.32}$$

which is the celebrated *Navier–Stokes equation*, in which $\mu > 0$ measures *viscosity* of the fluid.

Below, we return to using the convention $c = 1$ for the speed of light.

Newtonian limit and determination of coupling constant

We now consider the weak field limit of the Einstein gravitational equations (10.1.54) and show how to determine the coupling constant κ. For this purpose, we assume that the gravitational metric (10.1.1) is close to that of the standard Minkowski metric and static, and the motion of a test particle of mass m is slow such that the proper time may be replaced by the cosmic time and the 4-velocity (u^μ) of the particle is approximated by the vector $(1, 0, 0, 0)$. Thus, keeping leading orders, the geodesic equation or the equation of motion (10.1.19) of the particle assumes the approximate form

$$\frac{dv^i}{dt} + \Gamma^i_{00} = 0, \quad v^i = \frac{dx^i}{dt}. \tag{10.2.33}$$

On the other hand, by the definition of the Christoffel symbols, we have

$$\Gamma^i_{00} = -\frac{1}{2}g^{i\mu}(\partial_\mu g_{00}) \approx \frac{1}{2}\delta^{ij}(\partial_j g_{00}). \tag{10.2.34}$$

Inserting (10.2.34) into (10.2.33), we arrive at the equation

$$\frac{\mathrm{d}\mathbf{v}}{\mathrm{d}t} = -\nabla V, \quad V = \frac{g_{00}}{2}. \tag{10.2.35}$$

In other words, the quantity $\frac{g_{00}}{2}$ plays the role of the potential function in Newtonian theory of gravitation. Furthermore, from (10.1.27) and (10.1.37), we have

$$\begin{aligned} R_{00} &= R^\mu_{00\mu} \\ &= \Gamma^\mu_{0\mu,0} - \Gamma^\mu_{00,\mu} + \Gamma^\mu_{0\nu}\Gamma^\nu_{\mu 0} - \Gamma^\mu_{00}\Gamma^\nu_{\mu\nu} \\ &\approx -\Gamma^\mu_{00,\mu}, \end{aligned} \tag{10.2.36}$$

after neglecting higher-order terms. Assume the zero-pressure limit $p = 0$ such that both T_{00} and the trace of (10.2.18) are given by the mass density ρ. Insert these into the $\mu = \nu = 0$ component of (10.1.57) and use (10.2.36). We get

$$\Gamma^i_{00,i} = \frac{\kappa}{2}\rho. \tag{10.2.37}$$

Combining (10.2.34), (10.2.35), and (10.2.37), we obtain

$$\nabla^2 V = \frac{\kappa}{2}\rho. \tag{10.2.38}$$

Comparing (10.2.38) with the classical Poisson equation

$$\nabla^2 V = 4\pi G\rho, \tag{10.2.39}$$

for the Newton gravitational potential, V, we arrive at the relation (10.1.59), for the determination of κ, stated earlier.

Friedmann equation

We now consider some possible consequences of a homogeneous and isotropic universe in view of the Einstein theory. The cosmological principle requires that ρ and p in (10.2.16) depend on time t only. From (10.1.27) and (10.1.37), we can represent the Ricci tensor in terms of the Christoffel symbols by

$$R_{\mu\nu} = \Gamma^\alpha_{\mu\alpha,\nu} - \Gamma^\alpha_{\mu\nu,\alpha} + \Gamma^\alpha_{\mu\beta}\Gamma^\beta_{\alpha\nu} - \Gamma^\alpha_{\mu\nu}\Gamma^\beta_{\alpha\beta}. \tag{10.2.40}$$

Naturally, we label our coordinates according to $x^0 = t, x^1 = r, x^2 = \theta$, $x^3 = \phi$, in the Robertson–Walker metric (10.2.15). Then, its nonzero metric components and Christoffel symbols are

$$g_{00} = 1, \quad g_{11} = -\frac{a^2(t)}{1-kr^2}, \quad g_{22} = -a^2(t)r^2, \quad g_{33} = -a^2(t)r^2\sin^2\theta, \tag{10.2.41}$$

and

$$
\begin{aligned}
&\Gamma^0_{11} = \frac{a(t)\dot{a}(t)}{(1-kr^2)}, \quad \Gamma^0_{22} = a(t)\dot{a}(t)r^2, \quad \Gamma^0_{33} = a(t)\dot{a}(t)r^2\sin^2\theta,\\
&\Gamma^1_{01} = \frac{\dot{a}(t)}{a(t)}, \quad \Gamma^1_{11} = \frac{kr}{(1-kr^2)},\\
&\Gamma^1_{22} = -r(1-kr^2), \quad \Gamma^1_{33} = -r(1-kr^2)\sin^2\theta,\\
&\Gamma^2_{02} = \frac{\dot{a}(t)}{a(t)}, \quad \Gamma^2_{12} = \frac{1}{r}, \quad \Gamma^2_{33} = -\sin\theta\cos\theta,\\
&\Gamma^3_{03} = \frac{\dot{a}(t)}{a(t)}, \quad \Gamma^3_{13} = \frac{1}{r}, \quad \Gamma^3_{23} = \cot\theta, \qquad (10.2.42)
\end{aligned}
$$

where $\dot{a}(t) = \frac{\mathrm{d}a(t)}{\mathrm{d}t}$, respectively. Inserting (10.2.42) into (10.2.40), we see that the Ricci tensor $R_{\mu\nu}$ becomes diagonal with

$$R_{00} = \frac{3\ddot{a}}{a}, \qquad (10.2.43)$$

$$R_{11} = -\frac{a\ddot{a} + 2\dot{a}^2 + 2k}{1 - kr^2}, \qquad (10.2.44)$$

$$R_{22} = -(a\ddot{a} + 2\dot{a}^2 + 2k)r^2, \qquad (10.2.45)$$

$$R_{33} = -(a\ddot{a} + 2\dot{a}^2 + 2k)r^2\sin^2\theta. \qquad (10.2.46)$$

Note that, currently, the computation of Riemannian tensors has been facilitated enormously by available symbolic packages. See [84, 383, 457] and references therein. Hence, the scalar curvature (10.1.38) becomes

$$R = \frac{6}{a^2}(a\ddot{a} + \dot{a}^2 + k). \qquad (10.2.47)$$

Note that, in the static limit situation, the formula (10.2.47) coincides with the scalar curvature of the underlying 3-space when $k = 1$ or -1.

On the other hand, from (10.1.19) and (10.2.42), we see that the geodesics of the metric (10.2.15), which are the trajectories of moving stars and galaxies when net local interactions are neglected, are given by $r, \theta, \phi =$ constant. Thus, in (10.2.16) we have $u_0 = 1$ and $u_i = 0$, $i = 1, 2, 3$. Therefore, $T_{\mu\nu}$ is also diagonal with

$$T_{00} = \rho, \quad T_{11} = \frac{pa^2}{1-kr^2}, \quad T_{22} = pa^2r^2, \quad T_{33} = pa^2r^2\sin^2\theta. \qquad (10.2.48)$$

Substituting (10.2.43)–(10.2.46), (10.2.47), and (10.2.48) into the Einstein equations (10.1.60), we arrive at the following two equations,

$$3\ddot{a} = -4\pi G(\rho + 3p)a, \qquad (10.2.49)$$

$$a\ddot{a} + 2\dot{a}^2 + 2k = 4\pi G(\rho - p)a^2. \qquad (10.2.50)$$

Thus, eliminating $\ddot{a}$ from these equations, we arrive at the celebrated *Friedmann equation*

$$\dot{a}^2 + k = \frac{8\pi}{3} G\rho a^2, \tag{10.2.51}$$

which occupies a fundamental position in modern cosmology [396, 563, 569].

We can show that, in the category of time-dependent solutions, the Einstein cosmological equations, (10.2.49) and (10.2.50), are in fact equivalent to the single Friedmann equation (10.2.51). To this end, recall that both systems are to be subject to the conservation law for the energy-momentum tensor, namely, $T^{\mu\nu}{}_{;\nu} = 0$ or

$$\dot{\rho} + 3(\rho + p)\frac{\dot{a}}{a} = 0. \tag{10.2.52}$$

Differentiating (10.2.51), using (10.2.52), and assuming $\dot{a} \neq 0$ in observation of time dependence, we get (10.2.49). Inserting (10.2.51) into (10.2.49), we get (10.2.50).

Hubble constant, expansion of universe, and beginning of time

The relative rate of change of the radius of the universe is recognized as *Hubble's "constant," $H(t)$*, which is given by

$$H(t) = \frac{\dot{a}(t)}{a(t)}. \tag{10.2.53}$$

Recent estimates for Hubble's constant put it at about $(18 \times 10^9 \text{ years})^{-1}$. In particular, $\dot{a} > 0$ at present. However, since (10.2.49) indicates that $\ddot{a} < 0$ everywhere with the assumption $p \geq 0$, we conclude that $\dot{a} > 0$ for all time in the past. In other words, *the universe has undergone a process of expansion in the past.*

We now investigate whether the universe has a beginning time. For this purpose, let t_0 denote the present time and t denote any past time, $t < t_0$. The property $\ddot{a} < 0$ again gives us $\dot{a}(t) > \dot{a}(t_0)$, which implies that

$$a(t_0) - a(t) > \dot{a}(t_0)(t_0 - t). \tag{10.2.54}$$

Thus, there must be a finite time t in the past, $t < t_0$, when a vanishes. Such a time may be defined as the time when the universe begins. For convenience, we may assume that the universe begins at $t = 0$, namely, $a(0) = 0$. Hence, we arrive at the general picture of the *Big Bang cosmology*, which says that the universe started at a finite time in the past from a singular point when $a = 0$ and has been expanding in all its history of evolution.

Inserting (10.2.53) into (10.2.47), we obtain

$$R(t) = 6\left(\dot{H}(t) + 2H^2(t) + K(t)\right), \quad K = \frac{k}{a^2}, \tag{10.2.55}$$

which relates the scalar curvature of the spacetime to the Hubble constant and the space curvature.

Special solutions

We now present some special solutions of the Friedmann equation (10.2.51) in cosmology. For conciseness, we only consider the flat-space case $k = 0$, known as the *Einstein–de Sitter universe* [191], and thus rewrite the equation as

$$\dot{a}^2 = \frac{8\pi}{3} G\rho a^2. \tag{10.2.56}$$

This equation is to be supplemented with an equation that relates the fluid pressure p to the fluid mass density ρ generically by

$$p = f(\rho), \tag{10.2.57}$$

called the *equation of state*, which specifies the physical characteristics of the fluid. For example, the simplest situation

$$p = w\rho, \tag{10.2.58}$$

spells out a state called the *barotropic state*, where w is a constant, such that $w = 0$ gives rise to the *dust model* of vanishing pressure and $w = \frac{1}{3}$ the *radiation-dominated model*. Inserting (10.2.58) into (10.2.52), we have

$$\dot{\rho} + 3(1+w)\rho\frac{\dot{a}}{a} = 0, \tag{10.2.59}$$

which may be integrated to give us the solution

$$\rho = \rho_0 a^{-3(1+w)}, \tag{10.2.60}$$

where $\rho_0 > 0$ is an integration constant. Substituting (10.2.60) into (10.2.56) and integrating, we get the solution

$$a(t) = (6\pi G\rho_0)^{\frac{1}{3(1+w)}}([1+w]t)^{\frac{2}{3(1+w)}}, \quad t \geq 0, \tag{10.2.61}$$

evolving from $a(0) = 0$. In view of (10.2.60) and (10.2.61), we have

$$\rho(t) = \frac{1}{6\pi G(1+w)^2 t^2}, \quad t > 0. \tag{10.2.62}$$

Note that the right-hand side of (10.2.62) is independent of the prescribable integration constant ρ_0 and that $\rho(t)$ blows up as $t \to 0$, simulating a universe with extremely high mass density initially.

Use $t_0 > 0$ to denote the present *age of the universe*. Then (10.2.62) relates t_0 to the present matter density, $\rho(t_0)$, by

$$t_0^2 = \frac{1}{6\pi G(1+w)^2 \rho(t_0)}. \tag{10.2.63}$$

On the other hand, let

$$H_0 = \frac{\dot{a}(t_0)}{a(t_0)} \tag{10.2.64}$$

denote the present measured value of the Hubble constant. So, in view of (10.2.64), the equation (10.2.56) renders us the relation

$$\rho(t_0) = \frac{3H_0^2}{8\pi G}. \tag{10.2.65}$$

Substituting (10.2.65) into (10.2.63), we obtain

$$t_0 = \frac{2}{3(1+w)H_0}, \tag{10.2.66}$$

which clearly indicates, in the context of the Friedmann equation, how the estimate of t_0 depends on the choice of a model or the equation of state (10.2.58), labeled by w. Furthermore, in terms of the *Hubble time*

$$t_{\mathrm{H}} = \frac{1}{H_0}, \tag{10.2.67}$$

the age of the universe t_0 reads

$$t_0 = \frac{2t_{\mathrm{H}}}{3(1+w)}. \tag{10.2.68}$$

The current measured value of t_{H} is about 14.4×10^9(or 14.4 billion) years [282].

Another interesting situation is the *dust and radiation model* incorporating or interpolating (10.2.60) as follows

$$\rho = \rho_{\mathrm{d}} a^{-3} + \rho_{\mathrm{r}} a^{-4}, \tag{10.2.69}$$

where $\rho_{\mathrm{d}} > 0$ and $\rho_{\mathrm{r}} > 0$ are parameters arising respectively from the dust- and radiation-dominated integration constants as before. From (10.2.69), the equation (10.2.56) becomes

$$\dot{a}^2 = \frac{8\pi G}{3}\left(\frac{\rho_{\mathrm{d}}}{a} + \frac{\rho_{\mathrm{r}}}{a^2}\right), \tag{10.2.70}$$

which may be integrated to yield the implicit solution

$$\begin{aligned}&(\rho_{\mathrm{d}} a(t) + \rho_{\mathrm{r}})^{\frac{3}{2}} - 3\rho_{\mathrm{r}}\left(\rho_{\mathrm{d}} a(t) + \rho_{\mathrm{r}}\right)^{\frac{1}{2}} + 2\rho_{\mathrm{r}}^{\frac{3}{2}}\\ &= \sqrt{6\pi G}\rho_{\mathrm{d}}^2\, t, \quad t \geq 0; \quad a(0) = 0.\end{aligned} \tag{10.2.71}$$

It is clear that the function $a(t)$ obeys the asymptotic estimate

$$a(t) = \mathrm{O}(t^{\frac{2}{3}}), \quad t \to \infty, \tag{10.2.72}$$

indicating that eventually the dust model overwhelms the radiation model.

Rise and applications of cosmological constant

It is easy to see that the equations (10.2.49) do not allow *static* (time-independent) solutions. When Einstein applied his gravitational equations to cosmology, he hoped to obtain a homogeneous, isotropic, static, and compact universe. As a consequence, he modified his equations into [168]

$$G_{\mu\nu} + \Lambda g_{\mu\nu} = -8\pi G\, T_{\mu\nu}, \tag{10.2.73}$$

where Λ is a constant called the *cosmological constant.* Note that the added cosmological term, $\Lambda g_{\mu\nu}$, does not violate the required divergence-free condition. Although static models of the universe have long been discarded since Hubble's 1929 discovery that the universe is expanding, a nonvanishing cosmological constant gives important implications in the theoretical studies of the early-universe cosmology. In fact, the equations (10.2.73) may also be rewritten as

$$G_{\mu\nu} = -8\pi G\tilde{T}_{\mu\nu} \equiv -8\pi G\left(T_{\mu\nu} + T_{\mu\nu}^{(\text{vac})}\right), \quad T_{\mu\nu}^{(\text{vac})} = \frac{\Lambda}{8\pi G} g_{\mu\nu}, \tag{10.2.74}$$

where $T_{\mu\nu}^{(\text{vac})}$ is interpreted as the energy-momentum tensor associated with the vacuum. The vacuum polarization of quantum field theory endows the vacuum with a nonzero energy-momentum tensor, which is completely unobservable except by its gravitational effects. In particular,

$$\rho^{(\text{vac})} = T_{00}^{(\text{vac})} = \frac{\Lambda}{8\pi G}\, g_{00} \tag{10.2.75}$$

may be viewed as the mass-energy density of the vacuum. This viewpoint imposes a natural restriction on the sign of the cosmological constant, $\Lambda \geq 0$.

Multiplying (10.2.73) by the metric $g^{\mu\nu}$ and summing over repeated indices, we find

$$R = 8\pi G T + 4\Lambda, \quad T = g^{\mu\nu} T_{\mu\nu}. \tag{10.2.76}$$

Inserting (10.2.76) into (10.2.73), we obtain the more elegant equations

$$R_{\mu\nu} - \Lambda g_{\mu\nu} = -8\pi G\left(T_{\mu\nu} - \frac{1}{2} g_{\mu\nu} T\right). \tag{10.2.77}$$

In particular, in the absence of matter, we have the *vacuum Einstein equations*

$$R_{\mu\nu} = \Lambda g_{\mu\nu}. \tag{10.2.78}$$

Any spacetime satisfying (10.2.78) is called an *Einstein space* and its metric $g_{\mu\nu}$ is called an *Einstein metric.*

We now reexamine the perfect fluid model given by (10.2.16) in the presence of a cosmological constant. Inserting (10.2.16) into (10.2.74), we have

$$\tilde{T}_{\mu\nu} = (\tilde{\rho} + \tilde{p}) u_\mu u_\nu - \tilde{p} g_{\mu\nu}, \tag{10.2.79}$$

where $\tilde{\rho}$ and $\tilde{p}$ are the updated *effective* mass density and fluid pressure given by

$$\tilde{\rho} = \rho + \frac{\Lambda}{8\pi G}, \quad \tilde{p} = p - \frac{\Lambda}{8\pi G}, \tag{10.2.80}$$

respectively. Note that a positive cosmological constant contributes to the effective mass density *positively*, but the pressure *negatively*. As a result, we arrive at the modified cosmological equations

$$3\ddot{a} = -4\pi G(\tilde{\rho} + 3\tilde{p})a, \tag{10.2.81}$$

$$a\ddot{a} + 2\dot{a}^2 + 2k = 4\pi G(\tilde{\rho} - \tilde{p})a^2, \tag{10.2.82}$$

replacing (10.2.49) and (10.2.50).

Static universe

For a static universe, the equation (10.2.81) leads to the critical condition $\tilde{\rho} + 3\tilde{p} = 0$ or

$$\rho = \frac{\Lambda}{4\pi G} - 3p, \tag{10.2.83}$$

in view of (10.2.80). Of course, the conservation law for $\tilde{T}_{\mu\nu}$ now is automatically satisfied. Furthermore, the equation (10.2.82) gives us

$$(\Lambda - 8\pi G p)a^2 = k. \tag{10.2.84}$$

Thus, for the dust model, $p = 0$, (10.2.83) spells out the critical mass density in terms of Λ,

$$\rho = \frac{\Lambda}{4\pi G}, \tag{10.2.85}$$

and (10.2.84) indicates the relation

$$a^2 = \frac{k}{\Lambda}. \tag{10.2.86}$$

These results imply that a static universe with a positive mass density must be closed, $k = 1$, a universe also known as the *Einstein static universe.*

For the radiation-dominated model, $p = \frac{1}{3}\rho$, (10.2.83) and (10.2.84) give us

$$\rho = \frac{\Lambda}{8\pi G}, \quad a^2 = \frac{3k}{2\Lambda}, \tag{10.2.87}$$

which still lead to a closed universe, but with a smaller mass density and a larger space radius.

Time-dependent universe

We now discuss time-dependent solutions. For simplicity, we assume the flat space, $k = 0$, and recast (10.2.81)–(10.2.82) into the correspondingly modified Friedmann equation

$$\begin{aligned} \dot{a}^2 &= \frac{8\pi}{3} G\tilde{\rho} a^2 \\ &= \frac{8\pi}{3} G\rho a^2 + \frac{\Lambda}{3} a^2, \quad \Lambda > 0. \end{aligned} \tag{10.2.88}$$

Thus, with the linear equation of state (10.2.58) so that the mass density is given by (10.2.60), we may integrate (10.2.88) subject to the initial condition $a(0) = 0$ to get the solution

$$a^{3(1+w)}(t) = \frac{8\pi G\rho_0}{\Lambda} \sinh^2\left(\frac{1}{2}\sqrt{3\Lambda}(1+w)t\right), \quad t \geq 0, \tag{10.2.89}$$

giving us the following exponential growth pattern for the asymptotic form of the solution:

$$a(t) \sim \left(\frac{8\pi G\rho_0}{\Lambda}\right)^{\frac{1}{3(1+w)}} \mathrm{e}^{\sqrt{\frac{\Lambda}{3}}\,t}, \quad t \to \infty. \tag{10.2.90}$$

This kind of exponential growth pattern of the radial factor $a(t)$ is an indication of the presence of *dark energy*. In other words, the presence of a positive cosmological constant gives rise to a mechanism for the appearance of dark energy. Such a picture is consistent with (10.2.75), which says that the onset of a positive cosmological constant puts forth a positive mass density for the vacuum background, which may be interpreted as dark energy.

Chaplygin fluid and dark energy

We saw in (10.2.80) that the presence of a positive cosmological constant contributes positively to the mass density, but negatively to the pressure of the cosmological fluid, resulting in an exponential growth law, (10.2.90), of the radial factor, an indication of the presence of dark energy. Thus, it may be possible to achieve the same growth behavior for the radial factor when the pressure is allowed to be *negative* even when the cosmological constant is *absent*. The *Chaplygin fluid* is such a model with the equation of state given by [26, 110, 519]

$$p = A\rho - \frac{B}{\rho}, \tag{10.2.91}$$

where $A > -1, B > 0$ are some parameters. Inserting (10.2.91) into (10.2.52), we obtain

$$(1 + A)\rho^2 = Ca^{-6(1+A)} + B, \tag{10.2.92}$$

where $C > 0$ is an integration constant. The borderline situation

$$A = -1 \tag{10.2.93}$$

stands out, which is known as the *phantom divide line*, an actively pursued topic [136, 226, 390, 411, 562] in the study on dark energy.

With $\Lambda = 0$ and inserting (10.2.92) into the flat-space Friedmann equation (10.2.56), we have

$$\dot{a}^2 = \frac{8\pi G}{3\sqrt{1+A}}\left(Ca^{-6(1+A)} + B\right)^{\frac{1}{2}} a^2. \tag{10.2.94}$$

Thus, let $\beta > 0$ be given by

$$\beta^2 = \frac{8\pi G}{3\sqrt{1+A}}. \tag{10.2.95}$$

Then we may integrate (10.2.94) subject to the Big Bang initial condition $a(0) = 0$ to get the implicit solution

$$a^{6(1+A)}(t) = \frac{C}{B\left(\coth^4\left[3(1+A)\beta B^{\frac{1}{4}}\, t + b(t)\right] - 1\right)}, \tag{10.2.96}$$

$$b(t) = \arctan\left(1 + \frac{C}{B}a^{-6(1+A)}(t)\right)^{\frac{1}{4}} - \frac{\pi}{2}, \tag{10.2.97}$$

which looks rather complicated. Nevertheless, since the function $b(t)$ in (10.2.97) is bounded, we see that the scale factor $a(t)$ obeys the asymptotic exponential growth law

$$a(t) \sim \left(\frac{C}{8B}\right)^{\frac{1}{6(1+A)}} \mathrm{e}^{\beta B^{\frac{1}{4}}\, t}, \quad t \to \infty. \tag{10.2.98}$$

It is interesting to note the manner how the exponential rate in the growth law (10.2.98) given as

$$\varepsilon = \beta B^{\frac{1}{4}} = 2\left(\frac{2\pi G}{3}\right)^{\frac{1}{2}}\left(\frac{B}{1+A}\right)^{\frac{1}{4}}, \tag{10.2.99}$$

sometimes loosely referred to as dark energy, depends on the parameters A, B in the Chaplygin equation of state. In particular, we see that $\varepsilon \to \infty$ as $A \to -1$. In other words, the dark energy could undergo an explosive creation in the vicinity of the phantom divide line.

See [124, 125, 126] and references therein for a presentation of a wide range of solutions of the Friedmann equation for all types of spaces and the values of the cosmological constant, as well as some discussion on the associated cosmological consequences.

10.3 Schwarzschild black-hole solution

In the situation when both the cosmological constant Λ and the matter energy-momentum tensor vanish, the Einstein equations (10.2.77) become

$$R_{\mu\nu} = 0, \tag{10.3.1}$$

which says that the vacuum spacetime is characterized by its Ricci tensor being trivial. In the context of the static spherically symmetric limit with the standard ordered coordinates (t, r, θ, ϕ) used to count for (x^μ), the metric element takes the form [168, 211, 569]:

$$\mathrm{d}s^2 = A(r)\,\mathrm{d}t^2 - B(r)\,\mathrm{d}r^2 - r^2(\mathrm{d}\theta^2 + \sin^2\theta\,\mathrm{d}\phi^2), \tag{10.3.2}$$

where $A(r)$ and $B(r)$ are some unknown functions of the radial variable r only to be determined. With (10.3.2), the nontrivial and independent components of the metric tensor, Christoffel symbols, and Ricci tensor are given by

$$g_{00} = A(r), \quad g_{11} = -B(r), \quad g_{22} = -r^2, \quad g_{33} = -r^2\sin^2\theta, \tag{10.3.3}$$

$$\Gamma^0_{01} = \frac{A'}{2A}, \quad \Gamma^1_{00} = \frac{A'}{2B}, \quad \Gamma^1_{11} = \frac{B'}{2B}, \tag{10.3.4}$$

$$\Gamma^1_{22} = -\frac{r}{B}, \quad \Gamma^1_{33} = -\frac{r\sin^2\theta}{B}, \quad \Gamma^2_{12} = \frac{1}{r}, \tag{10.3.5}$$

$$\Gamma^2_{33} = -\cos\theta\sin\theta, \quad \Gamma^3_{13} = \frac{1}{r}, \quad \Gamma^3_{23} = \cot\theta, \tag{10.3.6}$$

$$R_{00} = -\frac{A''}{2B} + \frac{A'}{4B}\left(\frac{A'}{A} + \frac{B'}{B}\right) - \frac{A'}{rB}, \tag{10.3.7}$$

$$R_{11} = \frac{A''}{2A} - \frac{A'}{4A}\left(\frac{A'}{A} + \frac{B'}{B}\right) - \frac{B'}{rB}, \tag{10.3.8}$$

$$R_{22} = -1 + \frac{1}{B} + \frac{r}{2B}\left(\frac{A'}{A} - \frac{B'}{B}\right), \tag{10.3.9}$$

$$R_{33} = \sin^2\theta\, R_{22}, \tag{10.3.10}$$

respectively, where and in the sequel we use the notation $A' = \frac{\mathrm{d}A}{\mathrm{d}r}$, and so on.

We use these expressions to solve (10.3.1) in terms of A, B.

Solution to vacuum equations

First, from (10.3.7) and (10.3.8), we have

$$\frac{B}{A}R_{00} + R_{11} = -\frac{1}{r}\left(\frac{A'}{A} + \frac{B'}{B}\right). \tag{10.3.11}$$

Thus, setting $R_{00} = 0, R_{11} = 0$, we get

$$\frac{A'}{A} + \frac{B'}{B} = 0 \quad \text{or} \quad (AB)' = 0, \tag{10.3.12}$$

giving $AB = K =$ constant. Recall that in spherical coordinates the flat Minkowski metric reads

$$\mathrm{d}s^2 = \mathrm{d}t^2 - \mathrm{d}r^2 - r^2(\mathrm{d}\theta^2 + \sin^2\theta\,\mathrm{d}\phi^2), \tag{10.3.13}$$

which should serve as the asymptotic limit of (10.3.2) in the spatial infinity $r \to \infty$. Hence, $A(\infty) = 1, B(\infty) = 1$. So $K = 1$.

Next, using $AB = 1$ in R_{22}, we have

$$rA' + A = 1, \tag{10.3.14}$$

which may be integrated to give us

$$A(r) = 1 + \frac{C}{r}, \tag{10.3.15}$$

where C is an integration constant.

Finally, we can examine that the pair of functions A, B where A is given in (10.3.15) and $B = \frac{1}{A}$ indeed makes (10.3.7)–(10.3.9) vanish identically. Therefore, the metric element (10.3.2) becomes

$$\mathrm{d}s^2 = \left(1 + \frac{C}{r}\right) \mathrm{d}t^2 - \left(1 + \frac{C}{r}\right)^{-1} \mathrm{d}r^2 - r^2(\mathrm{d}\theta^2 + \sin^2\theta\, \mathrm{d}\phi^2). \tag{10.3.16}$$

It is clear that, in the $r \to \infty$ limit, the metric (10.3.16) becomes the flat Minkowski spacetime metric (10.3.13) as desired.

Determination of integration constant

Since regularity of the metric element requires $A > 0$, we may be tempted to take $C > 0$ in (10.3.15). Unfortunately, or fortunately, the situation we are facing here is not so simple and a more elaborate consideration is required so that the solution is physically meaningful. Indeed, we may seek for a solution that recovers the Newton law of gravitation generated from a centrally localized mass, say M, in the region where r is sufficiently large. Thus, we are led to

$$\frac{1}{2}\left(1 + \frac{C}{r}\right) = \frac{g_{00}}{2} = V, \tag{10.3.17}$$

with the notation of (10.2.35), where V should up to a constant approximate the Newton gravitation potential, $-\frac{GM}{r}$, of the central mass M, as $r \to \infty$, resulting in

$$\nabla V = \nabla\left(-\frac{GM}{r}\right), \tag{10.3.18}$$

neglecting errors. Hence, we arrive at the conclusion

$$C = -2GM. \tag{10.3.19}$$

Schwarzschild solution

In summary, we have arrived at the solution represented by

$$\mathrm{d}s^2 = \left(1 - \frac{2GM}{r}\right) \mathrm{d}t^2 - \left(1 - \frac{2GM}{r}\right)^{-1} \mathrm{d}r^2 - r^2(\mathrm{d}\theta^2 + \sin^2\theta\, \mathrm{d}\phi^2), \tag{10.3.20}$$

which is the celebrated *Schwarzschild metric* or *Schwarzschild solution* of the Einstein equations. The coordinates (t, r, θ, ϕ) are sometimes specifically called the *Schwarzschild coordinates.* Schwarzschild obtained this solution in 1915, the same year Einstein published his work on general relativity. For this solution, there are coordinate singularities at the radius

$$r = r_S = 2GM, \tag{10.3.21}$$

referred to as the *Schwarzschild radius.* With r_S, we rewrite (10.3.20) as

$$\mathrm{d}s^2 = \left(1 - \frac{r_S}{r}\right) \mathrm{d}t^2 - \left(1 - \frac{r_S}{r}\right)^{-1} \mathrm{d}r^2 - r^2(\mathrm{d}\theta^2 + \sin^2\theta\, \mathrm{d}\phi^2). \tag{10.3.22}$$

The singular sphere, $r = r_S$, in space, also called the *Schwarzschild surface,* is an *event horizon.* The solution (10.3.20) represents an empty-space solution or *exterior solution* that is valid outside a spherically distributed massive body occupying the region given by $r \leq R$ (say). In other words, the solution (10.3.20) is valid for $r > R$. To get the *interior solution* of the Einstein equations in $r < R$, we need to consider the full equations (10.1.60), with suitably given energy-momentum or stress tensor (10.2.16) (say), which enables us to match the exterior solution. A detailed discussion of the interior solution is in [563].

Section A.6 in Appendices presents a determination of the bending angle of light deflection around a massive celestial body, such as the Sun, using the Schwarzschild metric (10.3.20), which provides a well-known observational confirmation of a precise quantitative prediction of Einstein's theory of general relativity.

Some consequences

Here we discuss some of the simplest consequences of the Schwarzschild solution.

First, we note that, since the Newton constant G is a tiny quantity, the Schwarzschild radius r_S given by (10.3.21) is usually very small compared with the radius R of the gravitating body of mass M under normal circumstances. For example, $R = 6371$ km and $r_S = 9$ mm for the Earth; $R = 696000$ km and $r_S = 3$ km for the Sun. Thus, normally the Schwarzschild surface is well hidden in the bulk of the gravitating body, $r_S < R$, and there is no singular gravitational effect because as the exterior solution the expression (10.3.22) is only valid for $r > R$.

Next, we elaborate on the condition

$$R \leq r_S \tag{10.3.23}$$

for a spherical massive body of the radius R, mass M, and constant mass density ρ, such that

$$M = \frac{4\pi}{3} R^3 \rho. \tag{10.3.24}$$

In fact, (10.3.24) reveals that R grows sublinearly with respect to M, or $R \sim M^{\frac{1}{3}}$, which suggests that R will stay below r_S when M is sufficiently large. More

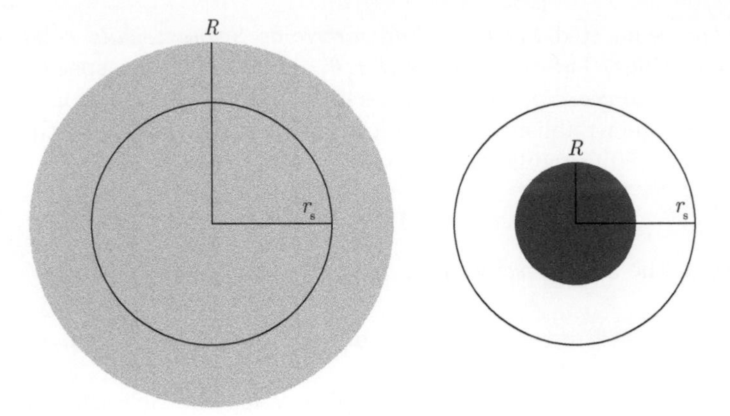

Figure 10.2 Two different situations of a spherically symmetric massive body with matter contained in a sphere of radius R relative to its associated Schwarzschild radius r_S.

precisely, in view of (10.3.21) and (10.3.24), we see that the condition (10.3.23) gives us

$$M^2 \geq \frac{3}{32\pi G^3 \rho}, \tag{10.3.25}$$

indicating a required mass amount for a fixed mass density $\rho > 0$, or

$$\rho \geq \frac{3}{32\pi G^3 M^2}, \tag{10.3.26}$$

spelling out a required mass density for a fixed total mass $M > 0$, to ensure (10.3.23).

Consequently, if the mass density or total mass of the gravitating body is so high that $R < r_S$, the singular surface $r = r_S$ gives rise to rich and complicated gravitational properties of the spacetime.

Figure 10.2 depicts these two distinct matter distribution situations with respect to the associated Schwarzschild radius.

To illustrate the situation with $R < r_S$, we may consider "light" propagation along the radial direction, characterized by the *null proper time* condition, $\mathrm{d}s^2 = 0$, subject to $\mathrm{d}\theta = 0$ and $\mathrm{d}\phi = 0$. Thus, we obtain

$$\left(\frac{\mathrm{d}r}{\mathrm{d}t}\right)^2 = \left(1 - \frac{r_S}{r}\right)^2. \tag{10.3.27}$$

Below, we split our study of the implications of (10.3.27) into two separate cases.

(i) Outside the event horizon, $r > r_S$, so (10.3.27) gives us the scenarios

$$\frac{\mathrm{d}r}{\mathrm{d}t} = \begin{cases} -\left(1 - \frac{r_S}{r}\right), & \text{for inbound light,} \\ \left(1 - \frac{r_S}{r}\right), & \text{for outbound light.} \end{cases} \tag{10.3.28}$$

Hence, inbound light is decelerated, but outbound accelerated. More precisely, solving (10.3.28), we have

$$r + r_S \ln(r - r_S) = \mp t + r_0 + r_S \ln(r_0 - r_S), \quad t \geq 0, \tag{10.3.29}$$

where $r_0 = r(0)$. Thus, it would take infinite *coordinate time* to reach the event horizon $r = r_S$ or spatial infinity outside the event horizon. For outbound light, it would take infinite coordinate time to reach spatial infinity and attain its vacuum speed.

(ii) Inside the event horizon, $R < r < r_S$, we have

$$\frac{dr}{dt} = \begin{cases} -\left(1 - \frac{r_S}{r}\right), & \text{for outbound light,} \\ \left(1 - \frac{r_S}{r}\right), & \text{for inbound light.} \end{cases} \tag{10.3.30}$$

So outbound light is decelerated, but inbound accelerated. Likewise, solving (10.3.30), we get

$$r + r_S \ln(r_S - r) = \mp t + r_0 + r_S \ln(r_S - r_0), \quad t \geq 0. \tag{10.3.31}$$

Hence, it would take infinite coordinate time to reach the event horizon $r = r_S$. Moreover, if the radius of the massive body is so small that

$$R < \frac{r_S}{2}, \tag{10.3.32}$$

then the speed of the inbound light exceeds the vacuum speed of light, 1, when

$$t > t_0 = \left(\frac{r_S}{2} - r_0\right) - r_S \ln 2 \left(1 - \frac{r_0}{r_S}\right). \tag{10.3.33}$$

Besides, the formula (10.3.31) indicates that the inbound light could even reach the spacetime singularity $r = 0$ in finite coordinate time,

$$t_1 = -r_0 + r_S \ln \frac{r_S}{(r_S - r_0)}, \tag{10.3.34}$$

if $r > 0$ were empty space. Thus we have observed that, inside the event horizon, it is "easier" to fall towards the center than deviate from it. Indeed, such a space region, enclosed by the event horizon and popularly called the *Schwarzschild black hole*, allows nothing, not even light, to escape from it, but rather tends to "collapse" everything towards its center.

Time lapses in other coordinate times

We have seen that it will take infinite coordinate time in the Schwarzschild coordinates for light to reach the event horizon $r = r_S$ either from outside or inside the event horizon. Such a picture indicates that the event horizon is unreachable. Here, we elaborate on this issue and we show that some modification of this picture is needed. We continue to assume $R < r_S$ as before.

First, we consider $r > r_S$. We note that the coordinate time and the proper time lapses for a resting observable at the space point (r, θ, ϕ) are related by

$$ds^2 = d\tau^2 = \left(1 - \frac{r_S}{r}\right) dt^2, \tag{10.3.35}$$

such that these two lapses, $d\tau$ and dt, are asymptotically the same as $r \to \infty$. However, for r near r_S, proper time contraction needs to be considered so that it may consume finite amount of proper time for a photon to reach the event horizon. In other words, in terms of location-adjusted proper time measurement, the event horizon may be reachable for a photon.

In fact, imagine we travel with an inbound photon such that we instantly use the location-dependent proper time lapse given by (10.3.35) or

$$d\tau = \left(1 - \frac{r_S}{r}\right)^{\frac{1}{2}} dt, \quad r > r_S, \tag{10.3.36}$$

to measure time. Thus, in view of (10.3.28) and (10.3.36), we have

$$\frac{dr}{d\tau} = -\left(1 - \frac{r_S}{r}\right)^{\frac{1}{2}}, \quad \tau \geq 0, \tag{10.3.37}$$

which may be integrated to render the solution for $r = r(\tau)$:

$$\begin{aligned}&(r^2 - r_S r)^{\frac{1}{2}} + \frac{r_S}{2} \ln\left(r - \frac{r_S}{2} + (r^2 - r_S r)^{\frac{1}{2}}\right) \\ &= -\tau + (r_0^2 - r_S r_0)^{\frac{1}{2}} + \frac{r_S}{2} \ln\left(r_0 - \frac{r_S}{2} + (r_0^2 - r_S r_0)^{\frac{1}{2}}\right), \ r_0 = r(0).\end{aligned} \tag{10.3.38}$$

In particular, $r(\tau)$ reaches r_S at finite proper time $\tau = \tau_0$ with

$$\tau_0 = (r_0^2 - r_S r_0)^{\frac{1}{2}} + \frac{r_S}{2} \ln\left(r_0 - \frac{r_S}{2} + (r_0^2 - r_S r_0)^{\frac{1}{2}}\right) - \frac{r_S}{2} \ln \frac{r_S}{2}. \tag{10.3.39}$$

Next, it may be tempting to use the same idea to consider the situation $r < r_S$. However, this changes the signs of g_{00} and g_{11} in (10.3.22) such that it now becomes necessary to swap the roles of coordinate time and radial variables [396, 563], namely, t and r, in the metric, which complicates the discussion, and which we avoid here.

To proceed, we collectively rewrite the left-hand side of (10.3.29) and (10.3.31) as

$$r^* = r + r_S \ln|r - r_S|, \quad r \neq r_S, \tag{10.3.40}$$

which is commonly referred to as the *tortoise coordinate* [396, 563]. Thus, from

$$\frac{dr^*}{dr} = \left(1 - \frac{r_S}{r}\right)^{-1}, \tag{10.3.41}$$

we convert (10.3.22) into

$$ds^2 = \left(1 - \frac{r_S}{r}\right) (dt^2 - (dr^*)^2) - r^2(d\theta^2 + \sin^2\theta d\phi^2). \tag{10.3.42}$$

Note that there is no more divergent metric coefficient at $r = r_S$ and that the singular surface $r = r_S$ in the Schwarzschild coordinates is pushed to $r^* = -\infty$.

Then, we introduce a new coordinate

$$v = t + r^* = t + r + r_S \ln |r - r_S|, \quad r \neq r_S, \tag{10.3.43}$$

resembling a new time coordinate. Collectively, v, r, θ, ϕ are called the *Eddington–Finkelstein coordinates* [211, 396, 563] in which the metric (10.3.22) reads

$$\mathrm{d}s^2 = \left(1 - \frac{r_S}{r}\right) \mathrm{d}v^2 - 2\mathrm{d}v\mathrm{d}r - r^2(\mathrm{d}\theta^2 + \sin^2\theta \mathrm{d}\phi^2). \tag{10.3.44}$$

As a consequence of using the Eddington–Finkelstein coordinates, we see from the results stated with the upper sign cases in (10.3.29) and (10.3.31) that inbound and outbound photons outside and inside the event horizon are jointly described by the equation

$$v = v_0. \tag{10.3.45}$$

Here, v_0 is a constant, which implies that, in terms of v, it is *instant* for a photon to reach the event horizon from outside or inside the horizon. Furthermore, in view of (10.3.41) and (10.3.43), we see that the equations (10.3.28) and (10.3.30) give us

$$\frac{\mathrm{d}v}{\mathrm{d}r} = 2\left(1 - \frac{r_S}{r}\right)^{-1}, \tag{10.3.46}$$

which may be more informatively rewritten as

$$\frac{\mathrm{d}r}{\mathrm{d}v} = \frac{1}{2}\left(1 - \frac{r_S}{r}\right), \tag{10.3.47}$$

for both outbound-photon outside and inbound-photon inside the event horizon situations. Hence, we obtain

$$\frac{\mathrm{d}r}{\mathrm{d}v} > 0, \quad r > r_S; \quad \frac{\mathrm{d}r}{\mathrm{d}v} < 0, \quad r < r_S, \tag{10.3.48}$$

for outbound and inbound photons in the respective regions that are consistent with the direction of events with respect to time. Moreover, by the uniqueness of a solution to the initial-value problem of ordinary differential equations, we know that the only solution of

$$\frac{\mathrm{d}r}{\mathrm{d}v} = \frac{1}{2}\left(1 - \frac{r_S}{r}\right), \quad r(v_0) = r_S, \tag{10.3.49}$$

is trivial, $r(v) \equiv r_S$. In other words, no solution is allowed to depart from the event horizon, $r = r_S$, in the Eddington–Finkelstein coordinates.

Regularity of gravitational metric at event horizon

The Schwarzschild metric (10.3.22) is singularly-behaved at $r = r_S$. Here, we show that the apparent singularity at $r = r_S$ may be removed by a suitable choice of coordinates.

First, recall that in the region $r > r_{\rm S}$, the inbound path of a photon is given by the upper line solution given in (10.3.29) or $v = t + r^* =$ constant. For this reason the coordinates v, r, θ, ϕ are also called the *inbound Eddington–Finkelstein coordinates.* Likewise, in the same region, the outbound path of a photon is given by the lower line solution given in (10.3.29) or $u \equiv t - r^* =$ constant, and the coordinates u, r, θ, ϕ are called the *outbound Eddington–Finkelstein coordinates.* These coordinates prompt the coordinate transformation

$$u = t - r^*, \quad v = t + r^*, \tag{10.3.50}$$

which recasts (10.3.42) into

$$\mathrm{d}s^2 = \left(1 - \frac{r_{\rm S}}{r}\right) \mathrm{d}u\, \mathrm{d}v - r^2(\mathrm{d}\theta^2 + \sin^2\theta\, \mathrm{d}\phi^2). \tag{10.3.51}$$

On the other hand, resolving r^* in (10.3.50) gives us the results

$$1 - \frac{r_{\rm S}}{r} = \pm\frac{1}{r}\mathrm{e}^{-\frac{1}{r_{\rm S}}\left(r + \frac{u-v}{2}\right)}, \quad r - r_{\rm S} = \pm|r - r_{\rm S}|. \tag{10.3.52}$$

Thus, (10.3.51) becomes

$$\begin{aligned} &\mathrm{d}s^2 = \pm\frac{1}{r}\mathrm{e}^{-\frac{1}{r_{\rm S}}\left(r + \frac{u-v}{2}\right)} \mathrm{d}u\, \mathrm{d}v - r^2(\mathrm{d}\theta^2 + \sin^2\theta\, \mathrm{d}\phi^2), \\ &r - r_{\rm S} = \pm|r - r_{\rm S}|. \end{aligned} \tag{10.3.53}$$

Furthermore, setting

$$U = -\mathrm{e}^{-\frac{u}{2r_{\rm S}}}, \quad V = \mathrm{e}^{\frac{v}{2r_{\rm S}}}, \tag{10.3.54}$$

we can rewrite (10.3.53) as

$$\begin{aligned} &\mathrm{d}s^2 = \pm\frac{4r_{\rm S}^2}{r}\,\mathrm{e}^{-\frac{r}{r_{\rm S}}}\, \mathrm{d}U\, \mathrm{d}V - r^2(\mathrm{d}\theta^2 + \sin^2\theta\, \mathrm{d}\phi^2), \\ &r - r_{\rm S} = \pm|r - r_{\rm S}|, \end{aligned} \tag{10.3.55}$$

as a further simplification. Consequently, in order to see the signature associated with the new coordinate variables, we introduce

$$\xi = \frac{1}{2}(U + V), \quad \eta = \frac{1}{2}(V - U). \tag{10.3.56}$$

Thus, (10.3.55) is updated into

$$\begin{aligned} &\mathrm{d}s^2 = \pm\frac{4r_{\rm S}^2}{r}\,\mathrm{e}^{-\frac{r}{r_{\rm S}}}\, \left(\mathrm{d}\xi^2 - \mathrm{d}\eta^2\right) - r^2(\mathrm{d}\theta^2 + \sin^2\theta\, \mathrm{d}\phi^2), \\ &r - r_{\rm S} = \pm|r - r_{\rm S}|. \end{aligned} \tag{10.3.57}$$

The advantages of using the coordinates ξ, η, θ, ϕ, known as the *Kruskal or Kruskal–Szekeres coordinates* [396, 563], are that they render the regularity of the gravitational metric across the event horizon $r = r_{\rm S}$, and that the interchange between time and space coordinates in terms of $r - r_S = \pm|r - r_S|$ becomes convenient.

In terms of the Schwarzschild time coordinate t and the tortoise coordinate r^* as defined in (10.3.40), we see that the coordinates ξ, η in the gravitational metric (10.3.57) follow the transformation rule

$$\xi = \frac{1}{2}\left(e^{\frac{1}{2r_S}(t+r^*)} - e^{-\frac{1}{2r_S}(t-r^*)}\right), \tag{10.3.58}$$

$$\eta = \frac{1}{2}\left(e^{\frac{1}{2r_S}(t+r^*)} + e^{-\frac{1}{2r_S}(t-r^*)}\right). \tag{10.3.59}$$

Inserting (10.3.40) into (10.3.58)–(10.3.59), we may express them in terms of t, r directly by

$$\xi = |r - r_S|^{\frac{1}{2}} e^{\frac{r}{2r_S}} \sinh\left(\frac{t}{2r_S}\right), \tag{10.3.60}$$

$$\eta = |r - r_S|^{\frac{1}{2}} e^{\frac{r}{2r_S}} \cosh\left(\frac{t}{2r_S}\right), \tag{10.3.61}$$

for $r \neq r_S$.

Spacetime singularity

The apparent event horizon singularity in the Schwarzschild metric is a coordinate singularity associated specifically with the Schwarzschild coordinates, which disappears in the Eddington–Finkelstein and Kruskal–Szekeres coordinates. In other words, such a singularity is coordinate-dependent but is not a spacetime or geometric singularity. Thus, it will be important to be able to recognize spacetime singularities from some quantities that are coordinate-independent. For this, the Ricci scalar curvature cannot serve our purpose because it vanishes identically as a result of the equation (10.3.1). However, another scalar curvature — the *Kretschmann invariant* — perfectly fits our needs.

In fact, we first form the *covariant Riemann tensor*

$$R_{\alpha\beta\mu\nu} = g_{\alpha\gamma} R^{\gamma}_{\beta\mu\nu}. \tag{10.3.62}$$

Then, the Kretschmann invariant or *Kretschmann scalar curvature* is defined to be [289, 396]

$$K = R_{\alpha\beta\mu\nu} R^{\alpha\beta\mu\nu}. \tag{10.3.63}$$

For the Schwarzschild metric (10.3.22), it may be calculated [92, 221, 289, 396] to yield

$$K = \frac{12\, r_S^2}{r^6} = \frac{48\, G^2 M^2}{r^6}. \tag{10.3.64}$$

In particular, this expression indicates that the center of the spherical symmetry, $r = 0$, appears intrinsically as a spacetime singularity, which is present in any coordinates. The Schwarzschild event horizon, on the other hand, makes no exhibition of its existence in (10.3.64) at all.

The next two sections present some of the well-known extensions of the Schwarzschild solution.

10.4 Reissner–Nordström solution

This section presents the Reissner–Nordström solution, which describes the Einstein gravity induced from static sources generated by centrally located mass, electric, and magnetic charges.

Static electromagnetism of central charges

Recall that the Maxwell equations allow a static centrally charged solution of the dually symmetric form

$$\mathbf{E} = \frac{Q\mathbf{x}}{r^3}, \quad \mathbf{B} = \frac{P\mathbf{x}}{r^3}, \quad \mathbf{x} \in \mathbb{R}^3, \tag{10.4.1}$$

where $\mathbf{E}, \mathbf{B}$ are the electric and magnetic fields generated by the prescribed electric and magnetic charges, Q, P, respectively, since now the Maxwell equations

$$\nabla \times \mathbf{E} = \mathbf{0}, \quad \nabla \times \mathbf{B} = \mathbf{0}, \quad \nabla \cdot \mathbf{E} = 0, \quad \nabla \cdot \mathbf{B} = 0, \tag{10.4.2}$$

are satisfied identically away from the space origin where the charges are located. The energy density of the solution is

$$\mathcal{E} = \frac{1}{2}\left(\mathbf{E}^2 + \mathbf{B}^2\right) = \frac{1}{2r^4}(Q^2 + P^2), \tag{10.4.3}$$

resulting in a divergent total energy as a consequence.

Maxwell equations in curved spacetime situation

Let (x^μ) denote a general collection of coordinates of the flat Minkowski spacetime. In absence of external sources, the Maxwell equations governing a skew-symmetric electromagnetic 2-tensor $F^{\mu\nu}$ read

$$F^{\mu\nu}{}_{,\mu} = 0, \tag{10.4.4}$$

$$F_{\mu\nu,\gamma} + F_{\nu\gamma,\mu} + F_{\gamma\mu,\nu} = 0, \tag{10.4.5}$$

where (10.4.5) is the Bianchi identity. In a curved spacetime setting, we replace the conventional derivatives in (10.4.4)–(10.4.5) by covariant derivatives to arrive at the updated Maxwell equations

$$F^{\mu\nu}{}_{;\mu} = 0, \tag{10.4.6}$$

$$F_{\mu\nu;\gamma} + F_{\nu\gamma;\mu} + F_{\gamma\mu;\nu} = 0. \tag{10.4.7}$$

From (10.1.21) and the skew-symmetry of $F_{\mu\nu}$, we see that (10.4.5) and (10.4.7) coincide. By the same reason and (10.1.23), we see that (10.4.6) simplifies itself into

$$F^{\mu\nu}{}_{,\mu} + \Gamma^{\mu}_{\alpha\mu} F^{\alpha\nu} = 0. \tag{10.4.8}$$

On the other hand, note that

$$\begin{aligned}\Gamma^{\mu}_{\alpha\mu} &= \frac{1}{2} g^{\mu\nu}(g_{\nu\alpha,\mu} + g_{\nu\mu,\alpha} - g_{\alpha\mu,\nu}) \\ &= \frac{1}{2}\, g^{\mu\nu} g_{\mu\nu,\alpha},\end{aligned} \tag{10.4.9}$$

where we have used the symmetry of $g^{\mu\nu}$ to cancel out the resulting first and third terms. Besides, using $C_{\mu\nu}$ to denote the cofactor of the matrix $(g_{\mu\nu})$ at the position (μ,ν), then we have the determinant expansion formula

$$g = \det(g_{\alpha\beta}) = \sum_{0\leq\mu\leq3} C_{\mu\nu} g_{\mu\nu}, \quad \nu = 0,1,2,3; \quad C_{\mu\nu} = g\, g^{\mu\nu}. \tag{10.4.10}$$

Hence, differentiating the determinant g and then expanding by cofactors, we get

$$\begin{aligned} g_{,\alpha} &= \sum_{0\leq\mu,\nu\leq3} C_{\mu\nu} g_{\mu\nu,\alpha} \\ &= g\, g^{\mu\nu} g_{\mu\nu,\alpha}. \end{aligned} \tag{10.4.11}$$

In view of (10.4.9) and (10.4.11), we see that (10.4.8) is converted into the following equations

$$\frac{1}{\sqrt{-g}} \partial_\mu \left(\sqrt{-g} F^{\mu\nu}\right) = 0, \tag{10.4.12}$$

which, when coupled with the Bianchi identity (10.4.5), are also recognized as the curved spacetime Maxwell equations.

Alternatively, we may also arrive at (10.4.12) by using the action principle

$$I = \int \mathcal{L} \sqrt{-g}\, \mathrm{d}x, \quad \mathcal{L} = -\frac{1}{4} F_{\mu\nu} F^{\mu\nu}. \tag{10.4.13}$$

Energy-momentum tensor

With (10.1.50) and (10.4.13), we have

$$\begin{aligned} T_{\mu\nu} &= 2\frac{\partial \mathcal{L}}{\partial g^{\mu\nu}} - g_{\mu\nu}\,\mathcal{L} \\ &= -F_{\mu\alpha} g^{\alpha\beta} F_{\nu\beta} + \frac{1}{4}\, g_{\mu\nu} \left(F_{\alpha\beta} F^{\alpha\beta}\right). \end{aligned} \tag{10.4.14}$$

In view of (10.4.14), we have $T = g^{\mu\nu} T_{\mu\nu} = 0$. Inserting this result into (10.2.77), with $\Lambda = 0$, we obtain the Einstein equations

$$R_{\mu\nu} = -8\pi G T_{\mu\nu}, \tag{10.4.15}$$

relating the Ricci tensor of the spacetime directly to the electromagnetic energy-momentum tensor.

Black hole metric

Guided by the Schwarzschild metric, we take the spacetime gravitational metric in the ordered spherical coordinates $(x^\mu) = (t, r, \theta, \phi)$ to be

$$\mathrm{d}s^2 = A(r)\,\mathrm{d}t^2 - \frac{\mathrm{d}r^2}{A(r)} - r^2(\mathrm{d}\theta^2 + \sin^2\theta\,\mathrm{d}\phi^2). \tag{10.4.16}$$

Then, the nontrivial components of the Ricci tensor following from (10.3.7)–(10.3.10) with $AB = 1$ are

$$R_{00} = -\frac{A}{2r^2}(r^2 A')', \tag{10.4.17}$$

$$R_{11} = \frac{1}{2r^2 A}(r^2 A')', \tag{10.4.18}$$

$$R_{22} = (rA)' - 1, \tag{10.4.19}$$

$$R_{33} = \sin^2\theta\, R_{22}. \tag{10.4.20}$$

Static radially symmetric electromagnetism

We consider the static electric and magnetic fields to be along the radial direction only. Then the only nontrivial components of $F^{\mu\nu}$ are $F^{01} = -F^{10} = -E$ (electric) and $F^{23} = -F^{32} = -B$ (magnetic). That is,

$$(F^{\mu\nu}) = \begin{pmatrix} 0 & -E & 0 & 0 \\ E & 0 & 0 & 0 \\ 0 & 0 & 0 & -B \\ 0 & 0 & B & 0 \end{pmatrix}. \tag{10.4.21}$$

Inserting $\sqrt{-g} = r^2 \sin\theta$ and (10.4.21) into (10.4.12), we see that the equation at $\nu = 0$ gives us the solution

$$E = E(r) = \frac{Q}{r^2}, \tag{10.4.22}$$

where Q is a constant. With this, (10.4.12) is automatically satisfied at $\nu = 1$. Moreover, at $\nu = 2, 3$, (10.4.21) and (10.4.12) lead to the equations

$$\frac{\partial}{\partial\phi}(r^2 \sin\theta\, B) = 0, \quad \frac{\partial}{\partial\theta}(r^2 \sin\theta\, B) = 0, \tag{10.4.23}$$

which are solved by

$$B = B(r, \theta) = \frac{C(r)}{\sin\theta}, \tag{10.4.24}$$

where $C(r)$ is a function of $r > 0$ to be determined. Furthermore, we may use

$$g_{00} = A, \quad g_{11} = -\frac{1}{A}, \quad g_{22} = -r^2, \quad g_{33} = -r^2 \sin^2\theta, \tag{10.4.25}$$

to get the nontrivial and independent components of $F_{\mu\nu}$ to be

$$F_{01} = g_{00}g_{11}F^{01} = -F^{01} = \frac{Q}{r^2}, \quad F_{23} = g_{22}g_{33}F^{23} = -r^4 \sin\theta\, C(r). \quad (10.4.26)$$

With (10.4.26) and setting $\mu = 1, \nu = 2, \gamma = 3$ in the Bianchi identity (10.4.5), we obtain

$$\frac{\partial}{\partial r}(r^4 \sin\theta\, C(r)) = F_{23,1} = 0, \quad (10.4.27)$$

which renders the solution

$$C(r) = \frac{P}{r^4}, \quad (10.4.28)$$

where P is a constant. Inserting (10.4.28) into (10.4.26), we arrive at the result

$$(F_{\mu\nu}) = \begin{pmatrix} 0 & \frac{Q}{r^2} & 0 & 0 \\ -\frac{Q}{r^2} & 0 & 0 & 0 \\ 0 & 0 & 0 & -P\sin\theta \\ 0 & 0 & P\sin\theta & 0 \end{pmatrix}. \quad (10.4.29)$$

For convenience of reference and comparison, we also express (10.4.21) explicitly as follows:

$$(F^{\mu\nu}) = \begin{pmatrix} 0 & -\frac{Q}{r^2} & 0 & 0 \\ \frac{Q}{r^2} & 0 & 0 & 0 \\ 0 & 0 & 0 & -\frac{P}{r^4\sin\theta} \\ 0 & 0 & \frac{P}{r^4\sin\theta} & 0 \end{pmatrix}. \quad (10.4.30)$$

In view of (10.4.29) and (10.4.30), we have

$$\mathcal{L} = -\frac{1}{4}F_{\mu\nu}F^{\mu\nu} = \frac{1}{2r^4}(Q^2 - P^2). \quad (10.4.31)$$

Furthermore, using (10.4.25), we get

$$F_{\mu\alpha}g^{\alpha\beta}F_{\nu\beta} = \frac{1}{A}F_{\mu 0}F_{\nu 0} - AF_{\mu 1}F_{\nu 1} - \frac{1}{r^2}F_{\mu 2}F_{\nu 2} - \frac{1}{r^2\sin^2\theta}F_{\mu 3}F_{\nu 3}. \quad (10.4.32)$$

Thus, in view of (10.4.14), (10.4.25), (10.4.31), and (10.4.32), we obtain the nontrivial components of the energy-momentum tensor $T_{\mu\nu}$ to be

$$T_{00} = \frac{A}{2r^4}(Q^2 + P^2), \quad (10.4.33)$$

$$T_{11} = -\frac{1}{2Ar^4}(Q^2 + P^2), \quad (10.4.34)$$

$$T_{22} = \frac{1}{2r^2}(Q^2 + P^2), \quad (10.4.35)$$

$$T_{33} = \sin^2\theta\, T_{22}. \quad (10.4.36)$$

Reissner–Nordström solution

Inserting (10.4.17)–(10.4.20) and (10.4.33)–(10.4.36) into (10.4.15), we arrive at the equations

$$(r^2A')' = \frac{8\pi G}{r^2}(Q^2 + P^2), \tag{10.4.37}$$

$$(rA)' = 1 - \frac{4\pi G}{r^2}(Q^2 + P^2), \tag{10.4.38}$$

which miraculously become linear in the unknown A but are seen to be overdetermined. Fortunately, (10.4.38) implies (10.4.37). Therefore, integrating (10.4.38), we obtain the solution

$$A(r) = 1 + \frac{C}{r} + \frac{4\pi G}{r^2}(Q^2 + P^2), \tag{10.4.39}$$

where C is an integration constant. Since in the special case of absence of electromagnetism, $Q = 0, P = 0$, we should recover the Schwarzschild metric (10.3.20), so we are led to the condition (10.3.19). That is, we conclude from (10.4.39) with

$$A(r) = 1 - \frac{2GM}{r} + \frac{4\pi G}{r^2}(Q^2 + P^2). \tag{10.4.40}$$

In summary, we see that the spherically symmetric gravitational metric induced from centrally distributed massive and electrically and magnetically charged static sources, parametrized by mass M, electric charge Q, and magnetic charge P, is found to be

$$\begin{aligned} ds^2 &= \left(1 - \frac{2GM}{r} + \frac{4\pi G}{r^2}(Q^2 + P^2)\right) dt^2 \\ &- \left(1 - \frac{2GM}{r} + \frac{4\pi G}{r^2}(Q^2 + P^2)\right)^{-1} dr^2 - r^2(d\theta^2 + \sin^2\theta\, d\phi^2), \end{aligned} \tag{10.4.41}$$

which is called the *Reissner–Nordström metric* or *Reissner–Nordström solution.*

Event horizons

An event horizon occurs when

$$1 - \frac{2GM}{r} + \frac{4\pi G}{r^2}(Q^2 + P^2) = 0, \quad r > 0, \tag{10.4.42}$$

which has no solution when $|Q|$ or $|P|$ is sufficiently large so that

$$4\pi\left(Q^2 + P^2\right) > GM^2. \tag{10.4.43}$$

In this case, the metric is regular everywhere in $r > 0$ and the curvature singularity, $r = 0$, stands out since there is no event horizon to "conceal" it, which is sometimes referred to be a *naked singularity.* On the other hand, when

$$4\pi\left(Q^2 + P^2\right) < GM^2, \tag{10.4.44}$$

the equation (10.4.42) gives rise to two concentric spherical event horizons with the radii

$$r_\pm = GM \pm \sqrt{G^2M^2 - 4\pi G(Q^2 + P^2)}, \tag{10.4.45}$$

which merge into a single spherical event horizon with the half of the Schwarzschild radius, $\frac{1}{2}r_S = GM$, in the degenerate limit

$$4\pi(Q^2 + P^2) = GM^2, \tag{10.4.46}$$

giving rise to a single-event-horizon black hole, as the Schwarzschild black hole, known as an *extremal black hole* [113]. In the non-degenerate situation, the radius r_+ gives rise to an *external horizon*, which has the same properties of the Schwarzschild horizon and is thus referred to as the event horizon. The radius r_- gives rise to an *internal horizon*, also called the *Cauchy horizon* [121], so that the metric factor $A(r)$ given in (10.4.40) changes its sign over the interval (r_-, r_+). In all situations, note that the presence of electromagnetism works to pull down the event horizon from that of Schwarzschild in absence of electromagnetism because there holds the general inequality

$$0 < r_- \leq r_+ < r_S, \quad 4\pi\left(Q^2 + P^2\right) \leq GM^2. \tag{10.4.47}$$

Thus, we rewrite the Reissner–Nordström metric as

$$\begin{aligned} ds^2 &= \left(1 - \frac{r_+}{r}\right)\left(1 - \frac{r_-}{r}\right) dt^2 \\ &- \left(1 - \frac{r_+}{r}\right)^{-1}\left(1 - \frac{r_-}{r}\right)^{-1} dr^2 - r^2(d\theta^2 + \sin^2\theta\, d\phi^2), \end{aligned} \tag{10.4.48}$$

in terms of $r_\pm$. In particular, in the extremal black hole situation, $r_- = r_+ = r_e \equiv \frac{r_S}{2}$, the metric (10.4.48) becomes

$$ds^2 = \left(1 - \frac{r_e}{r}\right)^2 dt^2 - \left(1 - \frac{r_e}{r}\right)^{-2} dr^2 - r^2(d\theta^2 + \sin^2\theta\, d\phi^2), \tag{10.4.49}$$

which is in an interesting contrast with (10.3.22). Moreover, in a general non-extremal situation, the radii r_+ and r_- are equally distant around r_e and approach r_e in the extremal limit.

Electric and magnetic charges

Use the Hamiltonian density $\mathcal{H} = T_0^0 = g^{\mu 0}T_{\mu 0}$ as the energy density. Then (10.4.25) and (10.4.33) give us the result

$$\mathcal{H} = \frac{1}{2r^4}(Q^2 + P^2), \tag{10.4.50}$$

which coincides with (10.4.3) expressed in terms of the electric charge Q and magnetic charge P, carried by spherically symmetric fields, $\mathbf{E}$ and $\mathbf{B}$, as in (10.4.1), given in the Cartesian coordinates.

We use $(\tilde{x}^i) = \mathbf{x} = (x, y, z)$ and $(x^i) = (r, \theta, \phi)$ to denote the Cartesian and spherical coordinates, respectively, related by (10.2.7), and $(\tilde{x}^\mu)$ and (x^μ) the corresponding coordinates for the Minkowski spacetime with $\tilde{x}^0 = x^0 = t$. Then, the electrostatic and magnetostatic fields $\mathbf{E}$ and $\mathbf{B}$ generated from the point electric and magnetic charges Q and P located at $\mathbf{x} = \mathbf{0}$ define the electromagnetic tensor

$$(\tilde{F}^{\mu\nu}) = \begin{pmatrix} 0 & -\frac{Qx}{r^3} & -\frac{Qy}{r^3} & -\frac{Qz}{r^3} \\ \frac{Qx}{r^3} & 0 & -\frac{Pz}{r^3} & \frac{Py}{r^3} \\ \frac{Qy}{r^3} & \frac{Pz}{r^3} & 0 & -\frac{Px}{r^3} \\ \frac{Qz}{r^3} & -\frac{Py}{r^3} & \frac{Px}{r^3} & 0 \end{pmatrix}. \tag{10.4.51}$$

Thus, from (10.4.51), we obtain

$$(\tilde{F}_{\mu\nu}) = \begin{pmatrix} 0 & \frac{Qx}{r^3} & \frac{Qy}{r^3} & \frac{Qz}{r^3} \\ -\frac{Qx}{r^3} & 0 & -\frac{Pz}{r^3} & \frac{Py}{r^3} \\ -\frac{Qy}{r^3} & \frac{Pz}{r^3} & 0 & -\frac{Px}{r^3} \\ -\frac{Qz}{r^3} & -\frac{Py}{r^3} & \frac{Px}{r^3} & 0 \end{pmatrix}. \tag{10.4.52}$$

As a consequence of the coordinate transformation (10.2.7), (10.4.52), and the covariant relation

$$F_{\mu\nu} = \frac{\partial \tilde{x}^\alpha}{\partial x^\mu} \frac{\partial \tilde{x}^\beta}{\partial x^\nu} \tilde{F}_{\alpha\beta}, \tag{10.4.53}$$

we obtain

$$(F_{\mu\nu}) = \begin{pmatrix} 0 & \frac{Q}{r^2} & 0 & 0 \\ -\frac{Q}{r^2} & 0 & 0 & 0 \\ 0 & 0 & 0 & -P\sin\theta \\ 0 & 0 & P\sin\theta & 0 \end{pmatrix}, \tag{10.4.54}$$

which recovers (10.4.29) as expected.

Kretschmann curvature

For convenience, we note that the Kretschmann curvature of the Reissner–Nordström metric (10.4.41) is

$$K = \frac{48G^2}{r^8}\left(M^2r^2 - 8\pi M(Q^2 + P^2)r + \frac{56\pi^2}{3}(Q^2 + P^2)^2\right), \tag{10.4.55}$$

which is seen to behave more singularly than the Schwarzschild metric at the curvature singularity, $r = 0$, as a result of the presence of charges, and contain a mixed "interaction" term between mass and charges.

The Reissner–Nordström solution allows us to develop the concept of *charged black holes*.

10.5 Kerr solution

Next, consider a rotating spherically symmetric gravitating body. Recall that, for a mass occupying a region V with mass density ρ and rotating about a certain axis ℓ, the moment of inertia may be calculated by

$$I = \int_V \delta^2(\mathbf{x})\rho(\mathbf{x})\,\mathrm{d}\mathbf{x}, \tag{10.5.1}$$

where $\delta(\mathbf{x})$ denotes the distance from the point $\mathbf{x}$ to the axis ℓ. If the angular velocity of the rotating motion of the mass about ℓ is ω, then the angular momentum is

$$J = I\omega. \tag{10.5.2}$$

In terms of J, the exterior gravitational metric, which is a solution to the Einstein equations, generated from a spherically distributed mass rotating around the vertical axis with angular velocity ω, is found to be

$$\begin{aligned}\mathrm{d}s^2 &= \left(1 - \frac{r_{\mathrm{S}}r}{\rho^2}\right)\mathrm{d}t^2 - \frac{\rho^2\,\mathrm{d}r^2}{r^2 - r_{\mathrm{S}}r + \alpha^2} - \rho^2\mathrm{d}\theta^2 \\ &\quad - \left(r^2 + \alpha^2 + \frac{r_{\mathrm{S}}r\alpha^2}{\rho^2}\sin^2\theta\right)\sin^2\theta\,\mathrm{d}\phi^2 + \frac{2r_{\mathrm{S}}r\alpha}{\rho^2}\sin^2\theta\,\mathrm{d}t\mathrm{d}\phi,\end{aligned} \tag{10.5.3}$$

where

$$r_{\mathrm{S}} = 2GM, \quad \alpha = \frac{J}{M}, \quad \rho^2 = r^2 + \alpha^2\cos^2\theta, \tag{10.5.4}$$

and M is the total mass. This metric is known as the *Kerr metric* [396, 563, 569] expressed in terms of the *Boyer–Lindquist coordinates* [78], with $x^0 = t, x^1 = r, x^2 = \theta, x^3 = \phi$. In the non-rotational limit, $\alpha = 0$, the metric (10.5.3) reduces itself into the Schwarzschild metric (10.3.20). The Kerr solution, originally derived in the Eddington–Finkelstein coordinates [325] and later reconstructed by Newman and Janis [414] using an algorithm involving a coordinate complexification technique, known as the *Newman–Janis algorithm*, enables the study of *rotating black holes* that exhibit some distinguishing features.

Consider again the conditions that $g_{00} = 0$ and the denominator of g_{11} vanishes. We see that the coordinate singularities revealing event horizons would occur at the radii

$$r_\pm^{\mathrm{E}} = \frac{1}{2}\left(r_{\mathrm{S}} \pm \sqrt{r_{\mathrm{S}}^2 - 4\alpha^2\cos^2\theta}\right), \tag{10.5.5}$$

$$r_\pm^{\mathrm{H}} = \frac{1}{2}\left(r_{\mathrm{S}} \pm \sqrt{r_{\mathrm{S}}^2 - 4\alpha^2}\right), \tag{10.5.6}$$

assuming $r_{\mathrm{S}}^2 \geq 4\alpha^2$. It is clear that

$$r_-^{\mathrm{E}} \leq r_-^{\mathrm{H}} \leq r_+^{\mathrm{H}} \leq r_+^{\mathrm{E}}, \tag{10.5.7}$$

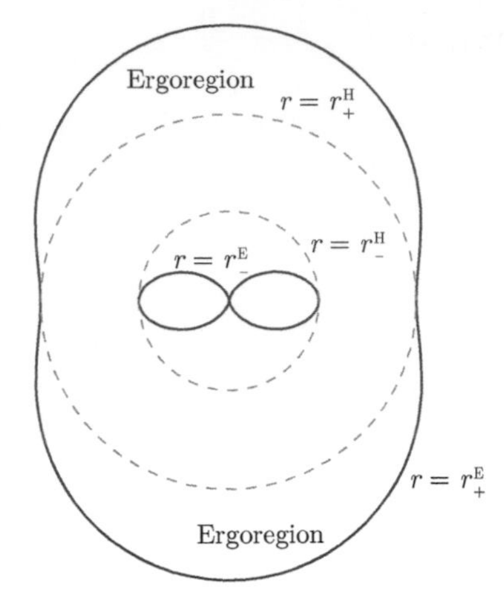

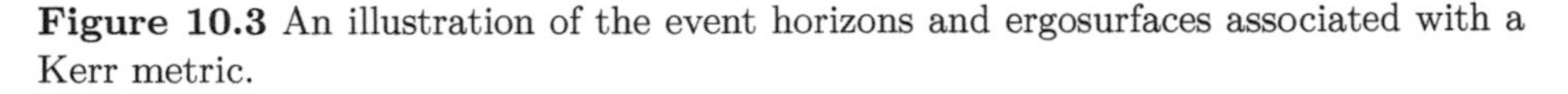

Figure 10.3 An illustration of the event horizons and ergosurfaces associated with a Kerr metric.

and that g_{00} changes its sign as r crosses $r_+^{\rm E}$ but the sign of g_{11} remains the same until r crosses $r_+^{\rm H}$ where both g_{00} and g_{11} flip their signs, as in the case of a Schwarzschild or Reissner–Nordström black hole. A similar situation occurs with the pair $r_-^{\rm E}$ and $r_-^{\rm H}$. As a consequence, $r_-^{\rm H}$ and $r_+^{\rm H}$ give rise to black hole event horizons but $r_-^{\rm E}$ and $r_+^{\rm E}$, which give rise to a pair of surfaces called *ergosurfaces*, do not. Instead, the sign change in g_{00} necessitates the interchange of the meaning of the time coordinate with that of an angular coordinate, which would enable a piece of matter caught in the region defined by

$$r_+^{\rm H} < r < r_+^{\rm E}, \tag{10.5.8}$$

known as the *ergoregion*, to split into two pieces, one entering the black hole, with $r = r_+^{\rm H}$, carrying negative energy, say $-\varepsilon$, and the other escaping from the ergosurface, with $r = r_+^{\rm E}$, carrying energy of an excessive amount ε. This mechanism [441] is known as the *Penrose process* [396, 563]. In particular, for a laboratory observer, the process is like seeing energy being *extracted* [443] from the black hole in the ergoregion.

Figure 10.3 plots the cross-sections of the two spherical event horizons, in dashed-line curves, and the two, respectively, externally and internally inscribed ergosurfaces, in solid-line curves, of a Kerr solution.

Note that the Kretschmann curvature of the Kerr metric (10.5.3) is

$$K = \frac{48G^2M^2(r^2 - \alpha^2\cos^2\theta)([r^2+\alpha^2\cos^2\theta]^2 - 16r^2\alpha^2\cos^2\theta)}{(r^2+\alpha^2\cos^2\theta)^6}, \tag{10.5.9}$$

such that the event horizons and ergosurfaces make no exhibition of their presence and the only curvature singularity shows up to be

$$r = 0, \quad \theta = \frac{\pi}{2}, \tag{10.5.10}$$

which is not quite a space point but the "equator" of a point when $\alpha \neq 0$. Thus it appears that curvature singularity is "reduced" by rotation.

The Reissner–Nordström and Kerr solutions show, in view of general relativity, that gravity, or rather spacetime curvature, can be generated from electromagnetism and spinning motion, as well as mass. Besides, the presence of event horizons or black holes is seen to be *a universal phenomenon* dictated by the Einstein equations.

The Schwarzschild solution may further be extended to incorporate both charged *and* rotating masses. In such a context, the solution is called the *Kerr–Newman solution* or *Kerr–Newman metric* [413, 414], which in terms of the Boyer–Lindquist coordinates reads

$$\begin{aligned} \mathrm{d}s^2 = & \frac{(\Delta - \alpha^2 \sin^2\theta)}{\rho^2}\,\mathrm{d}t^2 - \frac{\rho^2\,\mathrm{d}r^2}{\Delta} - \rho^2 \mathrm{d}\theta^2 \\ & - \frac{([r^2+\alpha^2]^2 - \Delta\,\alpha^2\sin^2\theta)}{\rho^2}\sin^2\theta\,\mathrm{d}\phi^2 + \frac{2([r^2+\alpha^2]-\Delta)\alpha}{\rho^2}\sin^2\theta\,\mathrm{d}t\mathrm{d}\phi, \end{aligned} \tag{10.5.11}$$

where

$$\Delta = r^2 - r_{\mathrm{S}} r + \alpha^2 + 4\pi G(Q^2 + P^2), \tag{10.5.12}$$

and Q, P are the electric and magnetic charges carried by the massive body. Thus, solving $\Delta = 0$, we obtain again a pair of event horizons at

$$r_\pm^{\mathrm{H}} = \frac{1}{2}\left(r_{\mathrm{S}} \pm \sqrt{r_{\mathrm{S}}^2 - 4(\alpha^2 + 4\pi G[Q^2 + P^2])} \right), \tag{10.5.13}$$

assuming $r_{\mathrm{S}}^2 \geq 4(\alpha^2 + 4\pi G[Q^2 + P^2])$. Likewise, solving $\Delta - \alpha^2\sin^2\theta = 0$ or $g_{00} = 0$, we arrive at a pair of *charged ergosurfaces* defined by

$$r_\pm^{\mathrm{E}} = \frac{1}{2}\left(r_{\mathrm{S}} \pm \sqrt{r_{\mathrm{S}}^2 - 4(\alpha^2\cos^2\theta + 4\pi G[Q^2 + P^2])} \right). \tag{10.5.14}$$

It will be instructive to see how (10.5.11), (10.5.13), (10.5.14) contain (10.4.41), (10.4.45), and (10.5.3), (10.5.5), (10.5.6), respectively, as special cases.

10.6 Gravitational mass and Penrose bounds

We saw that, through the Einstein equations, matters contained in a spacetime determine the geometric contents of the spacetime realized by its gravitational line element. Conversely, if a spacetime possesses a certain amount of nontrivial geometric properties resembling those brought forth by coupling matters to the geometry of the spacetime by the Einstein equations, it should contain a measurable amount of matter content represented as "energy" or "mass" of the matter contained in such a spacetime. In 1959, Arnowitt, Deser, and Misner [21] developed a Hamiltonian canonical variable formulation of general relativity, which was later proved to be crucial for quantum gravity formalism [163]. On the other hand, this formulation also enabled Arnowitt, Deser, and Misner to demonstrate a natural way for the computation of gravitational energy, or the ADM energy, which is simply given by the Hamiltonian in their canonical formalism [22, 23]. Interestingly, the ADM energy and ADM mass may all be represented as flux integrals near the infinity of an asymptotically Euclidean space, in a spirit similar to the calculation of the total electric charge based on the divergence theorem. Although there are other definitions of mass in general relativity, including the Bondi [69] mass and Komar mass [337], the ADM energy and mass are predominantly considered in the research on the problems discussed in this section, namely, a series of theorems and statements regarding energy, mass, and geometry of the spacetime of fundamental interests.

Thus, this section presents a brief discussion of the *positive energy theorem* [431, 491, 492, 493, 494, 586] and the *Penrose conjecture* [442] in general relativity, formulated over the concepts of the ADM energy and the ADM mass.

Dominant energy condition

In such a context, we are interested in an asymptotically Euclidean space-like hypersurface $(\mathcal{M}, g)$ embedded in a four-dimensional Lorentzian spacetime. The metric tensor of this spacetime satisfies the Einstein equations (10.1.60) for which the energy-momentum tensor $T_{\mu\nu}$ is subject to a physically motivated condition, called the *dominant energy condition*. In an orthonormal basis, known as a *tetrad* or *vielbein*, this condition reads [232, 278, 280]

$$T_{00} \geq |T_{\mu\nu}|, \quad \mu, \nu = 0, 1, 2, 3, \tag{10.6.1}$$

which implies that the pointwise energy density $\mathcal{H} = T_{00}$ of the system is nonnegative and dominates over all other components of the energy-momentum tensor. Sometimes (10.6.1) is also supplemented under the same basis with [431]

$$\mathcal{H} = T_{00} \geq \left(-T_{0i}T^{0i}\right)^{\frac{1}{2}} = \left(\sum_{i=1}^{3} T_{0i}^2\right)^{\frac{1}{2}}, \tag{10.6.2}$$

which states that the energy density is bounded from below by the magnitude of the linear momentum vector of the system. Of course, (10.6.2) automatically implies (10.6.1) when $\mu = 0$ and $\nu = i = 1, 2, 3$.

Examples fulfilling dominant energy condition

Consider two examples. With an orthonormal basis, it suffices to work over the flat Minkowski spacetime $(\mathbb{R}^{3,1}, \eta_{\mu\nu})$.

First, let

$$\mathcal{L} = -\frac{1}{4} F_{\mu\nu} F^{\mu\nu} \tag{10.6.3}$$

be the action density of the Maxwell theory such that

$$E^i = -F^{0i}, \quad B^i = -\frac{1}{2}\epsilon^{ijk}F_{jk}, \tag{10.6.4}$$

give rise to the associated electric and magnetic fields, respectively. Then, (10.4.14) leads to the well-known results

$$T_{00} = \frac{1}{2}(\mathbf{E}^2 + \mathbf{B}^2), \tag{10.6.5}$$

$$T_{0i} = -\epsilon_{ijk}E^j B^k, \tag{10.6.6}$$

in which $-(T_{0i}) = (T^{0i}) = \mathbf{E} \times \mathbf{B} = \mathbf{S}$ is the classical *Poynting vector field* [458], which determines the energy flux of the electromagnetic field [246, 256, 458]. Then, (10.6.2) is simply

$$|\mathbf{E} \times \mathbf{B}| \leq \frac{1}{2}(\mathbf{E}^2 + \mathbf{B}^2). \tag{10.6.7}$$

Next, let

$$\mathcal{L} = \frac{1}{2}\partial_\mu \phi \partial^\mu \overline{\phi} - V(|\phi|^2) \tag{10.6.8}$$

be the action density in the Klein–Gordon model governing a complex scalar field ϕ with $V \geq 0$. In this case, we have

$$T_{00} = \frac{1}{2}(|\partial_0 \phi|^2 + |\nabla \phi|^2) + V(|\phi|^2), \tag{10.6.9}$$

$$T_{0i} = \mathrm{Re}\{\partial_0 \phi \partial_i \overline{\phi}\}. \tag{10.6.10}$$

Thus

$$\sqrt{T_{0i}^2} \leq |\partial_0 \phi| \sqrt{|\partial_i \phi|^2}, \tag{10.6.11}$$

and (10.6.2) follows again by (10.6.9) and $V \geq 0$.

Similarly, the condition (10.6.1) when $\mu, \nu = i, j = 1, 2, 3$ may be examined as well (cf. Exercise 22).

Positive energy and positive mass theorems

The hypersurface $(\mathcal{M}, g)$ is said to be *asymptotically Euclidean* if there is a compact subset $\mathcal{K}$ in $\mathcal{M}$ such that $\mathcal{M} \setminus \mathcal{K}$ is the union of finitely many "*ends*" of $\mathcal{M}$ so that each end is diffeomorphic to $\mathbb{R}^3 \setminus B_r$ where B_r denotes the ball of radius $r > 0$ centered at the origin. Under this diffeomorphism, the metric g_{ij} of $\mathcal{M}$ at each end can be represented near infinity by

$$g_{ij}(x) = \delta_{ij} + a_{ij}(x), \quad a_{ij}(x) = \mathrm{O}(|x|^{-1}), \tag{10.6.12}$$

$$\partial_k a_{ij}(x) = \mathrm{O}(|x|^{-2}), \quad \partial_l \partial_k a_{ij}(x) = \mathrm{O}(|x|^{-3}), \tag{10.6.13}$$

for $x \in \mathbb{R}^3 \setminus B_r$ in Cartesian coordinates, and the second fundamental form (h_{jk}) there satisfies similar asymptotic estimates

$$h_{ij}(x) = \mathrm{O}(|x|^{-2}), \quad \partial_k h_{ij}(x) = \mathrm{O}(|x|^{-3}), \quad x \in \mathbb{R}^3 \setminus B_r. \tag{10.6.14}$$

(Recall that, if M is a submanifold of a Riemannian manifold N and ∇^M and ∇^N denote the covariant derivatives over M and N, respectively, then the "excessive" vector field

$$\alpha(X,Y) \equiv \nabla_X^N(Y) - \nabla_X^M(Y), \quad X, Y \in T(M), \tag{10.6.15}$$

valued in the tangent bundle $T(N)$ actually lies in the normal bundle $T(M)^\perp$. See [335]. Thus, if we use ν to denote the unit normal vector on M, or the Gauss map, in the co-dimension 1 situation, namely $\dim(N) = \dim(M) + 1$ such that M is regarded as a hypersurface in N, then (10.6.15) becomes

$$\alpha(X,Y) = h(X,Y)\nu, \tag{10.6.16}$$

where $h(X,Y)$ is a symmetric bilinear form called the *second fundamental form* of M, which also depends on the ambient manifold N. See also [223, 254, 520]. In (10.6.14), the matrix (h_{ij}) is the matrix representation of the second fundamental form h in (10.6.16) with respect to the standard coordinate basis given by $h_{ij} = h(\partial_i, \partial_j)$ $(i, j = 1, 2, 3)$, which, for the purposes here, is loosely referred to as the second fundamental form.) Without loss of generality, we assume only one end for convenience. According to Arnowitt, Deser, and Misner [22], the total energy E and the momentum P_ℓ of the gravitational system in space can be defined as the limits of the integral fluxes

$$E = \frac{1}{16\pi G} \lim_{r\to\infty} \int_{\partial B_r} (\partial_i g_{ij} - \partial_j g_{ii})\, n^j \mathrm{d}\sigma_r, \tag{10.6.17}$$

$$P_k = \frac{1}{8\pi G} \lim_{r\to\infty} \int_{\partial B_r} (h_{jk} - \delta_{jk} h_{ii})\, n^j \mathrm{d}\sigma_r, \tag{10.6.18}$$

where $\mathrm{d}\sigma_r$ is the area element of ∂B_r and $\mathbf{n} = (n^j)$ denotes the outnormal vector to ∂B_r. The Positive Energy Theorem [431, 491, 492, 493, 586] states that the total energy (10.6.17) is bounded from below by the total momentum (10.6.18) by

$$E \geq |P| \tag{10.6.19}$$

and that $E = 0$ if and only if $(\mathcal{M}, g)$ is the Euclidean space $(\mathbb{R}^3, \delta)$. When the second fundamental form (h_{ij}) vanishes identically,[3] $P \equiv 0$, the energy E is called the total mass or the *ADM mass*, M_{ADM}, which is always nonnegative, $M_{\text{ADM}} \geq 0$. The *positive mass theorem* [494] states that

$$M_{\text{ADM}} > 0, \tag{10.6.20}$$

unless the hypersurface $(\mathcal{M}, g)$ is the Euclidean space $(\mathbb{R}^3, \delta)$.

Simply put, these theorems imply that no energy or mass means no geometry or no gravitation.

---

[3] As a submanifold of an ambient manifold, with an induced metric, the condition of its second fundamental form vanishing is equivalent to the manifold being totally geodesic in its ambient manifold. That is, any geodesic in the manifold is also a geodesic in the ambient manifold, with respect to the metric of the ambient manifold. See [428].

Note that, using the Einstein equations (10.1.60), one may relate the scalar curvature R_g of $(\mathcal{M}, g)$ to the energy density T_{00} by

$$R_g + (h_i^i)^2 - h_j^i h_i^j = 16\pi G T_{00}. \tag{10.6.21}$$

Thus, as a consequence of the dominant energy condition (see (10.6.2)), the vanishing of the second fundamental form naturally leads to the positivity condition for the scalar curvature,

$$R_g \geq 0. \tag{10.6.22}$$

Penrose lower bounds

Naturally, one would hope to bound M_{ADM} away from zero by some physical information in a gravitational system. For example, one may start from considering an isolated black hole of mass $M > 0$ whose spacetime metric is known to be given by the Schwarzschild line element (10.3.20). We see that the spatial slice at any fixed t has the property that its second fundamental form vanishes and that its ADM mass is the same as the black hole mass M. To do so, we need to work within the framework described earlier. For example, we may consider the *isotropic coordinate* representation of (10.3.20) given as [211, 396, 563, 569]

$$\begin{aligned} ds^2 &= \frac{\left(1 - \frac{r_S}{4R}\right)^2}{\left(1 + \frac{r_S}{4R}\right)^2} dt^2 - \left(1 + \frac{r_S}{4R}\right)^4 (dx^2 + dy^2 + dz^2), \\ R &= \sqrt{x^2 + y^2 + z^2}. \end{aligned} \tag{10.6.23}$$

Here, we omit the details of calculation based on (10.6.17) and (10.6.23). Alternatively, Brewin's method [83] can be used as a shortcut to compute the ADM mass. Now, the singular surface or the event horizon, Σ_0, of the black hole is a sphere of radius $r_S = 2GM$ whose surface area has the value

$$|\Sigma_0| = 4\pi r_S^2 = 16\pi G^2 M^2. \tag{10.6.24}$$

The *Penrose conjecture* [442] states that the total energy E of the spacetime defined in (10.6.17) is bounded from below by the total surface area of its apparent horizon Σ, which coincides with the event horizon in the case of a Schwarzschild blackhole, by

$$16\pi G^2 E^2 \geq |\Sigma|. \tag{10.6.25}$$

When the second fundamental form of the hypersurface M vanishes, (10.6.25) becomes

$$16\pi G^2 M_{\text{ADM}}^2 \geq |\Sigma|, \tag{10.6.26}$$

which is referred to as the *Riemannian Penrose inequality*, for which the lower bound may be saturated only in the Schwarzschild limit [80, 81, 82, 300, 301].

Similarly, we may pursue an extension of (10.6.25) or (10.6.26) for charged black holes. For this purpose, use r_0 to denote a root of the event horizon equation (10.4.42) of the Reissner–Nordström metric (10.4.41), which is the radius of a spherically shaped event horizon, say Σ_0. Then we have

$$2GMr_0 = r_0^2 + 4\pi G(Q^2 + P^2), \tag{10.6.27}$$

which leads to

$$M = \frac{1}{G}\sqrt{\frac{|\Sigma_0|}{16\pi}} + 4\pi\sqrt{\frac{\pi}{|\Sigma_0|}}\,(Q^2 + P^2). \tag{10.6.28}$$

Therefore, with the same notation as in (10.6.25), we are naturally led to the following extended Penrose-like energy lower bound:

$$E \geq \frac{1}{G}\sqrt{\frac{|\Sigma|}{16\pi}} + 4\pi\sqrt{\frac{\pi}{|\Sigma|}}\,(Q^2 + P^2). \tag{10.6.29}$$

For study in this direction, see [170, 328, 389, 573] and references therein.

Brown–York quasilocal energy formula

Consider a general static spherically symmetric spacetime metric of the form

$$\mathrm{d}s^2 = N^2\mathrm{d}t^2 - h^2\mathrm{d}r^2 - r^2\left(\mathrm{d}\theta^2 + \sin^2\theta\,\mathrm{d}\phi^2\right), \tag{10.6.30}$$

where N and h are functions depending on the radial coordinate r only. In [89], Brown and York obtained a formulation of a *quasilocal energy* or *quasilocal mass* of a gravitational system based on the Hamilton–Jacobi formalism of the Einstein theory [21, 22, 23] so that, under (10.6.30), the quasilocal energy contained within the local region stretched to the "radial coordinate distance" $r > 0$ is expressed by the formula

$$E_{\mathrm{ql}}(r) = \frac{r}{G}\left(1 - \frac{1}{h(r)}\right), \tag{10.6.31}$$

and the limit

$$E_{\mathrm{ql}}(\infty) = \lim_{r\to\infty} E_{\mathrm{ql}}(r), \tag{10.6.32}$$

gives rise to the ADM energy or mass of the system in the full space. See also [88]. Thus, in terms of (10.4.16) and (10.6.31), we have

$$E_{\mathrm{ql}}(r) = \frac{r}{G}\left(1 - \sqrt{A(r)}\right). \tag{10.6.33}$$

This formula is convenient and useful for calculation in spherical coordinates. For example, for the Reissner–Nordström black hole metric defined by (10.4.40), we see that the associated quasilocal energy is

$$\begin{aligned} E_{\mathrm{ql}}(r) &= \frac{2\left(M - \frac{2\pi(Q^2+P^2)}{r}\right)}{1 + \sqrt{1 - \frac{2GM}{r} + \frac{4\pi G}{r^2}(Q^2 + P^2)}} \\ &= M + \frac{(GM^2 - 4\pi[Q^2 + P^2])}{2r}\left(1 + \frac{GM}{r}\right) + \mathrm{O}(r^{-3}), \quad r \gg 1. \end{aligned} \tag{10.6.34}$$

Consequently, in view of (10.6.32) and (10.6.34), we recover the ADM mass for a Reissner–Nordström black hole as well, $M_{\text{ADM}} = M$, which is independent of the electric and magnetic charges, in particular. In other words, the quasilocal energy or mass of a charged black hole is charge-dependent, but not the global one.

10.7 Gravitational waves

The wave nature of gravity in the context of the Einstein equations may be appreciated by considering a *weak gravity limit* such that the gravitational metric tensor assumes the form

$$g_{\mu\nu} = \eta_{\mu\nu} + h_{\mu\nu}, \tag{10.7.1}$$

where the tensor $h_{\mu\nu}$ is taken to represent some tiny fluctuations around the standard flat Minkowski metric tensor, $\eta_{\mu\nu}$.

Geometric quantities in leading-order approximation

Thus, keeping leading-order terms of $h_{\mu\nu}$, we have, for example,

$$h^{\mu\nu} = g^{\mu\alpha}g^{\nu\beta}h_{\alpha\beta} = \eta^{\mu\alpha}\eta^{\nu\beta}h_{\alpha\beta}, \tag{10.7.2}$$

$$h^{\nu}_{\mu} = g^{\nu\alpha}h_{\mu\alpha} = \eta^{\nu\alpha}h_{\mu\alpha}, \quad h = h^{\mu}_{\mu} = \eta^{\mu\nu}h_{\mu\nu}. \tag{10.7.3}$$

In this way, the Christoffel symbols are

$$\Gamma_{\mu\alpha\beta} = \frac{1}{2}(h_{\mu\alpha,\beta} + h_{\mu\beta,\alpha} - h_{\alpha\beta,\mu}), \tag{10.7.4}$$

$$\Gamma^{\mu}_{\alpha\beta} = \frac{1}{2}\eta^{\mu\nu}(h_{\nu\alpha,\beta} + h_{\nu\beta,\alpha} - h_{\alpha\beta,\nu}), \tag{10.7.5}$$

resulting in the Ricci tensor

$$\begin{aligned} R_{\mu\nu} &= \Gamma^{\alpha}_{\mu\alpha,\nu} - \Gamma^{\alpha}_{\mu\nu,\alpha} \\ &= \frac{1}{2}\eta^{\alpha\beta}(h_{\beta\mu,\alpha\nu} + h_{\beta\alpha,\mu\nu} - h_{\mu\alpha,\beta\nu}) - \frac{1}{2}\eta^{\alpha\beta}(h_{\beta\mu,\nu\alpha} + h_{\beta\nu,\mu\alpha} - h_{\mu\nu,\alpha\beta}). \end{aligned} \tag{10.7.6}$$

On the other hand, by symmetry, we have

$$\eta^{\alpha\beta}h_{\mu\alpha,\beta\nu} = \eta^{\alpha\beta}h_{\beta\mu,\alpha\nu}. \tag{10.7.7}$$

Thus, inserting (10.7.7) into (10.7.6) and using (10.7.2)–(10.7.3), we obtain

$$R_{\mu\nu} = \frac{1}{2}\left(\Box\, h_{\mu\nu} + h_{,\mu\nu} - h^{\alpha}_{\mu,\nu\alpha} - h^{\alpha}_{\nu,\mu\alpha}\right), \tag{10.7.8}$$

where $\Box = \eta^{\mu\nu}\partial_\mu\partial_\nu = \partial_0^2 - \nabla^2$ is the usual d'Alembertian wave operator. In view of (10.7.8), we arrive at the Ricci scalar curvature

$$R = \eta^{\mu\nu}R_{\mu\nu} = \Box\, h - h^{\mu\nu}{}_{,\mu\nu}. \tag{10.7.9}$$

Wave equations in general coordinates

Using (10.7.8) and (10.7.9), we see that, in leading-order approximation, the Einstein tensor is given by

$$\begin{aligned} G_{\mu\nu} &= R_{\mu\nu} - \frac{1}{2}\eta_{\mu\nu}R \\ &= \frac{1}{2}\left(\Box h_{\mu\nu} + h_{,\mu\nu} - h^{\alpha}_{\mu,\nu\alpha} - h^{\alpha}_{\nu,\mu\alpha}\right) - \frac{1}{2}\eta_{\mu\nu}(\Box h - h^{\alpha\beta}{}_{,\alpha\beta}). \end{aligned} \tag{10.7.10}$$

Thus, in view of (10.7.10), we arrive at the Einstein *gravitational wave equations*

$$\Box h_{\mu\nu} + h_{,\mu\nu} - h^{\alpha}_{\mu,\nu\alpha} - h^{\alpha}_{\nu,\mu\alpha} - \eta_{\mu\nu}\left(\Box h - h^{\alpha\beta}{}_{,\alpha\beta}\right) = -16\pi G T_{\mu\nu}, \tag{10.7.11}$$

in weak gravity limit.

Wave equations in harmonic coordinates

Let f be a scalar field. Then the divergence of the vector field $(f_{,\mu})$ is given by

$$\begin{aligned} \nabla^{\mu} f_{,\mu} &= g^{\mu\nu}\left(f_{,\mu\nu} - \Gamma^{\alpha}_{\mu\nu} f_{,\alpha}\right) \\ &= \frac{1}{\sqrt{-g}}\partial_{\mu}\left(\sqrt{-g}g^{\mu\nu}\partial_{\nu} f\right), \end{aligned} \tag{10.7.12}$$

which defines the Laplace–Beltrami operator induced from the metric $(g_{\mu\nu})$, Δ_g, and is also the curved spacetime d'Alembertian wave operator. Since a function f that nullifies (10.7.12) is called harmonic, a coordinate system (x^{α}) is a system of *harmonic coordinates* if it satisfies the condition

$$\nabla^{\mu} x^{\alpha}_{,\mu} = \Delta_g x^{\alpha} = 0, \quad \alpha = 0, 1, 2, 3. \tag{10.7.13}$$

As in [168], we see that the gravitational wave equations (10.7.11) may be recast into a much more transparent and familiar form in harmonic coordinates.

For this purpose, we first note that (x^{α}) is a system of harmonic coordinates if and only if

$$g^{\mu\nu}\Gamma^{\alpha}_{\mu\nu} = 0, \quad \alpha = 0, 1, 2, 3. \tag{10.7.14}$$

Next, multiplying (10.7.14) by $g_{\alpha\beta}$, we get $g^{\mu\nu}\Gamma_{\alpha\mu\nu} = 0$. Thus, using (10.1.12) in this equation, we have

$$g^{\mu\nu}\left(g_{\alpha\mu,\nu} - \frac{1}{2}g_{\mu\nu,\alpha}\right) = 0. \tag{10.7.15}$$

Differentiating (10.7.15) with respect to x^{β} and neglecting higher order terms involving $h_{\mu\nu}$, we get

$$\eta^{\mu\nu}\left(h_{\alpha\mu,\nu\beta} - \frac{1}{2}h_{\mu\nu,\alpha\beta}\right) = 0, \tag{10.7.16}$$

which implies

$$h_{,\alpha\beta} = \eta^{\mu\nu}(h_{\alpha\mu,\nu\beta} + h_{\beta\mu,\alpha\nu}) = h^{\nu}_{\alpha,\beta\nu} + h^{\nu}_{\beta,\alpha\nu}. \tag{10.7.17}$$

Now inserting (10.7.17) into (10.7.6), we arrive at

$$R_{\mu\nu} = \frac{1}{2}\Box h_{\mu\nu}. \tag{10.7.18}$$

Thus, in an empty spacetime where the energy-momentum tensor vanishes, the equations (10.2.77) with $\Lambda = 0$, in view of (10.7.18), become

$$\Box h_{\mu\nu} = 0. \tag{10.7.19}$$

Consequently, we may conclude that in weak gravity limit and empty spacetime gravitational waves propagate as light or electromagnetic fields. For this reason, such a phenomenon is also referred to as *gravitational radiation* [211, 396, 563, 569].

Finally, with (10.7.10) and (10.7.18), we see that, in harmonic coordinates, gravitational waves are governed by the equations

$$\Box h_{\mu\nu} - \eta_{\mu\nu}\left(\Box h - h^{\alpha\beta}{}_{,\alpha\beta}\right) = -16\pi G T_{\mu\nu}, \tag{10.7.20}$$

in weak gravity limit.

The detection of gravitational waves [63, 116, 291, 324, 340, 403], or gravitational radiation, long predicted by general relativity [111, 168, 211, 396, 563, 569], is a recent landmark progress in science.

10.8 Scalar-wave matters as quintessence

Section 10.2 considered the expansion of an isotropic and homogeneous universe filled with various cosmological fluids realized by their respective equations of state. Of the many patterns of expansion, an important one is the late-time exponential or accelerated growth dynamics of the scale factor of the universe, as observed by astrophysicists [444, 470] and attributed to dark energy [100, 318, 438], and may be achieved by the introduction of a positive cosmological constant or a suitable fluid model such as the Chaplygin fluid defined by the equation of state (10.2.91), say. Besides these hypothetical spacetime and cosmological fluid models, it is of physical interest to obtain the anticipated late-time exponential growth pattern, or dark energy, by coupling the Einstein gravitational equations with appropriate field-matter models. With regard to such exploration, a simple but successful formalism is that of a real scalar-wave matter governed canonically by a Klein–Gordon action density, popularly coined *quintessence* [101, 112, 179, 467, 553], believed to be responsible in propelling the accelerated expansion of the universe, which we now study.

Scalar-wave matter

Let φ be a real-valued scalar field governed by the Lagrangian action density

$$\mathcal{L} = \frac{1}{2}\partial_\mu\varphi\partial^\mu\varphi - V(\varphi), \tag{10.8.1}$$

of a Klein–Gordon theory type subject to a potential density function $V(\varphi) \geq 0$ and a gravitational metric $(g_{\mu\nu})$ with $\partial^\mu\varphi = g^{\mu\nu}\partial_\nu\varphi$. Then, the associated Euler–Lagrange equation and energy-momentum tensor are

$$\frac{1}{\sqrt{-g}}\partial_\mu\left(\sqrt{-g}\partial^\mu\varphi\right) + V'(\varphi) = 0, \tag{10.8.2}$$

and

$$T_{\mu\nu} = \partial_\mu\varphi\partial_\nu\varphi - g_{\mu\nu}\left(\frac{1}{2}\partial_\alpha\varphi\partial^\alpha\varphi - V(\varphi)\right), \tag{10.8.3}$$

respectively. Thus, the Hamiltonian energy density $\mathcal{H}$ and trace associated with $T_{\mu\nu}$ are

$$\mathcal{H} = T_0^0 = \frac{1}{2}\partial_0\varphi\partial^0\varphi - \frac{1}{2}\partial_i\varphi\partial^i\varphi + V(\varphi), \tag{10.8.4}$$

$$T = T_\mu^\mu = -\partial_\mu\varphi\partial^\mu\varphi + 4V(\varphi), \tag{10.8.5}$$

where $T_\nu^\mu = g^{\mu\alpha}T_{\nu\alpha}$, so that the Einstein equation (10.2.77) with $\Lambda = 0$ is simplified into

$$R_{\mu\nu} = -8\pi G\left(\partial_\mu\varphi\partial_\nu\varphi - g_{\mu\nu}V\right). \tag{10.8.6}$$

Perfect fluid interpretation

Comparing (10.8.4)–(10.8.5) with (10.2.19)–(10.2.20), we recognize the "fluid" density ρ and pressure P associated with the scalar-wave matter φ given by the expressions

$$\rho = \mathcal{H} = \frac{1}{2}\partial_0\varphi\partial^0\varphi - \frac{1}{2}\partial_i\varphi\partial^i\varphi + V(\varphi), \tag{10.8.7}$$

$$P = \frac{1}{3}(\mathcal{H} - T) = \frac{1}{2}\partial_0\varphi\partial^0\varphi + \frac{1}{6}\partial_i\varphi\partial^i\varphi - V(\varphi). \tag{10.8.8}$$

Robertson–Walker metric in Cartesian coordinates

The *Robertson–Walker gravitational metric in Cartesian coordinates* with $x^0 = t$ and $(x^i) = (x, y, z)$ assumes the form

$$g_{\mu\nu}\mathrm{d}x^\mu\mathrm{d}x^\nu = \mathrm{d}t^2 - a^2(t)\left(\mathrm{d}x^2 + \mathrm{d}y^2 + \mathrm{d}z^2\right), \tag{10.8.9}$$

giving rise to the nontrivial components of the Ricci tensor to be

$$R_{00} = \frac{3\ddot{a}}{a}, \quad R_{11} = R_{22} = R_{33} = -2\dot{a}^2 - a\ddot{a}. \tag{10.8.10}$$

Next, we assume that φ depends only on t. Hence, in view of (10.8.9), we see that (10.8.2) becomes

$$(a^3\dot{\varphi})^{\cdot} = -a^3 V'(\varphi), \tag{10.8.11}$$

and the nontrivial components of $T_{\mu\nu}$ are

$$T_{00} = \frac{1}{2}\dot{\varphi}^2 + V(\varphi), \tag{10.8.12}$$

$$T_{11} = T_{22} = T_{33} = a^2\left(\frac{1}{2}\dot{\varphi}^2 - V(\varphi)\right). \tag{10.8.13}$$

Inserting (10.8.9) and (10.8.10) into (10.8.6), we arrive at

$$\frac{\ddot{a}}{a} = -\frac{8\pi G}{3}(\dot{\varphi}^2 - V(\varphi)), \tag{10.8.14}$$

$$\frac{\ddot{a}}{a} + 2\left(\frac{\dot{a}}{a}\right)^2 = 8\pi G V(\varphi). \tag{10.8.15}$$

Combining these two equations, we get

$$\left(\frac{\dot{a}}{a}\right)^2 = \frac{8\pi G}{3}\left(\frac{1}{2}\dot{\varphi}^2 + V(\varphi)\right). \tag{10.8.16}$$

We see that (10.8.16), with (10.8.11), leads to (10.8.14) as well. In other words, the coupled system consisting of the equations (10.8.11), (10.8.14), and (10.8.15) and the system consisting of the equations (10.8.11) and (10.8.16) are equivalent.

Density, pressure, and equation of state

With $\varphi = \varphi(t)$, the expressions (10.8.7)–(10.8.8) lead to

$$\rho = \frac{1}{2}\dot{\varphi}^2 + V(\varphi), \tag{10.8.17}$$

$$P = \frac{1}{2}\dot{\varphi}^2 - V(\varphi). \tag{10.8.18}$$

Consequently, (10.8.16) assumes the same form of the Friedmann equation (10.2.56), expressed there in spherical coordinates, or

$$\left(\frac{\dot{a}}{a}\right)^2 = \frac{8\pi G}{3}\rho. \tag{10.8.19}$$

Moreover, in terms of (10.8.17)–(10.8.18) and (10.8.11), we see that there holds the conservation law

$$\dot{\rho} + 3(\rho + P)\frac{\dot{a}}{a} = 0, \tag{10.8.20}$$

which is formally identical to (10.2.52) and is a specialization of its general form $\nabla_\nu T^{\mu\nu} = 0$. Hence, the results obtained in Section 10.2 may be used here in the context of the Cartesian coordinates. Furthermore, extending (10.2.58), we find the associated equation of state

$$w = \frac{P}{\rho} = \frac{\frac{1}{2}\dot{\varphi}^2 - V(\varphi)}{\frac{1}{2}\dot{\varphi}^2 + V(\varphi)}. \tag{10.8.21}$$

Governing equations

Since we are interested in expansion, $\dot{a} \geq 0$, we may resolve (10.8.16) to obtain

$$H = \frac{\dot{a}}{a} = \sqrt{\frac{8\pi G}{3}\left(\frac{1}{2}\dot{\varphi}^2 + V(\varphi)\right)} \equiv H(\varphi, \dot{\varphi}), \tag{10.8.22}$$

for the Hubble constant. Hence, (10.8.11) becomes

$$\ddot{\varphi} + 3H(\varphi, \dot{\varphi})\dot{\varphi} + V'(\varphi) = 0, \tag{10.8.23}$$

which is a second-order quasilinear ordinary differential equation and hard to integrate in general. The system of the equations consisting of (10.8.22) and (10.8.23) governs the evolutionary dynamics of the isotropic and homogeneous universe propelled by a quintessence scalar-wave matter.

Special cases

The simplest situation is when the potential density function V is a constant,

$$V = V_0. \tag{10.8.24}$$

There are two interesting cases to be considered.

(i) $V_0 = 0$. Thus, $w = 1$ in (10.8.21) and the model is that of a stiff matter with its density and pressure specifically determined by

$$\rho_m = \frac{1}{2}\dot{\varphi}^2, \quad P_m = \frac{1}{2}\dot{\varphi}^2. \tag{10.8.25}$$

The Big Bang solution of the scale factor is given by the formula (10.2.61) so that there is no dark energy.

(ii) $V_0 > 0$. So (10.8.17) and (10.8.18) lead to the induced matter density and pressure

$$\rho = \rho_m + V_0, \quad P = P_m - V_0. \tag{10.8.26}$$

Comparing (10.8.26) with (10.2.80), we arrive at the recognition that V_0 now serves the role of an effective cosmological constant Λ with

$$V_0 = \frac{\Lambda}{8\pi G}. \tag{10.8.27}$$

In other words, the model may be interpreted as a universe filled with stiff matter and subject to a cosmological constant. Additionally, in this situation, it is consistent to assume $\varphi(t) = \text{constant}$ in (10.8.11) or (10.8.23). Hence, the Friedmann equation (10.8.16) becomes

$$\left(\frac{\dot{a}}{a}\right)^2 = \frac{8\pi G V_0}{3}, \tag{10.8.28}$$

corresponding to a pure cosmological-constant universe, in absence of matter, indeed, with $w = -1$.

Variable situation

In the general situation without assuming (10.8.24), the quantity w defined in (10.8.21) satisfies

$$-1 \leq w \leq 1. \tag{10.8.29}$$

Therefore, we see that in the general, variable situation, the quintessence model is viewed as an interpolation between the stiff matter and pure cosmological constant models. Besides, in view of (10.8.17)–(10.8.18) and (10.8.25)–(10.8.27), we see that the potential density function may be regarded as a *matter-dependent cosmological constant*,

$$\Lambda(\varphi) = 8\pi G\, V(\varphi), \tag{10.8.30}$$

so that, in order to obtain a dark-energy solution with

$$a(t) \sim \mathrm{e}^{E_0 t}, \quad E_0 > 0, \quad t \gg 1, \tag{10.8.31}$$

we aim to find a potential density function $V(\varphi)$ such that a solution to (10.8.23) will satisfy the condition

$$\lim_{t\to\infty} \rho(t) = \lim_{t\to\infty} \left(\frac{1}{2}\dot{\varphi}^2(t) + V(\varphi(t))\right) = \frac{3E_0^2}{8\pi G}, \tag{10.8.32}$$

in view of (10.8.17), (10.8.22), and (10.8.31). In particular, we see that dark energy enjoys an asymptotic matter density interpretation.

Some quintessence models

The work [467] considers the exponential model

$$V(\varphi) = \kappa \mathrm{e}^{-\alpha\varphi}, \tag{10.8.33}$$

See also [43, 143, 266, 284, 474, 553, 578] for subsequent studies. Another interesting model is given by the power function

$$V(\varphi) = \lambda \varphi^{\alpha}, \tag{10.8.34}$$

or its variants [45, 135, 446, 546, 553, 578, 616]. Besides, the sine-Gordon model given by

$$V(\varphi) = \lambda\left(1 + \cos(\alpha\varphi)\right), \tag{10.8.35}$$

with an infinite sequence of equilibria is also of interest in connection to thawing cosmology [135, 138, 218, 553]. Other potential density functions explored include [371, 372]

$$V(\varphi) = \lambda\varphi^2 \left(1 + \frac{\alpha}{\beta + (\varphi - \varphi_0)^2}\right), \tag{10.8.36}$$

$$V(\varphi) = \lambda \left(\varphi^2 + \frac{\alpha \sinh(\beta[\varphi - \varphi_0])}{\cosh^2(\beta[\varphi - \varphi_0])}\right), \tag{10.8.37}$$

in the study of *inflationary cosmology* in which the universe *initially* expands exponentially fast, in contrast to the dark-energy expansion picture, which concerns *ultimate* expansion.

In all these situations, the dynamical system consisting of the scalar-wave matter equation (10.8.23) and the Friedmann–Hubble scale-factor equation (10.8.22) is difficult to solve analytically and we need to rely on phase-space analysis and numerical computation in order to better understand the issues under investigation.

Realization of equation of state by quintessence

We now show how, theoretically, the equation of state of an underlying cosmological fluid may be realized by a quintessence. To see that, we assume that the pair of the scalar-wave density and pressure given by (10.8.17)–(10.8.18) are related by a general expression of the form (10.2.57) or

$$P = f(\rho). \tag{10.8.38}$$

Inserting (10.8.38) into (10.8.20), we subsequently obtain

$$\rho = \rho(a), \quad P = P(a). \tag{10.8.39}$$

Thus, in view of (10.8.17), (10.8.18), and (10.8.39), we have

$$\dot{\varphi}^2 = \rho(a) + P(a), \quad V(\varphi) = \frac{1}{2}(\rho(a) - P(a)). \tag{10.8.40}$$

So, combining (10.8.19) and the first relation in (10.8.40), we get

$$\frac{\mathrm{d}\varphi}{\mathrm{d}a} = \pm\sqrt{\frac{3}{8\pi G}\left(1 + \frac{P(a)}{\rho(a)}\right)}\,\frac{1}{a}, \tag{10.8.41}$$

which may be integrated to yield

$$\varphi = \varphi(a) \quad \text{or} \quad a = a(\varphi), \tag{10.8.42}$$

giving rise to the potential density function through the second equation in (10.8.40):

$$V(\varphi) = \frac{1}{2}(\rho(a(\varphi)) - P(a(\varphi))), \tag{10.8.43}$$

governing the scalar-wave matter field φ as quintessence.

Next, we work out two examples as illustrations of this formalism.

Linear fluids

With the linear equation of state (10.2.58) or $P = w\rho$ ($w > -1$) in the current context, the equation (10.8.41) becomes

$$\frac{\mathrm{d}\varphi}{\mathrm{d}a} = \pm\frac{\sigma}{a}, \quad \sigma = \sqrt{\frac{3(1+w)}{8\pi G}}, \tag{10.8.44}$$

which may be integrated to render

$$a = A_0 \mathrm{e}^{\pm\frac{\varphi}{\sigma}}, \tag{10.8.45}$$

where $A_0 > 0$ is an integration constant. In view of (10.2.60) and (10.8.45), we see that (10.8.43) leads us to [44]

$$V(\varphi) = (1-w)\lambda \mathrm{e}^{\mp 2\sqrt{6\pi G(1+w)}\varphi}, \quad \lambda > 0, \tag{10.8.46}$$

which is of the form of the exponential model (10.8.33). In particular, when $w = 1$, we are in the trivial potential situation, $V = 0$, which recovers the stiff-matter model.

Chaplygin fluids

We next consider the Chaplygin fluid model (10.2.91) with $A = 0$ for simplicity such that

$$P = f(\rho) = -\frac{B}{\rho}. \tag{10.8.47}$$

Hence, inserting (10.2.92) (observing $A = 0$) into (10.8.41), with (10.8.47), we arrive at

$$\frac{\mathrm{d}\varphi}{\mathrm{d}a} = \pm\frac{\sigma}{a}\sqrt{\frac{C}{Ba^6 + C}}, \quad \sigma = \sqrt{\frac{3}{8\pi G}}. \tag{10.8.48}$$

Integrating this equation, we obtain

$$\varphi = \pm\frac{\sigma}{3}\ln\left(\frac{A_0 a^3}{\sqrt{Ba^6 + C} + \sqrt{C}}\right), \tag{10.8.49}$$

where $A_0 > 0$ is an integration constant. Solving a^3 in terms of φ, we have

$$a^3 = \frac{2A_0\sqrt{C}\mathrm{e}^{\pm\frac{3\varphi}{\sigma}}}{A_0^2 - B\mathrm{e}^{\pm\frac{6\varphi}{\sigma}}}. \tag{10.8.50}$$

On the other hand, in view of (10.8.43), (10.8.47), and (10.2.92) (with $A = 0$), we have

$$\begin{aligned} V(\varphi) &= \frac{2Ba^6 + C}{2a^3\sqrt{Ba^6 + C}} \\ &= \frac{B^2\mathrm{e}^{\pm\frac{9\varphi}{\sigma}} + 6A_0^2 B\mathrm{e}^{\pm\frac{3\varphi}{\sigma}} + A_0^4\mathrm{e}^{\mp\frac{3\varphi}{\sigma}}}{4A_0\left(B\mathrm{e}^{\pm\frac{6\varphi}{\sigma}} + A_0^2\right)}, \end{aligned} \tag{10.8.51}$$

which looks rather complicated, although the integration constant C disappears. However, when setting $A_0^2 = B$, we see that (10.8.51) simplifies into

$$V(\varphi) = \frac{\sqrt{B}}{2}\left(\cosh\left[\pm\frac{3\varphi}{\sigma}\right] + \frac{1}{\cosh\left[\pm\frac{3\varphi}{\sigma}\right]}\right), \tag{10.8.52}$$

as obtained in [319].

Note that, in [44], a scalar-wave matter representation for the general equation of state

$$\rho + P = \gamma\rho^{\lambda}, \tag{10.8.53}$$

where γ, λ are constants with $\gamma \neq 0$, is systematically established.

Equation of state in terms of scale factor

If the model of the cosmological fluid is given in such a way that the matter density ρ and scale factor a are related by

$$\rho = h(a), \tag{10.8.54}$$

as, for example, by (10.2.69) describing a two-fluid model, then (10.8.20) leads to

$$P = -h(a) - \frac{1}{3}ah'(a). \tag{10.8.55}$$

Hence, the equation of state is now given in a parametric form in terms of the parameter a. Consequently, we may insert (10.8.54)–(10.8.55) into (10.8.41) to obtain (10.8.42) and, eventually, (10.8.43). Here we omit the details.

Chapter 14 discusses how the idea of quintessence may be extended in the context of the Born–Infeld nonlinear theory of electromagnetism to that of the so-called k-essence, which enables us to realize a broader range of equations of state of cosmic fluids of interests.

Exercises

1. Let $\mathbf{x} = \mathbf{x}(u, v)$ denote a parametrization of a surface S in $\mathbb{R}^3$ by two real parameters u and v.

   (a) Show that $\mathbf{x}_u$ and $\mathbf{x}_v$ are linearly independent for given u, v if and only if the *metric matrix*, also called the *Gram matrix*,

   $$M(u,v) = \begin{pmatrix} \mathbf{x}_u \cdot \mathbf{x}_u & \mathbf{x}_u \cdot \mathbf{x}_v \\ \mathbf{x}_u \cdot \mathbf{x}_v & \mathbf{x}_v \cdot \mathbf{x}_v \end{pmatrix} \equiv \begin{pmatrix} E & F \\ F & G \end{pmatrix} \tag{10.E.1}$$

   is positive definite. If $M(u, v)$ is positive definite for each pair u, v in the domain of definition of S, then the surface is called regular, which will be assumed throughout. (Hint: The following *Lagrange identity* involving vector dot and cross products,

   $$(\mathbf{a} \times \mathbf{b}) \cdot (\mathbf{a} \times \mathbf{b}) = \det\begin{pmatrix} \mathbf{a}\cdot\mathbf{a} & \mathbf{a}\cdot\mathbf{b} \\ \mathbf{a}\cdot\mathbf{b} & \mathbf{b}\cdot\mathbf{b} \end{pmatrix}, \quad \mathbf{a}, \mathbf{b} \in \mathbb{R}^3, \tag{10.E.2}$$

   may be useful for your study.)

(b) Show that the condition of being regular for the surface S is independent of the choice of parametrization. That is, if $(\tilde{u}, \tilde{v})$ is another parametrization of S which is related to (u, v) by a nonsingular transformation of the form

$$\tilde{u} = \tilde{u}(u, v), \quad \tilde{v} = \tilde{v}(u, v); \quad \frac{\partial(\tilde{u}, \tilde{v})}{\partial(u, v)} = \det \begin{pmatrix} \frac{\partial \tilde{u}}{\partial u} & \frac{\partial \tilde{u}}{\partial v} \\ \frac{\partial \tilde{v}}{\partial u} & \frac{\partial \tilde{v}}{\partial v} \end{pmatrix} \neq 0, \tag{10.E.3}$$

then $M(\tilde{u}, \tilde{v})$ is also positive definite.

(c) Show that the line element of S is given by

$$ds^2 = E du^2 + 2F du dv + G dv^2, \tag{10.E.4}$$

whose form is independent of the parametrization. That is, if we use $\tilde{u}, \tilde{v}$ as another parameter set for S and $\tilde{E}, \tilde{F}, \tilde{G}$ as the entries of the corresponding metric matrix, $M(\tilde{u}, \tilde{v})$, then

$$ds^2 = \tilde{E} d\tilde{u}^2 + 2\tilde{F} d\tilde{u} d\tilde{v} + \tilde{G} d\tilde{v}^2, \tag{10.E.5}$$

replacing (10.E.4).

(d) Show that the surface element

$$d\sigma = |\mathbf{x}_u \times \mathbf{x}_v| \, du dv \tag{10.E.6}$$

is parametrization independent as well. Then, use (10.E.2) to show that

$$d\sigma = \sqrt{\det M(u, v)} \, du dv = \sqrt{EG - F^2} \, du dv. \tag{10.E.7}$$

2. Equip the space $\mathbb{R}^3$ with the conformal metric

$$ds^2 = f(x, y, z)(dx^2 + dy^2 + dz^2), \tag{10.E.8}$$

$$f(x, y, z) = \frac{1}{\left(1 + \frac{K}{4}[x^2 + y^2 + z^2]\right)^2}, \tag{10.E.9}$$

where K is a constant. Show that the scalar curvature of such a space is K. The metric (10.E.8)–(10.E.9) is also known as the Robertson–Walker metric [396].

3. Derive the Friedmann equation (10.2.51) from (10.2.49)–(10.2.50).

4. From (10.2.48), derive the conservation law (10.2.52).

5. Integrate (10.2.70) to obtain the solution (10.2.71) and use it to arrive at the asymptotic estimate

$$a(t) = (6\pi G \rho_d)^{\frac{1}{3}} \, t^{\frac{2}{3}}, \quad t \to \infty, \tag{10.E.10}$$

which returns to (10.2.60) and shows that the dust model eventually dominates the evolution.

6. Derive the equation of state that leads to the dust and radiation model given in (10.2.69).
7. Assume $\Lambda > 0$ and a closed universe, $k = 1$. Show that the parameters $\rho_{\rm d}$ and $\rho_{\rm r}$ in (10.2.69) may be chosen such that a static dust and radiation model universe is permitted as a solution to the cosmological equations (10.2.81)–(10.2.82).
8. Present a study of a static universe given by the cosmological equations (10.2.81)–(10.2.82) when the cosmological constant Λ is negative concerning the possibility that the space is flat, $k = 0$, or open, $k = -1$.
9. Derive the solution (10.2.89).
10. Derive the Friedmann equation for the *de Sitter universe* governed by the vacuum Einstein equation

$$G_{\mu\nu} + \Lambda g_{\mu\nu} = 0, \tag{10.E.11}$$

with a positive cosmological constant, $\Lambda > 0$, and show that such an equation is contained in (10.2.88) and possesses the unique solution

$$a(t) = a_0 \mathrm{e}^{\sqrt{\frac{\Lambda}{3}}t}, \quad t \geq 0, \tag{10.E.12}$$

initiating from $a(0) = a_0$ at $t = 0$. In particular, it does not allow a Big Bang solution with $a(0) = 0$.
11. Let $a(t)$ be given by (10.2.96)–(10.2.97) for the Chaplygin fluid universe. Verify the initial condition $a(0) = 0$ and the exponential growth law (10.2.98).
12. Show that the quantities t_0 and t_1 given in (10.3.33) and (10.3.34), respectively, are both positive, and that $t_0 < t_1$.
13. Assume the radius R of the massive body is sufficiently small and consider an inbound travelling photon inside a black hole whose trajectory is given by the formula (10.3.31). Find a time when the speed of the photon reaches three times of the vacuum speed of light.
14. Use the outbound Eddington–Finkelstein coordinates to describe a photon that travels toward or away from the event horizon along the radial direction initiating outside and inside the event horizon.
15. Derive the transformation (10.3.60)–(10.3.61) for the Kruskal coordinates and find its inverse that expresses t, r in terms of ξ, η.
16. Establish the formula (10.3.64) for the Kretschmann curvature.
17. Note that, combining (10.4.9) and (10.4.11), we have

$$\Gamma^{\mu}_{\alpha\mu} = \frac{g_{,\alpha}}{2g}. \tag{10.E.13}$$

(a) Use (10.E.13) to establish the identity

$$\Gamma^{\mu}_{\mu\alpha,\beta} = \Gamma^{\mu}_{\mu\beta,\alpha}. \tag{10.E.14}$$

(b) Use (10.1.27) and (10.E.14) to derive the identity

$$R^{\mu}_{\mu\alpha\beta} = 0. \tag{10.E.15}$$

(c) By contracting $\mu = \nu$ in (10.1.29) and applying (10.1.28) and (10.E.15), show that $R_{\alpha\beta} = R_{\beta\alpha}$.

18. Let the electromagnetic tensor field $F_{\mu\nu}$ be generated from a gauge potential A_μ such that $F_{\mu\nu} = \partial_\mu A_\nu - \partial_\nu A_\mu$. Vary A_μ in (10.4.13) to derive the curved spacetime Maxwell equations (10.4.12).
19. For the Reissner–Nordström metric (10.4.41), obtain its Kretschmann curvature given in (10.4.55), generalizing (10.3.64).
20. Use (10.4.52) and (10.4.53) to derive (10.4.54).
21. Show that, formally, the Kerr–Newman metric (10.5.11) may be obtained by substituting $r_S r$ solved from (10.5.12) into the Kerr metric (10.5.3) in the following manner.

(a) First switch off the electric and magnetic charges in (10.5.12) by setting $Q, P = 0$ to get

$$r_S r = r^2 + \alpha^2 - \Delta. \tag{10.E.16}$$

(b) Next substitute (10.E.16) into (10.5.3) to arrive at (10.5.11) in the situation when $Q, P = 0$.

(c) Switch on Q, P to get (10.5.11).

22. Check to see that (10.6.1) remains valid when $\mu, \nu = i, j = 1, 2, 3$ for the examples (10.6.3) and (10.6.8).
23. For the Schwarzschild metric (10.3.20), use (10.6.17) to compute the gravitational energy.
24. For the Reissner–Nordström metric (10.4.41), use (10.6.17) to compute the gravitational energy.
25. Show that the extended inequality (10.6.29) may be recast into the form

$$E \geq \sqrt{\frac{|\Sigma|}{16\pi}\left(\frac{1}{G} + \frac{(4\pi)^2}{|\Sigma|}(Q^2 + P^2)\right)}, \tag{10.E.17}$$

which gives rise to a neater comparison with the Penrose inequality (10.6.25).

26. Show that, under the extremal charged black hole condition, (10.4.46), the Penrose-like energy lower bound (10.6.29) becomes

$$4\pi G^2 E^2 \geq |\Sigma|, \tag{10.E.18}$$

which sharpens (10.6.25). (Hint: Consider (10.6.27).)

27. Establish (10.6.34).
28. Derive the right-hand side of (10.7.12).
29. Use (10.8.11) and (10.8.17)–(10.8.18) to establish the conservation law (10.8.20).
30. With (10.8.33), formulate a phase-space dynamical system for the coupled system of the equations (10.8.22)–(10.8.23). Study whether the dark-energy behavior (10.8.31) may be achieved.
31. Derive (10.8.52).
32. For the two-fluid model (10.2.69), integrate (10.8.41) to find $\varphi = \varphi(a)$.
33. With (10.8.53), obtain (10.8.54) and (10.8.55) explicitly.
34. Consider the Chaplygin fluid model (10.2.91), that is,

$$P = A\rho - \frac{B}{\rho}, \quad A > -1, \quad B > 0, \tag{10.E.19}$$

which is not contained in (10.8.53). Show that, with (10.8.40), the scalar-wave matter field φ and scale factor a satisfy the equation

$$\frac{\mathrm{d}\varphi}{\mathrm{d}a} = \pm\frac{\sigma}{a}\sqrt{\frac{C}{Ba^{6(1+A)} + C}}, \quad \sigma = \sqrt{\frac{3(1+A)}{8\pi G}}, \tag{10.E.20}$$

which generalizes (10.8.48). For $A = -\frac{1}{2}$, integrate (10.E.20) and obtain the associated potential density function $V(\varphi)$ through (10.8.43).

11

Charged vortices and Chern–Simons equations

This chapter investigates whether it is possible to construct electrically and magnetically charged static solutions, known as dually charged vortices, of gauge field equations. It begins by discussing the Julia–Zee theorem, which says such solutions are not allowed in conventional situations. It next looks at the existence of dually charged vortices when a Chern–Simons topological term is added to the Lagrangian action density. It then considers a related problem in three spatial dimensions.

11.1 Julia–Zee theorem

Chapter 9 showed that gauge theory in three spatial dimensions allows the existence of finite-energy magnetically and electrically charged particle-like static solutions called dyons. We are curious as to whether there are dyons in two spatial dimensions. That is, whether there are finite-energy magnetically and electrically charged static vortices in gauge theory. In their now-classic 1975 paper [315], Julia and Zee studied the Abelian Higgs gauge field theory model. Using a radially symmetric field configuration ansatz and assuming a sufficiently fast decay rate at spatial infinity, they concluded that a finite-energy static solution of the equations of motion over the $(2+1)$-dimensional Minkowski spacetime must satisfy the temporal gauge condition, and thus, it is necessarily electrically neutral. This result, referred to as the Julia–Zee theorem, leads to many interesting consequences. For example, it makes it transparent that the static Abelian Higgs model is exactly the Ginzburg–Landau theory [237] which is purely magnetic [310, 416].

Since the work [315], it has been conceptualized [160, 161, 307, 496] and accepted [176, 295, 308, 346, 434, 559, 560] that, in order to obtain both

Mathematical Physics with Differential Equations. Yisong Yang, Oxford University Press.
 DOI: 10.1093/oso/9780192872616.003.0011

electrically and magnetically charged static vortices, it is necessary to introduce the Chern–Simons topological terms [131, 132] into the Lagrangian action density, which is an essential construct in anyon physics [581, 582] (see also [219]).

This chapter discusses these problems in the simplest Abelian situation.

Governing equations in conventional situation

Recall that the classical Abelian Higgs theory over the (2 + 1)-dimensional spacetime is governed by the Lagrangian action density (5.2.30) and the associated equations of motion are (5.2.31)–(5.2.33). In the static situation, the operator $\partial_0 = 0$ nullifies everything. Hence, the electric charge density ρ becomes

$$\rho = J^0 = \frac{\mathrm{i}}{2}(\overline{\phi} D^0 \phi - \phi \overline{D^0 \phi}) = -A_0 |\phi|^2, \tag{11.1.1}$$

where $D_\mu = \partial_\mu + \mathrm{i} A_\mu$ is the renormalized gauge-covariant derivative, and a nontrivial temporal component of the gauge field, A_0, is necessary for the presence of electric charge. On the other hand, the $\mu = 0$ component of the left-hand side of the Maxwell equation (5.2.32) is

$$\partial_\nu F^{0\nu} = \partial_i (F_{i0}) = \partial_i^2 A_0 = \Delta A_0. \tag{11.1.2}$$

Consequently, the static version of the equations of motion (5.2.31)–(5.2.33) may be written as

$$D_i^2 \phi = 2V'(|\phi|^2)\phi - A_0^2 \phi, \tag{11.1.3}$$

$$\partial_j F_{ij} = \frac{\mathrm{i}}{2}(\overline{\phi} D_i \phi - \phi \overline{D_i \phi}), \tag{11.1.4}$$

$$\Delta A_0 = |\phi|^2 A_0, \tag{11.1.5}$$

in which (11.1.5) is the Gauss law.

Energy-momentum tensor

It is clear that the energy-momentum or stress tensor of the action density (5.2.30) may be computed to be

$$\begin{aligned} T_{\mu\nu} &= 2\frac{\partial \mathcal{L}}{\partial \eta^{\mu\nu}} - \eta_{\mu\nu}\mathcal{L} \\ &= -\eta^{\alpha\beta} F_{\mu\alpha} F_{\nu\beta} + \frac{1}{2}(D_\mu \phi \overline{D_\nu \phi} + \overline{D_\mu \phi} D_\nu \phi) - \eta_{\mu\nu}\mathcal{L}, \end{aligned} \tag{11.1.6}$$

the associated Hamiltonian density is given by

$$\mathcal{H} = T_{00} = \frac{1}{2}|\partial_i A_0|^2 + \frac{1}{2}A_0^2|\phi|^2 + \frac{1}{4}F_{ij}^2 + \frac{1}{2}|D_i\phi|^2 + V(|\phi|^2), \tag{11.1.7}$$

so that the finite-energy condition reads

$$\int_{\mathbb{R}^2} \mathcal{H}\, \mathrm{d}x < \infty. \tag{11.1.8}$$

Julia–Zee theorem

With this formulation, the *Julia–Zee theorem* [315] may be stated as follows.

Suppose that (A_0, A_i, ϕ) is a finite-energy solution of the static Abelian Higgs equations (11.1.3)–(11.1.5) over $\mathbb{R}^2$. Then, either $A_0 = 0$ everywhere if ϕ is not identically zero or $A_0 \equiv$ constant and the solution is necessarily electrically neutral. In other words, the static Abelian Higgs model is exactly the Ginzburg–Landau theory.

Proof by cutoff

The work [525] provides a proof of the theorem, which relies on a crucial choice of a test function so that the argument is valid exactly in two dimensions.

Let $0 \leq \eta \leq 1$ be of compact support and define for $M > 0$ fixed the truncated function

$$A_0^M = \begin{cases} M & \text{if } A_0 > M, \\ A_0 & \text{if } |A_0| \leq M, \\ -M & \text{if } A_0 < -M. \end{cases} \tag{11.1.9}$$

Then, multiplying (11.1.5) by ηA_0^M and integrating, we have

$$\int_{\mathbb{R}^2} \left(\eta \nabla A_0 \cdot \nabla A_0^M + A_0^M \nabla A_0 \cdot \nabla \eta + \eta |\phi|^2 A_0^M A_0\right) \mathrm{d}x = 0. \tag{11.1.10}$$

Using (11.1.9) in (11.1.10), we find

$$\begin{aligned} &\int_{\{|A_0|<M\}\cap\, \mathrm{supp}(\eta)} \eta |\phi|^2 A_0^2 \, \mathrm{d}x + M^2 \int_{\{|A_0|>M\}\cap\, \mathrm{supp}(\eta)} \eta |\phi|^2 \, \mathrm{d}x \\ &+ \int_{\{|A_0|<M\}\cap\, \mathrm{supp}(\eta)} \eta |\nabla A_0|^2 \, \mathrm{d}x \\ &\leq M \left(\int_{\mathbb{R}^2} |\nabla A_0|^2 \, \mathrm{d}x\right)^{\frac{1}{2}} \left(\int_{\mathbb{R}^2} |\nabla \eta|^2 \, \mathrm{d}x\right)^{\frac{1}{2}}. \end{aligned} \tag{11.1.11}$$

For $R > 1$ (say), we now choose η to be a logarithmic cutoff function given as

$$\eta = \begin{cases} 1 & \text{if } |x| < R, \\ 2 - \frac{\ln |x|}{\ln R} & \text{if } R \leq |x| \leq R^2, \\ 0 & \text{if } |x| > R^2. \end{cases} \tag{11.1.12}$$

Then,

$$\int_{\mathbb{R}^2} |\nabla \eta|^2 \, \mathrm{d}x = \frac{2\pi}{\ln R}. \tag{11.1.13}$$

Using (11.1.13) in (11.1.11) gives

$$\begin{aligned}
&\int_{\{|A_0|<M\}\cap B_R} |\phi|^2 A_0^2 \,\mathrm{d}x + \int_{\{|A_0|<M\}\cap B_R} |\nabla A_0|^2 \,\mathrm{d}x \\
&\le \int_{\{|A_0|<M\}\cap B_R} |\phi|^2 A_0^2 \,\mathrm{d}x + M^2 \int_{\{|A_0|>M\}\cap B_R} |\phi|^2 \,\mathrm{d}x + \int_{\{|A_0|<M\}\cap B_R} |\nabla A_0|^2 \,\mathrm{d}x \\
&\le M \frac{\left(2\pi \int_{\mathbb{R}^2} |\nabla A_0|^2 \,\mathrm{d}x\right)^{\frac{1}{2}}}{(\ln R)^{\frac{1}{2}}}. \qquad (11.1.14)
\end{aligned}$$

The right-hand side of (11.1.14) tends to zero as R goes to infinity. Letting M tend to infinity proves the conclusion.

We note that the Julia–Zee theorem is also valid for non-Abelian gauge field theory. See [525] for detail.

11.2 Chern–Simons term

For simplicity, we again concentrate on the Abelian situation in $(2+1)$ dimensions. In this situation, with the gauge field A_μ and the induced field tensor $F_{\mu\nu} = \partial_\mu A_\nu - \partial_\nu A_\mu$, the *Chern–Simons term* reads [131, 132]

$$\frac{1}{4}\epsilon^{\alpha\mu\nu} A_\alpha F_{\mu\nu}, \qquad (11.2.1)$$

where $\epsilon^{\alpha\mu\nu}$ is the skew-symmetric Kronecker symbol with $\epsilon^{012} = 1$.

Gauge invariance

It is clear that the term (11.2.1) is *not invariant* under the gauge transformation

$$A_\mu \mapsto A_\mu - \mathrm{i}\partial_\mu \omega. \qquad (11.2.2)$$

However, such a violation of gauge symmetry is local, but not global. In other words, the integration of (11.2.1) over the full $(2+1)$-dimensional spacetime, that is,

$$\int \frac{1}{4}\epsilon^{\alpha\mu\nu} A_\alpha F_{\mu\nu} \,\mathrm{d}x, \qquad (11.2.3)$$

remains gauge invariant, which we can check through an integration by parts:

$$\begin{aligned}
\int \epsilon^{\alpha\mu\nu}(\partial_\alpha \omega) F_{\mu\nu} \,\mathrm{d}x &= \int \epsilon^{\alpha\mu\nu}([\partial_\nu \partial_\alpha \omega] A_\mu - [\partial_\mu \partial_\alpha \omega] A_\nu) \,\mathrm{d}x \\
&= \int \epsilon^{\alpha\mu\nu}(\partial_\nu \partial_\alpha \omega) A_\mu \,\mathrm{d}x - \int \epsilon^{\nu\alpha\mu}(\partial_\alpha \partial_\nu \omega) A_\mu \,\mathrm{d}x \\
&= \int \epsilon^{\alpha\mu\nu}(\partial_\nu \partial_\alpha \omega - \partial_\alpha \partial_\nu \omega) A_\mu \,\mathrm{d}x = 0. \qquad (11.2.4)
\end{aligned}$$

Consequently, we see that there is no violation of gauge symmetry when a Chern–Simons term is added into an action density.

11.3 Dually charged vortices

As a consequence of the preparation made in Section 11.2, the Abelian Chern–Simons–Higgs Lagrangian density introduced in [434, 559], which minimally extends the classical Abelian Higgs model [310, 416], as discussed in Chapter 5, defined over the Minkowski spacetime $\mathbb{R}^{2,1}$, may be written in the form

$$\mathcal{L} = -\frac{1}{4}F_{\mu\nu}F^{\mu\nu} + \frac{\kappa}{4}\varepsilon^{\mu\nu\alpha}A_\mu F_{\nu\alpha} + \frac{1}{2}D_\mu\phi\overline{D^\mu\phi} - \frac{\lambda}{8}(|\phi|^2-1)^2, \tag{11.3.1}$$

where κ is a constant referred to as the *Chern–Simons coupling parameter*. Sometimes, the quantity $k \equiv \pi\kappa$ is also called the *level of the Chern–Simons theory* so that the Chern–Simons term in the action density assumes the form

$$\frac{k}{4\pi}\epsilon^{\alpha\mu\nu}A_\alpha F_{\mu\nu}. \tag{11.3.2}$$

Governing equations

The extremals of the Lagrangian density (11.3.1) formally satisfy its Euler–Lagrange equations, or the Abelian *Chern–Simons–Higgs equations* [434],

$$D_\mu D^\mu\phi = \frac{\lambda}{2}\phi(1-|\phi|^2), \tag{11.3.3}$$

$$\partial_\nu F^{\mu\nu} - \frac{\kappa}{2}\varepsilon^{\mu\nu\alpha}F_{\nu\alpha} = -J^\mu, \tag{11.3.4}$$

in which (11.3.4) expresses the *modified Maxwell equations* so that the current density J^μ is given by

$$J^\mu = \frac{\mathrm{i}}{2}(\overline{\phi}D^\mu\phi - \phi\overline{D^\mu\phi}), \tag{11.3.5}$$

as before. Since we consider static configurations only so that all the fields are independent of the temporal coordinate, $t = x^0$, we have

$$\rho = J^0 = -A_0|\phi|^2. \tag{11.3.6}$$

Remember that the electric field $\mathbf{E} = (E^i)$, in the spatial plane, and magnetic fields H, perpendicular to the spatial plane, induced from the gauge field A_μ are

$$E^i = \partial_i A_0, \quad i = 1,2; \quad H = F_{12}, \tag{11.3.7}$$

respectively. The *static version of the Chern–Simons–Higgs equations* (11.3.3) and (11.3.4) take the explicit form

$$D_i^2\phi = \frac{\lambda}{2}(|\phi|^2-1)\phi - A_0^2\phi, \tag{11.3.8}$$

$$\partial_j F_{ij} - \kappa\,\varepsilon_{ij}\partial_j A_0 = \frac{\mathrm{i}}{2}(\overline{\phi}D_i\phi - \phi\overline{D_i\phi}), \tag{11.3.9}$$

$$\Delta A_0 = \kappa F_{12} + |\phi|^2 A_0. \tag{11.3.10}$$

Finite-energy condition

Since the Chern–Simons term gives rise to a *topological invariant* that is independent of the spacetime metric, $\eta_{\mu\nu}$, it makes no contribution to the energy-momentum tensor $T_{\mu\nu}$ of the action density (11.3.1), which may be calculated as

$$T_{\mu\nu} = -\eta^{\alpha\beta} F_{\mu\alpha} F_{\nu\beta} + \frac{1}{2}([D_\mu\phi]\overline{[D_\nu\phi]} + \overline{[D_\mu\phi]}[D_\nu\phi]) - \eta_{\mu\nu}\mathcal{L}_0, \tag{11.3.11}$$

where $\mathcal{L}_0$ is obtained from the Lagrangian (11.3.1) by setting $\kappa = 0$. Hence, it follows that the Hamiltonian $\mathcal{H} = T_{00}$ or the energy density of the theory is given by

$$\mathcal{H} = \frac{1}{2}|\nabla A_0|^2 + \frac{1}{2}|\phi|^2 A_0^2 + \frac{1}{2}F_{12}^2 + \frac{1}{2}(|D_1\phi|^2 + |D_2\phi|^2) + \frac{\lambda}{8}(|\phi|^2 - 1)^2, \tag{11.3.12}$$

which is contained as a special case of (11.1.7), and the terms in (11.3.12) not containing A_0 are exactly those appearing in the classical Abelian Higgs model in the temporal gauge. Thus, we arrive at the following finite-energy condition

$$E(\phi, A_0, A_i) = \int_{\mathbb{R}^2} \mathcal{H}(A_0, A_i, \phi)(x)\,\mathrm{d}x < \infty. \tag{11.3.13}$$

Boundary conditions

The condition (11.3.13) naturally gives rise to the asymptotic behavior of the fields A_0, A_i, and ϕ:

$$A_0,\ \partial_i A_0 \to 0, \tag{11.3.14}$$

$$F_{12} \to 0, \tag{11.3.15}$$

$$|\phi| \to 1,\ |D_1\phi|,\ |D_2\phi| \to 0, \tag{11.3.16}$$

as $|x| \to \infty$. In analogue to the Abelian Higgs model, we see that a finite-energy solution of the Chern–Simons–Higgs equations (11.3.8)–(11.3.10) should be classified by the winding number, say $N \in \mathbb{Z}$, of the complex scalar field ϕ near infinity, which is expected to give rise to the total quantized magnetic charge or magnetic flux.

Charged Chern–Simons vortices

The problem with the existence of charged vortices in the full Chern–Simons–Higgs theory is that we must prove that, for any integer N, the coupled nonlinear elliptic equations (11.3.8)–(11.3.10) over $\mathbb{R}^2$ possess a smooth solution (A_0, A_i, ϕ) satisfying the finite-energy condition (11.3.13) and natural boundary conditions (11.3.14)–(11.3.16) so that the winding number of ϕ near infinity is N.

Here, we provide an existence theorem [123] regarding such solutions.

For any given integer N, the Chern–Simons–Higgs equations (11.3.8)–(11.3.10) over $\mathbb{R}^2$ have a smooth finite-energy solution (ϕ, A_0, A_i) satisfying the asymptotic properties (11.3.14)–(11.3.16) as $|x| \to \infty$ such that the winding number of ϕ near infinity is N, which is also the algebraic multiplicity of zeros of ϕ in $\mathbb{R}^2$, and the total magnetic charge Q_m and electric charge Q_e are given by the quantization formulas

$$Q_m = \frac{1}{2\pi}\int_{\mathbb{R}^2} F_{12}\,\mathrm{d}x = N, \tag{11.3.17}$$

$$Q_e = \frac{1}{2\pi}\int_{\mathbb{R}^2} \rho\,\mathrm{d}x = \kappa N. \tag{11.3.18}$$

Such a solution represents an N-vortex soliton which is indeed both magnetically and electrically charged.

Note that the problem encountered here is that the static Chern–Simons–Higgs equations (11.3.8)–(11.3.10) are not the Euler–Lagrange equations of the energy functional (11.3.13) defined by the Hamiltonian density (11.3.12), but rather of the action functional defined by the Lagrangian density (11.3.1), which may be written explicitly as

$$\begin{aligned} I(\phi, A_0, A_i) &= \frac{1}{2}\int_{\mathbb{R}^2}\left(F_{12}^2 + |D_1\phi|^2 + |D_2\phi|^2 + \frac{\lambda}{4}(|\phi|^2-1)^2\right)\mathrm{d}x \\ &\quad - \frac{1}{2}\int_{\mathbb{R}^2}\left(|\nabla A_0|^2 + A_0^2|\phi|^2 + 2\kappa A_0 F_{12}\right)\mathrm{d}x. \end{aligned} \tag{11.3.19}$$

Thus, we need to find a critical point of the *indefinite action functional* (11.3.19) under the finite-energy condition (11.3.13) and the topological constraint $Q_m = N$ expressed in (11.3.17).

In order to tackle the difficulty arising from the negative part of the action functional, we introduce the constraint

$$\int_{\mathbb{R}^2}\left(\nabla A_0\cdot\nabla\tilde{A}_0 + |\phi|^2 A_0\tilde{A}_0 + \kappa F_{12}\tilde{A}_0\right)\mathrm{d}x = 0, \quad \forall\,\tilde{A}_0, \tag{11.3.20}$$

for each pair of fixed ϕ and A_i, maintaining the finite-energy condition. In particular, when $\tilde{A}_0 = A_0$, we have

$$\int_{\mathbb{R}^2}\kappa A_0 F_{12}\,\mathrm{d}x = -\int_{\mathbb{R}^2}\left(|\nabla A_0|^2 + A_0^2|\phi|^2\right)\mathrm{d}x. \tag{11.3.21}$$

Substituting (11.3.21) into (11.3.19), we see that the action functional becomes positive definite, which resolves the issue of dealing with an indefinite action functional in the original, unconstrained, setting.

Note that in the Chern–Simons–Higgs context, the electric charge Q_e is also quantized topologically. Such a property is naturally expected since the finite-energy condition (11.3.13) for the equations (11.3.8)–(11.3.10) implies the vanishing property

$$\int_{\mathbb{R}^2}\Delta A_0\,\mathrm{d}x = 0. \tag{11.3.22}$$

Thus, integrating (11.3.10) and using (11.3.6), we arrive at the following electric and magnetic charge relation

$$Q_e = \kappa Q_m, \tag{11.3.23}$$

which gives us (11.3.18).

Combining the established existence theorem with the Julia–Zee theorem, we can draw the conclusion that the Abelian static Chern–Simons–Higgs equations (11.3.8)–(11.3.10), which are the Euler–Lagrange equations of the minimally coupled action density (11.3.1), have an electrically nontrivial finite-energy solution if and only if the Chern–Simons term is present, which is characterized by the condition $\kappa \neq 0$. In such a situation, electricity and magnetism must co-exist.

For extensions of this study to the case of existence of *non-Abelian Chern–Simons–Higgs vortices*, see [123].

11.4 Rubakov–Tavkhelidze problem

This section presents an interesting partial differential equation problem at the juncture of the Ginzburg–Landau equations and the Chern–Simons equations formulated over $\mathbb{R}^3$ with the feature that the associated energy functional is similar to that of the Ginzburg–Landua theory which is then to be minimized subject to a Chern–Simons integral-type constraint, known as the Rubakov–Tavkhelidze soliton problem [463, 488]. The structure of the energy functional with respect to that of the Chern–Simons charge integral makes it possible to obtain some estimates of the energy in terms of the underlying Chern–Simons charge [488], which calls upon further exploration by analysts.

Energy functional

Closely related to the Chern–Simons theory and relevant in modeling electroweak interaction, in 1985 Rubakov and Tavkhelidze [473] introduced the following Ginzburg–Landau type energy functional

$$E(\mathbf{A}, u) = \int \left(\frac{1}{2}|\nabla \times \mathbf{A}|^2 + |\nabla u|^2 + g^2|\mathbf{A}|^2 u^2 + \lambda(u^2 - v^2)^2\right) \mathrm{d}x, \tag{11.4.1}$$

over $\mathbb{R}^3$, governing a vector field $\mathbf{A}$ and a real scalar field u, subject to the *prescribed Chern–Simons charge*

$$N_{\mathrm{CS}} = \frac{g^2}{16\pi^2} \int \mathbf{A} \cdot (\nabla \times \mathbf{A}) \, \mathrm{d}x, \tag{11.4.2}$$

where $g, v, \lambda > 0$ are coupling constants. See also [472, 504].

We are to minimize (11.4.1) subject to the integral constraint (11.4.2).

Governing equations

In order to derive the Euler–Lagrange equations of the problem, we recall the identity

$$\nabla\cdot(\mathbf{a}\times\mathbf{b})=(\nabla\times\mathbf{a})\cdot\mathbf{b}-\mathbf{a}\cdot(\nabla\times\mathbf{b}),\tag{11.4.3}$$

for two vector fields $\mathbf{a},\mathbf{b}$. Thus, with $\mathbf{A}_t=\mathbf{A}+t\mathbf{a}$ and applying (11.4.3), we have

$$\left(\frac{\mathrm{d}}{\mathrm{d}t}\left(\mathbf{A}_t\cdot[\nabla\times\mathbf{A}_t]\right)\right)_{t=0}=2\mathbf{a}\cdot(\nabla\times\mathbf{A})-\nabla\cdot(\mathbf{A}\times\mathbf{a}),\tag{11.4.4}$$

$$\begin{aligned}\left(\frac{\mathrm{d}}{\mathrm{d}t}|\nabla\times\mathbf{A}_t|^2\right)_{t=0}&=2(\nabla\times\mathbf{A})\cdot(\nabla\times\mathbf{a})\\&=2\mathbf{a}\cdot(\nabla\times\nabla\times\mathbf{A})+2\nabla\cdot(\mathbf{a}\times\nabla\times\mathbf{A}).\end{aligned}\tag{11.4.5}$$

In view of (11.4.4) and (11.4.5), we can vary the functional

$$I(\mathbf{A},u)=E(\mathbf{A},u)-\xi\left(\frac{8\pi^2}{g^2}\right)N_{\mathrm{CS}}(\mathbf{A},u),\tag{11.4.6}$$

where ξ is a Lagrange multiplier, to obtain the Euler–Lagrange equations

$$\Delta u-g^2|\mathbf{A}|^2u-2\lambda(u^2-v^2)u=0,\tag{11.4.7}$$

$$\nabla\times(\nabla\times\mathbf{A})-\xi(\nabla\times\mathbf{A})+2g^2u^2\mathbf{A}=\mathbf{0}.\tag{11.4.8}$$

Comparing with (11.3.1) and (11.3.3)–(11.3.4), the Lagrange multiplier ξ in (11.4.6) and (11.4.7)–(11.4.8) takes the role of the Chern–Simons parameter κ.

See [210, 488] for some subsequent work and [463] for a discussion in the context of a survey. So far, there is no rigorous mathematical study of such an *Euclidean three-dimensional Chern–Simons problem* regarding the existence of a solution to (11.4.7)–(11.4.8) with a prescribed value of the quantity N_{CS} in (11.4.2).

Restoration of gauge invariance

Note that the model (11.4.1) lacks gauge invariance. In order to recover its gauge invariance, we may replace the real scalar field u by a complex scalar field ϕ, use the gauge-covariant derivative

$$D_{\mathbf{A}}\phi=\nabla\phi-\mathrm{i}g\mathbf{A}\phi\tag{11.4.9}$$

as before, and reactivate the gauge transformation

$$\mathbf{A}\mapsto\mathbf{A}+\nabla\omega,\quad \phi\mapsto\mathrm{e}^{\mathrm{i}g\omega}\phi,\tag{11.4.10}$$

where ω is a real-valued function over $\mathbb{R}^3$. Since

$$|D_{\mathbf{A}}\phi|^2=|\nabla\phi|^2+g^2|\mathbf{A}|^2|\phi|^2+2g\mathbf{A}\cdot\mathrm{Im}(\phi\nabla\overline{\phi}),\tag{11.4.11}$$

we see that the *Rubakov–Tavkhelidze energy* (11.4.1) is the real-scalar-field version of the Ginzburg–Landau energy

$$E(\mathbf{A},\phi)=\int\left\{\frac{1}{2}|\nabla\times\mathbf{A}|^2+|D_{\mathbf{A}}\phi|^2+\lambda(|\phi|^2-v^2)^2\right\}\,\mathrm{d}x. \tag{11.4.12}$$

In other words, the model (11.4.1) is the Ginzburg–Landau model (11.4.12) stated in the *unitary gauge*. Hence, the equations of motion (11.4.7)–(11.4.8) are modified into

$$D_{\mathbf{A}}^2\phi=2\lambda(|\phi|^2-v^2)\phi, \tag{11.4.13}$$
$$\nabla\times\nabla\times\mathbf{A}=\mathrm{i}g(\phi\overline{D_{\mathbf{A}}\phi}-\overline{\phi}D_{\mathbf{A}}\phi)+\xi(\nabla\times\mathbf{A}). \tag{11.4.14}$$

Generalized London equation

Note that, in the *purely superconducting* limit, we may take $\phi=v$ and vary $\mathbf{A}$ in (11.4.12) to obtain the following single equation governing the gauge potential $\mathbf{A}$:

$$\nabla\times\nabla\times\mathbf{A}=-2g^2v^2\mathbf{A}+\xi(\nabla\times\mathbf{A}). \tag{11.4.15}$$

The magnetic field $\mathbf{B}$ induced from $\mathbf{A}$ is given by $\mathbf{B}=\nabla\times\mathbf{A}$. Thus, taking curl in (11.4.15), we have

$$\Delta\mathbf{B}=2g^2v^2\mathbf{B}-\xi(\nabla\times\mathbf{B}), \tag{11.4.16}$$

which is the London equation when the parameter ξ is absent. Thus, we arrive at a *generalized London equation* arising from the Rubakov–Tavkhelidze problem. It is clear that (11.4.15) is the Euler–Lagrange equation of the action functional

$$I(\mathbf{A})=\int\left(\frac{1}{2}|\nabla\times\mathbf{A}|^2+g^2v^2|\mathbf{A}|^2-\frac{\xi}{2}\mathbf{A}\cdot(\nabla\times\mathbf{A})\right)\,\mathrm{d}x, \tag{11.4.17}$$

which is *quadratic*. In view of the magnetic-field component of the Maxwell equations,

$$\nabla\times\mathbf{B}=\mathbf{J}+\frac{\partial\mathbf{E}}{\partial t}, \tag{11.4.18}$$

relating the magnetic field $\mathbf{B}$ to the electric current $\mathbf{J}$ and electric field $\mathbf{E}$, the parameter ξ-dependent term in (11.4.16) may naturally be interpreted as a steady-state superconductive-current contribution.

Constrained minimization

We now study the solution to (11.4.15) by considering the constrained minimization problem

$$\min\left\{E(\mathbf{A})\,\middle|\,Q(\mathbf{A})=\frac{16\pi^2}{g^2}N_{\mathrm{CS}}\right\}, \tag{11.4.19}$$

where the Chern–Simons charge $N_{\rm CS}$ is as prescribed in (11.4.2), and

$$E(\mathbf{A}) = \int \left(\frac{1}{2}|\nabla \times \mathbf{A}|^2 + g^2v^2|\mathbf{A}|^2\right) \mathrm{d}x, \tag{11.4.20}$$

$$Q(\mathbf{A}) = \int \mathbf{A} \cdot (\nabla \times \mathbf{A})\, \mathrm{d}x. \tag{11.4.21}$$

Bogomol'nyi lower bound for energy

From (11.4.20) and (11.4.21), we have

$$\begin{aligned} E(\mathbf{A}) &= \frac{1}{2}\int \left(\nabla \times \mathbf{A} \pm \sqrt{2}\, gv\mathbf{A}\right)^2 \mathrm{d}x \mp \sqrt{2}gvQ(\mathbf{A}) \\ &\geq \sqrt{2}gv|Q(\mathbf{A})|. \end{aligned} \tag{11.4.22}$$

Therefore, we arrive at the topological lower bound

$$E(\mathbf{A}) \geq \frac{16\sqrt{2}\,\pi^2 v}{g}|N_{\rm CS}|, \tag{11.4.23}$$

which is saturated when $\mathbf{A}$ satisfies the following *linear Bogomol'nyi equation*

$$\nabla \times \mathbf{A} = \pm\sqrt{2}\, gv\mathbf{A}, \quad N_{\rm CS} = \pm|N_{\rm CS}|. \tag{11.4.24}$$

Substituting (11.4.24) into (11.4.15), we find

$$\xi = \pm 2\sqrt{2}gv. \tag{11.4.25}$$

In other words, a finite-energy solution of (11.4.24) solves the problem (11.4.19), with the minimum energy given by the right-hand side of (11.4.23), which also determines the Lagrange multiplier ξ in the generalized London equation (11.4.15) through the expression (11.4.25).

Note that the equation (11.4.24) prompts an *eigenfunction problem for the curl operator*, which has been studied by many people [402, 529, 549] for its applications in electromagnetism and fluids.

This chapter has focused on the partial differential equation problems arising in the Abelian Chern–Simons–Higgs theory. For existence of dually charged vortices, we have exclusively only considered radially symmetric solutions in [123] for technical reasons, as in the situations of the Ginzburg–Landau vortices in Section 8.3 and the 't Hooft–Polyakov monopoles and Julia–Zee dyons in Section 9.3. Thus a natural question is whether multivortices may be constructed in the Chern–Simons gauge field theory as those in the Abelian Higgs theory considered in Section 8.2, by pursuing a BPS reduction under a similar critical coupling condition. Thanks to the pioneering investigations of Hong, Kim, and Pac [295] and Jackiw and Weinberg [308] made concurrently and independently in 1990, such a reduction is indeed available in the Chern–Simons case and has since then sparked a rich vista of mathematical studies [96, 117, 129, 523, 524, 538]. Subsequently, this reduction was successfully carried over to the situation of non-Abelian Chern–Simons theory [176, 177, 178], which unearthed a wealth of highly interesting systems of nonlinear elliptic equations of challenging structures governing non-Abelian Chern–Simons–Higgs multivortices. More recently, these ideas have also been further developed in supersymmetric gauge field theory in the context of the Bagger–Lambert–Gustavsson (BLG) model [34, 35, 36, 56, 119, 201, 263] and the Aharony–Bergman–Jafferis–Maldacena (ABJM) model [8, 331], both of which have been the focus of numerous activities in contemporary field-theoretical physics. See [128, 271, 272, 273, 275, 276, 418, 599] and references therein for some partial differential equation work on these lines of research.

Exercises

1. Obtain (11.1.6).
2. Verify the identity (11.1.13).
3. Subject to the Lorentz gauge $\partial_\mu A^\mu = 0$, recast the equations (11.3.3)–(11.3.4) into the form $\Box\phi = \cdots, \Box A^\mu = \cdots$.
4. Show that (11.3.8)–(11.3.10) are the Euler–Lagrange equations of the action functional (11.3.19).
5. Subject to the Coulomb gauge $\partial_i A_i = 0$, recast the equations (11.3.8)–(11.3.9) into the form $\Delta\phi = \cdots, \Delta A_i = \cdots$.
6. Impose suitable properties on ϕ and F_{12} compatible with the boundary conditions (11.3.15) and (11.3.16) to show that (11.3.20) has a unique solution that enjoys the boundary condition (11.3.14).
7. Elaborate on the boundary condition (11.3.14) to justify the identity (11.3.22).
8. Find all trivial or constant finite-energy solutions to the equations (11.4.7)–(11.4.8).
9. Replace (11.4.12) by

$$E(\mathbf{A},\phi) = \int \left\{\frac{1}{2}|\nabla\mathbf{A}|^2 + |D_{\mathbf{A}}\phi|^2 + \lambda(|\phi|^2 - v^2)^2\right\} \mathrm{d}x, \tag{11.E.1}$$

   where

$$|\nabla\mathbf{A}|^2 = \sum_{i=1}^{3} |\nabla A_i|^2, \quad \mathbf{A} = (A_i). \tag{11.E.2}$$

   (a) Derive the Euler–Lagrange equations of (11.E.1) by minimizing (11.E.1) subject to the constraint (11.4.2).

(b) Show that a finite-energy solution to the equations in (a) automatically lies in the Coulomb gauge, $\nabla \cdot \mathbf{A} = 0$, thus fulfilling the equations (11.4.7)–(11.4.8). (Hint: The identity

$$\nabla \times \nabla \times \mathbf{A} = -\Delta \mathbf{A} + \nabla(\nabla \cdot \mathbf{A}) \tag{11.E.3}$$

and the fact that a harmonic L^2-function must be zero will be useful.)

(c) Show that the energy (11.E.1) does not observe the gauge symmetry (11.4.10).

10. Replace (11.4.17) by the modified action functional

$$I(\mathbf{A}) = \int \left(\frac{1}{2}|\nabla \mathbf{A}|^2 + g^2 v^2 |\mathbf{A}|^2 - \frac{\xi}{2} \mathbf{A} \cdot (\nabla \times \mathbf{A}) \right) \mathrm{d}x. \tag{11.E.4}$$

(a) Derive the Euler–Lagrange equation of (11.E.4).

(b) Show that a finite-action and finite-charge solution of the equation in (a) automatically stays in the Coulomb gauge and thus solves (11.4.15).

11. Show that, when we set $(A_\mu) = (A_0, A_1, A_2) = \mathbf{A}$ and $\nabla = (\partial_0, \partial_1, \partial_2)$, then we have

$$\frac{1}{2}\epsilon^{\alpha\mu\nu} A_\alpha F_{\mu\nu} = \mathbf{A} \cdot (\nabla \times \mathbf{A}), \tag{11.E.5}$$

for the Chern–Simons density.

12. Consider the generalized London equation (11.4.15).

(a) Show that (11.4.15) permits a special family of solutions satisfying the equation

$$\nabla \times \mathbf{A} = a\mathbf{A}, \tag{11.E.6}$$

where a is a constant and ξ is prescribed.

(b) Find the range of ξ for which the only such solution must be trivial, $\mathbf{A} = \mathbf{0}$.

(c) In nontrivial situation, any solution of (11.E.6) must stay in the Coulomb gauge.

(d) In nontrivial situation, a solution of (11.E.6) must satisfy the elliptic "eigenvalue-eigenvector" equation

$$\Delta \mathbf{A} = -a^2 \mathbf{A}. \tag{11.E.7}$$

13. Let $\mathbf{A}$ be a solution to (11.4.19) and consider its one-parameter deformation defined by

$$\tilde{\mathbf{A}}(x) = \delta \mathbf{A}(\delta x) = \delta \mathbf{A}(\tilde{x}), \quad \tilde{x} = \delta x, \quad \delta > 0. \tag{11.E.8}$$

(a) Show that for $E(\mathbf{A})$ defined by (11.4.20) there holds

$$E(\tilde{\mathbf{A}}) = \int \left(\frac{\delta}{2}|\nabla \times \mathbf{A}|^2 + \frac{g^2 v^2}{\delta}|\mathbf{A}|^2 \right) \mathrm{d}x. \tag{11.E.9}$$

(b) Let $Q(\mathbf{A})$ be defined by (11.4.21). Show that $Q(\tilde{\mathbf{A}}) = Q(\mathbf{A})$. This invariant property is hardly surprising since Q is a topological quantity.

(c) Establish the equation

$$\left.\frac{\mathrm{d}E(\tilde{\mathbf{A}})}{\mathrm{d}\delta}\right|_{\delta=1} = 0, \tag{11.E.10}$$

which implies the *energy partition relation*

$$\int \frac{1}{2}|\nabla \times \mathbf{A}|^2 \,\mathrm{d}x = \int g^2 v^2 |\mathbf{A}|^2 \,\mathrm{d}x. \tag{11.E.11}$$

(d) Use the Cauchy–Schwarz inequality and (11.E.11) to establish the inequality

$$|Q(\mathbf{A})| \leq \frac{1}{\sqrt{2}gv} E(\mathbf{A}), \tag{11.E.12}$$

which recovers (11.4.22), with the equality valid if and only if $\mathbf{A}$ and $\nabla \times \mathbf{A}$ are linearly dependent. This latter feature further explains the linear Bogomol'nyi equation (11.4.24).

12

Skyrme model and related topics

The idea that elementary particles may be described by continuously distributed fields with localized energy concentrations, also called solitons, has a long history. In this formalism, it is of interest to construct static solitons describing particles at rest or in equilibrium. This chapter first revisits the classical Klein–Gordon scalar field theory and states an associated virial theorem known as the Derrick identity which indicates that there are no finite-energy static solutions in spatial dimensions beyond two. It discusses that such a dimensionality barrier can be overcome by an expansion of the theory to include gauge fields, and then shows that, the Skyrme model, on the other hand, allows static solutions in spatial dimension three without the need of adding gauge fields, but introduces extra quartic terms of derivatives. It also investigates the Faddeev model which is similar to the Skyrme model in this regard with a refined topological feature that its static solutions are characterized by the Hopf invariant and thus give rise to knotted solitons following a fractional-exponent energy-topology growth law. Afterwards, it explores the Q-ball model of Coleman that seeks energy minimizers among steady states with a prescribed conserved charge, leading to a constrained minimization problem, such that it also lifts the Derrick dimensionality barrier. These highly related problems offer rich analytic opportunities and challenges for future research.

12.1 Derrick theorem and Pohozaev identity

The central idea embedded in the 1961 Skyrme model [513, 514, 515, 516] is to use continuously extended, topologically characterized, relativistically invariant, locally concentrated, soliton-like fields to model elementary particles. One of the distinguishing features of the Skyrme model in comparison with the conventional

Mathematical Physics with Differential Equations. Yisong Yang, Oxford University Press.
 DOI: 10.1093/oso/9780192872616.003.0012

field-theoretical models is that it allows stable static solutions beyond spatial dimension two, without gauge fields, as imposed by the classical Derrick theorem [158]. At the quantized level, the Skyrme model gives rise to two types of hadronic particles that are essential for strong nuclear interactions, namely, the mesons, which are quantum fluctuations around the topologically trivial field configuration, and the baryons, which are effectively realized as (topologically nontrivial) solitons. In this way, the baryon–meson and baryon–baryon scattering [296, 585] come naturally into the picture. More recently, the Skyrme model and its various variations have been applied to many other areas, including the quantum Hall effect [200, 410], Bose–Einstein condensates [48, 476], and cosmology [58, 399, 508].

To appreciate some of the analytic features of the Skyme model, this section begins with a discussion on the dimensionality barrier dictated by the Derrick theorem regarding scalar field theory. Subsequent sections consider the Skyrme model, the Faddeev model, and the Q-balls, and demonstrate in particular how each of them manages to overcome the dimensionality barrier.

Scalar-field theory model

To start, we consider the standard Klein–Gordon field theory governing a complex scalar field ϕ over the Minkowski spacetime $\mathbb{R}^{n,1}$ whose Lagrangian action density reads

$$\mathcal{L} = \frac{1}{2}\partial_\mu\phi\overline{\partial^\mu\phi} - V(|\phi|^2), \tag{12.1.1}$$

where $V \geq 0$ is an arbitrary potential density function. The Euler–Lagrange equation of (12.1.1) is

$$\partial_\mu\partial^\mu\phi = -2V'(|\phi|^2)\phi, \tag{12.1.2}$$

with the associated energy-momentum tenor

$$T_{\mu\nu} = \frac{1}{2}(\partial_\mu\phi\overline{\partial_\nu\phi} + \partial_\nu\phi\overline{\partial_\mu\phi}) - \eta_{\mu\nu}\mathcal{L}. \tag{12.1.3}$$

Hence, the energy density or Hamiltonian is

$$\mathcal{H} = T_{00} = \frac{1}{2}|\partial_t\phi|^2 + \frac{1}{2}|\nabla\phi|^2 + V(|\phi|^2). \tag{12.1.4}$$

Static solution and Derrick theorem

In the static situation, the equation (12.1.2) becomes a semilinear elliptic equation,

$$\Delta\phi = 2V'(|\phi|^2)\phi, \tag{12.1.5}$$

which is the Euler–Lagrange equation of the Hamilton energy

$$E(\phi) = \int_{\mathbb{R}^n} \left(\frac{1}{2}|\nabla\phi|^2 + V(|\phi|^2)\right) \mathrm{d}x. \tag{12.1.6}$$

Thus, a solution of (12.1.5) is simply a critical point of the energy functional (12.1.6). Hence, for a finite-energy solution ϕ, we may use the same argument as in Section 8.1 to arrive at the virial identity

$$(2-n)\int_{\mathbb{R}^n}|\nabla\phi|^2\,\mathrm{d}x = 2n\int_{\mathbb{R}^n}V(|\phi|^2)\,\mathrm{d}x, \tag{12.1.7}$$

similar to (8.1.13). Although (12.1.7) is obtained for simplicity for a single-component complex-valued function, it is clear that it is also valid in the general situation of an arbitrary multi-component function.

Consequently, we see that there is no nontrivial solution if $n \geq 3$, which rules out the most physical dimension, $n = 3$. This statement is known as the *Derrick theorem* [158]. Besides, the case $n = 2$ is interesting only in the absence of potential energy, $V = 0$. Further, only when $n = 1$ is the potential density function V not subject to any restriction, and locally concentrated static solutions can indeed be constructed, which are often called kinks or domain walls [109, 369, 466, 556].

Pohozaev identity

The integral identity (12.1.7) is also called the *Pohozaev identity* or *Derrick–Pohozaev identity* in the study of partial differential equations [59, 199, 449], which we now elaborate on. For this purpose, we consider instead the elliptic boundary value problem

$$-\Delta\phi = \lambda|\phi|^{p-1}\phi \quad \text{in } \Omega, \quad \phi = 0 \quad \text{on } \partial\Omega, \tag{12.1.8}$$

for a complex-valued function ϕ, where Ω is a bounded domain in $\mathbb{R}^n$ with smooth boundary and $p > 1, \lambda > 0$ are constants.

Following [199], multiplying $\Delta\phi$ and $\Delta\overline{\phi}$ by $x\cdot\nabla\overline{\phi}$ and $x\cdot\nabla\phi$, respectively, integrating over Ω, using $\phi = 0$ on $\partial\Omega$, and observing the summation convention over repeated indices, we have

$$\begin{aligned}
&\int_\Omega x^j\left(\partial_j\overline{\phi}\partial_k\partial_k\phi + x^j\partial_j\phi\partial_k\partial_k\overline{\phi}\right)\mathrm{d}x\\
&= -n\int_\Omega(\overline{\phi}\Delta\phi+\phi\Delta\overline{\phi})\,\mathrm{d}x - \int_\Omega x^j(\overline{\phi}\partial_k\partial_k\partial_j\phi + \phi\partial_k\partial_k\partial_j\overline{\phi})\,\mathrm{d}x\\
&= -n\int_\Omega(\overline{\phi}\Delta\phi+\phi\Delta\overline{\phi})\,\mathrm{d}x + \int_\Omega \delta^j_k(\overline{\phi}\partial_k\partial_j\phi + \phi\partial_k\partial_j\overline{\phi})\,\mathrm{d}x\\
&\quad + \int_\Omega x^j(\partial_k\overline{\phi}\partial_k\partial_j\phi + \partial_k\phi\partial_k\partial_j\overline{\phi})\,\mathrm{d}x\\
&= (1-n)\int_\Omega(\overline{\phi}\Delta\phi+\phi\Delta\overline{\phi})\,\mathrm{d}x + \int_\Omega\partial_j(x^j|\nabla\phi|^2)\,\mathrm{d}x - n\int_\Omega|\nabla\phi|^2\,\mathrm{d}x\\
&= (1-n)\int_\Omega(\overline{\phi}\Delta\phi+\phi\Delta\overline{\phi})\,\mathrm{d}x + \int_{\partial\Omega}|\nabla\phi|^2(\nu(x)\cdot x)\,\mathrm{d}S - n\int_\Omega|\nabla\phi|^2\,\mathrm{d}x\\
&= (n-2)\int_\Omega|\nabla\phi|^2\,\mathrm{d}x + \int_{\partial\Omega}|\nabla\phi|^2(\nu(x)\cdot x)\,\mathrm{d}S,
\end{aligned} \tag{12.1.9}$$

where $\nu(x)$ denotes the unit outward normal at the boundary point $x \in \partial\Omega$. Further, multiplying $|\phi|^{p-1}\phi$ and $|\phi|^{p-1}\overline{\phi}$ by $x \cdot \nabla\overline{\phi}$ and $x \cdot \nabla\phi$, respectively, summing up, and integrating, we have

$$\begin{aligned}\int_\Omega |\phi|^{p-1} x \cdot \nabla|\phi|^2 \, \mathrm{d}x =& \frac{2}{p+1}\int_\Omega \nabla \cdot (x|\phi|^{p+1}) \, \mathrm{d}x - \frac{2n}{p+1}\int_\Omega |\phi|^{p+1} \, \mathrm{d}x \\ =& -\frac{2n}{p+1}\int_\Omega |\phi|^{p+1} \, \mathrm{d}x. \end{aligned} \tag{12.1.10}$$

Thus, multiplying the differential equation in (12.1.8) by $x \cdot \nabla\overline{\phi}$, adding it to its own complex conjugate, and using (12.1.9)–(12.1.10), we obtain

$$(n-2)\int_\Omega |\nabla\phi|^2 \, \mathrm{d}x + \int_{\partial\Omega} |\nabla\phi|^2 (\nu(x) \cdot x) \, \mathrm{d}S = \frac{2n\lambda}{(p+1)}\int_\Omega |\phi|^{p+1} \, \mathrm{d}x, \tag{12.1.11}$$

which is the *Pohozaev identity*.

If Ω is star-shaped about the origin, namely,

$$\tau x \in \overline{\Omega}, \quad x \in \overline{\Omega}, \quad \tau \in [0,1], \tag{12.1.12}$$

then $\nu(x) \cdot x \geq 0$ for $x \in \partial\Omega$, which says that the angle between the outward normal vector $\nu(x)$ and the radial vector x is never obtuse. This fact seems geometrically apparent and may be established following the ideas in [199]. Indeed, fix $x \in \partial\Omega$ and take $y \in \partial\Omega$ ($y \neq x$). Then,

$$\lim_{y \to x} \nu(x) \cdot \frac{y-x}{|y-x|} = 0, \tag{12.1.13}$$

since $\nu(x)$ is perpendicular to the tangent plane of $\partial\Omega$ at x. Thus, for any $\varepsilon > 0$, there is a $\delta > 0$ such that

$$\nu(x) \cdot \frac{y-x}{|y-x|} < \varepsilon, \quad |y-x| < \delta, \quad y \in \partial\Omega, \quad y \neq x. \tag{12.1.14}$$

Now consider

$$\Omega_\delta = \{z \in \overline{\Omega} \,|\, |z-x| < \delta\}. \tag{12.1.15}$$

For any $z \in \Omega_\delta$, $z \neq x$, we can find a point $y \in \Omega_\delta \cap \partial\Omega$ such that $|y-x| = |z-x|$ and that y lies in the plane spanned by $\nu(x)$ and $z-x$ in the nontrivial situation when $\nu(x)$ and $z-x$ are not colinear (in the trivial situation when $\nu(x)$ and $z-x$ are colinear we simply choose any $y \in \Omega_\delta \cap \partial\Omega$). Hence, using (12.1.14), we arrive at

$$\nu(x) \cdot \frac{z-x}{|z-x|} \leq \nu(x) \cdot \frac{y-x}{|y-x|} < \varepsilon, \quad |z-x| < \delta, \quad z \in \overline{\Omega}, \quad z \neq x. \tag{12.1.16}$$

Set $z = \tau x$ in (12.1.16). We have

$$-\nu(x) \cdot \frac{x}{|x|} = \nu(x) \cdot \frac{\tau x - x}{|\tau x - x|} < \varepsilon, \tag{12.1.17}$$

provided that $\tau \in (0,1)$ is sufficiently close to 1, which proves $\nu(x) \cdot x \geq 0$ as a result, since $\varepsilon > 0$ may be taken to be arbitrarily small.

Consequently, if Ω is star-shaped about the origin, we arrive from (12.1.11) at

$$\frac{(n-2)}{2} \int_\Omega |\nabla \phi|^2 \, dx \leq \frac{\lambda n}{(p+1)} \int_\Omega |\phi|^{p+1} \, dx. \tag{12.1.18}$$

On the other hand, multiplying the differential equation in (12.1.8) simply by $\overline{\phi}$, adding the result to its own complex conjugate, and integrating, we have

$$\int_\Omega |\nabla \phi|^2 \, dx = \lambda \int_\Omega |\phi|^{p+1} \, dx. \tag{12.1.19}$$

In view of (12.1.18)–(12.1.19), we have

$$\left(\frac{(n-2)}{2} - \frac{n}{(p+1)} \right) \int_\Omega |\phi|^{p+1} \, dx \leq 0, \tag{12.1.20}$$

which establishes that the only solution is the trivial solution $\phi = 0$ when

$$\frac{(n-2)}{2} - \frac{n}{(p+1)} > 0 \tag{12.1.21}$$

or

$$n \geq 3, \quad p > \frac{n+2}{n-2}. \tag{12.1.22}$$

In other words, nontrivial solutions may be possible only when $n = 1, 2$ or $n \geq 3$ and p satisfies

$$\text{the subcritical condition:} \quad 1 < p < \frac{n+2}{n-2}, \tag{12.1.23}$$

$$\text{or the critical condition:} \quad p = \frac{n+2}{n-2}. \tag{12.1.24}$$

Derrick theorem in presence of gauge field

We consider again the Derrick theorem over the Minkowski space $\mathbb{R}^{n,1}$. Of course, a finite-energy static solution (ϕ, A_μ) in the Yang–Mills–Higgs model, say, in the temporal gauge $A_0 = 0$ is a critical point of the energy functional

$$E(\phi, A) = \int_{\mathbb{R}^n} \left(\frac{1}{4} F_{ij}^2 + \frac{1}{2} |D_i \phi|^2 + V(|\phi|^2) \right) dx, \quad i, j = 1, \dots, n. \tag{12.1.25}$$

We may use the same argument as that in Section 8.1 to arrive at the new identity

$$(4-n) \int_{\mathbb{R}^n} F_{ij}^2 \, dx + 2(2-n) \int_{\mathbb{R}^n} |D_i \phi|^2 \, dx = 4n \int_{\mathbb{R}^n} V(|\phi|^2) \, dx, \tag{12.1.26}$$

which slightly extends (8.1.13). Therefore, with a gauge field, the allowance of spatial dimensions is extended to $n \leq 4$.

The interesting individual cases are listed as follows:

(i) $n = 4$: The matter field sector must be trivial,

$$V(|\phi|^2) = 0, \quad D_i\phi = 0, \quad i = 1, \dots, n, \tag{12.1.27}$$

and only a gauge field is allowed to be present (a pure gauge situation). Solutions in this situation are known as gauge-field instantons [466, 477, 614].

(ii) $n = 3, 2, 1$: All these are allowed with total freedom for choosing V.

Thus, we see that the presence of a gauge field component enhances dimensionality for the existence of static finite-energy solutions of field equations.

The rest of the chapter focuses on pure scalar field theory in absence of a gauge field, unless otherwise stated.

12.2 Skyrme model

This section presents the Skyrme model and focuses on static field configurations and their associated topological characterization.

Harmonic map model

To place our discussion in perspective, we first consider $n = 2$ in (12.1.7). Thus, we are led to $V(|\phi|^2) = 0$. In the typical normalized situation, we may let V vanish at $|\phi|^2 = 1$. This implies that the range of ϕ is compactified to satisfy $|\phi|^2 = 1$. Consequently, if ϕ is a m-component scalar field, $\phi = (\phi^1, \dots, \phi^m) \in \mathbb{R}^m$, say, then $\phi \in S^{m-1}$.

If $m = 2$, then $\phi \in S^1 \subset \mathbb{R}^2 = \mathbb{C}$, which may be represented by a real-valued function u such that $\phi = \mathrm{e}^{\mathrm{i}u}$. In the static situation, we see that u is harmonic over $\mathbb{R}^2$ and $\partial_i u \in L^2(\mathbb{R}^2)$, $i = 1, 2$. Hence, $u =$ constant, which trivializes the problem.

If $m = 3$, then $\phi \in S^2$, which may be regarded as a spin vector describing the magnetization of a 2-dimensional *Heisenberg ferromagnet* [466]. Since the potential density disappears, the Lagrangian density (12.1.1) assumes the form

$$\mathcal{L} = \frac{1}{2}\partial_\mu\phi \cdot \partial^\mu\phi, \quad \phi = (\phi^1, \phi^2, \phi^2), \quad |\phi|^2 = 1. \tag{12.2.1}$$

This model is known as the classical *σ-model* [466] or *harmonic map model* [187, 188, 189].

Equation of motion for harmonic map model

Take λ to be a coordinate-dependent Lagrange multiplier and rewrite (12.2.1) as

$$\mathcal{L} = \frac{1}{2}\partial_\mu\phi \cdot \partial^\mu\phi + \lambda(|\phi|^2 - 1). \tag{12.2.2}$$

Such a device allows us to vary ϕ as a free variable to get

$$\Box\phi = 2\lambda\phi. \tag{12.2.3}$$

Taking the scalar product of this equation with ϕ, we can solve for λ to arrive at

$$\Box\phi - (\phi \cdot \Box\phi)\phi = 0. \tag{12.2.4}$$

This is the simplest *wave-map equation.* Of course, in the static situation, (12.2.4) becomes

$$\Delta\phi - (\phi \cdot \Delta\phi)\phi = 0, \tag{12.2.5}$$

with the associated energy

$$E(\phi) = \frac{1}{2}\int_{\mathbb{R}^2} |\nabla\phi|^2 \,\mathrm{d}x, \quad |\nabla\phi|^2 = \sum_{a=1}^{3} |\nabla\phi^a|^2. \tag{12.2.6}$$

Topological characterization

A solution ϕ to (12.2.5), or a harmonic map, carrying a finite-energy, goes to a limiting vector in S^2 at infinity of $\mathbb{R}^2$. Thus, ϕ may be viewed as a map from $\mathbb{R}^2 \cup \{\infty\}$, which is S^2, into S^2. Consequently, as a map from S^2 into itself, ϕ represents a homotopy class in the *homotopy group*

$$\pi_2(S^2) = \mathbb{Z}, \tag{12.2.7}$$

given by an integer N, called the *topological degree*, or *Brouwer degree*, of ϕ, denoted by $\deg(\phi)$, which may also be expressed by the integral

$$\begin{aligned} \deg(\phi) &= \frac{1}{4\pi}\int_{\mathbb{R}^2} \phi \cdot (\partial_1\phi \times \partial_2\phi)\,\mathrm{d}x \\ &= \frac{1}{4\pi}\int_{\mathbb{R}^2} \det(\phi, \partial_1\phi, \partial_2\phi)\,\mathrm{d}x, \end{aligned} \tag{12.2.8}$$

in which the determined is evaluated for the matrix with $\phi, \partial_1\phi, \partial_2\phi$, respectively, as row vectors in a self-evidently arranged way. On the other hand, by the Cauchy–Schwarz inequality, there holds

$$|\phi \cdot (\partial_1\phi \times \partial_2\phi)| \le \frac{1}{2}\left(|\partial_1\phi|^2 + |\partial_2\phi|^2\right). \tag{12.2.9}$$

In view of (12.2.6), (12.2.8), and (12.2.9), we obtain the topological lower bound

$$E(\phi) \ge 4\pi|\deg(\phi)|. \tag{12.2.10}$$

We now study whether (12.2.10) is attainable. To this end, using $\phi \cdot \phi = 1$ and $\phi \cdot (\partial_i\phi) = 0$, we have

$$\begin{aligned} (\phi \times \partial_i\phi) \cdot (\phi \times \partial_i\phi) &= ([\phi \times \partial_i\phi] \times \phi) \cdot (\partial_i\phi) \\ &= (\partial_i\phi - \phi[\partial_i\phi \cdot \phi]) \cdot (\partial_i\phi) \\ &= |\partial_i\phi|^2, \quad i = 1, 2. \end{aligned} \tag{12.2.11}$$

From (12.2.11), we get

$$\begin{aligned}&|\partial_1\phi \pm \phi \times \partial_2\phi|^2 + |\partial_2\phi \mp \phi \times \partial_1\phi|^2 \\ &= 2|\partial_1\phi|^2 + 2|\partial_2\phi|^2 \mp 4\phi \cdot (\partial_1\phi \times \partial_2\phi).\end{aligned} \tag{12.2.12}$$

Integrating (12.2.12) over $\mathbb{R}^2$, we still arrive at (12.2.10), but with the additional insight that the lower bound there is attained if and only if ϕ satisfies the Bogomol'nyi equations

$$\partial_i\phi = \mp\epsilon_{ij}(\phi \times [\partial_j\phi]), \quad i, j = 1, 2, \tag{12.2.13}$$

for $\deg(\phi) = \pm|\deg(\phi)|$. We see that (12.2.13) implies (12.2.5). Belavin and Polyakov's classical work [55] establishes that, for any integer N, the equations (12.2.13) may be solved with an explicit construction of the solutions based on using meromorphic functions and that the solutions obtained are of the minimum energy

$$E_N = 4\pi|N|. \tag{12.2.14}$$

See also [466, 606] for details of such a construction.

An interesting development of the σ-model is the formulation of a *gauged σ-model* or *gauged harmonic map model* by Schroers [498, 499], whose BPS structure yields a rich range of topological solitons accommodating vortices and antivortices [270, 601, 602] dually characterized by the first Chern and Thom classes [510] of the underlying complex line bundles over which the model is defined.

If $n = 3$, the Derrick identity (12.1.7) indicates nonexistence of a static finite-energy nontrivial solution. This obstruction prompts us to consider the Skyrme model.

Skyrme model

The *Skyrme model* [513, 514, 515, 516] is an important particle-physics model in three spatial dimensions for which the Lagrangian action density assumes the normalized form

$$\mathcal{L} = \frac{1}{2}\partial_\mu\phi \cdot \partial^\mu\phi + \frac{1}{4}\left([\partial_\mu\phi \cdot \partial_\nu\phi][\partial^\mu\phi \cdot \partial^\nu\phi] - [\partial_\mu\phi \cdot \partial^\mu\phi]^2\right), \tag{12.2.15}$$

governing a scalar field ϕ with $S^3 \subset \mathbb{R}^4$ as its range space. The quantity in (12.2.15) that contains quartic terms of derivatives is commonly referred to as the *Skyrme term*. As in the harmonic map model, we may vary ϕ as a free variable in the Lagrange-multiplier-modified action density, $\mathcal{L} + \lambda(|\phi|^2 - 1)$, and then determine λ, to arrive at the following associated Euler–Lagrange equation of the Skyrme model (12.2.15):

$$\begin{aligned}&\Box\phi + \partial_\mu\left(\partial_\nu\phi[\partial^\mu\phi \cdot \partial^\nu\phi] - \partial^\mu\phi[\partial_\nu\phi \cdot \partial^\nu\phi]\right) \\ &- \left(\phi \cdot \Box\phi + \phi \cdot \partial_\mu\left(\partial_\nu\phi[\partial^\mu\phi \cdot \partial^\nu\phi] - \partial^\mu\phi[\partial_\nu\phi \cdot \partial^\nu\phi]\right)\right)\phi = 0.\end{aligned} \tag{12.2.16}$$

This equation is the wave-map equation arising from the Skyrme model.

In the static limit, the equation (12.2.16) becomes

$$\begin{aligned}&\Delta\phi-\partial_i\left(\partial_j\phi[\partial_i\phi\cdot\partial_j\phi]-\partial_i\phi|\nabla\phi|^2\right)\\&-\left(\phi\cdot\Delta\phi-\phi\cdot\partial_i\left(\partial_j\phi[\partial_i\phi\cdot\partial_j\phi]-\partial_i\phi|\nabla\phi|^2\right)\right)\phi=0.\end{aligned}\qquad(12.2.17)$$

Alternatively, in view of the *generalized Lagrange identity*

$$(p\wedge q)\cdot(p\wedge q)=(p\cdot p)(q\cdot q)-(p\cdot q)^2,\qquad(12.2.18)$$

involving dot and wedge products for two vectors p,q in $\mathbb{R}^m$, we may rewrite (12.2.15) as

$$\mathcal{L}=\frac{1}{2}\partial_\mu\phi\cdot\partial^\mu\phi-\frac{1}{4}(\partial_\mu\phi\wedge\partial_\nu\phi)\cdot(\partial^\mu\phi\wedge\partial^\nu\phi),\qquad(12.2.19)$$

such that the associated energy-momentum tensor reads

$$T_{\mu\nu}=\partial_\mu\phi\cdot\partial_\nu\phi-(\partial_\mu\phi\wedge\partial_\alpha\phi)\cdot(\partial_\nu\phi\wedge\partial^\alpha\phi)-\eta_{\mu\nu}\mathcal{L}.\qquad(12.2.20)$$

Therefore, in the static situation, the energy density or Hamiltonian of the Skyrme model is

$$\begin{aligned}\mathcal{H}&=T_{00}=-\mathcal{L}\\&=\frac{1}{2}|\nabla\phi|^2+\frac{1}{4}|\partial_i\phi\wedge\partial_j\phi|^2,\end{aligned}\qquad(12.2.21)$$

which leads to the following familiar expression for the associated *Skyrme energy*

$$E(\phi)=\int_{\mathbb{R}^3}\left(\frac{1}{2}|\nabla\phi|^2+\frac{1}{4}|\partial_i\phi\wedge\partial_j\phi|^2\right)\mathrm{d}x.\qquad(12.2.22)$$

The equation (12.2.17) is the Euler–Lagrange equation of (12.2.22). Further, it shows as before that energy (12.2.22) enjoys the scaling property

$$E(\phi^\lambda)=\int_{\mathbb{R}^3}\left(\frac{1}{2\lambda}|\nabla\phi|^2+\frac{\lambda}{4}|\partial_i\phi\wedge\partial_j\phi|^2\right)\mathrm{d}x,\qquad(12.2.23)$$

for $\phi^\lambda(x)=\phi(\lambda x)$ $(\lambda>0)$, which leads to the corresponding Derrick identity

$$\int_{\mathbb{R}^3}\frac{1}{2}|\nabla\phi|^2\,\mathrm{d}x=\int_{\mathbb{R}^3}\frac{1}{4}|\partial_i\phi\wedge\partial_j\phi|^2\,\mathrm{d}x,\qquad(12.2.24)$$

which renders no objection to the existence of nontrivial critical points.

Topological characterization

Let ϕ be a finite-energy map. Again, ϕ has a definitive limit at infinity so that ϕ can be viewed as a map from $\mathbb{R}^3\cup\{\infty\}\approx S^3$ into S^3. So now ϕ may be represented by an integer, say N, in the homotopy group

$$\pi_3(S^3)=\mathbb{Z},\qquad(12.2.25)$$

which is the topological degree of ϕ, $\deg(\phi)$, and has the integral representation

$$\deg(\phi) = \frac{1}{2\pi^2}\int_{\mathbb{R}^3} \det(\phi, \partial_1\phi, \partial_2\phi, \partial_3\phi)(x)\,\mathrm{d}x, \tag{12.2.26}$$

similar to that given by (12.2.8). Note that here, $2\pi^2$ is the total surface volume of the unit sphere S^3.

By a cofactor expansion along the row occupied by $\partial_1\phi$, using $|\phi| = 1$, and applying the Cauchy–Schwarz inequality, we have

$$\begin{aligned}|\det(\phi, \partial_1\phi, \partial_2\phi, \partial_3\phi)| &\leq \sum_{a=1}^{4} |\partial_1\phi^a| \sum_{1\leq b<c\leq 4}^{b,c\neq a} |\partial_2\phi^b\partial_3\phi^c - \partial_2\phi^c\partial_3\phi^b| \\ &\leq \frac{1}{2}|\partial_1\phi|^2 + |\partial_2\phi\wedge\partial_3\phi|^2.\end{aligned} \tag{12.2.27}$$

Thus, cycling such an expansion along the rows occupied by $\partial_2\phi$ and $\partial_3\phi$, similarly and respectively, and adding the results back to (12.2.27), we obtain

$$3|\det(\phi, \partial_1\phi, \partial_2\phi, \partial_3\phi)| \leq \frac{1}{2}|\nabla\phi|^2 + \frac{1}{2}|\partial_i\phi\wedge\partial_j\phi|^2. \tag{12.2.28}$$

In view of (12.2.26) and (12.2.28), we get

$$6\pi^2|\deg(\phi)| \leq \int_{\mathbb{R}^3}\left(\frac{1}{2}|\nabla\phi|^2 + \frac{1}{2}|\partial_i\phi\wedge\partial_j\phi|^2\right)\mathrm{d}x. \tag{12.2.29}$$

On the other hand, if ϕ is a finite-energy critical point of the Skyrme energy (12.2.22), then (12.2.24) implies that the quantities given there are both $\frac{1}{2}E(\phi)$. Using this property and (12.2.29), we arrive at the following topological lower bound

$$E(\phi) \geq 4\pi^2|\deg(\phi)|, \tag{12.2.30}$$

which is in the same spirit of (12.2.10) for harmonic maps.

Sharper lower bound

With some additional labor, the lower bound (12.2.30) may further be improved.

First, since $|\phi|^2 = 1$ and $\phi\cdot\partial_i\phi = 0$, we get from (12.2.18) the result

$$|\phi\wedge\partial_i\phi|^2 = |\partial_i\phi|^2, \quad i = 1, 2, 3. \tag{12.2.31}$$

Next, for $p, q \in \mathbb{R}^4$, the Hodge dual $*(p\wedge q)$ defined by

$$(*[p\wedge q])_{ab} = \frac{1}{2}\,\epsilon_{abcd}\,(p\wedge q)_{cd}, \quad a, b, c, d = 1, 2, 3, 4, \tag{12.2.32}$$

is an isometry over $\bigwedge^2(\mathbb{R}^4)$. With (12.2.32) and $p, q, s, t \in \mathbb{R}^4$, we can verify that the determinant of the 4×4 matrix comprised of the vectors $p, q, s, t \in \mathbb{R}^4$ is given by the identity

$$\det(p, q, s, t) = (p\wedge q)\cdot(*[s\wedge t]). \tag{12.2.33}$$

(See Exercise 7.)

Then, in view of (12.2.31) and (12.2.33), we have

$$|\partial_1\phi|^2 + |\partial_2\phi\wedge\partial_3\phi|^2$$
$$= (\phi\wedge\partial_1\phi \mp *[\partial_2\phi\wedge\partial_3\phi])^2 \pm 2\det(\phi,\partial_1\phi,\partial_2\phi,\partial_3\phi). \qquad (12.2.34)$$

By virtue of this, we can shuffle the coordinate indices to obtain

$$|\nabla\phi|^2 + \frac{1}{2}|\partial_i\phi\wedge\partial_j\phi|^2$$
$$= (\phi\wedge\partial_1\phi \mp *[\partial_2\phi\wedge\partial_3\phi])^2 + (\phi\wedge\partial_2\phi \mp *[\partial_3\phi\wedge\partial_1\phi])^2$$
$$+ (\phi\wedge\partial_3\phi \mp *[\partial_1\phi\wedge\partial_2\phi])^2 \pm 6\det(\phi,\partial_1\phi,\partial_2\phi,\partial_3\phi). \qquad (12.2.35)$$

Consequently, in view of (12.2.26) and (12.2.35), we see that the energy (12.2.22) enjoys the sharper topological lower bound

$$E(\phi) \geq 6\pi^2|\deg(\phi)|, \qquad (12.2.36)$$

with equality attained only when ϕ satisfies the Bogomol'nyi equations

$$\phi\wedge\partial_1\phi = \pm*(\partial_2\phi\wedge\partial_3\phi), \qquad (12.2.37)$$
$$\phi\wedge\partial_2\phi = \pm*(\partial_3\phi\wedge\partial_1\phi), \qquad (12.2.38)$$
$$\phi\wedge\partial_3\phi = \pm*(\partial_1\phi\wedge\partial_2\phi), \qquad (12.2.39)$$

or in short,

$$\phi\wedge\partial_i\phi = \pm\frac{1}{2}*(\epsilon_{ijk}\partial_j\phi\wedge\partial_k\phi), \qquad (12.2.40)$$

for $\deg(\phi) = \pm|\deg(\phi)|$. These equations generalize (12.2.13).

Note that, in arriving at (12.2.36), there is no need to assume that ϕ is a critical point of the Skyrme energy.

The derivation of (12.2.36) and (12.2.40) in the original $SU(2)$-valued field configuration setting of the Skyrme model is due to Manton and Ruback [385]. The derivation here in the current context of the S^3-valued field configuration formulation of the model was shown to the author by Bjarke S. Gudnason.

Minimization problem

An important unsolved question asks if the Skyrme energy has a minimizer among the prescribed homotopy or topological class

$$\mathcal{C}_N = \{\phi\,|\,E(\phi)<\infty,\ \deg(\phi)=N\}, \qquad (12.2.41)$$

defined by each integer N. This problem is only solved when $N=\pm1$ [366]. A relaxed question asks if the Skyrme energy has a finite-energy critical point in each given homotopy class. This latter problem is solved for radially symmetric maps [197, 598]. Problems of this type are numerous in field theory physics.

See [238, 296, 516, 613] for some introductory reviews on the Skyrme model.

12.3 Knots in Faddeev model

In 1997, Faddeev and Niemi [205] published their seminal work on knotted solitons arising in a quantum field theory model, known as the Faddeev model [203], which may be regarded as a refined formalism of the Skyrme model. However, in [49, 50], Battye and Sutcliffe point out that the original work [205] suffers from some deficiencies, such as an incorrect imposition of boundary and regularity conditions. Moreover, in [49, 50], truly three-dimensional knotted solitons with knot charges from one to eight were obtained numerically. This work suggests that knots may indeed be used as candidates to model elementary particles, first proposed by Lord Kelvin in 1860s. Today, we know that the concept of knots has important applications in science. In the past 100 years, mathematicians have made great progress in topological and combinatorial classifications of knots. In turn, the development of knot theory has also facilitated the advancement of mathematics in several of its frontiers, especially low-dimensional topology. In knot theory, an interesting problem concerns the existence of "ideal knots," which promises to provide a natural link between the geometric and topological contents of knotted structures. This problem originates in theoretical physics and aims to prove the existence and predict the properties of knots "based on a first principle approach" [417]. This approach seeks to determine the detailed physical characteristics of a knot, such as its energy (mass), geometric conformation, and topological identification, via conditions expressed in terms of temperature, viscosity, electromagnetic, nuclear, and possibly gravitational, interactions, which is also known as an Hamiltonian approach to knots as field-theoretical stable solitons. The Faddeev knots may provide such structures based on a first-principle [49, 50, 204, 205, 417] formalism.

Faddeev model

In normalized form, the action density of the *Faddeev model* over the standard Minkowski spacetime $\mathbb{R}^{3,1}$ reads

$$\mathcal{L} = \frac{1}{2}\partial_\mu \phi \cdot \partial^\mu \phi - \frac{1}{4}F_{\mu\nu}(\phi)F^{\mu\nu}(\phi), \tag{12.3.1}$$

where the field ϕ assumes its values in the unit 2-sphere in $\mathbb{R}^3$ and

$$F_{\mu\nu}(\phi) = \phi \cdot (\partial_\mu \phi \wedge \partial_\nu \phi). \tag{12.3.2}$$

Since ϕ lies in $\mathbb{R}^3$, it is more convenient to replace the wedge product $\wedge$ by the cross product $\times$ in (12.3.2) in our subsequent discussion. With this note and the fact that ϕ is parallel to $\partial_\mu \phi \times \partial_\nu \phi$, we have

$$F_{\mu\nu}(\phi)F^{\mu\nu}(\phi) = (\partial_\mu \phi \times \partial_\nu \phi) \cdot (\partial^\mu \phi \times \partial^\nu \phi), \tag{12.3.3}$$

such that we may rewrite (12.3.1) as

$$\mathcal{L} = \frac{1}{2}\partial_\mu \phi \cdot \partial^\mu \phi - \frac{1}{4}(\partial_\mu \phi \times \partial_\nu \phi) \cdot (\partial^\mu \phi \times \partial^\nu \phi), \tag{12.3.4}$$

which may be identified with the Skyrme model Lagrangian density (12.2.19), except that ϕ now assumes its values in S^2 instead of S^3. So the Faddeev model (12.3.4) may be viewed as a range-space restricted Skyrme model. As before and using the identity

$$(\mathbf{a} \times \mathbf{b}) \cdot \mathbf{c} = \mathbf{a} \cdot (\mathbf{b} \times \mathbf{c}), \quad \mathbf{a}, \mathbf{b}, \mathbf{c} \in \mathbb{R}^3, \tag{12.3.5}$$

we obtain the Euler–Lagrange equation of (12.3.4) to be

$$\Box\phi - \partial_\mu(\partial_\nu\phi \times [\partial^\mu\phi \times \partial^\nu\phi]) - (\phi \cdot \Box\phi - \phi \cdot \partial_\mu(\partial_\nu\phi \times [\partial^\mu\phi \times \partial^\nu\phi]))\,\phi = 0, \tag{12.3.6}$$

which is the wave-map equation arising in the Faddeev model. Besides, using the vector identity

$$\mathbf{a} \times (\mathbf{b} \times \mathbf{c}) = \mathbf{b}(\mathbf{a} \cdot \mathbf{c}) - \mathbf{c}(\mathbf{a} \cdot \mathbf{b}), \tag{12.3.7}$$

we may rewrite (12.3.6) as

$$\begin{aligned} &\Box\phi - \partial_\mu(\partial^\mu\phi[\partial_\nu\phi \cdot \partial^\nu\phi] - \partial^\nu\phi[\partial_\nu\phi \cdot \partial^\mu\phi]) \\ &- (\phi \cdot \Box\phi - \phi \cdot \partial_\mu(\partial^\mu\phi[\partial_\nu\phi \cdot \partial^\nu\phi] - \partial^\nu\phi[\partial_\nu\phi \cdot \partial^\mu\phi]))\,\phi = 0. \end{aligned} \tag{12.3.8}$$

Static solutions

The rest of this section considers static solutions such that the wave-map equation (12.3.8) becomes

$$\begin{aligned} &\Delta\phi + \partial_i(\partial_i\phi|\nabla\phi|^2 - \partial_j\phi[\partial_j\phi \cdot \partial_i\phi]) \\ &- (\phi \cdot \Delta\phi + \phi \cdot \partial_i(\partial_i\phi|\nabla\phi|^2 - \partial_j\phi[\partial_j\phi \cdot \partial_i\phi]))\,\phi = 0. \end{aligned} \tag{12.3.9}$$

As in the Skyrme model situation, the energy of a static field configuration in the Faddeev model (12.3.4) is found to be

$$E(\phi) = \int_{\mathbb{R}^3} \left(\frac{1}{2}|\nabla\phi|^2 + \frac{1}{4}|\partial_i\phi \times \partial_j\phi|^2 \right) \mathrm{d}x, \tag{12.3.10}$$

such that (12.3.9) is the Euler–Lagrange equation of (12.3.10).

Topological characterization

As in the Skyrme model situation, the finite-energy condition imposed on (12.3.10) implies that ϕ approaches a constant vector at spatial infinity of $\mathbb{R}^3$ such that ϕ may be viewed as a map from S^3 to S^2. As a consequence, we see that each finite-energy field configuration ϕ is associated with an integer, $Q(\phi)$, in $\pi_3(S^2) = \mathbb{Z}$. In fact, $Q(\phi)$ is known as the *Hopf invariant*, which has an integral characterization by Whitehead [579], as do the topological degrees in the harmonic map and Skyrme model situations discussed earlier.

In fact, from the skewsymmetric 2-tensor

$$F_{ij}(\phi) = \phi \cdot (\partial_i\phi \times \partial_j\phi), \quad i, j = 1, 2, 3, \tag{12.3.11}$$

we may define the associated magnetic field

$$\mathbf{B} = \mathbf{B}(\phi) = (B^i(\phi)), \quad B^i(\phi) = \frac{1}{2}\epsilon^{ijk} F_{jk}(\phi), \quad i, j, k = 1, 2, 3, \tag{12.3.12}$$

induced from ϕ. Since ϕ and $\partial_i \phi$ are perpendicular, we see that $\mathbf{B}$ is divergence-free, $\nabla \cdot \mathbf{B} = 0$. Hence, there is a vector field $\mathbf{A}$ such that $\mathbf{B} = \nabla \times \mathbf{A}$. In terms of $\mathbf{A}$ and $\mathbf{B}$, the Hopf invariant $Q(\phi)$ of the map ϕ is given by the integral [579]

$$Q(\phi) = \frac{1}{16\pi^2} \int_{\mathbb{R}^3} \mathbf{A} \cdot \mathbf{B} \, \mathrm{d}x, \tag{12.3.13}$$

also known as the *Whitehead integral*, which is a special form of the *Chern–Simons invariant* [131, 132].

Topological lower bound

For static solutions of the Faddeev model, the energy (12.3.10) enjoys a fractionally-powered topological lower bound, known as the *Vakulenko–Kapitanski inequality* [557], of the form

$$E(\phi) \geq C|Q(\phi)|^{\frac{3}{4}}, \tag{12.3.14}$$

where $C > 0$ is a universal constant.

We may aim to obtain the constant C in (12.3.14) as optimal as possible [367]. To this goal, we first recall the *sharp Sobolev inequality* [24, 537] for a scalar function $f \in W^{1,p}(\mathbb{R}^n)$: For $1 < p < n$ and $\frac{1}{q} = \frac{1}{p} - \frac{1}{n}$, there holds

$$C_0 \|f\|_q \leq \left(\int_{\mathbb{R}^n} |\nabla f|^p \, \mathrm{d}x \right)^{\frac{1}{p}}, \tag{12.3.15}$$

where the best constant C_0 is

$$C_0 = n^{\frac{1}{p}} \left(\frac{n-p}{p-1} \right)^{1-\frac{1}{p}} \left(\omega_n \frac{\Gamma(\frac{n}{p})\Gamma(n+1-\frac{n}{p})}{\Gamma(n)} \right)^{\frac{1}{n}}, \tag{12.3.16}$$

with ω_n the n-dimensional volume enclosed by the unit sphere S^{n-1} in $\mathbb{R}^n$, and $\|f\|_q$ denotes the standard $L^q(\mathbb{R}^n)$-norm. For our problem, we specialize with $n = 3$ and $p = 2$, resulting in $q = 6$, such that (12.3.15) and (12.3.16) give us

$$\left(\int_{\mathbb{R}^3} |f|^6 \, \mathrm{d}x \right)^{\frac{1}{6}} \leq \left(\frac{4}{3\sqrt{3}\,\pi^2} \right)^{\frac{1}{3}} \left(\int_{\mathbb{R}^3} |\nabla f|^2 \, \mathrm{d}x \right)^{\frac{1}{2}}. \tag{12.3.17}$$

We now derive (12.3.14). As in [557], we first bound the right-hand side of (12.3.13) in view of the Hölder inequality by

$$\left| \int_{\mathbb{R}^3} \mathbf{A} \cdot \mathbf{B} \, \mathrm{d}x \right| \leq \|\mathbf{A}\|_6 \|\mathbf{B}\|_{\frac{6}{5}} \leq \|\mathbf{A}\|_6 \|\mathbf{B}\|_1^{\frac{2}{3}} \|\mathbf{B}\|_2^{\frac{1}{3}}. \tag{12.3.18}$$

Here, and in the sequel, we use $\|\mathbf{A}\|_p$ to denote the $L^p(\mathbb{R}^3)$-norm for the amplitude $|\mathbf{A}|$ of a vector field $\mathbf{A}$. Thus, with this notation and using (12.3.17), we have

$$\|\mathbf{A}\|_6 = \|\,|\mathbf{A}|\,\|_6 \leq \left(\frac{4}{3\sqrt{3}\,\pi^2}\right)^{\frac{1}{3}} \left(\int_{\mathbb{R}^3} |\nabla|\mathbf{A}|\,|^2\,\mathrm{d}x\right)^{\frac{1}{2}}$$
$$\leq \left(\frac{4}{3\sqrt{3}\,\pi^2}\right)^{\frac{1}{3}} \left(\int_{\mathbb{R}^3} |\nabla\mathbf{A}|^2\,\mathrm{d}x\right)^{\frac{1}{2}}. \tag{12.3.19}$$

On the other hand, applying the identity

$$\int_{\mathbb{R}^3} |\nabla\mathbf{A}|^2\,\mathrm{d}x = \int_{\mathbb{R}^3} (\nabla\cdot\mathbf{A})^2\,\mathrm{d}x + \int_{\mathbb{R}^3} |\nabla\times\mathbf{A}|^2\,\mathrm{d}x, \tag{12.3.20}$$

and imposing the Coulomb gauge condition, $\nabla\cdot\mathbf{A} = 0$, we have in view of (12.3.19) the bound

$$\|\mathbf{A}\|_6 \leq \left(\frac{4}{3\sqrt{3}\,\pi^2}\right)^{\frac{1}{3}} \|\mathbf{B}\|_2. \tag{12.3.21}$$

Thus, inserting (12.3.21) into (12.3.18), we get

$$\left|\int_{\mathbb{R}^3} \mathbf{A}\cdot\mathbf{B}\,\mathrm{d}x\right| \leq \left(\frac{4}{3\sqrt{3}\,\pi^2}\right)^{\frac{1}{3}} \|\mathbf{B}\|_1^{\frac{2}{3}}\|\mathbf{B}\|_2^{\frac{4}{3}}. \tag{12.3.22}$$

We next estimate $\|\mathbf{B}\|_1$ and $\|\mathbf{B}\|_2$ in (12.3.22) in terms of the Faddeev energy (12.3.10). In fact, by (12.3.11) and (12.3.12), we have

$$\|\mathbf{B}\|_2^2 = \int_{\mathbb{R}^3} \frac{1}{2}|\partial_i\phi\times\partial_j\phi|^2\,\mathrm{d}x. \tag{12.3.23}$$

Furthermore, specializing Ward's argument [566] based on a paper by Manton [384] using symmetric polynomials, we have

$$|\mathbf{B}| \leq \frac{1}{2}\,|\nabla\phi|^2. \tag{12.3.24}$$

Indeed, let $\lambda_1, \lambda_2, \lambda_3$ be the eigenvalues of the symmetric matrix

$$M = (m_{ab}) = \big(\nabla\phi^a\cdot\nabla\phi^b\big). \tag{12.3.25}$$

Then, we have the relation

$$
\begin{aligned}
\lambda_1\lambda_2 + \lambda_2\lambda_3 + \lambda_1\lambda_3 &= \sigma_2(M) \\
&= \sum_{1\leq a<b\leq 3} \det\begin{pmatrix} \nabla\phi^a\cdot\nabla\phi^a & \nabla\phi^a\cdot\nabla\phi^b \\ \nabla\phi^a\cdot\nabla\phi^b & \nabla\phi^b\cdot\nabla\phi^b \end{pmatrix} \\
&= \sum_{1\leq a<b\leq 3} |\nabla\phi^a\times\nabla\phi^b|^2 \quad \text{(cf. (12.2.18))} \\
&= \sum_{1\leq a<b\leq 3}\ \sum_{1\leq i<j\leq 3} |\partial_i\phi^a\partial_j\phi^b - \partial_j\phi^a\partial_i\phi^b|^2 \\
&= \sum_{1\leq i<j\leq 3}\ \sum_{1\leq a<b\leq 3} |\partial_i\phi^a\partial_j\phi^b - \partial_i\phi^b\partial_j\phi^a|^2 \\
&= \sum_{1\leq i<j\leq 3} |\partial_i\phi\times\partial_j\phi|^2 = |\mathbf{B}|^2.
\end{aligned} \tag{12.3.26}
$$

On the other hand, since $|\phi|^2=1$, we have

$$
(\nabla\phi^a\cdot\nabla\phi^b)\phi^b = \nabla\phi^a\cdot([\nabla\phi^b]\phi^b) = 0, \quad a=1,2,3, \tag{12.3.27}
$$

which means $M\phi=0$ such that M has a zero eigenvalue (which we may assume to be λ_3). Inserting this information into (12.3.26) and using (12.3.25), we obtain

$$
|\mathbf{B}| = \sqrt{\lambda_1\lambda_2} \leq \frac{1}{2}(\lambda_1+\lambda_2) = \frac{1}{2}\mathrm{Tr}(M) = \frac{1}{2}|\nabla\phi|^2, \tag{12.3.28}
$$

as asserted in (12.3.24). Thus,

$$
\|\mathbf{B}\|_1 \leq \int_{\mathbb{R}^3} \frac{1}{2}|\nabla\phi|^2\,\mathrm{d}x. \tag{12.3.29}
$$

Finally, substituting (12.3.23) and (12.3.29) into (12.3.22), we obtain

$$
\begin{aligned}
&\left|\int_{\mathbb{R}^3} \mathbf{A}\cdot\mathbf{B}\,\mathrm{d}x\right| \\
&\leq \left(\frac{4}{3\sqrt{3}\,\pi^2}\right)^{\frac{1}{3}} 2^{\frac{2}{3}} \left(\int_{\mathbb{R}^3}\frac{1}{2}|\nabla\phi|^2\,\mathrm{d}x\right)^{\frac{2}{3}} \left(\int_{\mathbb{R}^3}\frac{1}{4}|\partial_i\phi\times\partial_j\phi|^2\,\mathrm{d}x\right)^{\frac{2}{3}} \\
&\leq \left(\frac{4}{3\sqrt{3}\,\pi^2}\right)^{\frac{1}{3}} 2^{\frac{2}{3}} \left(\frac{1}{2}E(\phi)\right)^{\frac{4}{3}}.
\end{aligned} \tag{12.3.30}
$$

In view of (12.3.13) and (12.3.30), we get

$$
E(\phi) \geq 8\pi^2 3^{\frac{3}{8}} |Q(\phi)|^{\frac{3}{4}}, \tag{12.3.31}
$$

which is a concrete realization of (12.3.14).

Faddeev knots

The geometric meaning of the Hopf invariant, or *Hopf charge*, (12.3.13), of a map $\phi : \mathbb{R}^3 \to S^2$, may be described as follows [175]: Let $s_1, s_2 \in S^2$ be an arbitrary pair of regular values of ϕ where $\mathrm{d}\phi$ is of full rank, namely, the Jacobian matrix of ϕ is of rank 2. So the preimages $\gamma_1 = \phi^{-1}(s_1)$ and $\gamma_2 = \phi^{-1}(s_2)$ are two closed curves in $\mathbb{R}^3$. Then, the integer $N = Q(\phi)$ is the linking number of γ_1 and γ_2. In [49, 50, 204, 205, 417], a static Hopf-charge N map ϕ minimizing the Faddeev energy (12.3.10) gives rise to a locally energetically concentrated soliton-like knotted field configuration, commonly referred to as the *Faddeev knot*. Thus, such a formalism naturally leads to the following minimization problem

$$E_N \equiv \inf\{E(\phi) \mid E(\phi) < \infty,\ Q(\phi) = N\}, \qquad N \in \mathbb{Z}, \tag{12.3.32}$$

also known as the *Faddeev knot problem*. Thus, we encounter a direct minimization problem over the full space $\mathbb{R}^3$. Here, a typical difficulty is that a minimizing sequence may fail to concentrate in a local region. This reminds us to look at what Lions' *concentration-compactness principle* [373, 374] can offer. A careful examination of the Faddeev knot problem indicates that we cannot make direct use of this method due to the lack of several key ingredients in the Faddeev energy (12.3.10) and in the *Hopf–Whitehead topological integral* (12.3.13).

The Faddeev knot problem was partially solved in [366] based on the sublinear growth upper bound

$$E_N \leq C|N|^{\frac{3}{4}}, \tag{12.3.33}$$

where $C > 0$ is an irrelevant constant. We also used a key tool (later called the *substantial inequality* [367]), which may well be explained by what happens in a *nuclear fission process*. When a nucleus fissions, it splits into several smaller fragments. The sum of the masses of these fragments is less than the original mass. The "missing" mass has been converted into energy according to Einstein's equation. For a survey on this problem in the context of the Skyrme model and its various extensions, see [369].

From (12.3.14) or (12.3.31) and (12.3.33), we see that the Faddeev energy E_N of a map ϕ of the Hopf charge N enjoys the fractional-exponent growth law

$$C_1|N|^{\frac{3}{4}} \leq E_N \leq C_2|N|^{\frac{3}{4}}, \tag{12.3.34}$$

where $C_1, C_2 > 0$ are two universal constants.

12.4 Other fractional-exponent growth laws and knot energies

The previous section showed that the Faddeev energy for a map from $\mathbb{R}^3$ into S^2 with Hopf invariant N grows sublinearly in N when N becomes large. This property implies that certain "particles" with large topological charges may be energetically preferred and that these large particles are prevented from splitting into particles with smaller topological charges. In other words, when the Hopf charge is large, the Faddeev solitons prefer to *stay knotted.* In order to achieve a deeper understanding of this rather peculiar property, we now place our discussion into a wider range of context.

Ropelength energy and growth law

The appearance of knotted structures may also be relevant to the existence and stability of large molecular conformation in polymers and gel electrophoresis of DNA. In these latter problems, a crucial geometric quantity that measures the "energy" of a physical knot of knot (or link) type K (or simply knot) is the "*rope length energy*" $L(K)$, of the knot K. To define it, we consider a uniform tube centered along a space curve Γ. The "rope length" $L(\Gamma)$ of Γ is the ratio of the arclength of Γ over the radius of the largest uniform tube centered along Γ. Then,

$$L(K) = \inf\{L(\Gamma) \mid \Gamma \in K\}. \tag{12.4.1}$$

A curve Γ achieving the infimum carries the minimum energy in K and gives rise to an "*ideal*" or "*physically preferred*" knot [322, 351], also called a *tight knot* [107]. Clearly, this ideal configuration determines the shortest piece of tube that can be closed to form the knot. Similarly, another crucial quantity that measures the geometric complexity of Γ is the average number of crossings in planar projections of the space curve Γ denoted by $N(\Gamma)$ (say). The *crossing number* $N(K)$ of the knot K is defined as

$$N(K) = \inf\{N(\Gamma) \mid \Gamma \in K\}, \tag{12.4.2}$$

which is a knot invariant. Naturally one expects the energy and the geometric complexity of the knot K to be closely related. Indeed, the combined results in [91, 107] lead to the relation

$$C_1 N(K)^p \leq L(K) \leq C_2 N(K)^p, \tag{12.4.3}$$

where $C_1, C_2 > 0$ are two universal constants and the exponent p satisfies $3/4 \leq p < 1$ so that, in truly three-dimensional situations, the preferred value of p is sharply at $p = 3/4$. This relation strikingly resembles the fractional-exponent growth law (12.3.34) for the Faddeev knots just discussed and reminds us that a sublinear energy growth law with regard to the topological content involved is essential for knotted structures to occur.

Faddeev model in general Hopf dimensions

In the Faddeev knot problem, it is the underlying property and structure of the homotopy group $\pi_3(S^2)$ and the Faddeev energy functional formula that guarantee the validity of the associated sublinear growth law. Generally, it seems that such a property may be related to the notion of quantitative homotopy introduced by Gromov [258]. For example, we may consider the Whitehead integral representation of the Hopf invariant and the "associated" knot energy à la Faddeev. More precisely, let

$$u:\mathbb{R}^{4n-1}\to S^{2n},\quad n\geq 1, \tag{12.4.4}$$

be a differentiable map that approaches a constant sufficiently fast at infinity. Denote by Ω the volume element of S^{2n} and

$$|S^{2n}|=\int_{S^{2n}}\Omega. \tag{12.4.5}$$

Then, the integral representation of u in the homotopy group $\pi_{4n-1}(S^{2n})$, say $Q(u)$, which is the Hopf invariant of u, is given by [77]

$$Q(u)=\frac{1}{|S^{2n}|}\int_{\mathbb{R}^{4n-1}}v\wedge u^*(\Omega),\quad \mathrm{d}v=u^*(\Omega). \tag{12.4.6}$$

We can introduce a generalized Faddeev knot energy for such a map u, in the *Hopf dimension* situation, characterized by the dimension match expressed in (12.4.4) between the domain and range spaces,

$$E(u)=\int_{\mathbb{R}^{4n-1}}\left(\frac{1}{2}|\mathrm{d}u|^2+\frac{1}{2}|u^*(\Omega)|^2\right)\,\mathrm{d}x. \tag{12.4.7}$$

For this energy functional, we are able to establish the following generalized sublinear energy growth estimate [368, 370]

$$C_1|N|^{\frac{4n-1}{4n}}\leq E_N\leq C_2|N|^{\frac{4n-1}{4n}}, \tag{12.4.8}$$

where $C_1, C_2>0$ are universal constants, and as before

$$E_N=\inf\{E(u)\,|\,E(u)<\infty,\,Q(u)=N\}. \tag{12.4.9}$$

In particular, we can see clearly that the fractional exponent in the generalized growth law is the ratio of the dimension of the domain space and twice of the dimension of the target space.

Knot energy and fractional growth law in general odd dimensions

The following paragraphs elaborate on the fractional-exponent topological lower bound in the spirit of the Faddeev knot problem and the Skyrme model, by presenting the work [462] on an Abelian Chern–Simons knot model over the space $\mathbb{R}^{2n+1}$. For this purpose, use $A=(A_i)$ $(i=1,\dots,2n+1)$ to denote a real-valued vector field and

$$F_{ij}=\partial_i A_j-\partial_j A_i,\quad i,j=1,\dots,2n+1, \tag{12.4.10}$$

the induced 2-form curvature. In general, use $F_{i_1 i_2 \cdots i_{2n}}$ to denote the $2n$-form curvature which is formed from taking the totally antisymmetric n-fold product of the 2-form curvature F_{ij}. Then, take

$$\Omega_{\mathrm{CS}} = \epsilon^{i j_1 j_2 \cdots j_{2n}} A_i F_{j_1 j_2 \cdots j_{2n}} \tag{12.4.11}$$

to be a Chern–Simons type topological invariant density so that

$$Q = \int \Omega_{\mathrm{CS}} \tag{12.4.12}$$

gives rise to the topological charge of the model, where and in the sequel the integral is understood to be evaluated over the domain space $\mathbb{R}^{2n+1}$, with the omission of the Lebesgue measure. The construction of $F_{i_1 i_2 \cdots i_{2n}}$ leads to the point-wise bound

$$|F_{i_1 i_2 \cdots i_{2n}}| \leq C |F_{ij}|^n, \tag{12.4.13}$$

where and in the sequel we use C to denote a generic positive constant.

The energy functional of a Skyrme–Faddeev type knot theory considered in [462] is proposed to be

$$E = \int \left(F_{ij}^2 + F_{i_1 i_2 \cdots i_{2n}}^2 \right). \tag{12.4.14}$$

With (12.4.12) and (12.4.14), we establish the energy lower bound

$$E \geq C |Q|^{\frac{2n+1}{2n+2}}, \tag{12.4.15}$$

which is again of a fractional-exponent feature.

First, in view of the Cauchy–Schwarz inequality, we have

$$|Q| \leq C \left(\int |A_i|^p \right)^{\frac{1}{p}} \left(\int |F_{j_1 j_2 \cdots j_{2n}}|^q \right)^{\frac{1}{q}}, \tag{12.4.16}$$

where $p, q > 1$ satisfy

$$\frac{1}{p} + \frac{1}{q} = 1. \tag{12.4.17}$$

Next, we recall the *Sobolev inequality* over $\mathbb{R}^m$ of the form

$$\|f\|_p \leq C \|\nabla f\|_2, \quad m \geq 3, \tag{12.4.18}$$

where p satisfies the relation

$$\frac{1}{p} = \frac{1}{2} - \frac{1}{m} = \frac{m-2}{2m}. \tag{12.4.19}$$

For our purpose, we need $m = 2n+1$ in (12.4.19), which gives us

$$p = \frac{2(2n+1)}{2n-1}, \tag{12.4.20}$$

which then yields by (12.4.18) the inequality

$$\|f\|_{\frac{2(2n+1)}{2n-1}} \leq C\|\nabla f\|_2, \tag{12.4.21}$$

for our application.

Then, with the Coulomb gauge, $\partial_i A_i = 0$, we have

$$\int F_{ij}^2 = \int (\partial_i A_j - \partial_j A_i)^2 = \int |\nabla A_i|^2. \tag{12.4.22}$$

Therefore, in view of (12.4.21) and (12.4.22), we obtain

$$\|A_i\|_{\frac{2(2n+1)}{2n-1}} \leq C\left(\int F_{ij}^2\right)^{\frac{1}{2}}. \tag{12.4.23}$$

On the other hand, by virtue of (12.4.17) and (12.4.20), we get

$$q = \frac{2(2n+1)}{2n+3}. \tag{12.4.24}$$

Thus, inserting (12.4.23) into (12.4.16), we arrive at the bound

$$|Q| \leq C\left(\int F_{ij}^2\right)^{\frac{1}{2}}\left(\int |F_{j_1 j_2 \cdots j_{2n}}|^{\frac{2(2n+1)}{2n+3}}\right)^{\frac{2n+3}{2(2n+1)}}. \tag{12.4.25}$$

To proceed, we seek constants $\alpha, \beta > 0$ and $s, t > 1$ such that

$$\alpha + \beta = \frac{2(2n+1)}{2n+3}, \quad \frac{1}{s} + \frac{1}{t} = 1, \quad \alpha s = \frac{2}{n}, \quad \beta t = 2. \tag{12.4.26}$$

If the simultaneous system of equations (12.4.26) has a solution, then we may use the Cauchy–Schwarz inequality to recast the second integral on the right-hand side of (12.4.25) into

$$\begin{aligned}
\int |F_{j_1 j_2 \cdots j_{2n}}|^{\frac{2(2n+1)}{2n+3}} &= \int |F_{j_1 j_2 \cdots j_{2n}}|^{\alpha+\beta} \\
&\leq \left(\int |F_{j_1 j_2 \cdots j_{2n}}|^{\alpha s}\right)^{\frac{1}{s}}\left(\int |F_{j_1 j_2 \cdots j_{2n}}|^{\beta t}\right)^{\frac{1}{t}} \\
&= \left(\int |F_{j_1 j_2 \cdots j_{2n}}|^{\frac{2}{n}}\right)^{\frac{1}{s}}\left(\int |F_{j_1 j_2 \cdots j_{2n}}|^{2}\right)^{\frac{1}{t}}.
\end{aligned} \tag{12.4.27}$$

Fortunately, when $n \neq 1$, the system (12.4.26) has a unique solution given by the expressions

$$\alpha = \frac{4}{(n-1)(2n+3)}, \quad \beta = \frac{2(2n^2-n-3)}{(n-1)(2n+3)}, \tag{12.4.28}$$

$$s = \frac{(n-1)(2n+3)}{2n}, \quad t = \frac{(n-1)(2n+3)}{2n^2-n-3}. \tag{12.4.29}$$

Substituting (12.4.28)–(12.4.29) into (12.4.27) and applying (12.4.13), we get

$$\int |F_{j_1 j_2\cdots j_{2n}}|^{\frac{2(2n+1)}{2n+3}} \leq \left(\int F_{ij}^2\right)^{\frac{2n}{(n-1)(2n+3)}} \left(\int F_{j_1 j_2\cdots j_{2n}}^2\right)^{\frac{2n^2-n-3}{(n-1)(2n+3)}}. \quad (12.4.30)$$

Finally, inserting (12.4.30) into (12.4.25) and applying (12.4.14), we find

$$\begin{aligned} |Q| &\leq C\left(\int F_{ij}^2\right)^{\frac{1}{2}+\frac{n}{(n-1)(2n+1)}} \left(\int F_{j_1 j_2\cdots j_{2n}}^2\right)^{\frac{2n^2-n-3}{2(n-1)(2n+1)}} \\ &\leq CE^{\frac{1}{2}+\frac{n}{(n-1)(2n+1)}+\frac{2n^2-n-3}{2(n-1)(2n+1)}} \\ &= CE^{\frac{2(n+1)}{2n+1}}, \quad n \neq 1. \end{aligned} \quad (12.4.31)$$

As a consequence, we have established the energy-topological charge inequality (12.4.15) when $n \neq 1$. In other words, for the topological charge (12.4.12) and the Skyrme–Faddeev type energy (12.4.14), the fractional-exponent energy lower bound (12.4.15) is valid over $\mathbb{R}^{2n+1}$ for $2n+1 \geq 5$. This result complements that found in the Hopf dimension situation for maps from $\mathbb{R}^{4n-1}$ into S^{2n}.

12.5 Q-balls

We saw that solitons in the Skyrme model and the Faddeev model are topological in that the field configurations are energy minimizers in stratified appropriate topological classes. In contrast, Q-balls are *nontopological solitons* in that they are stratified by a prescribed Q-charge of a nontopological nature. Such a feature leads us to consider a constrained minimization problem and thus bypass the dimensionality barrier set forth by the Derrick theorem (Section 12.1). This section presents a concise introduction to the Q-balls in the simplest field-theoretical model, the Klein–Gordon scalar field theory, as an illustration, to supplement our discussions on the Skyrme and Faddeev topological solitons (Sections 12.2–12.4).

Q-charge

Recall that the Klein–Gordon scalar field theory is governed by the Lagrangian action density (12.1.1). From the equation of motion, (12.1.2), of the theory, we can write down the conserved current density

$$j^\mu = \frac{1}{2\mathrm{i}}(\overline{\phi}\partial^\mu\phi - \phi\partial^\mu\overline{\phi}), \quad (12.5.1)$$

which gives rise to the associated charge density

$$\rho = j^0 = \frac{1}{2\mathrm{i}}(\overline{\phi}\partial^0\phi - \phi\partial^0\overline{\phi}) = \frac{1}{2\mathrm{i}}(\overline{\phi}\partial_t\phi - \phi\partial_t\overline{\phi}), \quad x^0 = t, \quad (12.5.2)$$

and the total charge

$$Q = \int_{\mathbb{R}^n} \rho \, dx = \frac{1}{2i} \int_{\mathbb{R}^n} (\overline{\phi} \partial_t \phi - \phi \partial_t \overline{\phi}) \, dx, \tag{12.5.3}$$

or the *Q-charge*. This charge is not a topological charge, but rather a charge associated with the global $U(1)$ symmetry of the model and may be thought of as a count of total particle number in the system.

Q-ball as constrained energy minimizer

From the Hamiltonian (12.1.4), the energy associated with the action density (12.1.1) is

$$E(\phi) = \int_{\mathbb{R}^n} \left(\frac{1}{2} |\partial_t \phi|^2 + \frac{1}{2} |\nabla \phi|^2 + V(|\phi|^2) \right) dx. \tag{12.5.4}$$

A *Q-ball* is described by a field configuration ϕ that minimizes the energy (12.5.4) with a *prescribed* non-topological charge Q given in (12.5.3). Therefore, Q-balls, first envisioned by Coleman [140] and based on an earlier study of Friedberg, Lee, and Sirlin [216], to describe blobs of bosonic particles, are also referred to as non-topological solitons [216, 353].

Therefore, we encounter a constrained minimization problem consisting of the objective functional (12.5.4) and the constraint functional (12.5.3) and so are led to the Lagrange functional

$$E_\omega(\phi) = E(\phi) + \omega \left(Q - \frac{1}{2i} \int_{\mathbb{R}^n} (\overline{\phi} \partial_t \phi - \phi \partial_t \overline{\phi}) \, dx \right), \tag{12.5.5}$$

where the parameter ω serves as a Lagrange multiplier.

Modified energy functional

To proceed, we collect the terms involving ∂_t in the integrand of (12.5.5) as

$$\frac{1}{2} |\partial_t \phi|^2 - \frac{\omega}{2i} (\overline{\phi} \partial_t \phi - \phi \partial_t \overline{\phi}) = \frac{1}{2} |\partial_t \phi - i\omega\phi|^2 - \frac{1}{2} \omega^2 |\phi|^2. \tag{12.5.6}$$

Inserting (12.5.6) into (12.5.5), we have

$$E_\omega(\phi) = \int_{\mathbb{R}^n} \left(\frac{1}{2} |\partial_t \phi - i\omega\phi|^2 + \frac{1}{2} |\nabla \phi|^2 + V(|\phi|^2) - \frac{1}{2} \omega^2 |\phi|^2 \right) dx + \omega Q. \tag{12.5.7}$$

We minimize the modified energy (12.5.7) by certain steady-state field configuration. For such a purpose, we see from (12.5.7) that we should choose

$$\phi(x,t) = e^{i\omega t} \phi(x) \tag{12.5.8}$$

such that (12.5.7) becomes a steady-state energy functional:

$$E_\omega(\phi) = \int_{\mathbb{R}^n} \left(\frac{1}{2} |\nabla \phi|^2 + V(|\phi|^2) - \frac{1}{2} \omega^2 |\phi|^2 \right) dx + \omega Q. \tag{12.5.9}$$

Furthermore, in view of (12.5.8), the constraint (12.5.3) reads

$$\omega \int_{\mathbb{R}^n} |\phi|^2 \, \mathrm{d}x = Q. \tag{12.5.10}$$

As a consequence, we see that the problem of the existence of Q-balls is recast into the existence of an energy minimizer of (12.5.9) subject to the constraint (12.5.10). Without loss of generality, we may assume $\omega > 0$ and $Q > 0$ in the subsequent discussion.

Solution by differential equation

As an initial approach, we *fix* $\omega > 0$ and minimize (12.5.9) subject to (12.5.10), or

$$\int_{\mathbb{R}^n} |\phi|^2 \, \mathrm{d}x = \frac{Q}{\omega}. \tag{12.5.11}$$

Thus, we arrive at solving the associated equation

$$\Delta\phi = 2V'(|\phi|^2)\phi - \omega^2\phi - \lambda\phi, \tag{12.5.12}$$

where λ is a Lagrange multiplier arising from the constraint (12.5.11), which appears as an eigenvalue in this context. Now, let ϕ_ω be a solution of (12.5.12) satisfying (12.5.11). Next, insert $\phi = \phi_\omega$ into (12.5.9) to get

$$F(\omega) \equiv E_\omega(\phi_\omega) = \int_{\mathbb{R}^n} \left(\frac{1}{2}|\nabla\phi_\omega|^2 + V(|\phi_\omega|^2) \right) \mathrm{d}x + \frac{1}{2}\omega Q. \tag{12.5.13}$$

Then, minimize $F(\omega)$ in $\omega > 0$.

Solution by nonlocal equation

As another approach, we follow [141] to substitute the constraint (12.5.11) into (12.5.9), eliminating ω, to obtain

$$\begin{aligned} E_Q(\phi) &\equiv E_\omega(\phi) \\ &= \int_{\mathbb{R}^n} \left(\frac{1}{2}|\nabla\phi|^2 + V(|\phi|^2) \right) \mathrm{d}x + \frac{Q^2}{2\int_{\mathbb{R}^n} |\phi|^2 \, \mathrm{d}x} \\ &= E(\phi) + \frac{Q^2}{2I(\phi)}, \quad I(\phi) = \int_{\mathbb{R}^n} |\phi|^2 \, \mathrm{d}x. \end{aligned} \tag{12.5.14}$$

In other words, the constraint (12.5.11) is absorbed into the parameter ω, which leads us to the unconstrained energy functional (12.5.14). In view of this formalism, we may vary ϕ in (12.5.14) to get the associated Euler–Lagrange equation to be

$$\Delta\phi = 2V'(|\phi|^2)\phi - \frac{Q^2}{2\left(\int_{\mathbb{R}^n} |\phi|^2 \, \mathrm{d}x\right)^2}\phi, \tag{12.5.15}$$

which is nonlocal.

Applying the same approach as in the derivation of the Derrick identity (12.1.7), we see that a finite-energy solution of the equation (12.5.15) fulfills the identity

$$(n-2)\int_{\mathbb{R}^n}\frac{1}{2}|\nabla\phi|^2\,\mathrm{d}x + n\int_{\mathbb{R}^n}V(|\phi|^2)\,\mathrm{d}x = \frac{nQ^2}{2\int_{\mathbb{R}^n}|\phi|^2\,\mathrm{d}x}, \tag{12.5.16}$$

as obtained in [347] when $n = 3$. In particular, we see that, in contrast to the classical Klein–Gordon model, there is no dimensionality obstruction to the existence of a finite-energy solution of (12.5.15) for the Q-ball problem as far as $Q \neq 0$. In this regard, as in the Skyrme and Faddeev models, the Q-ball model removes the dimensionality constraint in the Klein–Gordon model. An obvious mathematical advantage of the Q-ball model is that it does not involve non-quadratic terms of the derivatives of the field in the Lagrange action density.

We also note that non-Abelian Q-balls are formulated in [478] and [27] and gauged Q-balls are considered in [15, 261, 430]. Such non-topological solitons are found to play some roles in the theoretical study of early-universe cosmology [217] and dark matter [171, 195, 196, 348, 550]. The work [144] constructs charge-swapping Q-balls and points out their possible new applications.

Exercises

1. Let (ϕ, A_i) be a finite-energy critical point of the energy functional (12.1.25) for the static Abelian Higgs model over $\mathbb{R}^n$. Establish the Derrick identity (12.1.26).
2. Let the potential density $V(s)$ given in (12.1.1) be differentiable and attain its minimum at $s = 1$. Show that if $\phi = \mathrm{e}^{\mathrm{i}u}$ is a static solution of (12.1.2) where u is real valued, then $\Delta u = 0$ and $\partial_i u \in L^2(\mathbb{R}^n)$, $i = 1, \dots, n$, in view of the finite-energy condition. Thus, $u =$ constant.
3. Let ϕ be a solution of (12.2.13) and consider the stereographic projection from the south pole $(0, 0, -1)$ of the unit sphere S^2 onto the complex plane given by

$$\psi = \psi_1 + \mathrm{i}\psi_2, \quad \psi_1 = \frac{\phi^1}{1+\phi^3}, \quad \psi_2 = \frac{\phi^2}{1+\phi^3}. \tag{12.E.1}$$

   (a) Show that, with (12.E.1), the equations (12.2.13) are recast into the linear equations

$$\partial_1\psi_1 = \pm\partial_2\psi_2, \quad \partial_1\psi_2 = \mp\partial_2\psi_1, \tag{12.E.2}$$

   which are the Cauchy–Riemann equations for the complex meromorphic function ψ in $z = x^1 \pm \mathrm{i}x^2$, respectively.

   (b) Show that the inverse of the transformation (12.E.1) is given by

$$\phi^1 = \frac{2\psi_1}{1+|\psi|^2}, \quad \phi^2 = \frac{2\psi_2}{1+|\psi|^2}, \quad \phi^3 = \frac{1-|\psi|^2}{1+|\psi|^2}, \tag{12.E.3}$$

and use this to show that the static energy (12.2.6) becomes

$$E(\phi) = E(\psi) = 2\int_{\mathbb{R}^2} \frac{|\partial_1\psi|^2 + |\partial_2\psi|^2}{(1+|\psi|^2)^2}\,\mathrm{d}x. \tag{12.E.4}$$

(c) With (12.E.3), derive the expression

$$\phi\cdot(\partial_1\phi\times\partial_2\phi) = \frac{2\mathrm{i}}{(1+|\psi|^2)^2}(\partial_1\psi\partial_2\overline{\psi} - \partial_1\overline{\psi}\partial_2\psi). \tag{12.E.5}$$

(d) Combine (12.2.8), (12.E.4), and (12.E.5) to obtain the relation

$$E(\psi) = 2\int_{\mathbb{R}^2} \frac{|\partial_1\psi \pm \mathrm{i}\partial_2\psi|^2}{(1+|\psi|^2)^2}\,\mathrm{d}x + 4\pi|\deg(\phi)|, \tag{12.E.6}$$

which leads to the same energy lower bound (12.2.10), with equality saturated if and only if ψ satisfies the equation

$$\partial_1\psi \pm \mathrm{i}\partial_2\psi = 0, \tag{12.E.7}$$

which is the same as (12.E.2).

4. Use the expression

$$T_{\mu\nu} = 2\frac{\partial\mathcal{L}}{\partial\eta^{\mu\nu}} - \eta_{\mu\nu}\mathcal{L} \tag{12.E.8}$$

and (12.2.19) to derive (12.2.20).

5. Show that the equation (12.2.17) is the Euler–Lagrange equation of the energy functional (12.2.22).

6. Verify (12.2.27).

7. Prove the identity (12.2.33). (Such an identity has an elegant n-dimensional extension which may be stated as follows. Let $* : \bigwedge^k(\mathbb{R}^n) \to \bigwedge^{n-k}(\mathbb{R}^n)$ be the Hodge dual [407, 576, 577]. Then, there holds

$$\begin{aligned}\det(u_1,\dots,u_n) &= (*[u_1\wedge\cdots\wedge u_k])\cdot(u_{k+1}\wedge\cdots\wedge u_n)\\ &= (-1)^{k(n-k)}(u_1\wedge\cdots\wedge u_k)\cdot(*[u_{k+1}\wedge\cdots\wedge u_n]),\end{aligned} \tag{12.E.9}$$

where $u_1,\dots,u_n\in\mathbb{R}^n$. Hint for a proof: Use $\dim(\bigwedge^n(\mathbb{R}^n)) = 1$).

8. Show that the (12.3.6) is the Euler–Lagrange equation associated with the Faddeev action density (12.3.4).

9. Formulate the energy-momentum tensor of the Faddeev action density (12.3.4) and use it to derive the Faddeev energy (12.3.10).

10. Show that the equation (12.3.9) is the Euler–Lagrange equation of the energy (12.3.10).

11. Let the differential forms A and F be defined by

$$A = A_i\mathrm{d}x^i,\quad F = F_{ij}\,\mathrm{d}x^i\wedge\mathrm{d}x^j,\quad i,j = 1,2,3. \tag{12.E.10}$$

(a) Show that

$$A \wedge F = \left(\epsilon^{ijk} A_i F_{jk}\right) \mathrm{d}x^1 \wedge \mathrm{d}x^2 \wedge \mathrm{d}x^3. \tag{12.E.11}$$

(b) Show that

$$A \wedge \mathrm{d}A = \frac{1}{2} \left(\epsilon^{ijk} A_i [\partial_j A_k - \partial_k A_j]\right) \mathrm{d}x^1 \wedge \mathrm{d}x^2 \wedge \mathrm{d}x^3. \tag{12.E.12}$$

The simple relations in this problem explain how, in some different ways, the Hopf or Chern–Simons invariant induced from a vector field or 1-form A over $\mathbb{R}^3$ may be expressed:

$$Q(A) = \frac{1}{(4\pi)^2} \int A \wedge \mathrm{d}A = \frac{1}{32\pi^2} \int \epsilon^{ijk} A_i F_{jk} \,\mathrm{d}x, \quad F_{ij} = \partial_i A_j - \partial_j A_i. \tag{12.E.13}$$

12. Verify that the magnetic field $\mathbf{B}$ defined by the expressions (12.3.11)–(12.3.12) is indeed divergence-free, $\nabla \cdot \mathbf{B} = 0$.
13. Assume that $\mathbf{A}$ and its derivatives vanish at infinity sufficiently rapidly. Establish the identity (12.3.20). Formulate its extension in n dimensions as stated in (12.4.22).
14. Use (12.1.2) to verify that the current density given in (12.5.1) satisfies the conservation law $\partial_\mu j^\mu = 0$.
15. Derive the Derrick identity (12.5.16) for the Q-ball problem.
16. Formulate the equation (12.5.15) and the virial theorem (12.5.16) when the potential density V is of a *running mass potential* type [144, 195]:

$$V(|\phi|^2) = \frac{1}{2} m^2 |\phi|^2 \left(1 - K \ln\left[\frac{|\phi|^2}{2M^2}\right]\right), \tag{12.E.14}$$

where $m, M, K > 0$ are constants.

13

Strings and branes

This chapter presents a brief introduction to some classical aspects of relativistic strings and branes, including their action principles and equations of motion. It first reviews the relativistic motion of a free point particle and next generalizes the discussion to consider the Nambu–Goto strings and branes. It then discusses the Polyakov strings and branes. Afterwards, it considers the motion of a relativistic particle interacting with background fields as an extension for the problems studied.

13.1 Motivation and relativistic motion of free particle as initial setup

The central idea of string theory is to use strings instead of point-like particles as models to describe matter and force particles in interactions in the physical world, especially in strong nuclear interactions, originally in order to overcome the so-called ultraviolet divergence problem associated with infinitesimal distances [54, 587]. At the quantum level, various physical particles could be imagined or realized as a basic string occupying its various characteristic oscillatory modes [618]. In such a formalism, it is *inevitable* to encounter a massless spin-2 particle [486, 487], the *graviton*, which gives rise to gravity [486, 487, 614], and it is possible *at the same time* to accommodate all other particles in quantum field theory [316, 317]. For this reason, string theory is a promising theory for the unification of all four fundamental forces in nature — gravitational, electromagnetic, weak, and strong — at quantum level. Additionally, string theory also finds applications in other important subject areas, including condensed-matter physics [150, 408], black-hole thermodynamics [54, 530], and cosmology [153, 186, 215, 561]. The study of strings, and more generally, branes, brings forth a wealth of highly interesting and challenging nonlinear partial differential equation problems. This chapter discusses some of these problems.

Mathematical Physics with Differential Equations. Yisong Yang, Oxford University Press.
 DOI: 10.1093/oso/9780192872616.003.0013

The first two sections consider the action principle and equation of motion of the Nambu–Goto strings. Next, the chapter studies branes. It then considers the Polyakov strings and branes. The chapter ends with a short discussion on some further extensions.

To begin, let m be the mass of a free particle with coordinates (x^i) in the space $\mathbb{R}^n$, depending on time t. Recall that the Newtonian action and Lagrangian of the moving particle are

$$S = \int L\,\mathrm{d}t, \quad L = \frac{1}{2}m\sum_{i=1}^{n}(\dot{x}^i)^2, \quad \dot{f} = \frac{\mathrm{d}f}{\mathrm{d}t}. \tag{13.1.1}$$

Action of relativistically moving particle

On the other hand, relativistically, the free motion of the particle follows the trajectory that extremizes the action

$$S = \kappa \int \mathrm{d}s, \tag{13.1.2}$$

where κ is an undetermined constant and $\mathrm{d}s^2$ is the metric element given by

$$\mathrm{d}s^2 = c^2\mathrm{d}t^2 - \sum_{i=1}^{n}(\mathrm{d}x^i)^2 = \eta_{\mu\nu}\mathrm{d}x^\mu \mathrm{d}x^\nu, \tag{13.1.3}$$

with $c > 0$ the speed of light, $x^0 = ct$, $\eta_{\mu\nu} = \mathrm{diag}(1, -1, \dots, -1)$ the standard Minkowskian metric tensor, and assuming the timelike condition $\mathrm{d}s^2 > 0$. In view of (13.1.2) and (13.1.3), if we consider the motion of the particle in terms of time t, we have

$$S = \kappa c \int \sqrt{1 - \sum_{i=1}^{n} \frac{(\dot{x}^i)^2}{c^2}}\,\mathrm{d}t. \tag{13.1.4}$$

We see that (13.1.2) gives the dynamics of (13.1.1) in low speed when $\kappa = -mc$. Thus, we arrive at the relativistic action and Lagrangian

$$S = \int L\,\mathrm{d}t, \quad L = -mc^2\sqrt{1 - \sum_{i=1}^{n} \frac{(\dot{x}^i)^2}{c^2}}, \tag{13.1.5}$$

respectively. Note that we have focused on the special situation with $n = 3$ in Chapter 4 (but see (4.3.32) for a comparison).

Energy, momentum, and equations of motion

As a consequence, we can compute the associated momentum vector

$$p_i = \frac{\partial L}{\partial \dot{x}^i} = \frac{m\dot{x}^i}{\sqrt{1 - \sum_{i=1}^{n} \frac{(\dot{x}^i)^2}{c^2}}}, \quad i = 1, \dots, n, \tag{13.1.6}$$

which leads to the Hamiltonian or energy

$$H = \sum_{i=1}^{n} p_i \dot{x}^i - L = \frac{mc^2}{\sqrt{1 - \sum_{i=1}^{n} \frac{(\dot{x}^i)^2}{c^2}}}, \tag{13.1.7}$$

giving rise to the energy-momentum relation

$$H^2 = m^2c^4 + c^2 \sum_{i=1}^{n} p_i^2. \tag{13.1.8}$$

In particular, when the particle is at rest, we obtain the popular-science energy formula $E = H = mc^2$. The equations of motion of the particle governed by (13.1.5) are

$$\frac{\mathrm{d}}{\mathrm{d}t} \frac{m\dot{x}^i}{\sqrt{1 - \sum_{j=1}^{n} \frac{(\dot{x}^j)^2}{c^2}}} = 0, \quad i = 1, \dots, n, \tag{13.1.9}$$

which simply say that the momentum vector defined by (13.1.6) is conserved, $\dot{p}_i = 0$, $i = 1, \dots, n$.

13.2 Nambu–Goto strings

Nambu and Goto's 1971 formalism [244] for the motion of a relativistic string is based on a geometric extension of the point-particle action principle (13.1.5).

For simplicity, we set the speed of light to unity, $c = 1$, use the notation $\mathbf{x}^2 = \sum_{i=1}^{n} (x^i)^2$, etc., and $\mathbf{v}(t) = \dot{\mathbf{x}}$. Then, by (13.1.5), the action of a free particle of mass m over a time span $[t_1, t_2]$ reads

$$S = -m \int_{t_1}^{t_2} \sqrt{1 - \mathbf{v}^2}\, \mathrm{d}t. \tag{13.2.1}$$

String action

We now consider the motion of a free string of a uniform mass density, ρ_0, parametrized by a real parameter, s, with the spatial coordinates given as a parametrized curve,

$$\mathbf{x} = \mathbf{x}(s, t), \quad s_1 \leq s \leq s_2, \tag{13.2.2}$$

at any fixed time t. Following (13.2.1), the action for the motion of the infinitesimal portion

$$\mathrm{d}\ell = \left| \frac{\partial \mathbf{x}}{\partial s} \right| \mathrm{d}s \equiv |\mathbf{x}'|\, \mathrm{d}s, \quad s_1 \leq s \leq s_2, \tag{13.2.3}$$

is given by

$$-\rho_0\, \mathrm{d}\ell \int_{t_1}^{t_2} \sqrt{1 - \mathbf{v}^2}\, \mathrm{d}t. \tag{13.2.4}$$

Integrating (13.2.4), we obtain the total action for a *Nambu–Goto string* as follows:

$$S = -\rho_0 \int_{t_1}^{t_2} \int_{s_1}^{s_2} \sqrt{\left(\frac{\partial \mathbf{x}}{\partial s}\right)^2 (1 - \mathbf{v}^2)}\, \mathrm{d}s\mathrm{d}t. \tag{13.2.5}$$

Furthermore, to appreciate the geometric meaning of the action (13.2.5), we recall that, for two vectors, $x = (x^\mu) = (x^0, \mathbf{x})$ and $y = (y^\mu) = (y^0, \mathbf{y})$, xy stands for the inner product

$$xy = x^\mu y_\mu = x^\mu \eta_{\mu\nu} y^\nu = x^0 y^0 - \sum_{i=1}^{n} x^i y^i = x^0 y^0 - \mathbf{x} \cdot \mathbf{y}, \tag{13.2.6}$$

in the Minkowski spacetime $\mathbb{R}^{n,1}$. Thus, with the notation

$$\dot{f} = \frac{\partial f}{\partial t}, \quad f' = \frac{\partial f}{\partial s}, \tag{13.2.7}$$

we have

$$\dot{x}x' = \left(1, \frac{\partial \mathbf{x}}{\partial t}\right)\left(0, \frac{\partial \mathbf{x}}{\partial s}\right) = -\frac{\partial \mathbf{x}}{\partial t} \cdot \frac{\partial \mathbf{x}}{\partial s}, \tag{13.2.8}$$

$$\dot{x}^2 = 1 - \left(\frac{\partial \mathbf{x}}{\partial t}\right)^2, \quad x'^2 = -\left(\frac{\partial \mathbf{x}}{\partial s}\right)^2. \tag{13.2.9}$$

In the *Nambu–Goto theory* [244], the internal forces between neighboring points along a string do not contribute to the action so that the velocity vector $\mathbf{v}$ is perpendicular to the tangent of the string curve. Thus, we have

$$\mathbf{v} = \frac{\mathrm{d}\mathbf{x}}{\mathrm{d}t} = \frac{\partial \mathbf{x}}{\partial t} + a(s,t)\frac{\partial \mathbf{x}}{\partial s}, \quad a(s,t) = \frac{\mathrm{d}s}{\mathrm{d}t}, \quad \mathbf{v} \cdot \frac{\partial \mathbf{x}}{\partial s} = 0. \tag{13.2.10}$$

From (13.2.10), we can determine the scalar factor $a(s,t)$ and thereby obtain $\mathbf{v}$ to be

$$\mathbf{v} = \frac{\partial \mathbf{x}}{\partial t} - \frac{\left(\frac{\partial \mathbf{x}}{\partial t} \cdot \frac{\partial \mathbf{x}}{\partial s}\right)}{\left(\frac{\partial \mathbf{x}}{\partial s}\right)^2} \frac{\partial \mathbf{x}}{\partial s}. \tag{13.2.11}$$

As a consequence, it follows from (13.2.11) and (13.2.8)–(13.2.9) that

$$\begin{aligned}\left(\frac{\partial \mathbf{x}}{\partial s}\right)^2 (1 - \mathbf{v}^2) &= \left(\frac{\partial \mathbf{x}}{\partial s}\right)^2 - \left(\frac{\partial \mathbf{x}}{\partial t}\right)^2 \left(\frac{\partial \mathbf{x}}{\partial s}\right)^2 + \left(\frac{\partial \mathbf{x}}{\partial t} \cdot \frac{\partial \mathbf{x}}{\partial s}\right)^2 \\ &= (\dot{x}x')^2 - \dot{x}^2 x'^2. \end{aligned} \tag{13.2.12}$$

Hence, the Nambu–Goto action (13.2.5) becomes

$$\mathcal{A} = -\rho_0 \int \sqrt{(\dot{x}x')^2 - \dot{x}^2 x'^2}\, \mathrm{d}s\mathrm{d}t. \tag{13.2.13}$$

On the other hand, we calculate the line element of the embedded 2-surface $x^\mu = x^\mu(s,t)$, in the flat Minkowski spacetime $\mathbb{R}^{n,1}$, parametrized by the parameters s and t, by

$$\begin{aligned} ds^2 &= dx^\mu \eta_{\mu\nu} dx^\nu \\ &= (\dot{x}^\mu dt + x'^\mu ds)\eta_{\mu\nu}(\dot{x}^\nu dt + x'^\nu ds) \\ &= \dot{x}^2 dt^2 + 2\dot{x}x' dt ds + x'^2 ds^2 \\ &= h_{ab} du^a du^b, \quad a,b = 0,1, \end{aligned} \tag{13.2.14}$$

where $u^0 = t, u^1 = s$, and

$$(h_{ab}) = \begin{pmatrix} \dot{x}^2 & \dot{x}x' \\ \dot{x}x' & x'^2 \end{pmatrix}. \tag{13.2.15}$$

From (13.2.13) and (13.2.15), we see that the Nambu–Goto string action is simply a surface integral

$$\mathcal{A} = -\rho_0 \int_\Omega \sqrt{|h|}\, dt ds = -\rho_0 \int_S dS, \tag{13.2.16}$$

where $|h|$ is the absolute value of the determinant of the matrix (13.2.15) and dS is the canonical area element of the embedded 2-surface, $(S, \{h_{ab}\})$, in the Minkowski spacetime.

As a comparison, the general form of the action (13.1.5) for a point particle is simply a path integral,

$$\mathcal{A} = -m \int_{\tau_1}^{\tau_2} \sqrt{\dot{x}^2}\, d\tau = -m \int_C ds, \tag{13.2.17}$$

where ds is the line element of the path C parametrized by $x^\mu = x^\mu(\tau)$, $\tau_1 \le \tau \le \tau_2$, in the Minkowski spacetime.

Therefore, since the motion of a point particle follows an extremized path (the *world line*), the motion of a Nambu–Goto string is to follow an extremized surface (the *world sheet*).

Of course, both (13.2.16) and (13.2.17) are parametrization invariant.

Equations of motion of string

Returning to (13.2.13), using the generalized time and string coordinates, τ and σ, with

$$t = t(\tau,\sigma), \quad s = s(\tau,\sigma), \quad f_\tau = \partial_\tau f = \frac{\partial f}{\partial \tau}, \quad f_\sigma = \partial_\sigma f = \frac{\partial f}{\partial \sigma}, \tag{13.2.18}$$

and setting the mass density ρ_0 to unity, we see that the Nambu–Goto string action assumes the form

$$S = -\int \sqrt{(x_\tau x_\sigma)^2 - (x_\tau)^2 (x_\sigma)^2}\, d\tau d\sigma. \tag{13.2.19}$$

Use P^τ_μ and P^σ_μ to denote the generalized "momenta" where

$$P^\tau_\mu = \frac{\partial L}{\partial x^\mu_\tau}, \quad P^\sigma_\mu = \frac{\partial L}{\partial x^\mu_\sigma}, \quad L = -\sqrt{(x_\tau x_\sigma)^2 - (x_\tau)^2(x_\sigma)^2}. \tag{13.2.20}$$

Then, the equations of motion of the Nambu–Goto string obtained from varying the action (13.2.19) may be written in the form of the conservation laws

$$\frac{\partial P^\tau_\mu}{\partial \tau} + \frac{\partial P^\sigma_\mu}{\partial \sigma} = 0, \quad \mu = 0, 1, \ldots, n, \tag{13.2.21}$$

or, more explicitly,

$$\partial_\tau \left(\frac{(x_\tau x_\sigma)\partial_\sigma x_\mu - (x_\sigma)^2 \partial_\tau x_\mu}{\sqrt{(x_\tau x_\sigma)^2 - (x_\tau)^2(x_\sigma)^2}} \right) + \partial_\sigma \left(\frac{(x_\tau x_\sigma)\partial_\tau x_\mu - (x_\tau)^2 \partial_\sigma x_\mu}{\sqrt{(x_\tau x_\sigma)^2 - (x_\tau)^2(x_\sigma)^2}} \right)$$
$$= 0, \quad \mu = 0, 1, \ldots, n. \tag{13.2.22}$$

Solutions

Although the equations (13.2.22) appear rather complicated, they may be simplified by choosing an appropriate set of coordinates satisfying the condition

$$x_\tau x_\sigma = 0, \quad (x_\tau)^2 + (x_\sigma)^2 = 0, \tag{13.2.23}$$

known as the *orthonormal gauge condition* [486]. Indeed, inserting (13.2.23) into (13.2.22), we arrive at

$$\left(\partial^2_\tau - \partial^2_\sigma\right) x = 0, \tag{13.2.24}$$

which says the string oscillates following the d'Alembertian wave equation. If x is periodic in the string parameter σ, the string is called *closed*; if σ varies in a finite interval I such that $x_\sigma = 0$ at the end points of I, the string is called *Neumann*; if $x_\tau = 0$ at the end points of the interval I, the string is called *Dirichlet*. This latter case is the same as imposing x to assume constant values at the end points of I.

Note that the orthonormal gauge condition (13.2.23) makes the string metric (13.2.15) conformally flat,

$$(h_{ab}) = \Lambda \begin{pmatrix} 1 & 0 \\ 0 & -1 \end{pmatrix}, \quad \Lambda = \Lambda(\tau, \sigma) = (x_\tau)^2, \tag{13.2.25}$$

which motivates the terminology used. Note also that, alternatively, the condition (13.2.23) may be recast into its equivalent form:

$$(x_\tau \pm x_\sigma)^2 = 0. \tag{13.2.26}$$

13.3 p-branes

More generally, consider an embedded $(p+1)$-dimensional hypersurface, or a membrane, parametrized by the coordinates (u^a) $(a = 0, 1, \ldots, p)$, so that the induced metric element over the hypersurface is given by

$$\mathrm{d}s^2 = h_{ab}\mathrm{d}u^a\mathrm{d}u^b, \quad h_{ab} = \eta_{\mu\nu}\frac{\partial x^\mu}{\partial u^a}\frac{\partial x^\nu}{\partial u^b}. \tag{13.3.1}$$

Action

In analogy to (13.2.16), the action of a *p-brane* is given by the volume integral of the hypersurface as,

$$S = -T_p \int \mathrm{d}V_{p+1} = -T_p \int \sqrt{|h|}\,\mathrm{d}u^0\mathrm{d}u^1\cdots\mathrm{d}u^p, \tag{13.3.2}$$

where T_p is a positive constant referred to as the *tension of the p-brane* and $h = \det(h_{ab})$. In particular, a point particle is a 0-brane, and a string is a 1-brane.

Equations of motion

Let C_{ab} denote the cofactor of the matrix (h_{ab}) at the position (a, b) and δ a variation resulting from varying x^μ in (13.3.1). Then, we have

$$\delta h = \sum_{0\leq a,b\leq p} C_{ab}\,\delta h_{ab}, \tag{13.3.3}$$

$$\delta h_{ab} = \frac{\partial x_\mu}{\partial u^a}\,\delta\left(\frac{\partial x^\mu}{\partial u^b}\right) + \frac{\partial x_\mu}{\partial u^b}\,\delta\left(\frac{\partial x^\mu}{\partial u^a}\right). \tag{13.3.4}$$

Thus, applying (13.3.3)–(13.3.4) and the symmetry of (C_{ab}), we obtain from (13.3.2) the associated equations of motion to be

$$\sum_{0\leq a,b\leq p} \frac{\partial}{\partial u^a}\left(\frac{C_{ab}}{\sqrt{|h|}}\frac{\partial x_\mu}{\partial u^b}\right) = 0, \quad \mu = 0, 1, \ldots, n. \tag{13.3.5}$$

It is clear that the Nambu–Goto string equations (13.2.22) are contained in (13.3.5) as a special case when $p = 1$.

Moreover, if we use (h^{ab}) to denote the inverse of (h_{ab}), then

$$C_{ab} = hh^{ab}. \tag{13.3.6}$$

Inserting (13.3.6) into (13.3.5), we arrive at

$$\frac{1}{\sqrt{|h|}}\partial_a\left(\sqrt{|h|}h^{ab}\,\partial_b x_\mu\right) = 0, \quad \mu = 0, 1, \ldots, n, \tag{13.3.7}$$

with the summation convention observed over the repeated indices, $a, b = 0, 1, \ldots, p$, indicating that a free p-brane oscillates as *d'Alembertian waves*.

Static equations

For convenience and simplicity, we now consider a static n-brane, M, in the Minkowski spacetime $\mathbb{R}^{n,1}$, which is realized as a graph of a function depending on the spatial coordinates only, given by

$$x^0 = f(x^1, \dots, x^n). \tag{13.3.8}$$

Using (13.3.1), we see that the metric tensor (h_{ij}) of M is

$$h_{ij} = \eta_{\mu\nu}\partial_i x^\mu \partial_j x^\nu = \partial_i f \partial_j f - \delta_{ij}, \quad i, j = 1, \dots, n. \tag{13.3.9}$$

We see that

$$h = \det(\partial_i f \partial_j f - \delta_{ij}) = (-1)^n(1 - |\nabla f|^2). \tag{13.3.10}$$

In fact, recall that, if A and B are $m \times n$ and $n \times m$ matrices, then AB and BA share the same nonzero eigenvalues [607]. Applying this fact to the matrix $(\partial_i f \partial_j f) = (\nabla f)^t(\nabla f)$ where ∇f is taken to be a row vector, we see that the only possible nonzero eigenvalue of $(\partial_i f \partial_j f)$ is

$$(\nabla f)(\nabla f)^t = |\nabla f|^2. \tag{13.3.11}$$

As a consequence, the eigenvalues of (h_{ij}) are $|\nabla f|^2 - 1$ and -1 with multiplicity $n-1$ if $\nabla f \neq 0$. Therefore, (13.3.10) is established.

We assume that M is time-like, $|\nabla f| < 1$. Hence, ignoring the coupling constant, the action of an n-brane, which happens to be a graph, is

$$S = -\int \sqrt{1 - |\nabla f|^2}\, \mathrm{d}x, \tag{13.3.12}$$

whose Euler–Lagrange equation reads

$$\nabla \cdot \left(\frac{\nabla f}{\sqrt{1 - |\nabla f|^2}} \right) = 0. \tag{13.3.13}$$

A well-known theorem of Cheng and Yau [130] states that all solutions of (13.3.13) over the full space $\mathbb{R}^n$ satisfying $|\nabla f| < 1$ must be affine linear,

$$f(x^1, \dots, x^n) = \sum_{i=1}^n a_i x^i + b, \tag{13.3.14}$$

where a_i's and b are constants.

Minimal hypersurfaces

It may be instructive to compare this study of an n-brane with its Euclidean space counterpart where we replace the Minkowski spacetime $\mathbb{R}^{n,1}$ with the Euclidean space $\mathbb{R}^{n+1}$ so that the inherited metric of the embedded n-hypersurface M defined by the graph of the function (13.3.8) is given by

$$h_{ij} = \delta_{\mu\nu}\partial_i x^\mu \partial_j x^\nu = \partial_i f \partial_j f + \delta_{ij}, \quad i, j = 1, \dots, n. \tag{13.3.15}$$

Consequently,

$$h = \det(h_{ij}) = 1 + |\nabla f|^2, \tag{13.3.16}$$

such that the canonical volume of the hypersurface M reads

$$\mathcal{V}_M = \int \sqrt{h}\,\mathrm{d}x = \int \sqrt{1 + |\nabla f|^2}\,\mathrm{d}x. \tag{13.3.17}$$

Minimizing (13.3.17) gives us the classical equation

$$\nabla \cdot \left(\frac{\nabla f}{\sqrt{1 + |\nabla f|^2}} \right) = 0, \quad x \in \mathbb{R}^n, \tag{13.3.18}$$

known as the *minimal hypersurface equation* for *non-parametric minimal hypersurfaces*, defined as the graph of a function. The Bernstein theorem for this equation states that all entire solutions are affine linear for $n \leq 7$ (cf. [426] and references therein).

Equivalence theorem

Calabi first observed that the equations (13.3.13) and (13.3.18) are equivalent [99] when $n = 2$. A proof of this fact is as follows.

Let u be a solution of (13.3.18) and

$$p = \partial_1 f, \quad q = \partial_2 f. \tag{13.3.19}$$

Set $w = \sqrt{1 + p^2 + q^2}$. Then, (13.3.18) reads

$$\partial_1 \left(\frac{p}{w} \right) + \partial_2 \left(\frac{q}{w} \right) = 0. \tag{13.3.20}$$

Hence, there is a real-valued function U such that

$$\partial_1 U = P \equiv -\frac{q}{w}, \quad \partial_2 U = Q \equiv \frac{p}{w}. \tag{13.3.21}$$

Therefore, we have

$$1 - P^2 - Q^2 = \frac{1}{w^2} > 0, \tag{13.3.22}$$

and U is time-like. Inserting the relations $p = Qw$, $q = -Pw$, and $w = \frac{1}{W}$ where $W = \sqrt{1 - P^2 - Q^2}$ into the identity $\partial_2 p = \partial_1 q$, we arrive at

$$\partial_1 \left(\frac{P}{W} \right) + \partial_2 \left(\frac{Q}{W} \right) = 0. \tag{13.3.23}$$

Thus, U solves (13.3.13). The inverse correspondence from (13.3.13) to (13.3.18) may be established similarly.

Calabi's equivalence theorem extends into arbitrary n-dimensional settings [606], which give rise to a rich range of open problems.

13.4 Polyakov strings and branes

This section considers the Polyakov strings and branes, which are characterized by being governed by quadratic action principles so that they are in general more tractable mathematically than the Nambu–Goto strings and branes. It first conducts a discussion in general terms. It then specializes on some specific interests.

Brane action

To start, consider a map

$$\phi : (M, h_{ab}) \to (N, g_{\mu\nu}), \quad \phi(u^0, u^1, \dots, u^m) = (x^0, x^1, \dots, x^n), \tag{13.4.1}$$

where the domain and target spaces (M, h_{ab}) and $(N, g_{\mu\nu})$ are $(m+1)$- and $(n+1)$-dimensional Minkowski manifolds parametrized with the coordinates (u^a) and (x^μ), respectively. The *Polyakov action* is simply the *harmonic map functional* defined as

$$S = -\int (D\phi)^2 \, \mathrm{d}V_h, \quad (D\phi)^2 = g_{\mu\nu} h^{ab} \partial_a x^\mu \partial_b x^\nu, \quad a, b = 0, 1, \dots, m, \tag{13.4.2}$$

where $(h^{ab}) = (h_{ab})^{-1}$ and $\mathrm{d}V_h$ is the canonical volume element of (M, h_{ab}) given by

$$\mathrm{d}V_h = \sqrt{|h|} \mathrm{d}u^0 \mathrm{d}u^1 \cdots \mathrm{d}u^m, \tag{13.4.3}$$

or customarily, for a p-brane,

$$\begin{aligned} S &= -\tau_p \int \mathcal{L} \, \mathrm{d}V_h \\ &= -\tau_p \int \sqrt{|\det(h_{ab})|} \, g_{\mu\nu} h^{ab} \partial_a x^\mu \partial_b x^\nu \, \mathrm{d}u^0 \mathrm{d}u^1 \cdots \mathrm{d}u^p, \end{aligned} \tag{13.4.4}$$

where we have attached the constant $\tau_p > 0$ to account for the *Polyakov p-brane* tension. Note that the metrics h_{ab} and $g_{\mu\nu}$ depend on $u = (u^0, u^1, \dots, u^p)$ and $x = (x^0, x^1, \dots, x^n)$, respectively.

Action of Polyakov strings

When $p = 1$, the action (13.4.4) defines the *Polyakov string* [455] action, which is conformally invariant (i.e. the action is invariant under the conformal transformation of the metric, $h_{ab} \mapsto \Lambda h_{ab}$). Such an invariance property is also called the *Weyl invariance*. An obvious technical advantage of the action (13.4.4) over (13.3.2) is that the former is quadratic in the derivatives of x^μs and gives rise to linear equations of motion for the branes when the target spacetime is flat.

Action of Nambu–Goto branes

In the special case when (M, h_{ab}) is regarded as a submanifold of the flat Minkowski spacetime $(\mathbb{R}^{n,1}, \eta_{\mu\nu})$, which is the target space, so that the metric h_{ab} is induced from the map (13.4.1), we have

$$h_{ab} = \eta_{\mu\nu}\partial_a x^\mu \partial_b x^\nu, \tag{13.4.5}$$

which leads us to

$$h^{ab}\eta_{\mu\nu}\partial_a x^\mu \partial_b x^\nu = p + 1. \tag{13.4.6}$$

Consequently, we see that the Polyakov p-brane action (13.4.4) reduces into the Nambu–Goto p-brane action (13.3.2) when $T_p = (p+1)\tau_p$.

Another point of view regarding the relationship between the Nambu–Goto strings and the Polyakov strings follows from extremizing the metric tensor h_{ab} in the Polyakov p-brane action (13.4.4). Specifically, by using a cofactor expansion, we have

$$\frac{\partial h}{\partial h^{ab}} = \frac{\partial}{\partial h^{ab}}\left(\frac{1}{\det(h^{ab})}\right) = -h h_{ab}. \tag{13.4.7}$$

Consequently, varying h^{ab} in (13.4.4) and using (13.4.7), we obtain

$$\delta S = -\frac{1}{2}\tau_p \int T_{ab}\,\delta(h^{ab})\,\sqrt{|h|}\,\mathrm{d}u, \tag{13.4.8}$$

where

$$T_{ab} = 2\frac{\partial \mathcal{L}}{\partial h^{ab}} - h_{ab}\mathcal{L} = 2g_{\mu\nu}\partial_a x^\mu \partial_b x^\nu - h_{ab}\mathcal{L} \tag{13.4.9}$$

is the metric stress tensor. Thus, extremizing the action (13.4.4) by setting $\delta S = 0$, we get from (13.4.8)–(13.4.9) the *vanishing stress tensor condition*

$$T_{ab} = 2g_{\mu\nu}\partial_a x^\mu \partial_b x^\nu - h_{ab}\mathcal{L} = 0, \quad \forall a, b, \tag{13.4.10}$$

which gives us the solution for h_{ab}:

$$h_{ab} = \frac{2g_{\mu\nu}\partial_a x^\mu \partial_b x^\nu}{\mathcal{L}} \equiv \frac{2}{\mathcal{L}}(\partial_a x \cdot \partial_b x), \tag{13.4.11}$$

assuming

$$\mathcal{L} \neq 0. \tag{13.4.12}$$

Therefore, inserting (13.4.11) into (13.4.4), we have

$$S = -2\tau_p \int \sqrt{|\det(\partial_a x \cdot \partial_b x)|}\left(\frac{2}{\mathcal{L}}\right)^{\frac{(p-1)}{2}} \mathrm{d}u, \tag{13.4.13}$$

which becomes a pure volume integral exactly when $p = 1$. In other words, in such a context, the Nambu–Goto and Polyakov string actions are equivalent.

Action of moving particle revisited

The geometric elegance of the Polyakov p-brane action (13.4.4) and the equivalence proof for string actions prompts us to consider the actions for the motion of a massive particle, or a 0-brane. In such a situation, formally, we obtain from the condition (13.4.10) that

$$g_{\mu\nu}\partial_0 x^\mu \partial_0 x^\nu = 0. \tag{13.4.14}$$

Hence, the condition (13.4.12) fails and the particle is actually a massless photon. In order to overcome this difficulty, we use $t = u^0$ to denote the coordinate time and τ the proper time such that

$$\mathrm{d}\tau^2 = h_{00}(t)\mathrm{d}t^2. \tag{13.4.15}$$

In view of (13.4.15), we see that the action (13.4.4) in the 0-brane situation becomes

$$S_0 = -\tau_0 \int g_{\mu\nu} \frac{\mathrm{d}x^\mu}{\mathrm{d}\tau}\frac{\mathrm{d}x^\nu}{\mathrm{d}\tau}\,\mathrm{d}\tau. \tag{13.4.16}$$

To proceed, we expand slightly into

$$S_\alpha = -\tau_0 \int \left(g_{\mu\nu} \frac{\mathrm{d}x^\mu}{\mathrm{d}\tau}\frac{\mathrm{d}x^\nu}{\mathrm{d}\tau} + \alpha \right) \mathrm{d}\tau, \tag{13.4.17}$$

where α is an adjustable constant. On the other hand, we may rewrite (13.4.15) as

$$\frac{\mathrm{d}\tau}{\mathrm{d}t} = e(t), \tag{13.4.18}$$

where $e(t) = \sqrt{h_{00}(t)}$ is sometimes referred to as the *einbein field* of the model. Thus, substituting (13.4.18) into (13.4.17), we have

$$S_\alpha = -\tau_0 \int \left(\frac{1}{e(t)} g_{\mu\nu}\dot{x}^\mu \dot{x}^\nu + \alpha e(t) \right) \mathrm{d}t. \tag{13.4.19}$$

Now, varying the einbein $e(t)$ in (13.4.19), we get

$$\frac{1}{e^2(t)} g_{\mu\nu}\dot{x}^\mu \dot{x}^\nu = \alpha. \tag{13.4.20}$$

Thus, when we consider the motion of a massive particle such that its trajectory is time-like, $g_{\mu\nu}\dot{x}^\mu \dot{x}^\nu > 0$, then $\alpha > 0$. Hence, (13.4.20) gives us

$$e(t) = \sqrt{\frac{1}{\alpha} g_{\mu\nu}\dot{x}^\mu \dot{x}^\nu}. \tag{13.4.21}$$

Inserting (13.4.21) into (13.4.19), we arrive at

$$S = -2\tau_0\sqrt{\alpha} \int \sqrt{g_{\mu\nu}\dot{x}^\mu \dot{x}^\nu}\,\mathrm{d}t, \tag{13.4.22}$$

which miraculously reduces to the action (13.1.5) when the target space is flat, in particular.

Critical dimensions of spacetime

Quantization of the Polyakov string leads to the *string partition function* [455]

$$Z = \int \mathcal{D}\varphi \exp\left(-\frac{(26-D)}{48\pi}\int \left[\frac{1}{2}(\partial_a\varphi)^2 + \kappa^2 \mathrm{e}^{\varphi}\right] \mathrm{d}^2u\right), \tag{13.4.23}$$

where $D = n+1$ is the spacetime dimension, the metric h_{ab} is Euclideanized into $\mathrm{e}^{\varphi}\delta_{ab}$ through a *Wick rotation* $u^0 \mapsto \mathrm{i}u^0$, $\kappa > 0$ is constant, and $\int \mathcal{D}\varphi$ denotes the *path integral* over the space of all possible conformal exponents. Note that the Polyakov string action is conformally invariant. However, the partition function (13.4.23) clearly shows that such a conformal invariance is no longer valid when $D \neq 26$. Such a phenomenon is called "*conformal anomaly*" and the vanishing of conformal anomaly gives us the unique condition

$$D = 26, \tag{13.4.24}$$

known as the *critical dimension* of *bosonic string theory*. Furthermore, when *fermions* are present, Polyakov's computation [456] of the partition function of quantized supersymmetric strings, or *superstrings*, gives us the unique condition

$$D = 10, \tag{13.4.25}$$

to avoid conformal anomaly, again. These results about the *critical dimensions of spacetime* are now standard facts in string theory [618]. Physicists [106, 317, 618] further conjectured that our ten-dimensional universe, $\mathcal{M}_{10}$, is a product of a four-dimensional spacetime, $\mathcal{M}_4$, and a six-dimensional compact manifold, $\mathcal{K}_6$, curled up in a tiny but highly sophisticated way, following a formalism called the *string compactification* [172, 245], so that the spacetime $\mathcal{M}_4$ is *maximally symmetric* (which implies that $\mathcal{M}_4$ can either be Minkowski, de Sitter, or anti-de Sitter)[1] and $\mathcal{K}_6$ is a *Calabi–Yau manifold* [97, 98, 610].[2]

Note also that, the action stemming out (cf. [614]) from (13.4.23), given by

$$L = \int \left(\frac{1}{2}\partial_a\varphi\partial^a\varphi - \kappa^2\mathrm{e}^{\varphi}\right) \mathrm{d}^2u, \tag{13.4.26}$$

and the associated wave equation

$$\varphi_{\tau\tau} - \varphi_{\sigma\sigma} = -\kappa^2 \mathrm{e}^{\varphi}, \quad u^0 = \tau, \quad u^1 = \sigma, \tag{13.4.27}$$

as the equation of motion, are jointly known to define the *Liouville field theory* [152, 230], which is integrable [606] and of independent interest as a toy model.

[1] A manifold is maximally symmetric if it has the same number of symmetries as ordinary Euclidean space. More precisely, a Riemannian manifold is maximally symmetric if it has $\frac{1}{2}n(n+1)$ (n = dimension of the manifold) linearly independent Killing vector fields that generate isometric flows on the manifold.

[2] A Calabi–Yau manifold is a compact Kähler manifold with a vanishing first Chern class. A Kähler manifold is an even-dimensional Riemannian manifold accommodating a complex or symplectic structure in a certain compatible manner [175, 257, 285, 335, 352].

13.5 Equations of motion with interactions

The previous sections considered the problems of freely moving particles, strings, and branes. This section extends our discussion to the situation when the objects are subject to interactions. For conciseness and simplicity, we focus on moving particles since an extension of discussion to the cases of strings and branes is straightforward and immediate.

Let the motion of a point particle of rest mass m be subject to a potential V depending on the coordinate and velocity vectors $x = (x^i)$ and $\dot{x} = (\dot{x}^i)$, respectively, of the particle at time t, given by $V = V(x, \dot{x}, t)$. Then, the free Lagrangian L in (13.1.5) is expanded into

$$L = -mc^2\sqrt{1 - \sum_{i=1}^{n} \frac{(\dot{x}^i)^2}{c^2}} - V(x, \dot{x}, t). \tag{13.5.1}$$

Hence, the equations of motion of the particle are

$$\frac{\mathrm{d}}{\mathrm{d}t}\left(\frac{\partial L}{\partial \dot{x}^i}\right) = \frac{\partial L}{\partial x^i}, \quad i = 1, \dots, n, \tag{13.5.2}$$

or explicitly,

$$\frac{\mathrm{d}}{\mathrm{d}t}\frac{m\dot{x}^i}{\sqrt{1 - \sum_{j=1}^{n} \frac{(\dot{x}^j)^2}{c^2}}} = \frac{\mathrm{d}}{\mathrm{d}t}\left(\frac{\partial V}{\partial \dot{x}^i}\right) - \frac{\partial V}{\partial x^i}, \quad i = 1, \dots, n. \tag{13.5.3}$$

If the moving particle is subject to Einstein's gravity described by a gravitational metric $(g_{\mu\nu})$, then the free part of the action function is given by that defined in (13.4.22) so that the full Lagrangian reads

$$L = -\gamma\sqrt{g_{\mu\nu}(x)\dot{x}^\mu\dot{x}^\nu} - V, \tag{13.5.4}$$

where $\gamma > 0$ is a constant and V a potential function of interest.

The literature of string theory is vast and enormous. We recommend the following well-referenced texts that cover both introductory materials and research frontiers of the subject, [54, 247, 248, 450, 451, 618]. See also [333, 423, 495].

Exercises

1. Show that the Hamiltonian defined by (13.1.7) is conserved in view of (13.1.6) and (13.1.9).
2. Show that the Nambu–Goto string equations (13.2.22) are contained as a special case of the p-brane equations (13.3.5) when $p = 1$.
3. Establish (13.3.10) directly when $n = 2, 3$.

4. Show that (13.3.13) is contained as a special case of (13.3.7) when the p-brane is an n-brane represented by (13.3.8).
5. Establish (13.3.16).
6. Use the Cheng–Yau theorem for (13.3.13) and Calabi's equivalence theorem to prove the Bernstein theorem for (13.3.18) when $n = 2$.
7. Establish (13.4.7) and use it to derive (13.4.8)–(13.4.9).
8. For the relativistic Lagrangian (13.5.1), formally formulate the Hamiltonian
$$H = p_i\dot{x}^i - L = \frac{\partial L}{\partial \dot{x}^i}\dot{x}^i - L \tag{13.E.1}$$
and use (13.5.3) to establish the energy conservation law $\frac{\mathrm{d}}{\mathrm{dt}}H = 0$ when V is conserved, that is, $V = V(x)$. What happens if V is not conserved?
9. Derive the equations of motion when the potential V in (13.5.3) is that of a harmonic oscillator or a constant force of the form
$$V(x) = \frac{\kappa}{2}x^2 \quad \text{or} \quad V(x) = a_i x^i, \tag{13.E.2}$$
respectively, or that of a Helmholtz–Kirchhoff vortex or Newton gravity type,
$$V(x) = \begin{cases} -GmM \ln|x|, & x \in \mathbb{R}^2, \quad x \neq 0, \\ -G\frac{mM}{|x|^{n-2}}, & x \in \mathbb{R}^n, \quad n \geq 3, \quad x \neq 0, \end{cases} \tag{13.E.3}$$
respectively, where κ, a_i's, G, M are constants.
10. Consider the relativistic motion of a particle of rest mass m and electric charge q in an electric field $\mathbf{E}$ and magnetic field $\mathbf{B}$ subject to the Lorentz force such that its position vector $\mathbf{x}$ in $\mathbb{R}^3$ follows the equation of motion
$$\frac{\mathrm{d}}{\mathrm{d}t}\frac{m\mathbf{v}}{\sqrt{1-\frac{\mathbf{v}^2}{c^2}}} = q\mathbf{E} + q\mathbf{v}\times\mathbf{B}, \quad \mathbf{v} = \dot{\mathbf{x}}. \tag{13.E.4}$$
Find the potential function $V = V(\mathbf{x}, \dot{\mathbf{x}})$ for this problem so that (13.E.4) assumes the form (13.5.3).
11. Use $x^0 = ct$ in (13.5.1) to rewrite (13.5.1) in the form (13.5.4) with $g_{\mu\nu} = \eta_{\mu\nu}$. Thus, (13.5.4) may be regarded as being *geometrized* from (13.5.1).
12. Derive the equations of motion associated with (13.5.4).
13. Extend (13.5.1) to the situation of the motion of a Nambu–Goto string, subject to a potential function V, and derive the associated equations of motion for the string.
14. Extend (13.5.1) to the situation of the motion of a Nambu–Goto p-brane, subject to a potential function V, and derive the associated equations of motion for the p-brane.
15. Formulate the action principle and derive the equations of motion for a Polyakov string subject to a potential function as in Exercise 13.

14

Born–Infeld theory of electromagnetism

As a natural development of the topics covered in Chapter 13, this chapter introduces the Born–Infeld theory of electromagnetism originally formulated to tackle the energy divergence problem associated with a point charge in the Maxwell theory. This highly nonlinear (but geometrically elegant) theory of electromagnetism is relevant and applicable to other areas of field theory, including gravitation and superstrings. This chapter begins with the Born formalism based on a modification of the action density inspired by special relativity. Next, it considers the systematic formalism of Born and Infeld motivated by invariance. It then discusses several mathematical aspects of the Born–Infeld theory, including generalized Bernstein problems, associated virial identities, and an integer-squared law for global vortex solutions of the Born–Infeld wave equations. Moreover, it presents an extensive discussion of charged black hole solutions of the Reissner–Nordström type arising from coupling the Einstein equations with the Born–Infeld nonlinear electromagnetism. Afterwards, it considers generalized Born–Infeld theories with a rich array of applications, including a nonlinear mechanism for monopole exclusion, relegation of curvature singularities of charged black holes, and realization of various cosmological expansion patterns through appropriate Born–Infeld scalar-wave matters in the form of k-essence.

14.1 Resolution of energy divergence problem of point charges

In 1933–1934, Born and Infeld introduced their nonlinear geometric theory of electromagnetism in a series of articles with a goal to overcome the *infinity problem* or *energy divergence problem* associated with a *point charge* source in the original Maxwell theory, then as a model for electron. Initiated by Born

Mathematical Physics with Differential Equations. Yisong Yang, Oxford University Press.
 DOI: 10.1093/oso/9780192872616.003.0014

in [73, 74] and fully developed by Born and Infeld in [75, 76], their idea came from considering the action principle of a particle in free relativistic motion that naturally imposes an upper bound to the velocity of the particle, such that, when an analogous formalism is introduced in the context of theory of electromagnetism, one is led to their theory, which succeeds in resulting in a bounded electric field so that an electric point charge miraculously carries a finite energy as desired. In contemporary theoretical physics, the Born–Infeld theory also plays an active and important role. For example, it is shown to arise in string and brane theories [102, 213, 231, 551, 552], and its idea is applied to modify the Einstein–Hilbert action in an attempt [159] to regularize gravity theory. See [311] for a recent comprehensive survey on subsequent development and progress of the subject. This chapter presents an extensive discussion of a rich array of subjects arising from the Born–Infeld theory centered around insights provided by differential equations. The first four sections introduce the basic ideas and present some illustrative calculations of the theory. Section 14.5 considers the interaction of static electromagnetic fields and relates the problem to a new coupled system of nonlinear maximal hypersurface type equations. Section 14.6 obtains a virial theorem that implies that the Derrick theorem dimensionality barrier (Section 12.1) is eliminated in the Born–Infeld theory. Section 14.7 derives a universal "integer-squared" law, relating energy to topology, for global vortices in the Born–Infeld theory, and Section 14.9 obtains black hole solutions carrying electric and magnetic charges produced from coupling the Born–Infeld equations with the Einstein equations. The rest of the chapter covers more advanced and technical topics, and addresses the following four main problems.

(i) Based on the density theorem of Stone–Weierstrass, it is shown that a nonlinear theory of electrodynamics generically permits a finite-energy electric point charge but rejects a magnetic one, which spells out an exclusion mechanism for magnetic monopoles.

(ii) It is demonstrated that the black hole of a finite electromagnetic energy (*viz* the Born–Infeld theory) enjoys reduced singularity or enhanced regularity in comparison with the classical Reissner–Nordström charged black hole, whose energy necessarily diverges.

(iii) It is established that the Born–Infeld type scalar-wave matters may be used as k-essence in cosmology to realize in principle any *prescribed* equation of state of a hypothetical cosmic fluid.

(iv) It is shown that the formalism of Born and Infeld may be extended to construct finite-energy *dyonically charged* black holes with relegated curvature singularities.

Conceptually and technically, in their original investigations [73, 74, 75, 76], Born and Infeld developed two different formulations of nonlinear electrodynamics. As described, the first is based on a modification of the action principle of the Maxwell theory of electromagnetism in view of the action principle of the free motion of a point mass in the context of special relativity. The more geometric second one is based on a consideration of invariance principle. In weak field limit, the latter returns to the former. In this sense, the latter contains the former. Besides, we shall see that the latter is crucial for the construction of finite-energy dyonically charged point particles and black holes.

This section concentrates on the first model of Born and Infeld, motivated from special relativity, which is mathematically simpler and more direct. This discussion lays the foundation and sparks interest for the study of the other subjects in the chapter.

Motivation from special relativity

Section 3.2 showed how, in the Maxwell theory of electromagnetism, a charged point source necessarily carries an infinite energy. Such a property prevents us from classically modeling the electron as a finite-energy electric point charge. To overcome this difficulty, Born and Infeld [73, 74, 75, 76] observed that, since the mechanics of Einstein's special relativity governing the free motion of a particle of mass m and velocity v may be obtained from the Newton mechanics by replacing the classical action function $\mathcal{L} = \frac{1}{2}mv^2$ by the relativistic one,

$$\begin{aligned}\mathcal{L} &= mc^2\left(1 - \sqrt{1 - \frac{v^2}{c^2}}\right) \\ &= b^2\left(1 - \sqrt{1 - \frac{1}{b^2}mv^2}\right), \quad b^2 = mc^2, \end{aligned} \tag{14.1.1}$$

where c is the velocity of light in vacuum (cf. (4.3.32)), we can replace the action density of the Maxwell theory,

$$\mathcal{L} = \frac{1}{2}(\mathbf{E}^2 - \mathbf{B}^2), \tag{14.1.2}$$

where $\mathbf{E}$ and $\mathbf{B}$ are electric and magnetic fields, respectively, by a correspondingly modified expression of the form

$$\mathcal{L} = b^2\left(1 - \sqrt{1 - \frac{1}{b^2}(\mathbf{E}^2 - \mathbf{B}^2)}\right), \tag{14.1.3}$$

where $b > 0$ is a suitable scaling parameter, often called the *Born–Infeld parameter*, or simply, the *Born parameter*. It is clear that (14.1.3) defines a nonlinear theory of electromagnetism and the Maxwell theory, (14.1.2), may be recovered in the weak field limit $\mathbf{E}, \mathbf{B} \to \mathbf{0}$. Note that the choice of sign in front of the Lagrangian density (14.1.2) is the opposite of that of Born and Infeld [76] and is widely adopted in contemporary literature. This convention will be observed throughout this chapter.

Born–Infeld theory as gauge field theory

Let A_μ be a real-valued gauge vector potential and

$$F_{\mu\nu} = \partial_\mu A_\nu - \partial_\nu A_\mu, \tag{14.1.4}$$

the electromagnetic field induced from A_μ. Following the sign convention in [299, 477, 606] as earlier, we use

$$\mathbf{E} = (E^1, E^2, E^3), \quad E^i = -F^{0i};$$
$$\mathbf{B} = (B^1, B^2, B^3), \quad B^i = -\frac{1}{2}\epsilon^{ijk}F_{jk}, \tag{14.1.5}$$

to denote the electric and magnetic fields, respectively, where the standard Minkowski metric tensor $(\eta_{\mu\nu})$ or $(\eta^{\mu\nu})$ is taken to lower or raise coordinate indices. Thus, (14.1.2) is replaced by $\mathcal{L} = -\frac{1}{4}F_{\mu\nu}F^{\mu\nu}$ and (14.1.3) becomes

$$\mathcal{L} = b^2\left(1 - \sqrt{1 + \frac{1}{2b^2}F_{\mu\nu}F^{\mu\nu}}\right). \tag{14.1.6}$$

Now, we rewrite (14.1.5) as

$$(F^{\mu\nu}) = \begin{pmatrix} 0 & -E^1 & -E^2 & -E^3 \\ E^1 & 0 & -B^3 & B^2 \\ E^2 & B^3 & 0 & -B^1 \\ E^3 & -B^2 & B^1 & 0 \end{pmatrix}, \tag{14.1.7}$$

as before such that the dual of $F^{\mu\nu}$ reads

$$\tilde{F}^{\mu\nu} = \frac{1}{2}\epsilon^{\mu\nu\alpha\beta}F_{\alpha\beta} = \begin{pmatrix} 0 & -B^1 & -B^2 & -B^3 \\ B^1 & 0 & E^3 & -E^2 \\ B^2 & -E^3 & 0 & E^1 \\ B^3 & E^2 & -E^1 & 0 \end{pmatrix}. \tag{14.1.8}$$

So, the Bianchi identity following from (14.1.4) is

$$\partial_\mu \tilde{F}^{\mu\nu} = 0. \tag{14.1.9}$$

Varying A_ν in (14.1.6), we see that the associated Euler–Lagrange equations are

$$\partial_\mu P^{\mu\nu} = 0, \tag{14.1.10}$$
$$P^{\mu\nu} = \frac{F^{\mu\nu}}{\sqrt{1 + \frac{1}{2b^2}F_{\alpha\beta}F^{\alpha\beta}}}. \tag{14.1.11}$$

Corresponding to the electric field $\mathbf{E}$ and magnetic field $\mathbf{B}$ given in (14.1.5), we introduce the *electric displacement* field $\mathbf{D}$ and *magnetic intensity* field $\mathbf{H}$ as follows:

$$\mathbf{D} = (D^1, D^2, D^3), \quad D^i = -P^{0i};$$
$$\mathbf{H} = (H^1, H^2, H^3), \quad H^i = -\frac{1}{2}\epsilon^{ijk}P_{jk}, \tag{14.1.12}$$

which may be suppressed into the matrix form

$$(P^{\mu\nu}) = \begin{pmatrix} 0 & -D^1 & -D^2 & -D^3 \\ D^1 & 0 & -H^3 & H^2 \\ D^2 & H^3 & 0 & -H^1 \\ D^3 & -H^2 & H^1 & 0 \end{pmatrix}. \tag{14.1.13}$$

Inserting (14.1.8) into (14.1.9) and (14.1.13) into (14.1.10), respectively, we obtain the governing *equations of the Born–Infeld electromagnetic theory*:

$$\frac{\partial \mathbf{B}}{\partial t} + \nabla \times \mathbf{E} = \mathbf{0}, \quad \nabla \cdot \mathbf{B} = 0, \tag{14.1.14}$$

$$-\frac{\partial \mathbf{D}}{\partial t} + \nabla \times \mathbf{H} = \mathbf{0}, \quad \nabla \cdot \mathbf{D} = 0, \tag{14.1.15}$$

which take the same form of the *vacuum Maxwell equations*, except that, in view of the relations (14.1.7), (14.1.11), and (14.1.13), the fields $\mathbf{E}, \mathbf{B}$ and $\mathbf{D}, \mathbf{H}$ are *related nonlinearly*,

$$\mathbf{D} = \frac{\mathbf{E}}{\sqrt{1 + \frac{1}{b^2}(\mathbf{B}^2 - \mathbf{E}^2)}}, \tag{14.1.16}$$

$$\mathbf{H} = \frac{\mathbf{B}}{\sqrt{1 + \frac{1}{b^2}(\mathbf{B}^2 - \mathbf{E}^2)}}. \tag{14.1.17}$$

In other words, the Born–Infeld electromagnetism introduces $\mathbf{E}$- and $\mathbf{B}$-dependent *dielectrics and permeability* "coefficients,"

$$\mathbf{D} = \varepsilon(\mathbf{E}, \mathbf{B})\mathbf{E}, \quad \mathbf{B} = \mu(\mathbf{E}, \mathbf{B})\mathbf{H}. \tag{14.1.18}$$

If there is an external current source, $(j^\mu) = (\rho, \mathbf{j})$, the equation (14.1.10) is replaced by

$$\partial_\mu P^{\mu\nu} = j^\nu \tag{14.1.19}$$

and equivalently, the equations in (14.1.15) become

$$-\frac{\partial \mathbf{D}}{\partial t} + \nabla \times \mathbf{H} = \mathbf{j}, \quad \nabla \cdot \mathbf{D} = \rho, \tag{14.1.20}$$

We now examine and solve the point charge problem mentioned at the beginning of the section.

Point charge problem

Consider the electrostatic field generated from a point particle of electric charge q placed at the origin. Then, $\mathbf{B} = \mathbf{0}$, $\mathbf{H} = \mathbf{0}$, and (14.1.20) becomes

$$\nabla \cdot \mathbf{D} = 4\pi q \delta(\mathbf{x}), \tag{14.1.21}$$

which can be solved to give us

$$\mathbf{D} = \frac{q\mathbf{x}}{|\mathbf{x}|^3}, \tag{14.1.22}$$

which is singular at the origin. However, from (14.1.16), we have

$$\mathbf{D} = \frac{\mathbf{E}}{\sqrt{1 - \frac{1}{b^2}\mathbf{E}^2}}, \tag{14.1.23}$$

which implies that

$$\begin{aligned}\mathbf{E} &= \frac{\mathbf{D}}{\sqrt{1+\frac{1}{b^2}\mathbf{D}^2}} \\ &= \frac{q\mathbf{x}}{|\mathbf{x}|\sqrt{|\mathbf{x}|^4+\left(\frac{q}{b}\right)^2}}.\end{aligned} \tag{14.1.24}$$

In particular, the electric field $\mathbf{E}$ is globally bounded. We see that $\nabla \times \mathbf{E} = \mathbf{0}$ for $\mathbf{x} \neq \mathbf{0}$, which is needed in (14.1.14).

Note that, when $|\mathbf{x}|$ is sufficiently large, $\mathbf{E}$ given in (14.1.24) approximates that given by the Coulomb law, a consequence of the Maxwell equations.

To compute the energy of the solution, we get from the Lagrange density (14.1.6) the energy-momentum tensor

$$\begin{aligned}T_{\mu\nu} &= 2\frac{\partial \mathcal{L}}{\partial \eta^{\mu\nu}} - \eta_{\mu\nu}\mathcal{L} \\ &= -\frac{F_{\mu\alpha}\eta^{\alpha\beta}F_{\nu\beta}}{\sqrt{1+\frac{1}{2b^2}F_{\alpha\beta}F^{\alpha\beta}}} - \eta_{\mu\nu}\mathcal{L}.\end{aligned} \tag{14.1.25}$$

Using (14.1.7) in (14.1.25), we see that the Hamiltonian energy density of the theory is

$$\mathcal{H} = T_{00} = b^2\left(\frac{1}{\sqrt{1-\frac{1}{b^2}(\mathbf{E}^2-\mathbf{B}^2)}} - 1\right) + \frac{\mathbf{B}^2}{\sqrt{1-\frac{1}{b^2}(\mathbf{E}^2-\mathbf{B}^2)}}, \tag{14.1.26}$$

which deviates away from the classical Maxwell energy density

$$\frac{1}{2}(\mathbf{E}^2+\mathbf{B}^2), \tag{14.1.27}$$

as it should, although we see that $\mathcal{H}$ given in (14.1.26) enjoys the property

$$\lim_{b\to\infty}\mathcal{H} = \frac{1}{2}(\mathbf{E}^2+\mathbf{B}^2). \tag{14.1.28}$$

Thus, in the electrostatic case, we have in particular

$$\begin{aligned}\mathcal{H} &= b^2\left(\frac{1}{\sqrt{1-\frac{1}{b^2}\mathbf{E}^2}} - 1\right) \\ &= b^2\left(\sqrt{1+\frac{1}{b^2}\mathbf{D}^2} - 1\right) \\ &= b^2\left(\sqrt{1+\left(\frac{q}{b}\right)^2\frac{1}{|\mathbf{x}|^4}} - 1\right),\end{aligned} \tag{14.1.29}$$

From (14.1.29), we see that the total energy of an electric point charge is now finite,

$$E = \int_{\mathbb{R}^3} \mathcal{H}\,\mathrm{d}\mathbf{x} = \int_{\mathbb{R}^3} b^2\left(\sqrt{1+\left(\frac{q}{b}\right)^2\frac{1}{|\mathbf{x}|^4}}-1\right)\mathrm{d}\mathbf{x} < \infty. \tag{14.1.30}$$

In other words, a point charge in the Born–Infeld theory is of a finite energy, which resolves the point charge problem. Furthermore, (14.1.30) shows that the energy of a point charge monotonically depends on the magnitude of the charge but is independent of its sign.

Similarly, we can consider the magnetostatic field generated from a *magnetic point charge* g placed at the origin of $\mathbb{R}^3$. In this case, $\mathbf{D}=\mathbf{0}$, $\mathbf{E}=\mathbf{0}$, and we are left with solving the equation

$$\nabla\cdot\mathbf{B} = 4\pi g\delta(\mathbf{x}). \tag{14.1.31}$$

From (14.1.31) and (14.1.17), we have as before,

$$\mathbf{B} = \frac{g\mathbf{x}}{|\mathbf{x}|^3}, \tag{14.1.32}$$

$$\mathbf{H} = \frac{g\mathbf{x}}{|\mathbf{x}|\sqrt{|\mathbf{x}|^4+\left(\frac{g}{b}\right)^2}}. \tag{14.1.33}$$

Thus, again $\mathbf{H}$ is a bounded vector field. In view of (14.1.26), the Hamiltonian density of a magnetostatic field takes the form,

$$\mathcal{H} = b^2\left(\sqrt{1+\frac{1}{b^2}\mathbf{B}^2}-1\right). \tag{14.1.34}$$

Inserting (14.1.32) into (14.1.34), we see that the total energy of a magnetic point charge is also finite in the Born–Infeld theory.

It is easy to extend the discussion to cover the situation of multiply distributed electric point charges or magnetic monopoles.

14.2 Some illustrative calculations

This section provides calculations aimed at gaining some concrete illustrative perspectives of the Born–Infeld theory about electric and magnetic point charges in static settings.

Uniformly charged ball

Although an electric point charge is not energetically permissible (subject to the Coulomb law), we may still construct a uniformly charged ball of radius $R>0$ and electric charge $q>0$ with a finite amount of energy. To see how, imagine that a ball, of radius $r > 0$ centered at the origin and constant electric charge density $\rho_0 > 0$, has been formed already, which carries the charge

$$Q(r) = \frac{4\pi r^3}{3}\rho_0. \tag{14.2.1}$$

Imagine also that we continue to build up the charge at the surface of the ball in such a way that the added charge, $\mathrm{d}Q$, is uniformly distributed in the spherical layer of thickness $\mathrm{d}r$ with the same charge density. Thus,

$$\mathrm{d}Q = (4\pi r^2 \mathrm{d}r)\rho_0. \tag{14.2.2}$$

Assume that this additional charge is brought to the surface of the ball from spatial infinity (Figure 14.1). Hence, the work needed to do so is

$$\begin{aligned} \mathrm{d}W &= \int_r^\infty \frac{Q\mathrm{d}Q}{\eta^2}\,\mathrm{d}\eta \\ &= \rho_0^2 \frac{(4\pi)^2 r^4}{3}\mathrm{d}r. \end{aligned} \tag{14.2.3}$$

Consequently, the total work needed to construct such a ball of radius R is

$$W = \int_0^R \rho_0^2 \frac{(4\pi)^2 r^4}{3}\mathrm{d}r = \rho_0^2 \frac{(4\pi)^2 R^5}{15}. \tag{14.2.4}$$

Inserting the total charge $q = \rho_0 \frac{4\pi R^3}{3}$ into (14.2.4), we obtain

$$W = \frac{3}{5}\frac{q^2}{R}, \tag{14.2.5}$$

which is the total energy of the charged ball.

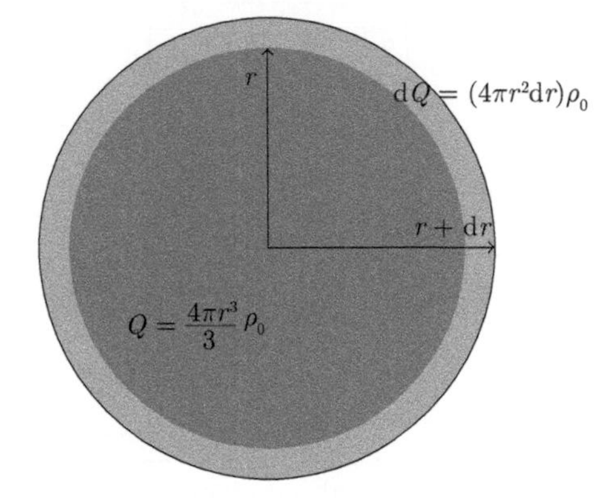

Figure 14.1 Illustration of the construction of a uniformly charged ball, which is carried out by accumulating an additional charge $\mathrm{d}Q$ to its surface layer of thickness $\mathrm{d}r$ and surface area $4\pi r^2$, brought from infinity, until its total charge reaches the desired amount, q.

We see from (14.2.5) that the energy blows up as $R \to 0$, which rules out the point-charge model for the electron.

If the electron were a uniformly charged ball, then we may use the data given in [75, 76] about the rest energy E_0 and electric charge e of the electron and identify E_0 with W in (14.2.5) with $q = e$ to obtain

$$r_0 = R = \frac{3}{5}\frac{e^2}{E_0} = \frac{3}{5 \times 1.2361} \times 2.28 \times 10^{-13} = 1.1067 \times 10^{-13} \text{ cm}, \quad (14.2.6)$$

as an estimate of the "*radius of the electron*" in the context of the Coulomb electrostatics.

Point charge in view of Born–Infeld electrostatics

To facilitate our calculation, we use the spherical coordinates (r, θ, ϕ) to represent $\mathbf{x} \in \mathbb{R}^3$. Hence, $\mathbf{x} = (r \sin\theta\cos\phi, r\sin\theta\sin\phi, r\cos\theta)$ and

$$\mathrm{d}\mathbf{x} = \mathbf{e}_r \mathrm{d}r + r\mathbf{e}_\theta \mathrm{d}\theta + r\sin\theta \mathbf{e}_\phi \mathrm{d}\phi, \quad (14.2.7)$$

where

$$\begin{aligned} \mathbf{e}_r &= (\sin\theta\cos\phi, \sin\theta\sin\phi, \cos\theta), \\ \mathbf{e}_\theta &= (\cos\theta\cos\phi, \cos\theta\sin\phi, -\sin\theta), \\ \mathbf{e}_\phi &= (-\sin\phi, \cos\phi, 0), \end{aligned} \quad (14.2.8)$$

form an ordered orthonormal moving frame at $\mathbf{x}$. We use this spherical representation for vector fields in general with the notation

$$\mathbf{A} = A^r \mathbf{e}_r + A^\theta \mathbf{e}_\theta + A^\phi \mathbf{e}_\phi = (A^r, A^\theta, A^\phi). \quad (14.2.9)$$

In view of (14.1.22), (14.1.24), and (14.2.9), we see that $\mathbf{D}$ and $\mathbf{E}$ only have nontrivial radial components, D^r and E^r, respectively, which are

$$D^r = \frac{q}{r^2}, \quad E^r = \frac{q}{\sqrt{r^4 + \left(\frac{q}{b}\right)^2}}. \quad (14.2.10)$$

So, the potential field, say ψ, associated with the electric field satisfying $\mathbf{E} = -\nabla\psi$ or

$$E^r = -\frac{\mathrm{d}\psi}{\mathrm{d}r}, \quad (14.2.11)$$

may be obtained to be

$$\psi(r) = \frac{q}{a} f\left(\frac{r}{a}\right), \quad f(x) = \int_x^\infty \frac{\mathrm{d}y}{\sqrt{y^4+1}}, \quad a = \sqrt{\frac{q}{b}}, \quad (14.2.12)$$

whose maximum, attained at $r = 0$, is given by

$$\psi(0) = \frac{1}{4} B\left(\frac{1}{4}, \frac{1}{4}\right) \frac{q}{a} \approx (1.85407)\, \frac{q}{a}, \quad (14.2.13)$$

where $B(x,y)$ is the beta function

$$B(x,y)=\int_0^1 t^{x-1}(1-t)^{y-1}\,\mathrm{d}t,\quad x,y>0. \tag{14.2.14}$$

In particular, the point charge has a finite potential energy as well at the spot where the charge resides.

The *free electric charge density*, ρ_{free}, induced from the electric field $\mathbf{E}$, and vice versa, is given by [76]:

$$\begin{aligned}\rho_{\text{free}}&=\frac{1}{4\pi}\nabla\cdot\mathbf{E}=\frac{1}{4\pi r^2}\frac{\mathrm{d}}{\mathrm{d}r}(r^2E^r)\\&=\frac{q}{2\pi a^3\,\frac{r}{a}\left(1+\left[\frac{r}{a}\right]^4\right)^{\frac{3}{2}}}.\end{aligned} \tag{14.2.15}$$

Note that such a charge density behaves like $\frac{1}{r}$ near $r=0$, like $\frac{1}{r^7}$ near infinity, and becomes insignificant beyond $r=a$ (Figure 14.2). For this reason, we may regard the free charge to be mainly distributed within the ball of radius a and refer to a as the *effective radius of the freely charged ball.* Using the first line in (14.2.15) and the asymptotic

$$E^r=\frac{q}{r^2}+\mathrm{O}\left(\frac{1}{r^6}\right),\quad r\to\infty, \tag{14.2.16}$$

as a result of (14.2.10), we get the total free electric charge

$$q_{\text{free}}=\int_{\mathbb{R}^3}\rho_{\text{free}}\,\mathrm{d}\mathbf{x}=\left(r^2E^r\right)_{r=0}^{r=\infty}=q, \tag{14.2.17}$$

which happens to coincide with the *prescribed* point charge q given in (14.1.21). The remarkable difference here is that the free charge is continuously distributed over space, but the prescribed charge is designated at a point.

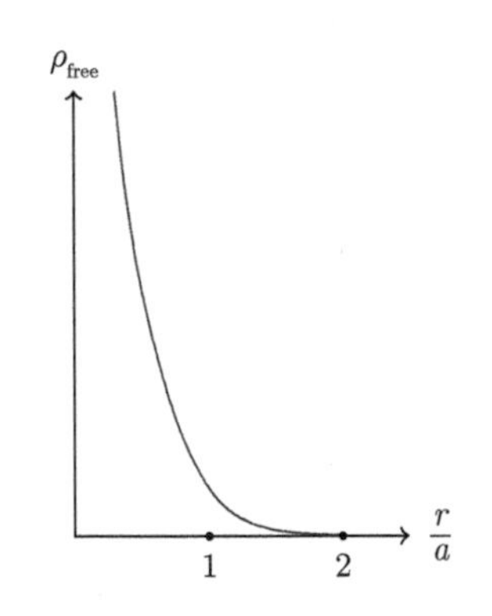

Figure 14.2 Distribution profile of the free charge density defined by the electric field generated from an electric point charge in the Born–Infeld theory, plotted against the scaled radial variable.

Furthermore, the expression (14.2.15) enables us to compute the free electric charge contained in the ball centered at the origin and of radius $r > 0$ explicitly as follows,

$$q_{\text{free}}(r) = \int_{|\mathbf{x}|\leq r} \rho_{\text{free}}\, \mathrm{d}\mathbf{x} = \frac{q}{\sqrt{1+\left(\frac{a}{r}\right)^4}}. \tag{14.2.18}$$

For example, we have

$$\begin{aligned} \frac{q_{\text{free}}(a)}{q} &= \frac{1}{\sqrt{2}} \approx 0.7071, \\ \frac{q_{\text{free}}(2a)}{q} &= \frac{4}{\sqrt{17}} \approx 0.97014, \\ \frac{q_{\text{free}}(3a)}{q} &= \frac{9}{\sqrt{82}} \approx 0.99388, \end{aligned} \tag{14.2.19}$$

showing a pattern of sharp concentration of charge.

From (14.1.30), the energy of the point charge is given by

$$\begin{aligned} \frac{E}{4\pi} &= \frac{q^2}{a}\int_0^\infty \frac{\mathrm{d}x}{\sqrt{x^4+1}+x^2} \\ &= \frac{\pi^{\frac{3}{2}}}{3\left(\Gamma\left(\frac{3}{4}\right)\right)^2}\frac{q^2}{a} \approx (1.2361)\,\frac{q^2}{a}. \end{aligned} \tag{14.2.20}$$

As in [76], when the left-hand side is replaced by the rest-mass energy and the charge q on the right-hand side by the electric charge of the electron, of the equation (14.2.20), respectively, the quantity a therein may be viewed as the "radius of the electron," denoted by r_0, which is calculated to be

$$r_0 = 2.28 \times 10^{-13}\ \text{cm}, \tag{14.2.21}$$

which is greater than that given in (14.2.6). In other words, the electron in the Born–Infeld theory has a bigger estimated "size" than that given in the Maxwell theory.

With (14.2.21) and $e = 4.803 \times 10^{-10}\ \text{cm}^{\frac{3}{2}}\text{g}^{\frac{1}{2}}\text{s}^{-1}$, we have

$$b = \frac{e}{r_0^2} \approx 9.24 \times 10^{15}\ \text{cm}^{-\frac{1}{2}}\text{g}^{\frac{1}{2}}\text{s}^{-1}, \tag{14.2.22}$$

as an estimate for the Born–Infeld parameter b, whose enormous magnitude suggests that the Maxwell theory may justifiably be regarded as a leading-order theory of that of Born–Infeld [75, 76].

Now, we explore how the energy of an electric point charge is distributed. For this purpose, we use (14.1.29) to compute the energy contained in the ball $|\mathbf{x}| \leq r$ as in (14.2.20):

$$E(r) = \int_{|\mathbf{x}|\leq r} \mathcal{H}\, \mathrm{d}\mathbf{x} = \frac{4\pi q^2}{a}\int_0^{\frac{r}{a}} \frac{\mathrm{d}x}{\sqrt{x^4+1}+x^2}. \tag{14.2.23}$$

As a result, we have, for example,

$$\frac{E(a)}{E} = 0.611703, \quad \frac{E(2a)}{E} = 0.798364, \quad \frac{E(3a)}{E} = 0.865245. \tag{14.2.24}$$

Therefore, in comparison with (14.2.19), we see that the energy is not as concentrated as the free charge of an electrically charged point particle in the Born–Infeld theory.

Magnetic point charge

For a magnetic point charge, the expressions (14.1.32), (14.1.33), and (14.1.34) replace their corresponding electric-field counterparts, (14.1.22), (14.1.24), and (14.1.29), respectively, with g replacing q. Thus, all the conclusions about an electric point charge are valid for a magnetic one. In particular, the total free magnetic charge is the same as the prescribed magnetic point charge,

$$g_{\text{free}} \equiv \frac{1}{4\pi} \int_{\mathbb{R}^3} \nabla \cdot \mathbf{H} \, d\mathbf{x} = g, \tag{14.2.25}$$

and the total energy of the magnetic charge is given by the formula (14.2.20) with setting $q = g$. In other words, as in the Maxwell theory, there is a perfect symmetry, qualitatively and quantitatively, between the electrostatic and magnetostatic descriptions of a point charge in the Born–Infeld theory.

14.3 Dyonic point charge

Section 3.5 presented the Schwinger dyons, which are point-source particles carrying both electric and magnetic charges. We saw there that the concept of a dyon is a direct extension of that of a magnetic monopole and gives rise to the elegant charge quantization formula of Schwinger for dyons [501], which generalizes the charge quantization formula of Dirac for monopoles [167]. However, like an electric or magnetic point charge, the Schwinger dyon carries an infinite energy, since it arises from the classical Maxwell electromagnetic theory. As a consequence of our earlier discussion on finite-energy electric and magnetic point charges, we explore whether a dyon in the Born–Infeld theory carries a finite energy as well.

Born–Infeld dyon

Thus, we now consider a dyonic point particle, residing at the origin, carrying both electric and magnetic charges, $q, g > 0$, whose electric displacement field $\mathbf{D}$ and magnetic field $\mathbf{B}$ are given by (14.1.22) and (14.1.32), respectively, or

$$D^r = \frac{q}{r^2}, \quad B^r = \frac{g}{r^2}, \quad r = |\mathbf{x}| > 0, \tag{14.3.1}$$

in terms of their nontrivial radial components. Inserting (14.3.1) into (14.1.16), we see that the nontrivial radial component of the electric field $\mathbf{E}$ is found to be

$$E^r = \frac{q}{r^2}\sqrt{\frac{r^4 + a_m^4}{r^4 + a_e^4}}, \quad a_e = \sqrt{\frac{q}{b}}, \quad a_m = \sqrt{\frac{g}{b}}, \tag{14.3.2}$$

giving rise to the asymptotics

$$E^r = \frac{g}{r^2} + \mathrm{O}(r^2) \quad \text{as } r \to 0; \quad E^r = \frac{q}{r^2} + \mathrm{O}\left(\frac{1}{r^6}\right) \quad \text{as } r \to \infty. \tag{14.3.3}$$

Thus, unlike in the electric point charge situation, the electric field E^r blows up at $r = 0$ as a consequence of the presence of the magnetic charge, in fact, it behaves like that of a magnetic point charge as expressed by B^r in (14.3.1), although it still obeys the Coulomb law near $r = \infty$. As in Section 14.2, we may calculate the free electric charge density by

$$\rho^e_{\text{free}} = \frac{1}{4\pi r^2}\frac{\mathrm{d}}{\mathrm{d}r}(r^2 E^r) = \frac{q(a_e^4 - a_m^4)\,r}{2\pi(r^4 + a_e^4)^2}\sqrt{\frac{r^4 + a_e^4}{r^4 + a_m^4}}. \tag{14.3.4}$$

So, we see that $\rho^e_{\text{free}} \sim r$ for $r \sim 0$, which is different from the behavior near an electric point charge, and $\rho^e_{\text{free}} \sim \frac{1}{r^7}$, which is the same as that of an electric charge. Besides, the free electric charge contained in the ball $\{|\mathbf{x}| \leq r\}$ is given in view of (14.3.4) and (14.3.2) by

$$q_{\text{free}}(r) = \int_{|\mathbf{x}|\leq r} \rho^e_{\text{free}}\,\mathrm{d}\mathbf{x} = r^2 E^r - (r^2 E^r)_{r=0} = q\sqrt{\frac{r^4 + a_m^4}{r^4 + a_e^4}} - g. \tag{14.3.5}$$

In particular, the total free electric charge is

$$q_{\text{free}} = q_{\text{free}}(\infty) = q - g, \tag{14.3.6}$$

which mixes the prescribed electric and magnetic charges. On the other hand, inserting B^r in (14.3.1) into (14.1.17), we get the nontrivial radial component of the magnetic intensity field $\mathbf{H}$ to be

$$H^r = \frac{g}{r^2}\sqrt{\frac{r^4 + a_e^4}{r^4 + a_m^4}}, \tag{14.3.7}$$

$$H^r = \frac{q}{r^2} + \mathrm{O}(r^2) \text{ as } r \to 0, \quad H^r = \frac{g}{r^2} + \mathrm{O}\left(\frac{1}{r^6}\right) \text{ as } r \to \infty, \tag{14.3.8}$$

which is completely analogous to (14.3.2). Thus, the associated magnetic intensity field H^r behaves like the electric displacement field D^r near $r = 0$ but resumes the behavior of the magnetic field B^r near infinity. The expressions (14.3.3) and (14.3.8) indicate that the dyonic point charge blends the electric and magnetic point charges symmetrically.

Furthermore, from (14.3.7), we obtain the free magnetic charge density, free magnetic charge contained in the ball $\{|\mathbf{x}| \leq r\}$, and total free magnetic charge,

$$\rho^m_{\text{free}} = \frac{g(a_m^4 - a_e^4)\,r}{2\pi(r^4 + a_m^4)^2}\sqrt{\frac{r^4 + a_m^4}{r^4 + a_e^4}}, \tag{14.3.9}$$

$$g_{\text{free}}(r) = g\sqrt{\frac{r^4 + a_e^4}{r^4 + a_m^4}} - q, \tag{14.3.10}$$

$$g_{\text{free}} = g_{\text{free}}(\infty) = g - q = -q_{\text{free}}, \tag{14.3.11}$$

respectively, such that q_{free} and g_{free} are mutually dependent quantities.

Dyon energy

To compute the energy of the dyon, we insert B^r and E^r given in (14.3.1) and (14.3.2) into (14.1.26) to get

$$\mathcal{H} = \frac{\left(\frac{qg}{b}\right)^2 + r^4(q^2 + g^2)}{r^4\left(r^4 + \sqrt{(r^4 + [\frac{q}{b}]^2)(r^4 + [\frac{g}{b}]^2)}\right)}. \tag{14.3.12}$$

Note that the two special cases of (14.3.12) we have already covered are

$$\mathcal{H}|_{g=0} = \frac{q^2}{r^2(\sqrt{r^4 + a_e^4} + r^2)}, \tag{14.3.13}$$

$$\mathcal{H}|_{q=0} = \frac{g^2}{r^2(\sqrt{r^4 + a_m^4} + r^2)}, \tag{14.3.14}$$

corresponding to the electric and magnetic point charge cases, respectively, are the *only* two situations when (14.3.12) gives rise to a finite energy, $\int_{\mathbb{R}^3} \mathcal{H}\,\mathrm{d}\mathbf{x} < \infty$. In other words, whenever $q \neq 0, g \neq 0$, the energy density (14.3.12) leads to the divergence of the total energy of the dyonic point charge.

Thus, although both electric and magnetic point charges are permitted in the Born–Infeld theory defined by the Lagrangian action density (14.1.6), dyonic point charges are not permitted energetically.

14.4 Formalism based on invariance

In [75, 76], Born and Infeld explored a nonlinear geometric theory of electromagnetism purely based on an invariance consideration, which elegantly extends the action density (14.1.6). Importantly, this model allows finite-energy dyonic point charges to exist, which do not exist in the first model of Born and Infeld based on a special-relativity action principle formalism introduced in Section 14.1.

Invariance consideration

We begin by considering an action functional

$$I = \int \mathcal{L}\,\mathrm{d}x \tag{14.4.1}$$

over the Minkowski spacetime $(\mathbb{R}^{3,1}, \eta_{\mu\nu})$. It is natural to postulate that the action is invariant with respect to change of coordinates. Thus, a simple and direct choice is to take

$$\mathcal{L} = \gamma\sqrt{|\det(a_{\mu\nu})|}, \tag{14.4.2}$$

where γ is a constant and $a_{\mu\nu}$ a certain covariant 2-tensor field, which may be decomposed as a sum of a symmetric and skewsymmetric parts, say $s_{\mu\nu}$ and $f_{\mu\nu}$ for the moment. Motivated by simplicity, we may choose $s_{\mu\nu} = \eta_{\mu\nu}$ and $f_{\mu\nu}$ to be in proportion of the electromagnetic field (14.1.4). That is, we set

$$a_{\mu\nu} = \eta_{\mu\nu} + f_{\mu\nu}, \quad f_{\mu\nu} = \kappa F_{\mu\nu}. \tag{14.4.3}$$

Besides, in order to maintain finiteness of the action, we need to subtract the "background" from (14.4.2), which pops up when $f_{\mu\nu}$ vanishes such that we modify (14.4.2) into

$$\mathcal{L} = \gamma(\sqrt{|\det(a_{\mu\nu})|} - \sqrt{|\det(\eta_{\mu\nu})|}) = \gamma(\sqrt{|\det(a_{\mu\nu})|} - 1). \tag{14.4.4}$$

For convenience of reference, from (14.1.7), here we write down

$$(F_{\mu\nu}) = \begin{pmatrix} 0 & E^1 & E^2 & E^3 \\ -E^1 & 0 & -B^3 & B^2 \\ -E^2 & B^3 & 0 & -B^1 \\ -E^3 & -B^2 & B^1 & 0 \end{pmatrix}, \tag{14.4.5}$$

since $F_{0i} = -F^{0i}, F_{ij} = F^{ij}$.

On the other hand, using the skewsymmetry of $f_{\mu\nu}$, (14.4.3), and (14.4.5) , we have the expansion

$$\begin{aligned} \det(a_{\mu\nu}) &= \det(\eta_{\mu\nu} + f_{\mu\nu}) \\ &= -1 + (f_{01}f_{23} - f_{02}f_{13} + f_{03}f_{12})^2 \\ &\quad + (f_{01}^2 + f_{02}^2 + f_{03}^2 - f_{12}^2 - f_{13}^2 - f_{23}^2) \\ &= -1 + \kappa^2(\mathbf{E}^2 - \mathbf{B}^2) + \kappa^4(\mathbf{E}\cdot\mathbf{B})^2. \end{aligned} \tag{14.4.6}$$

Inserting (14.4.6) into (14.4.4), we arrive at

$$\mathcal{L} = -\gamma\left(1 - \sqrt{1 - \kappa^2(\mathbf{E}^2 - \mathbf{B}^2) - \kappa^4(\mathbf{E}\cdot\mathbf{B})^2}\right), \tag{14.4.7}$$

where the choice of sign under the radical root is made in order to achieve consistency in the weak electromagnetic field limit. Consequently, comparing (14.4.7) with (14.1.3), we are led to the recognition

$$\kappa = \frac{1}{b}, \quad \gamma = -b^2, \tag{14.4.8}$$

such that (14.1.3) appears like a leading-order truncated theory model of (14.4.7) when b is large.

Born–Infeld action density

In view of (14.4.8), we have derived the *second Born–Infeld action density* to be

$$\mathcal{L} = b^2\left(1-\sqrt{1-\frac{1}{b^2}(\mathbf{E}^2-\mathbf{B}^2)-\frac{1}{b^4}(\mathbf{E}\cdot\mathbf{B})^2}\right)$$
$$= b^2\left(1-\sqrt{-\det\left[\eta_{\mu\nu}+\frac{1}{b}F_{\mu\nu}\right]}\right). \tag{14.4.9}$$

From (14.1.8) and (14.4.5), we have

$$F_{\mu\nu}\tilde{F}^{\mu\nu} = -4\,\mathbf{E}\cdot\mathbf{B}. \tag{14.4.10}$$

Hence, by virtue of (14.4.9) and (14.4.10), we obtain

$$\mathcal{L} = b^2\left(1-\sqrt{1+\frac{1}{2b^2}F_{\mu\nu}F^{\mu\nu}-\frac{1}{16b^4}\left(F_{\mu\nu}\tilde{F}^{\mu\nu}\right)^2}\right), \tag{14.4.11}$$

which exhibits from a different view angle how the extended model (14.4.9) deviates from the minimal one, (14.1.6).

Equations of motion

Varying A_ν (say) in (14.4.11) leads to the associated Euler–Lagrange equations

$$\partial_\mu P^{\mu\nu} = 0, \tag{14.4.12}$$

$$P^{\mu\nu} = \frac{\left(F^{\mu\nu}-\frac{1}{4b^2}[F_{\alpha\beta}\tilde{F}^{\alpha\beta}]\tilde{F}^{\mu\nu}\right)}{\sqrt{1+\frac{1}{2b^2}F_{\alpha\beta}F^{\alpha\beta}-\frac{1}{16b^4}\left(F_{\alpha\beta}\tilde{F}^{\alpha\beta}\right)^2}}. \tag{14.4.13}$$

Using the assignment (14.1.13) to obtain the electric displacement field $\mathbf{D}$ and magnetic intensity field $\mathbf{H}$ as before, we see that the Bianchi identity (14.1.9) and the Euler–Lagrange equations (14.4.12)–(14.4.13) are expressed as (14.1.14)–(14.1.15) for which

$$\mathbf{D} = \frac{\mathbf{E}+\frac{1}{b^2}(\mathbf{E}\cdot\mathbf{B})\,\mathbf{B}}{\sqrt{1-\frac{1}{b^2}(\mathbf{E}^2-\mathbf{B}^2)-\frac{1}{b^4}(\mathbf{E}\cdot\mathbf{B})^2}}, \tag{14.4.14}$$

$$\mathbf{H} = \frac{\mathbf{B}-\frac{1}{b^2}(\mathbf{E}\cdot\mathbf{B})\,\mathbf{E}}{\sqrt{1-\frac{1}{b^2}(\mathbf{E}^2-\mathbf{B}^2)-\frac{1}{b^4}(\mathbf{E}\cdot\mathbf{B})^2}}, \tag{14.4.15}$$

in view of (14.1.7), (14.1.8), and (14.4.10), which blend the electric and magnetic fields, $\mathbf{E}$ and $\mathbf{B}$. In either the electrostatic or magnetostatic situation, $\mathbf{B} = \mathbf{0}$ or $\mathbf{E} = \mathbf{0}$, or more generally, when $\mathbf{E} \cdot \mathbf{B} = 0$, the problem is identical to that discussed in the previous section. Thus, we have the same solution to the point electric and magnetic charge problems.

Note on relationship

Note that, in (14.4.6), we have

$$\begin{aligned} \det(f_{\mu\nu}) &= (f_{01}f_{23} - f_{02}f_{13} + f_{03}f_{12})^2 \\ &= \kappa^4 (\mathbf{E}\cdot\mathbf{B})^2. \end{aligned} \tag{14.4.16}$$

Therefore, in view of (14.4.8)–(14.4.9), we obtain

$$1 + \frac{1}{2b^2} F_{\mu\nu}F^{\mu\nu} = -\det\left(\eta_{\mu\nu} + \frac{1}{b}F_{\mu\nu}\right) + \det\left(\frac{1}{b}F_{\mu\nu}\right), \tag{14.4.17}$$

which relates (14.1.6) to (14.4.9), as noted in [76]. In fact, the theory of electromagnetism based on the action density (14.1.6) was originally formulated by Born himself in [73, 74], which is sometimes specifically referred to as the *Born theory.*

Dyonic point charge

Section 14.3 showed that the model (14.1.6) does not allow the existence of a finite-energy dyonic static point charge. Here, we show that such a point charge is allowed in the model (14.4.9).

To proceed, we first note that some suitable manipulation of (14.4.14) gives rise to the equations

$$\begin{aligned} &\left(1 + \frac{\mathbf{D}^2}{b^2}\right)\mathbf{E}^2 + \frac{1}{b^2}\left(2 + \frac{\mathbf{D}^2}{b^2} + \frac{\mathbf{B}^2}{b^2}\right)(\mathbf{E}\cdot\mathbf{B})^2 \\ &= \mathbf{D}^2\left(1 + \frac{\mathbf{B}^2}{b^2}\right), \end{aligned} \tag{14.4.18}$$

$$\begin{aligned} &\frac{(\mathbf{B}\cdot\mathbf{D})^2}{b^2}\mathbf{E}^2 + \left(\left[1 + \frac{\mathbf{B}^2}{b^2}\right]^2 + \frac{(\mathbf{B}\cdot\mathbf{D})^2}{b^4}\right)(\mathbf{E}\cdot\mathbf{B})^2 \\ &= (\mathbf{B}\cdot\mathbf{D})^2\left(1 + \frac{\mathbf{B}^2}{b^2}\right), \end{aligned} \tag{14.4.19}$$

with the solution

$$\mathbf{E}^2 = \frac{(\mathbf{B}^2 + 2b^2)(\mathbf{B}\times\mathbf{D})^2 + \mathbf{D}^2 b^4}{b^2(\mathbf{B}^2 + \mathbf{D}^2 + b^2) + (\mathbf{B}\times\mathbf{D})^2}, \tag{14.4.20}$$

$$(\mathbf{E}\cdot\mathbf{B})^2 = \frac{(\mathbf{B}\cdot\mathbf{D})^2 b^4}{b^2(\mathbf{B}^2 + \mathbf{D}^2 + b^2) + (\mathbf{B}\times\mathbf{D})^2}, \tag{14.4.21}$$

where we have used the vector identity

$$(\mathbf{B}\times\mathbf{D})^2 = \mathbf{B}^2\mathbf{D}^2 - (\mathbf{B}\cdot\mathbf{D})^2. \tag{14.4.22}$$

Inserting (14.3.1) into (14.4.20)–(14.4.21) and using the notation for a_e, a_m in (14.3.2), we obtain

$$\mathbf{E}^2 = \frac{q^2}{a_e^4 + a_m^4 + r^4}, \quad (\mathbf{E}\cdot\mathbf{B})^2 = \frac{q^2 g^2}{r^4(a_e^4 + a_m^4 + r^4)}, \tag{14.4.23}$$

which gives us the nontrivial radial component of $\mathbf{E}$ to be

$$E^r = \frac{q}{\sqrt{a_e^4 + a_m^4 + r^4}}, \tag{14.4.24}$$

which remains finite at $r = 0$ and is in sharp contrast with (14.3.2), although E^r still obeys the Coulomb law, $E^r \sim \frac{q}{r^2}$, for r near infinity. Besides, the free electric charge density reads

$$\rho^e_{\text{free}} = \frac{1}{4\pi r^2}\frac{\mathrm{d}}{\mathrm{d}r}(r^2 E^r) = \frac{q(a_e^4 + a_m^4)}{2\pi r(a_e^4 + a_m^4 + r^4)^{\frac{3}{2}}}, \tag{14.4.25}$$

such that $\rho^e_{\text{free}} \sim \frac{1}{r}$ and $\rho^e_{\text{free}} \sim \frac{1}{r^7}$ for r near zero and infinity, respectively. Furthermore, from (14.4.24), the free electric charge contained in $\{|\mathbf{x}| \le r\}$ is

$$q_{\text{free}}(r) = (r^2 E^r)_{r=0}^{r=r} = \frac{q r^2}{\sqrt{a_e^4 + a_m^4 + r^4}}, \tag{14.4.26}$$

such that the total free electric charge coincides with the prescribed one, $q_{\text{free}} = q_{\text{free}}(\infty) = q$.

On the other hand, using (14.3.1), (14.4.23), and (14.4.24) in (14.4.15), we get the nontrivial radial component of the magnetic intensity field

$$H^r = \frac{g}{\sqrt{a_e^4 + a_m^4 + r^4}}, \tag{14.4.27}$$

which is elegantly parallel to the expression for the electric field stated in (14.4.24). Consequently, the same conclusions drawn for the associated free magnetic charge density and charge holds. In particular, the total free magnetic charge coincides with the prescribed magnetic charge, $g_{\text{free}} = g$.

We now consider the energy carried by a dyonic point charge.

Similar to (14.1.25), we see that the energy-momentum tensor associated with (14.4.11) is

$$T_{\mu\nu} = -\frac{\left(F_{\mu\alpha}\eta^{\alpha\beta}F_{\nu\beta} - \frac{1}{4b^2}(F_{\mu'\nu'}\tilde{F}^{\mu'\nu'})F_{\mu\alpha}\eta^{\alpha\beta}\tilde{F}_{\nu\beta}\right)}{\sqrt{1 + \frac{1}{2b^2}F_{\mu\nu}F^{\mu\nu} - \frac{1}{16b^4}\left(F_{\mu\nu}\tilde{F}^{\mu\nu}\right)^2}} - \eta_{\mu\nu}\mathcal{L}. \tag{14.4.28}$$

Inserting (14.1.7), (14.1.8), (14.4.9), and (14.4.10) into (14.4.28), we can write the energy density as

$$\mathcal{H} = T_{00} = \frac{\mathbf{E}^2 + \frac{1}{b^2}(\mathbf{E}\cdot\mathbf{B})^2}{\mathcal{R}} - b^2(1-\mathcal{R})$$
$$= b^2\left(\frac{1}{\mathcal{R}} - 1\right) + \frac{\mathbf{B}^2}{\mathcal{R}}, \tag{14.4.29}$$
$$\mathcal{R} = \sqrt{1 - \frac{1}{b^2}(\mathbf{E}^2 - \mathbf{B}^2) - \frac{1}{b^4}(\mathbf{E}\cdot\mathbf{B})^2}. \tag{14.4.30}$$

Comparing (14.4.29) with (14.1.26) shows the similarity and difference of the two expressions.

Now, substituting (14.4.20)–(14.4.21) into (14.4.29)–(14.4.30), we arrive at

$$\mathcal{H} = b^2\left(\sqrt{1 + \frac{1}{b^2}(\mathbf{D}^2 + \mathbf{B}^2) + \frac{1}{b^4}(\mathbf{D}\times\mathbf{B})^2} - 1\right). \tag{14.4.31}$$

Inserting (14.3.1) into (14.4.31), we have

$$\mathcal{H} = b^2\left(\sqrt{1 + \frac{a_e^4 + a_m^4}{r^4}} - 1\right), \tag{14.4.32}$$

which renders the determination of the total energy of the dyonic point charge by

$$\frac{E}{4\pi} = \int_0^\infty r^2\mathcal{H}\,\mathrm{d}r = a^3b^2\int_0^\infty \frac{\mathrm{d}x}{\sqrt{x^4+1}+x^2} \quad (a^4 = a_e^4 + a_m^4)$$
$$= \frac{\pi^{\frac{3}{2}}}{3\left(\Gamma\left(\frac{3}{4}\right)\right)^2}\frac{Q^2}{a}, \tag{14.4.33}$$

where we set

$$Q = \sqrt{q^2 + g^2}, \tag{14.4.34}$$

to be viewed as a *combined dyonic charge* taking account of both electric and magnetic charges, q and g, jointly, residing at the same spot. The quantity a defined in (14.4.33) appears to be the "radius" of the dyonic point charge. Alternatively, if we use

$$b^2 = \frac{qg}{a_e^2 a_m^2} \tag{14.4.35}$$

in (14.4.33) instead, when $q, g > 0$, we obtain

$$\frac{E}{4\pi} = \frac{\pi^{\frac{3}{2}}}{3\left(\Gamma\left(\frac{3}{4}\right)\right)^2}\frac{Q_1^2}{a_1}, \quad Q_1 = \sqrt{qg}, \quad a_1 = \frac{a_e^2 a_m^2}{(a_e^4 + a_m^4)^{\frac{3}{4}}}, \tag{14.4.36}$$

where Q_1 and a_1 provide a different pair of combined charge and radius of the dyonic point charge.

Note that both (14.4.33) and (14.4.36) are of the same form as that for an electric point charge, (14.2.20). In other words, in terms of the combined charge defined in either (14.4.34) or (14.4.36), a dyonic point charge behaves like an electric point charge, energetically. Furthermore, the description of a dyonic point charge is symmetric with respect to the interchange of the roles of the electric and magnetic charges. Besides, the expression (14.4.33) indicates that a dyonic point charge possesses a larger radius as a result of the need to accommodate a greater joint charge.

14.5 Generalized Bernstein problem

As a nonlinear theory of electrodynamics, we explore whether there is self-sustained electromagnetism in the Born–Infeld theory as that in the Yang–Mills gauge field theory, in which the non-Abelian structure in the formalism serves to provide self-induced sources responsible for the existence of monopoles and dyons, for example, the Polyakov–'t Hooft monopole and Julia–Zee dyon (Section 9.3). Mathematically, this problem amounts to whether the source-free static Born–Infeld equations possess nontrivial solutions in a full space setting, a problem relating itself to the classical Bernstein problem in the study of the minimal surface equations [68, 392, 425, 426].

To proceed, we return to the Born–Infeld system consisting of the equations (14.1.14)–(14.1.15) and study its static solutions. Thus, (14.1.14) gives us

$$\nabla \times \mathbf{E} = \mathbf{0}, \quad \nabla \cdot \mathbf{B} = 0, \tag{14.5.1}$$

which leads to

$$\mathbf{E} = \nabla \phi, \quad \mathbf{B} = \nabla \times \mathbf{A}, \tag{14.5.2}$$

for a real-valued scalar field ϕ and a vector gauge field $\mathbf{A}$, both being time-independent.

Static equations

In terms of (14.5.2), the sourceless equations (14.1.14) and (14.1.15) become

$$\nabla \cdot \left(\frac{\nabla \phi}{\sqrt{1 + \frac{1}{b^2}(|\nabla \times \mathbf{A}|^2 - |\nabla \phi|^2)}} \right) = 0, \tag{14.5.3}$$

$$\nabla \times \left(\frac{\nabla \times \mathbf{A}}{\sqrt{1 + \frac{1}{b^2}(|\nabla \times \mathbf{A}|^2 - |\nabla \phi|^2)}} \right) = \mathbf{0}. \tag{14.5.4}$$

Reduction to scalar field equations

From (14.5.4), we see that there is a real scalar function ψ such that

$$\frac{\nabla \times \mathbf{A}}{\sqrt{1 + \frac{1}{b^2}(|\nabla \times \mathbf{A}|^2 - |\nabla \phi|^2)}} = \nabla \psi, \tag{14.5.5}$$

which leads us to the relation

$$|\nabla \times \mathbf{A}|^2 = \left(1 - \frac{1}{b^2}|\nabla\phi|^2\right)\frac{|\nabla\psi|^2}{\left(1 - \frac{1}{b^2}|\nabla\psi|^2\right)}. \tag{14.5.6}$$

Thus, we have

$$1 + \frac{1}{b^2}\left(|\nabla \times \mathbf{A}|^2 - |\nabla\phi|^2\right) = \frac{1 - \frac{1}{b^2}|\nabla\phi|^2}{1 - \frac{1}{b^2}|\nabla\psi|^2}. \tag{14.5.7}$$

Inserting (14.5.7) into (14.5.5), we obtain

$$\nabla \times \mathbf{A} = \nabla\psi\frac{\sqrt{1 - \frac{1}{b^2}|\nabla\phi|^2}}{\sqrt{1 - \frac{1}{b^2}|\nabla\psi|^2}}. \tag{14.5.8}$$

In view of (14.5.7) and (14.5.8), we see that the static Born–Infeld equations (14.5.3) and (14.5.4) are equivalent to the following coupled system of two scalar equations,

$$\nabla \cdot \left(\nabla\phi\frac{\sqrt{1 - \frac{1}{b^2}|\nabla\psi|^2}}{\sqrt{1 - \frac{1}{b^2}|\nabla\phi|^2}}\right) = 0, \tag{14.5.9}$$

$$\nabla \cdot \left(\nabla\psi\frac{\sqrt{1 - \frac{1}{b^2}|\nabla\phi|^2}}{\sqrt{1 - \frac{1}{b^2}|\nabla\psi|^2}}\right) = 0, \tag{14.5.10}$$

which are the the Euler–Lagrange equations of the action functional

$$\mathcal{A}(\phi,\psi) = \int_{\mathbb{R}^3}\left(1 - \sqrt{1 - \frac{1}{b^2}|\nabla\phi|^2}\sqrt{1 - \frac{1}{b^2}|\nabla\psi|^2}\right)\mathrm{d}x. \tag{14.5.11}$$

In the special situation when ϕ or ψ is a constant, we have $\mathbf{E} = \mathbf{0}$ or $\mathbf{B} = \mathbf{0}$, corresponding to the magnetostatic or electrostatic limit, and the system reduces into the single equation

$$\nabla \cdot \left(\frac{\nabla f}{\sqrt{1 - \frac{1}{b^2}|\nabla f|^2}}\right) = 0, \tag{14.5.12}$$

governing maximal hypersurfaces in the Minkowski spacetime [99, 130] as encountered earlier in the normalized form (13.3.13).

Generalized Bernstein problem

It is of interest to ask whether the Bernstein property holds for (14.5.9) and (14.5.10), over $\mathbb{R}^n$, or whether any entire solution (ϕ, ψ) of these equations over

$\mathbb{R}^n$ must be such that ϕ and ψ are affine linear functions. However, we see that an entire solution of (14.5.9) and (14.5.10) is not necessarily affine linear. In fact, a counterexample is obtained by setting $\phi = \psi = h$ in these equations where h is an arbitrary harmonic function over the full space. Thus, we have yet to identify the most general trivial solutions of these coupled equations.

Extending Calabi's equivalence theorem [99] about the minimal and maximal surface equations (Section 13.3), we see that the coupled system of equations, (14.5.9) and (14.5.10), is equivalent to the system of equations

$$\nabla\cdot\left(\nabla f\frac{\sqrt{1+|\nabla g|^2}}{\sqrt{1+|\nabla f|^2}}\right)=0, \tag{14.5.13}$$

$$\nabla\cdot\left(\nabla g\frac{\sqrt{1+|\nabla f|^2}}{\sqrt{1+|\nabla g|^2}}\right)=0, \tag{14.5.14}$$

over $\mathbb{R}^2$ (cf. Exercise 15), which are seen to be the Euler–Lagrange equations of the action functional

$$\mathcal{A}(f,g)=\int_{\mathbb{R}^2}\left(\sqrt{1+|\nabla f|^2}\sqrt{1+|\nabla g|^2}-1\right)\,\mathrm{d}x. \tag{14.5.15}$$

A warm-up question asks whether the solutions of (14.5.13)–(14.5.14) of finite action, $\mathcal{A}(f,g)<\infty$, are constant. See [511] for some further discussion.

It will be of interest to express the Lagrangian action density (14.1.3) and the Hamiltonian energy density (14.1.26) of the Born–Infeld theory in terms of the real scalar fields ϕ and ψ over $\mathbb{R}^3$ directly. In fact, with (14.5.2) and (14.5.5), we have (14.5.7) so that (14.1.3) becomes

$$\mathcal{L}=b^2\left(1-\sqrt{\frac{1-\frac{1}{b^2}|\nabla\phi|^2}{1-\frac{1}{b^2}|\nabla\psi|^2}}\right), \tag{14.5.16}$$

which, unlike (14.5.11), is not even symmetric with respect to ϕ and ψ. Moreover, in terms of ϕ and ψ again, we may rewrite (14.1.26) as

$$\begin{aligned}\mathcal{H}&=\frac{\mathbf{E}^2}{\sqrt{1-\frac{1}{b^2}(\mathbf{E}^2-\mathbf{B}^2)}}-\mathcal{L}\\&=b^2\left(\frac{1-\frac{1}{b^4}|\nabla\phi|^2|\nabla\psi|^2}{\sqrt{1-\frac{1}{b^2}|\nabla\phi|^2}\sqrt{1-\frac{1}{b^2}|\nabla\psi|^2}}-1\right),\end{aligned} \tag{14.5.17}$$

which is symmetric with respect to ϕ and ψ. We see that the right-hand side of (14.5.17) stays nonnegative (cf. Exercise 18).

Static field equations with sources

In the presence of an external source in the form of an electric or magnetic charge density distribution function, say ρ, the equation (14.5.12) becomes

$$\nabla \cdot \left(\frac{\nabla f}{\sqrt{1 - \frac{1}{b^2}|\nabla f|^2}} \right) = \rho. \tag{14.5.18}$$

See [70, 71, 72, 329, 330] for some analytic studies of this problem over $\mathbb{R}^n$ for $n \geq 3$. The work [612] considers a coupled system of the electrostatic Born–Infeld equation and a nonlinear Klein–Gordon equation. Here, we note that, when both electricity and magnetism are present, the full static equations of the Born–Infeld model (14.1.3) with charge density sources but free of current sources are

$$\nabla \cdot \left(\nabla\phi \frac{\sqrt{1 - \frac{1}{b^2}|\nabla\psi|^2}}{\sqrt{1 - \frac{1}{b^2}|\nabla\phi|^2}} \right) = \rho_e, \tag{14.5.19}$$

$$\nabla \cdot \left(\nabla\psi \frac{\sqrt{1 - \frac{1}{b^2}|\nabla\phi|^2}}{\sqrt{1 - \frac{1}{b^2}|\nabla\psi|^2}} \right) = \rho_m, \tag{14.5.20}$$

which are the Euler–Lagrange equations of the action functional

$$\mathcal{A}(\phi,\psi) = \int \left(b^2 \left[1 - \sqrt{1 - \frac{1}{b^2}|\nabla\phi|^2}\sqrt{1 - \frac{1}{b^2}|\nabla\psi|^2} \right] + \rho_e\phi + \rho_m\psi \right) \mathrm{d}x, \tag{14.5.21}$$

and may be derived from the static Born–Infeld equations

$$\nabla \cdot \mathbf{D} = \rho_e, \quad \nabla \cdot \mathbf{B} = \rho_m, \quad \nabla \times \mathbf{E} = \mathbf{0}, \quad \nabla \times \mathbf{H} = \mathbf{0}, \tag{14.5.22}$$

where ρ_e and ρ_m are time-independent functions representing electric and magnetic charge density distributions in space, respectively.

In fact, the relation $\mathbf{E} = \nabla\phi$ for some real-valued function ϕ is still valid, since $\mathbf{E}$ remains conservative. However, $\mathbf{B}$ is not guaranteed to have a vector potential since $\mathbf{B}$ may fail to be solenoidal. Nevertheless, since $\mathbf{H}$ is conservative, we have in view of (14.1.17) the relation

$$\frac{\mathbf{B}}{\sqrt{1 + \frac{1}{b^2}(\mathbf{B}^2 - \mathbf{E}^2)}} = \mathbf{H} = \nabla\psi, \tag{14.5.23}$$

for some real-valued function ψ again, which leads to

$$1 + \frac{1}{b^2}\left(\mathbf{B}^2 - \mathbf{E}^2\right) = 1 + \frac{1}{b^2}\left(\mathbf{B}^2 - |\nabla\phi|^2\right) = \frac{1 - \frac{1}{b^2}|\nabla\phi|^2}{1 - \frac{1}{b^2}|\nabla\psi|^2}, \tag{14.5.24}$$

replacing (14.5.7). Thus, substituting (14.5.24) into (14.1.16) and using the first relation in (14.5.22), we get (14.5.19), and substituting (14.5.24) into (14.5.23) and using the second relation in (14.5.22), we arrive at (14.5.20).

In the special situation where the dyonic charge distributions in (14.5.22) are of the type of point charge sources, $\rho_e = 4\pi q\delta(\mathbf{x}), \rho_m = 4\pi g\delta(\mathbf{x})$, we know that the electric field $\mathbf{E}$ is radial and given by (14.3.2), and the magnetic intensity field $\mathbf{H}$ is also radial and given by (14.3.7). As a consequence, the system of the equations (14.5.19) and (14.5.20) has the following radially symmetric solution

$$\phi = \phi(r) = -\int_r^\infty \frac{q}{\rho^2}\sqrt{\frac{\rho^4 + a_m^4}{\rho^4 + a_e^4}}\,\mathrm{d}\rho, \tag{14.5.25}$$

$$\psi = \psi(r) = -\int_r^\infty \frac{g}{\rho^2}\sqrt{\frac{\rho^4 + a_e^4}{\rho^4 + a_m^4}}\,\mathrm{d}\rho, \tag{14.5.26}$$

since the scalar potentials ϕ and ψ now satisfy $E^r = \phi'(r)$ and $H^r = \psi'(r)$.

Section 14.15 considers the static equations for dyons and associated point charge solutions in the most general setting, and where the equations (14.5.19)–(14.5.20) and the solution (14.5.25)–(14.5.26) rise as limiting situations.

14.6 Born–Infeld term and virial identities

Chapter 12 showed how added nonlinearity of a field-theoretical model inevitably leads to an adjusted virial identity, which often relaxes the usual dimensionality constraint imposed by the Derrick theorem on a conventional setting. As a nonlinear theory of electrodynamics, the Born–Infeld formalism naturally offers us correspondingly adjusted virial identities within its various settings as well, which consequently imply different dimensionality constraints or relaxations. As an illustration, this section considers the Abelian Higgs model case.

Born–Infeld modified electromagnetism

Consider a modified Abelian Higgs model for which the gauge field dynamical term is governed by the Born–Infeld electromagnetism such that the full action density reads

$$\mathcal{L} = b^2\left(1 - \sqrt{1 + \frac{1}{2b^2}F_{\mu\nu}F^{\mu\nu}}\right) + \frac{1}{2}D_\mu\phi\overline{D^\mu\phi} - V(|\phi|^2), \tag{14.6.1}$$

where ϕ is a complex-valued scalar field, $D_\mu\phi = \partial_\mu - \mathrm{i}A_\mu\phi$ the gauge-covariant derivative, and $V \geq 0$ a potential density function. For generality, the spacetime is taken to be $\mathbb{R}^{n,1}$.

We are to derive a virial identity of the model (14.6.1) for static solutions in the temporal gauge, $A_0 = 0$. In this situation, the Hamiltonian density of

(14.6.1) may be calculated to be

$$\mathcal{H} = b^2 \left(\sqrt{1 + \frac{1}{2b^2} F_{ij}^2} - 1 \right) + \frac{1}{2} |D_i \phi|^2 + V(|\phi|^2). \tag{14.6.2}$$

By exploring the rescaled fields, $\phi^\sigma(x) = \phi(\sigma x), A_i^\sigma(x) = \sigma A_i(\sigma x)$, as before, we arrive at for a critical point of the energy the anticipated virial identity

$$\int_{\mathbb{R}^n} \frac{F_{ij}^2}{\sqrt{1 + \frac{1}{2b^2} F_{kl}^2}} \, \mathrm{d}x + \left(1 - \frac{n}{2}\right) \int_{\mathbb{R}^n} |D_i \phi|^2 \, \mathrm{d}x$$

$$= n \int_{\mathbb{R}^n} \left(b^2 \left[\sqrt{1 + \frac{1}{2b^2} F_{ij}^2} - 1 \right] + V(|\phi|^2) \right) \mathrm{d}x, \tag{14.6.3}$$

which unlike (12.1.26) presents no restriction to n, although we may recover (12.1.26) by taking the $b \to \infty$ limit in (14.6.3).

Born–Infeld modified dynamics for scalar field

We may consider a similarly modified Abelian Higgs model for which the gauge field dynamics is the usual Maxwell type, but that of the scalar field instead follows the Born–Infeld formalism such that the action density assumes the form

$$\mathcal{L} = -\frac{1}{4} F_{\mu\nu} F^{\mu\nu} + b^2 \left(1 - \sqrt{1 - \frac{1}{b^2} D_\mu \phi \overline{D^\mu \phi}} \right) - V(|\phi|^2). \tag{14.6.4}$$

Focus again on the static situation within the temporal gauge. Then, the Hamiltonian of (14.6.4) is

$$\mathcal{H} = \frac{1}{4} F_{ij}^2 + b^2 \left(\sqrt{1 + \frac{1}{b^2} |D_i \phi|^2} - 1 \right) + V(|\phi|^2). \tag{14.6.5}$$

As a consequence, the associated virial identity now reads

$$\left(1 - \frac{n}{4}\right) \int_{\mathbb{R}^n} F_{ij}^2 \, \mathrm{d}x + \int_{\mathbb{R}^n} \frac{|D_i \phi|^2}{\sqrt{1 + \frac{1}{b^2} |D_j \phi|^2}} \, \mathrm{d}x$$

$$= n \int_{\mathbb{R}^n} \left(b^2 \left[\sqrt{1 + \frac{1}{b^2} |D_i \phi|^2} - 1 \right] + V(|\phi|^2) \right) \mathrm{d}x. \tag{14.6.6}$$

As before, we see that the dimensionality constraint spelled out in (12.1.26) disappears. Moreover, when $b \to \infty$ in (14.6.6), we return to (12.1.26), without surprise.

See [269, 365, 603] and references therein for some constructions of static multivortex solutions to the equations of motion of the model (14.6.1) and its extensions.

14.7 Integer-squared law for global Born–Infeld vortices

The Born–Infeld theory offers rich opportunities and challenging problems for various analytic aspects of field theories. The main ingredient of such a formalism is to "switch on" nonlinear electrodynamics by imposing a modified nonlinearity on the "dynamic terms" but leave the "potential terms" intact, as formulated in (4.3.32), (13.5.1), (13.5.4), (14.6.1), and (14.6.4). Following this concept, this section presents a study on a simplest Born–Infeld wave equation associated with the Lagrangian action density

$$\mathcal{L} = b^2\left(1-\sqrt{1-\frac{1}{b^2}\partial_\mu\phi\partial^\mu\overline{\phi}}\right) - V(|\phi|^2), \tag{14.7.1}$$

governing a complex scalar field ϕ over the Minkowski spacetime $\mathbb{R}^{n,1}$. Section 14.14 discusses how such a Born–Infeld wave-matter field theory may offer profound applications in cosmology.

Born–Infeld wave equation

The model (14.7.1) descends from (14.6.4) when the gauge field is absent, $A_\mu = 0$. The *Born–Infeld wave equation* or the Euler–Lagrange equation associated with (14.7.1) is

$$\frac{1}{2}\partial_\mu\left(\frac{\partial^\mu\phi}{\sqrt{1-\frac{1}{b^2}\partial_\nu\phi\partial^\nu\overline{\phi}}}\right) + V'(|\phi|^2)\phi = 0, \tag{14.7.2}$$

along with the induced energy-momentum tensor

$$T_{\mu\nu} = \frac{(\partial_\mu\phi\partial_\nu\overline{\phi} + \partial_\mu\overline{\phi}\partial_\nu\phi)}{2\sqrt{1-\frac{1}{b^2}\partial_\gamma\phi\partial^\gamma\overline{\phi}}} - \eta_{\mu\nu}\mathcal{L}. \tag{14.7.3}$$

Hence, with $x^0 = t$, the Hamiltonian energy density $\mathcal{H} = T_{00}$ of the model is

$$\begin{aligned}\mathcal{H} = {} & \frac{|\partial_t\phi|^2}{\sqrt{1+\frac{1}{b^2}\left(|\nabla\phi|^2 - |\partial_t\phi|^2\right)}} \\ & + b^2\left(\sqrt{1+\frac{1}{b^2}\left(|\nabla\phi|^2 - |\partial_t\phi|^2\right)} - 1\right) + V(|\phi|^2).\end{aligned} \tag{14.7.4}$$

Static wave limit

We are interested in static waves. So (14.7.2) and (14.7.4) become

$$\frac{1}{2}\nabla\cdot\left(\frac{\nabla\phi}{\sqrt{1+\frac{1}{b^2}|\nabla\phi|^2}}\right) = V'(|\phi|^2)\phi, \quad x \in \mathbb{R}^n, \tag{14.7.5}$$

$$\mathcal{H} = b^2\left(\sqrt{1+\frac{1}{b^2}|\nabla\phi|^2} - 1\right) + V(|\phi|^2), \tag{14.7.6}$$

respectively. The equation (14.7.5) is also the Euler–Lagrange equation of the energy functional

$$E(\phi) = \int_{\mathbb{R}^n} \mathcal{H}\,\mathrm{d}x = \int_{\mathbb{R}^n}\left(b^2\left[\sqrt{1+\frac{1}{b^2}|\nabla\phi|^2} - 1\right] + V(|\phi|^2)\right)\mathrm{d}^n x, \tag{14.7.7}$$

resulting in the virial identity

$$\int_{\mathbb{R}^n} \frac{|\nabla\phi|^2}{\sqrt{1+\frac{1}{b^2}|\nabla\phi|^2}}\,\mathrm{d}x = n\int_{\mathbb{R}^n}\left(b^2\left[\sqrt{1+\frac{1}{b^2}|\nabla\phi|^2} - 1\right] + V(|\phi|^2)\right)\mathrm{d}^n x, \tag{14.7.8}$$

for a finite-energy solution of (14.7.5). Thus, unlike in classical Klein–Gordon wave situation where the Derrick theorem excludes the spatial dimensions $n \neq 1$ (when $n = 2$, the potential density needs to be trivial, $V = 0$), we see that no dimension is excluded in the Born–Infeld wave equation formalism.

We now elaborate on the Born–Infeld vortex solutions in two dimensions with

$$V(|\phi|^2) = \frac{\lambda}{8}(|\phi|^2 - 1)^2, \quad \lambda > 0. \tag{14.7.9}$$

Ginzburg–Landau vortices

In the Klein–Gordon model case, the Lagrangian density is

$$\mathcal{L} = \frac{1}{2}\partial_\mu\phi\partial^\mu\overline{\phi} - \frac{\lambda}{8}(|\phi|^2 - 1)^2, \tag{14.7.10}$$

and the static equation of motion reads

$$\Delta\phi = \frac{\lambda}{2}(|\phi|^2 - 1)\phi, \tag{14.7.11}$$

which is the Ginzburg–Landau equation without a gauge field. By the Derrick theorem, (14.7.11) has no nontrivial finite-energy solution over $\mathbb{R}^n$ for any $n \geq 2$. Nevertheless, there are solutions of (14.7.11) over $\mathbb{R}^2$ that enjoy the *potential energy quantization* property [86, 87]:

$$\int_{\mathbb{R}^2} (|\phi|^2 - 1)^2\,\mathrm{d}^2 x = \frac{4\pi N^2}{\lambda}, \quad N = 0, 1, 2, \ldots, \infty, \tag{14.7.12}$$

indicating that the energy blow up of a nontrivial solution of (14.7.11) of finite potential energy must occur in its kinetic energy part:

$$\int_{\mathbb{R}^2} \frac{1}{2}|\nabla\phi|^2\,\mathrm{d}^2 x = \infty. \tag{14.7.13}$$

In fact, a prototype solution of these characteristics is given by the soliton solution of an N-vortex type of the radially symmetric "spiral" form [265]:

$$\phi(x) = u(r)\mathrm{e}^{\mathrm{i}N\theta}, \tag{14.7.14}$$

where r, θ are the polar coordinates of $\mathbb{R}^2$, N is an integer, and u a real-valued amplitude function, or profile function, satisfying the boundary condition

$$u(0) = 0, \quad u(\infty) = 1. \tag{14.7.15}$$

Asymptotic analysis also establishes the following precise properties

$$u(r) = \mathrm{O}(r^N) \quad \text{for } r \text{ small;} \quad u(r) = 1 + \mathrm{O}(\mathrm{e}^{-\sqrt{\lambda}r}) \quad \text{for } r \text{ large.} \tag{14.7.16}$$

Born–Infeld vortices

We now return to the Born–Infeld wave equation (14.7.5) over $\mathbb{R}^2$ with (14.7.9).

With the global vortex ansatz (14.7.14), we see that the energy (14.7.7) and the virial identity (14.7.8) become

$$E(u) = \pi \int_0^\infty \left(2b^2 \left[\sqrt{1 + \frac{1}{b^2}\left(u_r^2 + \frac{N^2}{r^2}u^2\right)} - 1 \right] + \frac{\lambda}{4}(u^2-1)^2 \right) r\mathrm{d}r, \tag{14.7.17}$$

and

$$\int_0^\infty \frac{\left(u_r^2 + \frac{N^2}{r^2}u^2\right)}{\sqrt{1 + \frac{1}{b^2}\left(u_r^2 + \frac{N^2}{r^2}u^2\right)}}\, r\,\mathrm{d}r$$

$$= \int_0^\infty \left(2b^2 \left[\sqrt{1 + \frac{1}{b^2}\left(u_r^2 + \frac{N^2}{r^2}u^2\right)} - 1 \right] + \frac{\lambda}{4}(u^2-1)^2 \right) r\mathrm{d}r, \tag{14.7.18}$$

respectively, so that the Born–Infeld wave equation (14.7.5) reads

$$\frac{\mathrm{d}}{\mathrm{d}r}\left(\frac{r u_r}{\sqrt{1 + \frac{1}{b^2}\left[u_r^2 + \frac{N^2}{r^2}u^2\right]}} \right)$$

$$= \frac{N^2 u}{r\sqrt{1 + \frac{1}{b^2}\left[u_r^2 + \frac{N^2}{r^2}u^2\right]}} + \frac{\lambda}{2} r(u^2-1)u, \quad r > 0, \tag{14.7.19}$$

which is also the Euler–Lagrange equation of the energy (14.7.17).

However, since $u(\infty) = 1$, the energy (14.7.17) diverges. So a Born–Infeld global vortex solution carries infinite energy.

Moreover, both sides of (14.7.18) diverge. Thus, (14.7.18) is only a formal, "infinity equals infinity," relation. Here, we investigate this issue and reveal a hidden quantization relation that roughly says that the difference of the left-hand and right-hand sides of (14.7.18) is exactly the half of the square of the vortex number. More precisely, we have

$$\lim_{R\to\infty}(I_R - J_R) = \frac{N^2}{2}, \tag{14.7.20}$$

where

$$I_R = \int_0^R \left(2b^2\left[\sqrt{1+\frac{1}{b^2}\left(u_r^2+\frac{N^2}{r^2}u^2\right)}-1\right]+\frac{\lambda}{4}(u^2-1)^2\right) r\mathrm{d}r, \tag{14.7.21}$$

$$J_R = \int_0^R \frac{\left(u_r^2+\frac{N^2}{r^2}u^2\right)}{\sqrt{1+\frac{1}{b^2}\left(u_r^2+\frac{N^2}{r^2}u^2\right)}} r\mathrm{d}r. \tag{14.7.22}$$

Proof of integer-squared law

We now prove (14.7.20). Indeed, multiplying (14.7.19) with ru_r, we have

$$\begin{aligned} &ru_r\left(\frac{ru_r}{\sqrt{1+\frac{1}{b^2}\left[u_r^2+\frac{N^2}{r^2}u^2\right]}}\right)_r \\ &= \frac{N^2 uu_r}{\sqrt{1+\frac{1}{b^2}\left[u_r^2+\frac{N^2}{r^2}u^2\right]}}+\frac{\lambda}{2}r^2(u^2-1)uu_r, \quad r>0. \end{aligned} \tag{14.7.23}$$

We first note that the left-hand side of (14.7.23) reads

$$\begin{aligned} ru_r\left(\frac{ru_r}{\sqrt{1+\frac{1}{b^2}\left[u_r^2+\frac{N^2}{r^2}u^2\right]}}\right)_r &= \left(\frac{r^2u_r^2}{\sqrt{1+\frac{1}{b^2}\left[u_r^2+\frac{N^2}{r^2}u^2\right]}}\right)_r \\ &\quad -\left(\frac{ru_r}{\sqrt{1+\frac{1}{b^2}\left[u_r^2+\frac{N^2}{r^2}u^2\right]}}\right)(u_r+ru_{rr}). \end{aligned} \tag{14.7.24}$$

Furthermore, we have

$$\begin{aligned} &b^2\left(r^2\left[\sqrt{1+\frac{1}{b^2}\left[u_r^2+\frac{N^2}{r^2}u^2\right]}-1\right]\right)_r \\ &= \frac{r^2}{\sqrt{1+\frac{1}{b^2}\left[u_r^2+\frac{N^2}{r^2}u^2\right]}}\left(u_ru_{rr}+\frac{N^2}{r^2}uu_r-\frac{N^2}{r^3}u^2\right) \\ &\quad +2b^2r\left(\sqrt{1+\frac{1}{b^2}\left[u_r^2+\frac{N^2}{r^2}u^2\right]}-1\right). \end{aligned} \tag{14.7.25}$$

Combining (14.7.23)–(14.7.25), we arrive at

$$\left(\frac{r^2u_r^2}{\sqrt{1+\frac{1}{b^2}\left[u_r^2+\frac{N^2}{r^2}u^2\right]}}\right)_r - \left(\frac{ru_r}{\sqrt{1+\frac{1}{b^2}\left[u_r^2+\frac{N^2}{r^2}u^2\right]}}\right)(u_r+ru_{rr})$$
$$= b^2\left(r^2\left[\sqrt{1+\frac{1}{b^2}\left[u_r^2+\frac{N^2}{r^2}u^2\right]}-1\right]\right)_r$$
$$- \frac{r^2}{\sqrt{1+\frac{1}{b^2}\left[u_r^2+\frac{N^2}{r^2}u^2\right]}}\left(u_r u_{rr} - \frac{N^2}{r^3}u^2\right)$$
$$- 2b^2 r\left(\sqrt{1+\frac{1}{b^2}\left[u_r^2+\frac{N^2}{r^2}u^2\right]}-1\right) + \frac{\lambda}{2}r^2(u^2-1)uu_r, \tag{14.7.26}$$

which may be simplified to give us the relation

$$\left(\frac{r^2u_r^2}{\sqrt{1+\frac{1}{b^2}\left[u_r^2+\frac{N^2}{r^2}u^2\right]}}\right)_r$$
$$= b^2\left(r^2\left[\sqrt{1+\frac{1}{b^2}\left[u_r^2+\frac{N^2}{r^2}u^2\right]}-1\right]\right)_r$$
$$+ \frac{r\left(u_r^2+\frac{N^2}{r^2}u^2\right)}{\sqrt{1+\frac{1}{b^2}\left[u_r^2+\frac{N^2}{r^2}u^2\right]}} - 2b^2 r\left(\sqrt{1+\frac{1}{b^2}\left[u_r^2+\frac{N^2}{r^2}u^2\right]}-1\right)$$
$$+ \frac{\lambda}{2}r^2(u^2-1)uu_r. \tag{14.7.27}$$

We now study the integration of (14.7.27) over the interval $(0, R)$ for $R > 0$ large.

First, we have

$$\int_0^R \left(\frac{r^2u_r^2}{\sqrt{1+\frac{1}{b^2}\left[u_r^2+\frac{N^2}{r^2}u^2\right]}}\right)_r \mathrm{d}r = \frac{R^2u_r^2(R)}{\sqrt{1+\frac{1}{b^2}\left[u_r^2(R)+\frac{N^2}{R^2}u^2(R)\right]}}, \tag{14.7.28}$$

which tends to zero as $R \to \infty$ since $u(r) \to 1$ sufficiently fast as $r \to \infty$.

Next we have

$$\int_0^R b^2\left(r^2\left[\sqrt{1+\frac{1}{b^2}\left[u_r^2+\frac{N^2}{r^2}u^2\right]}-1\right]\right)_r \mathrm{d}r$$
$$= b^2\left(R^2\left[\sqrt{1+\frac{1}{b^2}\left[u_r^2(R)+\frac{N^2}{R^2}u^2(R)\right]}-1\right]\right)$$
$$= \frac{N^2}{2} + \mathrm{O}(R^{-2}), \tag{14.7.29}$$

since $u(r) \to 1$ as $r \to \infty$ sufficiently fast.

Lastly, we have

$$\int_0^R r^2(u^2-1)uu_r\,\mathrm{d}r = \frac{R^2}{4}(u^2(R)-1)^2 - \frac{1}{2}\int_0^R (u^2-1)^2 r\,\mathrm{d}r. \tag{14.7.30}$$

Therefore, integrating (14.7.27) over $(0,R)$ and applying (14.7.28)–(14.7.30), we obtain

$$I_R - J_R = \frac{N^2}{2} + \mathrm{O}(R^{-2}), \tag{14.7.31}$$

for $R > 0$ sufficiently large.

It is worth noting that our quantization identity (14.7.20) or (14.7.31) is independent of the specific form of the potential function (14.7.9).

Thus, it is enlightening and worthwhile to recast the quantization identities derived in this section in general terms. For convenience, use $V(|\phi|^2)$ to denote a general potential density function with a spontaneous broken symmetry characterized by a normalized vacuum state, $|\phi| = 1$. That is, $V \geq 0, V(1) = 0$. Then it is clear that, without resorting to radial symmetry, for an N-vortex solution of the Born–Infeld wave equation (14.7.5) over $\mathbb{R}^2$ the quantization law (14.7.20) assumes the generalized form

$$\lim_{R\to\infty}\int_{|x|\leq R}\left(b^2\left[\sqrt{1+\frac{1}{b^2}|\nabla\phi|^2}-1\right] + V(|\phi|^2) - \frac{|\nabla\phi|^2}{2\sqrt{1+\frac{1}{b^2}|\nabla\phi|^2}}\right)\mathrm{d}^2x$$
$$= \frac{\pi N^2}{2}, \tag{14.7.32}$$

which refines the virial identity (14.7.8), when $n = 2$, for vortex waves.

See [225] and references therein for other models and quantization results for the Born–Infeld global vortices.

14.8 Electrically charged black hole solutions

Section 10.4 considered the Reissner–Nordström electromagnetically charged black holes. We saw that such black holes bring along some new concepts, including external and internal event horizons, extremal black holes, and naked singularities, which are not displayed in the context of the Schwarzschild black hole solutions (where electromagnetism is absent). Besides, another distinctive feature of the Reissner–Nordström black holes is their elevated singularity at the mass center, demonstrated by the Kretschmann scalar (10.4.55), which indicates that the presence of electromagnetism, no matter how weak it is, elevates the curvature singularity from that of the Schwarzschild of the type r^{-6} to r^{-8} in spherical coordinates. In fact, the Reissner–Nordström black holes are obtained from coupling the Einstein equations to the Maxwell equations in which the energy of an electromagnetic point charge diverges at $r = 0$, which is similar to an elevated curvature singularity at $r = 0$. In other words, the elevated singularity of the Reissner–Nordström black holes may well be regarded as resulting from an unbalanced treatment of their mass and electromagnetism aspects with respective to the finiteness issue: the Reissner–Nordström black hole has a finite mass but infinite electromagnetic energy. On the other hand, since the Born–Infeld theory allows finite-energy electromagnetic point charges, we would hope to obtain black holes with *relegated* curvature singularities, to the level of a Schwarzschild black hole, if possible. Thus, the next two sections aim to accomplish such an understanding. To make our discussion smoother and less technical, we begin with a study of electrically charged black holes. We then extend our study to a black hole carrying *both* electric and magnetic charges.

Einstein equations coupled with Born–Infeld equations

We consider electrically charged black hole solutions generated from the Einstein equations (10.2.77) with $\Lambda = 0$, or

$$R_{\mu\nu} = -8\pi G\left(T_{\mu\nu} - \frac{1}{2}g_{\mu\nu}T\right), \tag{14.8.1}$$

coupled with the Born–Infeld equations of the action density (14.1.6) in which the Minkowski metric $(\eta_{\mu\nu})$ is replaced by a gravitational one, $(g_{\mu\nu})$, which defines the Ricci tensor on the left-hand side of (14.8.1), namely,

$$\mathcal{L} = b^2\left(1 - \sqrt{1 + \frac{1}{2b^2}F_{\mu\nu}F^{\mu\nu}}\right), \quad F_{\mu\nu} = g_{\mu\alpha}g_{\nu\beta}F^{\alpha\beta}, \tag{14.8.2}$$

which gives rise to the associated curved-spacetime electromagnetic field equations

$$\frac{1}{\sqrt{-g}}\partial_\mu\left(\sqrt{-g}P^{\mu\nu}\right) = 0, \tag{14.8.3}$$

$$P^{\mu\nu} = \frac{F^{\mu\nu}}{\sqrt{1 + \frac{1}{2b^2}F_{\alpha\beta}F^{\alpha\beta}}}, \tag{14.8.4}$$

updating (14.1.10)–(14.1.11). Note that, in (14.8.1), the energy-momentum tensor $T_{\mu\nu}$ is determined from (14.8.2) by

$$\begin{aligned} T_{\mu\nu} &= 2\frac{\partial\mathcal{L}}{\partial g^{\mu\nu}} - g_{\mu\nu}\mathcal{L} \\ &= -\frac{F_{\mu\alpha}g^{\alpha\beta}F_{\nu\beta}}{\sqrt{1 + \frac{1}{2b^2}F_{\alpha\beta}F^{\alpha\beta}}} - g_{\mu\nu}\mathcal{L}, \end{aligned} \tag{14.8.5}$$

and, as before, $T = g^{\mu\nu}T_{\mu\nu}$ is the trace of $T_{\mu\nu}$.

Electric point charge

We consider an electrostatic situation for which the electric point charge is at the origin of space. Thus, we may impose the radially symmetric gravitational line element given by (10.4.16) such that the field $P^{\mu\nu}$ in (14.8.4) takes the form

$$(P^{\mu\nu}) = \begin{pmatrix} 0 & -D^r & 0 & 0 \\ D^r & 0 & 0 & 0 \\ 0 & 0 & 0 & 0 \\ 0 & 0 & 0 & 0 \end{pmatrix}, \tag{14.8.6}$$

where D^r depends on the radial variable r only. Accordingly, the electromagmetic field $F^{\mu\nu}$ assumes the same form of $P^{\mu\nu}$ as given in (14.8.6), or

$$(F^{\mu\nu}) = \begin{pmatrix} 0 & -E^r & 0 & 0 \\ E^r & 0 & 0 & 0 \\ 0 & 0 & 0 & 0 \\ 0 & 0 & 0 & 0 \end{pmatrix}, \tag{14.8.7}$$

where, as earlier, the scalar function E^r depends on the radial variable r only. With (14.8.7), we have

$$(F_{\mu\nu}) = (g_{\mu\alpha}g_{\nu\beta}F^{\alpha\beta}) = \begin{pmatrix} 0 & E^r & 0 & 0 \\ -E^r & 0 & 0 & 0 \\ 0 & 0 & 0 & 0 \\ 0 & 0 & 0 & 0 \end{pmatrix}. \tag{14.8.8}$$

In view of (14.8.6)–(14.8.8), we derive from (14.8.4) the relation

$$\frac{E^r}{\sqrt{1-\frac{1}{b^2}(E^r)^2}} = D^r, \tag{14.8.9}$$

and its conversion

$$E^r = \frac{D^r}{\sqrt{1+\frac{1}{b^2}(D^r)^2}}, \tag{14.8.10}$$

both as those in the flat-space situation.

Energy-momentum tensor

Substituting (14.8.7)–(14.8.8) into (14.8.5), we obtain the nontrivial components of $T_{\mu\nu}$ as follows,

$$T_{00} = A(r)\left(\frac{(E^r)^2}{\sqrt{1-\frac{1}{b^2}(E^r)^2}} - b^2\left[1-\sqrt{1-\frac{1}{b^2}(E^r)^2}\right]\right), \tag{14.8.11}$$

$$T_{11} = -\frac{1}{A(r)}\left(\frac{(E^r)^2}{\sqrt{1-\frac{1}{b^2}(E^r)^2}} - b^2\left[1-\sqrt{1-\frac{1}{b^2}(E^r)^2}\right]\right), \quad (14.8.12)$$

$$T_{22} = b^2 r^2 \left(1-\sqrt{1-\frac{1}{b^2}(E^r)^2}\right), \quad (14.8.13)$$

$$T_{33} = \sin^2\theta\, T_{22}. \quad (14.8.14)$$

Consequently, we have

$$T = \frac{2(E^r)^2}{\sqrt{1-\frac{1}{b^2}(E^r)^2}} - 4b^2\left(1-\sqrt{1-\frac{1}{b^2}(E^r)^2}\right), \quad (14.8.15)$$

for the trace of the energy-momentum tensor. Note that the Maxwell equation result $T = 0$ corresponds to the limit $T \to 0$ as $b \to \infty$, in (14.8.15), as anticipated.

Now, inserting (14.8.6) into (14.8.3) and using $\sqrt{-g} = r^2 \sin\theta$, we have

$$\frac{\mathrm{d}}{\mathrm{d}r}(r^2 D^r) = 0, \quad (14.8.16)$$

which renders the familiar point charge solution

$$D^r = \frac{q}{r^2}, \quad (14.8.17)$$

in which q arises as an integration constant and may be assumed to be positive for definiteness and convenience. Thus, we can use (14.8.10) to get

$$E^r = \frac{q}{\sqrt{r^4 + a^4}}, \quad a = \sqrt{\frac{q}{b}}. \quad (14.8.18)$$

Electrically charged black hole solutions

In view of (10.4.17)–(10.4.20) and (14.8.11)–(14.8.15), we see that the Einstein equations (14.8.1) become

$$(r^2 A')' = 16\pi G b^2 r^2 \left(1-\sqrt{1-\frac{1}{b^2}(E^r)^2}\right), \quad (14.8.19)$$

$$(rA)' = 1 - 8\pi G r^2 \left(\frac{(E^r)^2}{\sqrt{1-\frac{1}{b^2}(E^r)^2}} - b^2\left[1-\sqrt{1-\frac{1}{b^2}(E^r)^2}\right]\right), \quad (14.8.20)$$

which is an over-determined system of equations. However, with (14.8.18), the equation (14.8.20) leads to (14.8.19) by differentiating the former. Thus, inserting (14.8.18) into (14.8.20) and integrating, we obtain

$$A(r) = 1 + \frac{C}{r} + \frac{8\pi G a^4 b^2}{r} \int_r^\infty \frac{\mathrm{d}\eta}{\sqrt{\eta^4 + a^4} + \eta^2}, \tag{14.8.21}$$

where C is an integration constant. Of course, we should recover the Schwarzschild solution when $q = 0$. Thus, we have $C = -2GM$, where M is regarded as a mass parameter as before. In order to see the charge dependence explicitly, we now rewrite (14.8.21) as

$$A(r) = 1 - \frac{2GM}{r} + \frac{8\pi G q^2}{r} \int_r^\infty \frac{\mathrm{d}\eta}{\sqrt{\eta^4 + \frac{q^2}{b^2}} + \eta^2}. \tag{14.8.22}$$

We see that the Reissner–Nordström metric of an electrically charged black hole follows from taking the limit $b \to \infty$ in (14.8.22).

For any $b > 0$, we rewrite (14.8.22) in its asymptotic form

$$A(r) = 1 - \frac{2GM}{r} + \frac{4\pi G q^2}{r^2} - \frac{\pi G q^4}{5b^2 r^6} + \frac{\pi G q^6}{18 b^4 r^{10}} + \mathrm{O}\left(\frac{1}{r^{14}}\right), \quad r \gg 1. \tag{14.8.23}$$

Thus, in view of (14.8.23) and the methods in Section 10.6 or [83], we see that the ADM mass of the electrically charged back hole is still M.

Event horizons

An event horizon of the electrically charged black hole is at a root of (14.8.22), which is a positive solution to the equation $h(r) = rA(r) = 0$ where

$$h(r) \equiv r - 2GM + 8\pi G q^2 \int_r^\infty \frac{\mathrm{d}\eta}{\sqrt{\eta^4 + \frac{q^2}{b^2}} + \eta^2}. \tag{14.8.24}$$

At a first sight, the existence of a positive root of (14.8.24) is ensured by $h(0) < 0$ or

$$\begin{aligned} M &> 4\pi q^2 \int_0^\infty \frac{\mathrm{d}\eta}{\sqrt{\eta^4 + \frac{q^2}{b^2}} + \eta^2} \\ &= 4\pi q^{\frac{3}{2}} b^{\frac{1}{2}} \int_0^\infty \frac{\mathrm{d}x}{\sqrt{x^4+1} + x^2} = \frac{4\pi^{\frac{5}{2}} q^{\frac{3}{2}} b^{\frac{1}{2}}}{3\left(\Gamma\left(\frac{3}{4}\right)\right)^2}. \end{aligned} \tag{14.8.25}$$

However, since b is a large quantity, the opposite should be assumed to be valid in general, or

$$M < \frac{4\pi^{\frac{5}{2}} q^{\frac{3}{2}} b^{\frac{1}{2}}}{3\left(\Gamma\left(\frac{3}{4}\right)\right)^2}, \tag{14.8.26}$$

namely, $h(0) > 0$, which will be a condition we observe subsequently. Besides, since

$$h''(r) = \frac{16\pi G q^2 r}{\left(\sqrt{r^4 + \frac{q^2}{b^2}} + r^2\right)^2}\left(1 + \frac{r^2}{\sqrt{r^4 + \frac{q^2}{b^2}}}\right) > 0, \tag{14.8.27}$$

we see that the existence of a positive root of (14.8.24) is ensured by the condition $h(r_0) \leq 0$ where r_0 is the unique positive root (if any) of the equation $h'(r) = 0$ or

$$\sqrt{r^4 + \frac{q^2}{b^2}} + r^2 = 8\pi G q^2, \tag{14.8.28}$$

which has a solution if and only if

$$\frac{1}{b} < 8\pi G q. \tag{14.8.29}$$

This is another condition that is implied by the fact that b is large and will be assumed as well in addition to (14.8.26). Thus, solving (14.8.28), we get

$$r_0 = \frac{1}{4\sqrt{\pi G}}\sqrt{(8\pi G q)^2 - \frac{1}{b^2}}. \tag{14.8.30}$$

On the other hand, we can rewrite (14.8.24) as

$$h(r) = r - 2GM + 8\pi G\frac{q^2}{a}\,h_0\left(\frac{r}{a}\right), \quad h_0(x) = \int_x^\infty \frac{\mathrm{d}y}{\sqrt{y^4+1}+y^2}. \tag{14.8.31}$$

Hence, $h(r) = 0$ has exactly two positive roots, say r_- and r_+, if $h(r_0) < 0$ or

$$r_0 + 8\pi G\frac{q^2}{a}\,h_0\left(\frac{r_0}{a}\right) < 2GM, \tag{14.8.32}$$

and it is clear that

$$0 < r_- < r_0 < r_+ < 2GM = r_{\mathrm{S}}, \tag{14.8.33}$$

and a single positive root, which is actually r_0, if $h(r_0) = 0$ or

$$r_0 + 8\pi G\frac{q^2}{a}\,h_0\left(\frac{r_0}{a}\right) = 2GM. \tag{14.8.34}$$

In the former situation, $A(r) > 0$ for $r \in (0, r_-) \cup (r_+, \infty)$ and $A(r) < 0$ for $r \in (r_-, r_+)$. Thus, $r = r_-$ and $r = r_+$ correspond to an inner and outer event horizons, respectively. In the latter situation, $A(r) > 0$ for $r \neq r_0$. Therefore, (14.8.34) spells out the critical condition for the occurrence of an extremal black hole. In fact, rewriting (14.8.30) as

$$r_0 = r_0(a) = \frac{1}{4\sqrt{\pi G}\,q}\sqrt{(8\pi G q^2)^2 - a^4}, \tag{14.8.35}$$

and taking the limit $a \to 0$ or $b \to \infty$ in (14.8.34), we have

$$4\pi q^2 = GM^2, \tag{14.8.36}$$

as stated earlier in (10.4.46) for the extremal Reissner–Nordström black hole.

Kretschmann invariant

With (10.4.16), we may express the Kretschmann invariant (10.3.63) as

$$K = \frac{(r^2 A'')^2 + 4(rA')^2 + 4(A-1)^2}{r^4}. \tag{14.8.37}$$

From (14.8.22) and (14.8.37), we see that the only curvature singularity occurs at $r = 0$ where $K \sim \frac{1}{r^6}$ for $r \ll 1$ as that for the Schwarzschild black hole solution given by (10.3.64).

Naked singularity

Therefore, if $h(r_0) > 0$ or

$$r_0 + 8\pi G \frac{q^2}{a} h_0 \left(\frac{r_0}{a}\right) > 2GM, \tag{14.8.38}$$

there is no event horizon and the gravitational metric is everywhere regular. As a consequence, the singularity $r = 0$ becomes naked.

Gravitational mass versus electric energy

In view of (10.4.16) and (14.8.11), we may rewrite (14.8.21) in terms of the Hamiltonian energy density of the electric field, $\mathcal{H} = T_0^0$, which assumes the same form as that in the flat-space situation, as

$$(rA)' = 1 - 8\pi G r^2 \mathcal{H}. \tag{14.8.39}$$

Hence, the total static electric energy is

$$E = \int \mathcal{H}\sqrt{-g}\,\mathrm{d}r\mathrm{d}\theta\mathrm{d}\phi = 4\pi \int_0^\infty \mathcal{H}(r) r^2\,\mathrm{d}r, \tag{14.8.40}$$

since for (10.4.16) we have $\sqrt{-g} = r^2 \sin\theta$ as in the flat space. As a consequence, if we use the notation

$$E(r) = 4\pi \int_r^\infty \mathcal{H}(\eta)\eta^2\,\mathrm{d}\eta \tag{14.8.41}$$

to denote the electric energy contained in the spatial region outside the ball of radius r and centered at the origin, such that $E = E(0)$, then (14.8.39) leads us to the expression

$$A(r) = 1 - \frac{2GM}{r} + \frac{2GE(r)}{r}, \tag{14.8.42}$$

which places the gravitational mass and electric energy at somewhat *equal* footings. Besides, setting $h(r) = rA(r)$ as before, we see that the condition $h(0) < 0$, apparent for the occurrence of an event horizon, simply becomes

$$M > E, \tag{14.8.43}$$

which reinterprets (14.8.25) as an energy-mass condition.

14.9 Dyonic black hole solutions

This section extends our study of electrically charged black hole solutions of the Einstein equations coupled with the Born–Infeld electromagnetism to include a magnetic point charge as well. We first consider the model (14.1.6) and show that a dyonic solution carries divergent energy as in the setting without gravity. Taking this result as a motivation, we then consider the model (14.4.11) and obtain finite-energy solutions.

Radial reduction of electromagnetic field tensors

Expand (14.8.6) to contain both electric and magnetic fields in the radial direction such that the tensor field $P^{\mu\nu}$ defined by (14.4.13) assumes the form

$$(P^{\mu\nu}) = \begin{pmatrix} 0 & -D^r & 0 & 0 \\ D^r & 0 & 0 & 0 \\ 0 & 0 & 0 & -H^r \\ 0 & 0 & H^r & 0 \end{pmatrix}, \tag{14.9.1}$$

where D^r and H^r are the nontrivial radial components of the electric displacement and magnetic intensity fields, respectively, to be determined. Thus, by consistency in (14.8.4), we have

$$(F^{\mu\nu}) = \begin{pmatrix} 0 & -E^r & 0 & 0 \\ E^r & 0 & 0 & 0 \\ 0 & 0 & 0 & -B^r \\ 0 & 0 & B^r & 0 \end{pmatrix}. \tag{14.9.2}$$

Hence, with (14.9.2), (10.4.16), and $F_{\mu\nu} = g_{\mu\alpha}g_{\nu\beta}F^{\alpha\beta}$, we have

$$(F_{\mu\nu}) = \begin{pmatrix} 0 & E^r & 0 & 0 \\ -E^r & 0 & 0 & 0 \\ 0 & 0 & 0 & -r^4\sin^2\theta\, B^r \\ 0 & 0 & r^4\sin^2\theta\, B^r & 0 \end{pmatrix}. \tag{14.9.3}$$

Consequently, we have

$$F_{\mu\nu}F^{\mu\nu} = -2(E^r)^2 + 2r^4\sin^2\theta\,(B^r)^2. \tag{14.9.4}$$

Inserting (14.9.1), (14.9.2), and (14.9.4) into (14.8.4), we obtain

$$D^r = \frac{E^r}{\sqrt{1-\frac{1}{b^2}([E^r]^2 - r^4\sin^2\theta[B^r]^2)}}, \tag{14.9.5}$$

$$H^r = \frac{B^r}{\sqrt{1-\frac{1}{b^2}([E^r]^2 - r^4\sin^2\theta[B^r]^2)}}, \tag{14.9.6}$$

which may be converted to yield

$$E^r = \frac{D^r}{\sqrt{1+\frac{1}{b^2}([D^r]^2 - r^4\sin^2\theta[H^r]^2)}}, \tag{14.9.7}$$

$$B^r = \frac{H^r}{\sqrt{1+\frac{1}{b^2}([D^r]^2 - r^4\sin^2\theta[H^r]^2)}}. \tag{14.9.8}$$

Solution to Born–Infeld equations

Applying (14.9.1) to (14.8.3) and using $\sqrt{-g} = r^2\sin\theta$, we arrive at the equations

$$\partial_r(r^2\sin\theta D^r) = 0, \tag{14.9.9}$$

which is at $\nu = 0$ in (14.8.3), and

$$\partial_\theta(r^2\sin\theta H^r) = 0, \tag{14.9.10}$$

which is at $\nu = 3$ in (14.8.3). From (14.9.9), we get

$$D^r = \frac{q}{r^2}, \tag{14.9.11}$$

where q is an integration constant that gives rise to an electric charge as before. On the other hand, from (14.9.10), we get

$$H^r = \frac{H(r)}{\sin\theta}, \tag{14.9.12}$$

where $H(r)$ is a function of the radial variable r to be determined.

As a consequence of (14.9.11), (14.9.12), and (14.9.8), we have

$$B^r = \frac{H(r)}{\sin\theta\sqrt{1+\frac{1}{b^2}([\frac{q}{r^2}]^2 - r^4[H(r)]^2)}}. \tag{14.9.13}$$

Furthermore, we note that the Bianchi identity (10.4.5) with $F_{\mu\nu}$ assuming the form (14.9.3) has only one nontrivial component at $\mu = 1, \nu = 2, \gamma = 3$, which reads

$$-\partial_r(r^4\sin^2\theta B^r) = F_{23,1} = 0, \tag{14.9.14}$$

giving rise to the solution

$$B^r = \frac{g}{r^4\sin\theta}, \tag{14.9.15}$$

where g is an integration constant taken to be positive for convenience. Substituting (14.9.15) back into (14.9.13), we find

$$H(r) = \frac{g}{r^4}\sqrt{\frac{r^4+a_e^4}{r^4+a_m^4}}, \quad a_e = \sqrt{\frac{q}{b}}, \quad a_m = \sqrt{\frac{g}{b}}. \tag{14.9.16}$$

Applying (14.9.11), (14.9.12), and (14.9.16) to (14.9.7), we obtain

$$E^r = \frac{q}{r^2}\sqrt{\frac{r^4 + a_m^4}{r^4 + a_e^4}}. \tag{14.9.17}$$

Note that E^r given in (14.9.17) is the same as that obtained in (14.3.2). Besides, comparing (14.9.2) with (14.9.15) against (10.4.30), we see that the parameter g is recognized as the magnetic charge of the solution. For convenience of reference, we collect our results for the pair D^r, H^r here as well:

$$D^r = \frac{q}{r^2}, \quad H^r = \frac{g}{r^4 \sin\theta}\sqrt{\frac{r^4 + a_e^4}{r^4 + a_m^4}}. \tag{14.9.18}$$

Energy-momentum tensor

From (14.9.4), we have

$$\mathcal{R} \equiv \sqrt{1 + \frac{1}{2b^2}F_{\mu\nu}F^{\mu\nu}} = \sqrt{1 - \frac{1}{b^2}([E^r]^2 - r^4 \sin^2\theta [B^r]^2)}. \tag{14.9.19}$$

Thus, we may use (14.9.2), (14.9.3), and (10.4.16) to obtain the nontrivial components of (14.8.5) and its trace to be

$$T_{00} = A(r)\left(\frac{(E^r)^2}{\mathcal{R}} - b^2(1 - \mathcal{R})\right), \tag{14.9.20}$$

$$T_{11} = -\frac{1}{A(r)}\left(\frac{(E^r)^2}{\mathcal{R}} - b^2(1 - \mathcal{R})\right), \tag{14.9.21}$$

$$T_{22} = r^2\left(\frac{r^4 \sin^2\theta (B^r)^2}{\mathcal{R}} + b^2(1 - \mathcal{R})\right), \quad T_{33} = \sin^2\theta\, T_{22}, \tag{14.9.22}$$

$$T = \frac{2}{\mathcal{R}}\left([E^r]^2 - r^4 \sin^2\theta [B^r]^2\right) - 4b^2(1 - \mathcal{R}). \tag{14.9.23}$$

Solution to Einstein equations

Using (10.4.17)–(10.4.20) and (14.9.20)–(14.9.23), the Einstein equations (14.8.1) are reduced into

$$(r^2 A')' = 16\pi G r^2\left(\frac{r^4 \sin^2\theta (B^r)^2}{\mathcal{R}} + b^2(1 - \mathcal{R})\right), \tag{14.9.24}$$

$$(rA)' = 1 - 8\pi G r^2\left(\frac{(E^r)^2}{\mathcal{R}} - b^2(1 - \mathcal{R})\right). \tag{14.9.25}$$

Moreover, inserting (14.9.15) and (14.9.17) into (14.9.19), we have

$$\mathcal{R} = \sqrt{\frac{r^4 + a_m^4}{r^4 + a_e^4}}. \tag{14.9.26}$$

As a consequence of (14.9.15), (14.9.17), and (14.9.26), note that (14.9.25) implies (14.9.24). Thus, we integrate (14.9.25) with (14.9.17) and (14.9.26) to obtain the solution

$$A(r) = 1 - \frac{2GM}{r} + \frac{8\pi G}{r}\int_r^\infty \frac{\frac{q^2g^2}{b^2} + \eta^4(q^2+g^2)}{\eta^2\left(\eta^4 + \sqrt{(\eta^4 + [\frac{q}{b}]^2)(\eta^4 + [\frac{g}{b}]^2)}\right)}\,d\eta. \quad (14.9.27)$$

At large r, we have the asymptotic expression

$$A(r) = 1 - \frac{2GM}{r} + \frac{4\pi G(q^2+g^2)}{r^2} + \mathrm{O}\left(\frac{1}{r^6}\right), \quad (14.9.28)$$

which extends the Reissner–Nordström black hole metric (10.4.40).

By the methods in Section 10.6 or [83], the ADM mass of the solution is again M.

Divergence of energy of electromagnetic fields

From (10.4.16), (14.9.20)–(14.9.22), (14.9.17), and (14.9.26), we see that the Hamiltonian energy density $\mathcal{H} = T_0^0$ assumes the form

$$\begin{aligned}\mathcal{H} &= \frac{(E^r)^2}{\mathcal{R}} - b^2(1-\mathcal{R}) \\ &= \frac{\left(\frac{qg}{b}\right)^2 + r^4(q^2+g^2)}{r^4\left(r^4 + \sqrt{(r^4 + [\frac{q}{b}]^2)(r^4 + [\frac{g}{b}]^2)}\right)},\end{aligned} \quad (14.9.29)$$

as in the flat-space situation (14.3.12). To see how (14.9.29) leads to a divergent energy again at $r = 0$, we note that the line element of a spatial slice of the spacetime with the metric form (10.4.16) is

$$d\ell^2 = \frac{1}{A(r)}dr^2 + r^2(d\theta^2 + \sin^2\theta\, d\phi^2), \quad (14.9.30)$$

since $A(r) > 0$ for r small. Thus, correspondingly, the associated volume element reads

$$dv = \frac{r^2 \sin\theta}{\sqrt{A(r)}}\, dr d\theta d\phi. \quad (14.9.31)$$

On the other hand, from (14.9.27), we find

$$\lim_{r\to 0} r^2 A(r) = 8\pi G. \quad (14.9.32)$$

In view of (14.9.29) and (14.9.32), we arrive at the asserted energy divergence,

$$\int \mathcal{H}\, dv = \infty, \quad (14.9.33)$$

where the integration is over a small region around $r = 0$.

Gravitational mass versus dyonic energy

As in Section 14.8, the equation (14.9.25) is of the form (14.8.39) as well where the energy density $\mathcal{H}$ is now given by (14.9.29). Although the total static electric energy of the form (14.8.40) is again divergent in the dyonic context here, the partial energy of the form (14.8.41) is finite for all $r > 0$, such that the metric factor $A(r)$ follows the same expression as that given in (14.8.42). However, the apparent energy-mass condition of the form (14.8.43) for the occurrence of an event horizon is invalid since now E is infinite. Nevertheless, setting $h(r) = rA(r)$ for $A(r)$ given in (14.9.27), we have $h(r) \to \infty$ as $r \to 0$ and $r \to \infty$. Hence, $h(r)$ has a global minimum at some $r_0 > 0$ and the existence of an event horizon is equivalent to the condition $h(r_0) \leq 0$. Here, we omit the discussion.

Subsequently, we aim to obtain *finite-energy* dyonic black hole solutions in the context of the second Born–Infeld theory.

Dyonic black hole solutions based on invariant formalism

Consider the second Born–Infeld theory governed by the action density (14.4.11) where now the dual $\tilde{F}_{\mu\nu}$ of the electromagnetic field $F^{\mu\nu}$ reads

$$\tilde{F}_{\mu\nu} = \frac{\operatorname{sgn}(g)}{2}\sqrt{-g}\,\epsilon_{\mu\nu\alpha\beta}\,F^{\alpha\beta} = -\frac{\sqrt{-g}}{2}\,\epsilon_{\mu\nu\alpha\beta}\,F^{\alpha\beta}. \tag{14.9.34}$$

The equations of motion are still of the form (14.4.12)–(14.4.13), but the operations of lowering and raising indices on vector and tensor fields are done with the gravitational metric $(g_{\mu\nu})$. We again work within the context that the metric is of the Schwarzschild type given in (10.4.16).

Moreover, from (14.9.34), we get

$$(\tilde{F}_{\mu\nu}) = \begin{pmatrix} 0 & r^2\sin\theta B^r & 0 & 0 \\ -r^2\sin\theta B^r & 0 & 0 & 0 \\ 0 & 0 & 0 & r^2\sin\theta\, E^r \\ 0 & 0 & -r^2\sin\theta\, E^r & 0 \end{pmatrix}, \tag{14.9.35}$$

which results in

$$\begin{aligned}(\tilde{F}^{\mu\nu}) &= (g^{\mu\alpha}g^{\nu\beta}\tilde{F}_{\alpha\beta}) = \frac{1}{2\sqrt{-g}}\epsilon^{\mu\nu\alpha\beta}F_{\alpha\beta} \\ &= \begin{pmatrix} 0 & -r^2\sin\theta B^r & 0 & 0 \\ r^2\sin\theta B^r & 0 & 0 & 0 \\ 0 & 0 & 0 & \frac{E^r}{r^2\sin\theta} \\ 0 & 0 & -\frac{E^r}{r^2\sin\theta} & 0 \end{pmatrix}.\end{aligned} \tag{14.9.36}$$

As a consequence of (14.9.3) and (14.9.36), we obtain

$$F_{\mu\nu}\tilde{F}^{\mu\nu} = -4r^2\sin\theta\,(E^r B^r). \tag{14.9.37}$$

Hence, using (14.9.4) and (14.9.37), we have

$$\mathcal{R} \equiv \sqrt{1+\frac{1}{2b^2}F_{\mu\nu}F^{\mu\nu}-\frac{1}{16b^4}\left(F_{\mu\nu}\tilde{F}^{\mu\nu}\right)^2},$$
$$= \sqrt{1-\frac{1}{b^2}\left([E^r]^2-r^4\sin^2\theta\,[B^r]^2\right)-\frac{1}{b^4}\,r^4\sin^2\theta(E^rB^r)^2}. \qquad (14.9.38)$$

Therefore, inserting (14.9.37) and (14.9.38) into the curved-space version of (14.4.13), we arrive at the relations

$$D^r = \frac{E^r\left(1+\frac{1}{b^2}r^4\sin^2\theta\,[B^r]^2\right)}{\mathcal{R}}, \qquad (14.9.39)$$

$$H^r = \frac{B^r\left(1-\frac{1}{b^2}\,[E^r]^2\right)}{\mathcal{R}}. \qquad (14.9.40)$$

Solution to Born–Infeld equations

We now solve the Born–Infeld equations.

First, we may follow the same procedure to get the results (14.9.11) and (14.9.15). In fact, the results (14.9.11) and (14.9.12) are both clear. Using these in (14.9.39)–(14.9.40), we see that consistency requires both E^r and $(\sin\theta)B^r$ be functions of the radial variable r only. Since B^r also satisfies (14.9.14), we see that it is given by (14.9.15) indeed.

Next, substituting (14.9.11) and (14.9.15) into (14.9.39), (14.9.38), and (14.9.40), we obtain

$$E^r = \frac{q}{\sqrt{r^4+a_e^4+a_m^4}}, \qquad (14.9.41)$$

$$H^r = \frac{g}{r^2\,\sin\theta\sqrt{r^4+a_e^4+a_m^4}}, \qquad (14.9.42)$$

$$\mathcal{R} = \frac{r^4+a_m^4}{r^2\sqrt{r^4+a_e^4+a_m^4}}. \qquad (14.9.43)$$

Energy-momentum tensor

The curved-space version of the energy-momentum tensor (14.4.28) reads

$$T_{\mu\nu} = -\frac{\left(F_{\mu\alpha}g^{\alpha\beta}F_{\nu\beta}-\frac{1}{4b^2}(F_{\mu'\nu'}\tilde{F}^{\mu'\nu'})F_{\mu\alpha}g^{\alpha\beta}\tilde{F}_{\nu\beta}\right)}{\mathcal{R}} - g_{\mu\nu}\mathcal{L}, \qquad (14.9.44)$$

so that its nontrivial components and trace in our present context are

$$T_{00} = A(r)\left(\frac{(E^r)^2\left(1+\frac{1}{b^2}\,r^4\sin^2\theta\,(B^r)^2\right)}{\mathcal{R}} - b^2\,(1-\mathcal{R})\right), \qquad (14.9.45)$$

$$T_{11} = -\frac{1}{A(r)}\left(\frac{(E^r)^2\,(1+\frac{1}{b^2}\,r^4\sin^2\theta\,(B^r)^2)}{\mathcal{R}} - b^2\,(1-\mathcal{R})\right), \tag{14.9.46}$$

$$T_{22} = r^2\left(\frac{1}{\mathcal{R}}\,r^4\sin^2\theta(B^r)^2\left[1-\frac{(E^r)^2}{b^2}\right] + b^2(1-\mathcal{R})\right), \tag{14.9.47}$$

$$T_{33} = \sin^2\theta\,T_{22}, \tag{14.9.48}$$

$$T = \frac{2}{\mathcal{R}}\left((E^r)^2 - r^4\sin^2\theta(B^r)^2 + \frac{2}{b^2}r^4\sin^2\theta(E^rB^r)^2\right) - 4b^2(1-\mathcal{R}). \tag{14.9.49}$$

Solving Einstein equations

With (10.4.17)–(10.4.20) and (14.9.45)–(14.9.49), the Einstein equations (14.8.1) are reduced into

$$(r^2A')' = 16\pi G r^2\left(\frac{1}{\mathcal{R}}\,r^4\sin^2\theta(B^r)^2\left[1-\frac{(E^r)^2}{b^2}\right] + b^2(1-\mathcal{R})\right), \tag{14.9.50}$$

$$(rA)' = 1 - 8\pi G r^2\left(\frac{(E^r)^2}{\mathcal{R}}\left[1+\frac{1}{b^2}r^4\sin^2\theta(B^r)^2\right] - b^2(1-\mathcal{R})\right). \tag{14.9.51}$$

In view of (14.9.15), (14.9.41), and (14.9.42), the equation (14.9.50) is implied by (14.9.51), which may consequently be integrated to yield the solution [157, 233]

$$A(r) = 1 - \frac{2GM}{r} + \frac{8\pi G(q^2+g^2)}{r}\int_r^\infty \frac{\mathrm{d}\eta}{\sqrt{\eta^4+\frac{1}{b^2}(q^2+g^2)}+\eta^2}, \tag{14.9.52}$$

which elegantly extends the formula (14.8.22). Since (14.9.52) is of the same form of (14.8.22), we see that the curvature singularity at $r=0$ for the solution here is the same as that of a Schwarzschild black hole. Furthermore, asymptotically for r large, we have the expression

$$A(r) = 1 - \frac{2GM}{r} + \frac{4\pi G\,(q^2+g^2)}{r^2}$$
$$-\pi G(q^2+g^2)\left(\left[\frac{q^2+g^2}{b^2}\right]\frac{1}{5r^6} - \left[\frac{q^2+g^2}{b^2}\right]^2\frac{1}{18r^{10}}\right) + \mathrm{O}\left(\frac{1}{r^{14}}\right). \tag{14.9.53}$$

Convergence of energy of electromagnetic fields

Moreover, the Hamiltonian energy density $\mathcal{H} = T_0^0$ in view of (14.9.15), (14.9.41), and (14.9.43) is

$$\begin{aligned} \mathcal{H} &= \frac{(E^r)^2 + \frac{1}{b^2}\, r^4 \sin^2\theta\, (E^r B^r)^2}{\mathcal{R}} - b^2\,(1-\mathcal{R}) \\ &= \frac{q^2+g^2}{r^2(\sqrt{r^4+a_e^4+a_m^4}+r^2)}, \end{aligned} \tag{14.9.54}$$

which recovers (14.4.32).

For $h(r) \equiv rA(r)$, where $A(r)$ is given in (14.9.52), assume $h(0) > 0$ for convenience and simplicity. Thus, we have

$$M < \frac{4\pi^{\frac{5}{2}}(q^2+g^2)^{\frac{3}{4}} b^{\frac{1}{2}}}{3\left(\Gamma\left(\frac{3}{4}\right)\right)^2}, \tag{14.9.55}$$

as that in (14.8.26). In particular, $A(r) > 0$ for r small. Using

$$\lim_{r\to 0} rA(r) = h(0) > 0 \tag{14.9.56}$$

in the volume element (14.9.31), we see that (14.9.54) renders the finiteness of the local electromagnetic energy,

$$\int \mathcal{H}\, \mathrm{d}v < \infty, \tag{14.9.57}$$

where the integration is over a small region around $r = 0$.

Electrically and dyonically charged black hole solutions to the Einstein–Born–Infeld equations were first obtained in [157]. More recently, generalized Born–Infeld theory of electromagnetism is explored [28, 29, 311] in the context of general relativity to yield singularity-free charged black hole solutions, for the example, the *Bardeen black hole* [29, 471, 512, 555] defined by the metric factor [40]

$$A(r) = 1 - \frac{2GMr^2}{(r^2+Q^2)^{\frac{3}{2}}}, \quad r > 0, \tag{14.9.58}$$

with M being mass and $Q > 0$ a charge parameter. The unique global minimum of (14.9.58) is at

$$r_0 = \sqrt{2}Q. \tag{14.9.59}$$

Hence, we obtain

$$A(r_0) = 1 - \frac{4GM}{3\sqrt{3}Q}, \tag{14.9.60}$$

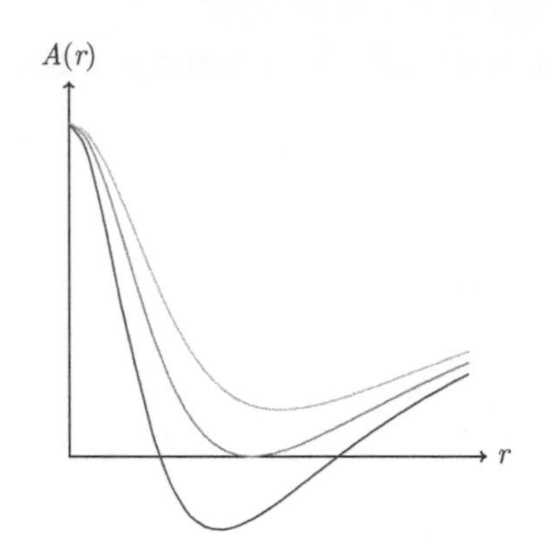

Figure 14.3 Profiles of the metric factor $A(r)$ of the Bardeen black hole solution free of singularity. As in the situation of the Reissner–Nordström black hole, there are appropriate ranges of the mass and charge parameters that give rise to inner and outer black hole horizons and extremal horizon, where the curve of $A(r)$ intersects the r-axis, as well as absence of horizon, when the curve of $A(r)$ does not intersect the r-axis.

which leads to the conditions $A(r_0) > 0, A(r_0) = 0, A(r_0) < 0$, or

$$4GM < 3\sqrt{3}Q, \quad 4GM = 3\sqrt{3}Q, \quad 4GM > 3\sqrt{3}Q, \tag{14.9.61}$$

giving rise to the absence of a black hole and occurrence of extremal and inner and outer black holes, respectively (Figure 14.3). Note that the extremal Bardeen black hole metric factor may be rewritten

$$\begin{aligned} A(r) &= 1 - \frac{3\sqrt{3}Qr^2}{2(r^2+Q^2)^{\frac{3}{2}}} \\ &= \frac{(Q^2+4r^2)(2Q^2-r^2)^2}{2(r^2+Q^2)^{\frac{3}{2}}(2(r^2+Q^2)^{\frac{3}{2}}+3\sqrt{3}Qr^2)}, \end{aligned} \tag{14.9.62}$$

which spells out the global minimum point (14.9.59) again and directly establishes the similarity between the extremal Bardeen black hole and the Reissner–Nordström one.

Another interesting and much-studied charged black hole solution free of curvature singularity is given by the metric factor [38, 283]

$$A(r) = 1 - \frac{2mr^2}{r^3+2l^2m}, \tag{14.9.63}$$

where $m, l > 0$ are constants, which shares the same spirit of the Bardeen black hole [40, 283] and is sometimes referred to as the *Hayward black hole* [220, 345].

Motivated by these and other developments, the subsequent sections of this chapter present a study of generalized Born–Infeld theories and some of their applications spanned over a much broader range of subjects.

14.10 Generalized Born–Infeld theories and applications

The rest of this chapter considers *generalized* Born–Infeld theories and some of their applications. It starts by discussing the formalism of generalized Born–Infeld theories and the associated governing equations, and goes on to elaborate on some of the rich applications of the generalized theories, including the asymmetry of electromagnetism regarding accommodating electric and magnetic point charges, relegation of curvature singularities of charged black holes, and the k-essence cosmology propelled by a scalar-wave matter of the Born–Infeld type.

Electromagnetic symmetry and asymmetry

Firstly, we recall that the Maxwell theory enjoys a perfect symmetry between its electric and magnetic sectors through the so-called electromagnetic duality. Such a symmetry is carried over to the situations of electric and magnetic point charges so that both suffer the same energy divergence. Such a symmetry is preserved by the Born–Infeld theory, in which electric and magnetic point charges both enjoy the same energy convergence or finiteness. In other words, in both the Maxwell and Born–Infeld theories, electric and magnetic point charges are given equal footing energetically, and the theories offer no mechanism to rule out the occurrence of a magnetic point charge, that is, a monopole [151, 167]. Although the notion of monopoles is fascinating and important in field theory physics [240, 460, 466, 567], such magnetically charged point particles have never been observed in nature or isolated in a laboratory, except for some condensed-matter-system simulations [235]. The Born–Infeld theory provides an opportunity to modify the Maxwell theory to break the electromagnetic symmetry so that a finite-energy electric point charge is maintained, but also a finite-energy magnetic point charge is excluded. Here, we show that such a *breakdown* of electric and magnetic point charge symmetry, referred to as *electromagnetic asymmetry*, may be regarded as a *generic* property of *nonlinear electrodynamics*. More precisely, we establish that, for any nonlinear electrodynamics governed by a polynomial function, the theory always accommodates finite-energy electric point charges but excludes magnetic ones. However, unlike what is seen in the classical Born–Infeld model, no upper bound for an electric field may be imposed in the current context. The word "generic" indicates that the set of polynomials is dense in the function space of nonlinear Lagrangian functions in view of the Stone–Weierstrass theorem [528, 611] such that any model of nonlinear electrodynamics may be approximated in a suitable sense by a sequence of models governed by polynomials. As a consequence, we could conclude that monopoles are generically ruled out with regard to the finite-energy condition.

Curvature singularities of charged black holes

Next, we expand our study in Sections 14.8 and 14.9 to explore the implications of finite-energy electric and magnetic point charges to the construction of charged black holes with regard to the associated curvature singularities. We saw that, in terms of the spherical coordinates, the Kretschmann invariant, K, of the Schwarzschild solution, near the center of the black hole, is of the type $K \sim r^{-6}$. For the Reissner–Nordström charged black hole solution, on the other hand, the curvature singularity is elevated to the type $K \sim r^{-8}$ due to the divergence of energy of an electric or magnetic point charge in the linear Maxwell electrodynamics. In the Born–Infeld theory formalism, however, such an elevation has been pulled back to $K \sim r^{-6}$ as a consequence of the energy convergence of point charges (Sections 14.8 and 14.9). In other words, in view of curvature singularities, finite mass and finite electromagnetic energy present themselves at an equal footing. In fact, such a singularity relegation phenomenon is universally valid in generalized Born–Infeld theories. Moreover, when a critical mass-energy condition is fulfilled, the Schwarzschild type curvature singularity may further be relegated or even eliminated in a systematic way. In this regard, the Bardeen black hole [28, 29, 37, 435] and the Hayward black hole [220, 283, 345] belong to this category of the regular black hole solutions of the Einstein equations coupled with the Born–Infeld type nonlinear electrodynamics, for which the critical mass-energy condition is embedded into the specialized forms of the Lagrangian or Hamiltonian densities. On the other hand, however, a common feature shared by the Bardeen and Hayward black holes is that the form of their Lagrangian action density can only accommodate one sector of electromagnetism, that is, either electric field or magnetic field is allowed to be present to model a point charge, but not both. This is due to the sign restriction under the radical root operation, in a sharp contrast to the classical Born–Infeld theory, which accommodates both electricity and magnetism at an equal footing. Hence, it will be of interest to find by taking advantage of the general formalism a model of nonlinear electrodynamics that accommodates both electric and magnetic point charges and at the same time gives rise to regular charged black hole solutions as in the Bardeen and Hayward models. Indeed, such a model may be obtained by taking the large n limit in a naturally formulated binomial model, which is a special situation of the polynomial model to be considered in the setting of the monopole exclusion problem. That is, the binomial model does not allow a finite-energy magnetic point charge but its large n limit, which assumes the form of an exponential model considered earlier by Hendi in [287, 288] in other contexts, accommodates both electric and magnetic point charges, and is free of sign restriction.

Cosmological expansion by Born–Infeld scalar-wave matters

Thirdly, Section 10.8 showed that, to obtain a theoretical interpretation of the observed accelerated expansion of the universe usually attributed to the existence of dark energy, a real scalar-wave field, referred to as quintessence, may be introduced to explain the rise of a hidden propelling force [101, 112, 179, 467, 553]. There, the quintessence is assumed to be governed canonically by the Klein–Gordon model such that the kinetic energy density term is minimal and the potential energy density may be adjusted to give rise to the desired dynamic evolution of the universe. When the kinetic density term is taken to be an adjustable nonlinear function of the canonical kinetic energy density, the scalar-wave matter is referred to as the *k-essence* [145, 165, 314, 461]. This nonlinear kinetic dynamics is of the form of the Born–Infeld type theory and conveniently provides a field-theoretical interpretation of a given equation of state relating the pressure and density of a cosmological fluid. Over such a meeting ground, we may examine the cosmic fluid contents or interpretations of the Born–Infeld type models in general settings. As an illustration, we present a full description of the dynamics of the cosmological evolution driven by the exponential model, as a kind of k-essence, used to generate regular charged black hole solutions earlier. In this situation, the equation of state is determined in terms of the Lambert W function, which gives rise to a Big Bang cosmology characterized by infinite curvature, density, and pressure at the start of time, and vanishing density and pressure at time infinity, such that the ratio of the pressure and density monotonically interpolates the dust and stiff-matter fluid models. Besides, the adiabatic squared speed of sound of the fluid of the exponential model remains within a physically anticipated range.

Section 14.9 showed that finite-energy dyonically charged black hole solutions generated from the Born–Infeld theory [73, 74, 75, 76], based on an invariance principle formalism as presented in Section 14.4, can be constructed. On the other hand, however, the classical Born–Infeld model defined by the action density (14.1.6) and its various generalizations do not support dyonic point charges with a finite energy. Thus, the last two sections of this chapter aim to construct finite-energy dyonically charged point particles and black holes based on the invariance principle formalism with a general nonlinear density profile function. In particular, we present some results and examples centered around the issues of relegated curvature singularities of charged black holes in such a general setting.

14.11 Electromagnetic asymmetry by virtue of point charges

Motivated by our study in Section 14.1, we now take the point of view that the Maxwell action density

$$\mathcal{L}_{\mathrm{M}} = -\frac{1}{4}F_{\mu\nu}F^{\mu\nu}, \tag{14.11.1}$$

is the leading-order approximation of the general action density

$$\mathcal{L} = f(\mathcal{L}_{\mathrm{M}}), \tag{14.11.2}$$

where $f(s)$ is a differentiable function of the real variable s over a suitable domain of definition containing $s = 0$ and satisfying the normalization condition

$$f(0) = 0, \quad f'(0) = 1. \tag{14.11.3}$$

Thus, subject to an applied source current (j^μ), the associated equations of motion are

$$\partial_\mu P^{\mu\nu} = j^\nu, \tag{14.11.4}$$

$$(P^{\mu\nu}) = (f'(\mathcal{L}_{\rm M})F^{\mu\nu}) = \begin{pmatrix} 0 & -D^1 & -D^2 & -D^3 \\ D^1 & 0 & -H^3 & H^2 \\ D^2 & H^3 & 0 & -H^1 \\ D^3 & -H^2 & H^1 & 0 \end{pmatrix}, \tag{14.11.5}$$

with $\mathbf{D} = (D^i)$ and $\mathbf{H} = (H^i)$ being the electric displacement and magnetic intensity fields, respectively, given by

$$D^i = P^{i0}, \quad H^i = -\frac{1}{2}\varepsilon^{ijk}P^{jk}, \tag{14.11.6}$$

as before, so that (14.11.4) assumes the Maxwell equation form

$$\nabla \cdot \mathbf{D} = \rho, \quad \frac{\partial \mathbf{D}}{\partial t} - \nabla \times \mathbf{H} = -\mathbf{j}, \tag{14.11.7}$$

where $\rho = j^0$ and $\mathbf{j} = (j^i)$ are the charge and current densities, respectively. On the other hand, the Bianchi identity remains intact, $\partial_\mu \tilde{F}^{\mu\nu} = 0$, where

$$\tilde{F}^{\mu\nu} = \frac{1}{2}\varepsilon^{\mu\nu\alpha\beta}F_{\alpha\beta}, \quad (\tilde{F}^{\mu\nu}) = \begin{pmatrix} 0 & -B^1 & -B^2 & -B^3 \\ B^1 & 0 & E^3 & -E^2 \\ B^2 & -E^3 & 0 & E^1 \\ B^3 & E^2 & -E^1 & 0 \end{pmatrix}, \tag{14.11.8}$$

is again the dual of $F_{\mu\nu}$, giving rise to the other two Maxwell equations,

$$\nabla \cdot \mathbf{B} = 0, \quad \frac{\partial \mathbf{B}}{\partial t} + \nabla \times \mathbf{E} = \mathbf{0}. \tag{14.11.9}$$

Furthermore, in this generalized setting, the energy-momentum tensor of the theory may be calculated as

$$T_{\mu\nu} = -f'(\mathcal{L}_{\rm M})F_{\mu\alpha}\eta^{\alpha\beta}F_{\nu\beta} - \eta_{\mu\nu}f(\mathcal{L}_{\rm M}), \tag{14.11.10}$$

such that the Hamiltonian energy density reads

$$\mathcal{H} = T_{00} = f'(\mathcal{L}_{\rm M})\mathbf{E}^2 - f(\mathcal{L}_{\rm M}), \quad \mathcal{L}_{\rm M} = \frac{1}{2}\left(\mathbf{E}^2 - \mathbf{B}^2\right). \tag{14.11.11}$$

Note that, when

$$f(s) = b^2\left(1 - \sqrt{1 - \frac{2}{b^2}s}\right), \quad b > 0, \tag{14.11.12}$$

we recover the classical Born–Infeld theory. In view of (14.11.5), we obtain the equations relating the pairs $\mathbf{D}, \mathbf{H}$ and $\mathbf{E}, \mathbf{B}$:

$$\mathbf{D} = \varepsilon(\mathbf{E}, \mathbf{B})\mathbf{E}, \quad \mathbf{B} = \mu(\mathbf{E}, \mathbf{B})\mathbf{H}, \tag{14.11.13}$$

where $\varepsilon(\mathbf{E}, \mathbf{B}), \mu(\mathbf{E}, \mathbf{B})$ are the field-dependent dielectrics and permeability coefficients given by

$$\varepsilon(\mathbf{E}, \mathbf{B}) = f'(\mathcal{L}_{\rm M}), \quad \mu(\mathbf{E}, \mathbf{B}) = \frac{1}{f'(\mathcal{L}_{\rm M})}. \tag{14.11.14}$$

Governing equations of generalized Born–Infeld electromagnetism

The governing equations describing the nonlinear electromagnetism defined by the generalized Lagrangian action density are comprised of (14.11.7), (14.11.9), and (14.11.13). The first equation in (14.11.9) indicates that $\mathbf{B}$ is solenoidal such that there is a vector field $\mathbf{A}$ serving as a potential for $\mathbf{B}$. That is, $\mathbf{B} = \nabla \times \mathbf{A}$. Inserting this into the second equation in (14.11.9), we get

$$\nabla \times \left(\mathbf{E} + \frac{\partial \mathbf{A}}{\partial t} \right) = \mathbf{0}. \tag{14.11.15}$$

Hence, there is a real scalar field ϕ such that

$$\mathbf{E} + \frac{\partial \mathbf{A}}{\partial t} = -\nabla \phi. \tag{14.11.16}$$

In view of (14.11.16), we see that the remaining governing equations, given in (14.11.7), become

$$\nabla \cdot \left(f'\left(\frac{1}{2} \left[\left| \nabla\phi + \frac{\partial \mathbf{A}}{\partial t} \right|^2 - |\nabla \times \mathbf{A}|^2 \right] \right) \left(\nabla\phi + \frac{\partial \mathbf{A}}{\partial t} \right) \right) = -\rho, \tag{14.11.17}$$

$$\frac{\partial}{\partial t} \left(f'\left(\frac{1}{2} \left[\left| \nabla\phi + \frac{\partial \mathbf{A}}{\partial t} \right|^2 - |\nabla \times \mathbf{A}|^2 \right] \right) \left(\nabla\phi + \frac{\partial \mathbf{A}}{\partial t} \right) \right)$$
$$+ \nabla \times \left(f'\left(\frac{1}{2} \left[\left| \nabla\phi + \frac{\partial \mathbf{A}}{\partial t} \right|^2 - |\nabla \times \mathbf{A}|^2 \right] \right) \nabla \times \mathbf{A} \right) = \mathbf{j}, \tag{14.11.18}$$

which is a closed coupled system of two non-homogeneous equations with unknowns ϕ and $\mathbf{A}$.

Polynomial model

We may consider the case that f in (14.11.2) is a polynomial of the form

$$f(s) = s + \sum_{k=2}^{n} a_k s^k, \quad a_2, \dots, a_n \in \mathbb{R}, \quad a_n \neq 0. \tag{14.11.19}$$

We regard (14.11.19) as an important general model because, by the celebrated Stone–Weierstrass density theorem [528, 611], any continuous function over a closed interval may be uniformly approximated by polynomials.

Electric point charge

Consider the electrostatic field generated from a point charge placed at the origin such that in (14.11.7) we have

$$\rho = 4\pi q \delta(\mathbf{x}), \quad q > 0, \quad \mathbf{j} = \mathbf{0}. \tag{14.11.20}$$

Thus, we have $\mathbf{H} = \mathbf{0}$ and

$$\mathbf{D} = \frac{q\mathbf{x}}{r^3}, \quad \mathbf{x} \neq \mathbf{0}, \quad r = |\mathbf{x}|. \tag{14.11.21}$$

On the other hand, from (14.11.13), (14.11.14), and (14.11.19), we have

$$\begin{aligned} \mathbf{D}^2 &= (f'(s))^2 \mathbf{E}^2, \quad s = \frac{\mathbf{E}^2}{2} \\ &= \left(1 + \sum_{k=2}^{n} \frac{k a_k}{2^{k-1}} \mathbf{E}^{2(k-1)}\right)^2 \mathbf{E}^2, \quad a_n \neq 0. \end{aligned} \tag{14.11.22}$$

In view of (14.11.21) and (14.11.22), we have

$$\mathbf{E}^2 = \mathrm{O}\left(r^{-\frac{4}{2n-1}}\right), \quad r \to 0. \tag{14.11.23}$$

Thus, by virtue of (14.11.23), we see that the Hamiltonian density (14.11.11) enjoys the property

$$\mathcal{H} = \mathrm{O}\left(r^{-\frac{4n}{2n-1}}\right), \quad r \to 0. \tag{14.11.24}$$

Moreover, using (14.11.21) in (14.11.22) again, we have

$$\mathbf{E}^2 = \mathrm{O}\left(r^{-4}\right), \quad r \to \infty. \tag{14.11.25}$$

Therefore, (14.11.11) leads to

$$\mathcal{H} = \mathrm{O}\left(r^{-4}\right), \quad r \to \infty. \tag{14.11.26}$$

Combining (14.11.24) and (14.11.26), we arrive at the finite energy conclusion $\int_{\mathbb{R}^3} \mathcal{H}\, \mathrm{d}\mathbf{x} < \infty$ for an electric point charge as anticipated as far as $n \geq 2$. Hence, we conclude that the existence of a finite-energy electric point charge naturally spells out the nonlinear "correction" terms to the action density generating function $f(s)$ as expressed in (14.11.19).

Magnetic point charge

We now consider a magnetic monopole. In this situation, the electric sector is trivial but the magnetic field $\mathbf{B}$ is generated from a magnetic point charge $g > 0$ resting at the origin so that

$$\nabla \cdot \mathbf{B} = 4\pi g \delta(\mathbf{x}), \quad g > 0, \tag{14.11.27}$$

which renders the solution

$$\mathbf{B} = \frac{g\mathbf{x}}{r^3}, \quad \mathbf{x} \neq \mathbf{0}. \tag{14.11.28}$$

Hence, inserting $\mathbf{E} = \mathbf{0}$ and (14.11.28) into (14.11.11) and observing (14.11.19), we see that $\mathcal{H} \sim r^{-4}$ regardless of the details of the model. In other words, for a magnetic point charge, we always get a divergent energy around the origin. Thus, we see that, energetically, the simple model (14.11.19) breaks the electric and magnetic symmetry and accepts an electric point charge but rejects a magnetic point charge.

Quadratic model

Next, we work out explicitly the simplest nonlinear situation when (14.11.19) assumes the quadratic form

$$f(s) = s + as^2, \quad a > 0, \tag{14.11.29}$$

which has also been studied earlier in the context of cosmology [156, 227] and is of independent interest. To this end, we insert (14.11.29) into (14.11.22) or

$$\mathbf{D}^2 = (1 + a\mathbf{E}^2)^2\mathbf{E}^2 \tag{14.11.30}$$

to obtain

$$\mathbf{E}^2 = \frac{1}{a}\left(\frac{h(\mathbf{D}^2)}{6} + \frac{2}{3h(\mathbf{D}^2)} - \frac{2}{3}\right), \tag{14.11.31}$$

$$h(\mathbf{D}^2) = \left(8 + 108a\mathbf{D}^2 + 12\sqrt{81a^2\mathbf{D}^4 + 12a\mathbf{D}^2}\right)^{\frac{1}{3}}. \tag{14.11.32}$$

Using (14.11.21), we see that (14.11.31)–(14.11.32) render

$$\mathbf{E}^2 = \mathrm{O}(r^{-\frac{4}{3}}), \quad r \to 0, \tag{14.11.33}$$

which in view of (14.11.11) and (14.11.29) gives us

$$\mathcal{H} = \mathrm{O}(r^{-\frac{8}{3}}), \quad r \to 0. \tag{14.11.34}$$

The asymptotics (14.11.33) and (14.11.34) are special cases of (14.11.23) and (14.11.24), respectively. The asymptotics of $\mathbf{E}$ and $\mathcal{H}$ at spatial infinity are clearly given by (14.11.25) and (14.11.26).

From (14.11.11) and (14.11.29), we have

$$a\mathcal{H} = \frac{1}{2}a\mathbf{E}^2 + \frac{3}{4}(a\mathbf{E}^2)^2, \tag{14.11.35}$$

where

$$a\mathbf{E}^2 = \frac{p(r)}{6} + \frac{2}{3p(r)} - \frac{2}{3}, \tag{14.11.36}$$

$$p(r) = h(\mathbf{D}^2) = \left(8 + \frac{108aq^2}{r^4} + 12\sqrt{\frac{81a^2q^4}{r^8} + \frac{12aq^2}{r^4}}\right)^{\frac{1}{3}}, \tag{14.11.37}$$

by using (14.11.21). These expressions appear complicated. Fortunately, the dependence of the function $p(r)$ on the parameters a and q may be scaled away by setting

$$r = (2a)^{\frac{1}{4}}q^{\frac{1}{2}}\rho, \quad \rho > 0, \tag{14.11.38}$$

where the parameter a is comparable to the quantity $\frac{1}{2b^2}$ in the classical Born–Infeld theory, as shown in (14.11.12), such that the energy of the point charge may be computed to yield

$$E = 4\pi \int_0^\infty \mathcal{H}\, r^2 \mathrm{d}r = 8\pi\,(2a)^{-\frac{1}{4}} q^{\frac{3}{2}} \int_0^\infty a\mathcal{H}\,\rho^2 \mathrm{d}\rho$$
$$\approx 4\pi (2a)^{-\frac{1}{4}} q^{\frac{3}{2}} (2.939283), \tag{14.11.39}$$

in view of (14.11.35)–(14.11.38), by utilizing a MAPLE 10 integration package.

In spherical coordinates, $\mathbf{E}$ is radially given by $\mathbf{E} = (E^r, 0, 0)$ with

$$E^r = -\frac{\mathrm{d}\psi}{\mathrm{d}r} = -\psi'(r). \tag{14.11.40}$$

Thus, inserting (14.11.36)–(14.11.37), we obtain

$$\psi(r) = \frac{q}{r_0}\, h\left(\frac{r}{r_0}\right), \quad r_0 = (2a)^{\frac{1}{4}} q^{\frac{1}{2}}, \tag{14.11.41}$$

$$h(\rho) = \sqrt{2}\int_\rho^\infty \left(\frac{p_0(\eta)}{6} + \frac{2}{3p_0(\eta)} - \frac{2}{3}\right)^{\frac{1}{2}} \mathrm{d}\eta, \quad \rho = \frac{r}{r_0}, \tag{14.11.42}$$

$$p_0(\rho) = \left(8 + \frac{54}{\rho^4} + 6\sqrt{\frac{81}{\rho^8} + \frac{24}{\rho^4}}\right)^{\frac{1}{3}}, \tag{14.11.43}$$

which resembles the parallel result derived for the classical Born–Infeld model since the integrand in (14.11.42) has the property

$$\sqrt{2}\left(\frac{p_0(\eta)}{6} + \frac{2}{3p_0(\eta)} - \frac{2}{3}\right)^{\frac{1}{2}}$$
$$= \frac{1}{\sqrt{1+\eta^4}} + \frac{3}{8\eta^{10}} + \mathrm{O}\left(\eta^{-14}\right), \quad \text{as } \eta \to \infty, \tag{14.11.44}$$

such that

$$\psi(r) = \frac{q}{r} - \frac{q r_0^4}{10 r^5} + \mathrm{O}\left(r^{-9}\right), \quad \text{as } r \to \infty, \tag{14.11.45}$$

which says the theory coincides with that given by the Coulomb law asymptotically. Moreover, in terms of the quantity r_0 in (14.11.41), we have

$$\frac{E}{4\pi} = (2.9393)\,\frac{q^2}{r_0}. \tag{14.11.46}$$

Recall that, in the classical Born–Infeld model, the numerical factor 2.9393 shown here is 1.2361 instead.

On the other hand, however, the expressions (14.11.41) and (14.11.42) lead to

$$\psi(0) \approx (4.408757)\frac{q}{r_0}. \tag{14.11.47}$$

This result deviates from that of the Coulomb potential, of course, which diverges like $\frac{1}{r}$ as $r \to 0$, and that of the Born–Infeld potential, which assumes the value:

$$\frac{q}{r_0}\int_0^\infty \frac{\mathrm{d}\eta}{\sqrt{1+\eta^4}} = \frac{1}{4}B\left(\frac{1}{4},\frac{1}{4}\right)\frac{q}{r_0} \approx (1.854075)\frac{q}{r_0}. \tag{14.11.48}$$

In view of (14.11.40)–(14.11.43), we obtain after a lengthy computation the asymptotics of the free charge density

$$\begin{aligned}\rho_{\text{free}} &= \frac{1}{4\pi}\nabla\cdot\mathbf{E} = \frac{1}{4\pi r^2}\frac{\mathrm{d}}{\mathrm{d}r}(r^2E^r)\\ &= \frac{qr_0^4}{2\pi r^7} + \mathrm{O}\left(r^{-11}\right), \quad \text{as } r\to\infty, &(14.11.49)\\ &= \frac{2^{\frac{1}{3}}q}{3\pi r_0^{\frac{4}{3}} r^{\frac{5}{3}}} + \mathrm{O}(r^{-\frac{1}{3}}), \quad \text{as } r\to 0. &(14.11.50)\end{aligned}$$

The result (14.11.49) coincides with that, but the result (14.11.50) deviates from that, which finds $\rho_{\text{free}} \sim r^{-1}$ as $r \to 0$, in the classical Born–Infeld model.

To compute the total free charge, we need to evaluate

$$q_{\text{free}} = \int_{\mathbb{R}^3} \rho_{\text{free}}\,\mathrm{d}\mathbf{x}, \tag{14.11.51}$$

in which the integrand ρ_{free} appears complicated. Fortunately, the asymptotic behavior of E^r is readily determined through (14.11.41)–(14.11.43) to yield

$$E^r = \frac{2^{\frac{1}{3}}q}{r_0^{\frac{4}{3}} r^{\frac{2}{3}}} + \mathrm{O}\left(r^{\frac{2}{3}}\right), \quad \text{as } r\to 0; \tag{14.11.52}$$

$$E^r = \frac{q}{r^2} + \mathrm{O}\left(r^{-6}\right), \quad \text{as } r\to\infty. \tag{14.11.53}$$

In view of the left-hand side of (14.11.49) and the estimates in (14.11.52)–(14.11.53), we get

$$q_{\text{free}} = (r^2E^r)_{r=0}^{r=\infty} = q. \tag{14.11.54}$$

Thus, as in the classical Born–Infeld theory, the total free charge q_{free} generated from the electric field induced from a point charge coincides with the prescribed value of the point charge q, although the former is given by a continuously distributed charge density while the latter by a measure concentrated at a single point. This fact is an interesting feature of the theory.

Although the free electric charge density ρ_{free} appears too complicated to present here, we may solve (14.11.13) with (14.11.21) and (14.11.29) to get

$$E^r = \frac{\sqrt{2}q}{r_0^2}\left(\frac{p(r)^{\frac{1}{3}}}{6} - \frac{2}{p(r)^{\frac{1}{3}}}\right), \tag{14.11.55}$$

$$p(r) = \frac{108}{\sqrt{2}}\left(\frac{r_0}{r}\right)^2 + 12\sqrt{\frac{81}{2}\left(\frac{r_0}{r}\right)^4 + 12}. \tag{14.11.56}$$

Hence, the free charge contained in $\{|\mathbf{x}| \leq r\}$ is

$$q_{\text{free}}(r) = \int_{|\mathbf{x}|\leq r} \rho_{\text{free}}\,\mathrm{d}\mathbf{x} = r^2 E^r = \sqrt{2}q\left(\frac{r}{r_0}\right)^2 \left(\frac{p(r)^{\frac{1}{3}}}{6} - \frac{2}{p(r)^{\frac{1}{3}}}\right), \quad (14.11.57)$$

where $p(r)$ is given in (14.11.56). As a consequence of (14.11.57), we have

$$q_{\text{free}}(r) = 2^{\frac{1}{3}} q \left(\frac{r}{r_0}\right)^{\frac{4}{3}} + \mathrm{O}(r^{\frac{8}{3}}), \quad \text{as } r \to 0; \quad (14.11.58)$$

$$q_{\text{free}}(r) = q - \frac{q}{2}\left(\frac{r_0}{r}\right)^4 + \mathrm{O}\left(r^{-8}\right), \quad \text{as } r \to \infty, \quad (14.11.59)$$

which are consistent with (14.11.52)–(14.11.53) and (14.11.54). At $r = r_0$ and $r = 2r_0$, the free charges given by (14.11.57) are

$$q_{\text{free}}(r_0) = (0.7709)\, q, \quad q_{\text{free}}(2r_0) = (0.9714)\, q, \quad (14.11.60)$$

respectively, within four decimal places. Figure 14.4 plots the quantity $\frac{q_{\text{free}}(r)}{q}$ against $\rho = \frac{r}{r_0}$, which shows that the free charge contained in $\{|\mathbf{x}| \leq r\}$ converges to its limiting value q rapidly as $r \to \infty$.

As noted earlier, the numerical factor on the right-hand side of (14.11.46) and that in the same calculation of the point-charge energy of the classical model lead to the ratio

$$\tau_0 = \frac{2.9391}{1.2361} = 2.37772. \quad (14.11.61)$$

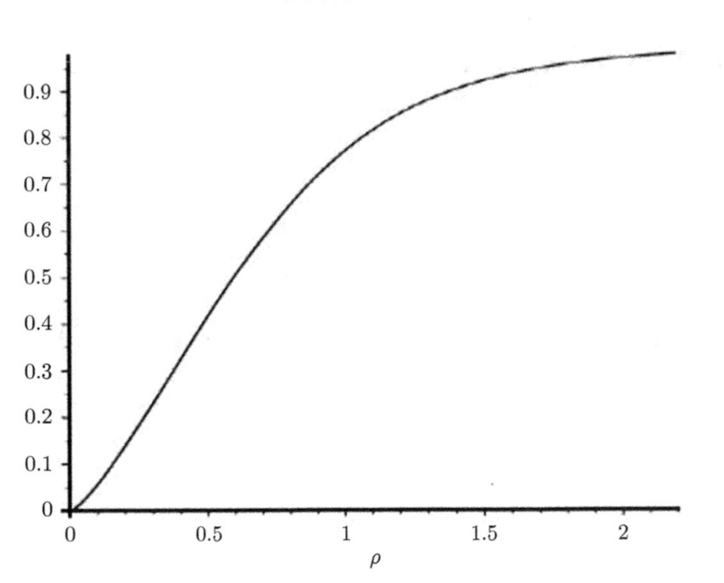

Figure 14.4 A plot of the free electric charge $q_{\text{free}}(r)$ in the ball of radius r around the point charge given as in (14.11.57) relative to its limit value q against the ratio $\rho = \frac{r}{r_0}$. Note that about 80% of the free charge is contained in the ball of radius r_0 and the charge quickly approaches its full-space limiting value when the ball of radius is beyond the threshold r_0.

Hence, inserting the electron energy and charge values, we obtain

$$r_0 = \tau_0(2.28 \times 10^{-13}\text{cm}) = 5.4212 \times 10^{-13}\text{cm}, \tag{14.11.62}$$

as an estimate for the electron radius. Consequently, we may use (14.11.41) to find

$$a = \frac{r_0^4}{2e^2} = \frac{r_0^4}{2 \times (4.8032 \times 10^{-10})^2} = 1.8719 \times 10^{-31}, \tag{14.11.63}$$

in the electrostatic system of units, or esu, which is a tiny quantity. This result suggests how the Maxwell electromagnetism may be viewed as the leading-order approximation of the nonlinear theory considered here as in the Born–Infeld theory. As a comparison, we have

$$\frac{e}{r_0^2} = \frac{1}{\sqrt{2a}} = 1.634345 \times 10^{15}, \tag{14.11.64}$$

which is of a similar magnitude of the Born–Infeld or Born parameter b.

Magnetic point charge

On the other hand, in view of (14.11.28) with (14.11.29) and (14.11.13), we see that the magnetic intensity field $\mathbf{H}$ of a magnetic point charge is

$$\mathbf{H} = \left(1 - \frac{ag^2}{r^4}\right)\frac{g\mathbf{x}}{r^3}, \quad \mathbf{x} \neq \mathbf{0}; \quad H^r = \frac{\mathbf{x}}{r}\cdot\mathbf{H} = \left(1 - \frac{ag^2}{r^4}\right)\frac{g}{r^2}, \tag{14.11.65}$$

which leads to a divergent total free magnetic charge as well, in addition to its divergent energy, since

$$\begin{aligned} g_{\text{free}} &= \frac{1}{4\pi}\int_{\mathbb{R}^3} \nabla\cdot\mathbf{H}\,\mathrm{d}x = \int_0^\infty \frac{\mathrm{d}}{\mathrm{d}r}\left(r^2 H^r\right)\,\mathrm{d}r \\ &= g\left(1 - \frac{ag^2}{r^4}\right)_{r=0}^{r=\infty} = \infty. \end{aligned} \tag{14.11.66}$$

Consequently, we see that, both energetically and in terms of free magnetic charge, parallel to the notion of free electric charge, the theory (14.11.29), or more generally, (14.11.19), rejects a magnetic monopole. This simple feature — the *exclusion of monopoles* — is not present in the classical Born–Infeld theory, in which a magnetic point charge is of finite energy as an electric point charge.

Thus, we conclude that a nonlinear theory of electromagnetism of an arbitrary polynomial type (14.11.19) always accommodates a finite-energy electric point charge but rejects a finite-energy magnetic point charge. Therefore, in view of the Stone–Weierstrass density theorem (that any continuous function over a closed interval may be approximated by a sequence of polynomials), we state that, generically, a nonlinear theory of electromagnetism allows a finite-energy electric point charge but does not allow a magnetic monopole, thereby establishing an electromagnetic asymmetry. This statement may be referred to as an *electromagnetic asymmetry theorem* for *nonlinear electrodynamics*.

Static equations in quadratic model

It is of interest to consider static solutions of the equations in the quadratic model. In this case, (14.11.9) indicates that $\mathbf{E}$ is conservative and $\mathbf{B}$ is solenoidal. Hence there are time-independent real scalar field ϕ and vector field $\mathbf{A}$ such that

$$\mathbf{E} = -\nabla\phi, \quad \mathbf{B} = \nabla \times \mathbf{A}. \tag{14.11.67}$$

Thus, using these in (14.11.13) and observing (14.11.29), we see that (14.11.7) assumes the form

$$\nabla \cdot \left([1 + a(|\nabla\phi|^2 - |\nabla \times \mathbf{A}|^2)]\nabla\phi\right) = -\rho, \tag{14.11.68}$$

$$\nabla \times \left([1 + a(|\nabla\phi|^2 - |\nabla \times \mathbf{A}|^2)]\nabla \times \mathbf{A}\right) = \mathbf{j}, \tag{14.11.69}$$

which are highly nonlinearly coupled. Of course, these equations are a limiting case of (14.11.17)–(14.11.18). In the special situation either $\mathbf{j} = \mathbf{0}$ or $\rho = 0$, it is consistent to set $\mathbf{A} = \mathbf{0}$ or $\phi = 0$, respectively, such that the system reduces into

$$\nabla \cdot \left([1 + a|\nabla\phi|^2]\nabla\phi\right) = -\rho, \tag{14.11.70}$$

which is a generalized Poisson equation, or its vector field version,

$$\nabla \times \left([1 - a|\nabla \times \mathbf{A}|^2]\nabla \times \mathbf{A}\right) = \mathbf{j}, \tag{14.11.71}$$

respectively, both of independent analytic interest.

14.12 Charged black holes

Subject to a gravitational background given by the metric tensor $g_{\mu\nu}$, the generalized Born–Infeld model is governed by the Lagrangian action density

$$\mathcal{L} = f(\mathcal{L}_\mathrm{M}), \quad \mathcal{L}_\mathrm{M} = -\frac{1}{4}F_{\mu\nu}F^{\mu\nu}, \quad F_{\mu\nu} = g_{\mu\alpha}g_{\nu\beta}F^{\alpha\beta}, \tag{14.12.1}$$

with the associated energy-momentum tenor

$$T_{\mu\nu} = -f'(\mathcal{L}_\mathrm{M})F_{\mu\alpha}g^{\alpha\beta}F_{\nu\beta} - g_{\mu\nu}f(\mathcal{L}_\mathrm{M}). \tag{14.12.2}$$

In this situation, the generalized Born–Infeld equations associated with (14.12.1) are

$$\frac{1}{\sqrt{-g}}\partial_\mu\left(\sqrt{-g}P^{\mu\nu}\right) = 0, \quad P^{\mu\nu} = f'(\mathcal{L}_\mathrm{M})F^{\mu\nu}. \tag{14.12.3}$$

Subsequently, we are interested in charged black hole solutions to the coupled system of the zero cosmological constant Einstein equations (14.8.1) and (14.12.3) for which, as before, the spacetime line element in the ordered spherical coordinates $(x^\mu) = (t, r, \theta, \phi)$ assumes the form (10.4.16) such that the nontrivial components of the Ricci tensor are given by (10.4.17)–(10.4.20). Note also that we use the prime $'$ to denote differentiation either with respect to the radial variable r or with respect to $s = \mathcal{L}_\mathrm{M}$ in (14.11.3), which should not cause confusion in various contexts.

Electrically charged black holes

In the spherically symmetric electrostatic situation, the field tensors $P^{\mu\nu}$, $F^{\mu\nu}$, and $F_{\mu\nu}$ assume the same forms (14.8.6), (14.8.7), and (14.8.8), respectively. Hence, we see that the nontrivial components of (14.12.2) are

$$T_{00} = A(f'(\mathcal{L}_\mathrm{M})(E^r)^2 - f(\mathcal{L}_\mathrm{M})), \tag{14.12.4}$$

$$T_{11} = -\frac{1}{A}(f'(\mathcal{L}_\mathrm{M})(E^r)^2 - f(\mathcal{L}_\mathrm{M})), \tag{14.12.5}$$

$$T_{22} = r^2 f(\mathcal{L}_\mathrm{M}), \quad T_{33} = \sin^2\theta T_{22}, \tag{14.12.6}$$

giving rise to

$$T = 2f'(\mathcal{L}_\mathrm{M})(E^r)^2 - 4f(\mathcal{L}_\mathrm{M}). \tag{14.12.7}$$

Thus, it follows from (10.4.17)–(10.4.20), (14.12.4)–(14.12.6), and (14.12.7) that (14.8.1) becomes

$$(r^2A')' = 16\pi G r^2 f(\mathcal{L}_\mathrm{M}), \tag{14.12.8}$$

$$(rA)' = 1 - 8\pi G r^2(f'(\mathcal{L}_\mathrm{M})(E^r)^2 - f(\mathcal{L}_\mathrm{M})). \tag{14.12.9}$$

On the other hand, from $\sqrt{-g} = r^2\sin\theta$ and (14.8.7)–(14.8.8), we see that the generalized Born–Infeld equations (14.12.3) are reduced into

$$(r^2 f'(\mathcal{L}_\mathrm{M})E^r)' = 0. \tag{14.12.10}$$

In view of (14.12.9) and (14.12.10), we have

$$\begin{aligned}(r^2A')' &= r((rA)'') \\ &= -8\pi G r\left((r^2 f'(\mathcal{L}_\mathrm{M})E^r)'E^r + r^2 f'(\mathcal{L}_\mathrm{M})E^r(E^r)'\right. \\ &\quad \left.-r^2 f'(\mathcal{L}_\mathrm{M})\mathcal{L}_\mathrm{M}' - 2rf(\mathcal{L}_\mathrm{M})\right) \\ &= 16\pi G r^2 f(\mathcal{L}_\mathrm{M}),\end{aligned} \tag{14.12.11}$$

which recovers (14.12.8). Hence, the reduced form of (14.8.1) is actually (14.12.9). Besides, note that, in the present context, the Hamiltonian energy density is $\mathcal{H} = T_0^0 = g^{00}T_{00}$, which assumes the form

$$\mathcal{H} = f'(\mathcal{L}_\mathrm{M})(E^r)^2 - f(\mathcal{L}_\mathrm{M}), \tag{14.12.12}$$

which happens to coincide with the flat-spacetime Hamiltonian energy density (14.11.11). Hence, with (14.12.12), we can integrate (14.12.9) to obtain the solution

$$A(r) = 1 - \frac{2GM}{r} + \frac{8\pi G}{r}\int_r^\infty \rho^2\mathcal{H}(\rho)\,\mathrm{d}\rho, \tag{14.12.13}$$

where $M > 0$ is an integration constant.

To understand the meaning of the integral on the right-hand side of (14.12.13), we recall that

$$E = \int \mathcal{H}\sqrt{-g}\,\mathrm{d}r\mathrm{d}\theta\mathrm{d}\phi = 4\pi\int_0^\infty \mathcal{H}(r)r^2\,\mathrm{d}r \tag{14.12.14}$$

is the total static energy of the electric field, which is the same as the flat-space Born–Infeld energy. Thus, if we denote by $E(r)$ the static energy of the electric field distributed in the region $\{|\mathbf{x}| \geq r\}$ such that $E = E(0)$, then (14.12.13) gives us

$$A(r) = 1 - \frac{2GM}{r} + \frac{2GE(r)}{r}. \tag{14.12.15}$$

Finiteness of energy implies $E(r) \to 0$ rapidly as $r \to \infty$, in general, such that M may be identified with the mass of the gravitational field [83]. An event horizon of the solution occurs at a positive root of (14.12.15) whose existence is ensured by the condition $a(0) < 0$, where $a(r) \equiv rA(r)$, namely,

$$M > E. \tag{14.12.16}$$

Since the Kretschmann invariant of (10.4.16) is as given in (14.8.37), which implies that the only curvature singularity is at $r = 0$, we find that a naked singularity may occur when (14.12.16) is violated.

It is interesting to see that in (14.12.15) and (14.12.16) the gravitational mass M and electric energy E are placed at "equal" footings.

In the electric point charge situation, when $\mathbf{D}$ is given by (14.11.21), the electric field $\mathbf{E}$ follows the relation (14.11.22) if $f(s)$ is defined by (14.11.19). So, we have

$$E^r = \frac{q}{r^2} + \mathrm{O}(r^{-4}), \quad r \to \infty. \tag{14.12.17}$$

Inserting (14.12.17) into (14.12.12), we get

$$\mathcal{H}(r) = \frac{q^2}{2r^4} + \mathrm{O}(r^{-8}), \quad r \to \infty. \tag{14.12.18}$$

Thus, we can apply (14.12.18) in the formula (14.12.13) to arrive at

$$A(r) = 1 - \frac{2GM}{r} + \frac{4\pi G q^2}{r^2} + \mathrm{O}(r^{-6}), \quad r \to \infty, \tag{14.12.19}$$

extending the Reissner–Nordström electrically charged black hole solution.

Relegation of curvature singularities

As an initial comparison, we recall that for the classical Reissner–Nordström black hole of mass M and electric charge q (Section 10.4), the metric factor reads

$$A(r) = 1 - \frac{2GM}{r} + \frac{4\pi G q^2}{r^2}, \tag{14.12.20}$$

such that the Kretschmann scalar (14.8.37) assumes the form

$$K = \frac{48G^2}{r^8}\left(M^2 r^2 - 8\pi M q^2 r + \frac{56\pi^2 q^4}{3}\right). \tag{14.12.21}$$

In other words, the presence of electricity elevates the Schwarzschild curvature singularity of the blow-up type r^{-6}, in absence of electricity, to the type r^{-8}, as is clearly implicated in (14.12.20) by the appearance of an inverse square term of the radial variable.

Interestingly, note that in the situation of the Born–Infeld theory we may rewrite (14.12.15) as

$$A(r) = 1 - \frac{2GM}{r} + \frac{2GE}{r} - \frac{8\pi G}{r}\int_0^r \mathcal{H}(\rho)\rho^2\,\mathrm{d}\rho, \tag{14.12.22}$$

where the integral on the right-hand side of (14.12.22) gives rise to the energy of the electric charge in $\{|\mathbf{x}| \leq r\}$, which is usually bounded by the structure of the theory. Thus, $A(r)$ is of the type r^{-1} as that for the Schwarzschild metric. In other words, convergence of electric energy "relegates" the curvature singularity of the Reissner–Nordström metric to the level of the Schwarzschild metric, in the general setting here.

As a concrete illustration, we reconsider and reexamine the classical Born–Infeld theory (Section 14.1) so that we have

$$D^r = \frac{q}{r^2}, \quad E^r = \frac{q}{\sqrt{r^4 + a^4}}, \quad a = \sqrt{\frac{q}{b}}, \tag{14.12.23}$$

$$\mathcal{H} = \frac{(E^r)^2}{\sqrt{1 - \frac{1}{b^2}(E^r)^2}} - b^2\left(1 - \sqrt{1 - \frac{1}{b^2}(E^r)^2}\right), \tag{14.12.24}$$

resulting in

$$\mathcal{H} = \frac{q^2}{r^2(\sqrt{r^4 + a^4} + r^2)}. \tag{14.12.25}$$

Inserting (14.12.25) into (14.12.14) and (14.12.22), we have

$$E = \frac{4\pi q^2}{a}\int_0^\infty \frac{\mathrm{d}x}{\sqrt{x^4 + 1} + x^2} = \frac{4\pi^{\frac{5}{2}} q^2}{3\Gamma^2(\frac{3}{4})\,a}, \tag{14.12.26}$$

$$A(r) = 1 - \frac{2G}{r}(M - E) - \frac{8\pi G q^2}{ar}\int_0^{\frac{r}{a}} \frac{\mathrm{d}x}{\sqrt{x^4 + 1} + x^2}, \tag{14.12.27}$$

such that we see clearly that $A(r)$ behaves like that for the Schwarzschild metric, $A(r) \sim r^{-1}$, for r near zero, as asserted.

In order to examine (14.8.37), we use (14.12.8) and (14.12.9) to obtain the general relations

$$r^2 A'' = 2(A - 1) + 16\pi G r^2 f'(\mathcal{L}_{\mathrm{M}})(E^r)^2, \tag{14.12.28}$$

$$rA' = 1 - A - 8\pi G r^2 \mathcal{H}. \tag{14.12.29}$$

As a consequence of (14.12.25) and (14.12.27)–(14.12.29), we see that (14.8.37) gives us

$$K = \frac{16G^2}{r^4}\left(\frac{3(M-E)^2}{r^2} + \frac{8\pi(M-E)qb}{r} + 16\pi^2 q^2 b^2 - \frac{32\pi^2 q b^3 r^2}{3} + \mathrm{O}(r^4)\right), \tag{14.12.30}$$

as $r \to 0$. This expression indicates that, in general, the Kretschmann curvature of the charged black hole solution behaves like that of the Schwarzschild solution, of the same blow-up type r^{-6}, at the singularity, and it recovers the formula for the Kretschmann curvature in the zero electric charge limit, $q = 0$ and $E = 0$, of the Schwarzschild black hole,

$$K = \frac{48G^2M^2}{r^6}. \tag{14.12.31}$$

More interestingly, in the critical situation when the mass M is equal to the electric energy,

$$M = E \tag{14.12.32}$$

subsequently referred to as the *critical mass-energy condition*, the expression (14.12.30) becomes

$$K = \frac{2^8\pi^2G^2qb^2}{r^4}\left(q - \frac{2br^2}{3} + \mathrm{O}(r^4)\right), \quad r \to 0. \tag{14.12.33}$$

Thus, the order of the blow-up of the curvature drops from 6 to 4, so that the singularity is considerably "relegated" or "regularized".

Dyonic black hole solutions

Consider a prescribed dyonic point charge at the origin giving rise to a radially symmetric electric and magnetic field distribution so that $P^{\mu\nu}$, $F^{\mu\nu}$, and $F_{\mu\nu}$ follow (14.9.1), (14.9.2), and (14.9.3), respectively, and

$$\mathcal{L}_{\mathrm{M}} = \frac{1}{2}\left((E^r)^2 - r^4\sin^2\theta\,(B^r)^2\right). \tag{14.12.34}$$

From (10.4.16), we see that (14.12.3) at $\nu = 0$ gives us $\partial_r\left(r^2\sin\theta D^r\right) = 0$, resulting in

$$D^r = \frac{q}{r^2}, \tag{14.12.35}$$

which is an electric point charge. Besides, by the same reason, the equation (14.12.3) at $\nu = 2$ reads $\partial_\theta(r^2\sin\theta H^r) = 0$, rendering

$$H^r = \frac{H(r)}{\sin\theta}, \tag{14.12.36}$$

where $H(r)$ is a function to be determined. Thus, consistency in (14.12.3) with (14.9.1) and (14.9.2) leads to the relations

$$\begin{aligned} B^r = \frac{B(r)}{\sin\theta}, \quad \mathcal{L}_{\mathrm{M}} = \frac{1}{2}\left((E^r)^2 - r^4B^2(r)\right), \\ f'(\mathcal{L}_{\mathrm{M}})E^r = D^r, \quad f'(\mathcal{L}_{\mathrm{M}})B = H. \end{aligned} \tag{14.12.37}$$

In particular, we see that H relates to B but is otherwise arbitrary. This structure offers convenience in generating solutions for our purposes.

Now consider the Einstein equations (14.8.1). Note that, among the nontrivial components of the energy-momentum tensor $T_{\mu\nu}$, the quantities T_{00} and T_{11} are as given in (14.12.4) and (14.12.5) with $\mathcal{L}_{\mathrm{M}}$ as given in (14.12.37), and the other quantities are

$$T_{22} = r^6 f'(\mathcal{L}_{\mathrm{M}})B^2 + r^2 f(\mathcal{L}_{\mathrm{M}}), \quad T_{33} = \sin^2\theta T_{22}. \tag{14.12.38}$$

Thus

$$\begin{aligned} T &= 2f'(\mathcal{L}_{\mathrm{M}})\left((E^r)^2 - r^4 B^2\right) - 4f(\mathcal{L}_{\mathrm{M}}) \\ &= 4\left(f'(\mathcal{L}_{\mathrm{M}})\mathcal{L}_{\mathrm{M}} - f(\mathcal{L}_{\mathrm{M}})\right), \end{aligned} \tag{14.12.39}$$

which extends (14.12.7). In view of these quantities and (10.4.17)–(10.4.20), we see that the system (14.8.1) becomes

$$(r^2 A')' = 16\pi G r^2\left(r^4 f'(\mathcal{L}_{\mathrm{M}})B^2 + f(\mathcal{L}_{\mathrm{M}})\right), \tag{14.12.40}$$

$$(rA)' = 1 - 8\pi G r^2 (f'(\mathcal{L}_{\mathrm{M}})(E^r)^2 - f(\mathcal{L}_{\mathrm{M}})). \tag{14.12.41}$$

To establish the compatibility of (14.12.40) and (14.12.41), we differentiate (14.12.41) with respect to r to get

$$\begin{aligned} (rA)'' &= -8\pi G\left((r^2 f'(\mathcal{L}_{\mathrm{M}})E^r)'E^r + r^2 f'(\mathcal{L}_{\mathrm{M}})E^r(E^r)'\right) \\ &\quad + 8\pi G\left(2rf(\mathcal{L}_{\mathrm{M}}) + r^2 f'(\mathcal{L}_{\mathrm{M}})\left[E^r(E^r)' - 2r^3B^2 - r^4BB'\right]\right) \\ &= 16\pi G r f(\mathcal{L}_{\mathrm{M}}) - 8\pi G r^5 f'(\mathcal{L}_{\mathrm{M}})(2B^2 + rBB'), \end{aligned} \tag{14.12.42}$$

where we have used the condition $\partial_r(r^2 f'(\mathcal{L}_{\mathrm{M}})E^r) = \partial_r(r^2 D^r) = 0$. Consequently, it follows from (14.12.42) that

$$(r^2A')' = r(rA)'' = 16\pi G r^2 f(\mathcal{L}_{\mathrm{M}}) - 8\pi G r^6 f'(\mathcal{L}_{\mathrm{M}})(2B^2 + rBB'), \tag{14.12.43}$$

which is compatible with (14.12.40) if and only if B satisfies the equation

$$B' = -\frac{4B}{r}. \tag{14.12.44}$$

This equation has the solution

$$B(r) = \frac{g}{r^4}, \tag{14.12.45}$$

where g is an arbitrary constant resembling the magnetic charge, which may be taken to be positive for convenience. In fact, the expression (14.12.45) also follows from the Bianchi identity for the electromagnetic field $F_{\mu\nu}$, which in the current context reads

$$F_{23,1} = 0 \quad \text{or} \quad \partial_r(r^4 \sin^2\theta\, B^r) = \partial_r(r^4 \sin\theta\, B) = 0. \tag{14.12.46}$$

Note that (14.12.45) is universally valid to ensure both the compatibility of (14.12.40) and (14.12.41) and the fulfillment of the Bianchi identity, and thus, model independent.

In summary, we see that the gravitational metric factor A for a black hole of a dyonic point charge is given by the general formula

$$
\begin{aligned}
A(r) &= 1 - \frac{2GM}{r} + \frac{8\pi G}{r}\int_r^\infty \mathcal{H}(\rho)\rho^2\,\mathrm{d}\rho \\
&= 1 - \frac{2GM}{r} \\
&+ \frac{8\pi G}{r}\int_r^\infty \left(f'\left[\frac{1}{2}\left((E^\rho)^2 - \frac{g^2}{\rho^4}\right)\right](E^\rho)^2 - f\left[\frac{1}{2}\left((E^\rho)^2 - \frac{g^2}{\rho^4}\right)\right]\right)\rho^2\,\mathrm{d}\rho,
\end{aligned}
\tag{14.12.47}
$$

following from an integration of (14.12.41), which is of the same form of (14.12.13) when magnetism is absent.

Formally, the expression (14.12.47) gives the asymptotic flatness of the metric, $A(r) \to 1$ as $r \to \infty$. Besides, under the general condition

$$
M < 4\pi \int_0^\infty \mathcal{H}(r) r^2\,\mathrm{d}r, \tag{14.12.48}
$$

which is valid automatically if the energy on the right-hand side of (14.12.48) diverges, we have $A(r) \to \infty$ as $r \to 0$. Hence, if $A(r)$ has a global minimum at some $r_0 > 0$ where $A(r_0) \leq 0$, there exist event horizons: if $A(r_0) = 0$, we have an extremal black hole metric, and if $A(r_0) < 0$, there are inner and outer event horizons at r_- and r_+, respectively, with $r_- < r_0 < r_+$. So, from $A'(r_0) = 0$, we have

$$
M = 4\pi \left(\mathcal{H}(r_0) r_0^3 + \int_{r_0}^\infty \mathcal{H}(\rho)\rho^2\,\mathrm{d}\rho \right). \tag{14.12.49}
$$

Assume $A(r_0) < 0$ such that the inner and outer horizons are present. Then, the behavior of $A(r)$ described implies $A'(r_+) \geq 0$. This property allows us to calculate the *Hawking black hole temperature* [208, 344, 404]

$$
T_{\mathrm{H}} = \frac{A'(r_+)}{4\pi} = \frac{1}{4\pi r_+} - 2G r_+ \mathcal{H}(r_+), \tag{14.12.50}
$$

where we have used the condition $A(r_+) = 0$ or

$$
r_+ = 2GM - 8\pi G \int_{r_+}^\infty \mathcal{H}(\rho)\rho^2\,\mathrm{d}\rho \tag{14.12.51}
$$

to eliminate the integral. In the limiting situation when the electromagnetic sector is absent, $\mathcal{H} = 0$, r_+ becomes the Schwarzschild radius $r_{\mathrm{s}} = 2GM$, and (14.12.50) recovers the classical Hawking temperature $T_{\mathrm{H}} = \frac{1}{8\pi GM}$ [279].

Relegation of curvature singularities with magnetically charged black holes

We may also relegate the curvature singularities of black hole solutions induced from finite-energy magnetic point charges or monopoles in the same spirit of those carrying electric point charges. In the present context, though, the magnetic field is universally given by (14.12.45), which gives rise to the Hamiltonian density explicitly when electricity is switched off, so that (14.12.47) may be computed directly by the much reduced formula

$$A(r) = 1 - \frac{2GM}{r} - \frac{8\pi G}{r}\int_r^\infty f\left(-\frac{g^2}{2\rho^4}\right)\rho^2\,\mathrm{d}\rho. \tag{14.12.52}$$

Since now the energy carried by the monopole is

$$E = -4\pi\int_0^\infty f\left(-\frac{g^2}{2r^4}\right)r^2\,\mathrm{d}r, \tag{14.12.53}$$

we may rewrite (14.12.52) as

$$A(r) = 1 - \frac{2G}{r}(M-E) + \frac{8\pi G}{r}\int_0^r f\left(-\frac{g^2}{2\rho^4}\right)\rho^2\mathrm{d}\rho. \tag{14.12.54}$$

Thus, as in the situation of an electrically charged black hole, we may explore the structure of (14.12.54) to relegate or even remove the apparent curvature singularity at $r=0$ of a magnetically charged black hole.

14.13 Relegation of curvature singularities of charged black holes

First, consider a special case of the polynomial model (14.11.19) given by the binomial function

$$f(s) = \frac{1}{\beta}\left(\left[1+\frac{\beta s}{n}\right]^n - 1\right), \quad \beta > 0. \tag{14.13.1}$$

We have seen that this model allows a finite-energy electric point charge but rejects a finite-energy magnetic point charge.

Exponential model

Taking $n \to \infty$ in (14.13.1), we arrive at the simple but rich-in-content *exponential model*

$$f(s) = \frac{1}{\beta}(\mathrm{e}^{\beta s} - 1), \quad \beta > 0, \tag{14.13.2}$$

proposed in [287, 288] in connection to asymptotic Reissner–Nordström black hole solutions. We now show that, unlike the model (14.13.1), both electric and magnetic point charges of finite energies are allowed in the model (14.13.2).

Electric point charge

In the electric point charge situation so that the nontrivial radial components E^r and D^r of the electric field and displacement field are related by their constitutive equation

$$f'\left(\frac{1}{2}(E^r)^2\right)E^r = D^r, \tag{14.13.3}$$

or concretely in view of (14.13.2) by

$$\mathrm{e}^{\frac{\beta}{2}(E^r)^2}E^r = D^r, \tag{14.13.4}$$

which may be recast into an elegant nonlinear equation of the form

$$\mathrm{e}^W W = \delta, \quad W = \beta(E^r)^2, \quad \delta = \beta(D^r)^2, \tag{14.13.5}$$

so that $\beta(E^r)^2 = W(\delta) = W(\beta[D^r]^2)$ is determined by the classical Lambert W function [146]. For our purpose, we note that $W(x)$ is analytic for $x > -\frac{1}{\mathrm{e}}$ and has the Taylor expansion

$$W(x) = \sum_{n=1}^{\infty}\frac{(-n)^{n-1}}{n!}x^n, \tag{14.13.6}$$

about $x = 0$, and the asymptotic expansion

$$W(x) = \ln x - \ln\ln x + \frac{\ln\ln x}{\ln x} + \cdots, \quad x > 3. \tag{14.13.7}$$

Now, in terms of this W-function, the Hamiltonian energy density of the electric point charge with $D^r = \frac{q}{r^2}$ reads

$$\mathcal{H} = \frac{1}{\beta}\left(\mathrm{e}^{\frac{1}{2}W\left(\frac{\beta q^2}{r^4}\right)}\left[W\left(\frac{\beta q^2}{r^4}\right) - 1\right] + 1\right), \tag{14.13.8}$$

giving rise to the total electric energy

$$\begin{aligned} E &= \int \mathcal{H}\,\mathrm{d}\mathbf{x} = \frac{4\pi}{\beta}\int_0^\infty \left(\mathrm{e}^{\frac{1}{2}W\left(\frac{\beta q^2}{r^4}\right)}\left[W\left(\frac{\beta q^2}{r^4}\right) - 1\right] + 1\right) r^2\,\mathrm{d}r \\ &= \frac{\pi q^{\frac{3}{2}}}{\beta^{\frac{1}{4}}}\int_0^\infty \frac{\left(\mathrm{e}^{\frac{1}{2}W(x)}\left[W(x) - 1\right] + 1\right)}{x^{\frac{7}{4}}}\,\mathrm{d}x. \end{aligned} \tag{14.13.9}$$

Using the asymptotic expansions (14.13.6) and (14.13.7) in (14.13.9), we see that the finiteness of the electric energy follows. A numerical evaluation of the integral in (14.13.9) gives us the approximation

$$E \approx (6.570481)\,\frac{\pi q^{\frac{3}{2}}}{\beta^{\frac{1}{4}}}. \tag{14.13.10}$$

Besides, it is also useful to write down the estimates for the electric field E^r here,

$$(E^r)^2 = \frac{1}{\beta}\ln\left(\frac{\beta q^2}{r^4}\right) + \mathrm{O}\left(\ln\ln\frac{1}{r}\right), \quad r \to 0; \tag{14.13.11}$$

$$E^r = \frac{q}{r^2} + \mathrm{O}\left(r^{-4}\right), \quad r \to \infty. \tag{14.13.12}$$

Consequently, we may calculate the free electric charge contained in $\{|\mathbf{x}| \leq r\}$ and the total free electric charge in the full space as before:

$$q_{\text{free}}(r) = \left(r^2 E^r\right) = \frac{r^2}{\sqrt{\beta}}\sqrt{W\left(\frac{\beta q^2}{r^4}\right)}, \quad r > 0; \quad q_{\text{free}} = q_{\text{free}}(\infty) = q. \tag{14.13.13}$$

Electrically charged black holes

From (14.12.47), (14.13.8), and (14.13.9), the metric factor $A(r)$ for an electrically charged black hole is found to be

$$\begin{aligned} A(r) &= 1 - \frac{2GM}{r} + \frac{8\pi G}{\beta r}\int_r^\infty \left(\mathrm{e}^{\frac{1}{2}W\left(\frac{\beta q^2}{\rho^4}\right)}\left[W\left(\frac{\beta q^2}{\rho^4}\right) - 1\right] + 1\right)\rho^2\,\mathrm{d}\rho \\ &= 1 - \frac{2G}{r}(M - E) - \frac{8\pi G}{\beta r}\int_0^r \left(\mathrm{e}^{\frac{1}{2}W\left(\frac{\beta q^2}{\rho^4}\right)}\left[W\left(\frac{\beta q^2}{\rho^4}\right) - 1\right] + 1\right)\rho^2\,\mathrm{d}\rho. \end{aligned} \tag{14.13.14}$$

Thus, from (14.13.6) and the upper line in (14.13.14), we have

$$A(r) = 1 - \frac{2GM}{r} + \frac{4\pi G q^2}{r^2}\left(1 - \frac{\beta q^2}{20 r^4} + \mathrm{O}\left(r^{-8}\right)\right), \quad r \to \infty, \tag{14.13.15}$$

and, from (14.13.7) and the lower line in (14.13.14), we have

$$A(r) = 1 - \frac{2G}{r}(M - E) + \frac{32\pi G q}{\beta^{\frac{1}{2}}}\left(\ln r + \mathrm{O}(1)\right), \quad r \to 0. \tag{14.13.16}$$

In particular, under the critical mass-energy condition (14.12.32) or $M = E$, we see from (14.13.16) that there holds

$$K = \mathrm{O}\left(\frac{(\ln r)^2}{r^4}\right), \quad r \to 0, \tag{14.13.17}$$

for the Kretschmann curvature. So, an electric black hole singularity still persists but is seen to be relegated.

Magnetically charged black holes

For a magnetic point charge with $B^r = \frac{g}{r^2}$ $(g > 0)$ for the radial component of the magnetic field, we apply (14.13.2) to (14.12.52) and (14.12.54) to get

$$A(r) = 1 - \frac{2GM}{r} + \frac{8\pi G}{\beta r}\int_r^\infty \left(1 - \mathrm{e}^{-\frac{\beta g^2}{2\rho^4}}\right)\rho^2 \mathrm{d}\rho$$

$$= 1 - \frac{2G}{r}(M - E) - \frac{8\pi G}{\beta r}\int_0^r \left(1 - \mathrm{e}^{-\frac{\beta g^2}{2\rho^4}}\right)\rho^2 \mathrm{d}\rho, \tag{14.13.18}$$

$$E = \frac{4\pi}{\beta}\int_0^\infty \left(1 - \mathrm{e}^{-\frac{\beta g^2}{2r^4}}\right) r^2 \mathrm{d}r$$

$$= \frac{2^{\frac{5}{4}}\pi g^{\frac{3}{2}}}{\beta^{\frac{1}{4}}}\int_0^\infty \left(1 - \mathrm{e}^{-\frac{1}{\eta^4}}\right)\eta^2\, \mathrm{d}\eta. \tag{14.13.19}$$

The integral on the right-hand side of (14.13.19) may be expressed in terms of the Whittaker M function [580], which gives rise to the approximate value 1.208535. The upper line in (14.13.18) renders the expression

$$A(r) = 1 - \frac{2GM}{r} + \frac{4\pi G g^2}{r^2}\left(1 - \frac{\beta g^2}{20 r^4} + \mathrm{O}\left(r^{-8}\right)\right), \quad r \to \infty, \tag{14.13.20}$$

which assumes the identical form as (14.13.15) within an exchange of the electric and magnetic charges, q and g, although in the magnetic situation there is no involvement of the Lambert W function. The lower line in (14.13.18), on the other hand, yields

$$A(r) = 1 - \frac{2G}{r}(M - E) - \frac{8\pi G r^2}{3\beta} + \frac{2^{\frac{9}{4}}\pi G g^{\frac{3}{2}}}{\beta^{\frac{1}{4}} r} F\left(\frac{2^{\frac{1}{4}} r}{\beta^{\frac{1}{4}} g^{\frac{1}{2}}}\right), \tag{14.13.21}$$

$$F(\rho) = \int_0^\rho \mathrm{e}^{-\frac{1}{\eta^4}}\eta^2\, \mathrm{d}\eta. \tag{14.13.22}$$

It is clear that the function (14.13.22) vanishes at $\rho = 0$ faster than any power function of ρ. Thus, we may rewrite (14.13.21) as

$$A(r) = 1 - \frac{2G}{r}(M - E) - \frac{8\pi G r^2}{3\beta} + \mathrm{O}(r^n), \quad r \to 0, \tag{14.13.23}$$

for an arbitrarily large integer n. The radial component of the induced magnetic intensity field, H^r, follows the constitutive equation

$$H^r = \mathrm{e}^{-\frac{\beta g^2}{2r^4}}\frac{g}{r^2}, \tag{14.13.24}$$

such that the free total magnetic charge contained in the region $\{|\mathbf{x}| \leq r\}$ and the full space are

$$g_{\text{free}}(r) = r^2 H^r = g\,\mathrm{e}^{-\frac{\beta g^2}{2r^4}}, \quad g_{\text{free}} = g_{\text{free}}(\infty) = g, \tag{14.13.25}$$

respectively. It is interesting that the magnetic intensity field and the induced free magnetic charge, generated from the prescribed magnetic point charge, vanish rapidly near the point charge and spread out in space.

Removal of curvature singularity

Assuming $M = E$ in (14.13.23) and inserting the resulting $A(r)$ into (14.8.37), we have

$$K = \frac{512\pi^2 G^2}{3\beta^2} + \mathrm{O}(r^m), \quad r \to 0, \tag{14.13.26}$$

where m is an arbitrarily large integer. Hence, the curvature singularity at $r = 0$ now disappears.

Dyonic charge

For a dyonic point charge with D^r and B^r given in (14.3.1), we have the constitutive equations

$$\mathrm{e}^{\frac{\beta}{2}([E^r]^2 - [B^r]^2)}\, E^r = D^r, \quad \mathrm{e}^{\frac{\beta}{2}([E^r]^2 - [B^r]^2)}\, B^r = H^r. \tag{14.13.27}$$

Again, we can use the Lambert W function to solve the first equation in (14.13.27) to obtain

$$\beta(E^r)^2 = W\left(\beta(D^r)^2 \mathrm{e}^{\beta(B^r)^2}\right) = W\left(\frac{\beta q^2}{r^4}\,\mathrm{e}^{\frac{\beta g^2}{r^4}}\right). \tag{14.13.28}$$

Thus, for r small, we apply (14.13.7) to get

$$\beta(E^r)^2 = \frac{\beta g^2}{r^4} + \ln\left(\frac{q^2}{g^2}\right) - \frac{r^4}{\beta g^2}\ln\left(\frac{\beta q^2}{r^4}\right) + \mathrm{O}(r^8). \tag{14.13.29}$$

Consequently, inserting (14.13.29) into the associated Hamiltonian energy density given as

$$\mathcal{H} = \frac{1}{\beta}\left(\mathrm{e}^{\frac{\beta}{2}([E^r]^2 - [B^r]^2)}\left(\beta[E^r]^2 - 1\right) + 1\right), \quad B^r = \frac{g}{r^2}, \tag{14.13.30}$$

we find

$$\mathcal{H} = \frac{qg}{r^4} + \frac{1}{2\beta g}\left(2(g-q) + q\ln\left[\frac{q^2 r^4}{\beta g^4}\right]\right) + \mathrm{O}(r^4), \quad r \to 0, \tag{14.13.31}$$

neglecting irrelevant truncation error terms. The dominant term in (14.13.31) is

$$\frac{qg}{r^4}, \tag{14.13.32}$$

which leads to the divergence of the dyon energy near $r = 0$. This term is a clear indication that the energy divergence is a consequence of a dyonic point charge, or both electric and magnetic charges assigned at a given point, characterized with $q, g \neq 0$.

Dyonic black holes

Nevertheless, we may obtain a dyonic black hole solution of the model (14.13.2) as follows. In fact, inserting (14.13.30) into (14.12.47), we have the explicit and exact result

$$A(r) = 1 - \frac{2GM}{r} + \frac{8\pi G}{\beta r}\int_r^\infty \left(\mathrm{e}^{\frac{1}{2}\left(W\left(\frac{\beta q^2}{\rho^4}\mathrm{e}^{\frac{\beta g^2}{\rho^4}}\right) - \frac{\beta g^2}{\rho^4}\right)} \left(W\left(\frac{\beta q^2}{\rho^4}\mathrm{e}^{\frac{\beta g^2}{\rho^4}}\right) - 1\right) + 1\right)\rho^2 \mathrm{d}\rho. \tag{14.13.33}$$

Hence, using (14.13.6), we get the asymptotic expression

$$A(r) = 1 - \frac{2GM}{r} + \frac{4\pi G}{r^2}\left((q^2+g^2) - \frac{\beta}{20r^4}(q^2-g^2)^2 + \mathrm{O}\left(r^{-8}\right)\right), \quad r \to \infty, \tag{14.13.34}$$

which contains the Reissner–Nordström solution in the $\beta \to 0$ limit, is symmetric with respect to the electric and magnetic charges, and covers (14.13.15) and (14.13.20) as special cases. On the other hand, from (14.13.31) and (14.13.32), we see that such an electric and magnetic charge symmetry is broken near $r = 0$.

Electromagnetism

The free electric charge contained in the region $\{|\mathbf{x}| \leq r\}$ and total free electric charge in the full space are

$$q_{\text{free}}(r) = \int_0^r \frac{\mathrm{d}}{\mathrm{d}\rho}\left(\rho^2 E^\rho\right)\mathrm{d}\rho = \frac{r^2}{\sqrt{\beta}}\sqrt{W\left(\frac{\beta q^2}{r^4}\mathrm{e}^{\frac{\beta g^2}{r^4}}\right)} - g, \tag{14.13.35}$$

$$q_{\text{free}} = q_{\text{free}}(\infty) = q - g, \tag{14.13.36}$$

respectively in view of (14.13.28), (14.13.29), and then (14.13.6). Furthermore, we may use (14.13.6), (14.13.28), and (14.13.29) to get the asymptotic expansions

$$E^r = \frac{q}{r^2}\left(1 + \frac{\beta(g^2-q^2)}{2r^4}\right) + \mathrm{O}(r^{-10}), \quad r \to \infty, \tag{14.13.37}$$

$$E^r = \frac{g}{r^2} + \frac{r^2}{\beta g}\ln\left(\frac{q}{g}\right) + \mathrm{O}(r^6), \quad r \to 0. \tag{14.13.38}$$

Interestingly, (14.13.37) implies that E^r follows an electric Coulomb law for r large, and (14.13.38) indicates that E^r follows a magnetic Coulomb law for r small. Such a balance explains why the total free electric charge is a combination of both electric and magnetic charges, as given in (14.13.36).

Likewise, in view of the second equation in (14.13.27) and (14.13.29), we obtain

$$H^r = \frac{q}{r^2} - \frac{qr^2}{2\beta g^2}\ln\left(\frac{\beta q^2}{r^4}\right) + \mathrm{O}(r^6), \quad r \to 0. \tag{14.13.39}$$

Consequently, in leading orders, the magnetic intensity field of a dyonic point charge behaves like a pure electric point charge at the origin. Moreover, using (14.13.6) in the second equation in (14.13.27) again, we get

$$H^r = \frac{g}{r^2}\left(1 + \frac{\beta(q^2 - g^2)}{2r^4}\right) + \mathrm{O}(r^{-10}), \quad r \to \infty. \tag{14.13.40}$$

Thus, the magnetic intensity field H^r behaves like the magnetic field B^r for r large and follows the Coulomb law as well.

It is interesting to note from (14.13.37) and (14.13.40) that there is a symmetry with regard to an exchange of fields and charges, $E^r \leftrightarrows H^r, q \leftrightarrows g$, asymptotically. However, from (14.13.38) and (14.13.39), we see that such a symmetry is invalid away from the leading-order approximation.

As a consequence of (14.13.39) and (14.13.40), we see that the associated total free magnetic charge is

$$g_{\text{free}} = (r^2 H^r)_{r=0}^{r=\infty} = g - q. \tag{14.13.41}$$

Energy conditions

At this juncture, it will be of interest to discuss various *energy conditions* [280] related to the occurrence and disappearance of curvature singularities we have since seen. In fact, in the current spherically symmetric dyon situation, we have, in view of (14.12.4)–(14.12.5) with (14.12.37) and (14.12.38),

$$T_0^0 = T_1^1 = f'(\mathcal{L}_{\mathrm{M}})(E^r)^2 - f(\mathcal{L}_{\mathrm{M}}), \quad T_2^2 = T_3^3 = -r^4 f'(\mathcal{L}_{\mathrm{M}})B^2 - f(\mathcal{L}_{\mathrm{M}}), \tag{14.13.42}$$

giving rise to the density $\rho = T_0^0$, radial pressure $p_1 = p_r = -T_1^1$, and tangential pressure $p_2 = p_3 = p_\perp = -T_2^2 = -T_3^3$ [180, 181], such that the *weak energy condition reads* $\rho \geq 0, \rho + p_i \geq 0$ $(i = 1, 2, 3)$, which may be reduced into

$$f'(\mathcal{L}_{\mathrm{M}})(E^r)^2 - f(\mathcal{L}_{\mathrm{M}}) \geq 0, \quad f'(\mathcal{L}_{\mathrm{M}})\left((E^r)^2 + r^4 B^2\right) \geq 0, \tag{14.13.43}$$

the *dominant energy condition* becomes $\rho \geq |p_i|$ $(i = 1, 2, 3)$, which combines (14.13.43) with the additional requirement $\rho - p_i \geq 0$ $(i = 1, 2, 3)$, or

$$f'(\mathcal{L}_{\mathrm{M}})\left((E^r)^2 - r^4 B^2\right) - 2f(\mathcal{L}_{\mathrm{M}}) \geq 0, \tag{14.13.44}$$

and the *strong energy condition* imposes $\rho + \sum_{i=1}^3 p_i \geq 0$, which recasts itself in the form

$$r^4 f'(\mathcal{L}_{\mathrm{M}})B^2 + f(\mathcal{L}_{\mathrm{M}}) \geq 0. \tag{14.13.45}$$

Thus, we can readily check that, for the exponential model (14.13.2) in the magnetic point charge situation with B given by (14.12.45), $\mathcal{L}_{\mathrm{M}} = -\frac{g^2}{2r^4}$, and $E^r = 0$, both the weak and dominant energy conditions hold. For example, in this situation the left-hand side of (14.13.44) assumes the form

$$-f'(\mathcal{L}_{\mathrm{M}})r^4 B^2 - 2f(\mathcal{L}_{\mathrm{M}}) = \frac{2h(\tau)}{\beta}, \quad h(\tau) = 1 - \mathrm{e}^{-\tau}(1 + \tau), \quad \tau = \frac{\beta g^2}{2r^4}, \tag{14.13.46}$$

and it is clear that $h(\tau) > 0$ for all $\tau > 0$. On the other hand, it is evident that the condition (14.13.45) is violated. In fact, in this situation, we may rewrite the left-hand side of (14.13.45) as

$$r^4 f'(\mathcal{L}_{\mathrm{M}})B^2 + f(\mathcal{L}_{\mathrm{M}}) = \frac{h(\tau)}{\beta}, \quad h(\tau) = (2\tau+1)\mathrm{e}^{-\tau} - 1, \quad \tau = \frac{\beta g^2}{2r^4}. \tag{14.13.47}$$

The function $h(\tau)$ defined in (14.13.47) decreases for $\tau > \frac{1}{2}$ and $h(2) < 0$ (say). Thus, $h(\tau) < 0$ for $\tau > 2$. In other words, the strong energy condition is invalid in the region

$$r < \beta^{\frac{1}{4}}\sqrt{\frac{g}{2}} \quad \text{(say)}. \tag{14.13.48}$$

In contrast, the electric point charge situation with $\mathcal{L}_{\mathrm{M}} = \frac{1}{2}(E^r)^2$ fulfills the strong energy condition, on the other hand. Such a picture explains the disappearance and occurrence of the curvature singularity associated with the magnetic and electric point charges, respectively, in view of the *Hawking–Penrose singularity theorems* [277, 280, 281, 396, 405, 439, 440, 503, 563] for which the fulfillment of the strong energy condition is essential. More generally, the strong energy condition (14.13.45) holds for a dyonic point charge as long as the electric charge dominates over the magnetic charge, $q \geq g$, since now $\mathcal{L}_{\mathrm{M}} \geq 0$. To see this, we use (14.13.28) to get

$$2\beta\mathcal{L}_{\mathrm{M}} = W\left(\frac{\beta q^2}{r^4}\mathrm{e}^{\frac{\beta g^2}{r^4}}\right) - \frac{\beta g^2}{r^4} \geq 0, \tag{14.13.49}$$

in view of the condition $q \geq g$ and the identity $W(\tau\mathrm{e}^{\tau}) = \tau$ $(\tau \geq 0)$ for the Lambert W function. Hence, (14.13.45) follows.

Summarizing our results regarding finite-energy charged black hole solutions arising in the generalized Born–Infeld theory of electromagnetism, we conclude that in the context of the Born–Infeld type nonlinear theory of electromagnetism, the gravitational metric of a finite-energy charged black hole solution enjoys a relegated curvature singularity at the same level of that of the Schwarzschild black hole without charge. In particular, the well-known higher-order, deteriorated, curvature singularity of the classical Reissner–Nordström black hole metric arises as a consequence of the divergence of its electromagnetic energy. In the critical coupling situation when the black hole gravitational mass is equal to its electromagnetic energy in the presence of electricity or magnetism or both, the curvature singularity may be further relegated from that of the Schwarzschild black hole or even completely removed so that the gravitational metric is free of singularity. This statement may be referred to as a *curvature singularity relegation theorem* of a charged black hole with a finite electromagnetic energy.

14.14 Cosmology driven by scalar-wave matters as k-essence

This section studies the cosmological expansion of a universe propelled by a Born–Infeld type scalar field governed by the Lagrangian action density of the general form [4, 10, 18, 32, 33, 51, 52, 53, 225, 475]

$$\mathcal{L} = f(X) - V(\varphi), \tag{14.14.1}$$

where

$$X = \frac{1}{2}\partial_\mu\varphi\partial^\mu\varphi = \frac{1}{2}g^{\mu\nu}\partial_\mu\varphi\partial_\nu\varphi, \tag{14.14.2}$$

φ being a real-valued scalar field, $g_{\mu\nu}$ the gravitational metric tensor, and V is a potential density function. The wave equation or the Euler–Lagrange equation associated with (14.14.1) is

$$\frac{1}{\sqrt{-g}}\partial_\mu\left(\sqrt{-g}f'(X)\partial^\mu\varphi\right) + V'(\varphi) = 0, \tag{14.14.3}$$

with the associated energy-momentum tensor given by

$$T_{\mu\nu} = \partial_\mu\varphi\partial_\nu\varphi f'(X) - g_{\mu\nu}(f(X) - V(\varphi)). \tag{14.14.4}$$

Friedmann equation

For cosmology, we consider an isotropic and homogeneous universe governed by the *Robertson–Walker line element* or *gravitational metric*

$$\mathrm{d}s^2 = g_{\mu\nu}\mathrm{d}x^\mu\mathrm{d}x^\nu = \mathrm{d}t^2 - a^2(t)(\mathrm{d}x^2 + \mathrm{d}y^2 + \mathrm{d}z^2), \tag{14.14.5}$$

in Cartesian coordinates, where $a(t) > 0$ is the scale factor or radius of the universe to be determined. With (14.14.5), we write down the nontrivial components of the associated Ricci tensor:

$$R_{00} = \frac{3\ddot{a}}{a}, \quad R_{11} = R_{22} = R_{33} = -2\dot{a}^2 - a\ddot{a}, \tag{14.14.6}$$

where $\dot{a} = \frac{\mathrm{d}a}{\mathrm{d}t}$, etc. Correspondingly, we assume that the scalar field is also only time-dependent so that (14.14.3) becomes

$$(a^3 f'(X)\dot{\varphi})^{\cdot} = -a^3 V'(\varphi). \tag{14.14.7}$$

Besides, now, the nontrivial components of (14.14.4) are

$$\begin{aligned} T_{00} &= \dot{\varphi}^2 f'(X) - (f(X) - V(\varphi)), \\ T_{11} &= T_{22} = T_{33} = a^2\left(f(X) - V(\varphi)\right),\ X = \frac{1}{2}\dot{\varphi}^2. \end{aligned} \tag{14.14.8}$$

From (14.14.8), we get the trace of $T_{\mu\nu}$ to be

$$T = \dot{\varphi}^2 f'(X) - 4\,(f(X) - V(\varphi)). \tag{14.14.9}$$

Thus, in view of (14.14.6), (14.14.8), and (14.14.9), we see that the Einstein equation (14.8.1) becomes

$$\frac{\ddot{a}}{a} = -\frac{8\pi G}{3}\left(\frac{1}{2}\dot{\varphi}^2 f'(X) + f(X) - V(\varphi)\right), \tag{14.14.10}$$

$$\frac{\ddot{a}}{a} + 2\left(\frac{\dot{a}}{a}\right)^2 = 8\pi G\left(\frac{1}{2}\dot{\varphi}^2 f'(X) - f(X) + V(\varphi)\right). \tag{14.14.11}$$

Combining (14.14.10) and (14.14.11), we have

$$\left(\frac{\dot{a}}{a}\right)^2 = \frac{8\pi G}{3}\left(\dot{\varphi}^2 f'(X) - f(X) + V(\varphi)\right). \tag{14.14.12}$$

We see that (14.14.7) and (14.14.12) imply (14.14.10) and (14.14.11) as well. Hence, the coupled governing equations (14.8.1) and (14.14.3)–(14.14.4) are recast into (14.14.7) and (14.14.12), which are now the focus of attention.

Comparing (14.14.4) with that of a perfect fluid, we may recognize a pair of quantities resembling the energy density ρ and pressure P through $T^{00} = \rho$ and $T^{ii} = -Pg^{ii} = a^{-2}P$ $(i = 1, 2, 3)$ (cf. (10.2.16)), respectively, thus leading to

$$\rho = \dot{\varphi}^2 f'(X) - (f(X) - V(\varphi)), \quad P = f(X) - V(\varphi), \tag{14.14.13}$$

such that the wave equation (14.14.7) is equivalent to the energy conservation law

$$\dot{\rho} + 3(\rho + P)\frac{\dot{a}}{a} = 0, \tag{14.14.14}$$

or $\nabla_\nu T^{\mu\nu} = 0$, where ∇_ν is the covariant derivative. As a consequence, (14.14.12) assumes the form

$$\left(\frac{\dot{a}}{a}\right)^2 = \frac{8\pi G}{3}\rho, \tag{14.14.15}$$

which returns to the celebrated *Friedmann equation.*

For simplicity, we consider the special situation where the potential density function V is constant, say V_0. Then, (14.14.7) becomes

$$\left(a^3 f'(X)\dot{\varphi}\right)^{\cdot} = 0, \tag{14.14.16}$$

which renders the integral

$$f'\left(\frac{1}{2}\dot{\varphi}^2\right)\dot{\varphi} = \frac{c}{a^3}, \tag{14.14.17}$$

where c is taken to be a nonzero constant to avoid triviality, which may be set to be positive for convenience and should not be confused with the speed of light in vacuum denoted the same elsewhere. Express $\dot{\varphi}$ in (14.14.17) by $\dot{\varphi} = h\left(\frac{c}{a^3}\right)$. Then, (14.14.12) becomes

$$\begin{aligned}\dot{a}^2 &= \frac{8\pi G}{3}\left(\dot{\varphi}^2 f'\left(\frac{1}{2}\dot{\varphi}^2\right) - f\left(\frac{1}{2}\dot{\varphi}^2\right) + V_0\right)a^2 \\ &= \frac{8\pi G}{3}\left(\frac{c}{a}h\left(\frac{c}{a^3}\right) - a^2 f\left(\frac{1}{2}h^2\left(\frac{c}{a^3}\right)\right) + V_0 a^2\right),\end{aligned} \tag{14.14.18}$$

which governs the scale factor a and may be integrated at least *in theory* to yield a dynamical description of the cosmological evolution under consideration. Next we work out a few illustrative examples.

In the simplest situation where $f(X) = X$, we have $\dot{\varphi} = \frac{c}{a^3}$ and the equation (14.14.18) reads

$$\dot{a}^2 = \frac{8\pi G}{3}\left(\frac{c^2}{2a^4} + V_0 a^2\right). \tag{14.14.19}$$

Comparing (14.14.19) with the classical Friedmann equation with a cosmological constant, Λ, of the form

$$H^2 = \frac{8\pi G}{3}\rho = \frac{8\pi G}{3}\left(\rho_m + \frac{\Lambda}{8\pi G}\right), \quad H = \frac{\dot{a}}{a}, \tag{14.14.20}$$

with the material energy density ρ_m, which is related to the material pressure P_m, of the cosmological fluid, through the energy conservation law

$$\dot{\rho}_m + 3(\rho_m + P_m)H = 0, \tag{14.14.21}$$

and the *barotropic equation of state*

$$P_m = w\rho_m, \tag{14.14.22}$$

we arrive at the identification

$$\rho_m = \frac{c^2}{2a^6} = \frac{1}{2}\dot{\varphi}^2 = P_m, \quad V_0 = \frac{\Lambda}{8\pi G}. \tag{14.14.23}$$

Thus, V_0 gives rise to a cosmological constant. Since the material energy density and pressure are related to their effective counterparts by the standard relations

$$\rho = \rho_m + \frac{\Lambda}{8\pi G}, \quad P = P_m - \frac{\Lambda}{8\pi G}, \tag{14.14.24}$$

we find from (14.14.13), (14.14.23), and (14.14.24) that the effective material energy density, pressure, and cosmological constant should be defined by

$$\rho_m = \dot{\varphi}^2 f'(X) - f(X), \quad P_m = f(X), \quad \Lambda = \Lambda(\varphi) = 8\pi G V(\varphi). \tag{14.14.25}$$

Note that, in such a context, the cosmological constant is a field-dependent quantity. In the special situation where $f(X) = X$, we have $w = 1$ in (14.14.22).

Equations of state

Since we focus on a homogeneous and isotropic universe for which fields are only time dependent, we have $X = \frac{1}{2}\dot{\varphi}^2$ and we may rewrite the material energy density ρ_m in terms of pressure P_m as

$$\begin{aligned}\rho_m &= 2Xf'(X) - f(X)\\ &= 2f^{-1}(P_m)f'\left(f^{-1}(P_m)\right) - P_m, \quad X = f^{-1}(P_m),\end{aligned} \tag{14.14.26}$$

which is a nonlinear scalar-wave-matter version of the barotropic equation of state, which generally is expected to be nonlinear as well.

As a first nonlinear example, we consider the following *fractional-powered model*

$$f(X) = b^2\left(1 - \left[1 - \frac{X}{pb^2}\right]^p\right), \quad 0 < p < 1, \tag{14.14.27}$$

which recovers the classical model (14.11.12) when $p = \frac{1}{2}$. Hence

$$X = f^{-1}(P_m) = pb^2\left(1 - \left[1 - \frac{P_m}{b^2}\right]^{\frac{1}{p}}\right). \tag{14.14.28}$$

Inserting (14.14.27) and (14.14.28) into (14.14.26), we obtain

$$\begin{aligned}\rho_m &= 2pb^2\left(\frac{P_m}{b^2} + \left[1 - \frac{P_m}{b^2}\right]^{1-\frac{1}{p}} - 1\right) - P_m\\ &= (2p-1)P_m + 2pb^2\left(\left[1 - \frac{P_m}{b^2}\right]^{1-\frac{1}{p}} - 1\right).\end{aligned} \tag{14.14.29}$$

Thus, corresponding to the classical Born–Infeld theory, $p = \frac{1}{2}$, (14.14.29) becomes [125]

$$\rho_m = \frac{b^2 P_m}{b^2 - P_m} \quad \text{or} \quad P_m = \frac{b^2 \rho_m}{b^2 + \rho_m}. \tag{14.14.30}$$

An elegantly illustrative example is suggested by the model (14.11.19) with taking

$$f(X) = X^p, \tag{14.14.31}$$

where $p \geq 1$ is an integer. Now

$$X = P_m^{\frac{1}{p}}. \tag{14.14.32}$$

Inserting (14.14.31) and (14.14.32) into (14.14.26), we see that the following linear equation of state holds:

$$\rho_m = (2p-1)P_m, \quad p > 0. \tag{14.14.33}$$

In particular, for $p = 2$, we have $P_m = \frac{1}{3}\rho_m$ and we arrive at a radiation-dominated universe.

For the exponential model (14.13.2), we have

$$f(X) = \frac{1}{\beta}\left(e^{\beta X} - 1\right), \quad X = \frac{1}{\beta}\ln(1 + \beta P_m). \tag{14.14.34}$$

Thus, in view of (14.14.26) and (14.14.34), we get

$$\begin{aligned}\rho_m &= \frac{2}{\beta}(1 + \beta P_m)\ln(1 + \beta P_m) - P_m \\ &= P_m + \beta P_m^2 - \frac{1}{3}\beta^2 P_m^2 + \mathrm{O}\left(\beta^3 P^4\right), \quad \text{for } P_m \text{ small.}\end{aligned} \tag{14.14.35}$$

Conversely, we may also express P_m in terms of ρ_m by the explicit formula through inverting (14.14.35) via the Lambert W function again:

$$\begin{aligned}P_m &= \frac{1}{\beta}\left(e^{\frac{1}{2} + W\left(\frac{\beta\rho_m - 1}{2e^{\frac{1}{2}}}\right)} - 1\right) \\ &= \rho_m - \beta\rho_m^2 + \mathrm{O}\left(\beta^2\rho_m^3\right), \quad \rho_m \to 0.\end{aligned} \tag{14.14.36}$$

Using (14.14.34) in (14.14.17), we see that the Friedmann equation (14.14.18) governing the scale factor a reads

$$\begin{aligned}\dot{a}^2 &= \frac{8\pi G}{3}(\rho_m + V_0)a^2 \\ &= \frac{8\pi G}{3}\left(\frac{c}{a}\sqrt{\frac{1}{\beta}W\left(\frac{\beta c^2}{a^6}\right)} - \frac{a^2}{\beta}\left[e^{\frac{1}{2}W\left(\frac{\beta c^2}{a^6}\right)} - 1\right] + V_0 a^2\right).\end{aligned} \tag{14.14.37}$$

Moreover, from (14.14.17), we find

$$X = \frac{1}{2}\dot{\varphi}^2 = \frac{1}{2\beta}W\left(\frac{\beta c^2}{a^6}\right). \tag{14.14.38}$$

Inserting (14.14.38) into (14.14.25) or (14.14.26), we get

$$\rho_m = \frac{1}{\beta}\left(1 + \left[W\left(\frac{\beta c^2}{a^6}\right) - 1\right]e^{\frac{1}{2}W\left(\frac{\beta c^2}{a^6}\right)}\right), \tag{14.14.39}$$

$$P_m = \frac{1}{\beta}\left(e^{\frac{1}{2}W\left(\frac{\beta c^2}{a^6}\right)} - 1\right), \tag{14.14.40}$$

which are both positive-valued quantities.

Cosmological expansion

Consequently, although it may be impossible to obtain an explicit integration of the equation (14.14.37), we see that the Big Bang solution with $a(0) = 0$ is monotone and enjoys the asymptotic estimates

$$a(t) \sim t^{\frac{2}{3}}, \quad t \to 0; \quad a(t) \sim \begin{cases} t^{\frac{1}{3}}, \quad t \to \infty, \quad \text{if } V_0 = 0; \\ e^{\sqrt{\frac{8\pi G V_0}{3}}t}, \quad t \to \infty, \quad \text{if } V_0 > 0, \end{cases} \tag{14.14.41}$$

in view of the expansions (14.13.6) and (14.13.7). Since the Kretschmann scalar of the line element (14.14.5) is

$$K=\frac{3\left(\dot{a}^4-2a\dot{a}^2\ddot{a}+2a^2\ddot{a}^2\right)}{2a^4}, \tag{14.14.42}$$

we see that (14.14.41) and (14.14.42) give us

$$K(t)\sim t^{-4},\ t\to 0; \tag{14.14.43}$$

$$K(t)\sim t^{-4},\ t\to\infty,\quad \text{if } V_0=0; \tag{14.14.44}$$

$$\lim_{t\to\infty} K(t)=\frac{32\pi^2G^2V_0^2}{3},\quad \text{if } V_0>0. \tag{14.14.45}$$

In particular, the Big Bang moment, $t=0$, is a curvature singularity.

In view of (14.14.39)–(14.14.40), (14.14.41), and the monotonicity of the scale factor, we see that both $\rho_m(t)$ and $P_m(t)$ are decreasing functions of the cosmic time t, with the limiting behavior

$$\lim_{t\to 0}\rho_m=\infty,\quad \lim_{t\to 0}P_m=\infty,\quad \lim_{t\to\infty}\rho_m=0,\quad \lim_{t\to\infty}P_m=0. \tag{14.14.46}$$

The behavior (14.14.46) depicts clearly a Big Bang universe which starts from a high-density and high-pressure state and dilutes into a zero-density and zero-pressure state. Besides,

$$\lim_{t\to 0}\frac{P_m}{\rho_m}=0,\quad \lim_{t\to\infty}\frac{P_m}{\rho_m}=1. \tag{14.14.47}$$

On the other hand, rewriting ρ_m and P_m given in (14.14.39)–(14.14.40) as $\rho_m(W)$ and $P_m(W)$, we have

$$\frac{\mathrm{d}}{\mathrm{d}W}\left(\frac{P_m}{\rho_m}\right)=-\frac{\mathrm{e}^{\frac{W}{2}}\left(1+\frac{W}{2}-\mathrm{e}^{\frac{W}{2}}\right)}{\left(1+(W-1)\mathrm{e}^{\frac{W}{2}}\right)^2}<0,\quad W>0. \tag{14.14.48}$$

Hence, the ratio $\frac{P_m(t)}{\rho_m(t)}$ monotonically increases in $t>0$ and interpolates between its limits stated in (14.14.47) at $t=0$ and $t=\infty$, respectively. This property indicates that the universe starts as a dust gas and eventually approaches a stiff matter state (Figure 14.5).

Furthermore, from (14.14.10), we have

$$\frac{\ddot{a}}{a}=-\frac{4\pi G}{3}\left(\rho_m+3P_m-2V_0\right). \tag{14.14.49}$$

Hence, if $V_0=0$, then $a(t)$ is globally concave down, and so the expansion of the universe undergoes a decelerated process; if $V_0>0$, $a(t)$ is initially concave down and then concave up, and so the expansion of the universe first experiences a

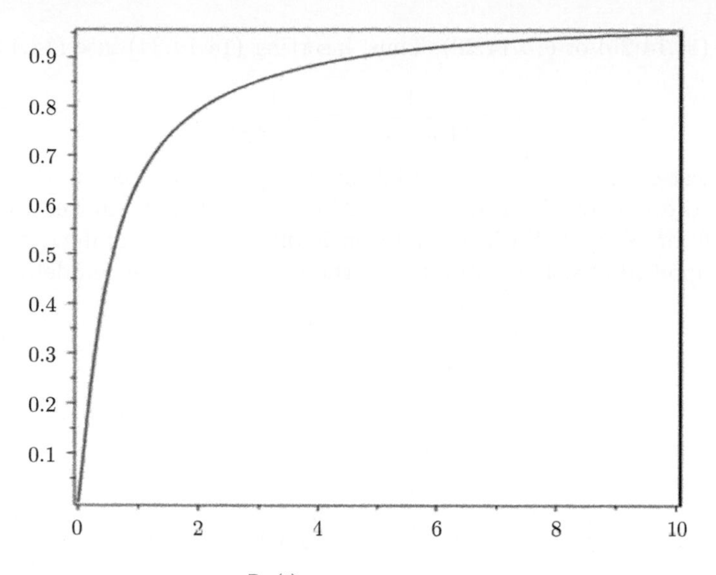

Figure 14.5 A plot of the ratio $\frac{P_m(t)}{\rho_m(t)}$ against the time variable t through $1/W\left(\frac{\beta c^2}{a^6(t)}\right)$ for the exponential model (14.14.34), which indicates that the scalar wave in such a context monotonically interpolates the dust model at $t = 0$ and the stiff matter model at $t = \infty$.

decelerated process but then an accelerated process after passing through a point of inflection at the moment $t = t_0$ where $t_0 > 0$ satisfies

$$\rho_m(t_0) + 3P_m(t_0) = 2V_0, \tag{14.14.50}$$

or

$$\left(W\left(\frac{\beta c^2}{a^6(t_0)}\right) + 2\right) \mathrm{e}^{\frac{1}{2}W\left(\frac{\beta c^2}{a^6(t_0)}\right)} = 2(1 + \beta V_0), \tag{14.14.51}$$

by (14.14.39)–(14.14.40). It is clear that (14.14.51) has a unique positive solution t_0, which may more explicitly be expressed in term of the Lambert W function as

$$\frac{\beta c^2}{a^6(t_0)} = W^{-1}\left(2\left[W([1 + \beta V_0]\mathrm{e}) - 1\right]\right). \tag{14.14.52}$$

Adiabatic squared speed of sound

Note that an important checkpoint [227, 343] for accepting or rejecting a cosmological model is to examine whether the resulting *adiabatic squared speed of sound*, c_s^2, would stay in the correct range $0 \leq c_s^2 < 1$. Recall that it follows from the classical Newton mechanics that c_s^2 is the rate of change of the fluid pressure with respect to density. Hence, we have

$$c_s^2 = \frac{\mathrm{d}P_m}{\mathrm{d}\rho_m} = \frac{P_m'(X)}{\rho_m'(X)} = \frac{f'(X)}{f'(X) + 2Xf''(X)}, \tag{14.14.53}$$

in view of (14.14.25) or (14.14.26). Thus, inserting (14.14.34) into (14.14.53), we find

$$c_s^2 = \frac{1}{1+2\beta X} = \frac{1}{1+\beta\dot{\varphi}^2}, \tag{14.14.54}$$

which indicates that the exponential model (14.14.34) meets the required relevance criterion for the range of c_s^2. More generally, it suffices to demand $f''(X) > 0$ for $X > 0$. Such a condition is often easily examined for various models in applications. For example, for the fractional-power model (14.14.27), we have

$$c_s^2 = \frac{pb^2 - X}{(pb^2 - X) + 2(1-p)X}, \tag{14.14.55}$$

which indicates that c_s^2 lies in the anticipated correct range as well.

In terms of (14.14.26), the ratio

$$w_m(X) \equiv \frac{P_m(X)}{\rho_m(X)} = \frac{f(X)}{2Xf'(X) - f(X)} \tag{14.14.56}$$

is also an informative quantity regarding the fluid nature of the scalar-wave matter concerned which is easier to calculate. For example, for the quadratic model (14.11.29), we have $f(X) = X + \alpha X^2$ in which $\alpha > 0$ is a constant coefficient (note that this coefficient is updated in order to avoid confusion with the scale factor in this section) such that

$$w_m(X) = \frac{1+\alpha X}{1+3\alpha X}, \tag{14.14.57}$$

which indicates that the model interpolates between stiff and radiation-dominated matters, characterized by $w_m = 1$ and $w_m = \frac{1}{3}$, respectively. For the fractional-powered model (14.14.27), we see that $\rho_m(X)$ stays positive and $w_m(X)$ assumes its values between stiff and dust matters as well.

Realization of equation of state of cosmological fluid

With these examples presented, we are interested in whether we are able to identify a Born–Infeld model to realize an *arbitrarily prescribed* equation of state,

$$P_m = w(\rho_m) \quad \text{or} \quad \rho_m = v(P_m), \tag{14.14.58}$$

at least in theory. In fact, by virtue of (14.14.25), (14.14.26), and (14.14.58), we see that the function f in (14.14.1) satisfies the differential equation

$$2X\frac{\mathrm{d}f}{\mathrm{d}X} - f(X) = v(f(X)) \quad \text{or} \quad 2X\frac{\mathrm{d}f}{\mathrm{d}X} = v(f) + f, \tag{14.14.59}$$

which is a first-order separable equation and may readily be solved in general. Note also that this equation is invariant under the rescaling $X \mapsto \gamma X$ $(\gamma > 0)$.

In conclusion, we see that, adapting the Born–Infeld type action principle to describe scalar-wave matters or k-essences, theoretically, the equation of state of *any* cosmological fluid model may be realized by obtaining the nonlinearity profile function of a k-essence model by solving a first-order separable ordinary differential equation defined by the equation of state.

As an example, we consider a *generalized Chaplygin fluid model* [26, 112, 125, 126, 519] with the equation of state

$$P_m = P_0 + w\rho_m - \frac{A}{\rho_m^{\frac{1}{k}}}, \quad k = 1, 2, \dots, \tag{14.14.60}$$

where P_0, $w > -1$, and $A > 0$ are constants and $k = 1$ is the classical case. In the linear equation of state limit, $A = 0$ and $w \neq 0$, we can use $v(f) = \frac{f - P_0}{w}$ in (14.14.59) to get

$$f(X) = \frac{P_0}{1 + w^2} + \alpha X^{\frac{1}{2}(1+\frac{1}{w})}, \tag{14.14.61}$$

where $\alpha > 0$ is an integration constant. Hence, P_0 serves the role of a cosmological constant. In the purely nonlinear limit, $w = 0$, from (14.14.60), we have

$$\rho_m = \frac{(-A)^k}{(P_m - P_0)^k}, \tag{14.14.62}$$

or $v(f) = \frac{(-A)^k}{(f - P_0)^k}$ within our formalism. Substituting this into (14.14.59), we obtain

$$2X\frac{\mathrm{d}f}{\mathrm{d}X} = \frac{(-A)^k}{(f - P_0)^k} + f. \tag{14.14.63}$$

An explicit integration of (14.14.63) may be made in the limit when $P_0 = 0$ to yield

$$f^{k+1} + (-A)^k = \alpha X^{\frac{k+1}{2}}, \tag{14.14.64}$$

where $\alpha \neq 0$ is an integration constant. Thus we have

$$f = f(X) = \left((-1)^{k+1}A^k + \alpha X^{\frac{k+1}{2}}\right)^{\frac{1}{k+1}}, \tag{14.14.65}$$

as obtained earlier in [461]. In general, an explicit expression for $f(X)$ is not available due to the difficulties associated with integrating (14.14.59) with (14.14.60). As an illustration, we take $k = 1$ in (14.14.63) to obtain

$$X\frac{\mathrm{d}(f - P_0)^2}{\mathrm{d}X} = f^2 - P_0 f - A. \tag{14.14.66}$$

For simplicity, we consider the region away from the equilibrium of the equation (14.14.66) where $f^2 - P_0 f - A > 0$. With this condition, we obtain the implicit solution

$$\frac{\alpha X}{(f^2 - P_0 f - A)}\left(\frac{\left(f - \frac{P_0}{2} - \sqrt{\left[\frac{P_0}{2}\right]^2 + A}\right)^2}{f^2 - P_0 f - A}\right)^{\frac{P_0}{\sqrt{P_0^2 + 4A}}} = 1, \quad \alpha > 0, \tag{14.14.67}$$

which is complicated. In the limit $P_0 = 0$, (14.14.67) returns to (14.14.65) with $k = 1$. Thus, we see that various generalized Chaplygin fluid models may be realized by their corresponding Born–Infeld scalar-wave matter models.

For the quadratic equation of state,

$$P_m = A\rho_m^2, \quad A > 0, \tag{14.14.68}$$

which is a special case of a general quadratic model covered in [16, 17, 126, 145, 148], we have

$$\rho_m = \sqrt{\frac{P_m}{A}}, \tag{14.14.69}$$

such that $v(f) = \sqrt{\frac{f}{A}}$ in (14.14.59). That is, f satisfies

$$2X\frac{\mathrm{d}f}{\mathrm{d}X} = \sqrt{\frac{f}{A}} + f, \tag{14.14.70}$$

whose solution is

$$f(X) = \frac{1}{A}\left(\alpha X^{\frac{1}{4}} - 1\right)^2, \tag{14.14.71}$$

where $\alpha > 0$ is an integration constant. As before, we see that the choice of the parameter α does not affect the equation of state, (14.14.68).

Sometimes the equation of state of a cosmological model is implicitly given in terms of the scale factor a. For example, the equation

$$\rho_m = \frac{A}{a^3} + \frac{B}{a^{\frac{3}{2}}}, \quad A, B > 0, \tag{14.14.72}$$

directly relating the matter density to the scale factor, defines a *two-fluid model* [125], which clearly states that $\rho_m \to \infty$ as $a \to 0$, as desired. On the other hand, we may rewrite the conservation law (14.14.21) as

$$P_m = -\rho_m - \frac{a}{3}\frac{\mathrm{d}\rho_m}{\mathrm{d}a}. \tag{14.14.73}$$

Hence, inserting (14.14.72) into (14.14.73), we find

$$P_m = -\frac{B}{2a^{\frac{3}{2}}}. \tag{14.14.74}$$

Combining (14.14.72) and (14.14.74), we arrive at the equation of state for the two-fluid model (14.14.72):

$$\rho_m = v(P_m) = -2P_m + \frac{4AP_m^2}{B^2}. \tag{14.14.75}$$

Using (14.14.75) in (14.14.59), we get the differential equation

$$2X\frac{\mathrm{d}f}{\mathrm{d}X} = f(\beta f - 1), \quad \beta = \frac{4A}{B^2}, \tag{14.14.76}$$

which may be solved to yield

$$f(X) = \frac{X^{\frac{1}{2}}}{\alpha + \beta X^{\frac{1}{2}}}, \tag{14.14.77}$$

where $\alpha > 0$ is an integration constant. In view of (14.14.26), we see that (14.14.77) gives rise to the same equation of state, (14.14.75), for any value of α.

14.15 Finite-energy dyonic point charge

The outcome of the generalized Born–Infeld models indicates that finite-energy point charges, either electric or magnetic, offer some distinctive features associated with black holes carrying such charges, including those with relegated and regularized curvature singularities. Thus, it is of interest to obtain black hole solutions carrying *both* electric *and* magnetic charges, or *dyonic black holes*, of a *finite* electromagnetic energy. However, note that a finite-energy dyonic point charge is generally not permitted in the model defined by (14.11.1)–(14.11.3), but that it is permitted in the model (14.4.11), based on an invariance formalism originally obtained by Born and Infeld [75, 76] as well. Here, we develop a generalized theory of dyonic point charges and black hole solutions, respectively, based on a formalism combining (14.4.11) and (14.11.1)–(14.11.3), in the next two sections.

Action density

As discussed, we now consider the action density

$$\mathcal{L} = f(s), \quad s = -\frac{1}{4}F_{\mu\nu}F^{\mu\nu} + \frac{\kappa^2}{32}\left(F_{\mu\nu}\tilde{F}^{\mu\nu}\right)^2, \tag{14.15.1}$$

where the nonlinear action density profile function f is assumed to satisfy the condition (14.11.3). Hence, the classical Born–Infeld theory (14.11.12) may slightly be extended to be given by

$$f(s) = \frac{1}{\beta}\left(1 - \sqrt{1 - 2\beta s}\right), \tag{14.15.2}$$

where and in the sequal we choose $\beta > 0$ to be a coupling parameter independent of the parameter κ for greater generality such that the limiting situation $\beta = \kappa^2 = \frac{1}{b^2}$ returns the model to classical theory (14.4.11) and reduces the computation considerably.

Equations of motion

Varying A_ν (say) in (14.15.1) leads to the associated Euler–Lagrange equations

$$\partial_\mu P^{\mu\nu} = 0, \tag{14.15.3}$$

$$P^{\mu\nu} = f'(s)\left(F^{\mu\nu} - \frac{\kappa^2}{4}\left(F_{\alpha\beta}\tilde{F}^{\alpha\beta}\right)\tilde{F}^{\mu\nu}\right). \tag{14.15.4}$$

In view of the formal associations with various electromagnetic fields, $\mathbf{E}, \mathbf{B}, \mathbf{D}, \mathbf{H}$, given in (14.1.5) or (14.1.7) and (14.1.12) or (14.1.13), we arrive at the same form of the vacuum Maxwell equations, (14.1.14)–(14.1.15). However, what is different here is that, in view of the relation (14.15.4), the electric displacement field $\mathbf{D}$ and magnetic intensity field $\mathbf{H}$ are now expressed as

$$\mathbf{D} = f'(s)\left(\mathbf{E} + \kappa^2(\mathbf{E}\cdot\mathbf{B})\,\mathbf{B}\right), \tag{14.15.5}$$

$$\mathbf{H} = f'(s)\left(\mathbf{B} - \kappa^2(\mathbf{E}\cdot\mathbf{B})\,\mathbf{E}\right), \tag{14.15.6}$$

in view of (14.1.7), (14.1.8), and (14.4.10). These relations blend the electric and magnetic fields, $\mathbf{E}$ and $\mathbf{B}$, in a complicated way.

Energy-momentum tensor

From (14.15.1), we see that the associated energy-momentum tensor is

$$T_{\mu\nu} = -f'(s)\left(F_{\mu\alpha}\eta^{\alpha\beta}F_{\nu\beta} - \frac{\kappa^2}{4}\left(F_{\mu'\nu'}\tilde{F}^{\mu'\nu'}\right)F_{\mu\alpha}\eta^{\alpha\beta}\tilde{F}_{\nu\beta}\right) - \eta_{\mu\nu}f(s). \tag{14.15.7}$$

Inserting (14.1.7), (14.1.8), (14.4.5), and (14.4.10) into (14.15.7), we can write the energy density as

$$\mathcal{H} = T_{00} = f'(s)\left(\mathbf{E}^2 + \kappa^2[\mathbf{E}\cdot\mathbf{B}]^2\right) - f(s), \tag{14.15.8}$$

$$s = \frac{1}{2}(\mathbf{E}^2 - \mathbf{B}^2) + \frac{\kappa^2}{2}(\mathbf{E}\cdot\mathbf{B})^2, \tag{14.15.9}$$

which will be useful later.

Dyonic point charge in Born–Infeld model

We first consider the classical Born–Infeld model defined by (14.15.2). With (14.15.9) and some suitable manipulation of (14.15.5), we obtain

$$\begin{aligned}&\left(1+\beta\mathbf{D}^2\right)\mathbf{E}^2 + \kappa^2\left(2+\beta\mathbf{D}^2+\kappa^2\mathbf{B}^2\right)(\mathbf{E}\cdot\mathbf{B})^2\\ &= \mathbf{D}^2\left(1+\beta\mathbf{B}^2\right),\end{aligned} \tag{14.15.10}$$

$$\begin{aligned}&\beta(\mathbf{B}\cdot\mathbf{D})^2\mathbf{E}^2 + \left(\left[1+\kappa^2\mathbf{B}^2\right]^2 + \beta\kappa^2(\mathbf{B}\cdot\mathbf{D})^2\right)(\mathbf{E}\cdot\mathbf{B})^2\\ &= (\mathbf{B}\cdot\mathbf{D})^2\left(1+\beta\mathbf{B}^2\right),\end{aligned} \tag{14.15.11}$$

with the solution

$$\mathbf{E}^2 = \frac{(\beta\mathbf{B}^2+1)(\mathbf{D}^2+\kappa^2[2+\kappa^2\mathbf{B}^2][\mathbf{B}\times\mathbf{D}]^2)}{(\kappa^2\mathbf{B}^2+1)(1+\kappa^2\mathbf{B}^2+\beta\mathbf{D}^2+\beta\kappa^2[\mathbf{B}\times\mathbf{D}]^2)}, \tag{14.15.12}$$

$$(\mathbf{E}\cdot\mathbf{B})^2 = \frac{(\beta\mathbf{B}^2+1)(\mathbf{B}\cdot\mathbf{D})^2}{(\kappa^2\mathbf{B}^2+1)(1+\kappa^2\mathbf{B}^2+\beta\mathbf{D}^2+\beta\kappa^2[\mathbf{B}\times\mathbf{D}]^2)}, \tag{14.15.13}$$

where we have used the identity (14.4.22). Inserting (14.11.21) and (14.11.28) into (14.15.12) and (14.15.13), we have

$$\mathbf{E}^2 = \frac{(\beta g^2 + r^4)q^2}{(\kappa^2 g^2 + r^4)(\beta q^2 + \kappa^2 g^2 + r^4)}, \tag{14.15.14}$$

$$(\mathbf{E}\cdot\mathbf{B})^2 = \frac{(\beta g^2 + r^4)q^2 g^2}{r^4(\kappa^2 g^2 + r^4)(\beta q^2 + \kappa^2 g^2 + r^4)}, \tag{14.15.15}$$

which in particular give us the nontrivial radial component of $\mathbf{E}$ to be

$$E^r = q\sqrt{\frac{\beta g^2 + r^4}{(\kappa^2 g^2 + r^4)(\beta q^2 + \kappa^2 g^2 + r^4)}}, \tag{14.15.16}$$

which remains finite at $r = 0$, although E^r still obeys the Coulomb law, $E^r \sim \frac{q}{r^2}$, for r near infinity. Besides, the free electric charge density reads

$$\begin{aligned}\rho^e_{\text{free}} &= \frac{1}{4\pi}\nabla\cdot\mathbf{E} = \frac{1}{4\pi r^2}\frac{\mathrm{d}}{\mathrm{d}r}(r^2 E^r) \\ &= \frac{q}{2\pi r}\frac{\left(\beta\kappa^2 g^4(\beta q^2 + \kappa^2 g^2) + 2\kappa^2 g^2(\beta q^2 + \kappa^2 g^2)r^4 + (2\kappa^2 g^2 + \beta[q^2 - g^2])r^8\right)}{(\beta g^2 + r^4)^{\frac{1}{2}}(\kappa^2 g^2 + r^4)^{\frac{3}{2}}(\beta q^2 + \kappa^2 g^2 + r^4)^{\frac{3}{2}}},\end{aligned} \tag{14.15.17}$$

such that

$$\rho^e_{\text{free}} = \frac{q}{2\pi\kappa r}\sqrt{\frac{\beta}{\beta q^2 + \kappa^2 g^2}}, \quad r \ll 1; \tag{14.15.18}$$

$$\rho^e_{\text{free}} = \frac{q}{2\pi r^7}\left(2\kappa^2 g^2 + \beta[q^2 - g^2]\right), \quad r \gg 1, \tag{14.15.19}$$

which clearly describe how electricity and magnetism mix and interact with each other through the induced free electric charge density, both locally and asymptotically. Furthermore, from (14.15.16) and (14.15.17), the free electric charge contained in $\{|\mathbf{x}| \leq r\}$ is

$$\begin{aligned}q_{\text{free}}(r) &= \int_{|\mathbf{x}|\leq r} \rho^e_{\text{free}}\,\mathrm{d}\mathbf{x} \\ &= (r^2 E^r)_{r=0}^{r=r} = qr^2\sqrt{\frac{\beta g^2 + r^4}{(\kappa^2 g^2 + r^4)(\beta q^2 + \kappa^2 g^2 + r^4)}},\end{aligned} \tag{14.15.20}$$

such that the total free electric charge coincides with the prescribed one, $q_{\text{free}} = q_{\text{free}}(\infty) = q$.

Furthermore, using (14.11.28), (14.15.14), (14.15.15), and (14.15.16) in (14.15.6) with (14.15.2) and (14.15.9), we get the nontrivial radial component of the magnetic intensity field

$$H^r = \frac{g\left(\kappa^4 g^4 + (2\kappa^2 g^2 + q^2[\beta - \kappa^2])r^4 + r^8\right)}{(\beta g^2 + r^4)^{\frac{1}{2}}(\beta q^2 + \kappa^2 g^2 + r^4)^{\frac{1}{2}}(\kappa^2 g^2 + r^4)^{\frac{3}{2}}}, \tag{14.15.21}$$

which is parallel to the expression for the electric field stated in (14.15.16) such that H^r remains finite at $r = 0$, and $H^r \sim \frac{g}{r^2}$ for $r \gg 1$, as anticipated for a magnetic monopole. Using (14.15.21), we may compute the associated free magnetic charge density to obtain

$$\rho^m_{\text{free}} = \frac{\kappa g}{2\pi r\sqrt{\beta(\beta q^2 + \kappa^2 g^2)}}, \quad r \ll 1; \tag{14.15.22}$$

$$\rho^m_{\text{free}} = \frac{g}{2\pi r^7}\left(2\kappa^2 q^2 + \beta[g^2 - q^2]\right), \quad r \gg 1, \tag{14.15.23}$$

similar to (14.15.18)–(14.15.19). Besides, the total free magnetic charge coincides with the prescribed magnetic charge as well, $g_{\text{free}} = g$.

In the classical Born–Infeld model limit with $\beta = \kappa^2$, the expressions (14.15.16) and (14.15.21) become

$$E^r = \frac{q}{\sqrt{\beta(q^2 + g^2) + r^4}}, \quad H^r = \frac{g}{\sqrt{\beta(q^2 + g^2) + r^4}}, \tag{14.15.24}$$

respectively, which recover the results (14.4.24) and (14.4.27) obtained earlier.

Energy of dyonic point charge

To compute the energy carried by a dyonic point charge, we insert (14.15.2) and (14.15.9) into (14.15.8) to get

$$\mathcal{H} = \frac{1}{\beta}\left(\frac{1}{\mathcal{R}} - 1\right) + \frac{\mathbf{B}^2}{\mathcal{R}}, \quad \mathcal{R} = \sqrt{1 - \beta(\mathbf{E}^2 - \mathbf{B}^2 + \kappa^2[\mathbf{E}\cdot\mathbf{B}]^2)}. \tag{14.15.25}$$

Thus, substituting (14.15.12) and (14.15.13) into (14.15.25), we arrive at

$$\mathcal{H} = \frac{1}{\beta}\left(\left[\frac{(1 + \beta\mathbf{B}^2)(1 + \beta(\mathbf{D}^2 + \kappa^2[\mathbf{D}\times\mathbf{B}]^2) + \kappa^2\mathbf{B}^2)}{1 + \kappa^2\mathbf{B}^2}\right]^{\frac{1}{2}} - 1\right), \tag{14.15.26}$$

given in terms of prescribable quantities. Inserting (14.11.21) and (14.11.28) into (14.15.26), we have

$$\begin{aligned}\mathcal{H} &= \frac{1}{\beta}\left(\sqrt{\frac{r^4 + \beta g^2}{r^4 + \kappa^2 g^2}}\sqrt{1 + \frac{\beta q^2 + \kappa^2 g^2}{r^4}} - 1\right) \\ &= \frac{(\beta q^2 + \kappa^2 g^2)g^2 + (q^2 + g^2)r^4}{r^2\sqrt{r^4 + \kappa^2 g^2}(\sqrt{r^4 + \beta g^2}\sqrt{r^4 + \beta q^2 + \kappa^2 g^2} + r^2\sqrt{r^4 + \kappa^2 g^2})},\end{aligned} \tag{14.15.27}$$

which renders the determination of the finite total energy E of the dyonic point particle charge by the usual expression

$$\frac{E}{4\pi} = \int_0^\infty r^2\mathcal{H}\,\mathrm{d}r. \tag{14.15.28}$$

First, we note that, when $\kappa = 0$, (14.15.27) becomes

$$\mathcal{H}_{\kappa=0} = \frac{\beta q^2 g^2 + (q^2 + g^2) r^4}{r^4(\sqrt{r^4 + \beta q^2}\sqrt{r^4 + \beta g^2} + r^4)}. \tag{14.15.29}$$

Hence, in the dyonic point charge situation, $q, g > 0$, the energy diverges, although in the pure electric or magnetic point charge situation with $q > 0, g = 0$ or $q = 0, g > 0$, respectively, the energy converges. In other words, the coupling parameter κ plays the role of an energy regulator for a dyonic point charge, so that when it is "switched off" by setting $\kappa = 0$ the energy blows up.

In contrast, when $\kappa > 0$ but the point charge is either electric, magnetic, or dyonic, with $q > 0$ and $g = 0$, $q = 0$ and $g > 0$, or $q > 0$ and $g > 0$, respectively, we see that the energy always converges.

Next, in the general dyonic situation with $\kappa > 0$, we note that it is impossible to obtain the exact value for the energy E given in (14.15.28). However, in order to estimate E, we observe that the quantity

$$\eta(r) = \sqrt{\frac{r^4 + \beta g^2}{r^4 + \kappa^2 g^2}} \tag{14.15.30}$$

is nondecreasing or nonincreasing in $r \geq 0$ according to whether $\beta \leq \kappa^2$ or $\beta \geq \kappa^2$. Consequently, we have the general bound

$$\min\left\{1, \frac{\sqrt{\beta}}{\kappa}\right\} \leq \eta(r) \leq \max\left\{1, \frac{\sqrt{\beta}}{\kappa}\right\}, \quad r \geq 0. \tag{14.15.31}$$

In particular, we have

$$\eta(r) \leq 1, \quad \beta \leq \kappa^2; \quad \eta(r) \geq 1, \quad \beta \geq \kappa^2. \tag{14.15.32}$$

Applying these bounds to (14.15.28), we get

$$E \leq E_0, \quad \beta \leq \kappa^2; \quad E \geq E_0, \quad \beta \geq \kappa^2, \tag{14.15.33}$$

where

$$\begin{aligned}\frac{E_0}{4\pi} &= \frac{1}{\beta}\int_0^\infty \left(\sqrt{1 + \frac{\beta q^2 + \kappa^2 g^2}{r^4}} - 1\right) r^2 \, \mathrm{d}r \\ &= \frac{a^3}{\beta}\int_0^\infty \frac{\mathrm{d}x}{\sqrt{x^4+1} + x^2} = \frac{\pi^{\frac{3}{2}}}{3\left(\Gamma\left(\frac{3}{4}\right)\right)^2}\frac{Q^2}{a},\end{aligned} \tag{14.15.34}$$

where we have set

$$Q = \left(q^2 + \left[\frac{\kappa^2}{\beta}\right] g^2\right)^{\frac{1}{2}}, \quad a = (\beta q^2 + \kappa^2 g^2)^{\frac{1}{4}}. \tag{14.15.35}$$

These results give some estimates of the energy of a dyonic point charge in terms of its *combined dyonic charge*, taking account of both electric and magnetic charges, q and g, jointly, imposed at the same spot, and of its "radius", respectively. Note that, when $\beta = \kappa^2$, we recover the classical result consisting of (14.4.33) and (14.4.34).

Note that (14.15.34) is of the same form as that for an electric point charge, as in the work of Born and Infeld [75, 76]. In other words, in terms of the combined charge defined in (14.15.35), a dyonic point charge behaves like an electric point charge, energetically. Furthermore, the description of a dyonic point charge is symmetric with respect to the interchange of the roles of the electric and magnetic charges in the classical model $\beta = \kappa^2$. Besides, the second expression in (14.15.35) indicates that a dyonic point charge possesses a larger radius as a result of the need to accommodate a greater joint charge, as given in the first expression in (14.15.35).

Although (14.15.27) appears complicated, it has the following simple expansion

$$\mathcal{H} = \frac{q^2+g^2}{2r^4} - \frac{(\beta(q^2-g^2)^2+4\kappa^2q^2g^2)}{8r^8} + \mathrm{O}(r^{-12}), \quad r \gg 1, \tag{14.15.36}$$

extending that of the Maxwell dyon solution, with a negative higher-order correction term.

Static equations for dyons

In general, consider the static equations for dyons, as given in (14.5.22). From (14.15.6), we get

$$\mathbf{E}\cdot\mathbf{B} = \frac{\mathbf{E}\cdot\mathbf{H}}{f'(s)(1-\kappa^2\mathbf{E}^2)}, \tag{14.15.37}$$

$$\mathbf{B} = \frac{\mathbf{H}}{f'(s)} + \kappa^2(\mathbf{E}\cdot\mathbf{B})\mathbf{E} = \frac{1}{f'(s)}\left(\mathbf{H} + \frac{\kappa^2(\mathbf{E}\cdot\mathbf{H})}{1-\kappa^2\mathbf{E}^2}\mathbf{E}\right). \tag{14.15.38}$$

For the Born–Infeld model (14.15.2), we have

$$\frac{1}{f'(s)} \equiv \mathcal{R} = \sqrt{1-2\beta s} = \sqrt{1-\beta(\mathbf{E}^2-\mathbf{B}^2+\kappa^2[\mathbf{E}\cdot\mathbf{B}]^2)}. \tag{14.15.39}$$

Thus, inserting (14.15.37)–(14.15.38) into (14.15.39), we obtain

$$\mathcal{R}^2 = 1-\beta\mathbf{E}^2 + \beta\mathcal{R}^2\left(\mathbf{H}+\frac{\kappa^2(\mathbf{E}\cdot\mathbf{H})}{1-\kappa^2\mathbf{E}^2}\mathbf{E}\right)^2 - \beta\kappa^2\mathcal{R}^2\left(\frac{\mathbf{E}\cdot\mathbf{H}}{1-\kappa^2\mathbf{E}^2}\right)^2, \tag{14.15.40}$$

which may be resolved to yield

$$\mathcal{R}^2 = \frac{1-\beta\mathbf{E}^2}{1-\beta\left(\mathbf{H}+\frac{\kappa^2(\mathbf{E}\cdot\mathbf{H})}{1-\kappa^2\mathbf{E}^2}\mathbf{E}\right)^2+\beta\kappa^2\left(\frac{\mathbf{E}\cdot\mathbf{H}}{1-\kappa^2\mathbf{E}^2}\right)^2}. \tag{14.15.41}$$

As a consequence of (14.15.5), (14.15.6), (14.15.37)–(14.15.39), and (14.15.41), it follows that

$$\mathbf{D} = \frac{\mathbf{E}}{\mathcal{R}} + \kappa^2 \mathcal{R} \frac{(\mathbf{E}\cdot\mathbf{H})}{1-\kappa^2\mathbf{E}^2}\left(\mathbf{H} + \frac{\kappa^2(\mathbf{E}\cdot\mathbf{H})}{1-\kappa^2\mathbf{E}^2}\mathbf{E}\right), \tag{14.15.42}$$

$$\mathbf{B} = \mathcal{R}\left(\mathbf{H} + \frac{\kappa^2(\mathbf{E}\cdot\mathbf{H})}{1-\kappa^2\mathbf{E}^2}\mathbf{E}\right). \tag{14.15.43}$$

The last two equations in (14.5.22) indicate that there are real-valued scalar fields, ϕ and ψ, such that $\mathbf{E} = \nabla\phi$ and $\mathbf{H} = \nabla\psi$. Thus, inserting these into (14.15.41), (14.15.42), and (14.15.43), we see that the first two equations in (14.5.22) become

$$\nabla\cdot\left(\frac{\nabla\phi}{\mathcal{R}} + \kappa^2\mathcal{R}\frac{\nabla\phi\cdot\nabla\psi}{1-\kappa^2|\nabla\phi|^2}\left[\nabla\psi + \frac{\kappa^2(\nabla\phi\cdot\nabla\psi)}{1-\kappa^2|\nabla\phi|^2}\nabla\phi\right]\right) = \rho_e, \tag{14.15.44}$$

$$\nabla\cdot\left(\mathcal{R}\left[\nabla\psi + \frac{\kappa^2(\nabla\phi\cdot\nabla\psi)}{1-\kappa^2|\nabla\phi|^2}\nabla\phi\right]\right) = \rho_m, \tag{14.15.45}$$

$$\mathcal{R} = \sqrt{\frac{1-\beta|\nabla\phi|^2}{1-\beta\left(\nabla\psi + \frac{\kappa^2(\nabla\phi\cdot\nabla\psi)}{1-\kappa^2|\nabla\phi|^2}\nabla\phi\right)^2 + \beta\kappa^2\left(\frac{\nabla\phi\cdot\nabla\psi}{1-\kappa^2|\nabla\phi|^2}\right)^2}}. \tag{14.15.46}$$

These are the static equations for dyons in the most general setting, and they appear highly complicated. However, in the point charge situation with $\rho_e = 4\pi q\delta(\mathbf{x})$ and $\rho_m = 4\pi g\delta(\mathbf{x})$, we have obtained a radially symmetric solution whose nontrivial radial components of the electric field and magnetic intensity field, E^r and H^r, are given by (14.15.16) and (14.15.21), respectively. Thus, we may assume that the scalar potentials ϕ and ψ for $\mathbf{E}$ and $\mathbf{H}$ are also radially symmetric with $\phi = \phi(r)$ and $\psi = \psi(r)$. Hence, $E^r = \phi'(r)$ and $H^r = \psi'(r)$ so that (14.15.16) and (14.15.21) give us the radially symmetric solution

$$\phi(r) = -\int_r^\infty \frac{q(\beta g^2+\rho^4)^{\frac{1}{2}}}{(\kappa^2 g^2+\rho^4)^{\frac{1}{2}}(\beta q^2+\kappa^2 g^2+\rho^4)^{\frac{1}{2}}}\,\mathrm{d}\rho, \tag{14.15.47}$$

$$\psi(r) = -\int_r^\infty \frac{g\left(\kappa^4 g^4 + (2\kappa^2 g^2 + q^2[\beta-\kappa^2])\rho^4 + \rho^8\right)}{(\beta g^2+\rho^4)^{\frac{1}{2}}(\beta q^2+\kappa^2 g^2+\rho^4)^{\frac{1}{2}}(\kappa^2 g^2+\rho^4)^{\frac{3}{2}}}\,\mathrm{d}\rho, \tag{14.15.48}$$

to (14.15.44)–(14.15.46) for the point charge situation.

In the classical Born–Infeld model situation with $\beta = \kappa^2$ the equations (14.15.44)–(14.15.46) become

$$\nabla\cdot\left(\frac{\mathcal{R}_0}{1-\beta|\nabla\phi|^2}\nabla\phi + \frac{\beta(\nabla\phi\cdot\nabla\psi)}{\mathcal{R}_0}\left[\nabla\psi + \frac{\beta(\nabla\phi\cdot\nabla\psi)}{1-\beta|\nabla\phi|^2}\nabla\phi\right]\right) = \rho_e, \tag{14.15.49}$$

$$\nabla\cdot\left(\frac{1}{\mathcal{R}_0}\left[(1-\beta|\nabla\phi|^2)\nabla\psi + \beta(\nabla\phi\cdot\nabla\psi)\nabla\phi\right]\right) = \rho_m, \tag{14.15.50}$$

$$\mathcal{R}_0 = \sqrt{1-\beta|\nabla\phi|^2 - \beta|\nabla\psi|^2 + \beta^2(\nabla\phi\times\nabla\psi)^2}, \tag{14.15.51}$$

in view of the vector identity (14.4.22), so that the point charge solution (14.15.47)–(14.15.48) assumes the simplified form

$$\phi(r) = -\int_r^\infty \frac{q}{\sqrt{\rho^4 + \beta q^2 + \beta g^2}}\,\mathrm{d}\rho, \tag{14.15.52}$$

$$\psi(r) = -\int_r^\infty \frac{g}{\sqrt{\rho^4 + \beta q^2 + \beta g^2}}\,\mathrm{d}\rho. \tag{14.15.53}$$

This result recovers (14.4.24) and (14.4.27), as anticipated.

In the situation when $\kappa = 0$, the equations (14.15.44)–(14.15.46) become (14.5.19)–(14.5.20) so that the point charge solution (14.15.47)–(14.15.48) reduces itself into

$$\phi(r) = -\int_r^\infty \frac{q}{\rho^2}\sqrt{\frac{\beta g^2 + \rho^4}{\beta q^2 + \rho^4}}\,\mathrm{d}\rho, \quad \psi(r) = -\int_r^\infty \frac{g}{\rho^2}\sqrt{\frac{\beta q^2 + \rho^4}{\beta g^2 + \rho^4}}\,\mathrm{d}\rho, \tag{14.15.54}$$

which recovers the result (14.5.25)–(14.5.26).

Thus, we have obtained a unified treatment of the static equations for dyons and associated point charge solutions in the Born–Infeld theory.

Exponential model

Here, we consider the exponential model given by

$$f(s) = \frac{1}{\beta}\left(\mathrm{e}^{\beta s} - 1\right), \tag{14.15.55}$$

where $\beta > 0$ is a parameter, with s defined in (14.15.1). In view of (14.15.55) and (14.15.9), the equations (14.15.5) and (14.15.6) become

$$\mathbf{D} = \mathrm{e}^{\frac{\beta}{2}\left(\mathbf{E}^2 - \mathbf{B}^2 + \kappa^2[\mathbf{E}\cdot\mathbf{B}]^2\right)}\left(\mathbf{E} + \kappa^2[\mathbf{E}\cdot\mathbf{B}]\mathbf{B}\right), \tag{14.15.56}$$

$$\mathbf{H} = \mathrm{e}^{\frac{\beta}{2}\left(\mathbf{E}^2 - \mathbf{B}^2 + \kappa^2[\mathbf{E}\cdot\mathbf{B}]^2\right)}\left(\mathbf{B} - \kappa^2[\mathbf{E}\cdot\mathbf{B}]\mathbf{E}\right). \tag{14.15.57}$$

From (14.15.56), we have

$$\mathbf{D}^2\mathrm{e}^{\beta\mathbf{B}^2} = \mathrm{e}^{\beta(\mathbf{E}^2 + \kappa^2[\mathbf{E}\cdot\mathbf{B}]^2)}\left(\mathbf{E}^2 + \kappa^2[2 + \kappa^2\mathbf{B}^2][\mathbf{E}\cdot\mathbf{B}]^2\right), \tag{14.15.58}$$

$$(\mathbf{B}\cdot\mathbf{D})^2\mathrm{e}^{\beta\mathbf{B}^2} = \mathrm{e}^{\beta(\mathbf{E}^2 + \kappa^2[\mathbf{E}\cdot\mathbf{B}]^2)}\left(1 + \kappa^2\mathbf{B}^2\right)^2(\mathbf{E}\cdot\mathbf{B})^2. \tag{14.15.59}$$

Note that the left-hand sides of these two equations are prescribable in view of (14.11.21) and (14.11.28) so that their combination renders the equation

$$\mathrm{e}^W W = \delta, \quad W = \beta\left(\mathbf{E}^2 + \kappa^2[\mathbf{E}\cdot\mathbf{B}]^2\right),$$
$$\delta = \beta\mathrm{e}^{\beta\mathbf{B}^2}\left(\mathbf{D}^2 - \frac{\kappa^2(\mathbf{B}\cdot\mathbf{D})^2}{1 + \kappa^2\mathbf{B}^2}\right), \tag{14.15.60}$$

which may be solved with the Lambert W function, resulting in

$$\beta\left(\mathbf{E}^2+\kappa^2[\mathbf{E}\cdot\mathbf{B}]^2\right)=W(\delta)=W\left(\beta \mathrm{e}^{\beta\mathbf{B}^2}\left[\mathbf{D}^2-\frac{\kappa^2(\mathbf{B}\cdot\mathbf{D})^2}{1+\kappa^2\mathbf{B}^2}\right]\right)$$

$$=W\left(\beta \mathrm{e}^{\beta\mathbf{B}^2}\left[\frac{\mathbf{D}^2+\kappa^2(\mathbf{B}\times\mathbf{D})^2}{1+\kappa^2\mathbf{B}^2}\right]\right), \tag{14.15.61}$$

where $W(x)$ satisfies (14.13.6) and (14.13.7). Furthermore, by (14.15.59), we get

$$(\mathbf{E}\cdot\mathbf{B})^2=\frac{(\mathbf{B}\cdot\mathbf{D})^2}{(1+\kappa^2\mathbf{B}^2)^2}\mathrm{e}^{\beta\mathbf{B}^2-W(\delta)}. \tag{14.15.62}$$

Combining (14.15.61) and (14.15.62), we arrive at

$$\mathbf{E}^2=\frac{W(\delta)}{\beta}-\frac{\kappa^2(\mathbf{B}\cdot\mathbf{D})^2}{(1+\kappa^2\mathbf{B}^2)^2}\mathrm{e}^{\beta\mathbf{B}^2-W(\delta)}, \tag{14.15.63}$$

where $W(\delta)$ is as stated in (14.15.61). This expression appears rather complicated. However, in the radially symmetric dyonically charged point particle situation when the electric and magnetic charges are given by (14.11.21) and (14.11.28), the quantity δ in (14.15.60) or (14.15.61) simplifies itself into

$$\delta=\frac{\beta q^2}{r^4+\kappa^2 g^2}\,\mathrm{e}^{\frac{\beta g^2}{r^4}}. \tag{14.15.64}$$

In view of (14.13.6) and (14.15.64), we have

$$W(\delta)=\frac{\beta q^2}{r^4}\left(1+\frac{1}{r^4}\left[(\beta-\kappa^2)g^2-\beta q^2\right]\right)+\mathrm{O}(r^{-12}),\quad r\gg 1, \tag{14.15.65}$$

such that the nontrivial radial component E^r of the electric field $\mathbf{E}$ given in (14.15.63) reads

$$E^r=\frac{q}{r^2}+\frac{q}{r^6}\left(\frac{\beta}{2}(g^2-q^2)-\kappa^2 g^2\right)+\mathrm{O}(r^{-10}),\quad r\gg 1, \tag{14.15.66}$$

asymptotically. In other words, the electric field of the dyonic point charge appears like that of an electric Coulomb charge asymptotically. On the other hand, for $\kappa>0$, in view of (14.13.7) and (14.15.64), we have

$$W(\delta)=\frac{\beta g^2}{r^4}+\ln r^4+\ln\left(\frac{q^2}{\kappa^2 g^4}\right)$$

$$-\frac{1}{g^2}\left(\frac{1}{\kappa^2}+\frac{1}{\beta}\ln\left[\frac{\beta q^2}{\kappa^2 g^2}\right]\right)r^4+\mathrm{o}(r^8),\quad r\ll 1, \tag{14.15.67}$$

in leading-order approximation, resulting in the behavior

$$E^r=\frac{g}{r^2}\sqrt{1-\frac{\kappa^4 g^4}{(r^4+\kappa^2 g^2)^2}},\quad r\ll 1, \tag{14.15.68}$$

by (14.15.63). In other words, locally, the electric field of the dyonic point charge depends only on its magnetic charge g. As a consequence of (14.15.66) and (14.15.68), we see that the total free electric charge of the particle is

$$q_{\text{free}} = \left(r^2 E^r\right)_{r=0}^{r=\infty} = q, \tag{14.15.69}$$

which coincides with the prescribed electric charge.

Furthermore, in view of (14.15.57) and (14.15.66), we see that the nontrivial radial component of the magnetic intensity field $\mathbf{H}$ is

$$H^r = \frac{g}{r^2} + \frac{g}{r^6}\left(\frac{\beta}{2}(q^2 - g^2) - \kappa^2 q^2\right) + \mathrm{O}(r^{-10}), \quad r \gg 1, \tag{14.15.70}$$

in leading-order approximation, which indicates that this field asymptotically behaves like that of a magnetic point charge. Moreover, there is a perfect symmetry, with $E^r \leftrightarrows H^r$, $q \leftrightarrows g$, between the two formulas, (14.15.66) and (14.15.70).

On the other hand, when $\kappa > 0$, inserting (14.15.67) and (14.15.68) into (14.15.57), we have, up to leading orders, the expression

$$\begin{aligned} H^r &= \frac{q}{\kappa g}\left(1 - \frac{\kappa^2 g^2}{r^4}\left[1 - \frac{\kappa^4 g^4}{(r^4 + \kappa^2 g^2)^2}\right]\right) \\ &= \frac{q}{\kappa g}\left(-1 + \frac{3r^4}{\kappa^2 g^2}\right) + \mathrm{O}(r^8), \quad r \ll 1. \end{aligned} \tag{14.15.71}$$

In particular, H^r stays finite near the origin and the associated total free magnetic charge is calculated to be

$$g_{\text{free}} = (r^2 H^r)_{r=0}^{r=\infty} = g, \tag{14.15.72}$$

which coincides with the prescribed magnetic point charge, by using (14.15.70) and (14.15.71) as before.

Energy density

To compute the energy of a dyonic point charge in the exponential model, we insert (14.15.55) into (14.15.8)–(14.15.9) to get the Hamiltonian density

$$\mathcal{H} = \frac{1}{\beta}\mathrm{e}^{-\frac{\beta}{2}\mathbf{B}^2 + \frac{1}{2}W(\delta)}\, W(\delta) - \frac{1}{\beta}\left(\mathrm{e}^{-\frac{\beta}{2}\mathbf{B}^2 + \frac{1}{2}W(\delta)} - 1\right). \tag{14.15.73}$$

Consequently, in view of (14.15.65), we have

$$\mathcal{H} = \frac{q^2 + g^2}{2r^4} - \frac{\left(\beta[q^2 - g^2]^2 + 4\kappa^2 q^2 g^2\right)}{8r^8} + \mathrm{O}(r^{-12}), \quad r \gg 1, \tag{14.15.74}$$

which, in leading order, is identical to that in the Maxwell electromagnetism theory case, and (incidentally) coincides with (14.15.36). Moreover, we have, with $\kappa > 0$,

$$\mathcal{H} = \frac{q}{\kappa r^2} + \frac{1}{\beta} - \frac{q r^2}{2\beta\kappa^3 g^2}\left(\beta + \kappa^2\left[2 + \ln\left(\frac{\beta\kappa^2 g^6}{q^2 r^6}\right)\right]\right) + \mathrm{O}(r^6), \quad r \ll 1, \tag{14.15.75}$$

and, when $\kappa = 0$, we have

$$\mathcal{H} = \frac{qg}{r^4}, \quad r \ll 1, \quad \kappa = 0, \tag{14.15.76}$$

which coincides with (14.13.32) whose more refined asymptotic expression is (14.13.31).

As a consequence of these results, we conclude that there is energy convergence when $\kappa > 0$, but divergence when $\kappa = 0$. In other words, the coupling parameter κ plays the role of an energy regulator for a dyonic point charge as before.

14.16 Dyonically charged black holes with relegated singularities

This section considers *finite-energy*, dyonically charged black hole solutions generated from the Einstein equations, with a vanishing cosmological constant, coupled with the Born–Infeld equations associated with a generalized action density, (14.15.1), in which the Minkowski metric $\eta_{\mu\nu}$ is replaced by a gravitational one, $g_{\mu\nu}$, used to raise or lower indices and construct the Hodge dual of the electromagnetic field as before. Thus, the curved-spacetime Born–Infeld electromagnetic field equations of (14.15.1) now read

$$\frac{1}{\sqrt{-g}}\partial_\mu \left(\sqrt{-g}\, P^{\mu\nu}\right) = 0, \tag{14.16.1}$$

$$P^{\mu\nu} = f'(s)\left(F^{\mu\nu} - \frac{\kappa^2}{4}\left[F_{\alpha\beta}\tilde{F}^{\alpha\beta}\right]\tilde{F}^{\mu\nu}\right), \tag{14.16.2}$$

which update (14.15.3)–(14.15.4). Likewise, the energy-momentum tensor $T_{\mu\nu}$ is now given by

$$\begin{aligned} T_{\mu\nu} &= 2\frac{\partial \mathcal{L}}{\partial g^{\mu\nu}} - g_{\mu\nu}\mathcal{L} \\ &= -f'(s)\left(F_{\mu\alpha}g^{\alpha\beta}F_{\nu\beta} - \frac{\kappa^2}{4}\left[F_{\mu'\nu'}\tilde{F}^{\mu'\nu'}\right]F_{\mu\alpha}g^{\alpha\beta}\tilde{F}_{\nu\beta}\right) - g_{\mu\nu}f(s). \end{aligned} \tag{14.16.3}$$

With radial symmetry

Imposing radial symmetry so that the gravitational line element is given by (10.4.16), we see that various associated electromagnetic fields, $P^{\mu\nu}, F^{\mu\nu}, F_{\mu\nu}, \tilde{F}^{\mu\nu}$, are still represented by the same matrices, (14.9.1)–(14.9.3) and (14.9.36), in terms of the radial quantities, D^r, H^r, E^r, B^r, respectively, and that (14.9.4) and (14.9.37) hold. As a consequence, we get

$$s = \frac{1}{2}\left([E^r]^2 - r^4\sin^2\theta[B^r]^2\right) + \frac{\kappa^2}{2}r^4\sin^2\theta(E^rB^r)^2. \tag{14.16.4}$$

Besides, by (14.16.2), we have

$$D^r = f'(s)\left(E^r + \kappa^2 r^4 \sin^2\theta\, E^r [B^r]^2\right), \tag{14.16.5}$$

$$H^r = f'(s)\left(B^r - \kappa^2 [E^r]^2 B^r\right). \tag{14.16.6}$$

Moreover, (14.16.1) yields the relations (14.9.9) and (14.9.10), resulting in (14.9.11) and (14.9.12) again. Thus, in view of consistency in (14.16.5) and (14.16.6) with (14.9.11) and (14.9.12), we arrive at

$$B^r = \frac{B(r)}{\sin\theta}, \tag{14.16.7}$$

which resembles (14.9.12) and also appears in (14.12.37), where the function $B(r)$ is to be determined. Hence, (14.9.4), (14.9.37), and (14.16.4) are simplified into

$$F_{\mu\nu}F^{\mu\nu} = -2(E^r)^2 + 2r^4 B^2, \tag{14.16.8}$$

$$F_{\mu\nu}\tilde{F}^{\mu\nu} = -4r^2 E^r B, \tag{14.16.9}$$

$$s = \frac{1}{2}\left([E^r]^2 - r^4 B^2\right) + \frac{\kappa^2}{2} r^4 (E^r B)^2. \tag{14.16.10}$$

Energy-momentum tensor

In view of these results, we see that the nontrivial components of the energy-momentum tensor given in (14.16.3) are

$$T_{00} = A f'(s)\left([E^r]^2 + \kappa^2 r^4 [E^r B]^2\right) - A f(s), \tag{14.16.11}$$

$$T_{11} = -\frac{f'(s)}{A}\left([E^r]^2 + \kappa^2 r^4 [E^r B]^2\right) + \frac{f(s)}{A}, \tag{14.16.12}$$

$$T_{22} = r^6 f'(s)\left(B^2 - \kappa^2 [E^r B]^2\right) + r^2 f(s), \quad T_{33} = \sin^2\theta\, T_{22}, \tag{14.16.13}$$

with the associated trace density

$$T = 2f'(s)\left([E^r]^2 - r^4 B^2 + 2\kappa^2 r^4 [E^r B]^2\right) - 4f(s). \tag{14.16.14}$$

Reduction of Einstein equations

Therefore, in view of (10.4.17)–(10.4.20) and (14.16.11)–(14.16.14), we see that the Einstein equations (14.8.1) are reduced into

$$(r^2 A')' = 16\pi G r^2 \left(r^4 f'(s)(B^2 - \kappa^2 [E^r B]^2) + f(s)\right), \tag{14.16.15}$$

$$(rA)' = 1 - 8\pi G r^2 \left(f'(s)([E^r]^2 + \kappa^2 r^4 [E^r B]^2) - f(s)\right), \tag{14.16.16}$$

which are over-determined as before. We show again that (14.16.15) is contained in (14.16.16) under a certain condition. In fact, inserting (14.16.5) into (14.9.9) and using (14.16.7), we have

$$\left(r^2 f'(s) E^r (1 + \kappa^2 r^4 B^2)\right)_r = 0. \tag{14.16.17}$$

Thus, applying (14.16.17), we see that (14.16.15) and (14.16.16) lead to

$$(r^2A')' - r(rA)'' = 16\pi G r^6 B f'(s)\left(1-\kappa^2[E^r]^2\right)\left(2B+\frac{rB'}{2}\right), \tag{14.16.18}$$

in view of

$$\frac{\mathrm{d}s}{\mathrm{d}r} = (1+\kappa^2 r^4 B^2)E^r(E^r)' + r^4(-1+\kappa^2[E^r]^2)BB' + 2r^3(-1+\kappa^2[E^r]^2)B^2, \tag{14.16.19}$$

from (14.16.10). Since consistency requires $(r^2A')' - r(rA)'' = 0$, we are led to impose the equation (14.12.44) and arrive at the same result (14.12.45).

Insight from Bianchi identity

We remark that the condition (14.12.45) is in fact a consequence of the Bianchi identity

$$\frac{1}{\sqrt{-g}}\partial_\mu\left(\sqrt{-g}\tilde{F}^{\mu\nu}\right) = 0, \tag{14.16.20}$$

under the radial symmetry assumption, as well. Indeed, inserting (14.9.36) and (14.16.7) into (14.16.20) and observing $\sqrt{-g} = r^2\sin\theta$, we have

$$(r^4\sin\theta\, B)_r = 0, \quad (E^r)_\theta = 0. \tag{14.16.21}$$

The first equation in (14.16.21) indicates that $r^4 B$ is a constant, implying (14.12.45), and the second equation in (14.16.21) simply reconfirms that E^r is a radial function only.

Metric factor

Consequently, it suffices to concentrate on (14.16.16). In order to understand the physical meaning of the right-hand side of (14.16.16), we recall that

$$\mathcal{H} = T_0^0 = g^{00}T_{00} = f'(s)\left([E^r]^2 + \kappa^2 r^4[E^r B]^2\right) - f(s), \tag{14.16.22}$$

in view of the line element (10.4.16) and (14.16.11). In other words, equation (14.16.16) simply reads

$$(rA)' = 1 - 8\pi G r^2\mathcal{H}(r), \tag{14.16.23}$$

which directly relates the metric factor A to the energy density of the electromagnetic sector through an integration, exactly as in (14.12.47),

$$A(r) = 1 - \frac{2GM}{r} + \frac{8\pi G}{r}\int_r^\infty \mathcal{H}(\rho)\rho^2\,\mathrm{d}\rho, \tag{14.16.24}$$

where M is an integration constant that may be taken to be positive to represent a mass. Since the energy of the dyonic point charge is given by

$$E = \int \mathcal{H}\sqrt{-g}\,\mathrm{d}r\mathrm{d}\theta\mathrm{d}\phi = 4\pi \int_0^\infty \mathcal{H}r^2\,\mathrm{d}r, \tag{14.16.25}$$

so, if this quantity is finite, we may rewrite (14.16.24) as

$$A(r) = 1 - \frac{2G(M-E)}{r} - \frac{8\pi G}{r}\int_0^r \mathcal{H}(\rho)\,\rho^2\,\mathrm{d}\rho, \tag{14.16.26}$$

where, now,

$$\mathcal{H} = f'(s)\left(1 + \frac{\kappa^2 g^2}{r^4}\right)(E^r)^2 - f(s), \tag{14.16.27}$$

$$s = \frac{1}{2}\left((E^r)^2 - \frac{g^2}{r^4}\right) + \frac{\kappa^2 g^2}{2r^4}(E^r)^2, \tag{14.16.28}$$

where E^r is determined by (14.16.5) or

$$D^r = \frac{q}{r^2} = f'\left(\frac{1}{2}\left[(E^r)^2 - \frac{g^2}{r^4}\right] + \frac{\kappa^2 g^2}{2r^4}(E^r)^2\right)\left(1 + \frac{\kappa^2 g^2}{r^4}\right)E^r, \tag{14.16.29}$$

implicitly, which is the radial version of the general equation (14.15.5) when gravity is absent. In other words, all the results found earlier to describe a dyonic point charge in the situation *without* gravity may directly be carried over to and used in the situation *with* gravity in our context of a spherically symmetric black hole solution.

Critical mass-energy condition

From the expression (14.16.26), we arrive as before at the critical mass-energy condition

$$M = E, \tag{14.16.30}$$

under which (14.16.26) simplifies itself into

$$A(r) = 1 - \frac{8\pi G}{r}\int_0^r \mathcal{H}(\rho)\,\rho^2\,\mathrm{d}\rho, \tag{14.16.31}$$

indicating clearly that the possible singularity at $r = 0$ is embodied in the local energy term represented by the integral on the right-hand side of the formula.

Asymptotic form of solution

Finally, we obtain the metric factor $A(r)$ asymptotically away from locally regions. For this purpose, note that (14.11.3) and (14.16.29) imply that

$$E^r = \frac{q}{r^2}, \quad r \gg 1, \tag{14.16.32}$$

in leading-order approximation. Thus, using (14.16.32) in (14.16.27) with (14.16.28), we get

$$\mathcal{H}(r) = \frac{1}{2r^4}\left(q^2+g^2\right), \quad r \gg 1, \tag{14.16.33}$$

within the same truncation errors. Therefore, substituting (14.16.33) into (14.16.24), we have

$$A(r) = 1 - \frac{2GM}{r} + \frac{4\pi G}{r^2}(q^2+g^2), \quad r \gg 1. \tag{14.16.34}$$

In other words, in leading-order approximation, we recover the classical Reissner–Nordström solution, as anticipated. Note that the spherically symmetric asymptotic form (14.16.34) is indifferent to the fine structure of the nonlinear profile function $f(s)$ in the general model defined by (14.15.1) with (14.11.3).

Quasilocal mass

Inserting (14.16.26) into (10.6.33), we obtain the quasilocal mass or energy of Brown–York [88] for a general dyonic black hole metric to be

$$E_{\mathrm{ql}}(r) = \frac{2}{1+\sqrt{A(r)}}\left([M-E] + 4\pi\int_0^r \mathcal{H}(\rho)\rho^2\,\mathrm{d}\rho\right). \tag{14.16.35}$$

Furthermore, for $r \gg 1$, since $A(r)$ is given by (14.16.34) asymptotically, we see that (10.6.34) is still valid. That is,

$$E_{\mathrm{ql}}(r) = M + \frac{(GM^2 - 4\pi[q^2+g^2])}{2r}\left(1+\frac{GM}{r}\right) + \mathrm{O}(r^{-3}), \quad r \gg 1. \tag{14.16.36}$$

Consequently, we obtain the same total ADM energy or mass, $E_{\mathrm{ql}}(\infty) = M$, from the gravitational mass of the black hole, which is independent of the fine electromagnetic structure of the generalized Born–Infeld model.

Born–Infeld black holes

Consider first the dyonically charged black holes in the classical Born–Infeld theory defined by (14.15.2). In this context, we insert (14.15.27) directly into (14.16.26) to obtain the exact solution

$$\begin{aligned} A(r) &= 1 - \frac{2G(M-E)}{r} \\ &\quad - \frac{8\pi G}{r}\int_0^r \frac{(\beta q^2 + \kappa^2 g^2)g^2 + (q^2+g^2)\rho^4}{\sqrt{\rho^4+\kappa^2 g^2}\left(\sqrt{\rho^4+\beta g^2}\sqrt{\rho^4+\beta q^2+\kappa^2 g^2} + \rho^2\sqrt{\rho^4+\kappa^2 g^2}\right)}\,\mathrm{d}\rho \\ &= 1 - \frac{2G(M-E)}{r} - 8\pi G\left(\frac{Q}{\kappa} - \frac{r^2}{3\beta} + \frac{(\kappa^2 Q^2 - \beta q^2)r^4}{10\beta\kappa^3 g^2 Q}\right) + \mathrm{O}(r^6),\ r \ll 1, \end{aligned} \tag{14.16.37}$$

where $\kappa > 0$, and we have used the joint charge Q defined in (14.15.35) in order to considerably simplify the expression for the asymptotic expansion.

Curvature singularity

With the Kretschmann invariant (14.8.37), we are led by (14.16.37) to draw the following conclusions for a finite-energy dyonically charged black hole solution just obtained:

(i) Under the non-critical mass-energy condition $M \neq E$, the curvature singularity at $r = 0$ is of the same type as that of the Schwarzschild black-hole solution, $K = \mathrm{O}(r^{-6})$.

(ii) Under the critical mass-energy condition $M = E$, the metric factor $A(r)$ is analytic at $r = 0$ and enjoys the expansion

$$A(r) = 1 - 8\pi G\left(\frac{Q}{\kappa} - \frac{r^2}{3\beta} + \frac{(\kappa^2 Q^2 - \beta q^2)r^4}{10\beta\kappa^3 g^2 Q}\right) + \mathrm{O}(r^6), \quad r \ll 1, \tag{14.16.38}$$

so that the curvature singularity at $r = 0$ is considerably relegated from that of the Schwarzschild type $K = \mathrm{O}(r^{-6})$ to the type $K = \mathrm{O}(r^{-4})$.

Similarly, we have

$$\begin{aligned} A(r) &= 1 - \frac{2GM}{r} \\ &+ \frac{8\pi G}{r}\int_r^\infty \frac{(\beta q^2 + \kappa^2 g^2)g^2 + (q^2+g^2)\rho^4}{\sqrt{\rho^4+\kappa^2 g^2}(\sqrt{\rho^4+\beta g^2}\sqrt{\rho^4+\beta q^2+\kappa^2 g^2} + \rho^2\sqrt{\rho^4+\kappa^2 g^2})}\,\mathrm{d}\rho \\ &= 1 - \frac{2GM}{r} + \frac{4\pi G(q^2+g^2)}{r^2} - \frac{\pi G(\beta[q^2-g^2]^2 + 4\kappa^2 q^2 g^2)}{5r^6} + \mathrm{O}(r^{-10}), \end{aligned} \tag{14.16.39}$$

for $r \gg 1$, in view of (14.15.36). Note that there is no need now to use the joint charge Q to arrive at a relatively simple asymptotic expression. This result refines the general formula (14.16.34) in the context of the Born–Infeld theory.

Charged black holes in exponential model

We next consider the exponential model (14.15.55).

Since finite energy requires $\kappa > 0$, we insert (14.15.75) into (14.16.26) to obtain

$$\begin{aligned} A(r) &= 1 - \frac{2G(M-E)}{r} \\ &- 8\pi G\left(\frac{q}{\kappa} + \frac{r^2}{3\beta} - \frac{qr^4}{150\beta\kappa^3 g^2}\left[15\beta + 48\kappa^2 + 15\kappa^2\ln\left(\frac{\beta\kappa^2 g^6}{q^2 r^6}\right)\right]\right) + \mathrm{O}(r^8), \end{aligned} \tag{14.16.40}$$

for $r \ll 1$, in leading order approximation. Thus, we draw similar conclusions regarding the curvature singularity of the metric factor at $r = 0$ as those for the Born–Infeld model (14.15.2).

It is interesting to note that the joint charge Q plays the same role as the electric charge q, in leading orders, in (14.16.38) and (14.16.40), around the center of mass of the black hole.

Furthermore, we insert (14.15.74) into (14.16.26) to obtain the metric factor $A(r)$ in the region $r \gg 1$, which happens to coincide with the asymptotic formula expressed in (14.16.39), and is a pleasant surprise, indicating that, within the sixth-order approximation, the black hole metric of the Born–Infeld theory (14.15.2) and that of the exponential model (14.15.55) are asymptotically nondistinguishable.

Under the critical mass-energy condition $M = E$, we see from (14.16.40) that curvature singularity at $r = 0$ disappears when electricity is switched off, $q = 0$. In this situation, the black hole is purely magnetically charged (Section 14.13).

On the other hand, in view of (14.16.38), we see that the curvature singularity at $r = 0$ can never be completely removed in all situations, unlike what happens in the exponential model.

Energy conditions

From (14.16.11)–(14.16.13) with (14.12.45), we have

$$T_0^0 = T_1^1 = f'(s)\left(1 + \frac{\kappa^2 g^2}{r^4}\right)(E^r)^2 - f(s), \tag{14.16.41}$$

$$T_2^2 = T_3^3 = -f'(s)\left(1 - \kappa^2 (E^r)^2\right)\frac{g^2}{r^4} - f(s), \tag{14.16.42}$$

which provide the energy density $\rho = \mathcal{H} = T_0^0$, radial pressure $p_1 = p_r = -T_1^1$, and tangential pressure $p_2 = p_3 = p_\perp = -T_2^2 = -T_3^3$, so that in view of (14.16.41) and (14.16.42) we may express various energy conditions as follows:

(i) The weak energy condition $\rho \geq 0, \rho + p_i \geq 0$ $(i = 1, 2, 3)$ now reads

$$\mathcal{H} = f'(s)\left(1 + \frac{\kappa^2 g^2}{r^4}\right)(E^r)^2 - f(s) \geq 0, \quad f'(s)\left((E^r)^2 + \frac{g^2}{r^4}\right) \geq 0. \tag{14.16.43}$$

(ii) The dominant energy condition $\rho \geq |p_i|$ $(i = 1, 2, 3)$ becomes (14.16.43) and $\rho \geq p_i$ $(i = 1, 2, 3)$ or

$$f'(s)\left((E^r)^2 - \frac{g^2}{r^4} + \frac{2\kappa^2 g^2}{r^4}(E^r)^2\right) - 2f(s) \geq 0. \tag{14.16.44}$$

(iii) The strong energy condition $\rho + \sum_{i=1}^3 p_i \geq 0$ is equivalent to the inequality

$$f'(s)\left(1 - \kappa^2 (E^r)^2\right)\frac{g^2}{r^4} + f(s) \geq 0. \tag{14.16.45}$$

Born–Infeld model case

It is clear that the classical Born–Infeld model satisfies (14.16.43).

To check (14.16.44) for the Born–Infeld model, we insert (14.15.2), (14.15.9), (14.15.14), and (14.15.15) into the left-hand side of (14.16.44), to obtain after a lengthy calculation, the result

$$\frac{1}{\sqrt{1-2\beta s}}\left((E^r)^2-\frac{g^2}{r^4}+\frac{2\kappa^2 g^2}{r^4}(E^r)^2\right)-\frac{2}{\beta}\left(1-\sqrt{1-2\beta s}\right)\equiv\frac{P(r)}{Q(r)}, \tag{14.16.46}$$

$$\begin{aligned}P(r)=&\frac{\beta}{2}(\kappa^2 g^4[\beta q^2+\kappa^2 g^2])^2+2\beta(\kappa^2 g^3)^2(q^2+g^2)(\beta q^2+\kappa^2 g^2)r^4\\&+(\kappa g^2)^2(2[\kappa^2 qg]^2+\beta\kappa^2[2q^4+3q^2g^2+3g^4]+\beta^2 q^2[q^2+g^2])r^8\\&+2(\kappa g)^2(2\kappa^2 q^2 g^2+\beta[q^4+g^4])r^{12}+\frac{1}{2}([2\kappa qg]^2+\beta[q^2-g^2]^2)r^{16},\end{aligned} \tag{14.16.47}$$

$$\begin{aligned}Q(r)=&R(r)\left(R(r)+\frac{1}{2}(\beta\kappa q g^2)^2+\frac{\beta}{2}([\kappa^2 g^3]^2+2[\kappa g]^2[q^2+g^2]r^4\right.\\&\left.+[q^2+g^2]r^8)+(\kappa^2 g^2+r^4)^2 r^4\right),\end{aligned} \tag{14.16.48}$$

$$R(r)=r^2\sqrt{r^4+\beta g^2}\,(r^4+\kappa^2 g^2)^{\frac{3}{2}}\,\sqrt{r^4+\beta q^2+\kappa^2 g^2}. \tag{14.16.49}$$

Hence, (14.16.46) is positive for $r>0$ and the dominant energy condition holds.

To examine (14.16.45), we insert (14.15.2), (14.15.9), (14.15.14), and (14.15.15) into the left-hand side of (14.16.45) to get

$$\begin{aligned}&\frac{1}{\sqrt{1-2\beta s}}\left(1-\kappa^2[E^r]^2\right)\frac{g^2}{r^4}+\frac{1}{\beta}\left(1-\sqrt{1-2\beta s}\right)\\&=\frac{1}{\beta}\left(1+\frac{\beta[E^r]^2-1}{\sqrt{1-2\beta s}}\right)\equiv\frac{P_1(r)}{Q_1(r)},\end{aligned} \tag{14.16.50}$$

$$\begin{aligned}P_1(r)=&\kappa^6 g^8(\beta q^2+\kappa^2 g^2)+g^4(\beta^2 q^4[2\kappa^2-\beta]+\kappa^4 q^2 g^2[5\beta-\kappa^2]\\&+\kappa^4[4\kappa^2 g^4-\beta q^4])r^4+\kappa^2 g^4(6\kappa^2 g^2+q^2[7\beta-\kappa^2])r^8+g^2([3\beta+\kappa^2]q^2\\&+4\kappa^2 g^2)r^{12}+(q^2+g^2)r^{16},\end{aligned} \tag{14.16.51}$$

$$Q_1(r)=R_1(r)\left(R_1(r)+r^2[r^8+g^2(\kappa^2[2r^4+\kappa^2 g^2]+\beta q^2[\kappa^2-\beta])]\right), \tag{14.16.52}$$

$$R_1(r)=\sqrt{r^4+\beta g^2}\,(r^4+\kappa^2 g^2)^{\frac{3}{2}}\,\sqrt{r^4+\beta q^2+\kappa^2 g^2}. \tag{14.16.53}$$

These expressions appear complicated. Fortunately, if we content ourselves with the classical Born–Infeld model where $\beta=\kappa^2$, we see clearly that

$$P_1(r)>0,\quad Q_1(r)>0,\quad r>0. \tag{14.16.54}$$

In other words, the dyonic black hole solution of the classical Born–Infeld theory also satisfies the strong energy condition.

Exponential model case

For the exponential model (14.15.55), the second inequality in (14.16.43) is self-evidently true and the left-hand side of the first inequality satisfies

$$\mathcal{H} \geq \frac{\mathrm{e}^{-\frac{\beta}{2}\mathbf{B}^2}}{\beta}\left(W\mathrm{e}^{\frac{W}{2}} - \mathrm{e}^{\frac{W}{2}} + 1\right), \quad W = W(\delta) \geq 0. \tag{14.16.55}$$

It is clear that $W\mathrm{e}^{\frac{W}{2}} - \mathrm{e}^{\frac{W}{2}} + 1 \geq 0$ for $W \geq 0$. As a consequence, the weak energy condition is valid for the exponential model.

To examine (14.16.44), we use (14.15.55) to rewrite its left-hand side as

$$\begin{aligned}
&\mathrm{e}^{\beta s}\left(2s + \frac{\kappa^2 g^2}{r^4}(E^r)^2\right) - \frac{2}{\beta}\left(\mathrm{e}^{\beta s} - 1\right) \\
&\geq 2\left(s\,\mathrm{e}^{\beta s} - \frac{\mathrm{e}^{\beta s}}{\beta} + \frac{1}{\beta}\right) \equiv 2h(s).
\end{aligned} \tag{14.16.56}$$

It is clear that $h(s) \to \frac{1}{\beta}$ as $s \to -\infty$, $h(s) \to \infty$ as $s \to \infty$, and the only root of $h'(s) = 0$ is at $s = 0$ where $h(s)$ attains its global minimum, $h(0) = 0$. Hence, $h(s) \geq 0$ for all s. Therefore, the dominant energy condition holds for the exponential model.

We then examine the strong energy condition for the exponential model when $\kappa > 0$. Inserting (14.15.55) into the left-hand side of (14.16.45) and applying (14.15.68), we see that it renders

$$\begin{aligned}
&\mathrm{e}^{\beta s}\left(1 - \kappa^2[E^r]^2\right)\frac{g^2}{r^4} + \frac{1}{\beta}\left(\mathrm{e}^{\beta s} - 1\right) \\
&= \mathrm{e}^{\beta s}\left(\frac{1}{\beta} + \frac{g^2}{r^4}\left(1 - \kappa^2[E^r]^2\right)\right) - \frac{1}{\beta} \\
&= \mathrm{e}^{\beta s}\left(\frac{1}{\beta} - \frac{g^2}{r^4} + \frac{3}{\kappa^2} + \mathrm{O}(r^4)\right) - \frac{1}{\beta} < 0, \quad r \ll 1.
\end{aligned} \tag{14.16.57}$$

Hence, (14.16.45) fails near $r = 0$. On the other hand, for $r \gg 1$, we may use (14.16.28) on the left-hand side of (14.16.45) to obtain

$$\begin{aligned}
&\mathrm{e}^{\beta s}\left(1 - \kappa^2[E^r]^2\right)\frac{g^2}{r^4} + \frac{1}{\beta}\left(\mathrm{e}^{\beta s} - 1\right) \\
&= \frac{(E^r)^2}{2} + \left(\mathrm{e}^{\beta s} - \frac{1}{2}\right)\frac{g^2}{r^4} + \left(\frac{1}{2} - \mathrm{e}^{\beta s}\right)\frac{\kappa^2 g^2}{r^4}(E^r)^2,
\end{aligned} \tag{14.16.58}$$

in leading orders, which is obviously positive. In other words, the strong energy condition does not hold globally for the exponential model, in the current situation.

When $\kappa = 0$, there is no finite-energy dyonic point charge. Thus, to observe finite energy, we need to consider the two singly charged situations, where a charged black hole carries either an electric charge or a magnetic charge, but not both. Section 14.13 showed that an electrically charged black hole solution fulfills the strong energy condition but a magnetically charged one violates such a condition. This explains why the curvature singularity at $r = 0$ is eliminated for a magnetically charged black hole solution in the exponential model, under the critical mass-energy condition.

In conclusion, the finite-energy dyonically charged black hole solutions to the Einstein equations, coupled with the generalized Born–Infeld type nonlinear electrodynamics constructed in this section are of the following specific features and characteristics:

(i) Being of finite electromagnetic energies, the curvature singularities of these solutions at the center of the matter sources are of the same type as that of the Schwarzschild black hole.

(ii) Asymptotically, all the solutions approach the Reissner–Nordström black hole. As a consequence, the Brown–York quasilocal energies of the solutions are asymptotically the same as that of the Reissner–Nordström solution. In particular, the ADM mass of such a charged black hole solution is of the Schwarzschild black hole mass, M, such that its electromagnetic energy, E, is seen to make no contribution to the gravitational energy or mass of the system.

(iii) The general formulation here provides a convenient framework for constructing dyonically charged black holes in concrete situations, as illustrated with two examples — one based on the classical Born–Infeld theory and another on an exponential model.

(iv) For the dyonically charged black hole obtained in the classical Born–Infeld model, the curvature singularity at the center of matter sources, $r = 0$, measured by the Kretschmann invariant, K, is of the same order as that of the Schwarzschild black hole, $K \sim r^{-6}$, when $M \neq E$, and is relegated to $K \sim r^{-4}$, when $M = E$, which stands out as a critical mass-energy condition. Such a black hole solution satisfies the weak, dominant, and strong energy conditions.

(v) The dyonically charged black hole obtained in the exponential model enjoys the same local and asymptotic properties as that in the classical Born–Infeld model. However, when its electric charge is set to zero and the black hole carries a magnetic charge only, its curvature singularity disappears and the black hole becomes regular under the critical mass-energy condition. Such a black hole satisfies the weak and dominant energy conditions but does not satisfy the strong energy condition.

(vi) The gravitational metric factors of the dyonically charged black holes in the classical Born–Infeld model and the exponential model are indistinguishable up to the order of r^{-6} for $r \gg 1$. However, they behave rather differently locally with respect to how they depend on electric and magnetic charges, in addition to the associated total electromagnetic energies. More precisely, the leading-order charge contribution in the former situation is truly dyonic, of the form $8\pi G\sqrt{\frac{q^2}{\kappa^2}+\frac{g^2}{\beta}}$, and that in the latter situation is purely electric, of the form $8\pi G\frac{q}{\kappa}$, which appears as if the magnetic charge were turned off by setting $g = 0$ in the former, where q, g are the electric and magnetic charges of the dyonic black hole. Such a picture explains why curvature singularity of a dyonically charged black hole in the former situation cannot be removed but that can be in the latter situation, by imposing the critical mass-energy condition, $M = E$, and switching off electricity with setting $q = 0$, while switching on magnetism with maintaining $g > 0$.

Exercises

1. Use the identity $\nabla\times(f\mathbf{F}) = \nabla f\times\mathbf{F}+f\nabla\times\mathbf{F}$ for the scalar and vector fields f and $\mathbf{F}$ to show that $\mathbf{E}$ defined in (14.1.24) is irrotational, $\nabla\times\mathbf{E} = \mathbf{0}$, in $\mathbb{R}^3\setminus\{\mathbf{0}\}$.
2. Let $\mathcal{H}$ be given by (14.1.26). Establish the limit (14.1.28).
3. Establish the gravitational analogue of (14.2.5) by showing that the work needed to "destroy" a ball of mass m and radius R of constant mass density, by gradually placing the mass its carries to infinity, is

$$W = \frac{3}{5}\frac{Gm^2}{R}, \tag{14.E.1}$$

where G is the gravitational constant.
4. Use (14.E.1) and the ideas in Section 14.2 to estimate the radius of a massive ball of total mass m with a constant mass density.
5. Derive the free charge formula (14.2.18).
6. Use (14.2.18) to obtain a radius with which the ball centered at the origin carries at least 90% of the total free electric charge.
7. Use (14.2.20) and (14.2.23) to obtain an estimate for a radius with which the ball centered at the origin carries at least 90% of the total energy.
8. Show that for the energy density given by (14.3.12) the finite-energy condition $\int_{\mathbb{R}^3}\mathcal{H}\,\mathrm{d}\mathbf{x} < \infty$ is valid only when $g = 0$ or $q = 0$.
9. Establish the expansion formula (14.4.6).
10. Verify the identity (14.4.10).
11. Derive the equations (14.4.12)–(14.4.13).
12. Show that the transformation

$$\tilde{\mathbf{D}} = \mathbf{D} - \frac{1}{b^2}(\mathbf{E}\cdot\mathbf{B})\mathbf{H}, \quad \tilde{\mathbf{H}} = \frac{1}{b^2}(\mathbf{E}\cdot\mathbf{B})\mathbf{D} + \mathbf{H}, \tag{14.E.2}$$

for $\mathbf{D}, \mathbf{H}$ given in (14.4.14)–(14.4.15), serves the purpose that $\tilde{\mathbf{D}}$ and $\tilde{\mathbf{H}}$ are now recast into the same directions of $\mathbf{E}$ and $\mathbf{B}$, respectively. Obtain also the corresponding dielectrics and permeability coefficients as defined in (14.1.18).

13. Derive the solution given in (14.4.20)–(14.4.21).
14. Obtain the expression for the Hamiltonian density $\mathcal{H}$ as stated in (14.4.29)–(14.4.30).
15. Show that in terms of the entire solutions satisfying $|\nabla u|^2 \neq 1$ and $|\nabla v|^2 \neq 1$ of the coupled equations

$$\nabla \cdot \left(\nabla u \sqrt{\frac{1 - |\nabla v|^2}{1 - |\nabla u|^2}} \right) = 0, \tag{14.E.3}$$

$$\nabla \cdot \left(\nabla v \sqrt{\frac{1 - |\nabla u|^2}{1 - |\nabla v|^2}} \right) = 0, \tag{14.E.4}$$

the systems of the equations (14.5.13)–(14.5.14) and (14.E.3)–(14.E.4) are equivalent over $\mathbb{R}^2$.

16. Derive the expression (14.5.16).
17. Derive the expression (14.5.17).
18. Show that, as an energy density, the right-hand side of (14.5.17) stays nonnegative by establishing the lower bound

$$\frac{1 - st}{\sqrt{1 - s}\sqrt{1 - t}} \geq 1, \quad s, t \in [0, 1). \tag{14.E.5}$$

19. Derive the Euler–Lagrange equations of (14.6.1).
20. Derive the virial identity (14.6.3) for a finite-energy critical point of the energy functional $\int_{\mathbb{R}^n} \mathcal{H}\, dx$ where $\mathcal{H}$ is the Born–Infeld type Abelian Higgs energy density given in (14.6.2).
21. Show that (14.6.3) becomes (12.1.26) when $b \to \infty$.
22. Establish (14.6.6) and show that, when $b \to \infty$ in (14.6.6), we recover (12.1.26).
23. Let ϕ be a solution to (14.7.11) over $\mathbb{R}^2$ satisfying the global vortex ansatz (14.7.14) with the boundary condition (14.7.15) fulfilled. Establish the quantization property (14.7.12).
24. With (14.8.35), establish the limit

$$\lim_{a \to 0} \frac{1}{a} h_0 \left(\frac{r_0(a)}{a} \right) = \frac{1}{2 r_0(0)} = \frac{1}{4\sqrt{\pi G}\, q}. \tag{14.E.6}$$

Then take the limit $a \to 0$ in the critical condition (14.8.34) and obtain (14.8.36).

25. Use (14.8.22) in (14.8.37) to show that $K \sim \frac{1}{r^6}$ for r small.
26. Establish the limit (14.9.32). What happens to this limit when setting $b = \infty$ in (14.9.27)?
27. For the metric factor $A(r)$ given in (14.9.52), identify the conditions for the occurrence of inner, outer, and extremal black holes, and naked singularity at $r = 0$, respectively.
28. Rewrite (14.9.51) in the form (14.8.39) with $\mathcal{H}$ given by (14.9.54) and represent $A(r)$ in the form (14.8.42). Derive the apparent energy-mass condition (14.8.43) for the occurrence of an event horizon in such a context.
29. For the Bardeen black hole defined by the metric factor (14.9.58), compute the Kretschmann invariant and show that it is free of singularity.
30. For the Hayward black hole given by the metric factor (14.9.63), find the conditions under which there are the situations of occurrence of inner and outer event horizons, an extremal horizon, and no horizon, respectively.
31. In the scalar-wave matter Lagrangian action density (14.14.1), consider a specific exponential model [342] given by the function

$$f(X) = X\mathrm{e}^{\beta X}, \quad \beta > 0. \tag{14.E.7}$$

Assume that the potential density function V in (14.14.1) is a nonnegative constant. In the isotropic and homogeneous universe situation in which the k-essence is realized by a spatially independent real-valued scalar-wave matter, compute the associated adiabatic squared speed of sound, $c_{\mathrm{s}}^2 = c_{\mathrm{s}}^2(\beta)$, as defined in (14.14.53), and show that it lies in the correct range, $0 \leq c_{\mathrm{s}}^2(\beta) < 1$. More precisely, show also that $c_{\mathrm{s}}^2(\beta)$ is a decreasing function of β such that $c_{\mathrm{s}}^2(\beta) \to 1$ as $\beta \to 0$ and $c_{\mathrm{s}}^2(\beta) \to 0$ as $\beta \to \infty$.
32. Consider an isotropic and homogeneous universe driven by a real scalar-wave matter as in Exercise 31 for which the k-essence nonlinearity function $f(X)$ in (14.14.1) is quadratic,

$$f(X) = X + \alpha X^2, \quad \alpha > 0, \tag{14.E.8}$$

and $V = V_0 \geq 0$ a constant.

(a) Derive the equations that govern the evolution of the scale factor $a(t)$ given in the Robertson–Walker line element (14.14.5) and the k-essence scalar field $\varphi(t)$ given in (14.14.1)–(14.14.2), in terms of time t, and show that the equations always permit a solution realizing the Big Bang scenario characterized by the behavior

$$\lim_{t\to 0} a(t) = 0, \quad \lim_{t\to\infty} a(t) = \infty. \tag{14.E.9}$$

(b) For the cosmic matter density ρ_m and fluid pressure P_m defined in (14.14.25), obtain them as functions of the scale factor a and use (14.E.9) to establish the asymptotic limits

$$\lim_{t\to 0}\rho_m(t)=\infty,\quad \lim_{t\to 0}P_m(t)=\infty;\quad \lim_{t\to\infty}\rho_m(t)=0,\quad \lim_{t\to\infty}P_m(t)=0. \tag{14.E.10}$$

(c) Derive the equation of state for the k-essence fluid that relates P_m to ρ_m.

(d) Establish the asymptotic limits

$$\lim_{t\to 0}w_m(t)=\frac{1}{3},\quad \lim_{t\to\infty}w_m(t)=1,\quad w_m(t)\equiv\frac{P_m(t)}{\rho_m(t)}, \tag{14.E.11}$$

and show that the pressure-density ratio $w_m(t)$ is an increasing function of t. In other words, the quadratic k-essence model monotonically interpolates two linear cosmic fluid models, namely, the radiation dominated universe with $w_m = \frac{1}{3}$ and the stiff-matter dominated universe with $w_m = 1$, respectively.

(e) For the adiabatic squared speed of sound c_s^2 defined by (14.14.53), obtain it as a function depending on ρ_m and α but independent of V_0. Use the result to establish the properties

$$\frac{1}{3}<c_s^2<1;\quad \lim_{t\to 0}c_s^2=\frac{1}{3},\quad \lim_{t\to\infty}c_s^2=1. \tag{14.E.12}$$

15

Canonical quantization of fields

This chapter presents a brief description on field quantization and the differential equations encountered and utilized, centered around harmonic oscillators, both classically and quantum mechanically. It first explores the classical and quantum aspects of harmonic oscillators and demonstrates a path to quantization through canonical formalism and particle creation and annihilation. It then uses harmonic oscillators as basic building blocks to show how to quantize the Klein–Gordon equation and Schrödinger equation. It distinguishes these two problems such that it treats the real Klein–Gordon equation as a natural extension of the many-degree-of-freedom harmonic oscillator system and explains the rise of a divergent zero-point energy and its removal by the concept of renormalization. For the Schrödinger equation, it needs to reformulate the Lagrange mechanics in terms of complex coordinates and consider its quantization in terms of complex canonical coordinates. Moreover, the chapter shows how to quantize the Maxwell equations in free space. As a consequence, it explains how the Planck–Einstein formula and Compton–Debye equation in the photoelectric effect arise in quantum field theory. The chapter concludes the discussion using the harmonic oscillator of one degree of freedom as an example to show how to derive some thermodynamic information of the system based on the acquired knowledge about the energy spectrum of the system, both classically and quantum-mechanically.

15.1 Quantum harmonic oscillator

The idea of canonical quantization may be explained clearly by a thorough discussion of one of the simplest mechanical systems: the harmonic oscillator.

Mathematical Physics with Differential Equations. Yisong Yang, Oxford University Press.
© Yisong Yang (2023). DOI: 10.1093/oso/9780192872616.003.0015

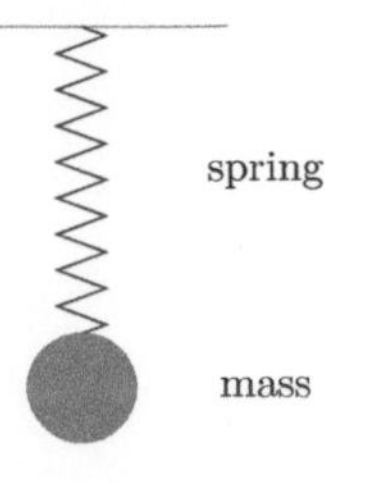

Figure 15.1 A harmonic oscillator of one degree of freedom modeled by a spring-mass system.

Harmonic oscillator

The motion of a one-dimensional spring-mass system described by the displacement coordinate, say q, from its equilibrium position $q = 0$, of the spring constant k and mass m, is governed by Hooke's law

$$m\ddot{q} = -kq \tag{15.1.1}$$

(see Figure 15.1). This simple mechanic system is also known as the *harmonic oscillator*, whose solution is periodic with the angular frequency

$$\omega = \sqrt{\frac{k}{m}}, \tag{15.1.2}$$

and easily available by solving (15.1.1) directly. Alternatively, we may complexify the equation (15.1.1) and rewrite it as

$$\left(\frac{\mathrm{d}}{\mathrm{d}t} - \mathrm{i}\omega\right)\left(\frac{\mathrm{d}}{\mathrm{d}t} + \mathrm{i}\omega\right) q = \ddot{q} + \omega^2 q = 0. \tag{15.1.3}$$

We may get the general solution of (15.1.3) by superimposing those of the equations

$$\left(\frac{\mathrm{d}}{\mathrm{d}t} - \mathrm{i}\omega\right) a_+ = 0, \quad \left(\frac{\mathrm{d}}{\mathrm{d}t} + \mathrm{i}\omega\right) a_- = 0, \tag{15.1.4}$$

or

$$a_+(t) = a_+(0)\mathrm{e}^{\mathrm{i}\omega t}, \quad a_-(t) = a_-(0)\mathrm{e}^{-\mathrm{i}\omega t}, \tag{15.1.5}$$

to obtain

$$q(t) = a_+(0)\mathrm{e}^{\mathrm{i}\omega t} + a_-(0)\mathrm{e}^{-\mathrm{i}\omega t}, \tag{15.1.6}$$

where $a_+(0)$ and $a_-(0)$ are arbitrary constants.

Quantization

Use p to denote the momentum of the system. Since the potential energy of the system is

$$V(q) = \frac{1}{2}kq^2, \tag{15.1.7}$$

the energy or Hamiltonian of it reads

$$E = H = H(q,p) = \frac{1}{2m}p^2 + \frac{1}{2}kq^2 = \frac{1}{2m}(p^2 + m^2\omega^2 q^2). \tag{15.1.8}$$

Thus, by the first quantization procedure (cf. Section 2.2),

$$E \mapsto \hat{E} = \mathrm{i}\hbar\frac{\partial}{\partial t}, \quad p \mapsto \hat{p} = -\mathrm{i}\hbar\frac{\partial}{\partial q}, \quad q \mapsto \hat{q} = q, \tag{15.1.9}$$

we arrive at the Schrödinger equation

$$\mathrm{i}\hbar\frac{\partial \phi}{\partial t} = \hat{H}\phi, \quad \phi = \phi(q,t), \tag{15.1.10}$$

describing the quantum mechanical motion of the harmonic oscillator, where

$$\hat{H} = \frac{1}{2m}(\hat{p}^2 + m^2\omega^2\hat{q}^2) = \frac{1}{2m}\left(\left[\frac{\hbar}{\mathrm{i}}\frac{\mathrm{d}}{\mathrm{d}q}\right]^2 + [m\omega q]^2\right). \tag{15.1.11}$$

The steady-state version of (15.1.10)–(15.1.11) is the eigenvalue problem

$$\hat{H}\psi = E\psi, \quad \psi = \psi(q), \tag{15.1.12}$$

which is a second-order and variable-coefficient ordinary differential equation and thus not easily solved. Nevertheless, its solution is known to be given by the following normalized sequence of eigenpairs [154]

$$\psi(q) = \psi_n(q) = \left(\frac{m\omega}{\pi\hbar}\right)^{\frac{1}{4}} \frac{1}{\sqrt{2^n n!}} H_n(\xi)\mathrm{e}^{-\frac{\xi^2}{2}}, \quad \xi = \sqrt{\frac{m\omega}{\hbar}}q, \tag{15.1.13}$$

$$E = E_n = \left(n + \frac{1}{2}\right)\hbar\omega, \tag{15.1.14}$$

for $n = 0, 1, 2, \dots$, where H_n are the *Hermite polynomials* satisfying

$$H_n(\xi) = (-1)^n \mathrm{e}^{\xi^2}\frac{\mathrm{d}^n}{\mathrm{d}\xi^n}\mathrm{e}^{-\xi^2}, \quad \int_{\mathbb{R}} H_n(\xi)H_{n'}(\xi)e^{-\xi^2}\mathrm{d}\xi = \sqrt{\pi}2^{\frac{n+n'}{2}}(n!n'!)^{\frac{1}{2}}\delta_{nn'}, \tag{15.1.15}$$

such that the set $\{\psi_n\}$ is an orthonormal basis of $L^2(\mathbb{R})$.

In order to lay a foundation for the formalism of canonical quantization, we now use the operator method of Dirac [169] to solve this same problem.

First, in terms of the Poisson bracket $\{,\}$ and (15.1.8), the equations of motion of the harmonic oscillator, classically, are

$$\dot{q} = \{q, H\} = \frac{p}{m}, \quad \dot{p} = \{p, H\} = -m\omega^2 q. \tag{15.1.16}$$

However, in general, if f, g are a pair of classical quantities and $\hat{f}, \hat{g}$ their corresponding quantum-mechanical operator representations in the Heisenberg picture, then

$$\dot{f} = \{f, H\} = g \tag{15.1.17}$$

is quantized into

$$\dot{\hat{f}} = \frac{1}{\mathrm{i}\hbar}[\hat{f}, \hat{H}] = \hat{g}. \tag{15.1.18}$$

This rule of correspondence is handy for computation. For example, since $\{q, p\} = 1$, we have immediately $[\hat{q}, \hat{p}] = \mathrm{i}\hbar$. Likewise, since $\{q^2, p\} = 2q$ and $\{p^2, q\} = -2p$, we have $[\hat{q}^2, \hat{p}] = 2\mathrm{i}\hbar\hat{q}$ and $[\hat{p}^2, \hat{q}] = -2\mathrm{i}\hbar\hat{p}$, as may also be established directly by definition.

Hence, we are led from (15.1.16) to the Heisenberg equations for the evolution of the quantum-mechanical observables or operators which read

$$\dot{\hat{q}} = \frac{1}{m}\hat{p}, \quad \dot{\hat{p}} = -m\omega^2\hat{q}. \tag{15.1.19}$$

In order to solve these equations, we naturally set

$$\hat{a}_+ = \frac{1}{\sqrt{2\hbar m\omega}}\left(\hat{p} + \mathrm{i}m\omega\hat{q}\right), \quad \hat{a}_- = \frac{1}{\sqrt{2\hbar m\omega}}\left(\hat{p} - \mathrm{i}m\omega\hat{q}\right), \tag{15.1.20}$$

to arrive at the *decoupled* system of equations

$$\dot{\hat{a}}_+ = \mathrm{i}\omega\hat{a}_+, \quad \dot{\hat{a}}_- = -\mathrm{i}\omega\hat{a}_-, \tag{15.1.21}$$

whose solution is

$$\hat{a}_+(t) = \mathrm{e}^{\mathrm{i}\omega t}\hat{a}_+(0), \quad \hat{a}_-(t) = \mathrm{e}^{-\mathrm{i}\omega t}\hat{a}_-(0), \tag{15.1.22}$$

generalizing (15.1.5), where $\hat{a}_+(0)$ and $\hat{a}_-(0)$ are initial operators.

From (15.1.20), we have

$$\hat{q} = \mathrm{i}\sqrt{\frac{\hbar}{2m\omega}}(\hat{a}_- - \hat{a}_+), \quad \hat{p} = \sqrt{\frac{\hbar m\omega}{2}}(\hat{a}_- + \hat{a}_+). \tag{15.1.23}$$

Inserting (15.1.23) into (15.1.11), we find

$$\hat{H} = \frac{\hbar\omega}{2}(\hat{a}_-\hat{a}_+ + \hat{a}_+\hat{a}_-). \tag{15.1.24}$$

Using the commutator relations stated for $\hat{q}$ and $\hat{p}$, we have

$$[\hat{a}_-, \hat{a}_+] = 1, \quad [a_-, \hat{H}] = \hbar\omega\hat{a}_-, \quad [\hat{a}_+, \hat{H}] = -\hbar\omega\hat{a}_+. \tag{15.1.25}$$

With (15.1.25), we get from (15.1.24) the useful relations

$$\hbar\omega\hat{a}_-\hat{a}_+ = \hat{H} + \frac{1}{2}\hbar\omega, \tag{15.1.26}$$

$$\hbar\omega\hat{a}_+\hat{a}_- = \hat{H} - \frac{1}{2}\hbar\omega. \tag{15.1.27}$$

We note that the operators $\hat{a}_+$ and $\hat{a}_-$ can be used to raise and lower energies in the amount of $\hbar\omega$. More precisely, if ψ is a solution of the steady-state Schrödinger equation (15.1.12), then $\psi_+ = \hat{a}_+\psi$, $\psi_- = \hat{a}_-\psi$ satisfy

$$\hat{H}\psi_+ = (E + \hbar\omega)\psi_+, \tag{15.1.28}$$

$$\hat{H}\psi_- = (E - \hbar\omega)\psi_-, \tag{15.1.29}$$

respectively, in view of (15.1.24)–(15.1.27). For this reason, the operators $\hat{a}_\pm$ are often called the *ladder operators* or the *raising* and *lowering operators*. They are also called the *creation and annihilation operators*.

On the other hand, since $\hat{H}$ is positive definite, all eigenvalues of it must be positive. Therefore, if we start from an eigenpair (ψ, E) of $\hat{H}$, then $E > 0$, and there must be an integer $n \geq 0$ such that $\hat{a}_-^n \psi \neq 0$, but $\hat{a}_-^{n+1}\psi = 0$. Now, set $\psi_0 = \hat{a}_-^n \psi$. We have $\psi_0 \neq 0$ and

$$\hat{a}_- \psi_0 = 0. \tag{15.1.30}$$

So, inserting (15.1.20), we see that ψ_0 satisfies the equation

$$\frac{\mathrm{d}\psi_0}{\mathrm{d}q} + \left(\frac{m\omega}{\hbar}\right) q\psi_0 = 0, \tag{15.1.31}$$

whose solution is of the Gauss type

$$\psi_0(q) = C_0 \mathrm{e}^{-\frac{m\omega}{2\hbar}q^2}, \quad C_0 \neq 0. \tag{15.1.32}$$

Substituting (15.1.30) into (15.1.12), we find

$$E = E_0 = \frac{1}{2}\hbar\omega. \tag{15.1.33}$$

However, this result may also be obtained directly without any computation if we apply (15.1.27) to ψ_0 and use (15.1.30). Thereafter, we use $|0\rangle$ to denote a normalized eigenstate, $\langle 0|0\rangle = 1$, associated with the lowest positive eigenvalue (15.1.33) of the system.

Define $|\psi_n\rangle = \hat{a}_+^n|0\rangle$ for $n = 1, 2, \ldots$. Then we can iterate (15.1.28) to get

$$\hat{H}|\psi_n\rangle = \left(\frac{1}{2}\hbar\omega + n\hbar\omega\right)|\psi_n\rangle. \tag{15.1.34}$$

Thus, we may conclude that the full eigenspectrum of $\hat{H}$ is given by

$$E_n = \left(n + \frac{1}{2}\right)\hbar\omega, \quad n = 0, 1, 2, \ldots, \tag{15.1.35}$$

as stated in (15.1.14).

Combining (15.1.27) and (15.1.34), we have

$$\hat{a}_+\hat{a}_-|\psi_n\rangle = n|\psi_n\rangle. \tag{15.1.36}$$

Likewise, we also have

$$\hat{a}_-\hat{a}_+|\psi_n\rangle = (n+1)|\psi_n\rangle. \tag{15.1.37}$$

Consequently, since it is clear that $\hat{a}_+^\dagger = \hat{a}_-$, we may use $|\psi_n\rangle = \hat{a}_+|\psi_{n-1}\rangle$ and (15.1.37) to get

$$\langle\psi_n|\psi_n\rangle = \langle\psi_{n-1}|\hat{a}_-\hat{a}_+|\psi_{n-1}\rangle = n\langle\psi_{n-1}|\psi_{n-1}\rangle. \tag{15.1.38}$$

From (15.1.38), we find

$$\langle\psi_n|\psi_n\rangle = n!|0\rangle, \tag{15.1.39}$$

which enables us to obtain the normalized nth eigenstate, associated with the eigenvalue E_n, to be, say,

$$|n\rangle = \frac{1}{\sqrt{n!}}\hat{a}_+^n|0\rangle, \quad n = 1, 2, \ldots. \tag{15.1.40}$$

Thus, for the quantized harmonic oscillator, the state $|n\rangle$ may be thought of an n-particle state representing n identical (bosonic) particles, each of the unit energy $\hbar\omega$, and the state $|0\rangle$ is the vacuum state in absence of a particle. Thus, the application of the creation operator $\hat{a}_+$ to the state $|n\rangle$ creates a particle, and that of the annihilation operator $\hat{a}_-$ annihilates a particle. Such a phenomenon justifies their names. Thus, the vacuum energy (15.1.33) is called the *zero-point energy* or *zero-particle energy.* Furthermore, in view of (15.1.36), we can define $\hat{N} = \hat{a}_+\hat{a}_-$ to be the *number operator* that satisfies

$$\hat{N}|n\rangle = n|n\rangle. \tag{15.1.41}$$

Namely, as an observable, its value describes the number of particles in the system. Besides, (15.1.27) gives us

$$\hat{N} = \frac{1}{\hbar\omega}\hat{H} - \frac{1}{2}. \tag{15.1.42}$$

In particular, $[\hat{N}, \hat{H}] = 0$. Hence, $\hat{N}$ is conserved or time-independent.

Therefore, we saw that the Heisenberg picture enables a direct quantization of a classical mechanical system from its Hamiltonian formalism, based on the Poisson brackets, and leads to a description of the quantized system in terms of creation and annihilation of particles, which is not possible in the Schrödinger picture in terms of a wave function. This quantization formulation is called canonical quantization.

15.2 Canonical quantization

The term *canonical quantization* comes from quantizing a classical mechanical system, within the Heisenberg picture, based on the canonical formulation of the system in terms of the Lagrangian action, Hamiltonian, and Poisson brackets.

Canonical formulation of mechanical system

Consider a mechanical system governed by the Lagrangian action function

$$L = L(q, \dot{q}, t), \tag{15.2.1}$$

where $q = (q^i)$ $(i = 1, \ldots, n)$ is the coordinate vector. The associated momentum vector $p = (p_i)$ then reads

$$p_i = \frac{\partial L}{\partial \dot{q}^i}, \quad i = 1, \ldots, n. \tag{15.2.2}$$

Thus, the Euler–Lagrange equations (or equations of motion of the system) are

$$\dot{p}_i = \frac{\partial L}{\partial q^i}, \quad i = 1, \ldots, n. \tag{15.2.3}$$

On the other hand, the Hamiltonian function of the system is

$$H(q, p, t) = \sum_{i=1}^{n} p_i \dot{q}^i - L(q, \dot{q}, t), \tag{15.2.4}$$

such that the equations of motion, (15.2.3), become a Hamiltonian system of equations

$$\dot{q}^i = \frac{\partial H}{\partial p_i}, \quad \dot{p}_i = -\frac{\partial H}{\partial q^i}, \quad i = 1, \ldots, n, \tag{15.2.5}$$

and that any mechanical quantity $f = f(q, p, t)$ follows the evolution equation

$$\frac{\mathrm{d}f}{\mathrm{d}t} = \{f, H\} + \frac{\partial f}{\partial t}, \tag{15.2.6}$$

where

$$\{f, g\} = \frac{\partial f}{\partial q^i}\frac{\partial g}{\partial p_i} - \frac{\partial f}{\partial p_i}\frac{\partial g}{\partial q^i}, \tag{15.2.7}$$

is the usual Poisson bracket.

Chapter 1 discussed much of the details of this formulation. Here we only remind ourselves of some key ingredients for convenience of presentation of canonical quantization.

Canonical quantization

As in the procedure of quantizing the harmonic oscillator, to quantize a mechanical system of canonical coordinates, (q^i) and (p_i), we first replace (q^i) and (p_i) by their quantum operators $(\hat{q}^i)$ and $(\hat{p}_i)$, or more generally,

$$f(q, p, t) \mapsto \hat{f} = f(\hat{q}, \hat{p}, t), \tag{15.2.8}$$

and then update all the associated Poisson brackets by the corresponding commutators following the rule

$$\{f, g\} = w \mapsto \frac{1}{\mathrm{i}\hbar}[\hat{f}, \hat{g}] = \hat{w} \quad \text{or} \quad [\hat{f}, \hat{g}] = \mathrm{i}\hbar\, \hat{w}. \tag{15.2.9}$$

Furthermore, the classical evolution equation (15.2.6) becomes the Heisenberg equation

$$\frac{\mathrm{d}\hat{f}}{\mathrm{d}t} = \frac{1}{\mathrm{i}\hbar}[\hat{f}, \hat{H}] + \frac{\partial \hat{f}}{\partial t}, \tag{15.2.10}$$

which may be solved to see how the observable $\hat{f}$ evolves with respect to time.

Commutators involving coordinates and momenta

Since by (15.2.7) there hold

$$\{q^i, q^j\} = 0, \quad \{p_i, p_j\} = 0, \quad \{q^i, p_j\} = \delta^i_j, \tag{15.2.11}$$

we have

$$[\hat{q}^i, \hat{q}^j] = 0, \quad [\hat{p}_i, \hat{p}_j] = 0, \quad [\hat{q}^i, \hat{p}_j] = \mathrm{i}\hbar\, \delta^i_j. \tag{15.2.12}$$

Harmonic oscillator of high degree of freedom

Consider the Lagrangian

$$L(q, \dot{q}) = \frac{1}{2}m\dot{q}^2 - \frac{1}{2}kq^2, \quad q = (q^i) \in \mathbb{R}^N, \tag{15.2.13}$$

governing a harmonic oscillator of mass m and spring constant k in $\mathbb{R}^N$. With canonical coordinates $(q, p) = (q^i, p_i) \in \mathbb{R}^{2N}$, the associated Hamiltonian is

$$H(q, p) = \frac{1}{2m}p^2 + \frac{1}{2}kq^2. \tag{15.2.14}$$

Let $\hat{q} = (\hat{q}^i), \hat{p} = (\hat{p}_i)$ be the quantum operators corresponding to q, p. Then, (15.2.12) holds. Define $(\hat{a}^i_\pm) = \hat{a}_\pm$ by (15.1.20). Then, by (15.2.12), we have

$$[\hat{a}^i_-, \hat{a}^j_-] = 0, \quad , [\hat{a}^i_+, \hat{a}^j_+] = 0, \quad [\hat{a}^i_-, \hat{a}^j_+] = \delta^{ij}. \tag{15.2.15}$$

Besides, since (15.1.23) is valid, we have

$$\begin{aligned} \hat{H} &= \frac{1}{2m}\hat{p}^2 + \frac{1}{2}k\hat{q}^2 \\ &= \frac{\hbar\omega}{2}\sum_{i=1}^N (\hat{a}^i_-\hat{a}^i_+ + \hat{a}^i_+\hat{a}^i_-), \end{aligned} \tag{15.2.16}$$

in view of the prescription (15.2.8). Thus, applying the last relation in (15.2.15), we have

$$\hbar\omega\sum_{i=1}^N \hat{a}^i_-\hat{a}^i_+ = \hat{H} + \frac{N}{2}\hbar\omega, \tag{15.2.17}$$

$$\hbar\omega\sum_{i=1}^N \hat{a}^i_+\hat{a}^i_- = \hat{H} - \frac{N}{2}\hbar\omega, \tag{15.2.18}$$

generalizing (15.1.26)–(15.1.27). Moreover, if ψ satisfies (15.1.12), then we may use (15.2.15) to see that $\psi^i_\pm = \hat{a}^i_\pm\psi$ enjoy the property

$$\hat{H}\psi^i_+ = (E + \hbar\omega)\psi^i_+, \tag{15.2.19}$$

$$\hat{H}\psi^i_- = (E - \hbar\omega)\psi^i_-, \tag{15.2.20}$$

which generalize (15.1.28)–(15.1.29). Hence, $\hat{a}_+^i$ and $\hat{a}_-^i$ are again creation and annihilation operators, for each $i = 1, \dots, N$. In particular, if (ψ, E) is an eigenpair satisfying the equation (15.1.12), then there is a nonnegative integer n such that $(\hat{a}_-^i)^{n+1}\psi = 0$ but $\psi_0 = (\hat{a}_-^i)^n\psi \neq 0$. Using this property in (15.2.20) iteratively, we obtain

$$\hat{H}(\hat{a}_-^i)^k\psi = (E - k\hbar\omega)(\hat{a}_-^i)^k\psi, \quad k = 0, 1, \dots, n, \quad i = 1, \dots, N. \tag{15.2.21}$$

Besides, inserting

$$\hat{a}_-^i\psi_0 = 0, \quad i = 1, \dots, N, \tag{15.2.22}$$

into (15.2.18), we have

$$\hat{H}(\hat{a}_-^i)^n\psi = \hat{H}\psi_0 = \frac{N}{2}\hbar\omega\,\psi_0. \tag{15.2.23}$$

Specifically, solving the differential equations (15.2.22), we obtain

$$\psi_0(q) = C_0 \mathrm{e}^{-\frac{m\omega}{2\hbar}\sum_{i=1}^N (q^i)^2}, \quad C_0 \neq 0. \tag{15.2.24}$$

In view of (15.2.21) and (15.2.23), we arrive at the result

$$E = E_n = n\hbar\omega + \frac{N}{2}\hbar\omega, \tag{15.2.25}$$

which is an n-particle energy. In particular, the vacuum or zero-point energy is found to be

$$E_0 = \frac{N}{2}\hbar\omega. \tag{15.2.26}$$

As before, we use $|0\rangle$ to denote the associated normalized zero-particle state.

Fixing $i = 1, \dots, N$ and $m = 1, 2, \dots$, we have

$$\hat{H}(\hat{a}_+^i)^m|0\rangle = \left(\frac{N}{2}\hbar\omega + m\hbar\omega\right)(\hat{a}_+^i)^m|0\rangle, \tag{15.2.27}$$

from (15.2.19). Thus, for an n-particle state, ψ_n, we can use (15.2.27), the commutativity stated in (15.2.15), and the combination equation

$$n = \sum_{i=1}^N n_i, \quad n_i \in \mathbb{N}, \quad i = 1, \dots, N, \tag{15.2.28}$$

to get

$$\psi_n = \prod_{i=1}^N (\hat{a}_+^i)^{n_i}|0\rangle, \quad n = \sum_{i=1}^N n_i. \tag{15.2.29}$$

In particular, we see that the nth energy level for each $n \geq 1$ is degenerate for any $N \geq 2$.

In fact, we see that the set of states

$$\left\{ \prod_{i=1}^{N} (\hat{a}_{+}^{i})^{n_i}|0\rangle \,\middle|\, n_1, \dots, n_N \in \mathbb{N} \right\}, \tag{15.2.30}$$

is a linearly independent set. Moreover, by the properties stated in (15.2.15), we have

$$\hat{a}_{-}^{j}(\hat{a}_{+}^{i})^{m}|0\rangle = (\hat{a}_{+}^{i})^{m}\hat{a}_{-}^{j}|0\rangle = 0, \quad i \neq j. \tag{15.2.31}$$

Hence, by (15.2.18) and (15.2.27), we have

$$\left(\sum_{j=1}^{N} \hat{a}_{+}^{j}\hat{a}_{-}^{j} \right)(\hat{a}_{+}^{i})^{m}|0\rangle = \left(\hat{H} - \frac{N}{2}\hbar\omega \right)(\hat{a}_{+}^{i})^{m}|0\rangle = m(\hat{a}_{+}^{i})^{m}|0\rangle, \tag{15.2.32}$$

resulting in

$$\hat{a}_{+}^{i}\hat{a}_{-}^{i}(\hat{a}_{+}^{i})^{m}|0\rangle = m(\hat{a}_{+}^{i})^{m}|0\rangle, \quad i = 1, \dots, N. \tag{15.2.33}$$

However, in view of (15.2.15) again, which says $\hat{a}_{\pm}^{i}$ commute with $\hat{a}_{+}^{j}$ for $i \neq j$, we deduce from (15.2.33) the result

$$\hat{a}_{+}^{i}\hat{a}_{-}^{i}\prod_{j=1}^{N}(\hat{a}_{+}^{j})^{n_j}|0\rangle = n_i \prod_{j=1}^{N}(\hat{a}_{+}^{j})^{n_j}|0\rangle. \tag{15.2.34}$$

That is, in the state ψ_n defined in (15.2.29), we have the number operator $\hat{N}_i$, which enjoys the property

$$\hat{N}_i\psi_n \equiv \hat{a}_{+}^{i}\hat{a}_{-}^{i}\psi_n = n_i\psi_n, \quad i = 1, \dots, N. \tag{15.2.35}$$

It is clear that each $\hat{N}$ is conserved,

$$[\hat{N}_i, \hat{H}] = 0, \quad i = 1, \dots, N. \tag{15.2.36}$$

Similarly, we can use the same method to establish

$$\hat{a}_{-}^{i}\hat{a}_{+}^{i}\psi_n = (n_i + 1)\psi_n, \quad i = 1, \dots, N. \tag{15.2.37}$$

For fixed $n_i \geq 1$, we may rewrite ψ_n as $\hat{a}_{+}^{i}\psi_{n-1}$ in an obvious manner. Thus, in view of (15.2.37), we have

$$\langle \psi_n | \psi_n \rangle = \langle \psi_{n-1} | \hat{a}_{-}^{i}\hat{a}_{+}^{i} | \psi_{n-1} \rangle = n_i \langle \psi_{n-1} | \psi_{n-1} \rangle. \tag{15.2.38}$$

Iterating this relation in descending order of n_i for all $i = 1, \dots, N$ until arriving at $|0\rangle$, we get

$$\langle \psi_n | \psi_n \rangle = n_1! \cdots n_N!. \tag{15.2.39}$$

Consequently, we obtain the normalized n-particle state, which we rewrite as

$$|n_1, \dots, n_N\rangle = \frac{1}{\sqrt{n_1! \cdots n_N!}} \prod_{i=1}^{N} (\hat{a}_{+}^{i})^{n_i}|0\rangle. \tag{15.2.40}$$

Such an n-particle state represents $n_1, \ldots, n_N$ particles in the corresponding particle states labeled by $1, \ldots, N$, which are created from the zero-particle state by the creation operators $\hat{a}_1, \ldots, \hat{a}_N$, respectively. We also say that the particle state i is occupied with n_i particles and n_i is the *occupation number* of the particle state i. In this way, canonical quantization is also called *occupation number representation.*

In the context of occupation number representation, the state space describing all possible numbers of particles in a system, that is, $n = 0, 1, 2, \ldots$, may be built up with the set of the states of the form (15.2.40), called the *Fock state* or *number state* basis, and the resulting state space the *Fock space.*

15.3 Field equation formalism

We now consider canonical quantization for fields following the ideas developed for quantizing harmonic oscillators in which we need to convert the Hamiltonian formalism of a system into its quantum-mechanical setting through a Heisenberg picture description. For convenience and simplicity, this section studies a real-valued scalar field, $\phi(\mathbf{x}, t)$, depending on the spatial coordinate vector $\mathbf{x} \in \mathbb{R}^3$ and time $t \in \mathbb{R}$. Then, $\phi(\mathbf{x}, t)$ may be viewed as $q^i(t)$ in which the coordinate index i is replaced by $\mathbf{x}$ such that the discrete system becomes a continuous system and summation over i is updated into an integration in the spatial domain of $\mathbf{x}$. This correspondence or analogy is followed in our subsequent formulation, as in [253].

Action principle and field equation

Let ϕ be governed by a Lagrangian action density

$$\mathcal{L} = \mathcal{L}(\phi(\mathbf{x}, t), \nabla\phi(\mathbf{x}, t), \dot{\phi}(\mathbf{x}, t), t), \quad \nabla\phi = \nabla_{\mathbf{x}}\phi, \quad \dot{\phi} = \frac{\partial \phi}{\partial t}. \tag{15.3.1}$$

Then, the associated action functional is

$$L(\phi, \dot{\phi}, t) = \int \mathcal{L}(\phi(\mathbf{x}, t), \nabla\phi(\mathbf{x}, t), \dot{\phi}(\mathbf{x}, t), t)\,\mathrm{d}\mathbf{x}, \tag{15.3.2}$$

where the integration is evaluated over the full spatial domain of interest, and $\dot{\phi}(\mathbf{x}, t)$ replaces the velocity coordinate $\dot{q}^i$ before. Thus, we are led to define the conjugate momentum to be

$$\pi(\mathbf{x}, t) = \frac{\partial \mathcal{L}}{\partial \dot{\phi}}(\mathbf{x}, t), \tag{15.3.3}$$

which takes the role of p_i. In other words, now $\phi(\mathbf{x}, t)$ and $\pi(\mathbf{x}, t)$ are canonical coordinates labeled by $\mathbf{x}$.

If we use δ to denote the Fréchet derivative [60, 164], also referred to as the functional derivative or first variation in calculus of variations [149, 212, 574], and A the action of the problem,

$$A = \int L(\phi, \dot{\phi}, t)\,\mathrm{d}t = \int\int (\phi(\mathbf{x}, t), \nabla\phi(\mathbf{x}, t), \dot{\phi}(\mathbf{x}, t), t)\,\mathrm{d}\mathbf{x}\mathrm{d}t, \tag{15.3.4}$$

then the Euler–Lagrange equation of the problem is simply $\delta A = 0$ or

$$\frac{\partial}{\partial t}\left(\frac{\partial \mathcal{L}}{\partial \dot{\phi}}\right) = \frac{\partial \mathcal{L}}{\partial \phi} - \nabla \cdot \left(\frac{\partial \mathcal{L}}{\partial(\nabla\phi)}\right), \quad \frac{\partial \mathcal{L}}{\partial(\nabla\phi)} = \frac{\partial \mathcal{L}}{\partial(\partial_i \phi)}. \tag{15.3.5}$$

More elegantly, we may also recast (15.3.5) following the same manner of (15.2.3) as

$$\dot{\pi}(\mathbf{x},t) = \frac{\delta L}{\delta \phi(\mathbf{x},t)}. \tag{15.3.6}$$

Hamiltonian system of equations

Using the action functional (15.3.2) and the conjugate momentum (15.3.3), we can express the associated Hamiltonian functional as

$$\begin{aligned} H(\phi,\pi,t) &= \int \pi(\mathbf{x},t)\dot{\phi}(\mathbf{x},t)\,\mathrm{d}\mathbf{x} - L(\phi,\dot{\phi},t) \\ &\equiv \int \mathcal{H}(\phi(\mathbf{x},t),\pi(\mathbf{x},t),t)\,\mathrm{d}\mathbf{x}, \end{aligned} \tag{15.3.7}$$

where

$$\mathcal{H}(\phi(\mathbf{x},t),\pi(\mathbf{x},t),t) = \pi(\mathbf{x},t)\dot{\phi}(\mathbf{x},t) - \mathcal{L}. \tag{15.3.8}$$

Note that, although $\mathcal{H}$ apparently depends on $\phi, \dot{\phi}$, and π, its dependence on $\dot{\phi}$ is actually absent as seen in

$$\frac{\partial \mathcal{H}}{\partial \dot{\phi}} = \pi - \frac{\partial \mathcal{L}}{\partial \dot{\phi}} = 0, \tag{15.3.9}$$

as a consequence of (15.3.3). This justifies our notation in (15.3.7) and (15.3.8).

On the other hand, the Fréchet derivatives of (15.3.7) may be computed to yield

$$\dot{\phi}(\mathbf{x},t) = \frac{\delta H}{\delta \pi(\mathbf{x},t)}, \tag{15.3.10}$$

$$\dot{\pi}(\mathbf{x},t) = -\frac{\delta H}{\delta \phi(\mathbf{x},t)}, \tag{15.3.11}$$

which is a Hamiltonian system of equations.

Poisson bracket

To proceed towards canonical quantization, we consider the Poisson bracket formalism of the equations (15.3.10)–(15.3.11). Note that, for a general functional F depending on the canonical coordinates ϕ and π and defined by spatially integrating a density function $\mathcal{F}$, we have

$$\frac{\delta F}{\delta \phi} = \frac{\partial \mathcal{F}}{\partial \phi} - \nabla \cdot \left(\frac{\partial \mathcal{F}}{\partial(\nabla\phi)}\right), \quad \frac{\delta F}{\delta \pi} = \frac{\partial \mathcal{F}}{\partial \pi} - \nabla \cdot \left(\frac{\partial \mathcal{F}}{\partial(\nabla\pi)}\right), \tag{15.3.12}$$

where, for applications, the second term in the second equation is usually absent. As a result, we define the Poisson bracket of the functionals F and G to be

$$\{F, G\} = \int \left(\frac{\delta F}{\delta \phi} \frac{\delta G}{\delta \pi} - \frac{\delta F}{\delta \pi} \frac{\delta G}{\delta \phi} \right) \mathrm{d}\mathbf{x}. \tag{15.3.13}$$

Therefore, along a solution to (15.3.10)–(15.3.11), we have

$$\begin{aligned} \frac{\mathrm{d}F}{\mathrm{d}t} &= \int \left(\frac{\delta F}{\delta \phi} \dot{\phi} + \frac{\delta F}{\delta \pi} \dot{\pi} \right) \mathrm{d}\mathbf{x} + \frac{\partial F}{\partial t} \\ &= \int \left(\frac{\delta F}{\delta \phi} \frac{\delta H}{\delta \pi} - \frac{\delta F}{\delta \pi} \frac{\delta H}{\delta \phi} \right) \mathrm{d}\mathbf{x} + \frac{\partial F}{\partial t} \\ &= \{F, H\} + \frac{\partial F}{\partial t}. \end{aligned} \tag{15.3.14}$$

It is clear that this single equation contains (15.3.10)–(15.3.11).

Examples

As illustrative examples, we show that (15.3.10) and (15.3.11) are covered as special cases of the evolution equation (15.3.14). For such a purpose, we need to represent $\phi(\mathbf{x}, t)$ and $\pi(\mathbf{x}, t)$ as functionals and compute their Fréchet derivatives.

In fact, using the Dirac delta function $\delta(\mathbf{x})$, we have

$$\phi(\mathbf{x}, t) = \int \phi(\mathbf{x}', t) \delta(\mathbf{x} - \mathbf{x}') \, \mathrm{d}\mathbf{x}'. \tag{15.3.15}$$

Similarly, we can represent $\pi(\mathbf{x}, t)$. Thus, from these, we have

$$\frac{\delta \phi(\mathbf{x}, t)}{\delta \phi(\mathbf{x}', t)} = \delta(\mathbf{x} - \mathbf{x}'), \quad \frac{\delta \phi(\mathbf{x}, t)}{\delta \pi(\mathbf{x}', t)} = 0, \quad \frac{\delta \pi(\mathbf{x}, t)}{\delta \phi(\mathbf{x}', t)} = 0, \quad \frac{\delta \pi(\mathbf{x}, t)}{\delta \pi(\mathbf{x}', t)} = \delta(\mathbf{x} - \mathbf{x}'). \tag{15.3.16}$$

Inserting (15.3.16) into (15.3.13), we arrive at

$$\{\phi(\mathbf{x}, t), H\} = \int \left(\frac{\delta \phi(\mathbf{x}, t)}{\delta \phi(\mathbf{x}', t)} \frac{\delta H}{\delta \pi(\mathbf{x}', t)} \right) \mathrm{d}\mathbf{x}' = \frac{\delta H}{\delta \pi(\mathbf{x}, t)}, \tag{15.3.17}$$

$$\{\pi(\mathbf{x}, t), H\} = -\int \left(\frac{\delta \pi(\mathbf{x}, t)}{\delta \pi(\mathbf{x}', t)} \frac{\delta H}{\delta \phi(\mathbf{x}', t)} \right) \mathrm{d}\mathbf{x}' = -\frac{\delta H}{\delta \phi(\mathbf{x}, t)}, \tag{15.3.18}$$

which return to (15.3.10)–(15.3.11) after using (15.3.14), as anticipated.

From the definitions of $\phi(\mathbf{x}, t)$ and $\pi(\mathbf{x}, t)$ as functionals, we also have the equal-time Poisson brackets

$$\begin{aligned} \{\phi(\mathbf{x}, t), \pi(\mathbf{x}', t)\} &= \int \left(\frac{\delta \phi(\mathbf{x}, t)}{\delta \phi(\mathbf{x}'', t)} \frac{\delta \pi(\mathbf{x}', t)}{\delta \pi(\mathbf{x}'', t)} \right) \mathrm{d}\mathbf{x}'' \\ &= \int \delta(\mathbf{x} - \mathbf{x}'') \delta(\mathbf{x}' - \mathbf{x}'') \, \mathrm{d}\mathbf{x}'' = \delta(\mathbf{x} - \mathbf{x}'), \end{aligned} \tag{15.3.19}$$

$$\{\phi(\mathbf{x}, t), \phi(\mathbf{x}', t)\} = \{\pi(\mathbf{x}, t), \pi(\mathbf{x}', t)\} = 0, \tag{15.3.20}$$

by virtue of (15.3.13) and (15.3.16).

Quantization

Following (15.2.8)–(15.2.12), we replace the classical canonical coordinate functionals ϕ and π by their quantum operator representatives and the Poisson brackets by the corresponding operator commutators according to the rule

$$\phi \mapsto \hat{\phi}, \quad \pi \mapsto \hat{\pi}, \quad f(\phi, \pi, t) \mapsto \hat{f} = f(\hat{\phi}, \hat{\pi}, t), \quad \{\,,\} \mapsto \frac{1}{\mathrm{i}\hbar}[\,,]. \tag{15.3.21}$$

As a consequence, (15.3.19)–(15.3.20) lead to the equal-time commutators

$$[\hat{\phi}(\mathbf{x}, t), \hat{\pi}(\mathbf{x}', t)] = \mathrm{i}\hbar\delta(\mathbf{x} - \mathbf{x}'), \tag{15.3.22}$$

$$[\hat{\phi}(\mathbf{x}, t), \hat{\phi}(\mathbf{x}', t)] = 0, \quad [\hat{\pi}(\mathbf{x}, t), \hat{\pi}(\mathbf{x}', t)] = 0. \tag{15.3.23}$$

Let $\hat{H} = H(\hat{\phi}, \hat{\pi}, t)$ be the quantum Hamiltonian of the system and $\hat{F}$ any observable. Then, in view of (15.3.14) and (15.3.21), $\hat{F}$ evolves according to the Heisenberg equation

$$\frac{\mathrm{d}\hat{F}}{\mathrm{d}t} = \frac{1}{\mathrm{i}\hbar}[\hat{F}, \hat{H}] + \frac{\partial \hat{F}}{\partial t}. \tag{15.3.24}$$

In particular, we can solve for $\hat{\phi}$ and $\hat{\pi}$ once they are known initially.

The subsequent sections show how to quantize the Klein–Gordon and Schrödinger equations as concrete illustrative examples.

15.4 Quantization of Klein–Gordon equation

To simplify the notation, we set $c = 1$. Then, the Lagrangian action density of the Klein–Gordon model governing a real-valued scalar field ϕ depending on the time and spatial coordinates, t and $\mathbf{x}$, reads

$$\mathcal{L} = \frac{1}{2}\dot{\phi}^2 - \frac{1}{2}|\nabla\phi|^2 - \frac{1}{2}m^2\phi^2, \tag{15.4.1}$$

where $m > 0$ is the mass. Since the canonically conjugate momentum field is

$$\pi = \frac{\partial \mathcal{L}}{\partial \dot{\phi}} = \dot{\phi}, \tag{15.4.2}$$

the Hamiltonian density is

$$\mathcal{H} = \pi\dot{\phi} - \mathcal{L} = \frac{1}{2}\left(\pi^2 + |\nabla\phi|^2 + m^2\phi^2\right). \tag{15.4.3}$$

Thus, the Hamiltonian energy functional assumes the form

$$\begin{aligned} H(\phi, \pi) &= \frac{1}{2}\int \left(\pi^2 + |\nabla\phi|^2 + m^2\phi^2\right)\,\mathrm{d}\mathbf{x} \\ &= \frac{1}{2}\int \left(\pi^2 + \phi[-\nabla^2 + m^2]\phi\right)(\mathbf{x}, t)\,\mathrm{d}\mathbf{x}. \end{aligned} \tag{15.4.4}$$

The equation of motion of (15.4.1), or the Klein–Gordon equation, is

$$\ddot{\phi} - \Delta\phi + m^2\phi = 0. \tag{15.4.5}$$

By separation of variables, $\phi(\mathbf{x}, t) = q(t)u(\mathbf{x})$, we recast (15.4.5) into its temporal and spatial parts, respectively,

$$\ddot{q} + (m^2 + \lambda)q = 0, \tag{15.4.6}$$

$$\Delta u + \lambda u = 0. \tag{15.4.7}$$

For simplicity, we assume that the eigenvalue problem (15.4.7) is such that the associated sequence of the eigenpairs $\{(u_i, \lambda_i)\}$ is countable, for which $\{u_i(\mathbf{x})\}$ is an orthonormal basis for the L^2 function space consisting of all spatially dependent functions of interest and

$$0 < \lambda_1 < \cdots < \lambda_i < \cdots, \quad \lambda_i \to \infty \quad \text{as} \quad i \to \infty. \tag{15.4.8}$$

Inserting this information into (15.4.6), we arrive at the following sequence of equations of harmonic oscillators:

$$\ddot{q}^i + \omega_i^2 q^i = 0, \quad \omega_i = \sqrt{m^2 + \lambda_i}, \tag{15.4.9}$$

for each fixed $i = 1, 2, \ldots$. Of course, for this system, the associated momentum is

$$p_i = \dot{q}^i. \tag{15.4.10}$$

Let $\{q^i(t)\}$ be a corresponding sequence of solutions to these equations. Then, the solution of (15.4.5) is

$$\phi(\mathbf{x}, t) = \sum_{i=1}^{\infty} q^i(t)u_i(\mathbf{x}). \tag{15.4.11}$$

Thus, in view of (15.4.2), (15.4.10), and (15.4.11), we have

$$\pi(\mathbf{x}, t) = \sum_{i=1}^{\infty} p_i(t)u_i(\mathbf{x}). \tag{15.4.12}$$

Therefore, we see that, classically, the Klein–Gordon waves result from *spatially weighted* superimposed harmonic oscillations such that the momentum of the waves is the sum of the momenta, p_i, of the oscillations associated with the phase coordinates, q^i, in the same manner, at any given spatial location $\mathbf{x}$.

Quantization

Following our earlier quantization formalism for the harmonic oscillator, we have

$$q^i \mapsto \hat{q}^i, \quad p_i \mapsto \hat{p}_i = -\mathrm{i}\hbar\frac{\partial}{\partial q^i}, \quad i = 1, 2, \ldots, \tag{15.4.13}$$

with

$$[\hat{q}^i, \hat{q}^j] = 0, \quad [\hat{p}_i, \hat{p}_j] = 0, \quad [\hat{q}^i, \hat{p}_j] = \mathrm{i}\hbar\, \delta^i_j, \tag{15.4.14}$$

since the phase coordinates q^i ($i = 1, 2, \dots$) are independent. There also hold $[(\hat{q}^i)^2, \hat{p}_i] = 2\mathrm{i}\hbar \hat{q}^i$ and $[(\hat{p}_i)^2, \hat{q}^i] = -2\mathrm{i}\hbar \hat{p}_i$. Consequently, if we use

$$\hat{a}^i_+ = \frac{1}{\sqrt{2\hbar\omega_i}} \left(\hat{p}_i + \mathrm{i}\omega_i \hat{q}^i\right), \quad \hat{a}_- = \frac{1}{\sqrt{2\hbar\omega_i}} \left(\hat{p}^i - \mathrm{i}\omega_i \hat{q}^i\right), \tag{15.4.15}$$

as creation and annihilation operators, then we have their commutators

$$[\hat{a}^i_-, \hat{a}^j_-] = 0, \quad , [\hat{a}^i_+, \hat{a}^j_+] = 0, \quad [\hat{a}^i_-, \hat{a}^j_+] = \delta^{ij}, \tag{15.4.16}$$

as before. Furthermore, we may invert (15.4.15) to get

$$\hat{q}^i = \mathrm{i}\sqrt{\frac{\hbar}{2\omega_i}}(\hat{a}^i_- - \hat{a}^i_+), \quad \hat{p}_i = \sqrt{\frac{\hbar\omega_i}{2}}(\hat{a}^i_- + \hat{a}^i_+). \tag{15.4.17}$$

On the other hand, inserting (15.4.11) and (15.4.12) into (15.4.4), we find

$$H(\phi, \pi) = \frac{1}{2}\sum_{i=1}^{\infty} \left(p_i^2 + \omega_i^2 [q^i]^2\right). \tag{15.4.18}$$

Hence, the quantized Hamiltonian is read off to be

$$\begin{aligned} \hat{H} &= \frac{1}{2}\sum_{i=1}^{\infty} \left(\hat{p}_i^2 + \omega_i^2 [\hat{q}^i]^2\right) \\ &= \frac{\hbar}{2}\sum_{i=1}^{\infty} \omega_i \left(\hat{a}^i_- \hat{a}^i_+ + \hat{a}^i_+ \hat{a}^i_-\right), \end{aligned} \tag{15.4.19}$$

which may be compared nicely with (15.2.16).

Quantized Klein–Gordon equation

In view of (15.4.19) and the Heisenberg picture, we have

$$\dot{\hat{q}}^i = [\hat{q}^i, \hat{H}] = \hat{p}_i, \quad \dot{\hat{p}}_i = [\hat{p}_i, \hat{H}] = -\omega_i^2 \hat{q}^i. \tag{15.4.20}$$

Thus, as an operator, $\hat{q}^i$ satisfies the same the same harmonic oscillator equation, (15.4.9). That is,

$$\ddot{\hat{q}}^i + \omega_i^2 \hat{q}^i = 0, \quad \omega_i = \sqrt{m^2 + \lambda_i}. \tag{15.4.21}$$

Moreover, the functional (15.3.15) and the quantization correspondence (15.2.8) lead to

$$\begin{aligned} \hat{\phi}(\mathbf{x}, t) &= \int \left(\sum_{i=1}^{\infty} \hat{q}^i(t) u_i(\mathbf{x}')\right) \delta(\mathbf{x} - \mathbf{x}')\, \mathrm{d}\mathbf{x}' \\ &= \sum_{i=1}^{\infty} \hat{q}^i(t) u_i(\mathbf{x}). \end{aligned} \tag{15.4.22}$$

In view of (15.4.21), we see that $\hat{\phi}$ given in (15.4.22) satisfies the *operator equation*

$$\ddot{\hat{\phi}} - \Delta\hat{\phi} + m^2\hat{\phi} = 0, \tag{15.4.23}$$

which is the quantum field version of the Klein–Gordon equation (15.4.5).

By virtue of (15.4.12) and (15.4.20), we also have

$$\hat{\pi}(\mathbf{x}, t) = \sum_{i=1}^{\infty} \hat{p}_i(t)u_i(\mathbf{x}) = \dot{\hat{\phi}}(\mathbf{x}, t), \tag{15.4.24}$$

which is consistent with (15.4.2).

Infinity problem

Besides, using (15.4.16), we have

$$\hat{a}^i_-\hat{a}^i_+ + \hat{a}^i_+\hat{a}^i_- = 2\hat{a}^i_+\hat{a}^i_- + 1. \tag{15.4.25}$$

Since the Hermitian conjugate of the operator $\hat{a}^i_\pm$ is $\hat{a}^i_\mp$, or $(\hat{a}^i_\pm)^\dagger = \hat{a}^i_\mp$, we have

$$\langle\psi|\hat{a}^i_+\hat{a}^i_-|\psi\rangle = \langle(\hat{a}^i_-|\psi\rangle)|(\hat{a}^i_-|\psi\rangle)\rangle \geq 0. \tag{15.4.26}$$

Applying (15.4.25) in (15.4.19), we have the formal result

$$\langle\psi|\hat{H}|\psi\rangle = \hbar\sum_{i=1}^{\infty}\omega_i\langle\psi|\hat{a}^i_+\hat{a}^i_-|\psi\rangle + \frac{\hbar}{2}\sum_{i=1}^{\infty}\omega_i, \tag{15.4.27}$$

which diverges in view of (15.4.26) and the divergent tail-part

$$E_0 = \frac{\hbar}{2}\sum_{i=1}^{\infty}\omega_i = \infty, \tag{15.4.28}$$

due to (15.4.8)–(15.4.9). This leads to an *infinity problem* arising from the zero-point energy E_0 of the Klein–Gordon model.

On the other hand, since we are interested in what happens in nonvacuum situations, we may subtract the zero-point energy away from the Hamiltonian to consider its "active" part only, via

$$\hat{H} \mapsto \hat{H} = \hbar\sum_{i=1}^{\infty}\omega_i\hat{a}^i_+\hat{a}^i_-, \tag{15.4.29}$$

free of the divergence problem. In other words, the quantum Hamiltonian is now a weighted sum of the number operators,

$$\hat{H} = \hbar\sum_{i=1}^{\infty}\omega_i\hat{N}_i, \tag{15.4.30}$$

superimposed from the full spectrum of underlying harmonic oscillators.

In quantum field theory, *renormalization* refers to any systematic technical process that enables one to bypass an infinity difficulty, such as by going from (15.4.19) to (15.4.29), through a reformulation of the problem.

Renormalization via normal ordering

As an illustration, we study how we can achieve a renormalization to resolve the infinity problem for the quantized Klein–Gordon model, by passing from (15.4.19) to (15.4.29).

To proceed, consider (15.4.19) expressed as a sum of the Hamiltonians

$$\hat{H}_i = \frac{\hbar}{2}\omega_i\left(\hat{a}^i_-\hat{a}^i_+ + \hat{a}^i_+\hat{a}^i_-\right), \quad i = 1, 2, \ldots, \tag{15.4.31}$$

in which both $\hat{a}^i_-\hat{a}^i_+$ and $\hat{a}^i_+\hat{a}^i_-$ are positive semi-definite. Furthermore, from (15.4.16), we have

$$\hat{a}^i_-\hat{a}^i_+ = \hat{a}^i_+\hat{a}^i_- + 1, \tag{15.4.32}$$

which indicates clearly that it is the term $\hat{a}^i_-\hat{a}^i_+$ that produces the quantity 1, which attributes to the divergence problem. In other words, if we come up with a mechanism such that in (15.4.31) the term $\hat{a}^i_-\hat{a}^i_+$ is replaced by $\hat{a}^i_+\hat{a}^i_-$, then the issue will disappear. Therefore, in a sense, the issue arises as a result of the order of product or noncommutativity of the operators $\hat{a}^i_-$ and $\hat{a}^i_+$.

On the other hand, note that the classical mechanics counterpart of (15.4.31) is the Hamiltonian

$$H_i = \frac{1}{2}p_i^2 + \omega_i^2(q^i)^2, \tag{15.4.33}$$

where there is no issue of noncommutativity. Thus, there is room to explore the process of quantization as we go from the classical commutative variables to noncommutative operators.

Here, let q, p be two classical quantities and $\hat{q}, \hat{p}$ their quantum-mechanical operator representatives, which may be expressed as linear combinations of annihilation and creation operators, $\hat{a}_-, \hat{a}_+$, respectively,

$$\hat{q} = c_1\hat{a}_- + c_2\hat{a}_+ \equiv \hat{q}_{(-)} + \hat{q}_{(+)}, \quad \hat{p} = c_3\hat{a}_- + c_4\hat{a}_+ \equiv \hat{p}_{(-)} + \hat{p}_{(+)}, \tag{15.4.34}$$

where $c_1, c_2, c_3, c_4 \in \mathbb{C}$. The *normally-ordered product* of $\hat{q}$ and $\hat{p}$ is defined to be

$$:\hat{q}\hat{p}: = \hat{q}_{(-)}\hat{p}_{(-)} + \hat{q}_{(+)}\hat{p}_{(-)} + \hat{p}_{(+)}\hat{q}_{(-)} + \hat{q}_{(+)}\hat{p}_{(+)}, \tag{15.4.35}$$

which is designed to make sure that the operators associated with creation, or subfix +, always appear to the right of those with annihilation, or subfix −, in a product, in order to bypass the problem caused by (15.4.32).

As a consequence, if the conventional operator products are replaced by normal products, wherever applicable, we see that quantum-mechanical version of (15.4.33) becomes

$$\begin{aligned}\hat{H}_i &= \frac{1}{2} :\hat{p}_i^2: + \omega_i^2 :(\hat{q}^i)^2: \\ &= \hbar\omega_i\hat{a}^i_+\hat{a}^i_-.\end{aligned} \tag{15.4.36}$$

Thus, by superposition, we arrive at (15.4.29) and obtain a renormalization of the model as anticipated. This simple renormalization method is known as *normal ordering*.

Back to canonical quantization

Note that the same quantization formalism may be made at the wave function level directly.

Using the notation in (15.4.11) and (15.4.12), we have their quantum-mechanical representatives

$$\hat{\phi}(\mathbf{x},t)=\sum_{i=1}^{\infty}\hat{q}^i(t)u_i(\mathbf{x}),\quad \hat{\pi}(\mathbf{x},t)=\sum_{i=1}^{\infty}\hat{p}_i(t)u_i(\mathbf{x}). \tag{15.4.37}$$

Thus, we have

$$\hat{q}^i(t)=\int\hat{\phi}(\mathbf{x},t)u_i(\mathbf{x})\,\mathrm{d}\mathbf{x},\quad \hat{p}_i(t)=\int\hat{\pi}(\mathbf{x},t)u_i(\mathbf{x})\,\mathrm{d}\mathbf{x}. \tag{15.4.38}$$

Therefore,

$$\begin{aligned}[\hat{q}^i(t),\hat{p}_j(t)]&=\int\int[\hat{\phi}(\mathbf{x},t),\hat{\pi}(\mathbf{x}',t)]u_i(\mathbf{x})u_j(\mathbf{x}')\,\mathrm{d}\mathbf{x}\mathrm{d}\mathbf{x}'\\&=\mathrm{i}\hbar\int\int\delta(\mathbf{x}-\mathbf{x}')u_i(\mathbf{x})u_j(\mathbf{x}')\,\mathrm{d}\mathbf{x}\mathrm{d}\mathbf{x}'\\&=\mathrm{i}\hbar\int u_i(\mathbf{x})u_j(\mathbf{x})\,\mathrm{d}\mathbf{x}=\mathrm{i}\hbar\delta^i_j,\end{aligned} \tag{15.4.39}$$

using (15.3.22). Similarly $[\hat{q}^i,\hat{q}^j]=0$ and $[\hat{p}_i,\hat{p}_j]=0$. These are as stated in (15.4.14).

Let $\hat{\phi},\hat{\psi}$ be the operator representatives of the classical wave quantities ϕ,ψ, respectively, and $\hat{\phi}=\hat{\phi}^{(-)}+\hat{\phi}^{(+)},\hat{\psi}=\hat{\psi}^{(-)}+\hat{\psi}^{(+)}$ are similarly decomposed. Then the normally ordered product of $\hat{\phi}$ and $\hat{\psi}$ is defined as

$$:\hat{\phi}\hat{\psi}:=\hat{\phi}_{(-)}\hat{\psi}_{(-)}+\hat{\phi}_{(+)}\hat{\psi}_{(-)}+\hat{\psi}_{(+)}\hat{\phi}_{(-)}+\hat{\phi}_{(+)}\hat{\psi}_{(+)}. \tag{15.4.40}$$

As a consequence, from (15.4.4), we arrive at the renormalized Hamiltonian

$$\begin{aligned}\hat{H}&=\frac{1}{2}\int:\left(\hat{\pi}^2+\hat{\phi}[-\nabla^2+m^2]\hat{\phi}\right)(\mathbf{x},t):\,\mathrm{d}\mathbf{x}\\&=\frac{1}{2}\sum_{i=1}^{\infty}:\left(\hat{p}_i^2+\omega_i^2(\hat{q}^i)^2\right):\\&=\hbar\sum_{i=1}^{\infty}\omega_i\hat{a}^i_+\hat{a}^i_-,\end{aligned} \tag{15.4.41}$$

in terms of creation and annihilation operators, as obtained earlier.

For the functional (15.4.4), its Fréchet derivative is given by

$$\frac{\delta H}{\delta\phi(\mathbf{x},t)}=\left(-\Delta+m^2\right)\phi(\mathbf{x},t). \tag{15.4.42}$$

In view of (15.3.18), (15.3.21), (15.4.2), and (15.4.42), we get the following evolution equation

$$\ddot{\hat{\phi}} = \dot{\hat{\pi}} = \left(\Delta - m^2\right)\hat{\phi}, \tag{15.4.43}$$

which recovers the operator-version Klein–Gordon equation (15.4.23), as expected.

Alternatively, from (15.4.4), we have

$$\hat{H} = H(\hat{\phi}, \hat{\pi}) = \frac{1}{2}\int \left(\hat{\pi}^2 + \hat{\phi}[-\nabla^2_{\mathbf{x}'} + m^2]\hat{\phi}\right)(\mathbf{x}', t)\,\mathrm{d}\mathbf{x}'. \tag{15.4.44}$$

Therefore, it follows from the Heisenberg equation (15.3.24) and the commutativity of $\hat{\pi}(\mathbf{x}, t)$ with $\hat{\pi}(\mathbf{x}', t)$ and $\nabla^2_{\mathbf{x}'}$ that

$$\begin{aligned}
\dot{\hat{\pi}}(\mathbf{x}, t) &= \frac{1}{\mathrm{i}\hbar}[\hat{\pi}(\mathbf{x}, t), \hat{H}] \\
&= \frac{1}{\mathrm{i}\hbar}\int [\hat{\pi}(\mathbf{x}, t), \hat{\phi}(\mathbf{x}', t)]\left(-\nabla^2_{\mathbf{x}'} + m^2\right)\hat{\phi}(\mathbf{x}', t)\,\mathrm{d}\mathbf{x}' \\
&= \int \delta(\mathbf{x} - \mathbf{x}')\left(\nabla^2_{\mathbf{x}'} - m^2\right)\hat{\phi}(\mathbf{x}', t)\,\mathrm{d}\mathbf{x}' \\
&= \left(\nabla^2_{\mathbf{x}} - m^2\right)\hat{\phi}(\mathbf{x}, t),
\end{aligned} \tag{15.4.45}$$

by virtue of (15.3.22), which renders (15.4.43) again.

15.5 Quantization of Schrödinger equation

In the previous section, the real-field version of the Klein–Gordon equation is quantized for simplicity and clarity. Since such a real situation is electrically neutral, it limits its range of applications. This section shows how to naturally extend the formalism to a complex setting. For this purpose, we start by considering a mechanical system of two degrees of freedom.

Mechanical system in complex coordinates

Consider a mechanical system of two degrees of freedom described in terms of the real coordinates x, y and governed by the Lagrange action function

$$L = L(x, y, \dot{x}, \dot{y}, t). \tag{15.5.1}$$

The equations of motion are

$$\dot{p}_x = \frac{\partial L}{\partial x}, \quad \dot{p}_y = \frac{\partial L}{\partial y}, \tag{15.5.2}$$

where

$$p_x = \frac{\partial L}{\partial \dot{x}}, \quad p_y = \frac{\partial L}{\partial \dot{y}}, \tag{15.5.3}$$

are the associated canonical momenta, respectively. Now introduce the complex variables

$$q = \frac{1}{\sqrt{2}}(x + \mathrm{i}y), \quad \overline{q} = \frac{1}{\sqrt{2}}(x - \mathrm{i}y). \tag{15.5.4}$$

Then,

$$\frac{\partial L}{\partial x} = \frac{1}{\sqrt{2}}\left(\frac{\partial L}{\partial q} + \frac{\partial L}{\partial \overline{q}}\right), \quad \mathrm{i}\frac{\partial L}{\partial y} = \frac{1}{\sqrt{2}}\left(-\frac{\partial L}{\partial q} + \frac{\partial L}{\partial \overline{q}}\right). \tag{15.5.5}$$

The same relations hold for partial derivatives involving $\dot{x}, \dot{y}, \dot{q}, \dot{\overline{q}}$, of course.

From (15.5.2)–(15.5.4), we see that the momenta $p, \overline{p}$ associated with $q, \overline{q}$, respectively, may similarly be defined as

$$p = \frac{\partial L}{\partial \dot{q}} = \frac{1}{\sqrt{2}}(p_x - \mathrm{i}p_y), \quad \overline{p} = \frac{\partial L}{\partial \dot{\overline{q}}} = \frac{1}{\sqrt{2}}(p_x + \mathrm{i}p_y). \tag{15.5.6}$$

Thus, (15.5.2) becomes

$$\dot{p} = \frac{\partial L}{\partial q}, \quad \dot{\overline{p}} = \frac{\partial L}{\partial \overline{q}}, \quad L = L(q, \overline{q}, \dot{q}, \dot{\overline{q}}, t). \tag{15.5.7}$$

Furthermore, define the Hamiltonian

$$H(q, \overline{q}, p, \overline{p}, t) = p\dot{q} + \overline{p}\dot{\overline{q}} - L(q, \overline{q}, \dot{q}, \dot{\overline{q}}, t). \tag{15.5.8}$$

Then, the equations of motion are recast into the Hamiltonian system

$$\dot{q} = \frac{\partial H}{\partial p}, \quad \dot{\overline{q}} = \frac{\partial H}{\partial \overline{p}}, \quad \dot{p} = -\frac{\partial H}{\partial q}, \quad \dot{\overline{p}} = -\frac{\partial H}{\partial \overline{q}}. \tag{15.5.9}$$

As a consequence, for any mechanical quantity $F = F(q, \overline{q}, p, \overline{p}, t)$, we deduce from (15.5.9) the evolution equation

$$\frac{\mathrm{d}F}{\mathrm{d}t} = \{F, H\} + \frac{\partial F}{\partial t}, \tag{15.5.10}$$

where the Poisson bracket $\{\,,\,\}$ is defined to be

$$\{F, H\} = \frac{\partial F}{\partial q}\frac{\partial H}{\partial p} + \frac{\partial F}{\partial \overline{q}}\frac{\partial H}{\partial \overline{p}} - \frac{\partial F}{\partial p}\frac{\partial H}{\partial q} - \frac{\partial F}{\partial \overline{p}}\frac{\partial H}{\partial \overline{q}}. \tag{15.5.11}$$

In particular, we have the useful results

$$\{q, \overline{q}\} = 0, \; \{p, \overline{p}\} = 0, \; \{q, p\} = 1, \; \{\overline{q}, \overline{p}\} = 1, \; \{q, \overline{p}\} = 0, \; \{\overline{q}, p\} = 0. \tag{15.5.12}$$

Quantization

With the quantization

$$x \mapsto \hat{x} = x, \quad y \mapsto \hat{y} = y, \quad p_x \mapsto \hat{p}_x = -\mathrm{i}\hbar\frac{\partial}{\partial x}, \quad p_y \mapsto \hat{p}_y = -\mathrm{i}\hbar\frac{\partial}{\partial y}, \tag{15.5.13}$$

and (15.5.6), we have

$$q \mapsto \hat{q} = q, \quad \overline{q} \mapsto \hat{\overline{q}} = \overline{q}, \quad p \mapsto \hat{p} = -\mathrm{i}\hbar\frac{\partial}{\partial q}, \quad \overline{p} \mapsto \hat{\overline{p}} = -\mathrm{i}\hbar\frac{\partial}{\partial \overline{q}}, \tag{15.5.14}$$

which are elegant. Note that the front factors in the last two relations remain intact in the correspondence. From (15.5.14), we have

$$[\hat{q}, \hat{\overline{q}}] = 0,\ [\hat{p}, \hat{\overline{p}}] = 0,\ [\hat{q}, \hat{p}] = \mathrm{i}\hbar,\ [\hat{\overline{q}}, \hat{\overline{p}}] = \mathrm{i}\hbar,\ [\hat{q}, \hat{\overline{p}}] = 0,\ [\hat{\overline{q}}, \hat{p}] = 0, \tag{15.5.15}$$

which may also be seen directly from quantizing (15.5.12) by using the rule stated in (15.2.9).

Extension of this formalism to the situation of higher degrees of freedom is immediate and is omitted.

Quantization of harmonic oscillator

For later applications, we now work out the quantization of a harmonic oscillator of two degrees of freedom governed by the Hamiltonian

$$H = \frac{1}{2}(\dot{x}^2 + \dot{y}^2) + \frac{\omega^2}{2}(x^2 + y^2), \quad (x, y) \in \mathbb{R}^2. \tag{15.5.16}$$

First, using (15.5.4) and (15.5.6), we have

$$H = \frac{1}{2}(p^2 + \overline{p}^2) + \frac{\omega^2}{2}(q^2 + \overline{q}^2). \tag{15.5.17}$$

Next, with

$$\hat{a}_\pm = \frac{1}{\sqrt{2\hbar\omega}}\,(\hat{p} \pm \mathrm{i}\omega\hat{q}), \quad \hat{\overline{a}}_\pm = \frac{1}{\sqrt{2\hbar\omega}}\,(\hat{\overline{p}} \pm \mathrm{i}\omega\hat{\overline{q}}), \tag{15.5.18}$$

and (15.5.15), we have

$$[\hat{a}_-, \hat{a}_+] = 1, \quad [\hat{\overline{a}}_-, \hat{\overline{a}}_+] = 1, \quad [\hat{a}_-, \hat{\overline{a}}_+] = 0, \quad [\hat{\overline{a}}_-, \hat{a}_+] = 0, \tag{15.5.19}$$

such that $\hat{a}_\pm$ and $\hat{\overline{a}}_\pm$ are two pairs of creation and annihilation operators.

Thus, from (15.5.18), we get

$$\hat{p} = \sqrt{\frac{\hbar\omega}{2}}\,(\hat{a}_- + \hat{a}_+), \quad \hat{q} = \mathrm{i}\sqrt{\frac{\hbar}{2\omega}}\,(\hat{a}_- - \hat{a}_+), \tag{15.5.20}$$

$$\hat{\overline{p}} = \sqrt{\frac{\hbar\omega}{2}}\,(\hat{\overline{a}}_- + \hat{\overline{a}}_+), \quad \hat{\overline{q}} = \mathrm{i}\sqrt{\frac{\hbar}{2\omega}}\,(\hat{\overline{a}}_- - \hat{\overline{a}}_+). \tag{15.5.21}$$

Note that $\hat{\overline{q}}$ in (15.5.21) is not simply the complex conjugate of $\hat{q}$ in (15.5.20).

In view of (15.5.17) and (15.5.20)–(15.5.21), we obtain the quantized Hamiltonian

$$\hat{H} = \frac{\hbar\omega}{2}\,(\hat{a}_-\hat{a}_+ + \hat{a}_+\hat{a}_- + \hat{\overline{a}}_-\hat{\overline{a}}_+ + \hat{\overline{a}}_+\hat{\overline{a}}_-). \tag{15.5.22}$$

Furthermore, using (15.5.19), we may rewrite (15.5.22) as

$$\begin{aligned}\hat{H} &= \hbar\omega\left(\hat{a}_-\hat{a}_+ + \hat{\overline{a}}_-\hat{\overline{a}}_+\right) - \hbar\omega \\ &= \hbar\omega\left(\hat{N}_1 + \hat{N}_2\right) + \hbar\omega,\end{aligned} \tag{15.5.23}$$

where $\hat{N}_1 = \hat{a}_+\hat{a}_-$ and $\hat{N}_2 = \hat{\overline{a}}_+\hat{\overline{a}}_-$ are the number operators associated with the particles corresponding to the classical dynamic variables (q, p) and $(\overline{q}, \overline{p})$, respectively, and $\hbar\omega$ is the zero-point energy of the quantized system.

Canonical description of Schrödinger equation

Consider the Schrödinger equation

$$\mathrm{i}\hbar\dot{\phi} = -\frac{\hbar^2}{2m}\nabla^2\phi + V(\mathbf{x}, t)\phi, \tag{15.5.24}$$

governing a complex scalar field, ϕ, in a potential field, V. The Lagrangian action density is

$$\mathcal{L}(\phi, \overline{\phi}, \dot{\phi}, \dot{\overline{\phi}}, t) = \mathrm{i}\hbar(\dot{\phi}\overline{\phi} - \phi\dot{\overline{\phi}}) - \frac{\hbar^2}{2m}\nabla\phi\cdot\nabla\overline{\phi} - V(\mathbf{x}, t)\phi\overline{\phi}, \tag{15.5.25}$$

where ϕ and $\overline{\phi}$ are treated as two independent variables. The canonical momenta associated with $\phi, \overline{\phi}$ are

$$\pi(\mathbf{x}, t) = \frac{\partial\mathcal{L}}{\partial\dot{\phi}} = \mathrm{i}\hbar\overline{\phi}(\mathbf{x}, t), \quad \overline{\pi}(\mathbf{x}, t) = \frac{\partial\mathcal{L}}{\partial\dot{\overline{\phi}}} = -\mathrm{i}\hbar\phi(\mathbf{x}, t). \tag{15.5.26}$$

Thus, the equations of motion of (15.5.25) may be represented as the Lagrange equations

$$\dot{\pi}(\mathbf{x}, t) = \frac{\delta L}{\delta\phi(\mathbf{x}, t)}, \quad \dot{\overline{\pi}}(\mathbf{x}, t) = \frac{\delta L}{\delta\overline{\phi}(\mathbf{x}, t)}, \tag{15.5.27}$$

where L is the Lagrange action functional

$$L(\phi, \overline{\phi}, \dot{\phi}, \dot{\overline{\phi}}, t) = \int \mathcal{L}(\phi, \overline{\phi}, \dot{\phi}, \dot{\overline{\phi}}, t)(\mathbf{x}, t)\,\mathrm{d}\mathbf{x}, \tag{15.5.28}$$

associated with (15.5.25).

With (15.5.25), we may follow (15.5.8) to form the Hamiltonian action density

$$\mathcal{H}(\phi, \overline{\phi}, \pi, \overline{\pi}, t) = \pi\dot{\phi} + \overline{\pi}\dot{\overline{\phi}} - \mathcal{L}(\phi, \overline{\phi}, \dot{\phi}, \dot{\overline{\phi}}, t). \tag{15.5.29}$$

Hence, using (15.5.26), we have

$$\mathcal{H} = \frac{\hbar^2}{2m}\nabla\phi\cdot\nabla\overline{\phi} + V(\mathbf{x}, t)\phi\overline{\phi}, \tag{15.5.30}$$

which does not contain the momenta π and $\overline{\pi}$ because they have been replaced by $\overline{\phi}$ and ϕ through (15.5.26). However, in practical computation, we can switch on the roles of π and $\overline{\pi}$ back and forth according to convenience. So, as in (15.4.4), we obtain the Hamiltonian energy functional

$$H(\phi, \overline{\phi}, \pi, \overline{\pi}, t) = \int \mathcal{H}\, \mathrm{d}\mathbf{x} = \int \left(\frac{\hbar^2}{2m} \nabla\phi \cdot \nabla\overline{\phi} + V(\mathbf{x}, t)\phi\overline{\phi} \right) \mathrm{d}\mathbf{x}$$
$$= \int \left(\overline{\phi} \left[-\frac{\hbar^2}{2m}\nabla^2 + V \right] \phi \right) (\mathbf{x}, t)\, \mathrm{d}\mathbf{x}, \tag{15.5.31}$$

with the associated Hamiltonian system of equations

$$\dot{\phi}(\mathbf{x}, t) = \frac{\delta H}{\delta \pi(\mathbf{x}, t)}, \quad \dot{\pi}(\mathbf{x}, t) = -\frac{\delta H}{\delta \phi(\mathbf{x}, t)}, \tag{15.5.32}$$
$$\dot{\overline{\phi}}(\mathbf{x}, t) = \frac{\delta H}{\delta \overline{\pi}(\mathbf{x}, t)}, \quad \dot{\overline{\pi}}(\mathbf{x}, t) = -\frac{\delta H}{\delta \overline{\phi}(\mathbf{x}, t)}. \tag{15.5.33}$$

As a consequence, we are led to define the Poisson bracket of the functionals F and G to be

$$\{F, G\} = \int \left(\frac{\delta F}{\delta \phi}\frac{\delta G}{\delta \pi} + \frac{\delta F}{\delta \overline{\phi}}\frac{\delta G}{\delta \overline{\pi}} - \frac{\delta F}{\delta \pi}\frac{\delta G}{\delta \phi} - \frac{\delta F}{\delta \overline{\pi}}\frac{\delta G}{\delta \overline{\phi}} \right) \mathrm{d}\mathbf{x}, \tag{15.5.34}$$

which generalizes (15.3.13) so that any dynamical functional $F(\phi, \overline{\phi}, \pi, \overline{\pi}, t)$ obeys the same evolution equation, (15.3.14).

We may define $\phi, \overline{\phi}, \pi, \overline{\pi}$ as functionals such as in (15.3.15). Thus, when treating these as independent quantities, we extend (15.3.19)–(15.3.20) to see that the only two non-vanishing Poisson brackets involving these are

$$\{\phi(\mathbf{x}, t), \pi(\mathbf{x}', t)\} = \delta(\mathbf{x} - \mathbf{x}'), \quad \{\overline{\phi}(\mathbf{x}, t), \overline{\pi}(\mathbf{x}', t)\} = \delta(\mathbf{x} - \mathbf{x}'). \tag{15.5.35}$$

Quantization

Use $\hat{\phi}, \hat{\overline{\phi}}, \hat{\pi}, \hat{\overline{\pi}}, \hat{H}$ to denote the quantum-mechanical operator representatives of $\phi, \overline{\phi}, \pi, \overline{\pi}, H$, respectively. Then, they all evolve following the Heisenberg equation (15.3.24). For example, $\hat{\phi}$ satisfies the equation

$$\dot{\hat{\phi}} = \frac{1}{\mathrm{i}\hbar}[\hat{\phi}, \hat{H}]. \tag{15.5.36}$$

On the other hand, the Poisson brackets in (15.5.35) give us the commutators

$$[\hat{\phi}(\mathbf{x}, t), \hat{\pi}(\mathbf{x}', t)] = \mathrm{i}\hbar\delta(\mathbf{x} - \mathbf{x}'), \quad [\hat{\overline{\phi}}(\mathbf{x}, t), \hat{\overline{\pi}}(\mathbf{x}', t)\} = \mathrm{i}\hbar\delta(\mathbf{x} - \mathbf{x}'). \tag{15.5.37}$$

Moreover, from (15.5.26) and (15.5.31), we have

$$\hat{H} = \frac{1}{\mathrm{i}\hbar} \int \left(\hat{\pi} \left[-\frac{\hbar^2}{2m}\nabla^2 + V \right] \hat{\phi} \right) (\mathbf{x}, t)\, \mathrm{d}\mathbf{x}. \tag{15.5.38}$$

Therefore, we obtain

$$\begin{aligned}
[\hat{\phi}(\mathbf{x},t),\hat{H}] &= \frac{1}{\mathrm{i}\hbar}\int \left[\hat{\phi}(\mathbf{x},t), \left(\hat{\pi}\left[-\frac{\hbar^2}{2m}\nabla^2_{\mathbf{x}'}+V\right]\hat{\phi}\right)(\mathbf{x}',t)\right]\mathrm{d}\mathbf{x}' \\
&= \frac{1}{\mathrm{i}\hbar}\int [\hat{\phi}(\mathbf{x},t),\hat{\pi}(\mathbf{x}',t)]\left(-\frac{\hbar^2}{2m}\nabla^2_{\mathbf{x}'}+V(\mathbf{x}',t)\right)\hat{\phi}(\mathbf{x}',t)\,\mathrm{d}\mathbf{x}' \\
&= \int \delta(\mathbf{x}-\mathbf{x}')\left(-\frac{\hbar^2}{2m}\nabla^2_{\mathbf{x}'}+V(\mathbf{x}',t)\right)\hat{\phi}(\mathbf{x}',t)\,\mathrm{d}\mathbf{x}' \\
&= \left(-\frac{\hbar^2}{2m}\nabla^2_{\mathbf{x}}+V(\mathbf{x},t)\right)\hat{\phi}(\mathbf{x},t),
\end{aligned} \tag{15.5.39}$$

where we have used the fact that $\hat{\phi}(\mathbf{x},t)$ commutes with

$$-\frac{\hbar^2}{2m}\nabla^2_{\mathbf{x}'}+V(\mathbf{x}',t), \quad \hat{\phi}(\mathbf{x}',t), \tag{15.5.40}$$

and (15.5.37). Inserting (15.5.39) into (15.5.36), we get

$$\mathrm{i}\hbar\dot{\hat{\phi}} = -\frac{\hbar^2}{2m}\nabla^2\hat{\phi}+V(\mathbf{x},t)\hat{\phi}, \tag{15.5.41}$$

which is the quantum-mechanical operator version of (15.5.24).

Since the Schrödinger equation is derived on a process called the first quantization based on the wave-particle duality, canonical quantization, which enables "further quantization" of the Schrödinger equation to get its operator version, is also commonly referred to as the *second quantization* [154, 253, 297].

We next consider a special situation [253] when the potential V is static, $V = V(\mathbf{x})$. For simplicity, we assume that the steady-state Schrödinger equation

$$\left(-\frac{\hbar^2}{2m}\nabla^2_{\mathbf{x}}+V(\mathbf{x},t)\right)u(\mathbf{x}) = Eu(\mathbf{x}) \tag{15.5.42}$$

has a sequence of solutions, $\{(u_i,E_i)\}$, such that $\{u_i\}$ is an orthonormal basis of the L^2 function space over the spatial domain and $0 < E_1 < \cdots < E_i < \cdots$. Thus, the solution of (15.5.24) reads

$$\phi(\mathbf{x},t) = \sum_{i=1}^{\infty} q^i(t)u_i(\mathbf{x}), \tag{15.5.43}$$

where q^i solves

$$\dot{q}^i = -\mathrm{i}\frac{E_i}{\hbar}q^i, \quad i = 1,2,\ldots, \tag{15.5.44}$$

governing a sequence of harmonic oscillators in complex coordinates. Besides, by the relation (15.5.26), if we set

$$\pi(\mathbf{x},t) = \sum_{i=1}^{\infty} p_i(t)\overline{u}_i(\mathbf{x}), \tag{15.5.45}$$

then we have

$$p_i(t) = \mathrm{i}\hbar \overline{q}^i(t), \quad i = 1, 2, \ldots. \tag{15.5.46}$$

From (15.5.43) and (15.5.45), we also have

$$q^i(t) = \int \phi(\mathbf{x}, t)\overline{u}_i(\mathbf{x})\, \mathrm{d}\mathbf{x}, \quad p_i(t) = \int \pi(\mathbf{x}, t)u_i(\mathbf{x})\, \mathrm{d}x. \tag{15.5.47}$$

Promoting these relations to the operator level, we have

$$\hat{\phi}(\mathbf{x}, t) = \sum_{i=1}^{\infty} \hat{q}^i(t)u_i(\mathbf{x}), \quad \hat{\pi}(\mathbf{x}, t) = \sum_{i=1}^{\infty} \hat{p}_i(t)\overline{u}_i(\mathbf{x}), \tag{15.5.48}$$

$$\hat{q}^i(t) = \int \hat{\phi}(\mathbf{x}, t)\overline{u}_i(\mathbf{x})\, \mathrm{d}\mathbf{x}, \quad \hat{p}_i(t) = \int \hat{\pi}(\mathbf{x}, t)u_i(\mathbf{x})\, \mathrm{d}x. \tag{15.5.49}$$

Thus, in view of (15.5.49) and then (15.5.37), we deduce the usual coordinate-momentum commutator:

$$\begin{aligned} [\hat{q}^i(t), \hat{p}_j(t)] &= \int\int [\hat{\phi}(\mathbf{x}, t), \hat{\pi}(\mathbf{x}', t)]\overline{u}_i(\mathbf{x})u_j(\mathbf{x}')\, \mathrm{d}\mathbf{x}\mathrm{d}\mathbf{x}' \\ &= \mathrm{i}\hbar \int\int \delta(\mathbf{x} - \mathbf{x}')\overline{u}_i(\mathbf{x})u_j(\mathbf{x}')\, \mathrm{d}\mathbf{x}\mathrm{d}\mathbf{x}' \\ &= \mathrm{i}\hbar \int \overline{u}_i(\mathbf{x})u_j(\mathbf{x})\, \mathrm{d}\mathbf{x} = \mathrm{i}\hbar\, \delta^i_j. \end{aligned} \tag{15.5.50}$$

On the other hand, inserting (15.5.48) into (15.5.38), we get

$$\hat{H} = \frac{1}{\mathrm{i}\hbar} \sum_{i=1}^{\infty} E_i \hat{p}_i \hat{q}^i. \tag{15.5.51}$$

Thus, by the Heisenberg equation, we have

$$\dot{\hat{q}}^i = \frac{1}{\mathrm{i}\hbar}[\hat{q}^i, \hat{H}] = -\mathrm{i}\frac{E_i}{\hbar}\hat{q}^i, \tag{15.5.52}$$

which is the operator realization of the harmonic oscillator equation (15.5.44). Moreover, since the operator realization of (15.5.46) is

$$(\hat{q}^i)^\dagger = \frac{1}{\mathrm{i}\hbar}\hat{p}_i, \tag{15.5.53}$$

then (15.5.51) becomes

$$\hat{H} = \sum_{i=1}^{\infty} E_i (\hat{q}^i)^\dagger \hat{q}^i. \tag{15.5.54}$$

In view of (15.5.50), we have

$$[\hat{q}^i, (\hat{q}^j)^\dagger] = \delta^{ij}, \quad i, j = 1, 2, \ldots, \tag{15.5.55}$$

such that $(\hat{q}^i)^\dagger$ and $\hat{q}^i$ play the roles of creation and annihilation operators. Hence, $\hat{N}_i = (\hat{q}^i)^\dagger \hat{q}^i$ is a number operator and

$$\hat{H} = \sum_{i=1}^{\infty} E_i \hat{N}_i \tag{15.5.56}$$

gives a clear picture about the energy of the system with regard to the numbers of particles across the entire energy spectrum.

Note that in (15.5.54) the creation operators all appear to the left of annihilation operators. In other words, normal ordering is automatically turned on in the formulation.

15.6 Quantization of electromagnetic fields

This section shows how to quantize source-free electromagnetic fields in free space. For convenience, we assume the speed of light is unity, $c = 1$.

Recall that, if we use A_μ to denote a vector gauge field, then $F_{\mu\nu} = \partial_\mu A_\nu - \partial_\nu A_\mu$ is the generated electromagnetic field tensor such that the electric field $\mathbf{E} = (E^i)$ and magnetic field $\mathbf{B} = (B^i)$ are given by

$$E^i = -F^{0i}, \quad B^i = B_i = -\frac{1}{2}\epsilon_{ijk}F^{jk}, \tag{15.6.1}$$

respectively. For simplicity, we work on the temporal gauge, $A_0 = 0$. Thus, in terms of the temporal and spatial coordinates, $t = x^0$ and $\mathbf{x} = (x^i)$, the fields $\mathbf{E}$ and $\mathbf{B}$ are now induced from $\mathbf{A} = (A^i) = -(A_i)$ by the relations

$$\mathbf{E} = -\frac{\partial \mathbf{A}}{\partial t} \equiv -\dot{\mathbf{A}}, \quad \mathbf{B} = \nabla \times \mathbf{A}. \tag{15.6.2}$$

Besides, the associated Lagrangian action and Hamiltonian energy densities assume the forms

$$\mathcal{L} = \frac{1}{2}(\mathbf{E}^2 - \mathbf{B}^2) = \frac{1}{2}(\dot{\mathbf{A}}^2 - |\nabla \times \mathbf{A}|^2), \tag{15.6.3}$$

$$\mathcal{H} = \frac{1}{2}(\mathbf{E}^2 + \mathbf{B}^2) = \frac{1}{2}(\dot{\mathbf{A}}^2 + |\nabla \times \mathbf{A}|^2), \tag{15.6.4}$$

so that $\mathbf{A}$ is seen to serve the role of a "dynamic variable" of a mechanical system with its momentum

$$\mathbf{P} = \frac{\partial \mathcal{L}}{\partial \dot{\mathbf{A}}} = \dot{\mathbf{A}} = -\mathbf{E}. \tag{15.6.5}$$

From (15.6.3), we have the equation of motion for $\mathbf{A}$:

$$\frac{\partial^2 \mathbf{A}}{\partial t^2} + \nabla \times \nabla \times \mathbf{A} = \mathbf{0}. \tag{15.6.6}$$

Hence, in view of (15.6.2), we get

$$\frac{\partial \mathbf{E}}{\partial t} = \nabla \times \mathbf{B}, \quad \frac{\partial \mathbf{B}}{\partial t} = -\nabla \times \mathbf{E}, \tag{15.6.7}$$

which are the time evolution sectors of the Maxwell equations in our context. Moreover, from (15.6.2) again, we see that the source-free condition consisting of $\nabla \cdot \mathbf{E} = 0$ and $\nabla \cdot \mathbf{B} = 0$ is implied by the Coulomb gauge condition

$$\nabla \cdot \mathbf{A} = 0, \tag{15.6.8}$$

such that (15.6.6) becomes the Klein–Gordon equation

$$\ddot{\mathbf{A}} - \Delta \mathbf{A} = \mathbf{0}. \tag{15.6.9}$$

In other words, the Maxwell equations are now converted into a constrained Klein–Gordon equation. Thus, quantum-mechanically, the propagation of electromagnetic fields may be described by a collection of harmonic oscillators, with constrained oscillations, as well.

Wave motion

As before, we first need to separate the temporal and spatial dependence of the solution of the governing equations (15.6.8)–(15.6.9) in order to isolate the dynamical part of the problem to be quantized. For simplicity and convenience, we take a large spatial cube of the side length $L > 0$ and volume $V = L^3$ and assume that the electromagnetic fields under consideration are L-periodic in all three spatial directions. Thus, we may expand $\mathbf{A}(\mathbf{x}, t)$ in terms of the orthonormal basis consisting of modes

$$\frac{\mathrm{e}^{\mathrm{i}(\mathbf{k}\cdot\mathbf{x})}}{\sqrt{V}}, \quad \mathbf{k} = (k_i), \quad k_i = \frac{2\pi n_i}{L}, \quad n_i \in \mathbb{Z}, \quad i = 1, 2, 3, \tag{15.6.10}$$

satisfying

$$\int \mathrm{e}^{\mathrm{i}(\mathbf{k}\cdot\mathbf{x})}\mathrm{e}^{-\mathrm{i}(\mathbf{k}'\cdot\mathbf{x})}\mathrm{d}\mathbf{x} = V\delta_{\mathbf{k}\mathbf{k}'}, \tag{15.6.11}$$

where (and in the sequel) the integral is evaluated over the spatial cube domain, such that

$$\mathbf{A}(\mathbf{x}, t) = \frac{1}{\sqrt{V}}\sum_{\mathbf{k}} \mathbf{A}_{\mathbf{k}}(t)\mathrm{e}^{\mathrm{i}(\mathbf{k}\cdot\mathbf{x})}. \tag{15.6.12}$$

Inserting (15.6.12) into (15.6.9), we see that $\mathbf{A}_k$ satisfies

$$\ddot{\mathbf{A}}_{\mathbf{k}} + \omega_{\mathbf{k}}^2\mathbf{A}_{\mathbf{k}} = 0, \quad \omega_{\mathbf{k}}^2 = \mathbf{k}^2, \tag{15.6.13}$$

whose solution renders (15.6.12) into

$$\mathbf{A}(\mathbf{x}, t) = \frac{1}{\sqrt{V}}\sum_{\mathbf{k}} \left(\mathbf{A}_{\mathbf{k}}^{(1)}\mathrm{e}^{\mathrm{i}(\mathbf{k}\cdot\mathbf{x}-\omega_{\mathbf{k}}t)} + \mathbf{A}_{\mathbf{k}}^{(2)}\mathrm{e}^{\mathrm{i}(\mathbf{k}\cdot\mathbf{x}+\omega_{\mathbf{k}}t)}\right). \tag{15.6.14}$$

So the solution is a superposition of plane waves, each propagating along the direction vectors $\pm\mathbf{k}$ with angular frequency $\omega_{\mathbf{k}}$.

Transverse mode decomposition

Let $\mathbf{C}$ be a constant vector. Then, we have the useful identities

$$\nabla \cdot \left(\mathbf{C}e^{i(\mathbf{k}\cdot\mathbf{x})}\right) = i(\mathbf{k}\cdot\mathbf{C})\, e^{i(\mathbf{k}\cdot\mathbf{x})}, \quad \nabla \times \left(\mathbf{C}e^{i(\mathbf{k}\cdot\mathbf{x})}\right) = i(\mathbf{k}\times\mathbf{C})\, e^{i(\mathbf{k}\cdot\mathbf{x})}. \quad (15.6.15)$$

Inserting (15.6.12) into (15.6.8) and applying the first identity in (15.6.15), we obtain

$$\mathbf{A}_{\mathbf{k}}(t)\cdot\mathbf{k} = 0. \quad (15.6.16)$$

In the nontrivial situation when $\mathbf{k} \neq \mathbf{0}$, which is the situation of interest for electromagnetism due to (15.6.2) and which is observed here and subsequently, we can find two unit vectors $\mathbf{e}_{\mathbf{k},1}$ and $\mathbf{e}_{\mathbf{k},2}$ in $\mathbb{R}^3$ such that the set $\{\mathbf{e}_{\mathbf{k},1}, \mathbf{e}_{\mathbf{k},2}, \mathbf{k}\}$ forms a right-handed system of orthogonal basis of $\mathbb{R}^3$ (Figure 15.2).

Thus, we have

$$\mathbf{A}_{\mathbf{k}}(t) = Z_{\mathbf{k},1}(t)\,\mathbf{e}_{\mathbf{k},1} + Z_{\mathbf{k},2}(t)\,\mathbf{e}_{\mathbf{k},2}, \quad (15.6.17)$$

where $Z_{\mathbf{k},1}(t), Z_{\mathbf{k},2}(t)$ are two complex-valued scalar functions describing the oscillation of the $\mathbf{k}$th mode of $\mathbf{A}(\mathbf{x},t)$, in (15.6.12), in the plane spanned by $\{\mathbf{e}_{\mathbf{k},1}, \mathbf{e}_{\mathbf{k},2}\}$, perpendicular or transverse to the vector $\mathbf{k}$.

Besides, since (15.6.12) is real, we have

$$\overline{\mathbf{A}}_{-\mathbf{k}}(t) = \mathbf{A}_{\mathbf{k}}(t) \quad \text{or} \quad \mathbf{A}_{-\mathbf{k}}(t) = \overline{\mathbf{A}}_{\mathbf{k}}(t). \quad (15.6.18)$$

As a consequence, we may rewrite (15.6.12) in its "symmetrized" form as

$$\begin{aligned}\mathbf{A}(\mathbf{x},t) &= \frac{1}{2\sqrt{V}}\sum_{\mathbf{k}}(\mathbf{A}_{\mathbf{k}}(t)e^{i(\mathbf{k}\cdot\mathbf{x})} + \mathbf{A}_{-\mathbf{k}}(t)e^{-i(\mathbf{k}\cdot\mathbf{x})}) \\ &= \frac{1}{2\sqrt{V}}\sum_{\mathbf{k}}(\mathbf{A}_{\mathbf{k}}(t)e^{i(\mathbf{k}\cdot\mathbf{x})} + \overline{\mathbf{A}}_{\mathbf{k}}(t)e^{-i(\mathbf{k}\cdot\mathbf{x})}), \end{aligned} \quad (15.6.19)$$

resulting in

$$\mathbf{P}(\mathbf{x},t) = \dot{\mathbf{A}}(\mathbf{x},t) = \frac{1}{2\sqrt{V}}\sum_{\mathbf{k}}(\dot{\mathbf{A}}_{\mathbf{k}}(t)e^{i(\mathbf{k}\cdot\mathbf{x})} + \dot{\overline{\mathbf{A}}}_{\mathbf{k}}(t)e^{-i(\mathbf{k}\cdot\mathbf{x})}), \quad (15.6.20)$$

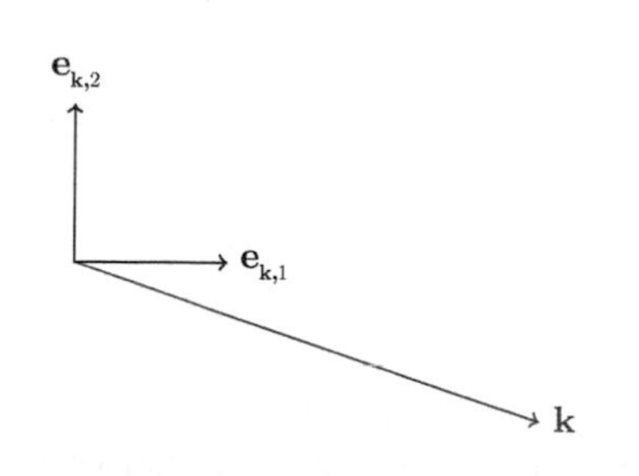

Figure 15.2 Wave propagation the $\mathbf{k}$th mode of a gauge vector field consisting of plane waves with transverse modes oscillating in the plane spanned by $\mathbf{e}_{\mathbf{k},1}, \mathbf{e}_{\mathbf{k},2}$, perpendicular to the direction $\mathbf{k}$ of the $\mathbf{k}$th mode, such that $\{\mathbf{e}_{\mathbf{k},1}, \mathbf{e}_{\mathbf{k},2}, \mathbf{k}\}$ forms a right-handed frame system.

for the canonical momentum vector (15.6.5), which also gives rise to the electric field $\mathbf{E}$. Furthermore, using the second identity in (15.6.15), we have

$$\mathbf{B} = \nabla \times \mathbf{A} = \frac{\mathrm{i}}{2\sqrt{V}} \sum_{\mathbf{k}} \left((\mathbf{k} \times \mathbf{A}_{\mathbf{k}}(t)) \mathrm{e}^{\mathrm{i}(\mathbf{k}\cdot\mathbf{x})} - (\mathbf{k} \times \overline{\mathbf{A}}_{\mathbf{k}}(t)) \mathrm{e}^{-\mathrm{i}(\mathbf{k}\cdot\mathbf{x})} \right). \quad (15.6.21)$$

In view of (15.6.17) and (15.6.18), we also have

$$\overline{Z}_{\mathbf{k},1} = Z_{-\mathbf{k},1}, \quad \overline{Z}_{\mathbf{k},2} = Z_{-\mathbf{k},2}. \quad (15.6.22)$$

Mode representation

From (15.6.11), (15.6.18), (15.6.20), and (15.6.22), we have

$$\begin{aligned} \int \mathbf{P}^2 \, \mathrm{d}\mathbf{x} = \int \dot{\mathbf{A}}^2 \, \mathrm{d}\mathbf{x} &= \sum_{\mathbf{k}} \dot{\mathbf{A}}_{\mathbf{k}}(t) \cdot \dot{\overline{\mathbf{A}}}_{\mathbf{k}}(t) \\ &= \sum_{\mathbf{k}} \left(|\dot{Z}_{\mathbf{k},1}|^2 + |\dot{Z}_{\mathbf{k},2}|^2 \right). \end{aligned} \quad (15.6.23)$$

Moreover, using the vector identity

$$(\mathbf{a} \times \mathbf{b}) \cdot (\mathbf{c} \times \mathbf{d}) = (\mathbf{a} \cdot \mathbf{c})(\mathbf{b} \cdot \mathbf{d}) - (\mathbf{a} \cdot \mathbf{d})(\mathbf{b} \cdot \mathbf{c}), \quad (15.6.24)$$

we obtain from (15.6.21) the result

$$\begin{aligned} \int \mathbf{B}^2 \, \mathrm{d}\mathbf{x} = \int |\nabla \times \mathbf{A}|^2 \, \mathrm{d}\mathbf{x} &= \sum_{\mathbf{k}} \mathbf{k}^2 \mathbf{A}_{\mathbf{k}}(t) \overline{\mathbf{A}}_{\mathbf{k}}(t) \\ &= \sum_{\mathbf{k}} \mathbf{k}^2 \left(|Z_{\mathbf{k},1}|^2 + |Z_{\mathbf{k},2}|^2 \right), \end{aligned} \quad (15.6.25)$$

by virtue of the constraint (15.6.16).

In view of (15.6.3), (15.6.4), (15.6.23), and (15.6.25), we arrive at the Lagrangian action and Hamiltonian energy

$$L = \frac{1}{2} \sum_{\mathbf{k}} \left(|\dot{Z}_{\mathbf{k},1}|^2 + |\dot{Z}_{\mathbf{k},2}|^2 \right) - \frac{1}{2} \sum_{\mathbf{k}} \mathbf{k}^2 \left(|Z_{\mathbf{k},1}|^2 + |Z_{\mathbf{k},2}|^2 \right), \quad (15.6.26)$$

$$H = \frac{1}{2} \sum_{\mathbf{k}} \left(|\dot{Z}_{\mathbf{k},1}|^2 + |\dot{Z}_{\mathbf{k},2}|^2 \right) + \frac{1}{2} \sum_{\mathbf{k}} \mathbf{k}^2 \left(|Z_{\mathbf{k},1}|^2 + |Z_{\mathbf{k},2}|^2 \right), \quad (15.6.27)$$

respectively, which clearly demonstrate how, classically, electromagnetic waves in free space may be regarded as a superposition of a collection of decoupled harmonic oscillators of two degrees of freedom, in terms of real coordinate variables $X_{\mathbf{k},a}, Y_{\mathbf{k},a}$, with

$$Z_{\mathbf{k},a} = X_{\mathbf{k},a} + \mathrm{i} Y_{\mathbf{k},a}, \quad a = 1, 2. \quad (15.6.28)$$

Quantization

Of course, we may quantize the system within such a real-variable setting. However, since we have used complex formalism to represent **E** and **B**, it is convenient to continue to work within the complex-variable setting. To do so, we adopt our formulation from the previous section for a single harmonic oscillator to set

$$q_{\mathbf{k},a} = \frac{1}{\sqrt{2}}(X_{\mathbf{k},a} + \mathrm{i}Y_{\mathbf{k},a}), \quad \overline{q}_{\mathbf{k},a} = \frac{1}{\sqrt{2}}(X_{\mathbf{k},a} - \mathrm{i}Y_{\mathbf{k},a}), \tag{15.6.29}$$

$$p_{\mathbf{k},a} = \frac{1}{\sqrt{2}}(\dot{X}_{\mathbf{k},a} - \mathrm{i}\dot{Y}_{\mathbf{k},a}), \quad \overline{p}_{\mathbf{k},a} = \frac{1}{\sqrt{2}}(\dot{X}_{\mathbf{k},a} + \mathrm{i}\dot{Y}_{\mathbf{k},a}). \tag{15.6.30}$$

Hence, (15.6.27) becomes

$$H = \frac{1}{2}\sum_{\mathbf{k}}\sum_{a=1}^{2}\left(p_{\mathbf{k},a}^2 + \overline{p}_{\mathbf{k},a}^2 + k^2[q_{\mathbf{k},a}^2 + \overline{q}_{\mathbf{k},a}^2]\right), \quad k^2 = \mathbf{k}^2. \tag{15.6.31}$$

Consequently, applying (15.5.14), with q being updated by $q_{\mathbf{k},a}$, etc., and using (15.5.18), with ω being replaced by k, we obtain two sequences of creation and annihilation operators, $\{\hat{a}_{\mathbf{k},a}^{\pm}\}$ and $\{\hat{\overline{a}}_{\mathbf{k},a}^{\pm}\}$, denoted now by the superscripts $\pm$, respectively, with the nontrivial commutators

$$[\hat{a}_{\mathbf{k},a}^{-}, \hat{a}_{\mathbf{k},a}^{+}] = 1, \quad [\hat{\overline{a}}_{\mathbf{k},a}^{-}, \hat{\overline{a}}_{\mathbf{k},a}^{+}] = 1, \tag{15.6.32}$$

such that the dynamical variables are quantized according to

$$\hat{p}_{\mathbf{k},a} = \sqrt{\frac{\hbar k}{2}}\left(\hat{a}_{\mathbf{k},a}^{-} + \hat{a}_{\mathbf{k},a}^{+}\right), \quad \hat{q}_{\mathbf{k},a} = \mathrm{i}\sqrt{\frac{\hbar}{2k}}\left(\hat{a}_{\mathbf{k},a}^{-} - \hat{a}_{\mathbf{k},a}^{+}\right), \tag{15.6.33}$$

$$\hat{\overline{p}}_{\mathbf{k},a} = \sqrt{\frac{\hbar k}{2}}\left(\hat{\overline{a}}_{\mathbf{k},a}^{-} + \hat{\overline{a}}_{\mathbf{k},a}^{+}\right), \quad \hat{\overline{q}}_{\mathbf{k},a} = \mathrm{i}\sqrt{\frac{\hbar}{2k}}\left(\hat{\overline{a}}_{\mathbf{k},a}^{-} - \hat{\overline{a}}_{\mathbf{k},a}^{+}\right), \tag{15.6.34}$$

resulting in the quantized Hamiltonian

$$\hat{H} = \frac{\hbar}{2}\sum_{\mathbf{k}}\sum_{a=1}^{2} k\left(\hat{a}_{\mathbf{k},a}^{-}\hat{a}_{\mathbf{k},a}^{+} + \hat{a}_{\mathbf{k},a}^{+}\hat{a}_{\mathbf{k},a}^{-} + \hat{\overline{a}}_{\mathbf{k},a}^{-}\hat{\overline{a}}_{\mathbf{k},a}^{+} + \hat{\overline{a}}_{\mathbf{k},a}^{+}\hat{\overline{a}}_{\mathbf{k},a}^{-}\right). \tag{15.6.35}$$

Formally, this expression may be rewritten as

$$\hat{H} = \hbar\sum_{\mathbf{k}}\sum_{a=1}^{2} k\left(\hat{a}_{\mathbf{k},a}^{+}\hat{a}_{\mathbf{k},a}^{-} + \hat{\overline{a}}_{\mathbf{k},a}^{+}\hat{\overline{a}}_{\mathbf{k},a}^{-}\right) + \hbar\sum_{\mathbf{k}}\sum_{a=1}^{2} k, \tag{15.6.36}$$

in which the second term on the right-hand side, the zero-point energy, diverges. However, this issue does not concern us because it may always be truncated off either by hand or through renormalization by normal ordering during the process of quantization:

$$\begin{aligned}\hat{H} &= \frac{1}{2}\sum_{\mathbf{k}}\sum_{a=1}^{2}\left(:\hat{p}_{\mathbf{k},a}^2: + :\hat{\overline{p}}_{\mathbf{k},a}^2: + k^2\left[:\hat{q}_{\mathbf{k},a}^2: + :\hat{\overline{q}}_{\mathbf{k},a}^2:\right]\right) \\ &= \hbar\sum_{\mathbf{k}}\sum_{a=1}^{2} k\left(\hat{a}_{\mathbf{k},a}^{+}\hat{a}_{\mathbf{k},a}^{-} + \hat{\overline{a}}_{\mathbf{k},a}^{+}\hat{\overline{a}}_{\mathbf{k},a}^{-}\right) \\ &= \hbar\sum_{\mathbf{k}}\sum_{a=1}^{2} k\left(\hat{N}_{\mathbf{k},a,1} + \hat{N}_{\mathbf{k},a,2}\right),\end{aligned} \tag{15.6.37}$$

where $\hat{N}_{\mathbf{k},a,1} = \hat{a}_{\mathbf{k},a}^{+}\hat{a}_{\mathbf{k},a}^{-}$ and $\hat{N}_{\mathbf{k},a,2} = \hat{\overline{a}}_{\mathbf{k},a}^{+}\hat{\overline{a}}_{\mathbf{k},a}^{-}$ are the number operators associated with the particles corresponding to the classical dynamic variables $(q_{\mathbf{k},a}, p_{\mathbf{k},a})$ and $(\overline{q}_{\mathbf{k},a}, \overline{p}_{\mathbf{k},a})$, respectively.

The equation (15.6.37) is a quantum-field theory confirmation of the Planck–Einstein formula (2.1.2) for the photoelectric effect.

Electromagnetic momentum

The electromagnetic momentum vector density $\mathbf{S}$ is given by $\mathbf{S} = (T^{0i}) = -(T_{0i})$ where

$$T_{\mu\nu} = -F_{\mu\alpha}\eta^{\alpha\beta}F_{\nu\beta} - \eta_{\mu\nu}\mathcal{L}, \tag{15.6.38}$$

is the usual energy-momentum tensor associated with the electromagnetic field $F_{\mu\nu}$ under discussion. Hence, inserting (15.6.1), we get

$$\mathbf{S} = \mathbf{E} \times \mathbf{B}, \tag{15.6.39}$$

which happens to be the Poynting vector. Thus, the total *electromagnetic momentum* is

$$\mathbf{P}_{\text{EM}} = \int \mathbf{S}\,\mathrm{d}\mathbf{x} = \int (\mathbf{E} \times \mathbf{B})\,\mathrm{d}\mathbf{x}. \tag{15.6.40}$$

In order to facilitate the calculation, we note that the general solution of (15.6.13) is spanned by the solutions of $\dot{\mathbf{A}}_{\mathbf{k}} = \pm\mathrm{i}\omega_{\mathbf{k}}\mathbf{A}_{\mathbf{k}} = \pm\mathrm{i}k\mathbf{A}_{\mathbf{k}}$. So we may take the lower sign case,

$$\dot{\mathbf{A}}_{\mathbf{k}} = -\mathrm{i}k\mathbf{A}_{\mathbf{k}}, \tag{15.6.41}$$

to rewrite electric field $\mathbf{E}$, in view of (15.6.20) and (15.6.41), as

$$\mathbf{E} = -\dot{\mathbf{A}} = \frac{\mathrm{i}}{2\sqrt{V}}\sum_{\mathbf{k}} k\left(\mathbf{A}_{\mathbf{k}}(t)\mathrm{e}^{\mathrm{i}(\mathbf{k}\cdot\mathbf{x})} - \overline{\mathbf{A}}_{\mathbf{k}}(t)\mathrm{e}^{-\mathrm{i}(\mathbf{k}\cdot\mathbf{x})}\right). \tag{15.6.42}$$

Consequently, in view of (15.6.5), (15.6.11), (15.6.18), (15.6.20), (15.6.21), and (15.6.42), we obtain from (15.6.40) the expression

$$\mathbf{P}_{\text{EM}} = \sum_{\mathbf{k}} k\,\mathbf{A}_{\mathbf{k}} \times (\mathbf{k} \times \overline{\mathbf{A}}_{\mathbf{k}}). \tag{15.6.43}$$

Furthermore, applying the vector identity

$$\mathbf{a} \times (\mathbf{b} \times \mathbf{c}) = (\mathbf{a} \cdot \mathbf{c})\mathbf{b} - (\mathbf{a} \cdot \mathbf{b})\mathbf{c}, \tag{15.6.44}$$

and using the transverse condition (15.6.16), we may recast (15.6.43) into

$$\begin{aligned} \mathbf{P}_{\text{EM}} &= \sum_{\mathbf{k}} k\mathbf{k}(\mathbf{A}_{\mathbf{k}} \cdot \overline{\mathbf{A}}_{\mathbf{k}}) \\ &= \sum_{\mathbf{k}} \sum_{a=1}^{2} k\mathbf{k} \left| Z_{\mathbf{k},a} \right|^2 = \sum_{\mathbf{k}} \sum_{a=1}^{2} k\mathbf{k} \left(X_{\mathbf{k},a}^2 + Y_{\mathbf{k},a}^2 \right). \end{aligned} \tag{15.6.45}$$

On the other hand, note that (15.6.41) leads to $\dot{X}_{\mathbf{k},a} = kY_{\mathbf{k},a}$, $\dot{Y}_{\mathbf{k},a} = -kX_{\mathbf{k},a}$, which renders (15.6.45) into the form

$$\begin{aligned} \mathbf{P}_{\text{EM}} &= \frac{1}{2} \sum_{\mathbf{k}} \sum_{a=1}^{2} \mathbf{k} \left(\frac{1}{k} \left[\dot{X}_{\mathbf{k},a}^2 + \dot{Y}_{\mathbf{k},a}^2 \right] + k \left[X_{\mathbf{k},a}^2 + Y_{\mathbf{k},a}^2 \right] \right) \\ &= \frac{1}{2} \sum_{\mathbf{k}} \sum_{a=1}^{2} \mathbf{k} \left(\frac{1}{k} \left[p_{\mathbf{k},a}^2 + \overline{p}_{\mathbf{k},a}^2 \right] + k \left[q_{\mathbf{k},a}^2 + \overline{q}_{\mathbf{k},a}^2 \right] \right), \end{aligned} \tag{15.6.46}$$

using (15.6.29)–(15.6.30). Therefore, from (15.6.33)–(15.6.34), we obtain the quantized version of (15.6.46):

$$\begin{aligned} \hat{\mathbf{P}}_{\text{EM}} &= \frac{1}{2} \sum_{\mathbf{k}} \sum_{a=1}^{2} \mathbf{k} \left(\frac{1}{k} \left[\hat{p}_{\mathbf{k},a}^2 + \hat{\overline{p}}_{\mathbf{k},a}^2 \right] + k \left[\hat{q}_{\mathbf{k},a}^2 + \hat{\overline{q}}_{\mathbf{k},a}^2 \right] \right) \\ &= \frac{\hbar}{2} \sum_{\mathbf{k}} \sum_{a=1}^{2} \mathbf{k} \left(\hat{a}_{\mathbf{k},a}^- \hat{a}_{\mathbf{k},a}^+ + \hat{a}_{\mathbf{k},a}^+ \hat{a}_{\mathbf{k},a}^- + \hat{\overline{a}}_{\mathbf{k},a}^- \hat{\overline{a}}_{\mathbf{k},a}^+ + \hat{\overline{a}}_{\mathbf{k},a}^+ \hat{\overline{a}}_{\mathbf{k},a}^- \right) \\ &= \hbar \sum_{\mathbf{k}} \sum_{a=1}^{2} \mathbf{k} \left(\hat{a}_{\mathbf{k},a}^+ \hat{a}_{\mathbf{k},a}^- + \hat{\overline{a}}_{\mathbf{k},a}^+ \hat{\overline{a}}_{\mathbf{k},a}^- + 1 \right) \\ &= \hbar \sum_{\mathbf{k}} \sum_{a=1}^{2} \mathbf{k} \left(\hat{N}_{\mathbf{k},a,1} + \hat{N}_{\mathbf{k},a,2} \right), \end{aligned} \tag{15.6.47}$$

when applying (15.6.32), where the constant term is dropped, since its collected contribution to the sum is zero, and $\hat{N}_{\mathbf{k},a,1}$ and $\hat{N}_{\mathbf{k},a,2}$ are the number operators encountered before when calculating the quantized Hamiltonian. Note that, here, there is no need to do renormalization by normal ordering in order to avoid divergence.

The equation (15.6.47) is a quantum-field theory confirmation of the Compton–Debye formula (2.1.7) for the Compton effect.

Spin

Similar to the *orbital angular momentum* about the origin of a moving particle of mass m and position vector $\mathbf{x}$ given by the expression $\mathbf{L} = m\mathbf{x} \times \dot{\mathbf{x}}$, the *spin angular momentum* of the electromagnetic fields under consideration is defined analogously by

$$\mathbf{L}_{\text{spin}} = \int (\mathbf{A} \times \dot{\mathbf{A}})\,\mathrm{d}\mathbf{x} = -\int (\mathbf{A} \times \mathbf{E})\,\mathrm{d}\mathbf{x}. \tag{15.6.48}$$

Thus, inserting (15.6.19) and (15.6.42) into (15.6.48), and applying (15.6.11), we have

$$\mathbf{L}_{\text{spin}} = \mathrm{i}\sum_{\mathbf{k}} \mathbf{k}\left(Z_{\mathbf{k},1}\overline{Z}_{\mathbf{k},2} - \overline{Z}_{\mathbf{k},1}Z_{\mathbf{k},2}\right), \tag{15.6.49}$$

where we have used the fact that $\{\mathbf{e}_{\mathbf{k},1}, \mathbf{e}_{\mathbf{k},2}, \mathbf{k}\}$ forms a right-handed orthogonal basis of $\mathbb{R}^3$ for $\mathbf{k} \neq \mathbf{0}$. On the other hand, in view of (15.6.29), (15.6.30), and (15.6.41), we have

$$Z_{\mathbf{k},a} = \sqrt{2}q_{\mathbf{k},a}, \quad \overline{Z}_{\mathbf{k},a} = -\mathrm{i}\frac{\sqrt{2}}{k}p_{\mathbf{k},a}, \quad \mathbf{k} \neq \mathbf{0}, \tag{15.6.50}$$

which recasts (15.6.49) into the form

$$\mathbf{L}_{\text{spin}} = \sum_{\mathbf{k}} \hat{\mathbf{k}}\left((q_{\mathbf{k},1}p_{\mathbf{k},2} - p_{\mathbf{k},1}q_{\mathbf{k},2}) + (\overline{q}_{\mathbf{k},1}\overline{p}_{\mathbf{k},2} - \overline{p}_{\mathbf{k},1}\overline{q}_{\mathbf{k},2})\right), \quad \mathbf{k} = k\hat{\mathbf{k}}. \tag{15.6.51}$$

Thus, we can now quantize (15.6.51), following (15.6.33)–(15.6.34), to obtain

$$\begin{aligned}
\hat{\mathbf{L}}_{\text{spin}} &= \sum_{\mathbf{k}} \hat{\mathbf{k}}\left((\hat{q}_{\mathbf{k},1}\hat{p}_{\mathbf{k},2} - \hat{p}_{\mathbf{k},1}\hat{q}_{\mathbf{k},2}) + (\hat{\overline{q}}_{\mathbf{k},1}\hat{\overline{p}}_{\mathbf{k},2} - \hat{\overline{p}}_{\mathbf{k},1}\hat{\overline{q}}_{\mathbf{k},2})\right) \\
&= \mathrm{i}\hbar\sum_{\mathbf{k}} \hat{\mathbf{k}}\left((\hat{a}^-_{\mathbf{k},1}\hat{a}^+_{\mathbf{k},2} - \hat{a}^+_{\mathbf{k},1}\hat{a}^-_{\mathbf{k},2}) + (\hat{\overline{a}}^-_{\mathbf{k},1}\hat{\overline{a}}^+_{\mathbf{k},2} - \hat{\overline{a}}^+_{\mathbf{k},1}\hat{\overline{a}}^-_{\mathbf{k},2})\right).
\end{aligned} \tag{15.6.52}$$

In order to relate (15.6.52) to creation and annihilation operators, we "diagonalize" the operators, $\{\hat{a}^\pm_{\mathbf{k},a}\}$, by the unitary transformations

$$\hat{\alpha}^-_{\mathbf{k}} = \frac{1}{\sqrt{2}}(\hat{a}^-_{\mathbf{k},1} - \mathrm{i}\hat{a}^-_{\mathbf{k},2}), \quad \hat{\alpha}^+_{\mathbf{k}} = \frac{1}{\sqrt{2}}(\hat{a}^+_{\mathbf{k},1} + \mathrm{i}\hat{a}^+_{\mathbf{k},2}), \tag{15.6.53}$$

$$\hat{\beta}^-_{\mathbf{k}} = \frac{1}{\sqrt{2}}(\hat{a}^-_{\mathbf{k},1} + \mathrm{i}\hat{a}^-_{\mathbf{k},2}), \quad \hat{\beta}^+_{\mathbf{k}} = \frac{1}{\sqrt{2}}(\hat{a}^+_{\mathbf{k},1} - \mathrm{i}\hat{a}^+_{\mathbf{k},2}), \tag{15.6.54}$$

and the correspondingly similar ones for $\{\hat{\overline{a}}^\pm_{\mathbf{k},a}\}$. Then, we see that the nontrivial commutators associated with these operators are

$$[\hat{\alpha}^-_{\mathbf{k}}, \hat{\alpha}^+_{\mathbf{k}}] = 1, \quad [\hat{\beta}^-_{\mathbf{k}}, \hat{\beta}^+_{\mathbf{k}}] = 1, \quad [\hat{\overline{\alpha}}^-_{\mathbf{k}}, \hat{\overline{\alpha}}^+_{\mathbf{k}}] = 1, \quad [\hat{\overline{\beta}}^-_{\mathbf{k}}, \hat{\overline{\beta}}^+_{\mathbf{k}}] = 1, \tag{15.6.55}$$

in view of (15.6.32), such that $\hat{\alpha}^+_{\mathbf{k}}, \hat{\beta}^+_{\mathbf{k}}, \hat{\overline{\alpha}}^+_{\mathbf{k}}, \hat{\overline{\beta}}^+_{\mathbf{k}}$ and $\hat{\alpha}^-_{\mathbf{k}}, \hat{\beta}^-_{\mathbf{k}}, \hat{\overline{\alpha}}^-_{\mathbf{k}}, \hat{\overline{\beta}}^-_{\mathbf{k}}$ are creation and annihilation operators, respectively, in pairs correspondingly labeled by the superscripts $\pm$.

Substituting (15.6.53)–(15.6.54) into (15.6.52), and applying (15.6.55), we obtain

$$\begin{aligned}\hat{\mathbf{L}}_{\text{spin}} &= \hbar \sum_{\mathbf{k}} \hat{\mathbf{k}} \left(\left[\hat{\alpha}^+_{\mathbf{k}} \hat{\alpha}^-_{\mathbf{k}} - \hat{\beta}^+_{\mathbf{k}} \hat{\beta}^-_{\mathbf{k}} \right] + \left[\hat{\overline{\alpha}}^+_{\mathbf{k}} \hat{\overline{\alpha}}^-_{\mathbf{k}} - \hat{\overline{\beta}}^+_{\mathbf{k}} \hat{\overline{\beta}}^-_{\mathbf{k}} \right] \right) \\ &= \hbar \sum_{\mathbf{k}} \hat{\mathbf{k}} \left(\hat{N}_{\mathbf{k},\alpha,1} - \hat{N}_{\mathbf{k},\beta,1} + \hat{N}_{\mathbf{k},\alpha,2} - \hat{N}_{\mathbf{k},\beta,2} \right), \end{aligned} \tag{15.6.56}$$

where $\hat{N}_{\mathbf{k},\alpha,1}, \hat{N}_{\mathbf{k},\beta,1}, \hat{N}_{\mathbf{k},\alpha,2}, \hat{N}_{\mathbf{k},\beta,2}$ are the number operators describing the α, β, unbarred, and barred quanta, or particles, respectively.

Thus, the expression (15.6.56) establishes that, when the quanta describing quantized electromagnetic fields are interpreted as photons, the spin angular momentum of a single photon is of the unit amplitude $\hbar$ along the directions $\pm\hat{\mathbf{k}}$. For this reason, photons are said to be spin-1 particles.

Field quantization

From (15.6.17), (15.6.19), and (15.6.29), we have

$$\mathbf{A}(\mathbf{x}, t) = \frac{1}{\sqrt{2V}} \sum_{\mathbf{k}} \left(q_{\mathbf{k},a} \mathrm{e}^{\mathrm{i}(\mathbf{x}\cdot\mathbf{k})} + \overline{q}_{\mathbf{k},a} \mathrm{e}^{-\mathrm{i}(\mathbf{k}\cdot\mathbf{x})} \right) \mathbf{e}_{\mathbf{k},a}. \tag{15.6.57}$$

Thus, following (15.6.33) and (15.6.34), we obtain

$$\hat{\mathbf{A}}(\mathbf{x}, t) = \frac{\mathrm{i}}{2} \sqrt{\frac{\hbar}{V}} \sum_{\mathbf{k}} \frac{1}{\sqrt{k}} \left((\hat{a}^-_{\mathbf{k},a} - \hat{a}^+_{\mathbf{k},a}) \mathrm{e}^{\mathrm{i}(\mathbf{k}\cdot\mathbf{x})} + (\hat{\overline{a}}^-_{\mathbf{k},a} - \hat{\overline{a}}^+_{\mathbf{k},a}) \mathrm{e}^{-\mathrm{i}(\mathbf{k}\cdot\mathbf{x})} \right) \mathbf{e}_{\mathbf{k},a}, \tag{15.6.58}$$

as the quantized gauge field.

Quantized electric and magnetic fields in terms of creation and annihilation operators may be similarly obtained, and are not discussed here.

For canonical quantization of the Dirac fields and massive electromagnetic fields, see [154, 169, 253, 299] for some treatments.

Note that, unlike traditional treatments [154, 169, 253, 299], in this section, our formulation is carried out as far as possible within classical formalism such that quantization is made directly on the quantities of interest, that is, energy, momentum, and spin angular momentum, without resorting to quantized gauge and electromagnetic fields.

15.7 Thermodynamics of harmonic oscillator

Previous sections showed how to quantize a system governed by a wave equation, which provided information about the energy spectrum of the system. Here, we show how to use the information obtained to understand some thermodynamic properties of the problem. For simplicity and conciseness, we concentrate on a one-degree of freedom harmonic oscillator because it is the basic building block

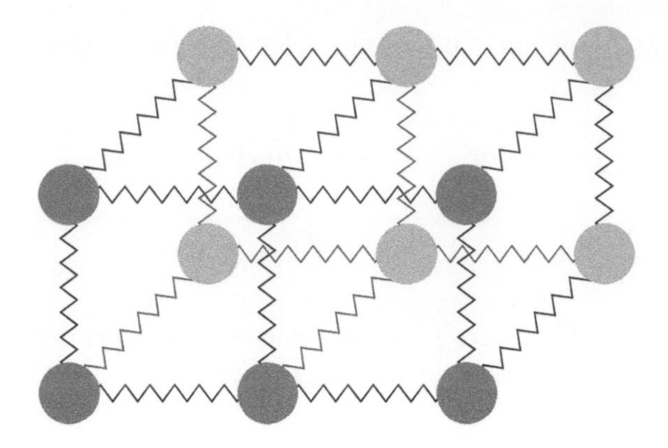

Figure 15.3 The Einstein solid modeled by a uniform array of independent but elastically identical harmonic oscillators of three degrees of freedom.

of all general wave motions. Furthermore, three-degrees of freedom harmonic oscillators are of fundamental importance in solid-state physics: an idealized solid called the *Einstein solid*, for example, is a solid model assuming that each atom in the solid crystal lattice behaves like a harmonic oscillator and that all atoms oscillate with the same angular frequency (Figure 15.3). Here, we compare the thermodynamics of a harmonic oscillator both classically and quantum-mechanically.

Classical system

For the Hamiltonian energy of a one-degree of freedom harmonic oscillator given in terms of canonical coordinates q, p by (15.1.8) or

$$H = H(q,p) = \frac{p^2}{2m} + \frac{m\omega^2}{2}q^2, \tag{15.7.1}$$

we use (1.5.20) to compute the associated partition function Z to get

$$\begin{aligned} Z &= \int\int \mathrm{e}^{-\beta\left(\frac{p^2}{2m}+\frac{m\omega^2}{2}q^2\right)}\,\mathrm{d}q\mathrm{d}p \\ &= \frac{2\pi}{\beta\omega}, \end{aligned} \tag{15.7.2}$$

where β is the inverse temperature as defined in (1.5.2). Hence, we may follow (1.5.21) to obtain the average or thermodynamic energy of the system:

$$\begin{aligned} U = \langle H \rangle &= \frac{1}{Z}\int\int H(q,p)\mathrm{e}^{-\beta\left(\frac{p^2}{2m}+\frac{m\omega^2}{2}q^2\right)}\,\mathrm{d}q\mathrm{d}p \\ &= -\frac{\partial \ln Z}{\partial \beta} = k_{\mathrm{B}}T, \end{aligned} \tag{15.7.3}$$

which relates the thermodynamic energy directly to the absolute temperature, T, but is independent of the frequency of the oscillation, ω. From (1.5.6) and (15.7.3), we obtain the associated heat capacity

$$C_v = \frac{\partial U}{\partial T} = k_{\mathrm{B}}, \tag{15.7.4}$$

which is the Boltzmann constant. Besides, from (1.5.8) and (15.7.2), we may write down the Helmholtz free energy

$$A = -k_{\mathrm{B}} T \ln Z = -k_{\mathrm{B}} T \ln \left(\frac{2\pi k_{\mathrm{B}} T}{\omega} \right). \tag{15.7.5}$$

Thus, from (1.5.7) , we see that the thermodynamic entropy of the system is

$$S = -\frac{\partial A}{\partial T} = k_{\mathrm{B}} (1 + \ln Z) = k_{\mathrm{B}} \left(1 + \ln \left[\frac{2\pi}{\beta\omega} \right] \right), \tag{15.7.6}$$

which depends on the absolute temperature and oscillation frequency in such a way that the inverse temperature and oscillation frequency play equivalent roles.

Quantum system

In this situation, the full energy spectrum is given by (15.1.35). Thus, by (1.5.1) the partition function is

$$\begin{aligned} Z &= \sum_{n=0}^{\infty} \mathrm{e}^{-\beta\left(n+\frac{1}{2}\right)\hbar\omega} \\ &= \frac{\mathrm{e}^{-\beta\frac{\hbar\omega}{2}}}{1 - \mathrm{e}^{-\beta\hbar\omega}} = \frac{1}{2\sinh\left(\beta\frac{\hbar\omega}{2}\right)}. \end{aligned} \tag{15.7.7}$$

Consequently, we obtain the thermodynamic energy

$$U = \langle E \rangle = -\frac{\partial \ln Z}{\partial \beta} = \hbar\omega \left(\frac{1}{2} + \frac{1}{\mathrm{e}^{\beta\hbar\omega} - 1} \right), \tag{15.7.8}$$

which looks complicated. However, an immediate result is

$$\langle E \rangle > E_0 = \frac{\hbar\omega}{2}. \tag{15.7.9}$$

Namely, the thermodynamic energy is always above the zero-point energy, which is naturally expected. Besides, we have the limiting values

$$\lim_{\beta\to\infty} \langle E \rangle = \frac{\hbar\omega}{2}, \tag{15.7.10}$$

$$\lim_{\beta\to 0} \beta \langle E \rangle = 1. \tag{15.7.11}$$

The first result, (15.7.10), indicates that the thermodynamic energy approaches the zero-point energy at zero-temperature limit, and the second result, (15.7.11), reveals that in high temperature the quantum thermodynamic energy depends on temperature following approximately the classical formula, (15.7.3).

By (1.5.6) and (15.7.8), we obtain the heat capacity to be

$$C_v(\beta) = \frac{\partial U}{\partial T} = -\frac{1}{k_{\mathrm{B}}T^2}\frac{\partial U}{\partial \beta} = k_{\mathrm{B}}\frac{(\beta\hbar\omega)^2 \mathrm{e}^{\beta\hbar\omega}}{(\mathrm{e}^{\beta\hbar\omega}-1)^2}. \tag{15.7.12}$$

Since there holds

$$\lim_{\beta\to 0} C_v(\beta) = k_{\mathrm{B}}, \tag{15.7.13}$$

so the quantum heat capacity approaches that of classical one at the high-temperature limit. It is also clear that

$$\lim_{\beta\to\infty} C_v(\beta) = 0. \tag{15.7.14}$$

Thus, we see that quantum theory departs sharply from classical theory at low temperatures.

Sometimes (15.7.12) is also rewritten as

$$C_v(\beta) = k_{\mathrm{B}}\left(\frac{T_{\mathrm{E}}}{T}\right)^2 \frac{\mathrm{e}^{\frac{T_{\mathrm{E}}}{T}}}{\left(\mathrm{e}^{\frac{T_{\mathrm{E}}}{T}}-1\right)^2} = k_{\mathrm{B}}\left(\frac{T_{\mathrm{E}}}{2T}\right)^2 \frac{1}{\sinh^2\left(\frac{T_{\mathrm{E}}}{2T}\right)}, \tag{15.7.15}$$

where the quantity

$$T_{\mathrm{E}} = \frac{\hbar\omega}{k_{\mathrm{B}}} \tag{15.7.16}$$

depends only on the oscillation frequency ω, has the dimension of temperature, and is referred to as the *Einstein temperature* [334, 391, 497], which serves as a basic unit of temperature and may characterize the underlying solid crystal in consideration. Note that, for T small, (15.7.15) gives the result

$$C_v \sim k_{\mathrm{B}}\left(\frac{T_{\mathrm{E}}}{T}\right)^2 \mathrm{e}^{-\frac{T_{\mathrm{E}}}{T}}, \tag{15.7.17}$$

which vanishes faster than any power of T as $T \to 0$ and significantly deviates from the experimental observation $C_v \sim T^3$, commonly known as the T^3-law in solid-state physics. A modification of the Einstein theory, called the *Debye theory*, successfully resolves this discrepancy [334, 391, 419].

Furthermore, solving (15.7.8), we get

$$\beta = \frac{1}{\hbar\omega}\ln\left(1 + \left[\frac{\langle E\rangle}{\hbar\omega} - \frac{1}{2}\right]^{-1}\right), \tag{15.7.18}$$

which clearly indicates that, at the quantum level, the temperature of the system depends on both its thermodynamic energy and oscillation frequency.

Moreover, inserting (15.7.7) into (1.5.8), we find

$$A = -\frac{1}{\beta}\ln Z = \frac{\hbar\omega}{2} + \frac{1}{\beta}\ln\left(1 - \mathrm{e}^{-\beta\hbar\omega}\right). \tag{15.7.19}$$

Thus, substituting (15.7.19) into (1.5.7), we obtain the entropy of the quantum harmonic oscillator to be

$$S = -\frac{\partial A}{\partial T} = k_{\mathrm{B}}\left(\frac{\beta\hbar\omega}{\mathrm{e}^{\beta\hbar\omega} - 1} - \ln\left[1 - \mathrm{e}^{-\beta\hbar\omega}\right]\right), \tag{15.7.20}$$

which is a monotone increasing function of T since

$$\frac{\partial S}{\partial \beta} = -k_{\mathrm{B}}\frac{\beta(\hbar\omega)^2\mathrm{e}^{\beta\hbar\omega}}{(\mathrm{e}^{\beta\hbar\omega} - 1)^2}. \tag{15.7.21}$$

Besides, we have the limits

$$\lim_{\beta\to 0} S = \infty, \quad \lim_{\beta\to\infty} S = 0. \tag{15.7.22}$$

These results clearly describe how the entropy of a quantum harmonic oscillator depends on its temperature. The same may be said about the entropy with respect to the oscillation frequency ω, since the roles of β and ω are interchangeable in (15.7.20).

It will be interesting to compare the classical entropy (15.7.6), now denoted by $S_{\mathrm{C}}(\beta)$, and the quantum one (15.7.20), now by $S_{\mathrm{Q}}(\beta)$. In fact, although these quantities look very different, there holds the limit

$$\lim_{\beta\to 0}\frac{S_{\mathrm{C}}(\beta)}{S_{\mathrm{Q}}(\beta)} = 1. \tag{15.7.23}$$

In other words, like thermodynamic energies and heat capacities, the classical and quantum entropies also agree at the high-temperature limit.

Exercises

1. Use the definition (15.1.9) to show directly that

$$[\hat{q}, \hat{p}] = \mathrm{i}\hbar, \quad [\hat{q}^2, \hat{p}] = 2\mathrm{i}\hbar\hat{q}, \quad [\hat{p}^2, \hat{q}] = -2\mathrm{i}\hbar\hat{p}. \tag{15.E.1}$$

2. Use (15.1.11) and (15.E.1) to obtain

$$[\hat{q}, \hat{H}] = \mathrm{i}\frac{\hbar}{m}\hat{p}, \quad [\hat{p}, \hat{H}] = -\mathrm{i}\hbar m\omega^2\hat{q}. \tag{15.E.2}$$

   Then use (15.E.2) and the Heisenberg equation to get (15.1.19).

3. Prove (15.1.25).

4. Check directly that (15.1.32) is a solution of (15.1.12) such that E is given by (15.1.33).
5. Show that the full eigenspectrum of the Hamiltonian (15.1.11) of the harmonic oscillator is given by (15.1.35).
6. Show that, for the creation and annihilation operators $\hat{a}_+$ and $\hat{a}_-$ defined in (15.1.20), there hold

$$\hat{a}^+|n\rangle = \sqrt{n+1}|n+1\rangle; \quad \hat{a}_-|n\rangle = \sqrt{n}|n-1\rangle, \quad n \geq 1. \tag{15.E.3}$$

7. Show that, for the creation and annihilation operators $\hat{a}_+$ and $\hat{a}_-$ defined in (15.1.20), there holds

$$\hat{a}_-\hat{a}_+ = \hat{N} + 1, \tag{15.E.4}$$

where $\hat{N}$ is the number operator of the harmonic oscillator.
8. Determine the constant factor C_0 in (15.2.24) and find a concrete expression for $|0\rangle$ for the N-dimensional harmonic oscillator.
9. Prove that the states given in (15.2.30) are linearly independent.
10. Verify (15.2.37).
11. For the number state given in (15.2.40), show that

$$\hat{a}^i_+|n_1, \ldots, n_i, \ldots, n_N\rangle = \sqrt{n_i+1}\,|n_1, \ldots, n_i+1, \ldots, n_N\rangle. \tag{15.E.5}$$

Besides, if $n_i \geq 1$, show that

$$\hat{a}^i_-|n_1, \ldots, n_i, \ldots, n_N\rangle = \sqrt{n_i}\,|n_1, \ldots, n_i-1, \ldots, n_N\rangle. \tag{15.E.6}$$

12. Use (15.3.22) to show that

$$[\hat{\phi}^2(\mathbf{x}, t), \hat{\pi}(\mathbf{x}', t)] = 2\mathrm{i}\hbar\delta(\mathbf{x} - \mathbf{x}')\hat{\phi}(\mathbf{x}, t), \tag{15.E.7}$$

$$[\hat{\phi}(\mathbf{x}, t), \hat{\pi}^2(\mathbf{x}', t)] = 2\mathrm{i}\hbar\delta(\mathbf{x} - \mathbf{x}')\hat{\pi}(\mathbf{x}', t). \tag{15.E.8}$$

13. Verify (15.4.41) concerning normal orderings in two different settings.
14. Explain why (15.5.8) is a real-valued quantity.
15. Show that (15.5.8) is independent of $\dot{q}$ and $\dot{\overline{q}}$. That is,

$$\frac{\partial H}{\partial \dot{q}} = 0, \quad \frac{\partial H}{\partial \dot{\overline{q}}} = 0. \tag{15.E.9}$$

16. Verify the quantization (15.5.14).
17. Explain why (15.5.24) and the Lagrange equations (15.5.27) are equivalent.
18. Explain why the Hamiltonian system (15.5.32)–(15.5.33) is equivalent to (15.5.24).
19. Show that the $\hat{q}^i$ and $(\hat{q}^i)^\dagger$ given in (15.5.49) and (15.5.53), respectively, are annihilation and creation operators, and determine the energy annihilated or created each time when applying these operators to a particle state.

20. Work out the canonical quantization for the complex Klein–Gordon equation along the lines for the Schrödinger equation.
21. Derive (15.6.39).
22. Establish (15.6.43).
23. Confirm (15.6.49).
24. Obtain (15.7.2) for the partition function of the harmonic oscillator (15.7.1).
25. Establish (15.7.23).
26. Consider a classical harmonic oscillator with N degrees of freedom and obtain its partition function, Helmholtz free energy, thermodynamic energy, entropy, and heat capacity.

Appendices

This Appendices chapter presents concise but self-contained introductions to a few important concepts encountered elsewhere in the main text of the book which may be of independent interests to some readers as well. These include the index of a vector field and topological degree of a map, linking number and the Hopf invariant, the Noether theorem relating conserved quantities to continuous symmetries of a mechanical or field-theoretical system, eigenvalues of quantum angular momentum operators, spins, particle statistics, and gravitational deflection of light.

A.1 Index of vector field and topological degree of map

This section discusses the notions of the index of a vector field, also known as the *Poincaré index*, and topological degree of a map, also known as the *Brouwer degree*. We start from a two-dimensional setting provided by the classical *argument principle* in the context of complex analysis [90, 224]. The formalism then extends beyond complex analysis and to higher dimensions. As we move along the way, we also see some direct applications of the concepts and results formulated and developed.

Argument principle

Let C be a simple closed curve in the complex plane coordinated by the complex variable $z = x + \mathrm{i}y$, $x, y \in \mathbb{R}$, parametrized by a real parameter t, and oriented counterclockwise, such that

$$z = z(t) = x(t) + \mathrm{i}y(t), \quad a \leq t \leq b, \quad a < b, \tag{A.1.1}$$

$z(t_1) \neq z(t_2)$ whenever $t_1 \neq t_2$, except $z(a) = z(b)$. Let $w = f(z)$ be a complex-valued function that is meromorphic in the domain D enclosed by the closed curve C and differentiable over C as well. Hence, through the function f, the closed curve C is mapped into a closed curve, Γ, in the complex plane coordinated in terms of the variable $w = u + \mathrm{i}v$, $u, v \in \mathbb{R}$ (Figure A.1).

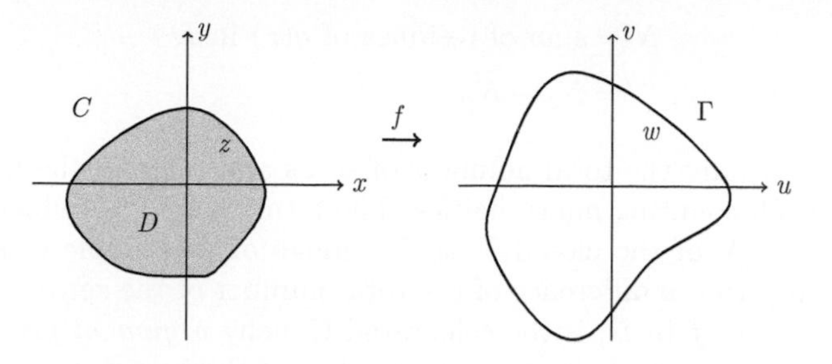

Figure A.1 A simple closed curve C is mapped into another closed curve Γ through the complex function $w = u + \mathrm{i}v = f(z)$.

Now assume $f(z) \neq 0$ for $z \in C$. Hence, we can represent the image $w = f(z)$ in its polar form

$$w = f(z(t)) = \rho(t)\mathrm{e}^{\mathrm{i}\Theta(t)}, \quad a \leq t \leq b, \tag{A.1.2}$$

along the curve C, where $\rho(t)$ and $\Theta(t)$ are the modulus and argument of the nonzero complex quantity $f(z(t))$, respectively, such that

$$\rho(a) = \rho(b), \quad \Theta(b) = \Theta(a) + 2\pi N, \tag{A.1.3}$$

for some integer N, which is the *winding number* of the curve Γ with respect to the curve C under the map f. In view of (A.1.2) and (A.1.3), we have

$$\begin{aligned} N &= \frac{1}{2\pi}\int_a^b \Theta'(t)\,\mathrm{d}t \\ &= \frac{1}{2\pi\mathrm{i}}\left(\int_a^b \frac{f'(z(t))}{f(z(t))}z'(t)\,\mathrm{d}t - \int_a^b \frac{\mathrm{d}}{\mathrm{d}t}\ln\rho(t)\,\mathrm{d}t\right) \\ &= \frac{1}{2\pi\mathrm{i}}\int_C \frac{f'(z)}{f(z)}\,\mathrm{d}z. \end{aligned} \tag{A.1.4}$$

On the other hand, if $z_0 \in D$ is a zero of $f(z)$ of multiplicity m, then it is clear that it is a simple pole of the function $q(z) = \frac{f'(z)}{f(z)}$ such that

$$q(z) = \frac{m}{z - z_0} + \phi(z), \tag{A.1.5}$$

where $\phi(z)$ is analytic for z in a neighborhood of z_0. So the residue of $q(z)$ at z_0 is m. Likewise, if $z_0 \in D$ is a pole of $f(z)$ of order n, then $q(z)$ has the local expression

$$q(z) = -\frac{n}{z - z_0} + \psi(z), \tag{A.1.6}$$

where $\psi(z)$ is analytic for z in a neighborhood of z_0. So the residue of $q(z)$ at z_0 is $-n$. Thus, as a consequence of the *Cauchy residue theorem*, we obtain from (A.1.4) the formula

$$\begin{aligned} N &= \text{sum of residues of } q(z) \text{ in } D \\ &= N_{\text{Z}} - N_{\text{P}}, \end{aligned} \tag{A.1.7}$$

where N_{Z} and N_{P} are the total numbers of zeros and poles of the function $f(z)$ in the domain D, counting multiplicities. The formula (A.1.7), which relates the winding number N of the closed Γ as the image of the simple closed curve C under the map f to the difference of the total number of the zeros, N_{Z}, and that of the poles, N_{P}, of f in D, is the celebrated Cauchy *argument principle.*

As an example, we find the winding number of the image

$$C_R = \{z \,|\, |z| = R\}, \quad R > 0, \tag{A.1.8}$$

under an arbitrary degree-n polynomial, $n \geq 1$,

$$f(z) = a_n z^n + a_{n-1} z^{n-1} + \cdots + a_1 z + a_0, \quad a_0, a_1, \ldots, a_n \in \mathbb{C}, \quad a_n \neq 0. \tag{A.1.9}$$

Thus, if $f(z) \neq 0$ for $|z| = R$, then N is the algebraic number of the zeros of $f(z)$ in $D_R = \{z \,|\, |z| < R\}$ (the algebraic number of zeros is the total number of zeros counting multiplicities). In particular, if R is sufficiently large so that all the zeros of $f(z)$ are in D_R, then $N = n$.

We now directly compute the winding number for R sufficiently large such that $f(z) \neq 0$ for any z satisfying $|z| \geq R$. In this situation, note that, if we replace C by C_R in (A.1.4), then the right-hand side of (A.1.4) depends on R continuously but the left-hand side of (A.1.4) stays constant since N is an integer. Consequently, we may parametrize C_R with $C_R : z = R\mathrm{e}^{\mathrm{i}t}$ $(0 \leq t \leq 2\pi)$ to obtain

$$N = \frac{1}{2\pi \mathrm{i}} \lim_{R \to \infty} \int_{C_R} \frac{f'(z)}{f(z)} \, \mathrm{d}z = n. \tag{A.1.10}$$

In other words, the polynomial (A.1.9) has exactly n zeros in the complex plane. This statement is known as the *fundamental theorem of algebra.*

Winding number in general

We now go beyond the context of complex functions and consider instead a real vector field of the form

$$\mathbf{A}(x, y) = (u(x, y), v(x, y)) \in \mathbb{R}^2, \quad (x, y) \in \mathbb{R}^2. \tag{A.1.11}$$

As before, we use C to denote a simple closed curve in $\mathbb{R}^2$ and assume $\mathbf{A}(x, y) \neq \mathbf{0}$ for $(x, y) \in C$. Thus, with $\mathbf{A} = (u, v) \neq \mathbf{0}$, we have the polar variable representation

$$u = \rho \cos \Theta, \quad v = \rho \sin \Theta, \quad \rho = \|\mathbf{A}\| > 0, \tag{A.1.12}$$

such that

$$\Theta = \tan^{-1} \frac{v}{u}, \quad u \neq 0; \quad \Theta = \cot^{-1} \frac{u}{v}, \quad v \neq 0. \tag{A.1.13}$$

Hence, analogously as in (A.1.4), we obtain the following expression for the winding number

$$N = \frac{1}{2\pi}\int_a^b \Theta'(t)\,\mathrm{d}t$$
$$= \frac{1}{2\pi}\int_a^b \left(\frac{uv' - u'v}{u^2+v^2}\right)\mathrm{d}t, \tag{A.1.14}$$

in view of (A.1.13), where $u = u(t) = u(x(t), y(t))$ and $v = v(t) = v(x(t), y(t))$ along the curve C.

Thus, we have produced a real vector-field version of the formula for the calculation of the winding number for the underlying map, $\mathbf{A} : C \to \Gamma$ (Figure A.1) replacing f there with $\mathbf{A}$ here.

From winding number to index of vector field

Subsequently, we only consider real-valued functions and quantities.

First, we show that the winding number formula (A.1.14) may be recast into a more elegant form. To do so, we use $\mathbf{A}_C$ to denote the *normalized* form of the vector field $\mathbf{A}$ along C:

$$\mathbf{A}_C = \frac{\mathbf{A}}{\|\mathbf{A}\|} = \left(\frac{u}{\sqrt{u^2+v^2}}, \frac{v}{\sqrt{u^2+v^2}}\right) \equiv (f, g). \tag{A.1.15}$$

After some computation, we see that, in terms of the normalized vector (A.1.15), the formula (A.1.14) assumes the following simplified form,

$$N = \frac{1}{2\pi}\int_a^b (f'g - fg')\,\mathrm{d}t = \frac{1}{2\pi}\int_a^b \begin{vmatrix} f & g \\ f' & g' \end{vmatrix}\,\mathrm{d}t. \tag{A.1.16}$$

This quantity is known as the *index*, or *Poincaré index*, of the vector field $\mathbf{A}$ along the simple closed curve C and often denoted by $\mathrm{ind}\,(\mathbf{A}|_C)$.

Geometric interpretation of index of vector field

The normalized vector field $\mathbf{A}_C$ may be regarded as a map from C into the unit circle S^1. Extracting this feature, we consider a general map $\mathbf{R} = (f, g) : C \to S^1$ where C is parametrized as before,

$$C: \quad x = x(t), \quad y = y(t), \quad a \le t \le b, \tag{A.1.17}$$

so that the quantity

$$\frac{1}{2\pi}\int_C \mathbf{T}\cdot\mathrm{d}\mathbf{R} \tag{A.1.18}$$

counts the number of times that the image of C under the map $\mathbf{R}$ covers S^1. Such an interpretation follows from noting that $|S^1| = 2\pi$ where S^1 is also oriented counterclockwise. Thus, its unit tangent vector $\mathbf{T}$ in the positive direction of the circle reads $\mathbf{T} = (-g, f)$. The quantity (A.1.18) is an integer known as the *topological degree* of the map $\mathbf{R} : C \to S^1$, and denoted $\deg(\mathbf{R})$. In view of the parametrization (A.1.17), we have $\mathrm{d}\mathbf{R} = (f', g')\,\mathrm{d}t$. Hence, (A.1.16) and (A.1.18) coincide. This leads us to write

$$\deg(\mathbf{R}) = \frac{1}{2\pi}\int_a^b \begin{vmatrix} f & g \\ f' & g' \end{vmatrix} \mathrm{d}t, \quad \mathbf{R} = (f, g) : C \to S^1. \tag{A.1.19}$$

With the notation (A.1.19), we arrive at the relation

$$\mathrm{ind}\,(\mathbf{A}|_C) = \deg(\mathbf{A}_C), \tag{A.1.20}$$

in particular. Namely, the index of a vector field along a simple closed curve is the degree of its normalized form regarded as a map from the curve into the unit circle.

Local representation of index

Consider the vector field (A.1.11), which is sufficiently differentiable in a domain containing a bounded domain D and its boundary C such that $\mathbf{A} \neq \mathbf{0}$ on C. Thus, we may rewrite the index formula (A.1.14) as a path integral

$$\begin{aligned}\mathrm{ind}\,(\mathbf{A}|_C) &= \frac{1}{2\pi}\int_\Gamma \frac{1}{u^2+v^2}\,(u\mathrm{d}v - v\mathrm{d}u) \\ &= \frac{1}{2\pi}\int_C \frac{1}{u^2+v^2}\,((uv_x - vu_x)\,\mathrm{d}x + (uv_y - vu_y)\,\mathrm{d}y),\end{aligned} \tag{A.1.21}$$

either over the (u, v)-plane or the (x, y)-plane. Note that, at $\mathbf{A} \neq 0$, we have

$$\frac{\partial}{\partial u}\left(\frac{u}{u^2+v^2}\right) = \frac{\partial}{\partial v}\left(\frac{-v}{u^2+v^2}\right), \tag{A.1.22}$$

$$\frac{\partial}{\partial x}\left(\frac{uv_y - vu_y}{u^2+v^2}\right) = \frac{\partial}{\partial y}\left(\frac{uv_x - vu_x}{u^2+v^2}\right). \tag{A.1.23}$$

Assume that $\mathbf{A}$ has finitely many zeros in D, say $p_1, \dots, p_n$. Let C_i be a small circle centered at p_i such that $C_i \subset D$ and $p_i \neq p_j$ and $C_i \cap C_j = \emptyset$ for $i \neq j$, $i, j = 1, \dots, n$. Then, using (A.1.21) to each C_i and applying the Green theorem, in view of (A.1.22) or (A.1.23), to the multiply connected domain with $C \cup C_1 \cup \cdots \cup C_n$ as its boundary, we obtain the relation

$$\mathrm{ind}\,(\mathbf{A}|_C) = \sum_{i=1}^n \mathrm{ind}\,(\mathbf{A}|_{C_i}), \tag{A.1.24}$$

(see Figure A.2). Of course, the circles C_i's may also be replaced by other simple closed curves around the points p_i's. Since $\mathrm{ind}\,(\mathbf{A}|_{C_i})$ is independent of C_i but p_i, we may denote it by $\mathrm{ind}\,(\mathbf{A}|_{p_i})$, instead, referred to as the index of the vector field $\mathbf{A}$ at its isolated zero p_i. Hence, we rewrite the formula (A.1.24) as

$$\mathrm{ind}\,(\mathbf{A}|_C) = \sum_{i=1}^n \mathrm{ind}\,(\mathbf{A}|_{p_i}), \tag{A.1.25}$$

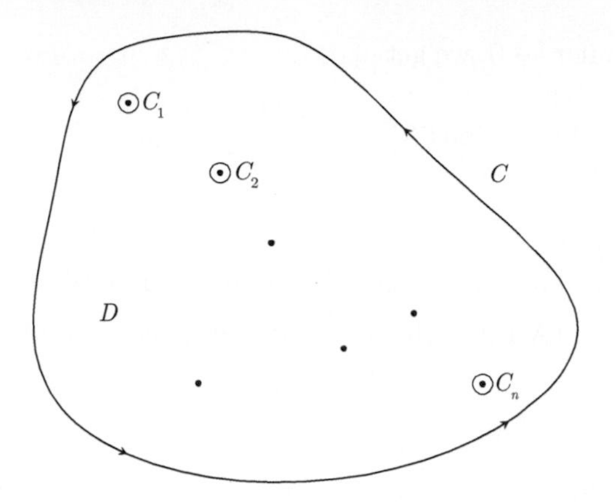

Figure A.2 The distinct zeros $p_1, \ldots, p_n$ of a vector field $\mathbf{A}$ are at the centers of some small non-intersecting circles $C_1, \ldots, C_n$ such that, as a result of the index formula and the Green theorem, the index of $\mathbf{A}$ along the boundary curve C of the domain D is seen to be equal to the sum of the indices of $\mathbf{A}$ along these small circles.

which clearly indicates how the global boundary information, expressed by the left-hand side of the equation, about a vector field relates to its local interior information, given by the right-hand side of the equation.

The equation (A.1.25) may well be regarded as a vector field analogue of the argument principle.

Index of vector field at isolated zero

Without loss of generality, let $p_0 = (0, 0)$ be an isolated zero of the vector field (A.1.11). We are to compute $\operatorname{ind}(\mathbf{A}|_{p_0})$. For convenience, we use the notation

$$a = u_x(0,0), \quad b = u_y(0,0), \quad c = v_x(0,0), \quad d = v_y(0,0). \tag{A.1.26}$$

Then, at $(0, 0)$, we have $\mathrm{d}u = a\mathrm{d}x + b\mathrm{d}y$ and $\mathrm{d}v = c\mathrm{d}x + d\mathrm{d}y$. We say that $(0, 0)$ is a non-degenerate zero of $\mathbf{A}$ if the differentials $\mathrm{d}u$ and $\mathrm{d}v$ there are not co-linear, namely, they are linearly independent, which amounts to imposing the condition

$$ad - bc \neq 0, \tag{A.1.27}$$

which is to be observed in subsequent discussion.

To compute $\operatorname{ind}(\mathbf{A}|_{p_0})$, we consider a circle, C_r, centered at p_0 and of a small radius $r > 0$, parametrized by t so that $x = r\cos t, y = r\sin t, 0 \leq t \leq 2\pi$. Hence,

$$u = r(a\cos t + b\sin t) + \mathrm{O}(r^2), \quad v = r(c\cos t + d\sin t) + \mathrm{O}(r^2), \tag{A.1.28}$$

along C_r. Therefore, inserting (A.1.28) into

$$\operatorname{ind}(\mathbf{A}|_{p_0}) = \operatorname{ind}(\mathbf{A}|_{C_r}) = \frac{1}{2\pi}\int_0^{2\pi} \frac{(uv' - u'v)}{u^2 + v^2}\,\mathrm{d}t, \tag{A.1.29}$$

and taking the limit $r \to 0$, we get

$$\operatorname{ind}(\mathbf{A}|_{p_0}) = \frac{(ad - bc)}{2\pi} I_0, \tag{A.1.30}$$

where

$$I_0 = \int_0^{2\pi} \frac{\mathrm{d}t}{(a\cos t + b\sin t)^2 + (c\cos t + d\sin t)^2}. \tag{A.1.31}$$

Under the condition (A.1.27), this integral can be calculated to yield the exact result

$$I_0 = \frac{2\pi}{|ad - bc|}. \tag{A.1.32}$$

Inserting (A.1.32) into (A.1.30), we obtain

$$\operatorname{ind}(\mathbf{A}|_{p_0}) = \operatorname{sgn}(ad - bc) = \operatorname{sgn}(J(\mathbf{A})_{p_0}), \tag{A.1.33}$$

where $\operatorname{sgn}(s) = 1$ or -1 depending on $s > 0$ or $s < 0$ and $J(\mathbf{A})_{p_0}$ denotes the Jacobian of the map $\mathbf{A}(x, y) = (u(x, y), v(x, y))$ at p_0 given by

$$J(\mathbf{A})_{p_0} = \frac{\partial(u, v)}{\partial(x, y)} = \begin{vmatrix} u_x & u_y \\ v_x & v_y \end{vmatrix}, \quad (x, y) = p_0. \tag{A.1.34}$$

In view of this calculation, if we assume all the zeros $p_1, \dots, p_n$ of $\mathbf{A}$ are non-degenerate so that

$$J(\mathbf{A})_{p_i} \neq 0, \quad i = 1, \dots, n, \tag{A.1.35}$$

then the formula (A.1.25) renders the explicit result

$$\operatorname{ind}(\mathbf{A}|_C) = \sum_{i=1}^{n} \operatorname{sgn}(J(\mathbf{A})_{p_i}), \tag{A.1.36}$$

which provides another interpretation or realization for the index of a vector field or the degree of a map.

Point index and stability of equilibrium

Next, associate the index of a vector field at an isolated zero with a physical interpretation. For this purpose, we consider the dynamical system

$$\frac{\mathrm{d}x}{\mathrm{d}t} = u(x, y), \quad \frac{\mathrm{d}y}{\mathrm{d}t} = v(x, y), \tag{A.1.37}$$

so that the zero $p_0 = (0, 0)$ of the vector field (A.1.11) is an equilibrium or critical point of the system (A.1.37). Hence, with the notation (A.1.26), the system (A.1.37) becomes

$$\frac{\mathrm{d}x}{\mathrm{d}t} = ax + by + \mathrm{O}(r^2), \quad \frac{\mathrm{d}y}{\mathrm{d}t} = cx + dy + \mathrm{O}(r^2), \quad r = \sqrt{x^2 + y^2}, \tag{A.1.38}$$

in a small neighborhood of p_0 so that the linear stability of the equilibrium p_0 is determined by the eigenvalues of the linear-coefficient matrix of (A.1.38) given by

$$\lambda_{1,2} = \frac{1}{2}\left((a+d) \pm \sqrt{(a+d)^2 - 4(ad-bc)}\right). \tag{A.1.39}$$

Thus, subject to the condition (A.1.27), we have the following alternatives:

(i) $ad - bc > 0$. If $a+d > 0$, then $\text{Re}(\lambda_{1,2}) > 0$ and p_0 is unstable. If $a+d < 0$, then $\text{Re}(\lambda_{1,2}) < 0$ and p_0 is stable. If $a+d = 0$, then $\text{Re}(\lambda_{1,2}) = 0$ and p_0 is a center, which is also (linearly) stable.

(ii) $ad - bc < 0$. Then λ_1 and λ_2 are both nonzero, real, and of opposite signs. So p_0 is a saddle point, which is sometimes referred to as half-stable.

Consequently, subject to the non-degeneracy condition and in terms of the dynamical system (A.1.37), we see that (A.1.36) becomes a number-counting formula for the associated equilibria of various types:

$$\text{ind}\,(\mathbf{A}|_C) = N_+ - N_-, \tag{A.1.40}$$

where N_+ denotes the number of zeros of $\mathbf{A}$ in D where the Jacobian is positive, which counts for the total number of non-saddle-point equilibria of (A.1.37) and N_- the number of zeros where the Jacobian is negative, which counts for the total number of saddle-point equilibria of (A.1.37).

Index of periodic orbit

In particular, if the simple closed curve C is a periodic orbit of the dynamical system (A.1.37), then at any point $(x, y) \in C$, the velocity vector $\mathbf{A}(x, y) = (u(x, y), v(x, y))$ is tangential to C. Thus, as one travels along C counterclockwise, one moves either following or against $\mathbf{A}(x, y)$ during the entire cycle of the trip. In either case, we have $\text{ind}\,(\mathbf{A}|_C) = 1$, as illustrated in Figure A.3 for the former case.

As a result, if we use n to denote the total number of saddle-point equilibria of (A.1.37) enclosed in a closed orbit C, then there must be $n + 1$ non-saddle-point equilibria of (A.1.37) also enclosed in C.

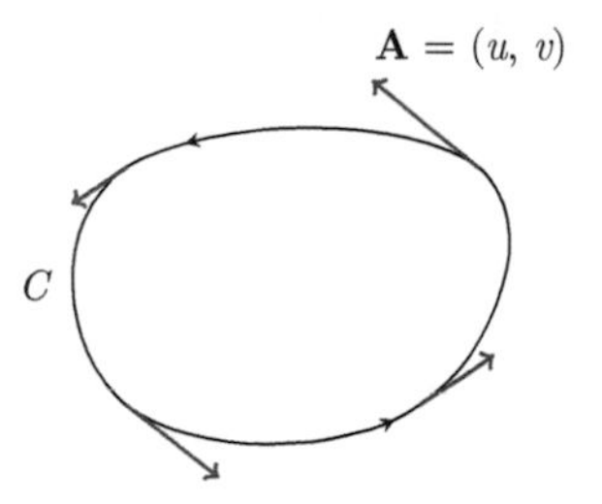

Figure A.3 A periodic orbit situation when its velocity vector $\mathbf{A} = (u, v)$ happens to coincide with the tangent vector of the orbit oriented counterclockwise, resulting in the reading $\text{ind}\,(\mathbf{A}|_C) = 1$.

Examples

As applications, we first consider the classical *predator-prey equations*, or the *Lotka–Volterra equations*, governing the interaction of a prey population, $x \geq 0$, and a predator population, $y \geq 0$, such as those of rabbits and wolves, respectively, in an isolated ecological system, of the form

$$\frac{\mathrm{d}x}{\mathrm{d}t} = x(r - ay), \quad \frac{\mathrm{d}y}{\mathrm{d}t} = y(bx - s), \tag{A.1.41}$$

where $a, b, r, s > 0$ are coupling parameters. When the predator is absent, $y = 0$, the prey population grows exponentially with rate r; when the predator is present, $y > 0$, it slows down the growth of the population of the prey by modifying its growth rate to $r - ay$. On the other hand, when the prey is absent, $x = 0$, the predator population decays exponentially with rate s; when the prey is present, $x > 0$, it enhances the maintenance of the population of the predator by adjusting its decay rate into a growth one, $bx - s$. There are two equilibria:

$$p = (0, 0), \quad q = \left(\frac{s}{b}, \frac{r}{a}\right), \tag{A.1.42}$$

in the (x, y)-phase plane. With $\mathbf{A}(x, y) = (x[r - ay], y[bx - s])$ and the notation (A.1.34), we have

$$J(\mathbf{A})_p = -rs < 0, \quad J(\mathbf{A})_q = rs > 0. \tag{A.1.43}$$

Thus, p, q are saddle and non-saddle equilibria of the system (A.1.41), respectively. As a consequence, it is possible to have periodic orbits in the first quadrant of the phase plane surrounding the equilibrium q given in (A.1.42). In fact, since (A.1.41) may be recast into the differential form

$$\frac{\mathrm{d}x}{x(r - ay)} - \frac{\mathrm{d}y}{y(bx - s)} = 0, \quad (x, y) \neq p, q, \tag{A.1.44}$$

whose integral is readily obtained to be

$$bx + ay - (s \ln x + r \ln y) = C, \tag{A.1.45}$$

where C is an integration constant. The expression (A.1.45) gives rise to a full family of periodic orbits of (A.1.41) clustered around the non-saddle equilibrium q. Figure A.4 provides a concrete example.

Next, we consider the classical *rabbits-versus-sheep equations*

$$\frac{\mathrm{d}x}{\mathrm{d}t} = x(r - ax - by), \quad \frac{\mathrm{d}y}{\mathrm{d}t} = y(s - cx - dy), \quad x, y \geq 0, \tag{A.1.46}$$

where x, y stand for the populations of rabbits and sheep, respectively, undergoing competitive interactions reflected by the coupling parameters, $r, s, a, b, c, d > 0$. Under the condition $ad - bc \neq 0$, we find the equilibria of (A.1.46) to be

$$p_0 = (0, 0), \quad p_1 = \left(0, \frac{s}{d}\right), \quad p_2 = \left(\frac{r}{a}, 0\right), \quad p_3 = \left(\frac{rd - sb}{ad - bc}, \frac{sa - rc}{ad - bc}\right). \tag{A.1.47}$$

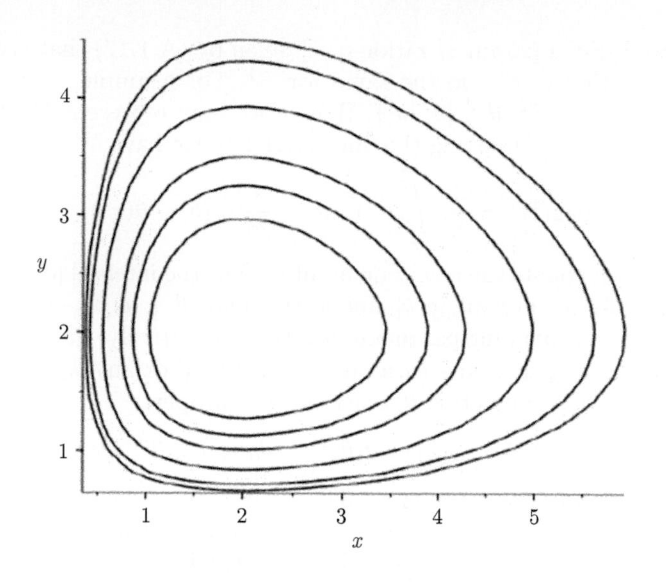

Figure A.4 A plot of a family of periodic orbits clustered around the non-saddle equilibrium q of the predator-prey equations (A.1.41) where $r=2, a=1, s=1, b=\frac{1}{2}$ such that $q=(2,2)$.

Thus, with $\mathbf{A}=(x[r-ax-by], y[s-cx-dy])$, we have

$$J_0 = rs,\ J_1 = -\frac{s}{d}(rd-sb),\ J_2 = -\frac{r}{a}(sa-rc),\ J_3 = \frac{(rd-sb)(sa-rc)}{ad-bc}, \tag{A.1.48}$$

where $J_i \equiv J(\mathbf{A})_{p_i}$ for $i=0,1,2,3$. For our interest, we assume that p_3 given in (A.1.47) lies in the first quadrant of the (x,y)-plane. Hence, $rd-sb$ and $sa-rc$ are of the same sign as $ad-bc$. So we have the following alternative situations:

(i) Assume $ad-bc<0$. Then, $rd-sb<0$ and $sa-rc<0$. Hence, $J_1>0, J_2>0, J_3<0$ in (A.1.48). Therefore, we conclude that the system (A.1.46) does not allow a periodic orbit in the first quadrant of the (x,y)-plane.

(ii) Assume $ad-bc>0$. Then, $rd-sb>0$ and $sa-rc>0$, leading to $J_1<0, J_2<0, J_3>0$ in (A.1.48). Under this circumstance, a periodic orbit of (A.1.46) in the first quadrant of the (x,y)-plane around the equilibrium point p_3 may exist.

Degree by virtue of number-counting

For the map $\mathbf{R}=(f,g): C\to S^1$ whose degree is calculated by (A.1.19) so that it counts the net algebraic number that the target curve S^1 is covered by the image of C under $\mathbf{R}$, we may analogously come up with a formula in the spirit of

(A.1.40) as well. Fix a parametrization of C given by (A.1.17) that is compatible with the orientation of C. Do the same for S^1. For example, a typical one is $S^1 = \{(\cos\theta, \sin\theta) \in \mathbb{R}^2 \,|\, \theta \in [0, 2\pi]\}$. Hence, we may represent $\mathbf{R} : C \to S^1$ as $\mathbf{R} = (\cos\theta(t), \sin\theta(t))$. Inserting this into (A.1.19), we have

$$\deg(\mathbf{R}) = \frac{1}{2\pi}\int_a^b \theta'(t)\,\mathrm{d}t = \frac{1}{2\pi}(\theta(b) - \theta(a)), \tag{A.1.49}$$

as (A.1.14), which counts the algebraic number that the interval $[0, 2\pi]$ is covered by the image of the interval $[a, b]$ under the map $\theta : [a, b] \to [0, 2\pi]$ at the associated and corresponding parameter levels. Alternatively, we may also reduce such a *global counting* task to a *local* one as follows. Pick any $\theta_0 \in [0, 2\pi]$, which is a *regular value* of the function θ satisfying the property

$$\theta'(t) \neq 0, \quad \text{whenever } \theta(t) = \theta_0 \text{ or } t \in \theta^{-1}(\theta_0). \tag{A.1.50}$$

(By Sard's theorem [175, 260, 293, 395], we know that regular values are abundant.) Thus, if $\theta'(t_0) > 0$ for $t_0 \in \theta^{-1}(\theta_0)$, then a neighborhood of θ_0 is covered by a neighborhood of t_0 with consistent interval orientations, and if $\theta'(t_0) < 0$, with opposite orientations, such that the net number of coverings is the difference of the number of consistent coverings and the number of opposite coverings, resulting in the expression

$$\begin{aligned}\deg(\mathbf{R}) &= \#\left\{t \in \theta^{-1}(\theta_0) \,|\, \theta'(t) > 0\right\} - \#\left\{t \in \theta^{-1}(\theta_0) \,|\, \theta'(t) < 0\right\} \\ &= \sum_{t \in \theta^{-1}(\theta_0)} \operatorname{sgn}(\theta'(t)),\end{aligned} \tag{A.1.51}$$

in analogue to (A.1.36) as well.

Index and zeros of vector field

Let $\mathbf{A}(x, y) = (u(x, y), v(x, y))$ be a vector field which is differentiable over a bounded domain Ω in $\mathbb{R}^2$. We are interested in acquiring zeros of $\mathbf{A}$ in Ω. For this purpose, take C to be a simple closed curve in Ω that encloses a subdomain D of Ω such that $\mathbf{A}$ has no zero on C. Then $\operatorname{ind}(\mathbf{A}|_C)$ is well defined. We now show that the condition $\operatorname{ind}(\mathbf{A}|_C) \neq 0$ ensures the existence of a zero of $\mathbf{A}$ in D if D is also contractible.

In fact, we assume otherwise that $\mathbf{A}$ has no zero in D. Let γ be another simple closed curve, say a circle, inside C. Since γ may be obtained from C through a continuous deformation and $\mathbf{A}$ has no zero in D, we have $\operatorname{ind}(\mathbf{A}|_\gamma) = \operatorname{ind}(\mathbf{A}|_C) \neq 0$ and

$$\operatorname{ind}(\mathbf{A}|_C) = \deg(\mathbf{A}_\gamma) = \frac{1}{2\pi}\int_\gamma \boldsymbol{\tau} \cdot \mathbf{A}'_\gamma(s)\,\mathrm{d}s, \quad \mathbf{A}_\gamma = \left.\left(\frac{1}{\|\mathbf{A}\|}\mathbf{A}\right)\right|_\gamma, \tag{A.1.52}$$

in view of (A.1.18), where s is the arclength parameter and $\boldsymbol{\tau}$ the unit tangent vector, respectively, along the curve γ. Hence, (A.1.52) leads to

$$1 \leq |\operatorname{ind}(\mathbf{A}|_\gamma)| \leq \frac{1}{2\pi}\int_\gamma \|\mathbf{A}'_\gamma(s)\|\,\mathrm{d}s \leq C_0|\gamma|, \tag{A.1.53}$$

where $C_0 > 0$ is a constant independent of γ and $|\gamma|$ denotes the total arclength of the curve γ. Thus, we arrive at a contradiction when we shrink γ to a point.

Three dimensions

We now extend our discussion on the notions of index and degree in their settings in two dimensions to three dimensions.

Let $\mathbf{A}(x,y,z) = (u,v,w)(x,y,z)$ be a real vector field defined over a domain Ω with a smooth boundary surface, say S. Assume also that $\mathbf{A}$ is differentiable on $\Omega \cup S$ and $\mathbf{A}(x,y,z) \neq \mathbf{0}$ for $(x,y,z) \in S$. Then, we obtain a map from S into S^2 by setting

$$\mathbf{R} = \left(\frac{1}{\|\mathbf{A}\|}\mathbf{A}\right)\Big|_S \equiv \mathbf{A}_S : S \to S^2. \tag{A.1.54}$$

On the other hand, denote by (s,t) a parametrization of the surface S such that a point $\mathbf{r} = (x,y,z)$ on S and the associated outnormal vector $\mathbf{n}$ there are given by the expressions

$$\mathbf{r} = \mathbf{r}(s,t) = (x,y,z)(s,t), \quad \mathbf{n}(x,y,z) = \frac{\mathbf{r}_s \times \mathbf{r}_t}{\|\mathbf{r}_s \times \mathbf{r}_t\|}, \quad (s,t) \in \mathcal{R}, \tag{A.1.55}$$

where $\mathcal{R}$ is an appropriate parameter region for the parameters s,t and $\mathbf{r}_s \times \mathbf{r}_t$ never vanishes. With such a parametrization, we rewrite the map (A.1.54) as

$$\mathbf{R}(s,t) = (f,g,h)(s,t), \quad (s,t) \in \mathcal{R}, \tag{A.1.56}$$

which may be viewed as a local reparametrization of S^2 with the induced area element

$$\mathrm{d}\mathbf{S} = (\mathbf{R}_s \times \mathbf{R}_t)\,\mathrm{d}s\mathrm{d}t. \tag{A.1.57}$$

Besides, the outnormal vector $\mathbf{N}$ of S^2 at the parameter point (s,t) is simply the radial position vector given by (A.1.56). Therefore, in analogue to (A.1.18) and replacing $\mathbf{T}$ and $\mathrm{d}\mathbf{R}$ there with $\mathbf{N}$ and $\mathrm{d}\mathbf{S}$ here, respectively, we arrive at the degree formula

$$\begin{aligned}\deg(\mathbf{R}) &= \frac{1}{4\pi}\int_S \mathbf{N}\cdot\mathrm{d}\mathbf{S}\\ &= \frac{1}{4\pi}\int_{\mathcal{R}} \mathbf{R}\cdot(\mathbf{R}_s \times \mathbf{R}_t)\,\mathrm{d}s\mathrm{d}t\\ &= \frac{1}{4\pi}\int_{\mathcal{R}} \begin{vmatrix} f & g & h\\ f_s & g_s & h_s\\ f_t & g_t & h_t\end{vmatrix}\,\mathrm{d}s\mathrm{d}t,\end{aligned} \tag{A.1.58}$$

since the total surface area of S^2 is 4π.

With (A.1.54), we may define the index of the vector field $\mathbf{A}(x,y,z)$ over the surface S by

$$\mathrm{ind}\,(\mathbf{A}|_S) = \deg(\mathbf{A}_S), \tag{A.1.59}$$

and establish the same result that if $\mathrm{ind}\,(\mathbf{A}|_S) \neq 0$ then $\mathbf{A}$ has a zero in the domain Ω enclosed by S, provided that Ω is contractible.

Spherical coordinate representation

Consider the standard spherical parametrization of S^2 in terms of the polar and azimuthal angles, θ and ϕ, respectively, such that

$$\mathbf{R} = (\cos\phi\sin\theta, \sin\phi\sin\theta, \cos\theta), \quad 0 \leq \theta \leq \pi, \quad 0 \leq \phi \leq 2\pi. \tag{A.1.60}$$

In view of (A.1.56) and (A.1.60), we may directly use $\theta = \theta(s,t), \phi = \phi(s,t)$ to represent the corresponding vector field $\mathbf{A}$ and the map $\mathbf{R} : S \to S^2$. Hence, we have

$$\mathbf{R} \cdot (\mathbf{R}_s \times \mathbf{R}_t) = \sin\theta \, \frac{\partial(\theta,\phi)}{\partial(s,t)} = \sin\theta \, J(\theta,\phi)(s,t), \tag{A.1.61}$$

expressed as the product of the usual factor $\sin\theta$ and the Jacobian of the transformation $(s,t) \mapsto (\theta,\phi)$ since the area element of the unit sphere S^2 is $\sin\theta \, \mathrm{d}\theta\mathrm{d}\phi$. Thus, as in the earlier situation $\mathbf{R} : C \to S^1$, if $\mathbf{R}_0 = \mathbf{R}(\theta_0, \phi_0)$ is a regular value of the map $\mathbf{R} : S \to S^2$ such that

$$J(\theta,\phi)(s,t) \neq 0, \quad \theta(s,t) = \theta_0, \quad \phi(s,t) = \phi_0, \tag{A.1.62}$$

or $(s,t) \in (\theta,\phi)^{-1}(\theta_0,\phi_0)$, which is a finite set, then the covering-number interpretation of the degree formula (A.1.58) as made before leads to

$$\deg(\mathbf{R}) = \sum_{(s,t)\in(\theta,\phi)^{-1}(\theta_0,\phi_0)} \mathrm{sgn}\,(J(\theta,\phi)(s,t)), \tag{A.1.63}$$

which is analogous to (A.1.51).

Index in terms of vector field directly

With (A.1.54) and the parametrization (A.1.55), we have

$$\mathbf{R}_{s,t} = \frac{1}{\|\mathbf{A}\|^3} \left(\|\mathbf{A}\|^2 \mathbf{A}_{s,t} - [\mathbf{A}_{s,t} \cdot \mathbf{A}]\mathbf{A} \right), \tag{A.1.64}$$

where the subscript $_{s,t}$ denotes differentiation with respect to s or t, separately. As a result, we get

$$\mathbf{R} \cdot (\mathbf{R}_s \times \mathbf{R}_t) = \frac{1}{\|\mathbf{A}\|^3} \mathbf{A} \cdot (\mathbf{A}_s \times \mathbf{A}_t). \tag{A.1.65}$$

Hence, inserting (A.1.65) into (A.1.58), we obtain the index formula

$$\mathrm{ind}\,(\mathbf{A}|_S) = \frac{1}{4\pi} \int_{\mathcal{R}} \frac{\mathbf{A} \cdot (\mathbf{A}_s \times \mathbf{A}_t)}{\|\mathbf{A}\|^3} \, \mathrm{d}s\mathrm{d}t, \tag{A.1.66}$$

for a vector field $\mathbf{A}$, which is assumed to be non-vanishing on the surface S.

General dimensions

Let Ω be a bounded domain in $\mathbb{R}^{n+1}$ with $n \geq 1$ and S the boundary of Ω which is assumed to be smooth. Use $\mathbf{A}(x^1, \dots, x^{n+1})$ to denote a vector field which is differentiable on $\overline{\Omega}$ and nonvanishing on S. As in (A.1.54), define

$$\mathbf{R} = \mathbf{A}_S \equiv \frac{1}{\|\mathbf{A}\|}\mathbf{A} : S \to S^n. \tag{A.1.67}$$

Take a parametrization of S with the parameters $s^1, \dots, s^n$:

$$S : x^1 = x^1(s^1, \dots, s^n), \dots, x^{n+1} = x^{n+1}(s^1, \dots, s^n),\ (s^1, \dots, s^n) \in \mathcal{R}, \tag{A.1.68}$$

where $\mathcal{R}$ is the parameter region. Then, the index of $\mathbf{A}$ over S, $\text{ind}(\mathbf{A}|_S)$, or the degree of the map $\mathbf{R} : S \to S^n$, is given by

$$\deg(\mathbf{R}) = \frac{1}{|S^n|}\int_{\mathcal{R}} \det(\mathbf{R}, \partial_1 \mathbf{R}, \dots, \partial_n \mathbf{R})\, \mathrm{d}s^1 \cdots \mathrm{d}s^n, \tag{A.1.69}$$

where

$$\partial_i \mathbf{R} = \frac{\partial \mathbf{R}}{\partial s^i}, \quad i = 1, \dots, n, \tag{A.1.70}$$

and

$$|S^n| = \frac{2\pi^{\frac{n+1}{2}}}{\Gamma\left(\frac{n+1}{2}\right)}, \tag{A.1.71}$$

is the total surface area of the unit sphere S^n.

As in the two-dimensional vector field situation, if $\text{ind}(\mathbf{A}|_S) \neq 0$, then $\mathbf{A}$ vanishes somewhere in Ω if Ω is contractible.

A simplest but perhaps most useful situation is the obvious result

$$\deg(I) = 1, \quad I : S^n \to S^n, \tag{A.1.72}$$

for the identity map I.

Brouwer fixed point theorem

As an immediate application of our discussion on topological degrees, we establish the celebrated *Brouwer fixed point theorem* which states that a continuous map from a closed ball of $\mathbb{R}^n$ into itself has a fixed point. Without loss of generality, we work out the case where the ball is the unit one,

$$B^n = \{(x^1, \dots, x^n) \in \mathbb{R}^n \mid (x^1)^2 + \cdots + (x^n)^2 \leq 1\}, \quad n \geq 2. \tag{A.1.73}$$

(The case $n = 1$ is trivial.)

We first assume that $F : B^n \to B^n$ is a differentiable map and that F has no fixed point in B^n. Then for any $\mathbf{x} \in B^n$ we have $F(\mathbf{x}) - \mathbf{x} \neq \mathbf{0}$. With this assumption, we consider the vector field

$$\mathbf{A}(\mathbf{x}; t) = \mathbf{x} - tF(\mathbf{x}), \quad \mathbf{x} \in B^n, \quad t \in [0, 1], \tag{A.1.74}$$

labeled by the parameter t. Then, for $\mathbf{x} \in S^{n-1}$ and $t \in [0,1)$ we have $\|\mathbf{A}(\mathbf{x};t)\| \geq 1-t\|F(\mathbf{x})\| \geq 1-t > 0$ since $\|F(\mathbf{x})\| \leq 1$. Furthermore, $\|\mathbf{A}(\mathbf{x};1)\| = \|\mathbf{x} - F(\mathbf{x})\| > 0$ for all $\mathbf{x} \in S^{n-1}$. Hence $\mathbf{A}(\mathbf{x};t) \neq \mathbf{0}$ for any $(\mathbf{x},t) \in S^{n-1} \times [0,1]$. Hence, we can construct the normalized map

$$\mathbf{R}(t) = \frac{\mathbf{A}(\cdot;t)}{\|\mathbf{A}(\cdot;t)\|} : S^{n-1} \to S^{n-1}, \tag{A.1.75}$$

as before, which leads to

$$\mathrm{ind}([I-F]|_{S^{n-1}}) = \deg(\mathbf{R}(1)) = \deg(\mathbf{R}(0)) = \deg(I) = 1, \tag{A.1.76}$$

indicating that the vector field $I - F = \mathbf{A}(\cdot;1)$ vanishes somewhere in B^n, which contradicts the assumption made. Therefore, F must have a fixed point in B^n as claimed.

We next take $F : B^n \to B^n$ to be a continuous map only. By the *Stone–Weierstrass theorem* [528, 611], for any $\varepsilon > 0$, there is a polynomial type approximation of F, say $P_\varepsilon : B^n \to \mathbb{R}^n$ such that

$$\|F(\mathbf{x}) - P_\varepsilon(\mathbf{x})\| < \varepsilon, \quad \mathbf{x} \in B^n. \tag{A.1.77}$$

Thus, we see that

$$\frac{P_\varepsilon}{1+\varepsilon} : B^n \to B^n, \tag{A.1.78}$$

has a fixed point, say $\mathbf{x}_\varepsilon$, in B^n. Without loss of generality, we may also assume $\mathbf{x}_\varepsilon \to \mathbf{x}_0 \in B^n$ as $\varepsilon \to 0$ by the compactness of B^n. As a result, we have

$$\lim_{\varepsilon\to 0} P_\varepsilon(\mathbf{x}_\varepsilon) = \lim_{\varepsilon\to 0}(1+\varepsilon)\mathbf{x}_\varepsilon = \mathbf{x}_0. \tag{A.1.79}$$

Therefore, substituting $\mathbf{x}$ with $\mathbf{x}_\varepsilon$ in (A.1.77), taking $\varepsilon \to 0$, and applying (A.1.79), we find $F(\mathbf{x}_0) = \mathbf{x}_0$ as anticipated.

A further extension of the Brouwer fixed point theorem, which is an important foundational theorem for the methods and theory of partial differential equations, is the *Schauder fixed-point theorem*, which states that a continuous map from a compact and convex subset into itself, or more generally, a continuous map from a convex and closed subset into itself whose image under the map is compact, of a Banach space, has a fixed point in the convex subset. See [199, 236] for details.

Using the Stone–Weierstrass theorem, the concept of topological degrees also may be established for continuous maps [175, 260, 395].

A.2 Linking number and Hopf invariant

As an important application of our discussion about topological degrees, we consider the concepts of linking number, the Hopf invariant, and other related constructions.

Let C_1 and C_2 be two non-intersecting simple closed curves in $\mathbb{R}^3$ with the parametrizations

$$C_1 : \mathbf{r} = \mathbf{r}_1(s), \quad a \leq s \leq b; \quad C_2 : \mathbf{r} = \mathbf{r}_2(t), \quad c \leq t \leq d, \tag{A.2.1}$$

for their respective position vectors. Then, with $\mathcal{R} = [a, b] \times [c, d]$, the position vector

$$\mathbf{R} = \mathbf{R}(s,t) = \frac{\mathbf{r}_1(s) - \mathbf{r}_2(t)}{\|\mathbf{r}_1(s) - \mathbf{r}_2(t)\|}, \quad (s,t) \in \mathcal{R}, \tag{A.2.2}$$

defines a map from the surface $S = C_1 \times C_2$ with its parametrization

$$\boldsymbol{\rho} = \boldsymbol{\rho}(s,t) = (\mathbf{r}_1(s), \mathbf{r}_2(t)), \quad (s,t) \in \mathcal{R}, \tag{A.2.3}$$

and induced product orientation, in $\mathbb{R}^6$, into the unit sphere S^2 in $\mathbb{R}^3$. Topologically, if C_1 and C_2 are viewed as two copies of the unit circle S^1, then $S = S^1 \times S^1 = T^2$, the 2-torus.

Linking number

The *linking number* or *linking coefficient* of the closed curves C_1 and C_2 is defined to be

$$\text{link}(C_1, C_2) = \deg(\mathbf{R}). \tag{A.2.4}$$

Gauss integral

In view of (A.1.54), we identify $\mathbf{A} = \mathbf{r}_1(s) - \mathbf{r}_2(t)$ for our calculation. Thus, by (A.1.58) or (A.1.66), and $\mathbf{A}_s = \dot{\mathbf{r}}_1(s)$, $\mathbf{A}_t = -\dot{\mathbf{r}}_2(t)$, we arrive at

$$\begin{aligned}\text{link}(C_1, C_2) &= \frac{1}{4\pi} \int_a^b \int_c^d \frac{(\mathbf{r}_2 - \mathbf{r}_1) \cdot (\dot{\mathbf{r}}_1 \times \dot{\mathbf{r}}_2)}{\|\mathbf{r}_2 - \mathbf{r}_1\|^3} \, \mathrm{d}s\mathrm{d}t \\ &= \frac{1}{4\pi} \oint_{C_1} \oint_{C_2} \frac{(\mathbf{r}_2 - \mathbf{r}_1) \cdot (\mathrm{d}\mathbf{r}_1 \times \mathrm{d}\mathbf{r}_2)}{\|\mathbf{r}_2 - \mathbf{r}_1\|^3},\end{aligned} \tag{A.2.5}$$

where we have flipped the positions of $\mathbf{r}_1$ and $\mathbf{r}_2$ to absorb the minus sign in the final expression. This integral representation of the linking number is known as the *Gauss integral* [5, 175, 469].

To get a pictorial grasp of the topological quantity (A.2.5), we use S to denote a 2-surface such that $\partial S = C_1$. The orientation of S is compatible with that of C_1 following the usual convention. In other words, we choose a positive unit normal vector at each point in the interior of S, say $\boldsymbol{\nu}$, so that the orientation of C_1 and $\boldsymbol{\nu}$ satisfy a right-handedness relation. Then, in a regular situation, the curve C_2 intersects S at finitely many points, say $p_1, \ldots, p_n$, and transversely, namely,

$$(\boldsymbol{\nu} \cdot \dot{\mathbf{r}}_2)(p_i) \neq 0, \quad i = 1, \ldots, n. \tag{A.2.6}$$

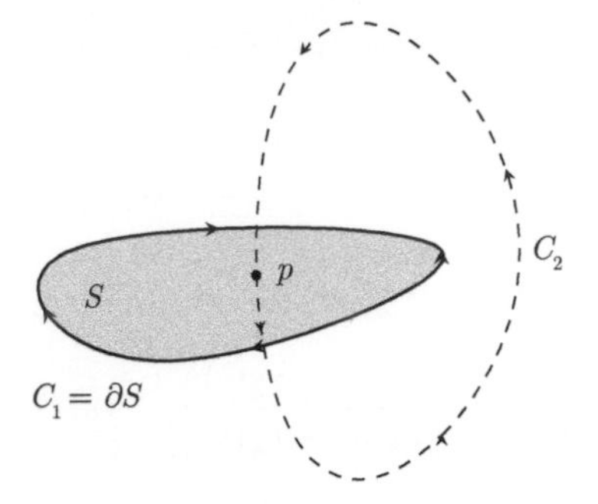

Figure A.5 Two linked closed curves, C_1 and C_2, so positioned that the orientation of C_1 gives rise to an orientation of the enclosed 2-surface S for which the positive unit normal vector of S, say $\boldsymbol{\nu}$, at the single intersection point p of the surface S and curve C_2, and the tangent vector of C_2 at p along the orientation of C_2, say $\boldsymbol{\tau}$, form a sharp angle, satisfying $\boldsymbol{\nu} \cdot \boldsymbol{\tau} > 0$, thereby leading to the result $\text{link}(C_1, C_2) = 1$ via (A.2.7).

Then, the linking number (A.2.5) is seen to be given by [77, 175, 260, 469]

$$\text{link}(C_1, C_2) = \sum_{i=1}^{n} \text{sgn}\,(\boldsymbol{\nu} \cdot \dot{\mathbf{r}}_2)\,(p_i). \tag{A.2.7}$$

In Figure A.5, we draw an example with $\text{link}(C_1, C_2) = 1$.

Writhe

In a series of papers [103, 104, 105], Calugareanu considered the case $C_1 = C_2 = C$ in (A.2.5) and established that the quantity, now called *writhe* or *writhing number*, written as

$$\text{wr}(C) = \frac{1}{4\pi} \oint_C \oint_C \frac{(\mathbf{r}_2 - \mathbf{r}_1) \cdot (\mathrm{d}\mathbf{r}_1 \times \mathrm{d}\mathbf{r}_2)}{\|\mathbf{r}_2 - \mathbf{r}_1\|^3}, \tag{A.2.8}$$

is well defined and describes how the curve twists and coils around itself [222]. Although writhe is not a topological invariant [398], it plays an important role in knot theory [5, 115].

Hopf invariant

For a map $\phi : S^3 \to S^2$, pick a pair of distinct points, $p_1, p_2 \in S^2$, which may be assumed to be two regular values of the map ϕ, based on Sard's theorem. Then $C_1 = \phi^{-1}(p_1)$ and $C_2 = \phi^{-1}(p_2)$ are two non-intersecting smooth simple closed curves in S^3, which may well be viewed as curves in $\mathbb{R}^3$ as well after disregarding a point in $S^3 \setminus (C_1 \cup C_2)$. Note that, in such a situation, the linking number (A.2.5) is independent of the choice of p_1, p_2 but only depends on ϕ, which is the classical Hopf invariant, $H(\phi)$.

Helicity

For a vector field $\mathbf{A}$ defined in a domain Ω in $\mathbb{R}^3$, the quantity

$$H(\mathbf{A}) = \int_\Omega \mathbf{A} \cdot (\nabla \times \mathbf{A})\, \mathrm{d}\mathbf{x}, \tag{A.2.9}$$

called the helicity of $\mathbf{A}$ in Ω, was studied by Woltjer [588] in the context of force-free magnetic fields (where $\mathbf{B} = \nabla \times \mathbf{A}$ is the magnetic field generated from the gauge field $\mathbf{A}$) and by Moffatt [397] and Moffatt and Ricca [398] as an integral invariant in hydrodynamics (where $\mathbf{A}$ represents the velocity distribution in a fluid and $\nabla \times \mathbf{A}$ the associated vorticity field). For simplicity and convenience in our discussion here, we assume that the fields are so behaved that *boundary terms are of no concern*. As a consequence, (A.2.9) is invariant under the gauge transformation $\mathbf{A} \mapsto \mathbf{A} + \nabla\omega$. We now follow [398] to show that the helicity (A.2.9) relates to the linking number (A.2.5).

To proceed, recall that, in the context of static magnetism in the presence of an external current source $\mathbf{j}$, the Ampère equation

$$\nabla \times \mathbf{B} = \mathbf{j}, \tag{A.2.10}$$

as one of the Maxwell equations governing the induced magnetic field $\mathbf{B}$, may be solved following the *Biot–Savart law*

$$\mathbf{B} = \frac{1}{4\pi} \int_\Omega \mathbf{j}(\mathbf{y}) \times \frac{(\mathbf{x} - \mathbf{y})}{\|\mathbf{x} - \mathbf{y}\|^3}\, \mathrm{d}\mathbf{y}. \tag{A.2.11}$$

In fact, let $\mathbf{A}$ be the gauge potential of $\mathbf{B}$ such that $\mathbf{B} = \nabla \times \mathbf{A}$. By a gauge transformation, we may also assume $\nabla \cdot \mathbf{A} \equiv 0$. Hence, (A.2.10) becomes

$$\nabla^2 \mathbf{A} = -\mathbf{j}, \quad \mathbf{x} \in \Omega. \tag{A.2.12}$$

On the other hand, applying the Green function $G(\mathbf{x})$ of the point-source problem

$$\nabla^2 G = \delta(\mathbf{x}), \tag{A.2.13}$$

namely,

$$G(\mathbf{x}) = -\frac{1}{4\pi\|\mathbf{x}\|}, \quad \mathbf{x} \neq \mathbf{0}, \tag{A.2.14}$$

to the general source problem (A.2.12), we have

$$\mathbf{A}(\mathbf{x}) = \frac{1}{4\pi} \int_\Omega \mathbf{j}(\mathbf{y}) \frac{1}{\|\mathbf{x} - \mathbf{y}\|}\, \mathrm{d}\mathbf{y}. \tag{A.2.15}$$

Therefore, we obtain

$$\begin{aligned}
\mathbf{B}(\mathbf{x}) &= \frac{1}{4\pi} \int_\Omega \nabla_{\mathbf{x}} \times \left(\frac{\mathbf{j}(\mathbf{y})}{\|\mathbf{x} - \mathbf{y}\|} \right) \mathrm{d}\mathbf{y} \\
&= -\frac{1}{4\pi} \int_\Omega \mathbf{j}(\mathbf{y}) \times \nabla_{\mathbf{x}} \left(\frac{1}{\|\mathbf{x} - \mathbf{y}\|} \right) \mathrm{d}\mathbf{y} \\
&= \frac{1}{4\pi} \int_\Omega \mathbf{j}(\mathbf{y}) \times \frac{(\mathbf{x} - \mathbf{y})}{\|\mathbf{x} - \mathbf{y}\|^3}\, \mathrm{d}\mathbf{y},
\end{aligned} \tag{A.2.16}$$

which renders (A.2.11).

Now applying (A.2.11) to the relation $\nabla \times \mathbf{A} = \mathbf{B}$, we have

$$\mathbf{A}(\mathbf{x}) = \frac{1}{4\pi}\int_\Omega \mathbf{B}(\mathbf{y}) \times \frac{(\mathbf{x}-\mathbf{y})}{\|\mathbf{x}-\mathbf{y}\|^3}\,\mathrm{d}\mathbf{y}. \tag{A.2.17}$$

Thus, inserting (A.2.17) into (A.2.9) and switching the scalar triple vector product, we get the elegant formula [398]

$$H(\mathbf{A}) = \frac{1}{4\pi}\int_\Omega\int_\Omega \frac{(\mathbf{B}(\mathbf{x}) \times \mathbf{B}(\mathbf{y}))\cdot(\mathbf{x}-\mathbf{y})}{\|\mathbf{x}-\mathbf{y}\|^3}\,\mathrm{d}\mathbf{x}\mathrm{d}\mathbf{y}, \quad \mathbf{B} = \nabla \times \mathbf{A}, \tag{A.2.18}$$

for the helicity of $\mathbf{A}$, which reminds us of the linking number formula (A.2.5). Indeed, when the field $\mathbf{B}$ assumes some idealized form and concentrates thinly along the flux tubes centered around two non-intersecting simple closed curves, C_1 and C_2, say, then (A.2.18) reduces itself into (A.2.5), up to a possible sign convention. See [398] for details regarding such a natural reduction.

Chern–Simons invariant

In fact, when considered over the full space $\mathbb{R}^3$ or generally over a three-dimensional closed manifold, the helicity (A.2.9) gives us the simplest Chern–Simons invariant [131, 132],

$$I_{\text{CS}}(\mathbf{A}) = \int \mathbf{A}\cdot(\nabla \times \mathbf{A})\,\mathrm{d}\mathbf{x}, \tag{A.2.19}$$

which explains why such an invariant is a knot or link invariant.

Whitehead integral and helicity

For a map $\phi : S^3 \to S^2$, after removing a point of S^3 such that S^3 is viewed as $\mathbb{R}^3$ in terms of an Euclidean coordinate system, the Hopf invariant of ϕ may be represented as an integral known as the Whitehead integral, obtained originally in [579], which relates itself to helicity as well. To appreciation this connection, we construct a vector field $\mathbf{B}$ from $\phi : \mathbb{R}^3 \to S^2$ by setting

$$\mathbf{B} = \mathbf{B}(\phi) = (B^i(\phi)), \quad B^i(\phi) = \frac{1}{2}\epsilon^{ijk}F_{jk}(\phi), \quad i,j,k = 1,2,3, \tag{A.2.20}$$

often interpreted as the "magnetic field" induced from the spin vector ϕ, where

$$F_{ij}(\phi) = \phi\cdot(\partial_i\phi \times \partial_j\phi), \quad \mathbf{x} = (x^i), \quad i,j = 1,2,3. \tag{A.2.21}$$

Since $|\phi|^2 = 1$ so that $\phi\cdot\partial_i\phi = 0$ for $i = 1,2,3$, we know that $\partial_1\phi, \partial_2\phi, \partial_3\phi$ are linearly dependent vectors. Hence, $\det(\partial_1\phi, \partial_2\phi, \partial_3\phi) = 0$. From this fact, we have

$$\nabla\cdot\mathbf{B} = \partial_i B^i(\phi) = 0. \tag{A.2.22}$$

That is, $\mathbf{B}$ given in (A.2.20) is solenoidal. Thus, there is a vector field $\mathbf{A} = \mathbf{A}(\phi)$ such that $\mathbf{B} = \nabla \times \mathbf{A}$. With such a ϕ-dependent pair of vector fields $\mathbf{A}(\phi)$ and $\mathbf{B}(\phi)$, Whitehead [579] found the following integral representation of the Hopf invariant of ϕ:

$$H(\phi) = \frac{1}{16\pi^2} \int_{\mathbb{R}^3} \mathbf{A}(\phi) \cdot \mathbf{B}(\phi) \, \mathrm{d}\mathbf{x}. \tag{A.2.23}$$

This expression is of the Chern–Simons type (A.2.19), in which $\mathbf{A}(\phi)$ is determined by the Biot–Savart law formula (A.2.17) in terms of $\mathbf{B}(\phi)$, or

$$\mathbf{A}(\phi)(\mathbf{x}) = \frac{1}{4\pi} \int_{\mathbb{R}^3} \mathbf{B}(\phi)(\mathbf{y}) \times \frac{(\mathbf{x} - \mathbf{y})}{\|\mathbf{x} - \mathbf{y}\|^3} \, \mathrm{d}\mathbf{y}. \tag{A.2.24}$$

Consequently, substituting (A.2.24) into (A.2.23), we arrive at

$$H(\phi) = \frac{1}{(4\pi)^3} \int_{\mathbb{R}^3} \int_{\mathbb{R}^3} \frac{(\mathbf{B}(\phi)(\mathbf{x}) \times \mathbf{B}(\phi)(\mathbf{y})) \cdot (\mathbf{x} - \mathbf{y})}{\|\mathbf{x} - \mathbf{y}\|^3} \, \mathrm{d}\mathbf{x}\mathrm{d}\mathbf{y}, \tag{A.2.25}$$

expressing the Hopf invariant of the map ϕ as the helicity of the magnetic field induced from ϕ.

A.3 Noether theorem

The celebrated *Noether theorem* is a general statement asserting that there is a time-conservative quantity associated to each continuous symmetry in a Lagrangian mechanical or field-theoretical system. This section presents a study of this foundational theorem. It discusses some basic constructions in Lagrangian mechanics concerning the concepts of energy and momenta arising from time- and space-symmetry structures as examples and motivations, and then develops such a formalism in the context of classical field theory as a natural extension.

Lagrangian mechanics

Let $q = (q^i) = (q^1, \dots, q^n)$ be the generalized coordinate vector of a moving particle at time t and

$$L = L(q, \dot{q}, t) \tag{A.3.1}$$

the Lagrangian action function. Recall that the equations of motion of the particle, which are the Euler–Lagrange equations of (A.3.1), are

$$\frac{\mathrm{d}}{\mathrm{d}t}\left(\frac{\partial L}{\partial \dot{q}^i}\right) = \frac{\partial L}{\partial q^i}, \quad i = 1, \dots, n. \tag{A.3.2}$$

Or in its vector form, we have

$$\frac{\mathrm{d}}{\mathrm{d}t} \nabla_{\dot{q}} L = \nabla_q L, \quad q, \dot{q} \in \mathbb{R}^n. \tag{A.3.3}$$

Linear momenta

Consider a uniform space-coordinate translation given by

$$q' = q + a, \quad a = (a^1, \ldots, a^n) \in \mathbb{R}^n, \tag{A.3.4}$$

which induces the transformation

$$q' = q + a, \quad \dot{q}' = \dot{q}, \quad t' = t. \tag{A.3.5}$$

Thus, if we impose the invariance or symmetry of the action function,

$$L(q', \dot{q}', t') = L(q, \dot{q}, t), \tag{A.3.6}$$

so that all the nontrivial components of a are allowed to change continuously near zero, then

$$0 = \left.\frac{\partial L}{\partial a^i}\right|_{a=0} = \frac{\partial L}{\partial q^i}, \tag{A.3.7}$$

if the ith component of a in (A.3.4) is nontrivial. In other words, L is independent of q^i. In view of (A.3.2) and (A.3.7), we get

$$\frac{\mathrm{d}}{\mathrm{d}t}\left(\frac{\partial L}{\partial \dot{q}^i}\right) = 0. \tag{A.3.8}$$

Therefore, we find that, if the Lagrangian action function enjoys a translation invariance in its ith coordinate q^i, then the function is independent of q^i and the quantity

$$p_i = \frac{\partial L}{\partial \dot{q}^i}, \tag{A.3.9}$$

called the *linear momentum* associated with the ith coordinate, is a conserved quantity with respect to the mechanical evolution of the system.

Hamiltonian energy

Likewise, we may consider the time translation or shift

$$t' = t + b, \quad b \in \mathbb{R}, \tag{A.3.10}$$

where $b \in \mathbb{R}$ is arbitrary, which induces the transformation

$$q' = q, \quad \dot{q}' = \dot{q}, \quad t' = t + b. \tag{A.3.11}$$

Hence, if we impose the symmetry (A.3.6) or

$$L(q, \dot{q}, t + b) = L(q, \dot{q}, t), \tag{A.3.12}$$

we have

$$\frac{\partial L}{\partial t} = 0, \tag{A.3.13}$$

as in (A.3.7).

On the other hand, if $q = q(t) = (q^i(t))$ is a solution of (A.3.2), then we have

$$\frac{\mathrm{d}L}{\mathrm{d}t} = \frac{\mathrm{d}}{\mathrm{d}t}\left(L(q(t), \dot{q}(t), t)\right) = \sum_{i=1}^{n}\left(\frac{\partial L}{\partial q^i}\dot{q}^i + \frac{\partial L}{\partial \dot{q}^i}\ddot{q}^i\right)(t) + \frac{\partial L}{\partial t}. \tag{A.3.14}$$

Therefore, in view of (A.3.2), (A.3.14) becomes

$$\frac{\mathrm{d}}{\mathrm{d}t}\left(\sum_{i=1}^{n}\frac{\partial L}{\partial \dot{q}^i}\dot{q}^i - L\right) + \frac{\partial L}{\partial t} = 0. \tag{A.3.15}$$

As a result of (A.3.9) and (A.3.15), if we introduce the quantity

$$H = \sum_{i=1}^{n} p_i \dot{q}^i - L, \tag{A.3.16}$$

then it is conserved,

$$\frac{\mathrm{d}H}{\mathrm{d}t} = 0, \tag{A.3.17}$$

provided that (A.3.13) holds. In other words, the quantity (A.3.16), called the Hamiltonian energy function, is conserved with respect to the evolution of the mechanical system if its Lagrangian action function L enjoys a time-translation invariance so that it is not explicitly dependent on the time variable t.

Angular momenta

In addition to the continuous symmetries associated with the translation invariance of space and time coordinates, we now turn our attention to rotational invariance. For this purpose, consider a one-parameter family, $\Omega(\theta)$ ($\theta \in \mathbb{R}$), in the orthogonal group $O(n)$, say, satisfying

$$\Omega(0) = I, \quad \Omega'(0) = \Gamma, \tag{A.3.18}$$

where Γ is a skew-symmetric $n \times n$ matrix that generates $\Omega(\theta)$. With $\Omega(\theta)$, we may introduce the transformed coordinate vector

$$q' = \Omega(\theta) q, \tag{A.3.19}$$

where q, q' are regarded as column vectors in $\mathbb{R}^n$ for notional convenience. If the Lagrangian action function (A.3.1) enjoys a one-parameter symmetry such that it is invariant under the transformation (A.3.19), with $\dot{q}' = \Omega(\theta)\dot{q}$ and $t' = t$, then (A.3.6) becomes

$$L(q', \dot{q}', t) = L(q, \dot{q}, t). \tag{A.3.20}$$

For example, if the motion of a particle of mass $m > 0$ is subject to a centripetal force, then its action function (A.3.1) assumes the form

$$L = \frac{1}{2} m \sum_{i=1}^{n} (\dot{q}^i)^2 - V(|q|^2, t), \tag{A.3.21}$$

which is invariant under the full orthogonal group $O(n)$ depending on $\frac{n(n-1)}{2}$ parameters.

With (A.3.19), (A.3.20), and setting

$$\left(\frac{\mathrm{d}L}{\mathrm{d}\theta}(q', \dot{q}', t)\right)\bigg|_{\theta=0} = 0, \tag{A.3.22}$$

we have

$$(\nabla_q L)^\tau \Gamma q + (\nabla_{\dot{q}} L)^\tau \Gamma \dot{q} = 0, \tag{A.3.23}$$

where τ denotes the transposition operation. Thus, in view of (A.3.2) and (A.3.9), we obtain

$$\frac{\mathrm{d}}{\mathrm{d}t}(p^\tau \Gamma q) = 0, \tag{A.3.24}$$

leading to the conserved quantity

$$a_\Gamma = p^\tau \Gamma q, \tag{A.3.25}$$

associated with the generator Γ. Hence, associated with $O(n)$, there are $\frac{n(n-1)}{2}$ angular momenta, possible.

Example

As an illustrative example, consider the motion of a particle of mass $m > 0$ with the space coordinates x, y, z, in $\mathbb{R}^3$, whose Lagrangian action function is given by

$$L = \frac{1}{2}m(\dot{x}^2 + \dot{y}^2 + \dot{z}^2) - V(\sqrt{x^2 + y^2}, z). \tag{A.3.26}$$

It is clear that the system is invariant under rotations about the z-axis, with

$$\Omega(\theta) = \begin{pmatrix} \cos\theta & -\sin\theta & 0 \\ \sin\theta & \cos\theta & 0 \\ 0 & 0 & 1 \end{pmatrix}. \tag{A.3.27}$$

Thus, using

$$\Omega'(0) = \begin{pmatrix} 0 & -1 & 0 \\ 1 & 0 & 0 \\ 0 & 0 & 0 \end{pmatrix} \equiv \Gamma_z, \quad p = m\dot{\mathbf{r}} \equiv \mathbf{p}, \quad \mathbf{r} = (x, y, z)^\tau, \tag{A.3.28}$$

in (A.3.25), we have

$$a_{\Gamma_z} = m\dot{\mathbf{r}} \cdot (\Gamma_z \mathbf{r}) = m(x\dot{y} - \dot{x}y) \equiv L_z, \tag{A.3.29}$$

which is the expression of the classical angular momentum of the motion about the z-axis.

Field theory situation

Consider now a field u governed by the Lagrangian action functional

$$S = \int \mathcal{L}(x; u, Du)\,\mathrm{d}x, \tag{A.3.30}$$

where $x = (x^\mu) = (x^0, x^1, \ldots, x^n)$ denotes the coordinate vector of a position point in an $(n+1)$-dimensional spacetime, which may either be Minkowskian or Euclidean,

$$u = u(x) = (u^a(x)) = (u^1(x), \ldots, u^m(x)) \tag{A.3.31}$$

stands for an m-component real-valued field, and $Du = (\partial_\mu u) = (\partial_0 u, \partial_1 u, \ldots, \partial_n u)$. The Euler–Lagrange equations of (A.3.30) read

$$\partial_\mu \left(\frac{\partial \mathcal{L}}{\partial(\partial_\mu u^a)} \right) = \frac{\partial \mathcal{L}}{\partial u^a}, \quad a = 1, \ldots, m. \tag{A.3.32}$$

Use δ to denote a generic infinitesimal variation initially caused by a spacetime coordinate change so that

$$\delta x = x' - x, \quad \delta u(x) = u'(x') - u(x), \quad \delta(\partial_\mu u(x)) = \partial'_\mu u'(x') - \partial_\mu u(x), \tag{A.3.33}$$

where $\partial'_\mu = \frac{\partial}{\partial x'^\mu}$. Here, $u'(x')$ denotes the "value" of the field at the same spacetime point with the coordinate vector x'. For a *scalar field*, for example, the temperature distribution function, we have $u'(x') = u(x)$. Thus, we have

$$\partial_\nu x'^\mu = \frac{\partial x'^\mu}{\partial x^\nu} = \delta^\mu_\nu + \partial_\nu(\delta x^\mu), \tag{A.3.34}$$

where (and in the sequel) δx^μ and $(\delta x)^\mu$ are used interchangeably without causing confusion. Conceptually and practically, it helps to regard the variation δ as to be associated with a parameter, $\theta \in \mathbb{R}$, such that

$$x' = \Omega(x; \theta), \quad \Omega(x; 0) = x, \tag{A.3.35}$$

which extends (A.3.19). Hence, δ corresponds to the first-order terms in the expansion in θ of (A.3.35). In this way, the quantity $\partial_\nu(\delta x)$ in (A.3.34), for example, is of the first order in δ. As an application of this fact, we see that (A.3.34) leads us to the Jacobian of the transformation (A.3.35),

$$J(x', x) = \det\left(\partial_\nu x'^\mu\right) = 1 + \partial_\mu(\delta x^\mu), \tag{A.3.36}$$

up to the first order of δ.

Assume now that the action (A.3.30) is invariant under the transformation (A.3.35). Then,

$$\begin{aligned} 0 &= \int_{\mathcal{R}'} \mathcal{L}(x'; u'(x'), D'u'(x'))\,\mathrm{d}x' - \int_{\mathcal{R}} \mathcal{L}(x; u(x), Du(x))\,\mathrm{d}x \\ &= \int_{\mathcal{R}} \left(\mathcal{L}(x'; u'(x'), D'u'(x'))J(x', x) - \mathcal{L}(x; u(x), Du(x))\right)\mathrm{d}x \\ &= \int_{\mathcal{R}} \left((\delta\mathcal{L})J(x', x) + \mathcal{L}(x; u(x), Du(x))(J(x', x) - 1)\right)\mathrm{d}x, \end{aligned} \tag{A.3.37}$$

where $\mathcal{R}$ and $\mathcal{R}'$ are the corresponding regions of integration, respectively, and

$$\delta\mathcal{L} = \mathcal{L}(x'; u'(x'), D'u'(x')) - \mathcal{L}(x; u(x), Du(x)). \tag{A.3.38}$$

In order to estimate (A.3.38) further, we need the *modified variation* $\tilde{\delta}$ [253] given as

$$\tilde{\delta}u(x) = u'(x) - u(x), \tag{A.3.39}$$

measuring the difference of the values of the field after and before the coordinate change at the same coordinate point. By virtue of this definition, the operations $\tilde{\delta}$ and ∂_μ clearly commute:

$$\partial_\mu(\tilde{\delta}u(x)) = \partial_\mu u'(x) - \partial_\mu u(x) = \tilde{\delta}(\partial_\mu u(x)). \tag{A.3.40}$$

Moreover, from

$$\begin{aligned}\tilde{\delta}u(x) &= (u'(x') - u(x)) - (u'(x') - u'(x))\\ &= \delta u(x) - (\partial_\mu u'(x))\delta x^\mu\\ &= \delta u(x) - (\partial_\mu u(x))\delta x^\mu,\end{aligned} \tag{A.3.41}$$

neglecting higher-order terms, we see that $\tilde{\delta}$ is also a first-order variation. On the other hand, the operations δ and ∂_μ do not commute:

$$\begin{aligned}\partial_\mu(\delta u(x)) &= \partial_\mu u'(x') - \partial_\mu u(x)\\ &= (\partial'_\mu u'(x') - \partial_\mu u(x)) + (\partial_\mu u'(x') - \partial'_\mu u'(x'))\\ &= \delta(\partial_\mu u(x)) + \partial_\mu(\delta x^\nu)\,\partial'_\nu u'(x')\\ &= \delta(\partial_\mu u(x)) + \partial_\mu(\delta x^\nu)\partial_\nu u(x),\end{aligned} \tag{A.3.42}$$

up to the first-order terms in δ, where we have applied the identity

$$\partial_\mu = \partial'_\mu + \partial_\mu(\delta x^\nu)\partial'_\nu, \tag{A.3.43}$$

in view of the chain rule and (A.3.34).

With (A.3.33) and (A.3.39), we can rewrite (A.3.38) as

$$\begin{aligned}\delta\mathcal{L} &= (\mathcal{L}(x; u'(x), Du'(x)) - \mathcal{L}(x; u(x), Du(x)))\\ &\quad + (\mathcal{L}(x'; u'(x'), D'u'(x')) - \mathcal{L}(x; u'(x), Du'(x)))\\ &= \frac{\partial\mathcal{L}}{\partial u^a}\tilde{\delta}u^a + \frac{\partial\mathcal{L}}{\partial(\partial_\mu u^a)}\tilde{\delta}(\partial_\mu u^a) + \partial_\mu(\mathcal{L}(x; u(x), Du(x)))(\delta x^\mu)\\ &= \partial_\mu\left(\frac{\partial\mathcal{L}}{\partial(\partial_\mu u^a)}\right)\tilde{\delta}u^a + \frac{\partial\mathcal{L}}{\partial(\partial_\mu u^a)}\partial_\mu(\tilde{\delta}u^a) + (\partial_\mu\mathcal{L})\delta x^\mu,\end{aligned} \tag{A.3.44}$$

keeping all first-order variation terms only and applying the equations of motion (A.3.32) and the commutativity property (A.3.40). Hence, in view of (A.3.36) and (A.3.44), we obtain from (A.3.37) the first-order result

$$\int_{\mathcal{R}}\left(\partial_\mu\left[\frac{\partial\mathcal{L}}{\partial(\partial_\mu u^a)}\tilde{\delta}u^a + \mathcal{L}\delta x^\mu\right]\right)\mathrm{d}x = 0. \tag{A.3.45}$$

Thus, inserting (A.3.41) into (A.3.45), we have

$$
\begin{aligned}
0 &= \int_{\mathcal{R}} \left(\partial_\mu \left[\frac{\partial \mathcal{L}}{\partial(\partial_\mu u^a)} (\delta u^a - \partial_\nu u^a \delta x^\nu) + \mathcal{L} \delta x^\mu \right] \right) \mathrm{d}x \\
&= \int_{\mathcal{R}} \left(\partial_\mu \left[\frac{\partial \mathcal{L}}{\partial(\partial_\mu u^a)} \delta u^a - \left(\frac{\partial \mathcal{L}}{\partial(\partial_\mu u^a)} \partial_\nu u^a - \delta^\mu_\nu \mathcal{L} \right) \delta x^\nu \right] \right) \mathrm{d}x. \qquad \text{(A.3.46)}
\end{aligned}
$$

To proceed, we assume that the transformations are parametrized by the parameters $\{\theta^1, \dots, \theta^k\}$ so that up to the first orders the variations δx^μ and δu^a may be expressed by

$$
\delta x^\mu = X^\mu_s(x) \theta^s, \quad \delta u^a(x) = U^a_s(x, u(x)) \theta^s, \quad \theta^s \sim 0, \quad s = 1, \dots, k. \qquad \text{(A.3.47)}
$$

Substituting (A.3.47) into (A.3.46), we arrive at

$$
\int_{\mathcal{R}} \partial_\mu \left(\frac{\partial \mathcal{L}}{\partial(\partial_\mu u^a)} U^a_s - T^\mu_\nu X^\nu_s \right) \theta^s \, \mathrm{d}x = 0, \qquad \text{(A.3.48)}
$$

where

$$
T^\nu_\mu = \frac{\partial \mathcal{L}}{\partial(\partial_\nu u^a)} \partial_\mu u^a - \delta^\nu_\mu \mathcal{L} \qquad \text{(A.3.49)}
$$

is often referred to as the *energy-momentum tensor*, which (through the spacetime metric tensor $g_{\mu\nu}$ or $g^{\mu\nu}$ satisfying $g_{\mu\alpha} g^{\nu\alpha} = \delta^\nu_\mu$) may be rewritten as

$$
T_{\mu\nu} = g_{\mu\alpha} T^\alpha_\nu = \frac{\partial \mathcal{L}}{\partial(\partial^\mu u^a)} \partial_\nu u^a - g_{\mu\nu} \mathcal{L}, \qquad \text{(A.3.50)}
$$

$$
T^{\mu\nu} = g^{\mu\alpha} T^\nu_\alpha = \frac{\partial \mathcal{L}}{\partial(\partial_\nu u^a)} \partial^\mu u^a - g^{\mu\nu} \mathcal{L}. \qquad \text{(A.3.51)}
$$

Noether currents and charges

Since the integration region $\mathcal{R}$ in (A.3.48) is arbitrary, we obtain immediately k conservation laws:

$$
\partial_\mu J^\mu_s = 0, \quad s = 1, \dots, k, \qquad \text{(A.3.52)}
$$

where the k currents labeled by s,

$$
J^\mu_s = \frac{\partial \mathcal{L}}{\partial(\partial_\mu u^a)} U^a_s - T^\mu_\nu X^\nu_s, \quad s = 1, \dots, k, \qquad \text{(A.3.53)}
$$

are referred to as the *Noether currents* over the $(n+1)$-dimensional spacetime with the spacetime coordinate indices $\mu, \nu = 0, 1, \dots, n$, field-component index $a = 1, \dots, m$, and transformation group index $s = 1, \dots, k$.

For any fixed $s = 1, \dots, k$, set

$$
(J^\mu_s) = (\rho_s, \mathbf{j}_s), \quad \rho_s = J^0_s, \quad \mathbf{j}_s = (J^1_s, \dots, J^n_s), \qquad \text{(A.3.54)}
$$

and take $t = x^0$ and $\mathbf{x} = (x^1, \dots, x^n)$ to be the time and space coordinates, respectively. Then, the conservation laws (A.3.52) assume the usual divergence form,

$$\frac{\partial \rho_s}{\partial t} + \nabla_{\mathbf{x}} \cdot \mathbf{j}_s = 0, \quad s = 1, \dots, k. \tag{A.3.55}$$

Consequently, if we introduce the total charges

$$Q_s = \int_{\mathbb{R}^n} \rho_s \, \mathrm{d}\mathbf{x}, \tag{A.3.56}$$

then using the divergence theorem, observing the laws (A.3.55), and neglecting boundary terms resulting from integration, we obtain

$$\frac{\mathrm{d}Q_s}{\mathrm{d}t} = \int_{\mathbb{R}^n} \frac{\partial \rho_s}{\partial t} \, \mathrm{d}\mathbf{x} = -\int_{\mathbb{R}^n} \nabla_{\mathbf{x}} \cdot \mathbf{j}_s \, \mathrm{d}\mathbf{x} = 0, \quad s = 1, \dots, k. \tag{A.3.57}$$

In other words, we have acquired k conservative quantities, (A.3.56), known as the *Noether charges*, associated with the k continuous parameters that parametrize the transformation group, which leaves the Lagrangian action functional of the field-theoretical model invariant. This is Noether's theorem.

Some examples

Let u be a scalar field. Then, $u'(x') = u(x)$ so that $U_s^a = 0$ in (A.3.47). Now we consider spacetime translations given by

$$x'^{\mu} = x^{\mu} - \theta^{\mu}, \quad \mu = 0, 1, \dots, n. \tag{A.3.58}$$

Hence, the transformation group index s in (A.3.47) is simply the spacetime coordinate index and the coefficient $X_s^{\mu}(x)$ in (A.3.47) becomes $X_{\nu}^{\mu} = -\delta_{\nu}^{\mu}$. Therefore, the Noether currents (A.3.53) assume the reduced form

$$J_{\nu}^{\mu} = T_{\nu}^{\mu}, \quad s = \nu = 0, 1, \dots, n. \tag{A.3.59}$$

As a consequence, the conservation laws (A.3.52) read

$$\partial_{\mu} T_{\nu}^{\mu} = 0, \quad \text{or equivalently,} \quad \partial_{\nu} T^{\mu\nu} = 0, \tag{A.3.60}$$

which is the conservation law of the energy-momentum tensor. Moreover, in view of (A.3.54), we find a total of $(n+1)$ charge densities

$$\rho_0 = J_0^0 = T_0^0, \quad \rho_i = J_i^0 = T_i^0, \quad i = 1, \dots, n, \tag{A.3.61}$$

for which $\rho_0 = T_0^0$ is referred to as the energy density and $\rho_i = T_i^0$ the ith component of the momentum vector density, which give rise to conserved total energy, E, and momentum vector $\mathbf{P} = (P_i)$, in view of (A.3.56) and (A.3.57), defined by the integrals

$$E = \int_{\mathbb{R}^n} T_0^0 \, \mathrm{d}\mathbf{x}, \tag{A.3.62}$$

$$P_i = \int_{\mathbb{R}^n} T_i^0 \, \mathrm{d}\mathbf{x}, \quad i = 1, \dots, n. \tag{A.3.63}$$

Note also that if we rewrite the Lagrangian action density in (A.3.30) in a relativistic manner such that

$$\mathcal{L} = \mathcal{L}\left(x; u, \partial_\mu u \partial^\mu u\right), \quad \partial_\mu u \partial^\mu u = \sum_{a=1}^{m} \partial_\mu u^a g^{\mu\nu} \partial_\nu u^a = \sum_{a=1}^{m} \partial^\mu u^a g_{\mu\nu} \partial^\nu u^a, \tag{A.3.64}$$

then the energy-momentum tensors (A.3.50) and (A.3.51) may also be recast into the forms

$$T_{\mu\nu} = 2\frac{\partial \mathcal{L}}{\partial g^{\mu\nu}} - g_{\mu\nu}\mathcal{L}, \quad T^{\mu\nu} = 2\frac{\partial \mathcal{L}}{\partial g_{\mu\nu}} - g^{\mu\nu}\mathcal{L}. \tag{A.3.65}$$

Another simple but interesting situation is $x' = x$, but $u' = \Omega u$ where $\Omega \in SO(m)$ and the $SO(m)$-invariant Lagrangian action density reads

$$\mathcal{L} = \frac{1}{2}\partial_\mu u \partial^\mu u - V(|u|^2). \tag{A.3.66}$$

Thus, $k = \dim(SO(m)) = \frac{1}{2}m(m-1)$. Use $\{\Gamma_1, \dots, \Gamma_k\}$ to denote a set of generators of $SO(m)$, which are skew-symmetric $m \times m$ real matrices such that each $\Omega \in SO(m)$ may be represented as

$$\Omega = \mathrm{e}^{\omega}, \quad \omega = \sum_{s=1}^{k} \theta^s \Gamma_s, \quad \theta^1, \dots, \theta^k \in \mathbb{R}. \tag{A.3.67}$$

As a result, we have

$$\delta u = \Omega u - u = \sum_{s=1}^{k} \theta^s \Gamma_s u, \tag{A.3.68}$$

within the first-order term approximation for $\theta^1, \dots, \theta^k$ near zero. Therefore, we see that the variation coefficients X_s^μ and U_s^a in (A.3.47) are determined to be

$$X_s^\mu = 0, \quad U_s^a = (\Gamma_s u)^a, \quad a = 1, \dots, m, \; s = 1, \dots, k, \; \mu = 0, 1, \dots, n. \tag{A.3.69}$$

As a consequence of the general construction (A.3.53) for the Noether currents and the expressions in (A.3.69) for the model (A.3.66), we arrive at the results

$$J_s^\mu = \frac{\partial \mathcal{L}}{\partial(\partial_\mu u^a)}(\Gamma_s u)^a = \sum_{a=1}^{m} \partial^\mu u^a (\Gamma_s u)^a = \partial^\mu u \cdot (\Gamma_s u) = (\partial^\mu u)^\tau \Gamma_s u, \tag{A.3.70}$$

as the $\frac{1}{2}m(m-1) = k$ Noether currents induced from the $SO(m)$-symmetry of the model over the $(n+1)$-dimensional spacetime, labeled by s, specifically. In particular, the conserved Noether charges are

$$Q_s = \int_{\mathbb{R}^n} (\partial^0 u)^\tau \Gamma_s u \,\mathrm{d}\mathbf{x}, \quad s = 1, \dots, k. \tag{A.3.71}$$

In addition to the Noether charges resulting from continuous symmetries of the system, other features of the system may lead to the appearance of some conserved charges of a *non-Noetherian* nature. For example, certain topological properties of the system may give birth to a distinctive family of conserved charges, called *topological charges*.

A.4 Spectra of angular momentum operators

Consider the motion of a particle of mass $m > 0$ in $\mathbb{R}^3$ with the space coordinates x, y, z. In terms of its position vector $\mathbf{r} = (x, y, z)$ and time variable t, the associated linear momentum vector $\mathbf{p}$ of the particle is

$$\mathbf{p} = (p_x, p_y, p_z) = m\frac{\mathrm{d}\mathbf{r}}{\mathrm{d}t} = m\dot{\mathbf{r}} = m(\dot{x}, \dot{y}, \dot{z}), \tag{A.4.1}$$

so that the induced orbital angular momentum vector for the motion reads

$$\mathbf{L} = (L_x, L_y, L_z) = \mathbf{r} \times \mathbf{p} = (yp_z - zp_y, zp_x - xp_z, xp_y - yp_x). \tag{A.4.2}$$

The square of the orbital angular momentum vector is

$$L^2 = \mathbf{L}^2 = L_x^2 + L_y^2 + L_z^2. \tag{A.4.3}$$

These quantities are the basis of the following discussion.

Quantized angular momenta

With the quantization of the coordinates and momenta of the system given by

$$\mathbf{r} \mapsto \hat{\mathbf{r}} = \mathbf{r}, \quad \mathbf{p} \mapsto \hat{\mathbf{p}} = -\mathrm{i}\hbar\nabla = -\mathrm{i}\hbar(\partial_x, \partial_y, \partial_z) = (\hat{p}_x, \hat{p}_y, \hat{p}_z), \tag{A.4.4}$$

we see that the quantum operator versions of various angular momenta are

$$\hat{L}_x = y\hat{p}_z - z\hat{p}_y, \quad \hat{L}_y = z\hat{p}_x - x\hat{p}_z, \quad \hat{L}_z = x\hat{p}_y - y\hat{p}_x, \tag{A.4.5}$$

$$\hat{L}^2 = \hat{L}_x^2 + \hat{L}_y^2 + \hat{L}_z^2, \tag{A.4.6}$$

which are all Hermitian or self-adjoint.

Commutators

Using a testing function, we check that the nontrivial and independent commutators among the operators $x, y, z, \hat{p}_x, \hat{p}_y, \hat{p}_z$ are

$$[x, \hat{p}_x] = [y, \hat{p}_y] = [z, \hat{p}_z] = \mathrm{i}\hbar. \tag{A.4.7}$$

In view of these, we have the cyclic relations

$$[\hat{L}_x, \hat{L}_y] = \mathrm{i}\hbar\hat{L}_z, \quad [\hat{L}_y, \hat{L}_z] = \mathrm{i}\hbar\hat{L}_x, \quad [\hat{L}_z, \hat{L}_x] = \mathrm{i}\hbar\hat{L}_y. \tag{A.4.8}$$

For example, we may examine the first relation in (A.4.8) with

$$\begin{aligned}[\hat{L}_x, \hat{L}_y] &= [y\hat{p}_z - z\hat{p}_y, z\hat{p}_x - x\hat{p}_z] \\ &= [y\hat{p}_z, z\hat{p}_x] + [z\hat{p}_y, x\hat{p}_z] \\ &= \mathrm{i}\hbar(x\hat{p}_y - y\hat{p}_x) \\ &= \mathrm{i}\hbar\hat{L}_z.\end{aligned} \tag{A.4.9}$$

As a consequence, we get

$$[\hat{L}^2, \hat{L}_x] = [\hat{L}_y^2, \hat{L}_x] + [\hat{L}_z^2, \hat{L}_x]$$
$$= \hat{L}_y[\hat{L}_y, \hat{L}_x] + [\hat{L}_y, \hat{L}_x]\hat{L}_y + \hat{L}_z[\hat{L}_z, \hat{L}_x] + [\hat{L}_z, \hat{L}_x]\hat{L}_z = 0, \quad \text{(A.4.10)}$$

after applying (A.4.8) again. Similarly, we also have $[\hat{L}^2, \hat{L}_y] = 0$ and $[\hat{L}^2, \hat{L}_z] = 0$.

Eigenvalues of angular momentum operators

We are to find the eigenvalues of the operators $\hat{L}_x, \hat{L}_y, \hat{L}_z, \hat{L}^2$. Since each of $\hat{L}_x, \hat{L}_y, \hat{L}_z$ commutes with $\hat{L}^2$, we know that each of $\hat{L}_x, \hat{L}_y, \hat{L}_z$, and $\hat{L}^2$ share simultaneous eigenfunctions. This property is essential for us to proceed. As an example, we consider the eigenvalues of the pair $\hat{L}_z$ and $\hat{L}^2$, subsequently. Thus, we compute λ and σ in the equations

$$\hat{L}_z f = \lambda f, \quad \hat{L}^2 f = \sigma f, \quad \text{(A.4.11)}$$

for a simultaneous eigenfunction f. Since L_z is Hermitian, we know that λ is real. Since $\hat{L}^2$ is positive semi-definite, we have $\sigma \geq 0$. Here, we focus on the nontrivial case, $\sigma > 0$.

Ladder operators

In order to simplify the calculation, we introduce the ladder operators

$$\hat{L}_+ = \hat{L}_x + \mathrm{i}\hat{L}_y, \quad \hat{L}_- = \hat{L}_x - \mathrm{i}\hat{L}_y, \quad \text{(A.4.12)}$$

resembling the creation and annihilation operators. Note that these operators are no longer Hermitian. We see that there hold the commutators

$$[\hat{L}_z, \hat{L}_\pm] = \pm\hbar\hat{L}_\pm, \quad [\hat{L}^2, \hat{L}_\pm] = 0, \quad [\hat{L}_-, \hat{L}_+] = -2\hbar\hat{L}_z. \quad \text{(A.4.13)}$$

Eigenvalues and eigenfunctions from ladder operators

Using (A.4.11) and (A.4.13), we have

$$\hat{L}_z(\hat{L}_\pm f) = \left(\hat{L}_z\hat{L}_\pm - \hat{L}_\pm\hat{L}_z + \hat{L}_\pm\hat{L}_z\right) f$$
$$= [\hat{L}_z, \hat{L}_\pm]f + \hat{L}_\pm\hat{L}_z f$$
$$= (\lambda \pm \hbar)(\hat{L}_\pm f). \quad \text{(A.4.14)}$$

In other words, if $\hat{L}_\pm f \neq 0$, then $\hat{L}_\pm f$ is an eigenfunction of $\hat{L}_z$ associated with the eigenvalue $\lambda \pm \hbar$. Thus, $\hat{L}_+$ and $\hat{L}_-$ are raising and lowering ladder operators, respectively. On the other hand, since $\hat{L}^2$ and $\hat{L}_\pm$ commute, $\hat{L}_\pm f$ lies in the same eigenspace of $\hat{L}^2$ associated with the eigenvalue σ:

$$\hat{L}^2(\hat{L}_\pm f) = \hat{L}_\pm(\hat{L}^2 f) = \sigma(\hat{L}_\pm f). \quad \text{(A.4.15)}$$

Bounds of eigenvalues

Use $\langle , \rangle$ to denote the inner product of the function space where the eigenfunctions reside. Then, by virtue of (A.4.6) and (A.4.11), we have

$$\begin{aligned}\sigma\langle f, f\rangle &= \langle f, \hat{L}^2 f\rangle \\ &= \langle \hat{L}_x f, \hat{L}_x f\rangle + \langle \hat{L}_y f, \hat{L}_y f\rangle + \langle f, \hat{L}_z^2 f\rangle \\ &\geq \lambda^2 \langle f, f\rangle,\end{aligned} \tag{A.4.16}$$

resulting in the bounds

$$-\sqrt{\sigma} \leq \lambda \leq \sqrt{\sigma}. \tag{A.4.17}$$

Climbing the ladder

Since $\hat{L}_+$ raises the eigenvalues, the upper bound in (A.4.17) indicates that there is an integer $k \geq 0$ such that

$$\hat{L}_+^{k+1} f = 0, \quad \hat{L}_+^k f \neq 0. \tag{A.4.18}$$

It is clear that $\hat{L}_+^k f$ is an eigenfunction that corresponds to the highest eigenvalue of $\hat{L}_z$, which is $\lambda + k\hbar$, accessible by applying the raising ladder operator. For this reason, we use the notation

$$f_{\text{top}} = \hat{L}_+^k f, \quad \lambda_{\text{top}} = \lambda + k\hbar. \tag{A.4.19}$$

Descending the ladder

Likewise, if we use $\hat{L}_-$ to lower eigenvalues, we may obtain an integer $l \geq 0$ such that

$$\hat{L}_-^{l+1} f = 0, \quad \hat{L}_-^l f \neq 0. \tag{A.4.20}$$

In a similar manner, we get the eigenpair

$$f_{\text{bottom}} = \hat{L}_-^l f, \quad \lambda_{\text{bottom}} = \lambda - l\hbar, \tag{A.4.21}$$

at the bottom level of the eigenpair ladder.

Determination of eigenvalues

From (A.4.8) and (A.4.12), we have

$$\hat{L}_- \hat{L}_+ = \hat{L}^2 - \hat{L}_z^2 - \hbar \hat{L}_z. \tag{A.4.22}$$

Thus, applying (A.4.22) to f_{top} and using $\hat{L}_+ f_{\text{top}} = 0$ and $\hat{L}_z f_{\text{top}} = \lambda_{\text{top}} f_{\text{top}}$, we obtain

$$\hat{L}^2 f_{\text{top}} = \left(\lambda_{\text{top}}^2 + \hbar\lambda_{\text{top}}\right) f_{\text{top}}, \tag{A.4.23}$$

which renders the result

$$\sigma = \lambda_{\text{top}}^2 + \hbar\lambda_{\text{top}}. \tag{A.4.24}$$

On the other hand, for the bottom eigenpair given in (A.4.21), we have

$$\hat{L}_z f_{\text{bottom}} = \lambda_{\text{bottom}} f_{\text{bottom}}, \quad \hat{L}_- f_{\text{bottom}} = 0. \tag{A.4.25}$$

Besides, from (A.4.13) and (A.4.22), we have the identity

$$\hat{L}_+ \hat{L}_- = \hat{L}^2 - \hat{L}_z^2 + \hbar \hat{L}_z. \tag{A.4.26}$$

Hence, applying (A.4.26) to f_{bottom} and using (A.4.25), we come up with

$$\hat{L}^2 f_{\text{bottom}} = \left(\lambda_{\text{bottom}}^2 - \hbar\lambda_{\text{bottom}}\right) f_{\text{bottom}}. \tag{A.4.27}$$

As a result, we get

$$\sigma = \lambda_{\text{bottom}}^2 - \hbar\lambda_{\text{bottom}}, \tag{A.4.28}$$

in analogue to (A.4.24).

Combining (A.4.24) and (A.4.28), we arrive at the factored equation

$$(\lambda_{\text{top}} + \lambda_{\text{bottom}})([\lambda_{\text{top}} - \lambda_{\text{bottom}}] + \hbar) = 0, \tag{A.4.29}$$

whose unique solution is

$$\lambda_{\text{top}} = -\lambda_{\text{bottom}}, \tag{A.4.30}$$

since the second factor on the left-hand side of (A.4.29) is positive.

Finally, as a consequence of (A.4.19), (A.4.21), and (A.4.30), we deduce

$$\lambda = \frac{\hbar}{2} N, \quad N \in \mathbb{Z}, \tag{A.4.31}$$

expressed as an integer multiple of *half* of the Planck constant or a *half-integer* multiple of the Planck constant.

Customarily, if we use ℓ to denote the largest allowed *half integer* in (A.4.31), then $\lambda_{\text{top}} = \hbar\ell$ so that (A.4.24) gives us

$$\sigma = \hbar^2 \ell(\ell + 1) \tag{A.4.32}$$

for the eigenvalue of $\hat{L}^2$. Correspondingly, the eigenvalues of $\hat{L}_z$ are listed accordingly in the expression

$$\lambda = \hbar m, \quad m = -\ell, -\ell + 1, \ldots, \ell - 1, \ell. \tag{A.4.33}$$

Therefore, if we denote by $\{|\ell, m\rangle\}$ the associated eigenfunctions or eigenstates, the results here are

$$\hat{L}^2 |\ell, m\rangle = \hbar^2 \ell(\ell + 1)|\ell, m\rangle, \quad \hat{L}_z|\ell, m\rangle = \hbar m|\ell, m\rangle, \tag{A.4.34}$$

where $m = -\ell, -\ell + 1, \ldots, \ell - 1, \ell$ and $\ell = \frac{1}{2}, 1, \frac{3}{2}, 2, \frac{5}{2}, \ldots$. In particular, for each ℓ, the spectrum point (A.4.32) of $\hat{L}^2$ has a $(2\ell + 1)$-dimensional degeneracy. When ℓ is an integer value, it is called the *orbital*, *azimuthal*, or *second quantum number*, and m the *magnetic quantum number*, correspondingly.

Charge quantization formulas

As an illustration, we first show how to derive Dirac's charge quantization formula considered in Chapter 3 by a method used by Saha [479, 480, 584] based on the knowledge on the eigenvalues of the angular momentum operators of an electric charge Q, of mass m, at $\mathbf{r}$ moving quantum-mechanically in the magnetic field of a monopole of magnetic charge g, residing at the origin. In such a situation, recall that the equation of motion reads

$$m\ddot{\mathbf{r}} = \frac{gQ}{|\mathbf{r}|^3}(\dot{\mathbf{r}} \times \mathbf{r}), \quad \mathbf{r} \neq \mathbf{0}, \tag{A.4.35}$$

so that the conserved angular momentum is

$$\mathbf{J} = m\mathbf{r} \times \dot{\mathbf{r}} - gQ\frac{\mathbf{r}}{|\mathbf{r}|}. \tag{A.4.36}$$

Hence, its radial component is constant,

$$J_r = \mathbf{J} \cdot \frac{\mathbf{r}}{|\mathbf{r}|} = -gQ. \tag{A.4.37}$$

On the other hand, like L_x, L_y, L_z, the eigenvalues of the quantized form of J_r are the multiples of the half of the Planck constant. Thus, we are led to the Dirac formula for monopoles:

$$gQ = \frac{\hbar}{2}n, \quad n \in \mathbb{Z}. \tag{A.4.38}$$

Likewise, if a dyon of electric charge Q and magnetic charge G is moving in the electromagnetic field of another dyon of respective charges q and g, residing at the origin, then the equation of motion becomes

$$m\ddot{\mathbf{r}} = \frac{(gG + qQ)}{|\mathbf{r}|^3}\mathbf{r} + \frac{(gQ - qG)}{|\mathbf{r}|^3}(\dot{\mathbf{r}} \times \mathbf{r}), \quad \mathbf{r} \neq \mathbf{0}, \tag{A.4.39}$$

such that the induced angular momentum is

$$\mathbf{J} = m\mathbf{r} \times \dot{\mathbf{r}} - (gQ - qG)\frac{\mathbf{r}}{|\mathbf{r}|}. \tag{A.4.40}$$

Since now $J_r = -(gQ - qG)$, we analogously arrive at the Schwinger quantization formula for dyons:

$$gQ - qG = \frac{\hbar}{2}n, \quad n \in \mathbb{Z}. \tag{A.4.41}$$

See also Zwanziger [617].

A significant feature of these quantum mechanical derivations is that the formalism is free of considering the Dirac strings.

A.5 Spins and spin-statistics theorem

The previous section considered all *possible* eigenvalues of the components of the orbital angular momentum $\hat{\mathbf{L}}$, which are seen to be the multiples of the half Planck constant. Here, we explore how these values may actually be *realized* by solutions of partial differential equations. In doing so, we encounter and understand the concept of *spins*. We focus on $\hat{L}_z$.

Spherical representations of angular momentum operators

To proceed, we use r, θ, ϕ to denote the spherical coordinates of a point away from the origin of $\mathbb{R}^3$, with r being the radial distance and θ and ϕ the polar and azimuthal angles, respectively, such that

$$x = r\sin\theta\cos\phi, \quad y = r\sin\theta\sin\phi, \quad z = r\cos\theta. \tag{A.5.1}$$

The inverse transformation of (A.5.1) has the local form (e.g.):

$$r^2 = x^2 + y^2 + z^2, \quad \cos\theta = \frac{z}{r}, \quad \tan\phi = \frac{y}{x}, \tag{A.5.2}$$

resulting in

$$\frac{\partial r}{\partial x} = \sin\theta\cos\phi, \quad \frac{\partial r}{\partial y} = \sin\theta\sin\phi, \quad \frac{\partial r}{\partial z} = \cos\theta, \tag{A.5.3}$$

$$\frac{\partial \theta}{\partial x} = \frac{\cos\theta\cos\phi}{r}, \quad \frac{\partial \theta}{\partial y} = \frac{\cos\theta\sin\phi}{r}, \quad \frac{\partial \theta}{\partial z} = -\frac{\sin\theta}{r}, \tag{A.5.4}$$

$$\frac{\partial \phi}{\partial x} = -\frac{\sin\phi}{r\sin\theta}, \quad \frac{\partial \phi}{\partial y} = \frac{\cos\phi}{r\sin\theta}, \quad \frac{\partial \phi}{\partial z} = 0. \tag{A.5.5}$$

In view of (A.4.4), (A.4.5), (A.5.1), and (A.5.3)–(A.5.5), we see that the chain rule gives us the expressions

$$\begin{aligned}
\hat{L}_x &= -\mathrm{i}\hbar(y\partial_z - z\partial_y) \\
&= -\mathrm{i}\hbar\left(r\sin\theta\sin\phi\left[\frac{\partial r}{\partial z}\partial_r + \frac{\partial\theta}{\partial z}\partial_\theta + \frac{\partial\phi}{\partial z}\partial_\phi\right]\right. \\
&\qquad \left. - r\cos\theta\left[\frac{\partial r}{\partial y}\partial_r + \frac{\partial\theta}{\partial y}\partial_\theta + \frac{\partial\phi}{\partial y}\partial_\phi\right]\right) \\
&= \mathrm{i}\hbar(\sin\phi\partial_\theta + \cot\theta\cos\phi\partial_\phi), \qquad (A.5.6) \\
\hat{L}_y &= -\mathrm{i}\hbar(\cos\phi\partial_\theta - \cot\theta\sin\phi\partial_\phi), \qquad (A.5.7) \\
\hat{L}_z &= -\mathrm{i}\hbar\partial_\phi. \qquad (A.5.8)
\end{aligned}$$

Inserting (A.5.6)–(A.5.8) into (A.4.6), we get

$$\hat{L}^2 = -\hbar^2\left(\frac{1}{\sin\theta}\partial_\theta(\sin\theta\partial_\theta) + \frac{1}{\sin^2\theta}\partial_\phi^2\right) = -\hbar^2\Delta_{S^2}, \tag{A.5.9}$$

which does not involve the radial variable r. Here, Δ_{S^2} is the Laplace–Beltrami operator on the standard unit sphere.

Eigenvalues

We proceed to find the eigenvalues of $\hat{L}_z$ given in (A.4.11) first. Hence, we solve the differential equation

$$-\mathrm{i}\hbar\partial_\phi f = \lambda f, \tag{A.5.10}$$

to get

$$f = K_m(\theta)\mathrm{e}^{\mathrm{i}m\phi} \text{ such that } \lambda = \hbar m \text{ with } m \in \mathbb{Z}, \tag{A.5.11}$$

where $K_m(\theta)$ is a function of θ only. In particular, we exclude all half-integer values of m.

Furthermore, with (A.5.9), the second equation in (A.4.11) may be resolved completely to render the following normalized solution [154, 251, 255, 388, 464]

$$\begin{aligned} f &= Y_{lm}(\theta, \phi) \\ &= (-1)^m \left(\frac{(2l+1)(l-|m|)!}{4\pi(l+|m|)!} \right)^{\frac{1}{2}} P_l^{|m|}(\cos\theta)\mathrm{e}^{\mathrm{i}m\phi}, \quad |m| \le l, \quad l \in \mathbb{N}, \end{aligned} \tag{A.5.12}$$

rightfully called the *spherical harmonics*, where the functions P_l^m's are the Legendre polynomials given by the formulas

$$P_l^m(x) = (1-x^2)^{\frac{m}{2}} \frac{\mathrm{d}^m P_l(x)}{\mathrm{d}x^m}, \quad m = 0, 1, \ldots, l, \tag{A.5.13}$$

with the "generating" ones produced by the *Rodrigues formula*:

$$P_l^0(x) \equiv P_l(x) = \frac{1}{2^l l!} \frac{\mathrm{d}^l}{\mathrm{d}x^l}(x^2-1)^l, \tag{A.5.14}$$

such that

$$\sigma = \hbar^2 l(l+1), \quad l \in \mathbb{N}. \tag{A.5.15}$$

Thus, again, we see that the possibility that ℓ taking half-integer values in (A.4.32) is ruled out, consistently.

In other words, all the integer cases of (A.4.34) may be realized by the solutions of the eigenvalue problems stated in (A.4.11) by virtue of partial differential equations, but not any of the half-integer cases. Such a "deficiency" prompts the introduction of the notion of "spins".

Generalized angular momentum operators

First, we note that the same general conclusion for the orbital angular momentum operator $\hat{\mathbf{L}}$ may be drawn for any Hermitian operator $\hat{\mathbf{J}}$ (say) satisfying the same commutator relations spelled out in (A.4.8). That is,

$$[\hat{J}_x, \hat{J}_y] = \mathrm{i}\hbar\hat{J}_z, \quad [\hat{J}_y, \hat{J}_z] = \mathrm{i}\hbar\hat{J}_x, \quad [\hat{J}_z, \hat{J}_x] = \mathrm{i}\hbar\hat{J}_y. \tag{A.5.16}$$

This observation creates extra space for us to manipulate the situation to open up new possibilities for the eigenvalues of the problem (A.4.11). In fact, for our purpose, we may take $\hat{\mathbf{J}}$ to assume the expanded form

$$\hat{\mathbf{J}} = \hat{\mathbf{L}} + \hat{\mathbf{S}}, \quad \hat{\mathbf{S}} = (\hat{S}_x, \hat{S}_y, \hat{S}_z), \tag{A.5.17}$$

where $\hat{S}_x, \hat{S}_y, \hat{S}_z$ are Hermitian operators as well, satisfying the same commutator relations stated in (A.5.16):

$$[\hat{S}_x, \hat{S}_y] = \mathrm{i}\hbar\hat{S}_z, \quad [\hat{S}_y, \hat{S}_z] = \mathrm{i}\hbar\hat{S}_x, \quad [\hat{S}_z, \hat{S}_x] = \mathrm{i}\hbar\hat{S}_y. \tag{A.5.18}$$

That is, $\hat{\mathbf{S}}$ resembles an angular momentum operator too. We should assume that $\hat{\mathbf{L}}$ and $\hat{\mathbf{S}}$ are independent observables such that their components are mutually commutative in pairs. With these conditions, it is clear that such an operator, (A.5.17), indeed fulfills (A.5.16) as well.

Let $\hbar s > 0$ be the largest eigenvalue of $\hat{S}_z$. Then, the corresponding eigenvalue of $\hat{S}^2$ is $\hbar^2 s(s+1)$, which is of a degeneracy of $(2s+1)$ dimensions associated with the $(2s+1)$ eigenvalues of $\hat{S}_z$ of the form

$$\hbar s_z, \quad s_z = -s, -s+1, \dots, s-1, s. \tag{A.5.19}$$

In particular, by the commutativity condition, we see that the largest eigenvalue of $\hat{J}_z = \hat{L}_z + \hat{S}_z$ is

$$j = \hbar(l+s). \tag{A.5.20}$$

Thus, the largest eigenvalue of $\hat{L}_z$ is boosted by the amount $s\hbar$. If s can assume a half-integer value, then j in (A.5.20) realizes the full possibilities of the quantum number ℓ, as described in (A.4.34).

When the additional angular momentum operator $\hat{\mathbf{S}}$ attached to the orbital angular momentum operator $\hat{\mathbf{L}}$ as expressed in (A.5.17) is taken to describe an *intrinsic angular momentum* associated with the particle under consideration, it is referred to as the *spin angular momentum* operator and s the *spin quantum number*, or simply *spin*, of the particle. The full operator $\hat{\mathbf{J}}$ given in (A.5.17) is called the *total angular momentum operator* and the sum $l+s$ in (A.5.20) the *total angular momentum quantum number*.

Spin matrices

An immediate realization of the spin angular momentum $\hat{\mathbf{S}}$ in (A.5.17) is to take it to be given by $n \times n$ Hermitian matrices of constant entries,

$$\hat{\mathbf{S}} = \mathbf{S} = (S_x, S_y, S_z), \tag{A.5.21}$$

where S_x, S_y, S_y are linearly independent matrices satisfying the condition

$$[S_x, S_y] = \mathrm{i}\hbar S_z, \quad [S_y, S_z] = \mathrm{i}\hbar S_x, \quad [S_z, S_x] = \mathrm{i}\hbar S_y. \tag{A.5.22}$$

Thus, $\{S_x, S_y, S_z\}$ is a basis of a three-dimensional (real) Lie algebra, often referred to as the *spin algebra*, due to its physical meaning. In order to understand the operator (A.5.17), it is clear that we need to consider a state space of functions taking values in $\mathbb{C}^n$, loosely called n-spinors. Here, $\mathbb{C}^n$ is a representation space of the spin Lie algebra at the fundamental level.

Pauli spin matrices

The most important example of spin matrices is at the bottom dimension of the representation space, $n = 2$. We study whether it is possible to realize $s = \frac{1}{2}$ in this context. To this end, without loss of generality, we take the eigenvectors associated with the eigenvalues $\pm\frac{1}{2}\hbar$ to be

$$u = \begin{pmatrix} 1 \\ 0 \end{pmatrix}, \quad v = \begin{pmatrix} 0 \\ 1 \end{pmatrix}, \tag{A.5.23}$$

otherwise we may use a suitable transformation. Hence, imposing

$$S_z u = \frac{\hbar}{2}\, u, \quad S_z v = -\frac{\hbar}{2}\, v, \tag{A.5.24}$$

and using $S_z = S_z^\dagger$, we obtain

$$S_z = \frac{\hbar}{2} \begin{pmatrix} 1 & 0 \\ 0 & -1 \end{pmatrix}. \tag{A.5.25}$$

On the other hand, since the associated ladder matrices

$$S_\pm = S_x \pm \mathrm{i} S_y, \tag{A.5.26}$$

are expected to raise and lower the eigenvalue readings of S_z, flipping between $-\frac{1}{2}\hbar$ and $\frac{1}{2}\hbar$ as described by (A.4.14), we may further impose the conditions

$$S_+ v = au, \qquad S_+ u = 0; \quad S_- u = bv, \quad S_- v = 0, \tag{A.5.27}$$

where $a, b \in \mathbb{C}$ are some nonzero numbers. Inserting (A.5.23) into (A.5.27), we obtain

$$S_+ = a \begin{pmatrix} 0 & 1 \\ 0 & 0 \end{pmatrix}, \quad S_- = b \begin{pmatrix} 0 & 0 \\ 1 & 0 \end{pmatrix}. \tag{A.5.28}$$

Hence, using (A.5.28) in (A.5.26), we arrive at

$$S_x = \frac{1}{2}(S_+ + S_-) = \frac{1}{2} \begin{pmatrix} 0 & a \\ b & 0 \end{pmatrix}, \; S_y = \frac{1}{2\mathrm{i}}(S_+ - S_-) = \frac{1}{2\mathrm{i}} \begin{pmatrix} 0 & a \\ -b & 0 \end{pmatrix}. \tag{A.5.29}$$

Applying the results (A.5.25) and (A.5.29) in (A.5.22), we get the general solution

$$a = \hbar \mathrm{e}^{\mathrm{i}\theta}, \quad b = \hbar \mathrm{e}^{-\mathrm{i}\theta}, \quad \theta \in \mathbb{R}. \tag{A.5.30}$$

For convenience, we may set $\theta = 0$ in (A.5.30) to have

$$\mathbf{S} = \frac{\hbar}{2}\,(\sigma_x, \sigma_y, \sigma_z), \tag{A.5.31}$$

in terms of the *Pauli spin matrices*

$$\sigma_x = \begin{pmatrix} 0 & 1 \\ 1 & 0 \end{pmatrix}, \quad \sigma_y = \begin{pmatrix} 0 & -\mathrm{i} \\ \mathrm{i} & 0 \end{pmatrix}, \quad \sigma_z = \begin{pmatrix} 1 & 0 \\ 0 & -1 \end{pmatrix}. \tag{A.5.32}$$

It is clear that the general result given by (A.5.29)–(A.5.30) may be obtained from the special solution (A.5.31) with a "rotation" of the "axis" S_+ given by (A.5.31) by an angle θ through $S_+ \mapsto e^{i\theta} S_+$. In fact, since we have

$$S_- = S_+^\dagger \mapsto e^{-i\theta} S_- = (e^{i\theta} S_+)^\dagger, \tag{A.5.33}$$

in correspondence, we recover (A.5.28) with (A.5.30) as a consequence.

Higher spin matrices

For spin $s = 1, \frac{3}{2}, 2, \frac{5}{2}, \ldots$, there are systematic ways to construct the associated spin matrices.

For example, for $s = 1$, we have

$$S_x = \frac{\hbar}{\sqrt{2}} \begin{pmatrix} 0 & 1 & 0 \\ 1 & 0 & 1 \\ 0 & 1 & 0 \end{pmatrix}, \; S_y = \frac{\hbar}{\sqrt{2}} \begin{pmatrix} 0 & -i & 0 \\ i & 0 & -i \\ 0 & i & 0 \end{pmatrix}, \; S_z = \hbar \begin{pmatrix} 1 & 0 & 0 \\ 0 & 0 & 0 \\ 0 & 0 & -1 \end{pmatrix}. \tag{A.5.34}$$

For $s = \frac{3}{2}$, we have

$$S_x = \frac{\hbar}{2} \begin{pmatrix} 0 & \sqrt{3} & 0 & 0 \\ \sqrt{3} & 0 & 2 & 0 \\ 0 & 2 & 0 & \sqrt{3} \\ 0 & 0 & \sqrt{3} & 0 \end{pmatrix}, \; S_y = \frac{\hbar}{2} \begin{pmatrix} 0 & -i\sqrt{3} & 0 & 0 \\ i\sqrt{3} & 0 & -2i & 0 \\ 0 & 2i & 0 & -i\sqrt{3} \\ 0 & 0 & i\sqrt{3} & 0 \end{pmatrix},$$

$$S_z = \frac{\hbar}{2} \begin{pmatrix} 3 & 0 & 0 & 0 \\ 0 & 1 & 0 & 0 \\ 0 & 0 & -1 & 0 \\ 0 & 0 & 0 & -3 \end{pmatrix}. \tag{A.5.35}$$

In both cases, the z-component spin matrices are taken as in the spin $s = \frac{1}{2}$ study carried out earlier so that the standard basis vectors of the representation spaces $\mathbb{C}^3$ and $\mathbb{C}^4$ are the eigenvectors associated with the eigenvalues $\hbar s_z$ for $s_z = 1, 0, -1$ and $s_z = \frac{3}{2}, \frac{1}{2}, -\frac{1}{2}, -\frac{3}{2}$, respectively.

Spins and particles

Elementary particles are subatomic particles without known substructures that are further categorized into matter particles such as electrons, quarks, leptons, and their anti-particles, and mediating or exchange particles as force carriers, such as photons for electromagnetic interaction, the W- and Z-particles for weak interaction, and gluons for strong interaction. All known matter particles are of spin $s = \frac{1}{2}$ and all known mediating particles are of spin $s = 1$ (except that the Higgs scalars responsible for the mechanism of spontaneous symmetry-breaking are of spin $s = 0$ and the hypothetically proposed gravity-carrying particles, known as gravitons, are of spin $s = 2$). Half-integer-spin particles with

$s = \frac{1}{2}, \frac{3}{2}, \frac{5}{2}, \dots$ satisfy the Pauli exclusion principle that a quantum state can only be occupied by one particle in a system of identical particles and are said to obey the Fermi–Dirac statistics. These particles are called fermions. In contrast, integer-spin particles with $s = 0, 1, 2, \dots$ do not satisfy the Pauli exclusion principle such that a quantum state can be taken by multiple particles in a system of identical particles and are said to follow the Bose–Einstein statistics. These particles are called bosons. For example, the W- and Z-particles are also called the weak bosons or intermediate vector bosons and the Higgs scalars Higgs bosons. Although at the levels of elementary particles, higher half-integer-spin fermionic particles (with $s \geq \frac{3}{2}$) and higher integer-spin bosonic particles (with $s \geq 2$) are yet to be observed in experiments, we have seen by virtue of (A.5.17) or its further extensions that such high-spin states in the forms of composite particles or excited states with nontrivial orbital momentum may theoretically be constructed.

In terms of a state vector description of a system of identical particles, the relation between particle spins and particle statistics may be spelled out by the symmetry of the state vector with respect to an exchange of any pair of the particles. To see this, consider n identical particles, either fermions or bosons, such that the ith particle occupies the state s_i, $i = 1, 2, \dots, n$, and that the state vector assumes the form

$$|\psi\rangle = |s_1, \dots, s_i, \dots, s_n\rangle. \tag{A.5.36}$$

Use $\hat{P}$ to denote the operation of swapping a pair of particles or states,

$$\hat{P}|\psi\rangle = \hat{P}|\dots, s_i, \dots, s_j, \dots\rangle = |\dots, s_j, \dots, s_i, \dots\rangle, \quad i \neq j. \tag{A.5.37}$$

Then, $\hat{P}^2|\psi\rangle = |\psi\rangle$ or $\hat{P}^2 = I$, the identity map. So the eigenvalues of $\hat{P}$ are ± 1. If $|\psi\rangle$ is an eigenstate associated with the eigenvalue 1, then $\hat{P}|\psi\rangle = |\psi\rangle$ or

$$|\dots, s_i, \dots, s_j, \dots\rangle = |\dots, s_j, \dots, s_i, \dots\rangle, \quad i \neq j. \tag{A.5.38}$$

If $|\psi\rangle$ is an eigenstate associated with the eigenvalue -1, then $\hat{P}|\psi\rangle = -|\psi\rangle$ or

$$|\dots, s_i, \dots, s_j, \dots\rangle = -|\dots, s_j, \dots, s_i, \dots\rangle, \quad i \neq j. \tag{A.5.39}$$

Bosons are those particles whose state vectors enjoy the symmetry property (A.5.38) and fermions the skew-symmetry property (A.5.39), both under arbitrary swapping operations stated.

Since identical particles are indistinguishable so that the measurement of an observable stays indifferent to the choice of swapped states, we have

$$\langle\psi|\hat{O}|\psi\rangle = \langle\psi|\hat{P}\hat{O}\hat{P}|\psi\rangle, \tag{A.5.40}$$

for any observable $\hat{O}$ and the state vector $|\psi\rangle$. This leads to $\hat{O} = \hat{P}\hat{O}\hat{P}$ or

$$\hat{O}\hat{P} = \hat{P}\hat{O}. \tag{A.5.41}$$

Indeed, the commutativity condition (A.5.41) is a necessary and sufficient condition for a system of identical particles to be indistinguishable quantum mechanically.

Consider a collection of identical fermions. Then the skew-symmetry condition (A.5.39) forbids the ith and jth particles to occupy the same quantum state, $s_i = s_j$, since we now have

$$|\dots, \underbrace{s_i,}_{i\text{th particle}} \dots, \underbrace{s_i,}_{j\text{th particle}} \dots\rangle = 0, \quad i \neq j, \tag{A.5.42}$$

which establishes the Pauli exclusion principle that a single quantum state cannot be occupied by more than one identical fermion.

In summary, half-integer-spin particles are fermions that follow the Fermi–Dirac statistics so that the state vector of a system of identical fermions is skew-symmetric with respect to exchange of particles and the Pauli exclusion principle holds. Integer-spin particles are bosons that follow the Bose–Einstein statistics so that the state vector of a system of identical bosons is symmetric with respect to exchange of particles and the Pauli exclusion principle does not apply. This collective statement is known as the *spin-statistics theorem* in quantum mechanics. See [209, 305, 436, 437, 464] for the conceptualization and development of this fundamental theorem.

Collectively, for the n-particle state (A.5.36) with $n \geq 2$, we may re-express the particle spin-statistics by the formula

$$P|\psi\rangle = e^{i\alpha\pi}|\psi\rangle, \tag{A.5.43}$$

so that

$$\alpha = 0, 2, 4, \dots, \quad \alpha = 1, 3, 5, \dots, \tag{A.5.44}$$

give rise to integer- and half-integer-spin particles, respectively. Thus, we arrive at the expression

$$s = \frac{1}{2}\alpha, \tag{A.5.45}$$

for the particle spin s under discussion. As a consequence, the parameter α in (A.5.43) is also called the *statistics* of the particle, with possible values listed in (A.5.44), which characterize point particles in three and higher space dimensions, as well. In two space dimensions, however, by constructing appropriate models, the statistics α may be made to take *any* designated real value [581]. Such "quasiparticles" of *fractional statistics* are called *anyons* [509] and of importance in condensed-matter physics [268, 327, 582, 583] and quantum computation [409]. See [377, 527] for reviews and further literature.

A.6 Deflection of light in gravitational field

The possibility that a light beam may be bent or deflected around a massive celestial body, a phenomenon known as *gravitational light deflection*, was suggested by Newton in 1704, and further explored by Cavendish in 1784 and expanded by Soldner in 1804. Such a phenomenon later became one of the three major observational confirmations (the other two being gravitational redshift and precession of the perihelion of Mercury around the Sun) of Einstein's theory of general relativity, which was firmly established in 1919 by Dyson, Eddington, and Davidson. This section presents a study of gravitational light deflection in the contexts of both Newton's gravity and Einstein's general relativity. Although the former falls short of yielding a correct prediction for the bending angle of light deflection, its mathematics illuminates the problem and inspires and prepares the study in the latter context.

Light deflection in general

Let the celestial body under consideration be the Sun, whose center is placed at the origin of $\mathbb{R}^3$ equipped with the Cartesian coordinates x, y, z and the spherical coordinates r, θ, ϕ, which are related by the usual repressions

$$x = r\cos\phi\sin\theta, \quad y = r\sin\phi\sin\theta, \quad z = r\cos\theta. \tag{A.6.1}$$

Imagine an incoming photon traveling towards the Sun, passing its outer region, and then continuing its journey outwards. For simplicity, assume that the trajectory of the photon is confined in the equatorial plane $\theta = \frac{\pi}{2}$. Let r_0 denote the *closest approach* of the traveling photon around the Sun which is attained when $\phi = 0$. The incoming photon starts its journey somewhere characterized by $y = -\infty$, $\phi = \phi_{-\infty}$ (say), which is below $-\frac{\pi}{2}$, or $\phi_{-\infty} < -\frac{\pi}{2}$, travels along a bent curve about the Sun as a result of the gravitational pulling of the Sun, reaches its point of the closest approach at $r = r_0, \phi = 0$, and finally escapes the gravity of the Sun at $y = \infty$, $\phi = \phi_\infty$ (say), which is above $\frac{\pi}{2}$, or $\phi_\infty > \frac{\pi}{2}$ (Figure A.6). As a consequence, the *angle of deflection* of the light beam consisting of such a single photon reads

$$\delta = (\phi_\infty - \phi_{-\infty}) - \pi. \tag{A.6.2}$$

Light deflection by Newton gravity

Consider a hypothetical particle of mass m outside the Sun at $\mathbf{r} \in \mathbb{R}^3$ and let M be the mass of the Sun. Then, the Newton gravity exerted on the particle is

$$\mathbf{f} = -\frac{GmM\mathbf{r}}{r^3}, \tag{A.6.3}$$

where G is Newton's gravitational constant. By the second law of Newton, the acceleration $\mathbf{a}$ of the particle resulting from the gravitational force (A.6.3) is

$$\mathbf{a} = -\frac{GM\mathbf{r}}{r^3}. \tag{A.6.4}$$

Since (A.6.4) is independent of m, we continue to use it for the gravitational acceleration of a photon, whose rest mass is actually zero, traveling about the Sun.

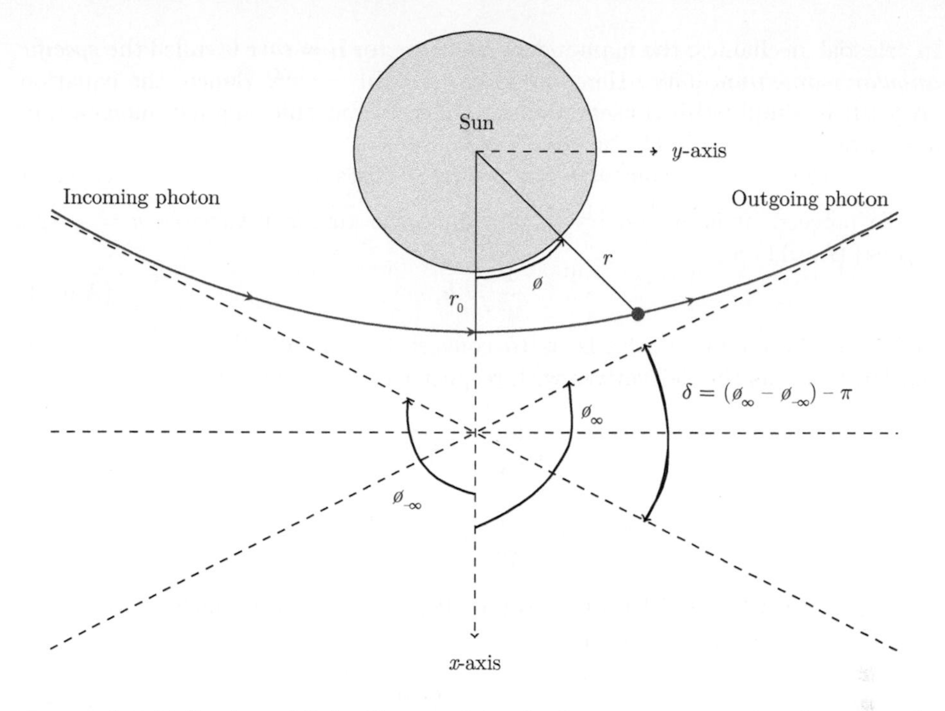

Figure A.6 Deflection of light illustrated by the bent trajectory of a photon in the gravitational field of the Sun confined in the equatorial plane of the Sun, $\theta = \frac{\pi}{2}$. The photon starts its journey at infinity at the limiting value of the azimuthal angle, $\phi = \phi_{-\infty}$ and ends its journey at $\phi = \phi_\infty$, resulting in the angle of deflection $\delta = (\phi_\infty - \phi_{-\infty}) - \pi$.

If we focus on when the photon travels in the equatorial plane $\theta = \frac{\pi}{2}$, then (A.6.1) indicates that its motion is described by the updated planar radial vector

$$\mathbf{r} = r(\cos\phi, \sin\phi, 0), \tag{A.6.5}$$

In view of (A.6.4) and (A.6.5), the gravitational acceleration becomes

$$\mathbf{a} = -\frac{GM}{r^2}(\cos\phi, \sin\phi, 0). \tag{A.6.6}$$

On the other hand, differentiating (A.6.5) with respect to the time variable t, we find the velocity and acceleration vectors of the photon to be

$$\mathbf{v} = \dot{\mathbf{r}} = \dot{r}(\cos\phi, \sin\phi, 0) + \omega r(-\sin\phi, \cos\phi, 0), \quad \omega = \dot{\phi}, \tag{A.6.7}$$

$$\mathbf{a} = \ddot{\mathbf{r}} = (\ddot{r} - \omega^2 r)(\cos\phi, \sin\phi, 0) + (\dot{\omega} r + 2\omega\dot{r})(-\sin\phi, \cos\phi, 0). \tag{A.6.8}$$

Thus, in view of (A.6.6) and (A.6.8), we arrive at the equations

$$\ddot{r} - \omega^2 r = -\frac{GM}{r^2}, \tag{A.6.9}$$

$$\dot{\omega} r + 2\omega\dot{r} = 0. \tag{A.6.10}$$

In celestial mechanics, the mass-independent vector $\mathbf{h} = \mathbf{r} \times \dot{\mathbf{r}}$ is called the *specific angular momentum* [558]. Here, we have $h = |\mathbf{h}| = \omega r^2$. Hence, the equation (A.6.10) is simply the conservation of the total specific angular momentum, $\dot{h} = 0$, or

$$\omega r^2 = h = \text{a positive constant.} \tag{A.6.11}$$

To proceed, it is customary to introduce the auxiliary variable $u = \frac{1}{r}$ and express (A.6.11) as

$$\frac{\omega}{u^2} = h, \tag{A.6.12}$$

such that the chain rule leads us to convert the radial velocity $\dot{r}$ and radial acceleration $\ddot{r}$ as the derivatives with respect to the angular variable ϕ only:

$$\dot{r} = -\frac{\dot{\phi}}{u^2}\frac{\mathrm{d}u}{\mathrm{d}\phi} = -h\frac{\mathrm{d}u}{\mathrm{d}\phi}, \tag{A.6.13}$$

$$\ddot{r} = -h\dot{\phi}\frac{\mathrm{d}^2 u}{\mathrm{d}\phi^2} = -h^2 u^2 \frac{\mathrm{d}^2 u}{\mathrm{d}\phi^2}. \tag{A.6.14}$$

Inserting (A.6.12) and (A.6.14) into (A.6.9), we obtain the following constant-coefficient differential equation

$$\frac{\mathrm{d}^2 u}{\mathrm{d}\phi^2} + u = \frac{GM}{h^2}, \tag{A.6.15}$$

which governs a "harmonic oscillator" subject to a constant "applied force," whose general solution is

$$u = A\cos\phi + B\sin\phi + \frac{GM}{h^2}, \tag{A.6.16}$$

where A, B are constants. Since $u = u(\phi)$ is maximized at the closest approach $\phi = 0$ where $u = \frac{1}{r_0}$, we get $B = 0$ and

$$A = \frac{1}{r_0} - \frac{GM}{h^2}. \tag{A.6.17}$$

On the other hand, since we expect that the incoming photon starts its journey from infinity with a limiting angle $-\pi < \phi_{-\infty} < -\frac{\pi}{2}$ (cf. Figure A.6) where $u = 0$, we have $A > 0$ or

$$e \equiv \frac{h^2}{GMr_0} - 1 > 0, \tag{A.6.18}$$

so that we may express the solution given in (A.6.16) as

$$u(\phi) = \frac{GM}{h^2}(1 + e\cos\phi). \tag{A.6.19}$$

Thus, using the correspondence relations

$$\lim_{\phi \to \phi_{-\infty}} u(\phi) = 0, \quad \lim_{\phi \to \phi_{\infty}} u(\phi) = 0, \tag{A.6.20}$$

in (A.6.19) and the symmetry of the cosine function, we find

$$\phi_\infty - \phi_{-\infty} = 2\phi_\infty = 2\cos^{-1}\left(-\frac{1}{e}\right). \tag{A.6.21}$$

Since ϕ_∞ is expected to be slightly above $\frac{\pi}{2}$, the quantity $\frac{1}{e}$ should be rather small. In fact, we have

$$\frac{1}{e} = \frac{GM}{\frac{h^2}{r_0} - GM}, \tag{A.6.22}$$

and $2GM$ is the Schwarzschild radius (which is indeed tiny). Hence, it suffices to use the asymptotic expansion

$$\cos^{-1}\left(-\frac{1}{e}\right) = \frac{\pi}{2} + \frac{1}{e} + \mathrm{O}\left(e^{-3}\right). \tag{A.6.23}$$

Using the results (A.6.21) and (A.6.23) with (A.6.22) in (A.6.2), we obtain the formula

$$\delta = \frac{2}{e} = \frac{2GM}{\frac{h^2}{r_0} - GM}, \tag{A.6.24}$$

for the angle of deflection after neglecting higher-order terms in $\frac{1}{e}$.

Alternatively, from (A.6.19), we obtain the *orbit equation*

$$r = r(\phi) = \frac{\frac{h^2}{GM}}{1 + e\cos\phi}, \tag{A.6.25}$$

where e appears to be the eccentricity of the orbit.

Note that this study is equally valid for the scattering of a fast-moving small massive particle as well and the formulas derived may further be simplified when the particle is a photon. In fact, in this situation, assume normalized units so that the speed of light in vacuum is unity, $c = 1$. Then, using (A.6.7) with $|\mathbf{v}|^2 = c^2 = 1$ and (A.6.12), we have

$$\dot{r}^2 + \frac{h^2}{r^2} \equiv 1. \tag{A.6.26}$$

Now, at the spot of the closest approach, we have $\dot{r} = 0$ and $r = r_0$. Thus, we can read off the value of the constant of motion h to be

$$h = r_0. \tag{A.6.27}$$

Substituting (A.6.27) into (A.6.24), we get

$$\delta = \frac{2GM}{r_0}\frac{1}{1 - \frac{GM}{r_0}} = \frac{2GM}{r_0}\left(1 + \frac{GM}{r_0} + \left[\frac{GM}{r_0}\right]^2 + \cdots\right). \tag{A.6.28}$$

The leading-order approximation of (A.6.28), or

$$\delta_{\mathrm{N}} = \frac{2GM}{r_0}, \tag{A.6.29}$$

is the *Newton deflection angle.* In the limiting situation that the light "grazes" the surface of the Sun, the closest approach r_0 may be the radius of the Sun, rendering a largest possible deflection angle. Thus, in this case, substituting the mass and radius of the Sun into (A.6.29), we arrive at the approximate value

$$\delta_{\mathrm{N}} = 0.875 \quad \text{arcsecond}, \tag{A.6.30}$$

when the photon or the light ray grazes the Sun. Note that an arcsecond is one 3600th of a degree or $\frac{\pi}{648000}$ radian. So this calculated value of deflection angle is really small.

Light deflection by Einstein gravity

To proceed, recall that the free motion of a particle in a spacetime equipped with the metric $(g_{\mu\nu})$ is governed by the geodesic equations (10.1.16) or

$$\ddot{x}^\mu + \Gamma^\mu_{\alpha\beta}\dot{x}^\alpha\dot{x}^\beta = 0, \tag{A.6.31}$$

where $\Gamma^\mu_{\alpha\beta}$ are the Christoffel symbols induced from $(g_{\mu\nu})$ and the overdot indicates differentiation with respect to a generic invariant parameter, say s. With the spherical coordinates, $(x^\mu) = (t, r, \theta, \phi)$, we see that the Schwarzschild metric (10.3.2) or

$$\mathrm{d}s^2 = A(r)\,\mathrm{d}t^2 - B(r)\,\mathrm{d}r^2 - r^2(\mathrm{d}\theta^2 + \sin^2\theta\,\mathrm{d}\phi^2), \tag{A.6.32}$$

leads to the concrete formulas (10.3.4)–(10.3.6), or

$$\Gamma^0_{01} = \frac{A'}{2A}, \quad \Gamma^1_{00} = \frac{A'}{2B}, \quad \Gamma^1_{11} = \frac{B'}{2B}, \tag{A.6.33}$$

$$\Gamma^1_{22} = -\frac{r}{B}, \quad \Gamma^1_{33} = -\frac{r\sin^2\theta}{B}, \quad \Gamma^2_{12} = \frac{1}{r}, \tag{A.6.34}$$

$$\Gamma^2_{33} = -\cos\theta\sin\theta, \quad \Gamma^3_{13} = \frac{1}{r}, \quad \Gamma^3_{23} = \cot\theta. \tag{A.6.35}$$

Substituting (A.6.33)–(A.6.35) into (A.6.31), we obtain the reduced geodesic equations

$$\ddot{t} + \frac{A'}{A}\dot{t}\dot{r} = 0, \tag{A.6.36}$$

$$\ddot{r} + \frac{A'}{2B}\dot{t}^2 + \frac{B'}{2B}\dot{r}^2 - \frac{r}{B}\dot{\theta}^2 - \frac{r\sin^2\theta}{B}\dot{\phi}^2 = 0, \tag{A.6.37}$$

$$\ddot{\theta} + \frac{2}{r}\dot{r}\dot{\theta} - \cos\theta\sin\theta\,\dot{\phi}^2 = 0, \tag{A.6.38}$$

$$\ddot{\phi} + \frac{2}{r}\dot{r}\dot{\phi} + 2\cot\theta\,\dot{\theta}\dot{\phi} = 0, \tag{A.6.39}$$

corresponding to $\mu = 0, 1, 2, 3$ in (A.6.31), respectively. We are interested in the motion of a photon only, which follows the null geodesic equation $g_{\mu\nu}\mathrm{d}x^\mu\mathrm{d}x^\nu = 0$ or $g_{\mu\nu}\dot{x}^\mu\dot{x}^\nu = 0$, which (in terms of (A.6.32)) reads simply

$$A\dot{t}^2 - B\dot{r}^2 - r^2\dot{\theta}^2 - r^2\sin^2\theta\,\dot{\phi}^2 = 0. \tag{A.6.40}$$

In fact, (A.6.40) is implied by (A.6.36)–(A.6.39) by observing that the left-hand side of (A.6.40) is a constant of motion that vanishes in outer regions where gravity is negligible and the spacetime metric becomes a flat Minkowskian metric.

Furthermore, if we confine the motion of the photon to the equatorial plane $\theta = \frac{\pi}{2}$ as before, (A.6.36)–(A.6.39) are reduced into

$$\ddot{t} + \frac{A'}{A}\dot{t}\dot{r} = 0, \tag{A.6.41}$$

$$\ddot{r} + \frac{A'}{2B}\dot{t}^2 + \frac{B'}{2B}\dot{r}^2 - \frac{r}{B}\dot{\phi}^2 = 0, \tag{A.6.42}$$

$$\ddot{\phi} + \frac{2}{r}\dot{r}\dot{\phi} = 0, \tag{A.6.43}$$

and (A.6.40) becomes

$$A\dot{t}^2 - B\dot{r}^2 - r^2\dot{\phi}^2 = 0. \tag{A.6.44}$$

Note that (A.6.41) yields the conservation law

$$A\dot{t} = k = \text{constant}, \tag{A.6.45}$$

and (A.6.43) renders

$$r^2\dot{\phi} = h = \text{constant}, \tag{A.6.46}$$

which is identical to its Newtonian version (A.6.11) with $\omega = \dot{\phi}$. Finally, with (A.6.41) and (A.6.43), the remaining equation (A.6.42) is equivalent to the null geodesic equation (A.6.44), which may be recast in view of (A.6.45) and (A.6.46) into the equation

$$AB\dot{r}^2 + \frac{A}{r^2}h^2 = k^2, \tag{A.6.47}$$

which (unsurprisingly) assumes the form of a conservation law, in analogy to (A.6.26).

For convenience, we introduce the variable $u = \frac{1}{r}$ as in the study of the light deflection subject to the Newton gravity. Thus, with (A.6.13) and (A.6.46), we have

$$AB\left(\frac{\mathrm{d}u}{\mathrm{d}\phi}\right)^2 + Au^2 = \frac{k^2}{h^2}. \tag{A.6.48}$$

In the context of the Schwarzschild-type black hole solutions, it suffices to consider the situation where $AB = 1$. Therefore, (A.6.48) is further simplified into the form

$$\left(\frac{\mathrm{d}u}{\mathrm{d}\phi}\right)^2 + Au^2 = \frac{k^2}{h^2}, \tag{A.6.49}$$

which is a kind of the Friedmann-type equations. Note that, unlike in the situation of light deflection subject to the Newton gravity where we encounter a linear differential equation, (A.6.15), here we face the challenge of a *fully nonlinear differential equation* (a differential equation is fully nonlinear if it is nonlinear in its highest-order derivative(s)), (A.6.49). Fortunately, to obtain the bending angle of light deflection from (A.6.49), we do not need to integrate it explicitly.

In fact, at the spot of the closest approach where $\phi = 0$, we have $\frac{du}{d\phi} = 0$. Hence, we obtain from (A.6.49) the value

$$\frac{k^2}{h^2} = A(r_0)u_0^2, \quad u_0 = \frac{1}{r_0}. \tag{A.6.50}$$

Inserting (A.6.50) into (A.6.49) and applying the same discussion as that in the study of the angle of light deflection under the Newton gravity, we find

$$\phi_\infty = \int_0^1 \frac{dv}{\sqrt{A\left(\frac{1}{u_0}\right) - A\left(\frac{1}{u_0 v}\right) v^2}}, \quad v = \frac{u}{u_0}, \tag{A.6.51}$$

for the limiting value of the azimuthal angle of the outgoing photon escaping to spatial infinity.

For a classical Schwarzschild black hole of mass M, the metric factor A is given by

$$A(r) = 1 - \frac{2GM}{r}. \tag{A.6.52}$$

Using (A.6.52) in (A.6.51), we get

$$\phi_\infty = f(\delta_{\rm N}) \equiv \int_0^1 \frac{dv}{\sqrt{1 - v^2 - \delta_{\rm N}(1 - v^3)}}, \tag{A.6.53}$$

where $\delta_{\rm N}$ is the Newton deflection angle given in (A.6.29) used here as a parameter. This integral is difficult. However, since $\delta_{\rm N}$ is small, it will be sufficient to find its leading-order terms. Indeed, we have

$$f(0) = \int_0^1 \frac{dv}{\sqrt{1 - v^2}} = \frac{\pi}{2}, \tag{A.6.54}$$

$$f'(0) = \frac{1}{2}\int_0^1 \frac{(1 - v^3)}{(1 - v^2)^{\frac{3}{2}}}\, dv = 1, \tag{A.6.55}$$

which lead to the expansion

$$\phi_\infty = \frac{\pi}{2} + \delta_{\rm N} + {\rm O}(\delta_{\rm N}^2). \tag{A.6.56}$$

Inserting (A.6.56) into (A.6.2), using $\phi_{-\infty} = -\phi_\infty$ following Figure A.6, and keeping only the leading-order term, we arrive at

$$\delta_{\rm E} = \delta = 2\delta_{\rm N} = \frac{4GM}{r_0}, \tag{A.6.57}$$

which is the celebrated *Einstein bending angle of light deflection* in general relativity and exactly doubles that of Newton. At the grazing limit for a light ray passing by the Sun, this angle is

$$\delta_{\rm E} = 1.75 \quad \text{arcsecond}. \tag{A.6.58}$$

Dyson, Eddington, and Davidson observed this value in their monumental 1919 work [182].

Angle of light deflection subject to Reissner–Nordström metric

Recall that the metric factor A of a spherically symmetric black hole of mass M, electric charge Q, and magnetic charge P is given by the Reissner–Nordström formula (10.4.40), or

$$\begin{aligned} A(r) &= 1 - \frac{2GM}{r} + \frac{4\pi G}{r^2}(Q^2 + P^2) \\ &= 1 - \delta_{\rm N} v + \delta_{\rm em} v^2, \quad u_0 = \frac{1}{r_0}, \quad r = \frac{r_0}{v}, \quad 0 < v \leq 1, \end{aligned} \tag{A.6.59}$$

where $\delta_{\rm N}$ is as given in (A.6.29) and

$$\delta_{\rm em} = \frac{4\pi G(Q^2 + P^2)}{r_0^2}, \tag{A.6.60}$$

denotes an arising electromagnetic adjustment to the deflection angle to be clarified shortly subsequently. Using (A.6.59) in (A.6.51), we get

$$\phi_\infty = f(\delta_{\rm N}, \delta_{\rm em}) \equiv \int_0^1 \frac{{\rm d}v}{\sqrt{1 - v^2 - \delta_{\rm N}(1 - v^3) + \delta_{\rm em}(1 - v^4)}}. \tag{A.6.61}$$

As in the Schwarzschild black hole case, we have $f(0,0) = \frac{\pi}{2}$ and $f_1(0,0) = 1$. Besides, we have

$$f_2(0,0) = -\frac{1}{2}\int_0^1 \frac{(1 - v^4)}{(1 - v^2)^{\frac{3}{2}}}\,{\rm d}v = -\frac{3\pi}{8}. \tag{A.6.62}$$

Here f_1 and f_2 denote the partial derivatives of f with respect to its first and second arguments, respectively.

Consequently, we get the following formula for the bending angle of light deflection about a Reissner–Nordström black hole:

$$\begin{aligned} \delta &= 2\phi_\infty - \pi = 2\delta_{\rm N} - \frac{3\pi}{4}\delta_{\rm em} \\ &= \frac{4GM}{r_0} - \frac{3\pi^2 G(Q^2 + P^2)}{r_0^2}, \end{aligned} \tag{A.6.63}$$

within leading orders terms. In particular, we see that the presence of electromagnetic charges contributes to reducing the value of the deflection angle. Moreover, in the context of an extreme Reissner–Nordström black hole satisfying (10.4.46), or

$$4\pi(Q^2 + P^2) = GM^2, \tag{A.6.64}$$

the formula (A.6.63) becomes

$$\delta = \left(2 - \frac{3\pi}{16}\delta_{\rm N}\right)\delta_{\rm N}, \tag{A.6.65}$$

and it is apparent that, in this situation, the correction from electromagnetism is a second-order quantity in $\delta_{\rm N}$.

Light deflection associated with nonlinear electrodynamics

The general formalism presented in this section is immediately applicable to the study of the gravitational light deflection phenomenon subject to nonlinear electrodynamics such as that of the Born–Infeld theory type discussed in Chapter 14, where a dually charged black hole metric is determined by the gravitational metric factor (14.12.47), say, so that we can rewrite the quantity under the radical root in the denominator of the integral in (A.6.51) as

$$A\left(\frac{1}{u_0}\right)-A\left(\frac{1}{u_0 v}\right)v^2 = 1-v^2-\frac{2GM}{r_0}(1-v^3) + \frac{8\pi G}{r_0}\left(\int_{r_0}^{\infty}\mathcal{H}(\rho)\rho^2\mathrm{d}\rho - v^3\int_{\frac{r_0}{v}}^{\infty}\mathcal{H}(\rho)\rho^2\mathrm{d}\rho\right). \tag{A.6.66}$$

Inserting (A.6.66) into (A.6.51), we deduce that, within the first-order approximation in the parameter G near $G=0$, the angle of light deflection has the expression

$$\delta = \frac{4GM}{r_0} - \delta_{\mathrm{em}} = \delta_{\mathrm{E}} - \delta_{\mathrm{em}}, \tag{A.6.67}$$

where δ_{em} is an electromagnetic charge-induced correction term to the gravitational deflection angle given by

$$\delta_{\mathrm{em}} = \frac{8\pi G}{r_0}\left(2\int_{r_0}^{\infty}\mathcal{H}(\rho)\rho^2\mathrm{d}\rho - \int_0^1\left[\frac{v^3}{(1-v^2)^{\frac{3}{2}}}\int_{r_0}^{\frac{r_0}{v}}\mathcal{H}(\rho)\rho^2\mathrm{d}\rho\right]\mathrm{d}v\right). \tag{A.6.68}$$

Although the expression of such a correction term appears complicated in general, it is clear that it is always positive in nontrivial situations, $\mathcal{H}>0$. That is,

$$\delta_{\mathrm{em}} > 0. \tag{A.6.69}$$

In other words, the presence of electromagnetism serves to reduce the bending angle of light deflection, which is consistent with that observed in the earlier Reissner–Nordström black hole calculation.

As an explicit example, we now work to obtain the bending angle of light deflection for the magnetically charged Hayward black hole metric [38, 220, 283, 345] given by (14.9.63), whose curvature regularity follows from a critical mass-magnetic energy condition [605], resulting in the specific form of the gravitational metric factor:

$$A(r) = 1 - \frac{2GMr^2}{r^3+a^3}, \tag{A.6.70}$$

where $a>0$ is a characteristic length parameter. Substituting (A.6.70) into the left-hand side of (A.6.66), we get

$$A\left(\frac{1}{u_0}\right)-A\left(\frac{1}{u_0 v}\right)v^2 = 1-v^2-\frac{2GM}{r_0}\left(\frac{1}{1+\left[\frac{a}{r_0}\right]^3} - \frac{v^3}{1+\left[\frac{a}{r_0}\right]^3 v^3}\right). \tag{A.6.71}$$

Using (A.6.71) in (A.6.51), we get the bending angle

$$\begin{aligned}\delta &= \frac{4GM}{r_0\left(1+\left[\frac{a}{r_0}\right]^3\right)} + \frac{2GM}{r_0}\int_0^1 \frac{v^3}{(1-v^2)^{\frac{3}{2}}}\left(\frac{1}{1+\left[\frac{a}{r_0}\right]^3} - \frac{1}{1+\left[\frac{a}{r_0}\right]^3 v^3}\right) dv \\ &= \frac{4GM}{r_0\left(1+\left[\frac{a}{r_0}\right]^3\right)} - \frac{2GM}{r_0}\frac{\left(\frac{a}{r_0}\right)^3}{1+\left(\frac{a}{r_0}\right)^3}\int_0^1 \frac{v^3(1-v^3)}{(1-v^2)^{\frac{3}{2}}} dv + \mathrm{O}\left(\left[\frac{a}{r_0}\right]^6\right), \\ &= \frac{4GM}{r_0\left(1+\left[\frac{a}{r_0}\right]^3\right)}\left(1-\left[\frac{15\pi}{32}-1\right]\left[\frac{a}{r_0}\right]^3\right) + \mathrm{O}\left(\left[\frac{a}{r_0}\right]^6\right), \end{aligned} \tag{A.6.72}$$

within the first-order approximation in G near $G = 0$. Consequently, a decrease in the bending angle in the Hayward black hole situation is explicit, and the Einstein angle remains as the dominant quantity.

See [591, 592] and the references therein for some explicit results of and approximation methods for the calculation of the bending angle of light deflection in the contexts of various extended gravity theories.

The well-established phenomenon of *gravitational lensing* [111, 190, 468, 484, 489, 490] attributes to gravitational light deflection.

Exercises

1. Derive (A.1.10).
2. Derive (A.1.14).
3. Establish (A.1.23) and use it to prove (A.1.24).
4. Show that the nondegeneracy condition (A.1.27) implies that the map $(x, y) \mapsto (u(x, y), v(x, y))$ is locally invertible at $(0, 0)$.
5. Use (A.1.28) in (A.1.29) to arrive at (A.1.30).
6. Although the result (A.1.32) may be obtained by calculating the integral (A.1.31) directly, it may also be obtained by topological means by carrying out the following steps:
   (a) Deform the parameters a, b, c, d while observing the condition (A.1.27) and use the fact that (A.1.30) is a constant (integer) independent of a, b, c, d subject to (A.1.27) to show that actually the value of (A.1.30) can only be ± 1.
   (b) Use (a) to verify (A.1.32).
7. If a simple closed curve C is a periodic orbit of the dynamical system (A.1.37) and, as one travels along C counterclockwise, one moves against

$\mathbf{A}(x,y)$ during the entire cycle of the orbital, draw a picture and use it to show that $\text{ind}\,(\mathbf{A}|_C) = 1$ still holds.

8. Solve (A.1.44) to obtain its integral (A.1.45).
9. Show that the integration constant C in (A.1.45) satisfies the condition

$$C > r\left(1 - \ln\left[\frac{r}{a}\right]\right) + s\left(1 - \ln\left[\frac{s}{b}\right]\right). \tag{A.E.1}$$

10. Consider the existence of a solution to the following simultaneous system of nonlinear equations:

$$x^3 - 3xy^2 = 7\cos(1+xy), \quad 3x^2y - y^3 = -5\mathrm{e}^{-x^2-y^4}, \quad (x,y) \in \mathbb{R}^2. \tag{A.E.2}$$

    (a) Reformulate the problem such that any solution to (A.E.2) is a zero of a certain vector field, say $\mathbf{A}$.
    (b) Let C_R be the circle centered at origin of $\mathbb{R}^2$ and of radius $R > 0$. Compute $\text{ind}\,(\mathbf{A}|_{C_R})$ when R is sufficiently large.
    (c) Use (b) to prove the existence of a solution to (A.E.2).

11. Directly verify (A.1.61).
12. Let $\mathbf{A}(x,y,z)$ be a real vector field defined over a contractible domain Ω that is assumed to be differentiable on $\overline{\Omega}$, $\mathbf{A} \neq \mathbf{0}$ on $\partial\Omega$, and $\text{ind}\,(\mathbf{A}|_{\partial\Omega}) \neq 0$. Show that $\mathbf{A}$ vanishes somewhere in Ω.
13. Extend the index formula (A.1.66) to its general dimensional setting.
14. Establish (A.1.72).
15. Consider the initial value problem

$$\frac{\mathrm{d}y}{\mathrm{d}x} = f(x,y), \quad y(x_0) = y_0, \tag{A.E.3}$$

of a differential equation, where $f(x,y)$ is a continuous function for (x,y) in a domain $\Omega \subset \mathbb{R}^2$ and $(x_0,y_0) \in \Omega$. The *Peano existence theorem* or *Cauchy–Peano theorem* states that (A.E.3) is always locally solvable [139]. In other words, there is some $\delta > 0$ such that there is a continuously differentiable function $y = y(x)$ defined over $[x_0 - \delta, x_0 + \delta]$ such that $(x, y(x)) \in \Omega$ and $y'(x) = f(x, y(x))$ for $x \in [x_0 - \delta, x_0 + \delta]$ and $y(x_0) = y_0$. In this problem, obtain a proof of this fundamental existence theorem by applying the Schauder fixed-point theorem using the following steps:

    (a) Convert the problem (A.E.3) into the *Volterra type integral equation*

$$y(x) = y_0 + \int_{x_0}^{x} f(s, y(s))\,\mathrm{d}s, \tag{A.E.4}$$

    so that its solution, if any, becomes a fixed point of a map T over a subset of $C(I_\delta)$, where $C(I_\delta)$ denotes the Banach space of continuous functions over $I_\delta = [x_0 - \delta, x_0 + \delta]$, $\delta > 0$ being small, equipped with the usual sup-norm, defined formally by the integral

$$T(u)(x) = y_0 + \int_{x_0}^{x} f(s, u(s))\, \mathrm{d}s, \quad u(x_0) = y_0, \quad u \in C(I_\delta). \quad \text{(A.E.5)}$$

(b) For $R > 0$, define

$$K_{\delta,R} = \left\{ u \in C(I_\delta) \,\middle|\, \sup_{x \in I_\delta} |u(x) - y_0| \leq R \right\}. \quad \text{(A.E.6)}$$

Show that $K_{\delta,R}$ is a closed convex subset of $C(I_\delta)$ and T given by (A.E.5) maps $K_{\delta,R}$ into itself if δ and R are small enough.

(c) If δ and R are small, show that $T : K_{\delta,R} \to K_{\delta,R}$ is continuous.

(d) Under the condition of (c), show that $T(K_{\delta,R})$ is compact by using the *Arzelá–Ascoli theorem* that if a sequence in $C[a, b]$ $(-\infty < a < b < \infty)$ is uniformly bounded and equicontinuous, then it contains a convergent subsequence in $C[a, b]$.

(e) Using (d) and the Schauder fixed-point theorem, show that $T : K_{\delta,R} \to K_{\delta,R}$ has a fixed point, thereby establishing the Peano theorem.

(f) Show by an example that the map (A.E.5) may not have a unique fixed point.

16. Use the formula (A.2.5) to show that the linking number is a symmetric invariant with respect to an exchange of the two closed curves, namely,

$$\mathrm{link}(C_1, C_2) = \mathrm{link}(C_2, C_1). \quad \text{(A.E.7)}$$

17. Use the formula (A.2.7) to draw some examples of two non-intersecting closed curves to obtain $0, -1, 2,$ in respective situations, as the linking numbers.

18. Show that the helicity (A.2.9) is invariant under the gauge transformation $\mathbf{A} \mapsto \mathbf{A} + \nabla\omega$ where ω is a real-valued smooth function that may be assumed to be compactly supported in Ω for convenience.

19. Check to see that the Biot–Savart law (A.2.11) indeed gives a solution of the static Ampére equation (A.2.10).

20. Show that any continuous function $f : [-1, 1] \to [-1, 1]$ has a fixed point. Construct an example $f : (-1, 1) \to (-1, 1)$ of a continuous function that has no fixed point in $(-1, 1)$.

21. Study the motion of a particle of mass $m > 0$ in the space $\mathbb{R}^3$ and use the construction of Noether's theorem to find the expressions of the angular momenta a_x and a_y about the x- and y-axises, respectively. Use your results and (A.3.29) to obtain the expression for the momentum vector of the general motion to be

$$\mathbf{L} = (L_x, L_y, L_z) = m\mathbf{r} \times \dot{\mathbf{r}} = \mathbf{r} \times \mathbf{p}. \quad \text{(A.E.8)}$$

22. Establish (A.3.36).
23. Verify (A.3.50).
24. Show that, with (A.3.64), the associated energy-momentum tensors, (A.3.50) and (A.3.51), are the same as those stated in (A.3.65).
25. For the Lagrangian action functional (A.3.30) subject to the symmetry

$$x'^0 = x^0, \quad \mathbf{x}' = \Omega\mathbf{x}, \quad \mathbf{x} = (x^1, \ldots, x^n) \in \mathbb{R}^n, \quad u'(x') = u(x), \quad \text{(A.E.9)}$$

where $\Omega \in SO(n)$, derive the associated Noether currents and charges.

26. Consider the Lagrangian action density

$$\mathcal{L} = \frac{1}{2}\partial_\mu\phi\partial^\mu\overline{\phi} - V(|\phi|^2), \quad \text{(A.E.10)}$$

governing a complex scalar field ϕ. It is clear that the model possesses a $U(1)$-symmetry, $\phi \mapsto \phi' = \mathrm{e}^{\mathrm{i}\theta}\phi$ ($\theta \in \mathbb{R}$). Use the idea of Noether's theorem to derive the energy-momentum tensor of the model associated with spacetime translation invariance and the $U(1)$-invariant current and charge. Confirm your results by reformulating the problem as a real-scalar field model such that the $U(1)$-symmetry is actually an $SO(2)$-symmetry for the underlying real-field formalism.

27. For the operators defined in (A.4.5)–(A.4.6), show that, if $\hat{L}^2 f = 0$, then

$$\hat{L}_x f = 0, \quad \hat{L}_y f = 0, \quad \hat{L}_z f = 0, \quad \text{(A.E.11)}$$

simultaneously. In other words, the zero-eigenvalue problem is actually trivial.

28. Establish (A.4.13).
29. Verify (A.4.22).
30. Show that, in terms of the polar and azimuthal angles θ and ϕ such that the line element of the standard unit sphere is

$$\mathrm{d}s^2 = \mathrm{d}\theta^2 + \sin^2\theta\,\mathrm{d}\phi^2, \quad \text{(A.E.12)}$$

the square of the angular momentum operator $\hat{L}^2$ is a multiple of the Laplace–Beltrami operator on the unit sphere, as given in (A.5.9).

31. For the spin matrices S_x, S_y, S_z given in (A.5.25), (A.5.29), and (A.5.30), compute $S^2 = S_x^2 + S_y^2 + S_y^2$ and show that S^2 indeed commutes with S_x, S_y, S_z.
32. For the spin matrices S_x and S_y given in (A.5.29)–(A.5.30), find all the eigenvectors associated with the eigenvalues $\pm\frac{\hbar}{2}$, respectively.
33. Using the properties of the spin algebra, derive the spin matrices (A.5.34) for the spin $s = 1$ situation, by taking

$$u = \begin{pmatrix} 1 \\ 0 \\ 0 \end{pmatrix}, \quad v = \begin{pmatrix} 0 \\ 1 \\ 0 \end{pmatrix}, \quad w = \begin{pmatrix} 0 \\ 0 \\ 1 \end{pmatrix}, \quad \text{(A.E.13)}$$

as the eigenvectors of S_z associated with the eigenvalues, $1, 0, -1$, respectively. Is your solution unique?

34. Derive (A.6.36)–(A.6.39) from (A.6.31) and (A.6.33)–(A.6.35).

35. Show that the quantity given on the left-hand side of (A.6.40) is a constant of motion. That is,

$$\frac{\mathrm{d}}{\mathrm{d}s}\left(A\dot{t}^2 - B\dot{r}^2 - r^2\dot{\theta}^2 - r^2\sin^2\theta\,\dot{\phi}^2\right) = 0, \tag{A.E.14}$$

in view of (A.6.36)–(A.6.39). Conversely, establish that any three equations among the equations (A.6.36)–(A.6.39) and (A.E.14) imply the validity of the full system of the equations (A.6.36)–(A.6.39).

36. Use (A.6.59) in (A.6.51) and vary the parameter G near $G = 0$ to derive the same expression (A.6.63) for the angle of deflection within the first-order approximation of G.

37. Show that (A.6.56) may be improved into the form

$$\phi_\infty = \frac{\pi}{2} + \delta_{\mathrm{N}} + \frac{3}{8}\left(\frac{5\pi}{4} - \frac{4}{3}\right)\delta_{\mathrm{N}}^2 + \mathrm{O}(\delta_{\mathrm{N}}^3), \tag{A.E.15}$$

which yields the following second-order Einstein bending angle for light deflection [66]

$$\delta = \frac{4GM}{r_0} + \left(\frac{15\pi}{4} - 4\right)\left(\frac{GM}{r_0}\right)^2. \tag{A.E.16}$$

38. Derive (A.6.68).

39. Establish (A.6.69).

40. Insert (A.6.71) into (A.6.51) and work out the details of the computation to obtain (A.6.72) for the bending angle of light deflection in the Hayward black hole situation.

Bibliography

[1] Abrikosov, AA. On the magnetic properties of superconductors of the second group. *Journal of Experimental and Theoretical Physics*. 1957; 5: 1174–1182.

[2] Acharya, PK, Bartolotti, LJ, Sears, SB, Parr, RG. An atomic kinetic energy functional with full Weizsäcker correction. *Proceedings of the National Academy of Sciences of the United States of America*. 1980; 77: 6978–6982.

[3] Actor, A. Classical solutions of $SU(2)$ Yang–Mills theories. *Reviews of Modern Physics*. 1979; 51: 461–525.

[4] Adam, C, Klimas, P, Sánchez-Guillén, J, Wereszczyński, A. Compact gauge K vortices. *Journal of Physics A: Mathematical and Theoretical*. 2009; 42(13): 135401.

[5] Adams, C. *The Knot Book: An Elementary Introduction to the Mathematical Theory of Knots*. American Mathematical Society: Providence; 2004.

[6] Aharonov, Y, Bohm, D. Significance of electromagnetic potentials in quantum theory. *Physical Review*. 1959; 115: 485–491.

[7] Aharonov, Y, Bohm, D. Further considerations on electromagnetic potentials in the quantum theory. *Physical Review*. 1961; 123: 1511–1524.

[8] Aharony, O, Bergman, O, Jaferis, DL, Maldacena, J. $\mathcal{N} = 6$ superconformal Chern–Simons-matter theories, M2-branes and their gravity duals. *Journal of High Energy Physics*. 2008; 0810: 091.

[9] Albert, VV, Glazman, LI, Jiang, L. Topological properties of linear circuit lattices. *Physical Review Letters*. 2015; 114: 173902.

[10] Almeida, CAG, Bazeia, D, Losano, L, Menezes, R. Scalar fields and defect structures: Perturbative procedure for generalized models. *Physical Review D*. 2013; 88: 025007.

[11] Ambjorn, J, Olesen, P. Anti-screening of large magnetic fields by vector bosons. *Physics Letters B*. 1988; 214: 565–569.

[12] Ambjorn, J, Olesen, P. On electroweak magnetism. *Nuclear Physics B*. 1989; 315: 606–614.

[13] Ambjorn, J, Olesen, P. A magnetic condensate solution of the classical electroweak theory. *Physics Letters B*. 1989; 218: 67–71.

[14] Ambjorn, J, Olesen, P. A condensate solution of the classical electroweak theory which interpolates between the broken and the symmetric phase. *Nuclear Physics B*. 1990; 330: 193–204.

[15] Anagnostopoulos, KN, Axenides, M, Floratos, EG, Tetradis, N. Large gauged Q balls. *Physical Review D*. 2001; 64: 125006.
[16] Ananda KN, Bruni, M. Cosmo-dynamics and dark energy with nonlinear equation of state: A quadratic model. *Physical Review D*. 2006; 74: 023523.
[17] Ananda, KN, Bruni, M. Cosmo-dynamics and dark energy with a quadratic EoS: Anisotropic models, large-scale perturbations and cosmological singularities. *Physical Review D*. 2006; 74: 023524.
[18] Andrews, M, Lewandowski, M, Trodden, M, Wesley, D. Doppelganger defects. *Physical Review D*. 2010; 82: 105006.
[19] Arafune, J, Freund, PGO, Goebel, CJ. Topology of Higgs fields. *Journal of Mathematical Physics*. 1975; 16: 433–437.
[20] Arnold, VI. *Mathematical Methods of Classical Mechanics*. Springer-Verlag: New York; 1978.
[21] Arnowitt, R, Deser, S, Misner, C. Dynamical structure and definition of energy in general relativity. *Physical Review*. 1959; 116: 1322–1330.
[22] Arnowitt, R, Deser, S, Misner, C. Coordinate invariance and energy expressions in general relativity. *Physical Review*. 1961; 122: 997–1006.
[23] Arnowitt, R, Deser, S, Misner, C. The dynamics of general relativity. In: Witten, L (ed.) *Gravitation: An Introduction to Current Research*. Wiley: London; 1962. pp. 227–265.
[24] Aubin, T. Problemes isoperimetriques et espaces de Sobolev. *Journal of Differential Geometry*. 1976; 11: 573–598.
[25] Auzzi, R, Bolognesi, S, Evslin, J, Konishi, K. Non-Abelian monopoles and the vortices that confine them. *Nuclear Physics B*. 2004; 686: 119–134.
[26] Avelino, PP, Bolejko, K, Lewis, GF. Nonlinear Chaplygin gas cosmologies. *Physical Review D*. 2014; 89: 103004.
[27] Axenides, M, Floratos, E, Kehagias, A. Non-Abelian Q-balls in supersymmetric theories. *Physics Letters B*. 1998; 444: 190–195.
[28] Ayon–Beato, E, Garcia, A. Regular black hole in general relativity coupled to nonlinear electrodynamics. *Physical Review Letters*. 1998; 80: 5056–5059.
[29] Ayon–Beato, E, Garcia, A. The Bardeen model as a nonlinear magnetic monopole. *Physics Letters B*. 2000; 493: 149–152.
[30] Babaev, E. Vortices with fractional flux in two-gap superconductors and in extended Faddeev model. *Physical Review Letters*. 2002; 89: 067001.
[31] Babaev, E. Phase diagram of planar $U(1) \times U(1)$ superconductors: Condensation of vortices with fractional flux and a superfluid state. *Nuclear Physics B*. 2004; 686: 397–412.
[32] Babichev, E. Global topological k-defects. *Physical Review D*. 2006; 74: 085004.
[33] Babichev, E. Gauge k-vortices. *Physical Review D*. 2008; 77: 065021.
[34] Bagger, J, Lambert, N. Modeling multiple M2-branes. *Physical Review D*. 2007; 75: 045020.
[35] Bagger, J, Lambert, N. Gauge symmetry and supersymmetry of multiple M2-branes. *Physical Review D*. 2008; 77: 065008.

[36] Bagger, J, Lambert, N. Comments on multiple M2-branes. *Journal of High Energy Physics.* 2008; 0802: 105.

[37] Balart, L, Vagenas, EC. Regular black hole metrics and the weak energy condition. *Physics Letters B.* 2014; 370: 14–17.

[38] Balbinot, R, Poisson, E. Stability of the Schwarzschild–de Sitter model. *Physical Review D.* 1990; 41: 395–402.

[39] Band, Y, Avishai, Y. *Quantum Mechanics with Applications to Nanotechnology and Information Science.* Elsevier: New York; 2013.

[40] Bardeen, J. Non-singular general-relativistic gravitational collapse. In: Fok, VA (ed.) *Abstracts/5th International Conference on Gravitation and the Theory of Relativity: Proceedings of International Conference GR5.* Publishing House of Tbilisi University: Tbilisi, USSR; 1968.

[41] Bardeen, J, Cooper, LN, Schrieffer, JR. Microscopic theory of superconductivity. *Physical Review.* 1957; 106: 162–164.

[42] Bardeen, J, Cooper, LN, Schrieffer, JR. Theory of superconductivity. *Physical Review.* 1957; 108: 1175–1204.

[43] Barreiro, T, Copeland, EJ, Nunes, NJ. Quintessence arising from exponential potentials. *Physical Review D.* 2000; 61: 127301.

[44] Barrow, JD. Graduated inflationary universes. *Physics Letters B.* 1990; 235: 40–43.

[45] Barrow, JD, Parsons, P. Inflationary models with logarithmic potentials. *Physical Review D.* 1995; 52: 5576–5587.

[46] von Barth, U. Basic density-functional theory–an overview. *Physica Scripta.* 2004; T109: 9–39.

[47] Bartolucci, D, Tarantello, G. Liouville type equations with singular data and their applications to periodic multivortices for the electroweak theory. *Communications in Mathematical Physics.* 2002; 229: 3–47.

[48] Battye, RA, Cooper, NR, Sutcliffe, PM. Stable skyrmions in two-component Bose–Einstein condensates. *Physical Review Letters.* 2002; 88: 080401.

[49] Battye, RA, Sutcliffe, PM. Knots as stable solutions in a three-dimensional classical field theory. *Physical Review Letters.* 1998; 81: 4798–4801.

[50] Battye, RA, Sutcliffe, PM. Solitons, links and knots. *Proceedings of the Royal Society A: Mathematical, Physical and Engineering Sciences.* 1999; 455: 4305–4331.

[51] Bazeia, D, Losano, L, Menezes, R, Oliveira, JCRE. Generalized global defect solutions. *European Physical Journal C.* 2007; 51: 953–962.

[52] Bazeia, D, da Hora, E, Rubiera-Garcia, D. Compact vortexlike solutions in a generalized Born–Infeld model. *Physical Review D.* 2011; 84: 125005.

[53] Bazeia, D, Lima, EEM, Losano, L. Kinklike structures in models of the Dirac–Born–Infeld type. *Annals of Physics.* 2018; 388: 408–427.

[54] Becker, K, Becker, M, Schwarz, J. *String Theory and M-Theory: A Modern Introduction.* Cambridge University Press: Cambridge; 2007.

[55] Belavin, AA, Polyakov, AM. Metastable states of two-dimensional isotropic ferromagnets. *Journal of Experimental and Theoretical Physics Letters.* 1975; 22: 245–247.

[56] Belyaev, D, Brink, L, Kim, S-S, Ramond, P. The BLG theory in light-cone superspace. *Journal of High Energy Physics.* 2010; 1004: 026.

[57] Benci, V, Rabinowitz, PH. Critical point theorems for indefinite functional. *Inventiones Mathematicae.* 1979; 52: 241–273.

[58] Benson, K, Bucher, M. Skyrmions and semilocal strings in cosmology. *Nuclear Physics B.* 1993; 406: 355–376.

[59] Berestycki, H, Lions, P-L. Nonlinear scalar field equations I: Existence of a ground state. *Archive for Rational Mechanics and Analysis.* 1983; 82: 313–345.

[60] Berger, MS. *Nonlinearity and Functional Analysis.* Academic Press: New York; 1977.

[61] Berger, MS, Chen, YY. Symmetric vortices for the Ginzburg–Landau equations of superconductivity and the nonlinear desingularization phenomenon. *Journal of Functional Analysis.* 1989; 82: 259–295.

[62] Bethuel, F, Brezis, H, Helein, F. *Ginzburg–Landau Vortices.* Birkhäuser: Boston; 1994.

[63] Bieri, L, Garfinkle, D, Yunes, N. Gravitational waves and their mathematics. *Notices of the American Mathematical Society.* 2017; 64: 693–707.

[64] Boatto, S, Cabral, HE. Nonlinear stability of a latitudinal ring of point vortices on a nonrotating sphere. *SIAM Journal on Applied Mathematics.* 2003; 64: 216–230.

[65] Boatto, S, Crowdy, D. Point-vortex dynamics. In: Françoise, J-P, Naber, GL, Tsun, TS (eds.) *Encyclopedia of Mathematical Physics.* Elsevier: Amsterdam; 2006. pp. 66–79.

[66] Bodenner, J, Will, CM. Deflection of light to second order: A tool for illustrating principles of general relativity. *American Journal of Physics.* 2003; 71: 770–773.

[67] Bogomol'nyi, EB. The stability of classical solutions. *Soviet Journal of Nuclear Physics.* 1976; 24: 449–454.

[68] Bombieri, E, De Giorgi, E, Giusti, E. Minimal cones and the Bernstein problem. *Inventiones Mathematicae.* 1969; 7: 243–268.

[69] Bondi, H, Van Der Burg, MGJ, Metzner, AWK. Gravitational waves in general relativity. VII. Waves from axi-symmetric isolated system. *Proceedings of the Royal Society A: Mathematical, Physical and Engineering Sciences.* 1962; 269: 21–52.

[70] Bonheure, D, d'Avenia, P, Pomponio, A. On the electrostatic Born–Infeld equation with extended charges. *Communications in Mathematical Physics.* 2016; 346: 877–906.

[71] Bonheure, D, d'Avenia, P, Pomponio, A, Reichel, W. Equilibrium measures and equilibrium potentials in the Born–Infeld model. *Journal de Mathématiques Pures et Appliquées.* 2020; 139: 35–62.

[72] Bonheure, D, Iacopetti, A. On the regularity of the minimizer of the electrostatic Born–Infeld energy. *Archive for Rational Mechanics and Analysis.* 2019; 232: 697–725.

[73] Born, M. Modified field equations with a finite radius of the electron. *Nature*. 1933; 132: 282.

[74] Born, M. On the quantum theory of the electromagnetic field. *Proceedings of the Royal Society A: Mathematical, Physical and Engineering Sciences*. 1934; 143: 410–437.

[75] Born, M, Infeld, L. Foundation of the new field theory. *Nature*. 1933; 132: 1004.

[76] Born, M, Infeld, L. Foundation of the new field theory. *Proceedings of the Royal Society A: Mathematical, Physical and Engineering Sciences*. 1934; 144: 425–451.

[77] Bott, R, Tu, LW. *Differential Forms in Algebraic Topology*. Springer-Verlag: New York; 1982.

[78] Boyer, RH, Lindquist, RW. Maximal analytic extension of the Kerr metric. *Journal of Mathematical Physics*. 1967; 8: 265–281.

[79] Bramwell, ST, Giblin, SR, Calder, R, Aldus, R, Prabhakaran, D, Fennell, T. Measurement of the charge and current of magnetic monopoles in spin ice. *Nature*. 2009; 461: 956–959.

[80] Bray, HL. Proof of the Riemannian Penrose inequality using the positive mass theorem. *Journal of Differential Geometry*. 2001; 59: 177–267.

[81] Bray, HL. Black holes, geometric flows, and the Penrose inequality in general relativity. *Notices of the American Mathematical Society*. 2002; 49: 1372–1381.

[82] Bray, HL, Schoen, RM. Recent proofs of the Riemannian Penrose conjecture. In: Jerison, D, Mazur, B, Schmid, W, Singer, I, Strook, DW, Yau, S-T (eds.) *Current Developments in Mathematics*. International Press of Boston: Somerville; 1999. pp. 1–36.

[83] Brewin, L. A simple expression for the ADM mass. *General Relativity and Gravitation*. 2007; 39: 521–528.

[84] Brewin, L. A brief introduction to Cadabra: A tool for tensor computations in general relativity. *Computer Physics Communications*. 2010; 181: 489–498.

[85] Brezis, H. The interplay between analysis and topology in some nonlinear PDE problems. *Bulletin of the American Mathematical Society*. 2003; 40: 179–201.

[86] Brezis, H, Merle, F, Rivière, T. Quantization effects for $-\Delta u = u(1-|u|^2)$ in R^2. *Comptes rendus de l'Académie des Sciences*. 1993; 317: 57–60.

[87] Brezis, H, Merle, F, Rivière, T. Quantization effects for $-\Delta u = u(1-|u|^2)$ in R^2. *Archive for Rational Mechanics and Analysis*. 1994; 126: 35–58.

[88] Brown, JD, Lau, SR, York, Jr., JW. Canonical quasilocal energy and small spheres. *Physical Review D*. 1999; 59: 064028.

[89] Brown, JD, York, Jr., JW. Quasilocal energy and conserved charges derived from the gravitational action. *Physical Review D*. 1993; 47: 1407–1419.

[90] Brown, JW, Churchill, RV. *Complex Variables and Applications*. 9th ed. McGraw-Hill: New York; 2014.

[91] Buck, G. Four-thirds power law for knots and links. *Nature.* 1998; 392: 238–239.

[92] Buoninfante, L, Koshelev, AS, Lambiase, G, Mazumdar, A. Classical properties of nonlocal, ghost-, and singularity-free gravity. *Journal of Cosmology and Astroparticle Physics.* 2018; 09: 034.

[93] Burzlaff, J. $SU(3)$ monopole with magnetic quantum numbers (0, 2). *Physical Revuew D.* 1981; 23: 1329–1334.

[94] Burzlaff, J, Navarro-Lerida, F. Nielsen–Olesen vortices for large Ginzburg–Landau parameter. *Physical Review D.* 2010; 82: 125033.

[95] Burzlaff, J, Moncrief, V. The global existence of time-dependent vortex solutions. *Journal of Mathematical Physics.* 1985; 26: 1368–1372.

[96] Caffarelli, LA, Yang, Y. Vortex condensation in the Chern–Simons Higgs model: An existence theorem. *Communications in Mathematical Physics.* 1995; 168: 321–336.

[97] Calabi, E. The space of Kähler metrics. In: Gerretsen, JCH, de Groot, J (eds.) *Proceedings of the International Congress of Mathematicians, 1954: Held at Amsterdam, September 2–9.* Noordhoff: Groningen; 1957. pp. 206–207.

[98] Calabi, E. On Kähler manifolds with vanishing canonical class. In: Fox, RH, Spencer, DC, Tucker, AW (eds.) *Algebraic Geometry and Topology. A Symposium in Honor of S. Lefschetz.* Princeton University Press: Princeton; 1957. pp. 78–89 (Princeton Mathematical Series; No. 12).

[99] Calabi, E. Examples of Bernstein Problems for some nonlinear equations. In: Chern, S-S, Smale, S (eds.) *Global Analysis, Proceedings of the Symposium in Pure Mathematics of the American Mathematical Society.* Vol. 15. American Mathematical Society: Berkeley; 1968. pp. 223–230.

[100] Caldwell, RR. Dark energy. *Physics World.* 2004; 17: 37–42.

[101] Caldwell, RR, Dave, R, Steinhardt, PJ. Cosmological imprint of an energy component with general equation-of-state. *Physical Review Letters.* 1998; 80: 1582–1585.

[102] Callan Jr., CG, Maldacena, JM. Brane dynamics from the Born–Infeld action. *Nuclear Physics B.* 1998; 513: 198–212.

[103] Calugareanu, G. L'integral de Gauss et l'analyse des nocuds tridimensionnels. *Romanian Journal of Pure and Applied Mathematics.* 1959; 4: 5–20.

[104] Calugareanu, G. Sur les classes d'isotopie des noeuds tridimensionnels et leurs invariants. *Czechoslovak Mathematical Journal.* 1961; 11: 588–625.

[105] Calugareanu, G. Sur les enlacements tridimensionnels des courbes fermees. *Comunicările Academiei Republicii Populare Române.* 1961; 11: 829–832.

[106] Candelas, P, Horowitz, GT, Strominger, A, Witten, E. Vacuum configurations for superstrings. *Nuclear Physics B.* 1985; 258: 46–74.

[107] Cantarella, J, Kusner, RB, Sullivan, JM. Tight knot values deviate from linear relations. *Nature.* 1998; 392: 237–238.

[108] Capelle, K. A bird's-eye view of density-functional theory. *Brazilian Journal of Physics.* 2006; 36: 1318–1343.

[109] Cao, L, Chen, S, Yang, Y. Domain wall solitons arising in classical gauge field theories. *Communications in Mathematical Physics.* 2019; 369: 317–349.

[110] Carneiro, S, Pigozzo, C. Observational tests for non-adiabatic Chaplygin gas. *Journal of Cosmology and Astroparticle Physics.* 2014; 1410: 060.

[111] Carroll, SM. *Spacetime and Geometry.* Addison Wesley: New York; 2004.

[112] Carroll, SM. Quintessence and the rest of the world: Suppressing long-range interactions. *Physical Review Letters.* 1998; 81: 3067–3070.

[113] Carroll, SM, Johnson, MC, Randall, L. Extremal limits and black hole entropy. *Journal of High Energy Physics.* 2009; 0911: 109.

[114] Castelnovo, C, Moessner, R, Sondhi, SL. Magnetic monopoles in spin ice. *Nature.* 2008; 451: 42–45.

[115] Cerf, C, Stasiak, A. A topological invariant to predict the three-dimensional writhe of ideal configurations of knots and links. *Proceedings of the National Academy of Sciences of the United States of America.* 2000; 97: 3795–3798.

[116] Cervantes-Cota, JL, Galindo-Uribarri, S, Smoot, GF. A brief history of gravitational waves. *Universe.* 2016; 2: 22.

[117] Chae, D, Imanuvilov, OY. The existence of non-topological multivortex solutions in the relativistic self-dual Chern–Simons theory. *Communications in Mathematical Physics.* 2000; 215: 119–142.

[118] Chaichian, M, Nelipa, NF. *Introduction to Gauge Field Theory.* Springer-Verlag: New York; 1984.

[119] Chakrabortty, S, Chowdhury, SP, Ray, K. Some BPS configurations of the BLG theory. *Physics Letters B.* 2011; 703: 172–179.

[120] Chambers, RG. Shift of an electron interference pattern by enclosed magnetic flux. *Physical Review Letters.* 1960; 5: 3–5.

[121] Chandrasekhar, S. *The Mathematical Theory of Black Holes.* Oxford University Press: Oxford; 1998.

[122] Chang, S-J, Ellis, SD, Lee, BW. Chiral confinement: An exact solution of the massive Thirring model. *Physical Review D.* 1975; 11: 3572–3582.

[123] Chen, RM, Guo, Y, Spirn, D, Yang, Y. Electrically and magnetically charged vortices in the Chern–Simons–Higgs theory. *Proceedings of the Royal Society A: Mathematical, Physical and Engineering Sciences.* 2009; 465: 3489–3516.

[124] Chen, S, Gibbons, GW, Li, Y, Yang, Y. Friedmann's equations in all dimensions and Chebyshev's theorem. *Journal of Cosmology and Astroparticle Physics.* 2014; 12: 035.

[125] Chen, S, Gibbons, GW, Yang, Y. Explicit integration of Friedmann's equation with nonlinear equations of state. *Journal of Cosmology and Astroparticle Physics.* 2015; 05: 020.

[126] Chen, S, Gibbons, GW, Yang, Y. Friedmann–Lemaitre cosmologies via roulettes and other analytic methods. *Journal of Cosmology and Astroparticle Physics.* 2015; 10: 056.

[127] Chen, S, Yang, Y. Existence of multiple vortices in supersymmetric gauge field theory. *Proceedings of the Royal Society A: Mathematical, Physical and Engineering Sciences.* 2012; 468: 3923–3946.

[128] Chen, S, Zhang, R, Zhu, M. Multiple vortices in the Aharony–Bergman–Jafferis–Maldacena model. *Annales Henri Poincaré*. 2013; 14: 1169–1192.

[129] Chen, X, Hastings, S, McLeod, JB, Yang, Y. A nonlinear elliptic equation arising from gauge field theory and cosmology. *Proceedings of the Royal Society A: Mathematical, Physical and Engineering Sciences*. 1994; 446: 453–478.

[130] Cheng, S-Y, Yau, S-T. Maximal spacelike hypersurfaces in the Lorentz–Minkowski spaces. *Annals of Mathematics*. 1976; 104: 407–419.

[131] Chern, SS, Simons, J. Some cohomology classes in principal fiber bundles and their application to Riemannian geometry. *Proceedings of the National Academy of Sciences of the United States of America*. 1971; 68: 791–794.

[132] Chern, SS, Simons, J. Characteristic forms and geometric invariants. *Annals of Mathematics*. 1974; 99: 48–69.

[133] Chiao, RY, Garmire, E, Townes, CH. Self-trapping of optical beams. *Physical Review Letters*. 1964; 13: 479–482.

[134] Chiba, T, De Felice, A, Tsujikawa, S. Observational constraints on quintessence: Thawing, tracker, and scaling models. *Physical Review D*. 2013; 87: 083505.

[135] Chiba, T, Sugiyama, N, Nakamura, T. Cosmology with x-matter. *Monthly Notices of the Royal Astronomical Society*. 1997; 289: L5–L9.

[136] Chimento, LP, Lazkoz, R, Maartens, R, Quiros, I. Crossing the phantom divide without phantom matter. *Journal of Cosmology and Astroparticle Physics*. 2006; 0609: 004.

[137] Cho, YM, Maison, D. Monopole configurations in Weinberg–Salam model. *Physics Letters B*. 1997; 391: 360–365.

[138] Clemson, TG, Liddle, AR. Observational constraints on thawing quintessence models. *Monthly Notices of the Royal Astronomical Society*. 2009; 395: 1585–1590.

[139] Coddington, EA, Levinson, N. *Theory of Ordinary Differential Equations*. McGraw-Hill: New York; 1955.

[140] Coleman, S. Quantum sine–Gordon equation as the massive Thirring model. *Physical Review D*. 1975; 11: 2088–2097.

[141] Coleman, S. Q-balls. *Nuclear Physics B*. 1985; 262: 263–283.

[142] Cooper, LN. Bound electron pairs in a degenerate Fermi gas. *Physical Review*. 1956; 104: 1189–1190.

[143] Copeland, EJ, Liddle, AR, Wands, D. Exponential potentials and cosmological scaling solutions. *Physical Review D*. 1998; 57: 4686–4690.

[144] Copeland, EJ, Saffin, PM, Zhou, S-Y. Charge-swapping Q-balls. *Physical Review Letters*. 2014; 113: 231603.

[145] Cordero, R, González, EL, Queijeiro, A. An equation of state for purely kinetic k-essence inspired by cosmic topological defects. *European Physical Journal C*. 2017; 77: 413.

[146] Corless, RM, Gonnet, GH, Hare, DEG, Jeffrey, DJ, Knuth, DE. On the Lambert W function. *Advances in Computational Mathematics*. 1996; 5: 329–359.

[147] Corrigan, E, Olive, DI, Fairlie, DB, Nuyts, J. Magnetic monopoles in $SU(3)$ gauge theories. *Nuclear Physics B*. 1976; 106: 475–492.

[148] Costa, SS. An entirely analytical cosmological model. *Modern Physics Letters A*. 2009; 24: 531–540.

[149] Courant, R, Hilbert, D. *Methods of Mathematical Physics*. Vol. 1. Wiley-Interscience: New York; 1989.

[150] Cubrovic, M, Zaanen, J, Schalm, K. String theory, quantum phase transitions, and the emergent Fermi liquid. *Science*. 2009; 325: 439–444.

[151] Curie, P. Sur la possibilité d'existence de la conductibilité magnétique et du magnétisme libre. *Séances de la Société Francaise de Physique (Paris)*. 1894: 76–77.

[152] Curtright, TL, Thorn, CB. Conformally invariant quantization of the Liouville theory. *Physical Review Letters*. 1982; 48: 1309–1313.

[153] Danielsson, UH. Lectures on string theory and cosmology. *Classical and Quantum Gravity*. 2005; 22: S1–S40.

[154] Davydov, AS. *Quantum Mechanics*. Pergamon Press: Oxford; 1985.

[155] de Gennes, PG. *Superconductivity of Metals and Alloys*. Benjamin: New York; 1966.

[156] De Lorenci, VA, Klippert, R, Novello, M, Salim, JM. Nonlinear electrodynamics and FRW cosmology. *Physical Review D*. 2002; 65: 063501.

[157] Demianski, M. Static electromagnetic geon. *Foundations of Physics*. 1986; 16: 187–190.

[158] Derrick, GH. Comments on nonlinear wave equations as models for elementary particles. *Journal of Mathematical Physics*. 1964; 5: 1252–1254.

[159] Deser, S, Gibbons, GW. Born–Infeld–Einstein actions. *Classical and Quantum Gravity*. 1998; 15: L35–L39.

[160] Deser, S, Jackiw, R, Templeton, S. Three-dimensional massive gauge theories. *Physical Review Letters*. 1982; 48: 975–978.

[161] Deser, S, Jackiw, R, Templeton, S. Topologically massive gauge theories. *Annals of Physics*. 1982; 140: 372–411.

[162] Desyatnikov, AS, Kivshar, YS, Torner, L. Optical vortices and vortex solitons. *Progress in Optics*. 2005; 47: 291–391.

[163] Dewitt, B. Quantum theory of gravity. I. The canonical theory. *Physical Review*. 1967; 160: 1113–1148.

[164] Dieudonné, J. *Foundations of Modern Analysis*. Academic Press: Boston; 1969.

[165] Diez-Tejedor, A, Feinstein, A. Relativistic hydrodynamics with sources for cosmological K-fluids. *International Journal of Modern Physics D*. 2005; 14: 1561–1576.

[166] Dirac, PAM. The quantum theory of the electron. *Proceedings of the Royal Society A: Mathematical, Physical and Engineering Sciences*. 1928; 117: 610–624.

[167] Dirac, PAM. Quantized singularities in the electromagnetic field. *Proceedings of the Royal Society A: Mathematical, Physical and Engineering Sciences*. 1931; 133: 60–72.

[168] Dirac, PAM. *General Theory of Relativity.* Princeton University Press: Princeton; 1996.

[169] Dirac, PAM. *The Principles of Quantum Mechanics.* 3rd ed. Oxford University Press: Oxford; 1947.

[170] Disconzi, MM, Khuri, MA. On the Penrose inequality for charged black holes. *Classical and Quantum Gravity.* 2012; 29: 245019.

[171] Dodelson, S, Widrow, L. Baryon symmetric baryogenesis. *Physical Review Letters.* 1990; 64: 340–343.

[172] Douglas, MR, Kachru, S. Flux compactification. *Reviews of Modern Physics.* 2007; 79: 733–796.

[173] Dreizler, RM, Gross, EKU. *Density Functional Theory: An Approach to the Quantum Many-Body Problem.* Springer-Verlag: Berlin; 1990.

[174] Du, Q, Gunzburger, M, Peterson, J. Analysis and approximation of the Ginzburg–Landau model of superconductivity. *SIAM Review.* 1992; 34: 54–81.

[175] Dubrovin, BA, Fomenko, AT, Novikov, SP. *Modern Geometry–Methods and Applications. Part II: The Geometry and Topology of Manifolds.* Springer-Verlag: New York; 1985.

[176] Dunne, G. *Self-Dual Chern–Simons Theories.* Springer-Verlag: Berlin; 1995 (Lecture Notes in Physics; No. 36).

[177] Dunne, G. Mass degeneracies in self-dual models. *Physics Letters B.* 1995; 345: 452–457.

[178] Dunne, G, Jackiw, R, Pi, S-Y, Trugenberger, C. Self-dual Chern–Simons solitons and two-dimensional nonlinear equations. *Physical Review D.* 1991; 43: 1332–1345.

[179] Dvali, G, Zaldarriaga, M. Changing α with time: Implications for fifth-force-type experiments and quintessence. *Physical Review Letters.* 2002; 88: 091303.

[180] Dymnikova, I. Regular electrically charged structures in nonlinear electrodynamics coupled to general relativity. *Classical and Quantum Gravity.* 2004; 21: 4417–4429.

[181] Dymnikova, I. Cosmological term as a source of mass. *Classical and Quantum Gravity.* 2002; 19: 725–740.

[182] Dyson, FW, Eddington, AS, Davidson, C. A determination of the deflection of light by the Sun's gravitational field, from observations made at the total eclipse of 29 May 1919. *Proceedings of the Royal Society A: Mathematical, Physical and Engineering Sciences.* 1920; 220: 291–333.

[183] Weinan, E. Dynamics of vortices in Ginzburg–Landau theories with applications to superconductivity. *Physica D: Nonlinear Phenomena.* 1994; 77: 383–404.

[184] Eardley, DM, Moncrief, V. The global existence of Yang–Mills–Higgs fields in 4-dimensional Minkowski space. I. Local existence and smoothness properties. *Communications in Mathematical Physics.* 1982; 83: 171–191.

[185] Eardley, DM, Moncrief, V. The global existence of Yang–Mills–Higgs fields in 4-dimensional Minkowski space. II. Completion of proof. *Communications in Mathematical Physics.* 1982; 83: 193–212.

[186] Easther, R, Maeda, K, Wands, D. Tree-level string cosmology. *Physical Review D*. 1996; 53: 4247–4256.

[187] Eells, Jr., J, Sampson, JH. Harmonic mappings of Riemannian manifolds. *American Journal of Mathematics*. 1964; 86: 109–160.

[188] Eells, Jr., J, Sampson, JH. A report on harmonic maps. *Bulletin of the London Mathematical Society*. 1978; 10: 1–68.

[189] Eells, Jr., J, Sampson, JH. Another report on harmonic maps. *Bulletin of the London Mathematical Society*. 1988; 20: 385–524.

[190] Einstein, A. Lens-like action of a star by the deviation of light in the gravitational field. *Science*. 1936; 84: 506–507.

[191] Einstein, A, de Sitter, W. On the relation between the expansion and the mean density of the universe. *Proceedings of the National Academy of Sciences of the United States of America*. 1932; 18: 213–214.

[192] Englander, SW, Kallenbach, NR, Heeger, AJ, Krumhansl, JA, Litwin, S. Nature of the open state in long polynucleotide double helices: Possibility of soliton excitations. *Proceedings of the National Academy of Sciences of the United States of America*. 1980; 77: 7222–7226.

[193] Engel, E, Dreizler, RM. Extension of the Thomas–Fermi–Dirac–Weizsäcker model: Four-order gradient corrections to the kinetic energy. *Journal of Physics B*. 1989; 22: 1901–1912.

[194] Englert, F, Brout, R. Broken symmetry and the mass of gauge vector mesons. *Physical Review Letters*. 1964; 13: 321–323.

[195] Enqvist, K, Mazumdar, A. Cosmological consequences of MSSM flat directions. *Physics Reports*. 2003; 380: 99–234.

[196] Enqvist, K, McDonald, J. Q-balls and baryogenesis in the MSSM. *Physics Letters B*. 1998; 425: 309–321.

[197] Esteban, MJ. Existence of symmetric solutions for the Skyrme's problem. *Annali di Matematica Pura ed Applicata*. 1987; 147: 187–195.

[198] Esteban, MJ, Lewin, M, Séré, E. Variational methods in relativistic quantum mechanics. *Bulletin of the American Mathematical Society*. 2008; 45: 535–593.

[199] Evans, LC. *Partial Differential Equations*. American Mathematical Society: Providence; 2002 (Graduate Studies in Mathematics; No. 19).

[200] Ezawa, ZF, Hasebe, K. Interlayer exchange interactions, $SU(4)$ soft waves and Skyrmions in bilayer quantum Hall ferromagnets. *Physical Review B*. 2002; 65: 075311.

[201] Ezhuthachan, B, Mukhi, S, Papageorgakis, C. The power of the Higgs mechanism: Higher-derivative BLG theories. *Journal of High Energy Physics*. 2009; 0904: 101.

[202] Faddeev, L. Some comments on the many-dimensional solitons. *Letters in Mathematical Physics*. 1976; 1: 289–293.

[203] Faddeev, L. Einstein and several contemporary tendencies in the theory of elementary particles. In: Pantaleo, M, de Finis, F (eds.) *Relativity, Quanta, and Cosmology*. Vol. 1. Johnson Reprint Corporation: New York; 1979. pp. 247–266.

[204] Faddeev, L. Knotted solitons. In: Li, T (ed.) *Proceedings of the International Congress of Mathematicians 2002, Beijing, China, 20–28 August 2002*. Vol. 1. Higher Education Press: Beijing; 2002. pp. 235–244.

[205] Faddeev, L, Niemi, AJ. Stable knot-like structures in classical field theory. *Nature*. 1997; 387: 58–61.

[206] Felsager, B. *Geometry, Particles, and Fields*. Springer-Verlag: New York; 1998 (Graduate Texts in Contemporary Physics).

[207] Fermi, E. Eine statistische methode zur bestimmung einiger eigenschaften des atoms und ihre anwendung auf die theorie des periodischen systems der elemente. *Zeitschrift für Physik A*. 1928; 48: 73–79.

[208] Fernando, S, Krug, D. Charged black hole solutions in Einstein–Born–Infeld gravity with a cosmological constant. *General Relativity and Gravitation*. 2003; 35: 129–137.

[209] Fierz, M. Über die relativistische theorie kräftefreier teilchen mit beliebigem spin. *Helvetica Physica Acta*. 1939; 12: 3–37.

[210] Figueroa, DG, Shaposhnikov, M. Anomalous non-conservation of fermion/chiral number in Abelian gauge theories at finite temperature. *Journal of High Energy Physics*. 2018; 2018: 26.

[211] Foster, J, Nightingale, JD. *A Short Course in General Relativity*. 3rd ed. Springer-Verlag: New York; 2006.

[212] Fox, C. *An Introduction to the Calculus of Variations*. Dover: New York; 1987.

[213] Fradkin, ES, Tseytlin, AA. Non-linear electrodynamics from quantized strings. *Physics Letters B*. 1985; 163: 123–130.

[214] Freifelder, D. *Molecular Biology*. Jones and Bartlett: Boston; 1987.

[215] Friedan, D. Nonlinear models in $2+\varepsilon$ dimensions. *Physical Review Letters*. 1980; 45: 1057–1060.

[216] Friedberg, R, Lee, TD, Sirlin, A. Class of scalar-field soliton solutions in three space dimensions. *Physical Review D*. 1976; 13: 2739–2761.

[217] Frieman, J, Gelmini, G, Gleiser, M, Kolb, E. Solitogenesis: Primordial origin of nontopological solitons. *Physical Review Letters*. 1988; 60: 2101–2104.

[218] Frieman, J, Hill, C, Stebbins, A, Waga, I. Cosmology with ultra-light pseudo-Nambu–Goldstone bosons. *Physical Review Letters*. 1995; 75: 2077–2080.

[219] Fröhlich, J, Marchetti, PA. Quantum field theories of vortices and anyons. *Communications in Mathematical Physics*. 1989; 121: 177–223.

[220] Frolov, VP. Notes on nonsingular models of black holes. *Physical Review D*. 2016; 94: 104056.

[221] Frolov, VP, Zelnikov, A, de Paula Netto, T. Spherical collapse of small masses in the ghost-free gravity. *Journal of High Energy Physics*. 2015; 09: 107.

[222] Fuller, FB. The writhing number of a space curve. *Proceedings of the National Academy of Sciences of the United States of America*. 1971; 68: 815–819.

[223] Gallot, S, Hulin, D, Lafontaine, L. *Riemannian Geometry.* Springer-Verlag: New York; 1987.

[224] Gamelin, TW. *Complex Analysis.* Springer-Verlag: New York; 2001.

[225] Gao, Z, Gudnason, SB, Yang, Y. Integer-squared laws for global vortices in the Born–Infeld wave equations. *Annals of Physics.* 2019; 400: 303–319.

[226] Garcia-Compean, H, Garcia-Jimenez, G, Obregon, O, Ramirez, C. Crossing the phantom divide in an interacting generalized Chaplygin gas. *Journal of Cosmology and Astroparticle Physics.* 2008; 0807: 016.

[227] Garcia-Salcedo, R, Gonzalez, T, Quiros, I. No compelling cosmological models come out of magnetic universes which are based in nonlinear electrodynamics. *Physical Review D.* 2014; 89: 084047.

[228] Georgi, H, Glashow, SL. Spontaneously broken gauge symmetry and elementary particle masses. *Physical Review D.* 1972; 6: 2977–2982.

[229] Georgi, H, Glashow, S. Unity of all elementary particle forces. *Physical Review Letters.* 1974; 32: 438–441.

[230] Gervais, J-L, Neveu, A. Dual string spectrum in Polyakov's quantization, parts I and II. *Nuclear Physics B.* 1982; 199: 59–76; *ibid* 1982; 209: 125–145.

[231] Gibbons, GW. Born–Infeld particles and Dirichlet p-branes. *Nuclear Physics B.* 1998; 514: 603–639.

[232] Gibbons, GW, Hawking, SW, Horowitz, GT, Perry, MJ. Positive mass theorems for black holes. *Communications in Mathematical Physics.* 1983; 88: 295–308.

[233] Gibbons, GW, Rasheed, DA. Electric-magnetic duality rotations in nonlinear electrodynamics. *Nuclear Physics B.* 1995; 454: 185–206.

[234] Giblin, SR, Bramwell, ST, Holdsworth, PCW, Prabhakaran, D, Terry, I. Creation and measurement of long-lived magnetic monopole currents in spin ice. *Nature Physics.* 2011; 7: 252–258.

[235] Gibney, E. Quantum cloud simulates magnetic monopole. *Nature.* 2014; 1038: 14612.

[236] Gilbarg, D, Trudinger, N. *Elliptic Partial Differential Equations of Second Order.* Springer-Verlag: New York; 1977.

[237] Ginzburg, VL, Landau, LD. On the theory of superconductivity. In: Ter Haar, D (ed.) *Collected Papers of L. D. Landau.* Pergamon: New York; 1965. pp. 546–568.

[238] Gisiger, T, Paranjape, MB. Recent mathematical developments in the Skyrme model. *Physics Reports.* 1998; 306: 109–211.

[239] Glashow, S. The renormalizability of vector meson interactions. *Nuclear Physics.* 1959; 10: 107–117.

[240] Goddard, P, Olive, DI. Magnetic monopoles in gauge field theories. *Reports on Progress in Physics.* 1978; 41: 1357–1437.

[241] Gorkov, L, Eliashberg, GM. Generalization of the Ginzburg–Landau equations for non-stationary problems in the case of alloys with paramagnetic impurities. *Journal of Experimental and Theoretical Physics.* 1968; 27: 328–334.

[242] Gorkov, LP. Microscopic derivation of the Ginzburg–Landau equations in the theory of superconductivity. *Journal of Experimental and Theoretical Physics.* 1959; 9: 1364–1367.

[243] Gorkov, LP, Melik-Barkhudarov, TK. Microscopic derivation of the Ginzburg–Landau equations for an anisotropic superconductor. *Journal of Experimental and Theoretical Physics.* 1964; 18: 1031–1034.

[244] Goto, T. Relativistic quantum mechanics of one-dimensional mechanical continuum and subsidiary condition of dual resonance model. *Progress of Theoretical Physics.* 1971; 46: 1560–1569.

[245] Grana, M. Flux compactifications in string theory: A comprehensive review. *Physics Reports.* 2006; 423: 91–158.

[246] Grant, IS, Phillips, WR. *Electromagnetism.* 2nd ed. John Wiley & Sons: New York; 1990.

[247] Green, M, Schwarz, J, Witten, E. *Superstring Theory. Volume 1: Introduction.* Cambridge University Press: Cambridge; 2012.

[248] Green, M, Schwarz, J, Witten, E. *Superstring Theory. Volume 2: Loop Amplitudes, Anomalies and Phenomenology.* Cambridge University Press: Cambridge; 2012.

[249] Greensite, J. *An Introduction to the Confinement Problem.* Springer-Verlag: New York; 2011 (Lecture Notes in Physics; No. 821).

[250] Greiner, W. *Relativistic Quantum Mechanics. Wave Equations.* 3rd ed. Springer-Verlag: New York; 1987.

[251] Greiner, W. *Quantum Mechanics: An Introduction.* 4th ed. Springer-Verlag: New York; 2001.

[252] Greiner, W, Neise, L, Stöcker, H. *Thermodynamics and Statistical Mechanics.* Springer-Verlag: New York; 1995.

[253] Greiner, W, Reinhart, J. *Field Quantization.* Springer-Verlag: New York; 1996.

[254] Greub, W, Halperin, S, Vanstone, R. *Connections, Curvature, and Cohomology.* Vol. II. Academic Press: New York; 1973.

[255] Griffiths, DJ. *Introduction to Quantum Mechanics.* Prentice-Hall: Englewood Cliffs; 1995.

[256] Griffiths, DJ. *Introduction to Electrodynamics.* 3rd ed. Addison-Wesley: Boston; 2012.

[257] Griffiths, P, Harris, J. *Principles of Algebraic Geometry.* John Wiley & Sons: New York; 1978.

[258] Gromov, M. *Quantitative Homotopy Theory.* Princeton University Press: Princeton; 1996 (Prospects in Mathematics).

[259] Gross, D, Neveu, A. Dynamical symmetry breaking in asymptotically free field theories. *Physical Review D.* 1974; 10: 3235–3253.

[260] Guillemin, V, Pollack, A. *Differential Topology.* Reprinted edition. American Mathematical Society: Providence; 2010.

[261] Gulamov, IE, Nugaev, EY, Smolyakov, MN. Theory of U(1) gauged Q-balls revisited. *Physical Review D.* 2014; 89: 085006.

[262] Guralnik, G, Hagen, CR, Kibble, TWB. Global conservation laws and massless particles. *Physical Review Letters.* 1964; 13: 585–587.

[263] Gustavsson, A. Algebraic structures on parallel M2-branes. *Nuclear Physics B.* 2009; 811: 66–76.

[264] Henyey, FS. Distinction between a perfect conductor and a superconductor. *Physical Review Letters*. 1982; 49: 416.

[265] Hagan, PS. Spiral waves in reaction-diffusion equations. *SIAM Journal on Applied Mathematics*. 1982; 42: 762–786.

[266] Halliwell, JJ. Scalar fields in cosmology with an exponential potential. *Physics Letters B*. 1987; 185: 341–344.

[267] Halmos, PR, *Finite-Dimensional Vector Spaces*. 2nd ed. Springer-Verlag: New York; 1987.

[268] Halperin, BI. Statistics of quasiparticles and the hierarchy of fractional quantized Hall states. *Physical Review Letters*. 1984; 52: 1583–1586.

[269] Han, X. The Born–Infeld vortices induced from a generalized Higgs mechanism. *Proceedings of the Royal Society A: Mathematical, Physical and Engineering Sciences*. 2016; 472: 20160012.

[270] Han, X, Huang, G, Yang, Y. Coexisting vortices and antivortices generated by dually gauged harmonic maps. *Journal of Mathematical Physics*. 2021; 62: 103503.

[271] Han, CS Lin, Tarantello, G, Yang, Y. Chern–Simons vortices in the Gudnason model. *Journal of Functional Analysis*. 2014; 267: 678–726.

[272] Han, X, Lin, CS, Yang, Y. Resolution of Chern–Simons–Higgs vortex equations. *Communications in Mathematical Physics*. 2016; 343: 701–724.

[273] Han, X, Tarantello, G. Doubly periodic self-dual vortices in a relativistic non-Abelian Chern–Simons model. *Calculus of Variations and Partial Differential Equations*. 2014; 49: 1149–1176.

[274] Han, X, Yang, Y. Magnetic impurity inspired Abelian Higgs vortices. *Journal of High Energy Physics*. 2016; 2: 046.

[275] Han, X, Yang, Y. Existence theorems for vortices in the Aharony–Bergman–Jaferis–Maldacena model. *Communications in Mathematical Physics*. 2015; 333: 229–259.

[276] Han, X, Yang, Y. Doubly periodic solutions of relativistic Chern–Simons–Higgs vortex equations. *Transactions of the American Mathematical Society*. 2016; 368: 3565–3590.

[277] Hawking, SW. The occurrence of singularities in cosmology. III. Causality and singularities. *Proceedings of the Royal Society A: Mathematical, Physical and Engineering Sciences*. 1967; 300: 187–201.

[278] Hawking, SW. The conservation of matter in general relativity. *Communications in Mathematical Physics*. 1970; 18: 301–306.

[279] Hawking, SW. Particle creation by black holes. *Communications in Mathematical Physics*. 1975; 43: 199–220.

[280] Hawking, SW, Ellis, GFR. *The Large Scale Structure of Spacetime*. Cambridge University Press: Cambridge; 1973.

[281] Hawking, SW, Penrose, R. The singularities of gravitational collapse and cosmology. *Proceedings of the Royal Society A: Mathematical, Physical and Engineering Sciences*. 1970; 314: 529–548.

[282] Hawley, JF, Holcomb, KA. *Foundations of Modern Cosmology*. 2nd ed. Oxford University Press: Oxford; 2005.

[283] Hayward, SA. Formation of evaporation of non-singular black holes. *Physical Review Letters*. 2006; 96: 031103.

[284] Heard, IPC, Wands, D. Cosmology with positive and negative exponential potentials. *Classical and Quantum Gravity*. 2002; 19: 5435–5447.

[285] Helgason, S. *Differential Geometry and Symmetric Spaces*. Academic Press: New York; 1962.

[286] von Helmholtz, H. On integrals of the hydrodynamical equations which express vortex motion [first published 1858]. *The London, Edinburgh, and Dublin Philosophical Magazine and Journal of Science*. 1867; 4: 485–512.

[287] Hendi, SH. Asymptotic charged BTZ black hole solutions. *Journal of High Energy Physics*. 2012; 03: 065.

[288] Hendi, SH. Asymptotic Reissner–Nordström black holes. *Annals of Physics*. 2013; 333: 282–289.

[289] Henry, RC. Kretschmann scalar for a Kerr–Newman black hole. *Astrophysics Journal*. 2000; 535: 350–353.

[290] Higgs, P. Broken symmetries and the masses of gauge bosons. *Physical Review Letters*. 1964; 13: 508–509.

[291] Hill, CD, Nurowski, P. How the green light was given for gravitational wave search. *Notices of the American Mathematical Society*. 2017; 64: 686–692.

[292] Hirayama, M, Yamakoshi, H, Yamashita, J. Estimation of the Lin-Yang bound of the least static energy of the Faddeev model. *Progress of Theoretical Physics*. 2006; 116: 273–283.

[293] Hirsch, MW. *Differential Topology*. Springer-Verlag: New York; 1976 (Graduate Texts in Mathematics).

[294] Hohenberg, P, Kohn, W. Inhomogeneous electron gas. *Physical Review*. 1964; 136: B864–B871.

[295] Hong, J, Kim, Y, Pac, P-Y. Multivortex solutions of the Abelian Chern–Simons–Higgs theory. *Physical Review Letters*. 1990; 64: 2330–2333.

[296] Holzwarth, G, Schwesinger, B. Baryons in the Skyrme model. *Reports on Progress in Physics*. 1986; 49: 825–872.

[297] Huang, K. *Statistical Mechanics*. John Wiley & Sons: New York; 1967.

[298] Huang, K. *Quarks, Leptons and Gauge Fields*. 2nd ed. World Scientific: Singapore; 1992.

[299] Huang, K. *Quantum Field Theory–From Operators to Path Integrals*. Wiley Interscience: New York; 1998.

[300] Huisken, G, Ilmanen, T. The Riemannian Penrose inequality. *International Mathematics Research Notices*. 1997; 20: 1045–1058.

[301] Huisken, G, Ilmanen, T. The inverse mean curvature flow and the Riemannian Penrose inequality. *Journal of Differential Geometry*. 2001; 59: 353–437.

[302] Humphreys, JE. *Introduction to Lie Algebras and Representation Theory*. Springer-Verlag: New York; 1972.

[303] Hunziker, W, Sigal, IM. The quantum N-body problem. *Journal of Mathematical Physics*. 2000; 41: 3448–3510.

[304] Imry, Y, Webb, RA. Quantum interference and the Aharonov–Bohm effect. *Scientific American*. 1989; 260: 56–62.

[305] Jabs, A. Connecting spin and statistics in quantum mechanics. *Foundations of Physics*. 2002; 40: 776–792.

[306] Jackiw, R. Quantum meaning of classical field theory. *Reviews of Modern Physics*. 1977; 49: 681–706.

[307] Jackiw, R, Templeton, S. How super-renormalizable interactions cure their infrared divergences. *Physical Reviews D*. 1981; 23: 2291–2304.

[308] Jackiw, R, Weinberg, EJ. Self-dual Chern–Simons vortices. *Physical Review Letters*. 1990; 64: 2334–2337.

[309] Jacobson, N. *Lie Algebras*. Dover: New York; 1979.

[310] Jaffe, A, Taubes, CH. *Vortices and Monopoles*. Birkhäuser: Boston; 1980.

[311] Jimenez, JB, Heisenberg, L, Olmo, GJ, Rubiera–Garcia, D. Born–Infeld inspired modifications of gravity. *Physics Reports*. 2018; 727: 1–129.

[312] Jones, RO. Density functional theory: Its origins, rise to prominence, and future. *Reviews of Modern Physics*. 2015; 87: 897–923.

[313] Jones, RO, Gunnarsson, O. The density functional formalism, its applications and prospects. *Reviews of Modern Physics*. 1989; 61: 689–746.

[314] Jorge, P, Mimoso, JP, Wands, D. On the dynamics and k-essence models. *Journal of Physics*. 2007; 66: 012031 (Conference Series; Open access).

[315] Julia B, Zee, A. Poles with both magnetic and electric charges in non-Abelian gauge theory. *Physical Review D*. 1975; 11: 2227–2232.

[316] Kaku, M. *Quantum Field Theory*. Oxford University Press: New York; 1993.

[317] Kaku, M. *Introduction to Superstrings and M-Theory*. 2nd ed. Springer-Verlag: New York; 1999.

[318] Kalita, R. Dark energy. *Journal of Modern Physics*. 2015; 6: 1007–1011.

[319] Kamenshchik, AY, Moschella, U, Pasquier, V. An alternative to quintessence. *Physics Letters B*. 2001; 511: 265–268.

[320] Kang, K, Pan, X-B. On a quasilinear parabolic curl system motivated by time evolution of Meissner states of superconductors. *SIAM Journal on Mathematical Analysis*. 2021; 53: 6471–6516.

[321] Kartashov, YV, Malomed, BA, Torner, L. Solitons in nonlinear lattices. *Reviews of Modern Physics*. 2011; 83: 247–305.

[322] Katritch, V, Bednar, J, Michoud, D, Scharein, RG, Dubochet, J, Stasiak, A. Geometry and physics of knots. *Nature*. 1996; 384: 142–145.

[323] Keel, M. Global existence for critical power Yang–Mills–Higgs equations in $\mathbb{R}^{3+1}$. *Communications in Partial Differential Equations*. 1997; 22: 1161–1225.

[324] Kennefick, D. *Traveling at the Speed of Thought: Einstein and the Quest for Gravitational Waves*. Princeton University Press: Princeton; 2016.

[325] Kerr, RP. Gravitational field of a spinning mass as an example of algebraically special metrics. *Physical Review Letters*. 1963; 11: 237–238.

[326] Ketterson, JB, Song, SN. *Superconductivity*. Cambridge University Press: Cambridge; 1999.

[327] Khare, A. *Fractional Statistics and Quantum Theory*. World Scientific: Singapore; 2005.

[328] Khuri, M, Weinstein, G, Yamada, S. Proof of the Riemannian–Penrose inequality with charge for multiple black holes. *Journal of Differential Geometry.* 2017; 106: 451–498.

[329] Kiessling, MK-H. On the quasi-linear elliptic PDE $-\nabla \cdot (\nabla u/\sqrt{1-|\nabla u|^2}) = 4\pi \sum_k a_k \delta_{s_k}$ in physics and geometry. *Communications in Mathematical Physics.* 2012; 314: 509–523.

[330] Kiessling, MK-H. Correction to: On the quasi-linear elliptic PDE $-\nabla \cdot (\nabla u/\sqrt{1-|\nabla u|^2}) = 4\pi \sum_k a_k \delta_{s_k}$ in physics and geometry. *Communications in Mathematical Physics.* 2018; 364: 825–833.

[331] Kim, C, Kim, Y, Kwon, OK, Nakajima, H. Vortex-type half-BPS solitons in ABJM theory, *Physical Reviews D.* 2009; 80: 045013.

[332] Kirchhoff, GR. *Vorlesungen über Mathematische Physik: Mechanik.* Vol. 1. Teubner: Leipzig; 1876.

[333] Kiritsis, E. *String Theory in a Nutshell.* 2nd ed. Princeton University Press: Princeton; 2019.

[334] Kittel, C. *Introduction to Solid State Physics.* 8th ed. John Wiley & Sons: Hoboken; 2004.

[335] Kobayashi, S, Nomizu, K. *Foundations of Differential Geometry.* Vol. II. Interscience: New York; 1969.

[336] Kohn, W, Sham, LJ. Self-consistent equations including exchange and correlation effects. *Physical Review Letters.* 1965; 140: 1133–1138.

[337] Komar, A. Covariant conservation laws in general relativity. *Physical Reviews.* 1959; 113: 934–936.

[338] Konishi, K. Advent of non-Abelian vortices and monopoles: Further thoughts about duality and confinement. *Progress of Theoretical Physics Supplements.* 2009; 177: 83–98.

[339] Konishi, K, Spanu, L. Non-Abelian vortex and confinement. *International Journal of Modern Physics A.* 2003; 18: 249–270.

[340] Krauss, LM, Dodelson, S, Meyer, S. Primordial gravitational waves and cosmology. *Science.* 2010; 328: 989–992.

[341] Krieger, J, Tataru, D. Global well-posedness for the Yang–Mills equation in 4 + 1 dimensions. Small energy. *Annals of Mathematics.* 2017; 185: 831–893.

[342] Kruglov, SI. Nonlinear electromagnetic fields as a source of universe acceleration. *International Journal of Modern Physics A.* 2016; 31: 1650058.

[343] Kruglov, SI. Acceleration of universe by nonlinear electromagnetic fields. *International Journal of Modern Physics D.* 2016; 25: 1640002.

[344] Kruglov, SI. Dyonic and magnetic black holes with nonlinear arcsin-electrodynamics. *Annals of Physics.* 2019; 409: 167937.

[345] Kumar, A, Singh, DV, Ghosh, SG. Hayward black holes in Einstein–Gauss–Bonnet gravity. *Annals of Physics.* 2020; 419: 168214.

[346] Kumar, CN, Khare, A. Charged vortex of finite energy in non-Abelian gauge theories with Chern–Simons term. *Physics Letters B.* 1986; 178: 395–399.

[347] Kusenko, A. Small Q balls. *Physics Letters B.* 1997; 404: 285–290.

[348] Kusenko, A, Shaposhnikov, M. Supersymmetric Q balls as dark matter. *Physics Letters B.* 1998; 418: 46–54.

[349] Lamb, H. *Hydrodynamics.* 6th ed. Cambridge University Press: Cambridge; 1932.

[350] Landau, LD, Lifshitz, EM. *Statistical Physics. Part 1.* 3rd ed. Butterworth-Heinemann: Oxford; 1996.

[351] Laurie, B, Katritch, V, Sogo, J, Koller, T, Dubochet, J, Stasiak, A. Geometry and physics of catenanes applied to the study of DNA replication. *Biophysical Journal.* 1998; 74: 2815–2822.

[352] Lawson, Jr., HB, Michelsohn, M-L. *Spin Geometry.* Princeton University Press: Princeton; 1989.

[353] Lee, TD, Pang, Y. Nontopological solitons. *Physics Reports.* 1992; 221: 251–350.

[354] Leggett, AJ. Number-phase fluctuations in two-band superconductors. *Progress of Theoretical Physics.* 1966; 36: 901–930.

[355] Lieb, EH. Thomas–Fermi and related theories of atoms and molecules. *Reviews of Modern Physics.* 1981; 53: 603–641.

[356] Lieb, EH. Thomas–Fermi theory. In: Hazewinkel, M (ed.) *Kluwer Encyclopedia of Mathematics.* Suppl. Vol. II. Springer-Verlag: Dordrecht; 2000. pp. 455–457.

[357] Lieb, EH, Seiringer, R. *The Stability of Matter in Quantum Mechanics.* Cambridge University Press: Cambridge; 2010.

[358] Lieb, EH, Simon, B. The Thomas–Fermi theory of atoms, molecules and solids. *Advances in Mathematics.* 1977; 23: 22–116.

[359] Lieb, EH, Yang, Y. Non-Abelian vortices in supersymmetric gauge field theory via direct methods. *Communications in Mathematical Physics.* 2012; 313: 445–478.

[360] Lin, CC. On the motion of vortices in two dimensions: I. Existence of the Kirchhoff–Routh function. *Proceedings of the National Academy of Sciences of the United States of America.* 1941; 27: 570–575.

[361] Lin, CC. On the motion of vortices in two dimensions: II. Some further investigations on the Kirchhoff–Routh function. *Proceedings of the National Academy of Sciences of the United States of America.* 1941; 27: 575–577.

[362] Lin, C-S, Yang, Y. Sharp existence and uniqueness theorems for non-Abelian multiple vortex solutions. *Nuclear Physics B.* 2011; 846: 650–676.

[363] Lin, F. Mixed vortex-antivortex solutions of Ginzburg–Landau equations. *Archive for Rational Mechanics and Analysis.* 1995; 133: 103–127.

[364] Lin, F. Complex Ginzburg–Landau equations and dynamics of vortices, filaments, and codimension-2 submanifolds. *Communications on Pure and Applied Mathematics.* 1998; 51: 385–441.

[365] Lin, F, Yang, Y. Gauged harmonic maps, Born–Infeld electromagnetism, and magnetic vortices. *Communications on Pure and Applied Mathematics.* 2003; 56: 1631–1665.

[366] Lin, F, Yang, Y. Existence of energy minimizers as stable knotted solitons in the Faddeev model. *Communications in Mathematical Physics.* 2004; 249: 273–303.

[367] Lin, F, Yang, Y. Energy splitting, substantial inequality, and minimization for the Faddeev and Skyrme models. *Communications in Mathematical Physics.* 2007; 269: 137–152.

[368] Lin, F, Yang, Y. Static knot energy, Hopf charge, and universal growth law. *Nuclear Physics B.* 2006; 747: 455–463.

[369] Lin, F, Yang, Y. Analysis on Faddeev knots and Skyrme solitons: Recent progress and open problems. *Contemporary Mathematics.* 2007; 446: 319–344.

[370] Lin, F, Yang, Y. Universal growth law for knot energy of Faddeev type in general dimensions. *Proceedings of the Royal Society A: Mathematical, Physical and Engineering Sciences.* 2008; 646: 2741–2757.

[371] Linde, A. A toy model for open inflation. *Physical Reviews D.* 1999; 59: 023503.

[372] Linde, A, Sasaki, M, Tanaka, T. CMB in open inflation. *Physical Reviews D.* 1999; 59: 123522.

[373] Lions, PL. The concentration-compactness principle in the calculus of variations. Part I and Part II. *Annales de l'Institut Henri Poincaré: Analyse Non Lineaire.* 1984; 1: 109–145; 223–283.

[374] Lions, PL. The concentration-compactness principle in the calculus of variations. Part I and Part II. *Revista Matemática Iberoamericana.* 1985; 1: 145–200; 2: 45–121.

[375] Liouville, J. Sur l'équation aux différences partielles $\frac{\mathrm{d}^2 \log \lambda}{\mathrm{d}u\mathrm{d}v} \pm \frac{\lambda}{2a^2} = 0$. *Journal de Mathématiques Pures et Appliquées.* 1853; 18: 71–72.

[376] London, F, London, H. The electromagnetic equations of the superconductor. *Proceedings of the Royal Society A: Mathematical, Physical and Engineering Sciences.* 1935; 149: 71–88.

[377] MacKenzie, R, Wilczek, F. Peculiar spin and statistics in two space dimensions. *International Journal of Modern Physics A.* 1988; 3: 2827–2853.

[378] Mackey, G. *Mathematical Foundations of Quantum Mechanics.* Benjamin: New York; 1963.

[379] Mahan, GD. *Many Particle Physics.* Springer-Verlag: New York; 1981.

[380] Maison, GD. Uniqueness of the Prasad–Sommerfield monopole solution. *Nuclear Physics B.* 1981; 182: 144–150.

[381] Mandelstam, S. Vortices and quark confinement in non-Abelian gauge theories. *Physics Letters B.* 1975; 53: 476–478.

[382] Mandelstam, S. General introduction to confinement. *Physical Reports C.* 1980; 67: 109–121.

[383] Manssur, LRU, Portugal, R. The canon package: A fast kernel for tensor manipulators. *Computer Physics Communications.* 2004; 157: 173–180.

[384] Manton, N. Geometry of Skyrmions. *Communications in Mathematical Physics.* 1987; 111: 469–478.

[385] Manton, N, Ruback, P. Skyrmions in flat space and curved space. *Physics Letters B.* 1986; 181: 137–140.

[386] Manton, N, Sutcliffe, P. *Topological Solitons.* Cambridge University Press: Cambridge; 2008.

[387] Marchisotto, EA, Zakeri, G-A. An invitation to integration in finite terms. *College Mathematics Journal.* 1994; 25: 295–308.

[388] Martin, JL. *Basic Quantum Mechanics.* Oxford University Press: Oxford; 1981.

[389] McCormick, S. On the charged Riemannian Penrose inequality with charged matter. *Classical and Quantum Gravity.* 2020; 37: 015007.

[390] McInnes, B. The phantom divide in string gas cosmology. *Nuclear Physics B.* 2005; 718: 55–82.

[391] McQuarrie, DA. *Statistical Mechanics.* Viva Books: New Delhi; 2003.

[392] Meeks, III, WH, Pérez, J. The classical theory of minimal surfaces. *Bulletin of the American Mathematical Society.* 2011; 48: 325–407.

[393] Messiah, A. *Quantum Mechanics.* Vols. I and II. Dover: New York; 1999.

[394] Meyer, KR, Hall, GR. *Introduction to Hamiltonian Dynamical Systems and the N-Body Problem.* Springer-Verlag: New York; 1992 (Applied Mathematical Sciences; No. 90).

[395] Milnor, JW. *Topology from the Differentiable Viewpoint.* Princeton University Press: Princeton; 1965.

[396] Misner, GW, Thorne, KS, Wheeler, JA. *Gravitation.* Freeman: New York; 1973.

[397] Moffatt, HK. The degree of knottedness of tangled vortex lines. *Journal of Fluid Mechanics.* 1969; 35: 117–129.

[398] Moffatt, HK, Ricca, RL. Helicity and the Calugareanu invariant. *Proceedings of the Royal Society A: Mathematical, Physical and Engineering Sciences.* 1992; 439: 411–429.

[399] Mornas, L. Neutron stars in a Skyrme model with hyperons. *European Physics Journal A.* 2005; 24: 293–312.

[400] Morris, DJP, Tennant, DA, Grigera, SA, Klemke, B, Castelnovo, C, Moessner, R, Czternasty, C, Meissner, M, Rule, KC, Hoffmann, J-U, Kiefer, K, Gerischer, S, Slobinsky, D, Perry, RS. Dirac strings and magnetic monopoles in the spin ice $Dy_2Ti_2O_7$. *Science.* 2009; 326: 411–414.

[401] Morse, PM. Diatomic molecules according to the wave mechanics. II. Vibrational levels. *Physical Review.* 1929; 34: 57–64.

[402] Moses, HE. Eigenfunctions of the curl operator, rotationally invariant Helmholtz theorem, and applications to electromagnetic theory and fluid mechanics. *SIAM Journal on Applied Mathematics.* 1971; 21: 114–144.

[403] Moskowitz, C. Gravity waves from big bang detected. *Scientific American* [online]. March 17, 2014. Available from: https://www.scientificamerican.com/article/gravity-waves-cmb-b-modepolarization/.

[404] Mukhanov, VF, Winitzki, S. *Introduction to Quantum Effects in Gravity.* Cambridge University Press: Cambridge; 2011.

[405] Naber, G. *Spacetime and Singularities – An Introduction.* Cambridge University Press: Cambridge; 1988.

[406] Nambu, Y. Strings, monopoles, and gauge fields. *Physical Review D.* 1974; 10: 4262–4268.

[407] Nash, C, Sen, S. *Topology and Geometry for Physicists.* Academic Press: New York; 1992.

[408] Nastase, H. *String Theory Methods for Condensed Matter Physics.* Cambridge University Press: Cambridge; 2017.

[409] Nayak, C, Simon, SH, Stern, A, Freedman, M, Sarma, SD. Non-Abelian anyons and topological quantum computation. *Reviews of Modern Physics.* 2008; 80: 1083–1155.

[410] Nayak, C, Wilczek, F. Quantum numbers of textured Hall effect quasiparticles. *Physical Review Letters.* 1996; 77: 4418–4421.

[411] Nesseris, S, Perivolaropoulos, L. Crossing the phantom divide: Theoretical implications and observational status. *Journal of Cosmology and Astroparticle Physics.* 2007; 0701: 018.

[412] von Neumann, J. *Mathematical Foundations of Quantum Mechanics.* Princeton University Press: Princeton; 1955.

[413] Newman, E, Couch, E, Chinnapared, K, Exton, A, Prakash, A, Torrence, R. Metric of a rotating, charged mass. *Journal of Mathematical Physics.* 1965; 6: 918–919.

[414] Newman, E, Janis, A. Note on the Kerr spinning-particle metric. *Journal of Mathematical Physics.* 1965; 6: 915–917.

[415] Newton, PK. *The N-Vortex Problem. Analytical Techniques.* Springer-Verlag: New York; 2001.

[416] Nielsen, HB, Olesen, P. Vortex-line models for dual strings. *Nuclear Physics B.* 1973; 61: 45–61.

[417] Niemi, AJ. Hamiltonian approach to knotted solitons. In: Stasiak, A, Katritch, V, Kauffman, LH (eds.) *Ideal Knots.* World Scientific: Singapore; 1998. pp. 274–287.

[418] Nolasco, M, Tarantello, G. Vortex condensates for the $SU(3)$ Chern–Simons theory. *Communications in Mathematical Physics.* 2000; 213: 599–639.

[419] Oberai, MM, Srikantiah, G. *A First Course in Thermodynamics.* Prentice-Hall: New Delhi; 1974.

[420] Oh, S-J. Finite energy global well-posedness of the Yang–Mills equations on $\mathbb{R}^{1+3}$: An approach using the Yang–Mills heat flow. *Duke Mathematical Journal.* 2015; 164: 1669–1732.

[421] Oh, S-J, Tataru, D. The threshold conjecture for the energy critical hyperbolic Yang–Mills equation. *Annals of Mathematics.* 2021; 194: 393–473.

[422] Onsager, L. Statistical hydrodynamics. *Il Nuovo Cimento.* 1949; 6: 279–287.

[423] Ortín, T. *Gravity and Strings.* Cambridge University Press: Cambridge; 2004.

[424] Osakabe, N, Matsuda, T, Kawasaki, T, Endo, J, Tonomura, A, Yano, S, Yamada, H. Experimental confirmation of Aharonov–Bohm effect using a toroidal magnetic field confined by a superconductor. *Physical Review A.* 1986; 34: 815–822.

[425] Osserman, R. Minimal varieties. *Bulletin of the American Mathematical Society*. 1969; 75: 1092–1120.
[426] Osserman, R. *A Survey of Minimal Surfaces*. Dover: New York; 1986.
[427] van Oudenaarden, A, Devoret, MH, Nazarov, YV, Mooij, JE. Magnetoelectric Aharonov–Bohm effect in metal rings. *Nature*. 1998; 391: 768–770.
[428] O'Neill, B. *Semi-Riemannian Geometry*. Academic Press: Cambridge (MA); 1983.
[429] Pacard, F, Riviére, T. *Linear and Nonlinear Aspects of Vortices: The Ginzburg–Landau Model*. Birkhäuser: Boston; 2012 (Progress in Nonlinear Differential Equations and Their Applications).
[430] Panin, AG, Smolyakov, MN. Problem with classical stability of $U(1)$ gauged Q-balls. *Physical Review D*. 2017; 95: 065006.
[431] Parker, TH, Taubes, CH. On Witten's proof of the positive energy theorem. *Communications in Mathematical Physics*. 1982; 84: 223–238.
[432] Parr, RG, Yang, W. *Density-Functional Theory of Atoms and Molecules*. Oxford University Press: New York; 1989.
[433] Parsons, P, Barrow, JD. Generalised scalar field potentials and inflation. *Physical Review D*. 1995; 51: 6757–6763.
[434] Paul, S, Khare, A. Charged vortices in an Abelian Higgs model with Chern–Simons term. *Physics Letters B*. 1986; 17: 420–422.
[435] Paula, M, Leite, L, Crispino, LCB. Electrically charged black holes in linear and nonlinear electrodynamics: Geodesic analysis and scalar absorption. *Physical Review D*. 2020; 102: 104033.
[436] Pauli, W. The connection between spin and statistics. *Physical Review*. 1940; 58: 716–722.
[437] Pauli, W. On the connection between spin and statistics. *Progress of Theoretical Physics*. 1950; 5: 526–543.
[438] Peebles, PJE, Ratra, B. The cosmological constant and dark energy. *Reviews of Modern Physics*. 2003; 75: 559–606.
[439] Penrose, R. Gravitational collapse and space-time singularities. *Physical Review Letters*. 1965; 14: 57–59.
[440] Penrose, R. *Techniques of Differential Topology in Relativity*. SIAM: Philadelphia; 1987.
[441] Penrose, R. Gravitational collapse: The role of general relativity. *La Rivista del Nuovo Cimento*. 1969; 1: 252–276; *General Relativity and Gravitation*. 2002; 34: 1141–1165.
[442] Penrose, R. Naked singularities. *Annals of the New York Academy of Science*. 1973; 224: 125–134.
[443] Penrose, R, Floyd, RM. Extraction of rotational energy from a black hole. *Nature Physical Science*. 1971; 229: 177–179.
[444] Perlmutter, S, Aldering, G, Goldhaber, G, Knop, RA, Nugent, P, Castro, PG, et al. Measurements of Ω and Λ from 42 high redshift supernovae. *The Astrophysical Journal*. 1999; 517: 565–586.
[445] Peyrard, M. Nonlinear dynamics and statistical physics of DNA. *Nonlinearity*. 2004; 17: R1–R40.

[446] Peyrard, M, Bishop, AR. Statistical mechanics of a nonlinear model for DNA denaturation. *Physical Review Letters.* 1989; 62: 2755–2757.

[447] Pippard, AB. An experimental and theoretical study of the relation between magnetic field and current in a superconductor. *Proceedings of the Royal Society A: Mathematical, Physical and Engineering Sciences.* 1953; 216: 547–568.

[448] Plohr, BJ. The existence, regularity, and behavior of isotropic solutions of classical gauge field theories. Thesis (PhD). Princeton University: Princeton (NJ); 1980.

[449] Pokhozhaev, SI. On the eigenfunctions of the equation $\Delta u + \lambda f(u) = 0$. *Proceedings of the USSR Academy of Sciences.* 1965; 165: 36–39.

[450] Polchinski, J. *String Theory. Volume 1: An Introduction to the Bosonic String.* Cambridge University Press: Cambridge; 1998.

[451] Polchinski, J. *String Theory. Volume 2: Superstring Theory and Beyond.* Cambridge University Press: Cambridge; 1998.

[452] Polchinski, J. Monopoles, duality, and string theory. *International Journal of Modern Physics A.* 2004; 19: 145–154.

[453] Pollard, H. *Mathematical Introduction to Celestial Mechanics.* Prentice-Hall: Englewood Cliffs; 1966.

[454] Polyakov, AM. Particle spectrum in quantum field theory. *Journal of Experimental and Theoretical Physics Letters.* 1974; 20: 194.

[455] Polyakov, AM. Quantum geometry of bosonic strings. *Physics Letters B.* 1981; 103: 207–210.

[456] Polyakov, AM. Quantum geometry of fermionic strings. *Physics Letters B.* 1981; 103: 211–213.

[457] Portugal, R, Sautu, SL. Applications of Maple to general relativity. *Computer Physics Communications.* 1997; 105: 233–253.

[458] Poynting, JH. On the transfer of energy in the electromagnetic field. *Philosophical Transactions of the Royal Society of London.* 1884; 175: 343–361.

[459] Prasad, MK, Sommerfield, CM. Exact classical solutions for the 't Hooft monopole and the Julia–Zee dyon. *Physical Review Letters.* 1975; 35: 760–762.

[460] Preskill, J. Magnetic monopoles. *Annual Review of Nuclear and Particle Science.* 1984; 34: 461–530.

[461] de Putter, R, Linder, EV. Kinetic k-essence and quintessence. *Astroparticle Physics.* 2007; 28: 263–272.

[462] Radu, E, Tchrakian, DH, Yang, Y. Abelian Hopfions of the CP^n model on $\mathbb{R}^{2n+1}$ and a fractionally powered topological lower bound. *Nuclear Physics B.* 2013; 875: 388–407.

[463] Radu, E, Volkov, MS. Stationary ring solitons in field theory: Knots and vortons. *Physics Reports.* 2008; 468: 101–151.

[464] Rae, AIM. *Quantum Mechanics.* Institute of Physics Publishing: Bristol; 2002.

[465] Rajantie, A. The search for magnetic monopoles. *Physics Today.* 2016; 69: 40–46.

[466] Rajaraman, R. *Solitons and Instantons.* Elsevier: Amsterdam; 1987.

[467] Ratra, B, Peebles, PJE. Cosmological consequences of a rolling homogeneous scalar field, *Physical Review D.* 1988; 37: 3406–3427.

[468] Renn, J, Sauer, T, Stachel, J. The origin of gravitational lensing: A postscript to Einstein's 1936 Science paper. *Science.* 1997; 275: 184–186.

[469] Ricca, RL, Nipoti, B. Gauss' linking number revisited. *The Journal of Knot Theory and Its Ramifications.* 2011; 20: 1325–1343.

[470] Riess, AG, Filippenko, AV, Challis, P, Clocchiatti, A, Diercks, A, Garnavich, PM, et al. Observational evidence from supernovae for an accelerating universe and a cosmological constant. *The Astronomical Journal.* 1998; 116: 1009–1038.

[471] Rodrigues, ME, de S. Silva, MV. Bardeen regular black hole with an electric source. *Journal of Cosmology and Astroparticle Physics.* 2018; 06: 025.

[472] Rubakov, VA. On the electroweak theory at high fermion density. *Progress of Theoretical Physics.* 1986; 75: 366–385.

[473] Rubakov, VA, Tavkhelidze, AN. Stable anomalous states of superdense matter in gauge theories. *Physics Letters B.* 1985; 165: 109–112.

[474] Rubano, C, Scudellaro, P. On some exponential potentials for a cosmological scalar field as quintessence. *General Relativity and Gravitation.* 2002; 34: 307–328.

[475] Rubiera-Garcia, D, dos Santos, C. BPS solitons in a Dirac–Born–Infeld action. *Journal of Physics A: Mathematical and Theoretical.* 2014; 47: 105402.

[476] Ruostekoski, J, Anglin, JR. Creating vortex rings and three-dimensional skyrmions in Bose–Einstein condensates. *Physical Review Letters.* 2001; 86: 3934–3937.

[477] Ryder, LH. *Quantum Field Theory.* 2nd ed. Cambridge University Press: Cambridge; 1996.

[478] Safian, AM, Coleman, S, Axenides, M. Some non-Abelian Q-balls. *Nuclear Physics B.* 1988; 297: 498–514.

[479] Saha, MN. The origin of mass in neutrons and protons. *Indian Journal of Physics.* 1937; 10: 141–153.

[480] Saha, MN. Note on Dirac's theory of magnetic poles. *Physical Review.* 1949; 75: 1968.

[481] Saint-James, D, Thomas, EJ, Sarma, G. *Type II Superconductivity.* Pergammon: New York; 1969.

[482] Salam, A, Ward, JC. Weak and electromagnetic interactions. *Nuovo Cimento.* 1959; 11: 568–577.

[483] Sandier, E, Serfaty, S. *Vortices in the Magnetic Ginzburg–Landau Model.* Birkhäuser: Boston; 2007 (Progress in Nonlinear Differential Equations and Their Applications; No. 70).

[484] Sauer, T. Nova Geminorum 1912 and the origin of the idea of gravitational lensing. *Archive for History of Exact Sciences.* 2008; 62: 1–22.

[485] Schechter, M, Weder, R. A theorem on the existence of dyon solutions. *Annals of Physics.* 1981; 132: 293–327.

[486] Scherk, J. An introduction to the theory of dual models and strings. *Reviews of Modern Physics.* 1975; 47: 123–164.

[487] Scherk, J, Schwarz, JH. Dual models for non-hadrons. *Nuclear Physics B.* 1974; 81: 118–144.

[488] Schmid, M, Shaposhnikov, M. Anomalous Abelian solitons. *Nuclear Physics B.* 2007; 775: 365–389.

[489] Schneider, P, Ehlers, J, Falco, EE. *Gravitational Lenses.* Springer-Verlag: Heidelberg; 1992.

[490] Schneider, P, Kochanek, C, Wambsganss, J. *Gravitational Lensing: Strong, Weak and Micro.* Springer-Verlag: Heidelberg; 2006.

[491] Schoen, R, Yau, S-T. Complete manifolds with nonnegative scalar curvature and the positive action conjecture in general relativity. *Proceedings of the National Academy of Sciences of the United States of America.* 1978; 75: 45–76.

[492] Schoen, R, Yau, S-T. On the proof of the positive mass conjecture in general relativity. *Communications in Mathematical Physics.* 1979; 65: 45–76.

[493] Schoen, R, Yau, S-T. Positivity of the total mass of a general space-time. *Physical Review Letters.* 1979; 43: 1457–1459.

[494] Schoen, R, Yau, S-T. Proof of the positive mass theorem II. *Communications in Mathematical Physics.* 1981; 79: 231–260.

[495] Schomerus, V. *A Primer on String Theory.* Cambridge University Press: Cambridge; 2017.

[496] Schonfeld, JS. A massive term for three-dimensional gauge fields. *Nuclear Physics B.* 1981; 185: 157–171.

[497] Schroeder, DV. *An Introduction to Thermal Physics.* Addison-Wesley: San Francisco; 2000.

[498] Schroers, BJ. Bogomol'nyi solitons in a gauged $O(3)$ sigma model. *Physics Letters B.* 1995; 356: 291–296.

[499] Schroers, BJ. The spectrum of Bogomol'nyi solitons in gauged linear sigma models. *Nuclear Physics B.* 1996; 475: 440–468.

[500] Schwarzschild, B. Currents in normal-metal rings exhibit Aharonov–Bohm effect. *Physics Today.* 1986; 39: 17–20.

[501] Schwinger, J. A magnetic model of matter. *Science.* 1969; 165: 757–761.

[502] Segal, IE. The Cauchy problem for the Yang–Mills equations. *Journal of Functional Analysis.* 1979; 33: 175–194.

[503] Senovilla, JMM, Garfinkle, D. The 1965 Penrose singularity theorem. *Classical and Quantum Gravity.* 2015; 32: 124008.

[504] Shaposhnikov, ME. Structure of the high temperature gauge ground state and electroweak production of the baryon asymmetry. *Nuclear Physics B.* 1988; 299: 797–817.

[505] Shekhter, RI, Gorelik, LY, Glazman, LI, Jonson, M. Electronic Aharonov-Bohm effect induced by quantum vibrations. *Physical Review Letters.* 2006; 97: 156801.

[506] Shifman, A, Yung, A. Supersymmetric solitons. *Reviews of Modern Physics.* 2007; 79: 1139–1196.

[507] Shifman, A, Yung, A. *Supersymmetric Solitons.* Cambridge University Press: Cambridge; 2009.

[508] Shiiki, N, Sawado, N. Regular and black hole solutions in the Einstein–Skyrme theory with negative cosmological constant. *Classical and Quantum Gravity.* 2005; 22: 3561–3574.

[509] Shtengel, K. A home for anyon? *Nature Physics.* 2007; 3: 763.

[510] Sibner, L, Sibner, R, Yang, Y. Abelian gauge theory on Riemann surfaces and new topological invariants. *Proceedings of the Royal Society A: Mathematical, Physical and Engineering Sciences.* 2000; 456: 593–613.

[511] Sibner, L, Sibner, R, Yang, Y. Generalized Bernstein property and gravitational strings in Born–Infeld theory. *Nonlinearity.* 2007; 20: 1193–1213.

[512] Singh, DV, Ghosh, SG, Maharaj, SD. Bardeen-like regular black holes in $5D$ Einstein–Gauss–Bonnet gravity. *Annals of Physics.* 2020; 412: 168025.

[513] Skyrme, THR. A nonlinear field theory. *Proceedings of the Royal Society A: Mathematical, Physical and Engineering Sciences.* 1961; 260: 127–138.

[514] Skyrme, THR. Particle states of a quantized meson field. *Proceedings of the Royal Society A: Mathematical, Physical and Engineering Sciences.* 1961; 262: 237–245.

[515] Skyrme, THR. A unified field theory of mesons and baryons. *Nuclear Physics.* 1962; 31: 556–569.

[516] Skyrme, THR. The origins of Skyrmions. *International Journal of Modern Physics A.* 1988; 3: 2745–2751.

[517] Slater, JC. The theory of complex spectra. *Physical Review.* 1929; 24: 1293–1322.

[518] Soler, M. Classical, stable, nonlinear spinor field with positive rest energy. *Physical Review D.* 1970; 1: 2766–2769.

[519] Sonego, S, Talamini, V. Qualitative study of perfect-fluid Friedmann–Lamaıtre–Robertson–Walker models with a cosmological constant. *American Journal of Physics.* 2012; 80: 670–679.

[520] Spivak, M. *A Comprehensive Introduction to Differential Geometry.* Volume III. Publish or Perish: Berkeley; 1975.

[521] Spruck, J, Yang, Y. On multivortices in the electroweak theory I: Existence of periodic solutions. *Communications in Mathematical Physics.* 1992; 144: 1–16.

[522] Spruck, J, Yang, Y. On multivortices in the electroweak theory II: Existence of Bogomol'nyi solutions in $\mathbb{R}^2$. *Communications in Mathematical Physics.* 1992; 144: 215–234.

[523] Spruck, J, Yang, Y. The existence of non-topological solitons in the self-dual Chern–Simons Theory. *Communications in Mathematical Physics.* 1992; 149: 361–376.

[524] Spruck, J, Yang, Y. Topological solutions in the self-dual Chern–Simons theory: Existence and approximation. *Annales de l'Institut Henri Poincaré, Analyse Non Linéaire.* 1995; 12: 75–97.

[525] Spruck, J, Yang, Y. Proof of the Julia–Zee theorem. *Communications in Mathematical Physics.* 2009; 291: 347–356.

[526] Srednicki, M. *Quantum Field Theory.* Cambridge University Press: Cambridge; 2015.
[527] Stern, A. Anyons and the quantum Hall effect — a pedagogical review. *Annals of Physics.* 2008; 323: 204–249.
[528] Stone, MH. The generalized Weierstrass approximation theorem. *Mathematics Magazine.* 1948; 21: 167–184.
[529] Ströhmer, G. About the eigenfunctions of the curl-operator. *Applicable Analysis.* 1993; 51: 187–196.
[530] Strominger, A, Vafa, C. Microscopic origin of the Bekenstein–Hawking entropy. *Physics Letters B.* 1996; 379: 99–104.
[531] Suhl, H, Matthias, BT, Walker, LR. Bardeen–Cooper–Schrieffer theory of superconductivity in the case of overlapping bands. *Physical Review Letters.* 1959; 3: 552–554.
[532] 't Hooft, G. Magnetic monopoles in unified gauge theories. *Nuclear Physics B.* 1974; 79: 276–284.
[533] 't Hooft, G. Computation of the quantum effects due to a four-dimensional pseudoparticle. *Physical Review D.* 1976; 14: 3432–3450.
[534] 't Hooft, G. On the phase transition towards permanent quark confinement. *Nuclear Physics B.* 1978; 138: 1–25.
[535] 't Hooft, G. A property of electric and magnetic flux in non-Abelian gauge theories. *Nuclear Physics B.* 1979; 153: 141–160.
[536] 't Hooft, G. Topology of the gauge condition and new confinement phases in non-Abelian gauge theories. *Nuclear Physics B.* 1981; 190: 455–478.
[537] Talenti, G. Best constant in Sobolev inequality. *Annali di Matematica Pura ed Applicata.* 1976; 110: 352–372.
[538] Tarantello, G. Multiple condensate solutions for the Chern–Simons–Higgs theory. *Journal of Mathematical Physics.* 1996; 37: 3769–3796.
[539] Tarantello, G. *Self-dual Gauge Field Vortices.* Birkhäuser: Boston; 2008 (Progress in Nonlinear Differential Equations and Their Applications; No. 72).
[540] Taubes, CH. Arbitrary N-vortex solutions to the first order Ginzburg–Landau equations. *Communications in Mathematical Physics.* 1980; 72: 277–292.
[541] Taubes, CH. On the equivalence of the first and second order equations for gauge theories. *Communications in Mathematical Physics.* 1980; 75: 207–227.
[542] Thirring, WE. A soluble relativistic field theory. *Annals of Physics.* 1958; 3: 91–112.
[543] Tilley, OR, Tilley, J. *Superfluidity and Superconductivity.* Adam Hilger: Bristol; 1986.
[544] Tinkham, M. *Introduction to Superconductivity.* McGraw-Hill: New York; 1975.
[545] Thomas, LH. The calculation of atomic fields. *Mathematical Proceedings of the Cambridge Philosophical Society.* 1927; 23: 542–548.
[546] Thompson, RI. Beta function quintessence cosmological parameters and fundamental constants I: Power and inverse power law dark energy

potentials. *Monthly Notices of the Royal Astronomical Society.* 2018; 477: 4104–4115.

[547] Tong, D, Wong, K. Vortices and impurities. *Journal of High Energy Physics.* 2014; 01: 090.

[548] Tonomura, A, Osakabe, N, Matsuda, T, Kawasaki, T, Endo, J. Evidence for Aharonov–Bohm effect with magnetic field completely shielded from electron wave. *Physical Review Letters.* 1986; 56: 792–795.

[549] Torres del Castillo, GF. Eigenfunctions of the curl operator in spherical coordinates. *Journal of Mathematical Physics.* 1994; 35: 499–507.

[550] Troitsky, S. Supermassive dark-matter Q-balls in galactic centers? *Journal of Cosmology and Astroparticle Physics.* 2016; 11: 027.

[551] Tseytlin, AA. Self-duality of Born–Infeld action and Dirichlet 3-brane of type IIB superstring theory. *Nuclear Physics B.* 1996; 469: 51–67.

[552] Tseytlin, AA. On non-Abelian generalisation of Born–Infeld action in string theory. *Nuclear Physics B.* 1997; 501: 41–52.

[553] Tsujikawa, S. Quintessence: a review. *Classical and Quantum Gravity.* 2013; 30: 214003.

[554] Tyupkin, YS, Fateev, VA, Shvarts, AS. Particle-like solutions of the equations of gauge theories. *Theoretical and Mathematical Physics.* 1976; 26: 270–273.

[555] Tzikas, AG. Bardeen black hole chemistry. *Physics Letters B.* 2019; 788: 219–224.

[556] Vachaspati, T. *Kinks and Domain Walls, an Introduction to Classical and Quantum Solitons.* Cambridge University Press: Cambridge; 2006.

[557] Vakulenko, AF, Kapitanski, LV. Stability of solitons in S^2 nonlinear σ-model. *Doklady Physics.* 1979; 24: 433–434.

[558] Vallado, DA. *Fundamentals of Astrodynamics and Applications.* 2nd ed. Kluwer Academic Publications: Dordrecht; 2001.

[559] de Vega, HJ, Schaposnik, F. Electrically charged vortices in non-Abelian gauge theories with Chern–Simons term. *Physical Review Letters.* 1986; 56: 2564–2566.

[560] de Vega, HJ, Schaposnik, F. Vortices and electrically charged vortices in non-Abelian gauge theories. *Physical Review D.* 1986; 34 3206–3213.

[561] Veneziano, G. Scale factor duality for classical and quantum strings. *Physics Letters B.* 1991; 265: 287–294.

[562] Vikman, A. Can dark energy evolve to the phantom? *Physical Review D.* 2005; 71: 023515.

[563] Wald, RM. *General Relativity,* University of Chicago Press: Chicago; 1984.

[564] Wang, Q. Relative periodic solutions of the N-vortex problem via the variational method. *Archive for Rational Mechanics and Analysis.* 2019; 231: 1401–1425.

[565] Wang, S, Yang, Y. Abrikosov's vortices in the critical coupling. *SIAM Journal on Mathematical Analysis.* 1992; 23: 1125–1140.

[566] Ward, RS. Hopf solitons on S^3 and $\mathbb{R}^3$. *Nonlinearity.* 1999; 12: 241–246.

[567] Weinberg, EJ. *Classical Solutions in Quantum Field Theory: Solitons and Instantons in High Energy Physics.* Cambridge University Press: Cambridge; 2012.

[568] Weinberg, S. A model of leptons. *Physical Review Letters.* 1967; 19: 1264–1266.

[569] Weinberg, S. *Gravitation and Cosmology: Principles and Applications of the General Theory of Relativity.* Wiley: New York; 1972.

[570] Weinberg, S. *The Quantum Theory of Fields. Volume 1: Foundations.* Cambridge University Press: Cambridge; 1995.

[571] Weinberg, S. *The Quantum Theory of Fields. Volume 2: Modern Applications.* Cambridge University Press: Cambridge; 1995.

[572] Weinberg, S. *The Quantum Theory of Fields. Volume 3: Supersymmetry.* Cambridge University Press: Cambridge; 1995.

[573] Weinstein, G, Yamada, S. On a Penrose inequality with charge. *Communications in Mathematical Physics.* 2005; 257: 703–723.

[574] Weinstock, R. *Calculus of Variations.* Dover: New York; 1974.

[575] von Weizsäcker, C. Zur theorie der Kernmassen. *Zeitschrift für Physik.* 1935; 96: 431–458.

[576] Wells, RO. *Differential Analysis on Complex Manifolds.* Springer-Verlag: New York; 1986.

[577] von Westenholz, C. *Differential Forms in Mathematical Physics.* North-Holland: Amsterdam; 1981.

[578] Wetterich, C. Cosmology and the fate of dilatation symmetry. *Nuclear Physics B.* 1988; 302: 668–696.

[579] Whitehead, JHC. An expression of Hopf's invariant as an integral. *Proceedings of the National Academy of Sciences of the United States of America.* 1947; 33: 117–123.

[580] Whittaker, ET. An expression of certain known functions as generalized hypergeometric functions. *Bulletin of the American Mathematical Society.* 1904; 10: 125–134.

[581] Wilczek, F. Quantum mechanics of fractional-spin particles. *Physical Review Letters.* 1982; 49: 957–959.

[582] Wilczek, F. *Fractional Statistics and Anyon Superconductors.* World Scientific: Singapore; 1990.

[583] Wilczek, F. From electronics to anyonics. *Physics World.* 2006; 19: 22–23.

[584] Wilson, HA. Note on Dirac's theory of magnetic poles. *Physical Review.* 1949; 75: 309.

[585] Witten, E. Baryons in $1/N$ expansion. *Nuclear Physics B.* 1979; 160: 57–115.

[586] Witten, E. A new proof of the positive energy theorem. *Communications in Mathematical Physics.* 1981; 80: 381–402.

[587] Witten, E. What every physicist should know about string theory. *Physics Today.* 2015; 68: 38–43.

[588] Woltjer, L. A theorem on force-free magnetic fields. *Proceedings of the National Academy of Sciences of the United States of America.* 1958; 44: 489–491.

[589] Wu, TT, Yang, CN. Concept of nonintegrable phase factors and global formulation of gauge fields. *Physical Review D.* 1975; 12: 3845–3857.
[590] Wu, TT, Yang, CN. Dirac monopole without strings: monopole harmonics. *Nuclear Physics B.* 1976; 107: 365–380.
[591] Xu, C, Yang, Y. Determination of angle of light deflection in higher-derivative gravity theories. *Journal of Mathematical Physics.* 2018; 59: 032501.
[592] Xu, C, Yang, Y. Determination of bending angle of light deflection subject to possible weak and strong quantum gravity effects. *International Journal of Modern Physics A.* 2020; 35: 2050188.
[593] Xu, L, Shu, Z, Wang, S. From BCS theory for isotropic homogeneous systems to the complete Ginzburg–Landau equations for anisotropic inhomogeneous systems. *Physical Review B.* 1998; 57: 11654–11660.
[594] Yakaboylu, E, Deuchert, A, Lemeshko, M. Emergence of non-Abelian magnetic monopoles in a quantum impurity problem. *Physical Review Letters.* 2017; 119: 235301.
[595] Yang, CN, Mills, RL. Conservation of isotopic spin and isotopic gauge invariance. *Physical Review.* 1954; 96: 191–195.
[596] Yang, Y, Peng, C, Zhu, D, Buljian, H, Joannopoulos, JD, Zhen, B, Soljacic, M. Synthesis and observation of non-Abelian gauge fields in real space. *Science.* 201; 365: 1021–1025.
[597] Yang, Y. Existence, regularity, and asymptotic behavior of the Ginzburg–Landau equations on $\mathbb{R}^3$. *Communications in Mathematical Physics.* 1989; 123: 147–161.
[598] Yang, Y. On the global behavior of symmetric Skyrmions. *Letters in Mathematical Physics.* 1990; 19: 25–33.
[599] Yang, Y. The relativistic non-Abelian Chern–Simons equations. *Communications in Mathematical Physics.* 1997; 186: 199–218.
[600] Yang, Y. Dually charged particle-like solutions in the Weinberg–Salam theory. *Proceedings of the Royal Society A: Mathematical, Physical and Engineering Sciences.* 1998; 454: 155–178.
[601] Yang, Y. Coexistence of vortices and anti-vortices in an Abelian gauge theory. *Physical Review Letters.* 1998; 80: 26–29.
[602] Yang, Y. Strings of opposite magnetic charges in a gauge field theory. *Proceedings of the Royal Society A: Mathematical, Physical and Engineering Sciences.* 1999; 455: 601–629.
[603] Yang, Y. Classical solutions in the Born–Infeld theory. *Proceedings of the Royal Society A: Mathematical, Physical and Engineering Sciences.* 2000; 456: 615–640.
[604] Yang, Y. Rigorous proof of isotope effect by Bardeen–Cooper–Schrieffer theory. *Journal of Mathematical Physics.* 2003; 44: 2009–2025.
[605] Yang, Y. Electromagnetic asymmetry, relegation of curvature singularities of charged black holes, and cosmological equations of state in view of the Born–Infeld theory. *Classical Quantum Gravity.* 2022; 39: 195007.
[606] Yang, Y. *Solitons in Field Theory and Nonlinear Analysis.* Springer-Verlag: New York; 2001 (Springer Monographs in Mathematics).

[607] Yang, Y. *A Concise Text on Advanced Linear Algebra.* Cambridge University Press: Cambridge; 2015.

[608] Yang, Y, Zhang, R. Existence of optical vortices. *SIAM Journal on Mathematical Analysis.* 2014; 46: 484–498.

[609] Yang, Y, Zhang, R, Zhao, L. Dynamics of electrostatic MEMS actuators. *Journal of Mathematical Physics.* 2012; 53: 022703.

[610] Yau, S-T. Calabi's conjecture and some new results in algebraic geometry. *Proceedings of the National Academy of Sciences of the United States of America.* 1977; 74: 1798–1799.

[611] Yosida, K. *Functional Analysis.* 5th ed. Springer-Verlag: New York; 1995.

[612] Yu, Y. Solitary waves for nonlinear Klein–Gordon equations coupled with Born–Infeld theory. *Annales de l'Institut Henri Poincaré, Analyse Non Linéaire.* 2010; 27: 351–376.

[613] Zahed, I, Brown, GE. The Skyrme model. *Physics Reports.* 1986; 142: 1–102.

[614] Zee, A. *Quantum Field Theory in a Nutshell.* Princeton University Press: Princeton; 2003.

[615] Zhang, X, Yang, Y. Solutions to the minimization problem arising in a dark monopole model in gauge field theory. *Nuclear Physics B.* 2020; 951: 114851.

[616] Zlatev, I, Wang, LM, Steinhardt, PJ. Quintessence, cosmic coincidence, and the cosmological constant. *Physical Review Letters.* 1999; 185: 896–899.

[617] Zwanziger, D. Quantum field theory of particles with both electric and magnetic charges. *Physical Review.* 1968; 176: 1489–1495.

[618] Zwiebach, B. *A First Course in String Theory.* Cambridge University Press: Cambridge; 2004.

Index